SOLITON EQUATIONS AND THEIR ALGEBRO-GEOMETRIC SOLUTIONS
Volume II: (1 + 1)-Dimensional Discrete Models

As a partner to Volume I: (1 + 1)-*Dimensional Continuous Models*, this monograph provides a self-contained introduction to algebro-geometric solutions of completely integrable, nonlinear, partial differential-difference equations, also known as soliton equations.

The systems studied in this volume include the Toda lattice hierarchy, the Kac–van Moerbeke hierarchy, and the Ablowitz–Ladik hierarchy. An extensive treatment of the class of algebro-geometric solutions in the stationary as well as time-dependent contexts is provided. The theory presented includes trace formulas, algebro-geometric initial value problems, Baker–Akhiezer functions, and theta function representations of all relevant quantities involved.

The book uses basic techniques from the theory of difference equations and spectral analysis, some elements of algebraic geometry and, especially, the theory of compact Riemann surfaces. The presentation is constructive and rigorous, with ample background material provided in various appendices. Detailed notes for each chapter, together with an exhaustive bibliography, enhance understanding of the main results.

Reviews of Volume I:

'...this is a book that I would recommend to any student of mine, for clarity and completeness of exposition...Any expert as well would enjoy the book and learn something stimulating from the sidenotes that point to alternative developments. We look forward to Volumes II and III!'
Mathematical Reviews

'The book is very well organized and carefully written. It could be particularly useful for analysts wanting to learn new methods coming from algebraic geometry.'
EMS Newsletter

CAMBRIDGE STUDIES IN ADVANCED MATHEMATICS

Editorial Board:

B. Bollobás, W. Fulton, A. Katok, F. Kirwan, P. Sarnak, B. Simon, B. Totaro

All the titles listed below can be obtained from good booksellers or from Cambridge University Press. For a complete series listing visit:
http://www.cambridge.org/series/sSeries.asp?code=CSAM

Already published

55 D. Bump *Automorphic forms and representations*
56 G. Laumon *Cohomology of Drinfeld modular varieties II*
57 D.M. Clark & B.A. Davey *Natural dualities for the working algebraist*
58 J. McCleary *A user's guide to spectral sequences II*
59 P. Taylor *Practical foundations of mathematics*
60 M. P. Brodmann & R. Y. Sharp *Local cohomology*
61 J. D. Dixon et al. *Analytic pro-p groups*
62 R. Stanley *Enumerative combinatorics II*
63 R. M. Dudley *Uniform central limit theorems*
64 J. Jost & X. Li-Jost *Calculus of variations*
65 A. J. Berrick & M. E. Keating *An introduction to rings and modules*
66 S. Morosawa *Holomorphic dynamics*
67 A. J. Berrick & M. E. Keating *Categories and modules with K-theory in view*
68 K. Sato *Levy processes and infinitely divisible distributions*
69 H. Hida *Modular forms and Galois cohomology*
70 R. Iorio & V. Iorio *Fourier analysis and partial differential equations*
71 R. Blei *Analysis in integer and fractional dimensions*
72 F. Borceaux & G. Janelidze *Galois theories*
73 B. Bollobás *Random graphs*
74 R. M. Dudley *Real analysis and probability*
75 T. Sheil-Small *Complex polynomials*
76 C. Voisin *Hodge theory and complex algebraic geometry, I*
77 C. Voisin *Hodge theory and complex algebraic geometry, II*
78 V. Paulsen *Completely bounded maps and operator algebras*
79 F. Gesztesy & H. Holden *Soliton Equations and Their Algebro-Geometric Solutions, I*
81 S. Mukai *An Introduction to Invariants and Moduli*
82 G. Tourlakis *Lectures in Logic and Set Theory, I*
83 G. Tourlakis *Lectures in Logic and Set Theory, II*
84 R. A. Bailey *Association Schemes*
85 J. Carlson, S. Müller-Stach & C. Peters *Period Mappings and Period Domains*
86 J. J. Duistermaat & J. A. C. Kolk *Multidimensional Real Analysis I*
87 J. J. Duistermaat & J. A. C. Kolk *Multidimensional Real Analysis II*
89 M. Golumbic & A. N. Trenk *Tolerance Graphs*
90 L. Harper *Global Methods for Combinatorial Isoperimetric Problems*
91 I. Moerdijk & J. Mrcun *Introduction to Foliations and Lie Groupoids*
92 J. Kollár, K. E. Smith & A. Corti *Rational and Nearly Rational Varieties*
93 D. Applebaum *Levy Processes and Stochastic Calculus*
94 B. Conrad *Modular Forms and the Ramanujan Conjecture*
95 M. Schechter *An Introduction to Nonlinear Analysis*
96 R. Carter *Lie Algebras of Finite and Affine Type*
97 H. L. Montgomery, R. C. Vaughan & M. Schechter *Multiplicative Number Theory I*
98 I. Chavel *Riemannian Geometry*
99 D. Goldfeld *Automorphic Forms and L-Functions for the Group GL(n,R)*
100 M. Marcus & J. Rosen *Markov Processes, Gaussian Processes, and Local Times*
101 P. Gille & T. Szamuely *Central Simple Algebras and Galois Cohomology*
102 J. Bertoin *Random Fragmentation and Coagulation Processes*
103 E. Frenkel *Langlands Correspondence for Loop Groups*
104 A. Ambrosetti & A. Malchiodi *Nonlinear Analysis and Semilinear Elliptic Problems*
105 T. Tao & V. H. Vu *Additive Combinatorics*
106 E. B. Davies *Linear Operators and their Spectra*
107 K. Kodaira *Complex Analysis*
108 T. Ceccherini-Silberstein, F. Scarabotti, F. Tolli *Harmonic Analysis on Finite Groups*
109 H. Geiges *An Introduction to Contact Topology*
110 J. Faraut *Analysis on Lie Groups*
111 E. Park *Complex Topological K-Theory*
112 D.W. Stroock *Partial Differential Equations for Probabilists*
113 A. Kirillov *An Introduction to Lie Groups and Lie Algebras*
114 F. Gesztesy, H. Holden, J. Michor & G. Teschl *Soliton Equations and Their Algebro-Geometric Solutions, II*

SOLITON EQUATIONS AND THEIR ALGEBRO-GEOMETRIC SOLUTIONS

Volume II: (1 + 1)-Dimensional Discrete Models

FRITZ GESZTESY
*University of Missouri, Columbia,
Missouri, USA*

HELGE HOLDEN
*Norwegian University of Science and
Technology, Trondheim, Norway*

JOHANNA MICHOR
*Universität Wien,
Wien, Austria*

GERALD TESCHL
*Universität Wien,
Wien, Austria*

CAMBRIDGE UNIVERSITY PRESS
Cambridge, New York, Melbourne, Madrid, Cape Town, Singapore, São Paulo, Delhi

Cambridge University Press
The Edinburgh Building, Cambridge CB2 8RU, UK

Published in the United States of America by Cambridge University Press, New York

www.cambridge.org
Information on this title: www.cambridge.org/9780521753081

© F. Gesztesy, H. Holden, J. Michor, and G. Teschl 2008

This publication is in copyright. Subject to statutory exception
and to the provisions of relevant collective licensing agreements,
no reproduction of any part may take place without
the written permission of Cambridge University Press.

First published 2008

Printed in the United Kingdom at the University Press, Cambridge

A catalog record for this publication is available from the British Library

Library of Congress Cataloging in Publication data
Gesztesy, Fritz, 1953–
Soliton equations and their algebro-geometric solutions / Fritz Gesztesy,
Helge Holden.
p. cm. – (Cambridge studies in advanced mathematics ; 79)
Includes bibliographical references and index.
Contents: v. 1. (1 + 1)-dimensional continuous models
ISBN 0-521-75307-4 (v. 1)
1. Differential equations, Nonlinear – Numerical solutions.   2. Solitons.
I. Holden, H. (Helge), 1956–   II. Title,   III. Series.
QC20.7.D5 G47   2003
530.15′5355 – dc21   2002074069
ISBN 0 521 75307 4 hardback

Helge Holden har mottatt støtte fra Det faglitterære fond

Cambridge University Press has no responsibility for the persistence or
accuracy of URLs for external or third-party internet websites referred to
in this publication, and does not guarantee that any content on such
websites is, or will remain, accurate or appropriate.

To
Gloria
Christian, Mads, Frederik, and Daniel
Elli, Peter, and Franziska
Susanne, Simon, and Jakob

Contents

Acknowledgments		*page* ix
Introduction		1
1	**The Toda Hierarchy**	**25**
	1.1 Contents	25
	1.2 The Toda Hierarchy, Recursion Relations, Lax Pairs, and Hyperelliptic Curves	26
	1.3 The Stationary Toda Formalism	41
	1.4 The Stationary Toda Algebro-Geometric Initial Value Problem	72
	1.5 The Time-Dependent Toda Formalism	84
	1.6 The Time-Dependent Toda Algebro-Geometric Initial Value Problem	103
	1.7 Toda Conservation Laws and the Hamiltonian Formalism	117
	1.8 Notes	145
2	**The Kac–van Moerbeke Hierarchy**	**161**
	2.1 Contents	161
	2.2 The KM Hierarchy and its Relation to the Toda Hierarchy	162
	2.3 The Stationary KM Formalism	172
	2.4 The Time-Dependent KM Formalism	178
	2.5 Notes	181
3	**The Ablowitz–Ladik Hierarchy**	**186**
	3.1 Contents	186
	3.2 The Ablowitz–Ladik Hierarchy, Recursion Relations, Zero-Curvature Pairs, and Hyperelliptic Curves	187
	3.3 Lax Pairs for the Ablowitz–Ladik Hierarchy	202
	3.4 The Stationary Ablowitz–Ladik Formalism	220

3.5	The Stationary Ablowitz–Ladik Algebro-Geometric Initial Value Problem	236
3.6	The Time-Dependent Ablowitz–Ladik Formalism	249
3.7	The Time-Dependent Ablowitz–Ladik Algebro-Geometric Initial Value Problem	267
3.8	Ablowitz–Ladik Conservation Laws and the Hamiltonian Formalism	281
3.9	Notes	314

Appendices

A	Algebraic Curves and Their Theta Functions in a Nutshell	324
B	Hyperelliptic Curves of the Toda-Type	353
C	Asymptotic Spectral Parameter Expansions and Nonlinear Recursion Relations	365
D	Lagrange Interpolation	385

List of Symbols 395

Bibliography 398

Index 423

Errata and Addenda for Volume I 426

Acknowledgments

It's been a hard day's night,
and I've been working like a dog.
It's been a hard day's night,
I should be sleeping like a log.

J. Lennon/P. McCartney[1]

This monograph is the second volume focusing on a certain class of solutions, namely the algebro-geometric solutions of hierarchies of soliton equations. While we studied nonlinear partial differential equations in one space and one time dimension in the first volume, with the Korteweg–de Vries (KdV) and AKNS hierarchies as the prime examples, we now discuss differential-difference equations, where the time variable is continuous, while the one-dimensional spatial variable is discretized in this second volume. The key examples treated here in great detail are the Toda and Ablowitz–Ladik lattice hierarchies.

As in the case of the previous volume, we have tried to make the presentation as detailed, explicit, and precise as possible. The text is aimed to be self-contained for graduate students with sufficient training in analysis. Ample background material is provided in the appendices. The notation is consistent with that of Volume I, whenever possible (but the present Volume II is independent of Volume I).

To a large extent this enterprise is the result of joint work with several colleagues and friends, in particular, Wolfgang Bulla and Jeff Geronimo.

The writing, and in particular the typesetting of a technical manuscript is no easy task. As was the case for Volume I we have had the great fortune to be assisted by Harald Hanche-Olsen whenever we got stuck, and we appreciate his unselfish assistance.

Parts of the manuscript have been read by Emma Previato and Maxim Zinchenko. We gratefully acknowledge their constructive comments. We are particularly indebted to Emma Previato for the comprehensive list of misprints we received and

---

[1] *A Hard Day's Night* (1964).

for her enthusiasm about this project in general. Her efforts required a considerable time commitment and we truly appreciate her help. We are also very grateful to Engui Fan for supplying us with a large number of corrections for Volume I.

The web-page with URL

    www.math.ntnu.no/~holden/solitons

contains an updated list of misprints and comments for Volume I and will include the same for this volume. Please send pertinent comments to the authors.

Our research in this area has been funded in part by the Research Council and the Office of Research of the University of Missouri, Columbia, the US National Science Foundation, the Research Council of Norway, and the Austrian Science Fund (FWF) under Grants No. P17762, Y330, and J2655.

Over the duration of this project we have enjoyed the very friendly hospitality of several institutions, including Imperial College, New York University, Institut Mittag-Leffler, University of Vienna, University of Missouri, Columbia, and the Norwegian University of Science and Technology, and we are grateful for their generous support.

July 7, 2008

Fritz Gesztesy
Department of Mathematics
University of Missouri
Columbia, MO 65211
USA
fritz@math.missouri.edu
www.math.missouri.edu/personnel/
faculty/gesztesyf.html

Helge Holden
Department of Mathematical Sciences
Norwegian University of
Science and Technology
NO-7491 Trondheim
Norway
holden@math.ntnu.no
www.math.ntnu.no/~holden/

Johanna Michor
Fakultät für Mathematik
Universität Wien
Nordbergstr. 15
1090 Wien
Austria
jmichor@esi.ac.at
www.mat.univie.ac.at/~jmichor/

Gerald Teschl
Fakultät für Mathematik
Universität Wien
Nordbergstr. 15
1090 Wien
Austria
Gerald.Teschl@univie.ac.at
www.mat.univie.ac.at/~gerald/

Introduction

> ...though I bestowed some time in writing the book, yet it cost
> me not half so much labor as this very preface.
>
> Miguel de Cervantes Saavedra[1]

Background: In the early 1950s, Fermi, Pasta, and Ulam[2] (FPU), in the unpublished report by Fermi et al. (1955), analyzed numerically, in one of the first computer simulations performed, the behavior of oscillations in certain nonlinear lattices. Expecting equipartition of energy among the various modes, they were highly surprised to discover that the energy did not equidistribute, but rather they observed that the system seemed to return periodically to its initial state. Motivated by the surprising findings in FPU, several researchers, including Ford (1961), Ford and Waters (1964), Waters and Ford (1964), Atlee Jackson et al. (1968), Payton and Visscher (1967a,b; 1968), Payton et al. (1967), studied lattice models with different nonlinear interactions, observing close to periodic and solitary behavior. It was Toda who in 1967 isolated the exponential interaction, see Toda (1967a,b) and hence introduced a model that supported an exact periodic and soliton solution. The model, now called the Toda lattice, is a nonlinear differential-difference system continuous in time and discrete in space,

$$x_{tt}(n,t) = e^{(x(n-1,t)-x(n,t))} - e^{(x(n,t)-x(n+1,t))}, \quad (n,t) \in \mathbb{Z} \times \mathbb{R}, \qquad (0.1)$$

where $x(n, t)$ denotes the displacement of the nth particle from its equilibrium position at time t. While nonlinear lattices are interesting objects of study and certainly

---

[1] *Don Quixote*, (1605) preface.
[2] "We [Fermi and Ulam]...decided to attempt to formulate a problem simple to state but such that a solution...could not be done with pencil and paper. ...Our problem turned out to be felicitously chosen. The results were entirely different from what even Fermi, with his great knowledge of wave motions, had expected. ...Fermi considered this to be, as he said, "a minor discovery." ...He intended to talk about this [at the Gibbs lecture; a lecture never given as Fermi became ill before the meeting]...", see Ulam (1991, p. 226f).

of fundamental importance in their own right, it should be mentioned that already in the paper Toda (1967b), it is also shown that the Korteweg–de Vries (KdV) equation emerges in a certain scaling, or continuum limit, from the Toda lattice, creating a link with the theory of the KdV equation. Indeed, the theory for the Toda lattice is closely intertwined with the corresponding theory for the KdV equation on several levels. Most notably, the Toda lattice shares many of the properties of the KdV equation and other completely integrable equations. This applies, in particular, to the Hamiltonian and algebro-geometric formalism treated in detail in the present monograph. While the developments for the KdV equation preceded those for the Toda lattice, in the context of algebro-geometric solutions the actual developments for the latter rapidly followed the former as described below.

Before turning to a description of the main contributors and their accomplishments in connection with the Hamiltonian and algebro-geometric formalism for the Toda lattice, we briefly recall a few milestones in the development leading up to soliton and algebro-geometric solutions of the KdV equation (for an in-depth presentation of that theory, we refer to the introduction of Volume I). In 1965, Kruskal and Zabusky (cf. Zabusky and Kruskal (1965)), while analyzing the numerical results of FPU on heat conductivity in solids, discovered that pulselike solitary wave solutions of the KdV equation, for which the name "solitons" was coined, interacted elastically. This was followed by the 1967 discovery of Gardner, Greene, Kruskal, and Miura (cf. Gardner et al. (1967; 1974)) that the inverse scattering method allowed one to solve initial value problems for the KdV equation with sufficiently fast decaying initial data. Soon after, Lax (1968) found the explanation of the isospectral nature of KdV solutions using the concept of Lax pairs and introduced a whole hierarchy of KdV equations. Subsequently, in the early 1970s, Zakharov and Shabat (1972; 1973; 1974), and Ablowitz et al. (1973a,b; 1974) extended the inverse scattering method to a wide class of nonlinear partial differential equations of relevance in various scientific contexts, ranging from nonlinear optics to condensed matter physics and elementary particle physics. In particular, soliton solutions found numerous applications in classical and quantum field theory, in connection with optical communication devices, etc.

Another decisive step forward in the development of completely integrable soliton equations was taken around 1974. Prior to that period, inverse spectral methods in the context of nonlinear evolution equations had been restricted to spatially decaying solutions to enable the applicability of inverse scattering techniques. From 1975 on, following some pioneering work of Novikov (1974), the arsenal of inverse spectral methods was extended considerably in scope to include periodic and certain classes of quasi-periodic and almost periodic KdV finite-band solutions. This new approach to constructing solutions of integrable nonlinear evolution equations, based on solutions of the inverse periodic spectral problem and on algebro-geometric methods

Introduction

and theta function representations, developed by pioneers such as Dubrovin, Its, Kac, Krichever, Marchenko, Matveev, McKean, Novikov, and van Moerbeke, to name just a few, was followed by very rapid development in the field and within a few years of intense activity worldwide, the landscape of integrable systems was changed forever. By the early 1980s the theory was extended to a large class of nonlinear (including certain multi-dimensional) evolution equations beyond the KdV equation, and the explicit theta function representation of quasi-periodic solutions of integrable equations (including soliton solutions as special limiting cases) had introduced new algebro-geometric techniques into this area of nonlinear partial differential equations. Subsequently, this led to an interesting cross-fertilization between the areas of integrable nonlinear partial differential equations and algebraic geometry, culminating, for instance, in a solution of Schottky's problem (Shiota (1986; 1990), see also Krichever (2006) and the references cited therein).

The present monograph is devoted to hierarchies of completely integrable differential-difference equations and their algebro-geometric solutions, treating, in particular, the Toda, Kac–van Moerbeke, and Ablowitz–Ladik hierarchies. For brevity we just recall the early historical development in connection with the Toda lattice and refer to the Notes for more recent literature on this topic and for the corresponding history of the Kac–van Moerbeke and Ablowitz–Ladik hierarchies. After Toda's introduction of the exponential lattice in 1967, it was Flaschka who in 1974 proved its integrability by establishing a Lax pair for it with Lax operator a tri-diagonal Jacobi operator on $\mathbb{Z}$ (a discrete Sturm–Liouville-type operator, cf. Flaschka (1974a)). He used the variable transformation

$$a(n, t) = \frac{1}{2} \exp\left(\tfrac{1}{2}(x(n, t) - x(n+1, t))\right),$$
$$b(n, t) = -\frac{1}{2} x_t(n, t), \quad (n, t) \in \mathbb{Z} \times \mathbb{R}, \tag{0.2}$$

which transforms (0.1) into a first-order system for a, b, the Toda lattice system, displayed in (0.3). Just within a few months, this was independently observed also by Manakov (1975). The corresponding integrability in the finite-dimensional periodic case had first been established by Hénon (1974) and shortly thereafter by Flaschka (1974b) (see also Flaschka (1975), Flaschka and McLaughlin (1976a), Kac and van Moerbeke (1975a), van Moerbeke (1976)). Soon after, integrability of the finite nonperiodic Toda lattice was established by Moser (1975a). Returning to the Toda lattice (0.2) on $\mathbb{Z}$, infinitely-many constants of motion (conservation laws) were derived by Flaschka (1974a) and Manakov (1975) (see also McLaughlin (1975)), moreover, the Hamiltonian formalism, Poisson brackets, etc., were also established by Manakov (1975) (see also Flaschka and McLaughlin (1976b)). The theta function representation of b in the periodic case was nearly simultaneously derived by Dubrovin et al.

(1976) and Date and Tanaka (1976a,b), following Its and Matveev (1975a,b) in their theta function derivation of the corresponding periodic finite-band KdV solution. An explicit theta function representation for a was derived a bit later by Krichever (1978) (see also Kričever (1982), Krichever (1982; 1983), and the appendix written by Krichever in Dubrovin (1981)). We also note that Dubrovin, Matveev, and Novikov as well as Date and Tanaka consider the special periodic case, but Krichever treats both the periodic and quasi-periodic cases.

Scope: We aim for an elementary, yet self-contained, and precise presentation of hierarchies of integrable soliton differential-difference equations and their algebro-geometric solutions. Our point of view is predominantly influenced by analytical methods. We hope this will make the presentation accessible and attractive to analysts working outside the traditional areas associated with soliton equations. Central to our approach is a simultaneous construction of all algebro-geometric solutions and their theta function representation of a given hierarchy. In this volume we focus on some of the key hierarchies in $(1+1)$-dimensions associated with differential-difference integrable models such as the Toda lattice hierarchy (Tl), the Kac–van Moerbeke hierarchy (KM), and the Ablowitz–Ladik hierarchy (AL). The key equations, defining the corresponding hierarchies, read[1]

$$\text{Tl:} \quad \begin{pmatrix} a_t - a(b^+ - b) \\ b_t - 2(a^2 - (a^-)^2) \end{pmatrix} = 0,$$

$$\text{KM:} \quad \rho_t - \rho\big((\rho^+)^2 - (\rho^-)^2\big) = 0, \tag{0.3}$$

$$\text{AL:} \quad \begin{pmatrix} -i\alpha_t - (1 - \alpha\beta)(\alpha^+ + \alpha^-) + 2\alpha \\ -i\beta_t + (1 - \alpha\beta)(\beta^+ + \beta^-) - 2\beta \end{pmatrix} = 0.$$

Our principal goal in this monograph is the construction of algebro-geometric solutions of the hierarchies associated with the equations listed in (0.3). Interest in the class of algebro-geometric solutions can be motivated in a variety of ways: It represents a natural extension of the classes of soliton solutions and similar to these, its elements can still be regarded as explicit solutions of the nonlinear integrable evolution equation in question (even though their complexity considerably increases compared to soliton solutions due to the underlying analysis on compact Riemann surfaces). Moreover, algebro-geometric solutions can be used to approximate more general solutions (such as almost periodic ones) although this is not a topic pursued in this monograph. Here we primarily focus on the construction of explicit solutions in terms of certain algebro-geometric data on a compact hyperelliptic Riemann surface and their representation in terms of theta functions. Solitons arise as the special case of solutions corresponding to an underlying singular hyperelliptic curve

---

[1] Here, and in the following, $\phi^\pm$ denotes the shift of a lattice function ϕ, that is, $\phi^\pm(n) = \phi(n \pm 1)$, $n \in \mathbb{Z}$.

Introduction

obtained by confluence of pairs of branch points. The theta function associated with the underlying singular curve then degenerates into appropriate determinants with exponential entries.

We use basic techniques from the theory of differential-difference equations, some spectral analysis, and elements of algebraic geometry (most notably, the basic theory of compact Riemann surfaces). In particular, we do not employ more advanced tools such as loop groups, Grassmanians, Lie algebraic considerations, formal pseudo-differential expressions, etc. Thus, this volume strays off the mainstream, but we hope it appeals to spectral theorists and their kin and convinces them of the beauty of the subject. In particular, we hope a reader interested in quickly reaching the fundamentals of the algebro-geometric approach of constructing solutions of hierarchies of completely integrable evolution equations will not be disappointed.

Completely integrable systems, and especially nonlinear evolution equations of soliton-type, are an integral part of modern mathematical and theoretical physics, with far-reaching implications from pure mathematics to the applied sciences. It is our intention to contribute to the dissemination of some of the beautiful techniques applied in this area.

Contents: In the present volume we provide an effective approach to the construction of algebro-geometric solutions of certain completely integrable nonlinear differential-difference evolution equations by developing a technique which simultaneously applies to all equations of the hierarchy in question.

Starting with a specific integrable differential-difference equation, one can build an infinite sequence of higher-order differential-difference equations, the so-called hierarchy of the original soliton equation, by developing an explicit recursive formalism that reduces the construction of the entire hierarchy to elementary manipulations with polynomials and defines the associated Lax pairs or zero-curvature equations. Using this recursive polynomial formalism, we simultaneously construct algebro-geometric solutions for the entire hierarchy of soliton equations at hand. On a more technical level, our point of departure for the construction of algebro-geometric solutions is not directly based on Baker–Akhiezer functions and axiomatizations of algebro-geometric data, but rather on a canonical meromorphic function ϕ on the underlying hyperelliptic Riemann surface $\mathcal{K}_p$ of genus $p \in \mathbb{N}_0$. More precisely, this fundamental meromorphic function ϕ carries the spectral information of the underlying Lax operator (such as the Jacobi operator in context of the Toda lattice) and in many instances represents a direct generalization of the Weyl–Titchmarsh m-function, a fundamental device in the spectral theory of difference operators. Riccati-type difference equations satisfied by ϕ separately in the discrete space and continuous time variables then govern the time evolutions of all quantities of interest (such as that of the associated Baker–Akhiezer function). The basic meromorphic function

ϕ on $\mathcal{K}_p$ is then linked with solutions of equations of the underlying hierarchy via trace formulas and Dubrovin-type equations for (projections of) the pole divisor of ϕ. Subsequently, the Riemann theta function representation of ϕ is then obtained more or less simultaneously with those of the Baker–Akhiezer function and the algebro-geometric solutions of the (stationary or time-dependent) equations of the hierarchy of evolution equations. This concisely summarizes our approach to all the $(1+1)$-dimensional discrete integrable models discussed in this volume.

In the following we will detail this verbal description of our approach to algebro-geometric solutions of integrable hierarchies with the help of the Toda hierarchy.

The Toda lattice, in Flaschka's variables, reads

$$a_t - a(b^+ - b) = 0,$$
$$b_t - 2((a^+)^2 - (a^-)^2) = 0,\qquad(0.4)$$

where $a = \{a(n,t)\}_{n\in\mathbb{Z}} \in \mathbb{C}^{\mathbb{Z}}, b = \{b(n,t)\}_{n\in\mathbb{Z}} \in \mathbb{C}^{\mathbb{Z}}, t \in \mathbb{R}$. The system (0.4) is equivalent to the Lax equation

$$L_t(t) - [P_2(t), L(t)] = 0.$$

Here L and P_2 are the difference expressions of the form

$$L = aS^+ + a^-S^- + b, \quad P_2 = aS^+ - a^-S^-,$$

and $S^\pm$ denote the shift operators

$$(S^\pm f)(n) = f(n \pm 1), \quad n \in \mathbb{Z}, \ f = \{f(m)\}_{m\in\mathbb{Z}} \in \mathbb{C}^{\mathbb{Z}},$$

with $\mathbb{C}^{\mathbb{Z}}$ abbreviating the set of complex-valued sequences indexed by $\mathbb{Z}$.

In this introduction we will indicate how to construct all real-valued algebro-geometric quasi-periodic finite-band solutions of a hierarchy of nonlinear evolution equations of which the first equation is the Toda lattice, abbreviated Tl. The approach is similar to the one advocated for the Korteweg–de Vries (KdV) and Zakharov–Shabat (ZS), or equivalently, Ablowitz–Kaup–Newell–Segur (AKNS), equations and their hierarchies in Chapters 1 and 3 of Volume I.

This means that we construct a hierarchy of difference operators P_{2p+2} such that the Lax relation

$$L_{t_p} - [P_{2p+2}, L] = 0,$$

defines a hierarchy of differential-difference equations where the time variation is continuous and space is considered discrete. We let each equation in this hierarchy run according to its own time variable t_p. The operators P_{2p+2} are defined recursively. In the stationary case, where we study

$$[P_{2p+2}, L] = 0,$$

Introduction

there is a hyperelliptic curve $\mathcal{K}_p$ of genus p which is associated with the equation in a natural way. This relation is established by introducing the analog of Burchnall–Chaundy polynomials, familiar from the KdV and ZS-AKNS theory. The basic relations for both the time-dependent and stationary Toda hierarchy as well as the construction of the Burchnall–Chaundy polynomials are contained in Section 1.2.

In Section 1.3 we discuss the stationary case in detail. We introduce the Baker–Akhiezer function ψ which is the common eigenfunction of the commuting difference operators L and P_{2p+2}. The main result of this section is the proof of theta function representations of $\phi = \psi^+/\psi$ and ψ, as well as the solutions a and b of the stationary Toda hierarchy.

In Section 1.4 we analyze the algebro-geometric initial value problem for the Toda hierarchy. By that we mean the following: Given a nonspecial Dirichlet divisor of degree p at one fixed lattice point, we explicitly construct an algebro-geometric solution, which equals the given data at the lattice point, of the qth stationary Toda lattice, $q \in \mathbb{N}$.

Section 1.5 parallels that of Section 1.3, but it discusses the time-dependent case. The goal of the section is to construct the solution of the rth equation in the Toda hierarchy with a given stationary solution of the pth equation in the Toda hierarchy as initial data. We construct the solution in terms of theta functions.

Section 1.6 treats the algebro-geometric time-dependent initial value problem for the Toda hierarchy. Given a stationary solution of an arbitrary equation in the Toda hierarchy and its associated nonsingular hyperelliptic curve as initial data, we construct explicitly the solution of any other time-dependent equation in the Toda hierarchy with the given stationary solution as initial data.

Finally, in Section 1.7 we construct an infinite sequence of local conservation laws for each of the equations in the Toda hierarchy. Moreover, we derive two Hamiltonian structures for the Toda hierarchy.

We now return to a more detailed survey of the results in this monograph for the Toda hierarchy. The Toda hierarchy is the simplest of the hierarchies of nonlinear differential-difference evolution equations studied in this volume, but the same strategy, with modifications to be discussed in the individual chapters, applies to the integrable systems treated in this monograph and is in fact typical for all $(1 + 1)$-dimensional integrable differential-difference hierarchies of soliton equations.

A discussion of the Toda case then proceeds as follows.[1] In order to define the Lax pairs and zero-curvature pairs for the Toda hierarchy, one assumes a, b to be bounded sequences in the stationary context and smooth functions in the time variable in the time-dependent case. Next, one introduces the recursion relation for some

---

[1] All details of the following construction are to be found in Chapter 1.

polynomial functions f_ℓ, g_ℓ of a, b and certain of its shifts by

$$f_0 = 1, \quad g_0 = -c_1,$$
$$2f_{\ell+1} + g_\ell + g_\ell^- - 2bf_\ell = 0, \quad \ell \in \mathbb{N}_0, \qquad (0.5)$$
$$g_{\ell+1} - g_{\ell+1}^- + 2(a^2 f_\ell^+ - (a^-)^2 f_\ell^-) - b(g_\ell - g_\ell^-) = 0, \quad \ell \in \mathbb{N}_0.$$

Here c_1 is a given constant. From the recursively defined sequences $\{f_\ell, g_\ell\}_{\ell \in \mathbb{N}_0}$ (whose elements turn out to be difference polynomials with respect to a, b, defined up to certain summation constants) one defines the *Lax pair* of the Toda hierarchy by

$$L = aS^+ + a^- S^- + b, \qquad (0.6)$$

$$P_{2p+2} = -L^{p+1} + \sum_{\ell=0}^{p}(g_{p-\ell} + 2af_{p-\ell}S^+)L^\ell + f_{p+1}. \qquad (0.7)$$

The commutator of P_{2p+2} and L then reads[1]

$$\begin{aligned}[P_{2p+2}, L] = &-a(g_p^+ + g_p + f_{p+1}^+ + f_{p+1} - 2b^+ f_p^+)S^+ \\ &+ 2(-b(g_p + f_{p+1}) + a^2 f_p^+ - (a^-)^2 f_p^- + b^2 f_p) \\ &- a^-(g_p + g_p^- + f_{p+1} + f_{p+1}^- - 2bf_p)S^-,\end{aligned} \qquad (0.8)$$

using the recursion (0.5). Introducing a deformation (time) parameter[2] $t_p \in \mathbb{R}$, $p \in \mathbb{N}_0$, into a, b, the *Toda hierarchy* of nonlinear evolution equations is then defined by imposing the *Lax commutator relation*

$$\frac{d}{dt_p}L - [P_{2p+2}, L] = 0, \qquad (0.9)$$

for each $p \in \mathbb{N}_0$. By (0.8), the latter are equivalent to the collection of evolution equations[3]

$$\mathrm{Tl}_p(a, b) = \begin{pmatrix} a_{t_p} - a(f_{p+1}^+(a, b) - f_{p+1}(a, b)) \\ b_{t_p} + g_{p+1}(a, b) - g_{p+1}^-(a, b) \end{pmatrix} = 0, \quad p \in \mathbb{N}_0. \qquad (0.10)$$

---

[1] The quantities P_{2p+2} and $\{f_\ell, g_\ell\}_{\ell=0,\ldots,p}$ are constructed such that all higher-order difference operators in the commutator (0.8) vanish. Observe that the factors multiplying $S^\pm$ are just shifts of one another.

[2] Here we follow Hirota's notation and introduce a separate time variable t_p for the pth level in the Toda hierarchy.

[3] In a slight abuse of notation we will occasionally stress the functional dependence of f_ℓ, g_ℓ on a, b, writing $f_\ell(a, b), g_\ell(a, b)$.

Explicitly,

$$\mathrm{Tl}_0(a,b) = \begin{pmatrix} a_{t_0} - a(b^+ - b) \\ b_{t_0} - 2(a^2 - (a^-)^2) \end{pmatrix} = 0,$$

$$\mathrm{Tl}_1(a,b) = \begin{pmatrix} a_{t_1} - a((a^+)^2 - (a^-)^2 + (b^+)^2 - b^2) \\ b_{t_1} + 2(a^-)^2(b + b^-) - 2a^2(b^+ + b) \end{pmatrix}$$
$$+ c_1 \begin{pmatrix} -a(b^+ - b) \\ -2(a^2 - (a^-)^2) \end{pmatrix} = 0,$$

$$\mathrm{Tl}_2(a,b) = \begin{pmatrix} a_{t_2} - a((b^+)^3 - b^3 + 2(a^+)^2 b^+ - 2(a^-)^2 b \\ \quad + a^2(b^+ - b) + (a^+)^2 b^{++} + (a^-)^2 b^-) \\ b_{t_2} - 2a^2(b^2 + bb^+ + (b^+)^2 + a^2 + (a^+)^2) \\ \quad + 2(a^-)^2(b^2 + bb^- + (b^-)^2 + (a^-)^2 + (a^{--})^2) \end{pmatrix}$$
$$+ c_1 \begin{pmatrix} -a((a^+)^2 - (a^-)^2 + (b^+)^2 - b^2) \\ 2(a^-)^2(b + b^-) - 2a^2(b^+ + b) \end{pmatrix}$$
$$+ c_2 \begin{pmatrix} -a(b^+ - b) \\ -2(a^2 - (a^-)^2) \end{pmatrix} = 0, \text{ etc.},$$

represent the first few equations of the time-dependent Toda hierarchy. For $p = 0$ one obtains *the* Toda lattice (0.4). Introducing the polynomials ($z \in \mathbb{C}$),

$$F_p(z) = \sum_{\ell=0}^{p} f_{p-\ell} z^\ell, \tag{0.11}$$

$$G_{p+1}(z) = -z^{p+1} + \sum_{\ell=0}^{p} g_{p-\ell} z^\ell + f_{p+1}, \tag{0.12}$$

one can alternatively introduce the Toda hierarchy as follows. One defines a pair of 2×2 matrices $(U(z), V_{p+1}(z))$ depending polynomially on z by

$$U(z) = \begin{pmatrix} 0 & 1 \\ -a^-/a & (z-b)/a \end{pmatrix}, \tag{0.13}$$

$$V_{p+1}(z) = \begin{pmatrix} G_{p+1}^-(z) & 2a^- F_p^-(z) \\ -2a^- F_p(z) & 2(z-b)F_p + G_{p+1}(z) \end{pmatrix}, \quad p \in \mathbb{N}_0, \tag{0.14}$$

and then postulates the discrete *zero-curvature equation*

$$0 = U_{t_p} + U V_{p+1} - V_{p+1}^+ U. \tag{0.15}$$

One verifies that both the Lax approach (0.10), as well as the zero-curvature approach

(0.15), reduce to the basic equations,

$$a_{t_p} = -a\big(2(z-b^+)F_p^+ + G_{p+1}^+ + G_{p+1}\big),$$
$$b_{t_p} = 2\big((z-b)^2 F_p + (z-b)G_{p+1} + a^2 F_p^+ - (a^-)^2 F_p^-\big). \tag{0.16}$$

Each one of (0.10), (0.15), and (0.16) defines the Toda hierarchy by varying $p \in \mathbb{N}_0$.

The strategy we will be using is then the following: First we assume the existence of a solution a, b, and derive several of its properties. In particular, we deduce explicit Riemann's theta function formulas for the solution a, b, the so-called Its–Matveev formulas (cf. (0.41) in the stationary case and (0.53) in the time-dependent case). As a second step we will provide an explicit algorithm to construct the solution given appropriate initial data.

The Lax and zero-curvature equations (0.9) and (0.15) imply a most remarkable isospectral deformation of L as will be discussed later in this introduction. At this point, however, we interrupt our time-dependent Toda considerations for a while and take a closer look at the special stationary Toda equations defined by

$$a_{t_p} = b_{t_p} = 0, \quad p \in \mathbb{N}_0. \tag{0.17}$$

By (0.8)–(0.10) and (0.15), (0.16), the condition (0.17) is then equivalent to each one of the following collection of equations, with p ranging in $\mathbb{N}_0$, defining the *stationary Toda hierarchy* in several ways,

$$[P_{2p+2}, L] = 0, \tag{0.18}$$
$$f_{p+1}^+ - f_{p+1} = 0, \quad g_{p+1} - g_{p+1}^- = 0, \tag{0.19}$$
$$UV_{p+1} - V_{p+1}^+ U = 0, \tag{0.20}$$
$$2(z-b^+)F_p^+ + G_{p+1}^+ + G_{p+1} = 0,$$
$$(z-b)^2 F_p + (z-b)G_{p+1} + a^2 F_p^+ - (a^-)^2 F_p^- = 0. \tag{0.21}$$

To set the stationary Toda hierarchy apart from the general time-dependent one, we will denote it by

$$\text{s-Tl}_p(a,b) = \begin{pmatrix} f_{p+1}^+(a,b) - f_{p+1}(a,b) \\ g_{p+1}(a,b) - g_{p+1}^-(a,b) \end{pmatrix} = 0, \quad p \in \mathbb{N}_0.$$

Explicitly, the first few equations of the stationary Toda hierarchy then read as follows

$$\text{s-Tl}_0(a,b) = \begin{pmatrix} b^+ - b \\ 2((a^-)^2 - a^2) \end{pmatrix} = 0,$$

$$\text{s-Tl}_1(a,b) = \begin{pmatrix} (a^+)^2 - (a^-)^2 + (b^+)^2 - b^2 \\ 2(a^-)^2(b+b^-) - 2a^2(b^+ + b) \end{pmatrix}$$

$$+ c_1 \begin{pmatrix} b^+ - b \\ 2((a^-)^2 - a^2) \end{pmatrix} = 0,$$

$$\text{s-Tl}_2(a,b) = \begin{pmatrix} (b^+)^3 - b^3 + 2(a^+)^2 b^+ - 2(a^-)^2 b \\ + a^2(b^+ - b) + (a^+)^2 b^{++} + (a^-)^2 b^- \\ 2(a^-)^2(b^2 + bb^- + (b^-)^2 + (a^-)^2 + (a^{--})^2) \\ -2a^2(b^2 + bb^+ + (b^+)^2 + a^2 + (a^+)^2) \end{pmatrix}$$

$$+ c_1 \begin{pmatrix} (a^+)^2 - (a^-)^2 + (b^+)^2 - b^2 \\ 2(a^-)^2(b+b^-) - 2a^2(b^+ + b) \end{pmatrix}$$

$$+ c_2 \begin{pmatrix} b^+ - b \\ 2((a^-)^2 - a^2) \end{pmatrix} = 0, \text{ etc.}$$

The class of *algebro-geometric* Toda potentials, by definition, equals the set of solutions a, b of the stationary Toda hierarchy. In the following analysis we fix the value of a, b in (0.18)–(0.21), and hence we now turn to the investigation of algebro-geometric solutions a, b of the pth equation within the stationary Toda hierarchy. Equation (0.18) is of special interest since by the discrete analog of a 1923 result of Burchnall and Chaundy, proven by Naiman in 1962, commuting difference expressions (due to a common eigenfunction, to be discussed below, cf. (0.34), (0.35)) give rise to an algebraic relationship between the two difference expressions. Similarly, (0.20) permits the important conclusion that

$$\det(yI_2 - V_{p+1}(z, n)) = \det(yI_2 - V_{p+1}(z, n+1)), \tag{0.22}$$

(with I_2 the identity matrix in $\mathbb{C}^2$) and hence

$$\begin{aligned} \det(yI_2 - V_{p+1}(z, n)) &= y^2 + \det(V_{p+1}(z, n)) \\ &= y^2 - G_{p+1}^-(z, n)^2 + 4a^-(n)^2 F_p^-(z, n) F_p(z, n) \\ &= y^2 - R_{2p+2}(z), \end{aligned} \tag{0.23}$$

for some n-independent monic polynomial R_{2p+2}, which we write as

$$R_{2p+2}(z) = \prod_{m=0}^{2p+1} (z - E_m) \text{ for some } \{E_m\}_{m=0,\ldots,2p+1} \subset \mathbb{C}.$$

In particular, the combination

$$G_{p+1}(z, n)^2 - 4a(n)^2 F_p(z, n) F_p^+(z, n) = R_{2p+2}(z) \tag{0.24}$$

is n-independent. Moreover, one can rewrite (0.21) to yield

$$\begin{aligned}
(z-b)^4 F_p^2 - 2a^2(z-b)^2 F_p F_p^+ &- 2(a^-)^2(z-b)^2 F_p F_p^- + a^4 (F_p^+)^2 \\
+ (a^-)^4 (F_p^-)^2 &- 2a^2(a^-)^2 F_p^+ F_p^- = (z-b)^2 R_{2p+2}(z), \\
(z-b)(z-b^+) G_{p+1}^2 &- a^2 (G_{p+1}^- + G_{p+1})(G_{p+1} + G_{p+1}^+) \\
&= (z-b)(z-b^+) R_{2p+2}(z),
\end{aligned} \quad (0.25)$$

with precisely the same integration constant $R_{2p+2}(z)$ as in (0.23). In fact, by (0.11) and (0.12), equations (0.24) and (0.25) are simply identical. Incidentally, the algebraic relationship between L and P_{2p+2} alluded to in connection with the vanishing of their commutator in (0.18) can be made precise as follows: Restricting P_{2p+2} to the (algebraic) kernel $\ker(L-z)$ of $L-z$, one computes, using (0.7) and (0.25),

$$\begin{aligned}
\left(P_{2p+2}\big|_{\ker(L-z)}\right)^2 &= \left((2aF_p S^+ + G_{p+1})\big|_{\ker(L-z)}\right)^2 \\
&= \left(2aF_p(G_{p+1}^+ + G_{p+1} + 2(z-b^+) F_p^+) S^+ \right. \\
&\quad \left. + G_{p+1}^2 - 4a^2 F_p F_p^+\right)\big|_{\ker(L-z)} \\
&= (G_{p+1}^2 - 4a^2 F_p F_p^+)\big|_{\ker(L-z)} = R_{2p+2}(L)\big|_{\ker(L-z)}.
\end{aligned}$$

Thus, P_{2p+2}^2 and $R_{2p+2}(L)$ coincide on the finite-dimensional nullspace of $L-z$. Since $z \in \mathbb{C}$ is arbitrary, one infers that

$$P_{2p+2}^2 - R_{2p+2}(L) = 0 \quad (0.26)$$

holds once again with the same polynomial R_{2p+2}. The characteristic equation of V_{p+1} (cf. (0.23)) and (0.26) naturally leads one to the introduction of the *hyperelliptic curve* $\mathcal{K}_p$ of genus $p \in \mathbb{N}_0$ defined by

$$\mathcal{K}_p: \mathcal{F}_p(z, y) = y^2 - R_{2p+2}(z) = 0, \quad R_{2p+2}(z) = \prod_{m=0}^{2p+1}(z - E_m). \quad (0.27)$$

One compactifies the curve by adding two distinct points $P_{\infty_-}, P_{\infty_+}$ (still denoting the curve by $\mathcal{K}_p$ for simplicity) and notes that points $P \neq P_{\infty_\pm}$ on the curve are denoted by $P = (z, y) \in \mathcal{K}_p \setminus \{P_{\infty_-}, P_{\infty_+}\}$, where $y(\cdot)$ is the meromorphic function on $\mathcal{K}_p$ satisfying[1] $y^2 - R_{2p+2}(z) = 0$. For simplicity, we will assume in the following that the (affine part of the) curve $\mathcal{K}_p$ is nonsingular, that is, the zeros E_m of R_{2p+2} are all simple. Remaining within the stationary framework a bit longer,

---

[1] For more details we refer to Appendix B and Chapter 1.

Introduction

one can now introduce the fundamental meromorphic function ϕ on $\mathcal{K}_p$ alluded to earlier, as follows,

$$\phi(P, n) = \frac{y - G_{p+1}(z, n)}{2a(n)F_p(z, n)} \tag{0.28}$$

$$= \frac{-2a(n)F_p^+(z, n)}{y + G_{p+1}(z, n)}, \quad P = (z, y) \in \mathcal{K}_p. \tag{0.29}$$

(We mention in passing that via (C.9) and (C.17), the two branches $\phi_\pm$ of ϕ are directly connected with the diagonal Green's function of the Lax operator L.) Equality of the two expressions (0.28) and (0.29) is an immediate consequence of the identity (0.24) and the fact $y^2 = R_{2p+2}(z)$. A comparison with (0.20) then readily reveals that ϕ satisfies the Riccati-type equation

$$a\phi(P) + a^-\phi^-(P)^{-1} = z - b. \tag{0.30}$$

The next step is crucial. It concerns the zeros and poles of ϕ and hence involves the zeros of $F_p(\,\cdot\,, n)$. Isolating the latter by introducing the factorization

$$F_p(z, n) = \prod_{j=1}^{p}(z - \mu_j(n)),$$

one can use the zeros of F_p and F_p^+ to define the following points $\hat{\mu}_j(n)$ and $\hat{\mu}_j^+(n)$ on $\mathcal{K}_p$,

$$\hat{\mu}_j(n) = (\mu_j(n), -G_{p+1}(\mu_j(n), n)), \quad j = 1, \ldots, p, \tag{0.31}$$

$$\hat{\mu}_j^+(n) = (\mu_j^+(n), G_{p+1}(\mu_j^+(n), n)), \quad j = 1, \ldots, p, \tag{0.32}$$

where μ_j^+, $j = 1, \ldots, p$, denote the zeros of F_p^+. The motivation for this choice stems from $y^2 = R_{2p+2}(z)$ by (0.23), the identity (0.24) (which combines to $G_{p+1}^2 - 4a^2 F_p F_p^+ = y^2$), and a comparison of (0.28) and (0.29). Given (0.28)–(0.32) one obtains for the divisor $(\phi(\,\cdot\,, n))$ of the meromorphic function ϕ,

$$(\phi(\,\cdot\,, n)) = \mathcal{D}_{P_{\infty+}\underline{\hat{\mu}}^+(n)} - \mathcal{D}_{P_{\infty-}\underline{\hat{\mu}}(n)}. \tag{0.33}$$

Here we abbreviated $\underline{\hat{\mu}} = \{\hat{\mu}_1, \ldots, \hat{\mu}_p\}$, $\underline{\hat{\mu}}^+ = \{\hat{\mu}_1^+, \ldots, \hat{\mu}_p^+\} \in \mathrm{Sym}^p(\mathcal{K}_p)$, with $\mathrm{Sym}^p(\mathcal{K}_p)$ the pth symmetric product of $\mathcal{K}_p$, and used our conventions[1] (A.39), (A.43), and (A.44) to denote positive divisors of degree p and $p+1$ on $\mathcal{K}_p$. Given

---

[1] $\mathcal{D}_{\underline{Q}}(P) = m$ if P occurs m times in $\{Q_1, \ldots, Q_p\}$ and zero otherwise, $\underline{Q} = \{Q_1, \ldots, Q_p\} \in \mathrm{Sym}^p(\mathcal{K}_p)$. Similarly, $\mathcal{D}_{Q_0\underline{Q}} = \mathcal{D}_{Q_0} + \mathcal{D}_{\underline{Q}}$, $\mathcal{D}_{\underline{Q}} = \mathcal{D}_{Q_1} + \cdots + \mathcal{D}_{Q_p}$, $Q_0 \in \mathcal{K}_p$, and $\mathcal{D}_Q(P) = 1$ for $P = Q$ and zero otherwise.

$\phi(\,\cdot\,, n)$ one defines the *stationary Baker–Akhiezer function* $\psi(\,\cdot\,, n, n_0)$ on $\mathcal{K}_p \setminus \{P_{\infty_\pm}\}$ by

$$\psi(P, n, n_0) = \begin{cases} \prod_{n'=n_0}^{n-1} \phi(P, n'), & n \geq n_0 + 1, \\ 1, & n = n_0, \\ \prod_{n'=n}^{n_0-1} \phi(P, n')^{-1}, & n \leq n_0 - 1. \end{cases}$$

In particular, this implies

$$\phi = \psi^+/\psi,$$

and the following normalization[1] of ψ, $\psi(P, n_0, n_0) = 1$, $P \in \mathcal{K}_p \setminus \{P_{\infty_\pm}\}$. The Riccati-type equation (0.30) satisfied by ϕ then shows that the Baker–Akhiezer function ψ is the common formal eigenfunction of the commuting pair of Lax difference expressions L and P_{2p+2},

$$L\psi(P) = z\psi(P), \tag{0.34}$$
$$P_{2p+2}\psi(P) = y\psi(P), \quad P = (z, y) \in \mathcal{K}_p \setminus \{P_{\infty_\pm}\}, \tag{0.35}$$

and at the same time the Baker–Akhiezer vector Ψ defined by

$$\Psi(P) = \begin{pmatrix} \psi(P) \\ \psi^+(P) \end{pmatrix}, \quad P \in \mathcal{K}_p \setminus \{P_{\infty_\pm}\}, \tag{0.36}$$

satisfies the zero-curvature equations,

$$\Psi(P) = U(z)\Psi^-(P), \tag{0.37}$$
$$y\Psi^-(P) = V_{p+1}(z)\Psi^-(P), \quad P = (z, y) \in \mathcal{K}_p \setminus \{P_{\infty_\pm}\}. \tag{0.38}$$

Moreover, one easily verifies that away from the branch points $(E_m, 0)$, $m = 0, \ldots, 2p+1$, of the two-sheeted Riemann surface $\mathcal{K}_p$, the two branches of ψ constitute a fundamental system of solutions of (0.34) and similarly, the two branches of ψ yield a fundamental system of solutions of (0.37). Since $\psi(\,\cdot\,, n, n_0)$ vanishes at $\hat{\mu}_j(n)$, $j = 1, \ldots, p$, and $\psi^+(\,\cdot\,, n, n_0)$ vanishes at $\hat{\mu}_j^+(n)$, $j = 1, \ldots, p$, we may call $\{\hat{\mu}_j(n)\}_{j=1,\ldots,p}$ and $\{\hat{\mu}_j^+(n)\}_{j=1,\ldots,p}$ the *Dirichlet* and *Neumann data* of L at the point $n \in \mathbb{Z}$, respectively.

Now the stationary formalism is almost complete; we only need to relate the solution a, b of the pth stationary Toda equation and $\mathcal{K}_p$-associated data. This can be accomplished as follows.

---

[1] This normalization is less innocent than it might appear at first sight. It implies that $\mathcal{D}_{\hat{\underline{\mu}}(n)}$ and $\mathcal{D}_{\hat{\underline{\mu}}(n_0)}$ are the divisors of zeros and poles of $\psi(\,\cdot\,, n, n_0)$ on $\mathcal{K}_p \setminus \{P_{\infty_\pm}\}$.

Introduction

First we relate a, b and the zeros μ_j of F_p. This is easily done by comparing the coefficients of the power z^{2p} in (0.25) and results in the *trace formulas*,[1]

$$a^2 = \frac{1}{2} \sum_{j=1}^{p} y(\hat{\mu}_j) \prod_{\substack{k=1 \\ k \neq j}}^{p} (\mu_j - \mu_k)^{-1} + \frac{1}{4}(b^{(2)} - b^2),$$

$$b = \frac{1}{2} \sum_{m=0}^{2p+1} E_m - \sum_{j=1}^{p} \mu_j,$$
(0.39)

where $b^{(2)} = \frac{1}{2} \sum_{m=0}^{2p+1} E_m^2 - \sum_{j=1}^{p} \mu_j^2$. However, the formula for a^2 is not useful for the algebro-geometric initial value problem as the quantities μ_j indeed may collide.[2] A more elaborate reconstruction algorithm, as described below, is required.

We will now indicate how to reconstruct a, b from $\mathcal{K}_p$ and given Dirichlet data at just one fixed point n_0. Due to the discrete spatial variation, this is considerably more involved than, say, for the KdV equation. Consider first the simplest case of self-adjoint Jacobi operators where a and b are real-valued and bounded sequences. In that case we are given Dirichlet divisors $\mathcal{D}_{\hat{\mu}(n_0)} \in \operatorname{Sym}^p(\mathcal{K}_p)$ with corresponding Dirichlet eigenvalues in appropriate spectral gaps of L (more precisely, in appropriate spectral gaps of a bounded operator realization of L in $\ell^2(\mathbb{Z})$, but for simplicity this aspect will be ignored in the introduction). Next one develops an algorithm that provides finite nonspecial divisors $\mathcal{D}_{\hat{\mu}(n)} \in \operatorname{Sym}^p(\mathcal{K}_p)$ in real position for all $n \in \mathbb{Z}$. In the self-adjoint case, the Dirichlet eigenvalues remain in distinct spectral gaps, and hence the expression for (0.39) for a^2 remains meaningful.

The self-adjoint situation is in sharp contrast to the general non-self-adjoint case in which the Dirichlet eigenvalues no longer are confined to distinct spectral gaps on the real axis. Moreover, Dirichlet eigenvalues are not necessarily separated and hence might coincide (i.e., collide) at particular lattice points. In addition, they may not remain finite and hit P_{∞_+} or P_{∞_-}. The algorithm has to take that into consideration; it is handled by further restricting the permissible set of initial data, which, however, remains a dense set of full measure even in this more involved setting. A key element

---

[1] Observe that only a^2 enters, and thus the sign of a is left undetermined.
[2] In the continuous case, e.g., for the Korteweg–de Vries equation, the situation is considerably simpler: The spatial variation of the μ_j, $j = 1, \ldots, p$, is determined by the Dubrovin equations, a first-order system of ordinary differential equations. Assuming the μ_j, $j = 1, \ldots, p$, are distinct at a given spatial point, there exists a small neighborhood around that point for which they remain distinct. This has no analog in the discrete case.

in the construction is the discrete dynamical system[1]

$$\underline{\alpha}_{Q_0}(\mathcal{D}_{\underline{\hat{\mu}}(n)}) = \underline{\alpha}_{Q_0}(\mathcal{D}_{\underline{\hat{\mu}}(n_0)}) - (n-n_0)\underline{A}_{P_{\infty_-}}(P_{\infty_+}),$$
$$\underline{\hat{\mu}}(n_0) = \{\hat{\mu}_1(n_0), \ldots, \hat{\mu}_p(n_0)\} \in \mathrm{Sym}^p(\mathcal{K}_p),$$

where $Q_0 \in \mathcal{K}_p \setminus \{P_{\infty_\pm}\}$ is a given base point. Starting from a nonspecial finite initial divisor $\mathcal{D}_{\underline{\hat{\mu}}(n_0)}$, we find that as n increases, $\mathcal{D}_{\underline{\hat{\mu}}(n)}$ stays nonspecial as long as it remains finite. If it becomes infinite, then it is still nonspecial and contains P_{∞_+} at least once (but not P_{∞_-}). Further increasing n, all instances of P_{∞_+} will be rendered into P_{∞_-} step by step, until we have again a nonspecial divisor that has the same number of P_{∞_-} as the first infinite one had P_{∞_+}. Generically, one expects the subsequent divisor to be finite and nonspecial again. A central part of the algorithm is to prove that for a full set of initial data, the iterates stay away from $P_{\infty_\pm}$. Summarizing, we solve the following inverse problem: Given $\mathcal{K}_p$ and appropriate initial data

$$\underline{\hat{\mu}}(n_0) = \{\hat{\mu}_1(n_0), \ldots, \hat{\mu}_p(n_0)\} \in \mathcal{M}_0,$$
$$\hat{\mu}_j(n_0) = (\mu_j(n_0), -G_{p+1}(\mu_j(n_0), n_0)), \quad j = 1, \ldots, p,$$

where $\mathcal{M}_0 \subset \mathrm{Sym}^p(\mathcal{K}_p)$ is the set of nonspecial Dirichlet divisors, we develop an algorithm that defines finite nonspecial divisors $\underline{\hat{\mu}}(n)$ for all $n \in \mathbb{Z}$.

Having constructed $\mu_j(n)$, $j = 1, \ldots, p$, $n \in \mathbb{Z}$, using an elaborate twelve-step procedure, one finds that the quantities a and b are given by

$$a(n)^2 = \frac{1}{2} \sum_{k=1}^{q(n)} \frac{(d^{p_k(n)-1} y(P)/d\zeta^{p_k(n)-1})|_{P=(\zeta,\eta)=\hat{\mu}_k(n)}}{(p_k(n)-1)!}$$
$$\times \prod_{k'=1,\, k' \neq k}^{q(n)} (\mu_k(n) - \mu_{k'}(n))^{-p_k(n)} + \frac{1}{4}(b^{(2)}(n) - b(n)^2), \quad (0.40)$$
$$b(n) = \frac{1}{2} \sum_{m=0}^{2p+1} E_m - \sum_{k=1}^{q(n)} p_k(n)\mu_k(n), \quad n \in \mathbb{Z},$$

where $p_k(n)$ are associated with degeneracies of the μ_j, $j = 1, \ldots, p$, and $\sum_{k=1}^{q(n)} p_k(n) = p$, see Theorem 1.32. We stress the resemblance between (0.40) and (0.39). Formulas (0.40) then yield a solution a, b of the pth stationary Toda equation.

An alternative reconstruction of a, b, nicely complementing the one just discussed, can be given with the help of the *Riemann theta function*[2] associated with

---

[1] Here $\underline{\alpha}_{Q_0}$ and $\underline{A}_{Q_0}$ denote Abel maps, see (A.30) and (A.29), respectively.
[2] For details on the p-dimensional theta function $\theta(\underline{z})$, $\underline{z} \in \mathbb{C}^p$, we refer to Appendices A and B.

Introduction

$\mathcal{K}_p$ and an appropriate homology basis of cycles on it. The known zeros and poles of ϕ (cf. (0.33)), and similarly, the set of zeros $\{P_{\infty_+}\} \cup \{\hat{\mu}_j(n)\}_{j=1,\ldots,p}$ and poles $\{P_{\infty_-}\} \cup \{\hat{\mu}_j(n_0)\}_{j=1,\ldots,p}$ of the Baker–Akhiezer function $\psi(\,\cdot\,, n, n_0)$, then permit one to find theta function representations for ϕ and ψ by referring to Riemann's vanishing theorem and the Riemann–Roch theorem. The corresponding theta function representation of the algebro-geometric solution a, b of the pth stationary Toda equation then can be obtained from that of ψ by an asymptotic expansion with respect to the spectral parameter near the point P_{∞_+}. The resulting final expression for a, b, the analog of the *Its–Matveev formula* in the KdV context, is of the type

$$a(n)^2 = \tilde{a}^2 \frac{\theta(\underline{A} - \underline{B} + \underline{B}n)\theta(\underline{A} + \underline{B} + \underline{B}n)}{\theta(\underline{A} + \underline{B}n)^2},$$

$$b(n) = \frac{1}{2}\sum_{m=0}^{2p+1} E_m - \sum_{j=1}^{p} \lambda_j \tag{0.41}$$

$$- \sum_{j=1}^{p} c_j(p) \frac{\partial}{\partial w_j} \ln\left(\frac{\theta(\underline{A} + \underline{B}n + \underline{w})}{\theta(\underline{A} - \underline{B} + \underline{B}n + \underline{w})}\right)\bigg|_{\underline{w}=0}.$$

Here the constants $\tilde{a}, \lambda_j, c_j(p) \in \mathbb{C}$, $j = 1, \ldots, p$, and the constant vector $\underline{B} \in \mathbb{C}^p$ are uniquely determined by $\mathcal{K}_p$ (and its homology basis), and the constant vector $\underline{A} \in \mathbb{C}^p$ is in one-to-one correspondence with the Dirichlet data $\underline{\hat{\mu}}(n_0) = (\hat{\mu}_1(n_0), \ldots, \hat{\mu}_p(n_0)) \in \mathrm{Sym}^p(\mathcal{K}_p)$ at the initial point n_0 as long as the divisor $\mathcal{D}_{\underline{\hat{\mu}}(n_0)}$ is assumed to be nonspecial.[1] Moreover, the theta function representation (0.41) remains valid as long as the divisor $\mathcal{D}_{\underline{\hat{\mu}}(n)}$ stays nonspecial. We emphasize the remarkable fact that the argument of the theta functions in (0.41) is linear with respect to n.

This completes our somewhat lengthy excursion into the stationary Toda hierarchy. In the following we return to the time-dependent Toda hierarchy and describe the analogous steps involved to construct solutions $a = a(n, t_r)$, $b = b(n, t_r)$ of the rth Toda equation with initial values being algebro-geometric solutions of the pth stationary Toda equation. More precisely, we are seeking a solution a, b of

$$\widetilde{\mathrm{Tl}}_r(a, b) = \begin{pmatrix} a_{t_r} - a\big(\tilde{f}^+_{p+1}(a, b) - \tilde{f}^-_{p+1}(a, b)\big) \\ b_{t_r} + \tilde{g}^-_{p+1}(a, b) - \tilde{g}^+_{p+1}(a, b) \end{pmatrix} = 0, \tag{0.42}$$

$$(a, b)\big|_{t_r=t_{0,r}} = (a^{(0)}, b^{(0)}),$$

$$\mathrm{s\text{-}Tl}_p\big(a^{(0)}, b^{(0)}\big) = \begin{pmatrix} f^+_{p+1}\big(a^{(0)}, b^{(0)}\big) - f^-_{p+1}\big(a^{(0)}, b^{(0)}\big) \\ g^+_{p+1}\big(a^{(0)}, b^{(0)}\big) - g^-_{p+1}\big(a^{(0)}, b^{(0)}\big) \end{pmatrix} = 0 \tag{0.43}$$

---

[1] If $\mathcal{D} = n_1 \mathcal{D}_{Q_1} + \cdots + n_k \mathcal{D}_{Q_k} \in \mathrm{Sym}^p(\mathcal{K}_p)$ for some $n_\ell \in \mathbb{N}$, $\ell = 1, \ldots, k$, with $n_1 + \cdots + n_k = p$, then $\mathcal{D}$ is called nonspecial if there is no nonconstant meromorphic function on $\mathcal{K}_p$ which is holomorphic on $\mathcal{K}_p \setminus \{Q_1, \ldots, Q_k\}$ with poles at most of order n_ℓ at Q_ℓ, $\ell = 1, \ldots, k$.

for some $t_{0,r} \in \mathbb{R}$, $p, r \in \mathbb{N}_0$, and a prescribed curve $\mathcal{K}_p$ associated with the stationary solution $a^{(0)}, b^{(0)}$ in (0.43).

We pause for a moment to reflect on the pair of equations (0.42), (0.43): As it turns out, it represents a dynamical system on the set of algebro-geometric solutions isospectral to the initial value $a^{(0)}, b^{(0)}$. By isospectral we here allude to the fact that for any fixed t_r, the solution $a(\,\cdot\,, t_r), b(\,\cdot\,, t_r)$ of (0.42), (0.43) is a stationary solution of (0.43),

$$\text{s-Tl}_p\bigl(a(\,\cdot\,,t_r), b(\,\cdot\,,t_r)\bigr)$$
$$= \begin{pmatrix} f_{p+1}^+(a(\,\cdot\,,t_r), b(\,\cdot\,,t_r)) - f_{p+1}^-(a(\,\cdot\,,t_r), b(\,\cdot\,,t_r)) \\ g_{p+1}^+(a(\,\cdot\,,t_r), b(\,\cdot\,,t_r)) - g_{p+1}^-(a(\,\cdot\,,t_r), b(\,\cdot\,,t_r)) \end{pmatrix} = 0$$

associated with the fixed underlying algebraic curve $\mathcal{K}_p$ (the latter being independent of t_r). Put differently, $a(\,\cdot\,, t_r), b(\,\cdot\,, t_r)$ is an isospectral deformation of $a^{(0)}, b^{(0)}$ with t_r the corresponding deformation parameter. In particular, $a(\,\cdot\,, t_r), b(\,\cdot\,, t_r)$ traces out a curve in the set of algebro-geometric solutions isospectral to $a^{(0)}, b^{(0)}$.

Since the summation constants in the functionals f_ℓ of a, b in the stationary and time-dependent contexts are independent of each other, we indicate this by adding a tilde on all the time-dependent quantities. Hence we shall employ the notation $\widetilde{P}_{2r+2}, \widetilde{V}_{r+1}, \widetilde{F}_r$, etc., in order to distinguish them from P_{2p+2}, V_{p+1}, F_p, etc. Thus $\widetilde{P}_{2r+2}, \widetilde{V}_{r+1}, \widetilde{F}_r, \widetilde{G}_{r+1}, \tilde{f}_s, \tilde{g}_s, \tilde{c}_s$ are constructed in the same way as $P_{2p+2}, V_{p+1}, F_p, G_p, f_\ell, g_\ell, c_\ell$ using the recursion (0.5) with the only difference being that the set of summation constants $\tilde{c}_r$ in $\tilde{f}_s$ is independent of the set c_k used in computing f_ℓ.

Our strategy will be the same as in the stationary case: Assuming existence of a solution a, b, we will deduce many of its properties which in the end will yield an explicit expression for the solution. In fact, we will go a step further, postulating the equations

$$a_{t_r} = -a\bigl(2(z-b^+)\widetilde{F}_r^+ + \widetilde{G}_{r+1}^+ + \widetilde{G}_{r+1}\bigr),$$
$$b_{t_r} = 2\bigl((z-b)^2 \widetilde{F}_r + (z-b)\widetilde{G}_{r+1} + a^2 \widetilde{F}_r^+ - (a^-)^2 \widetilde{F}_r^-\bigr), \qquad (0.44)$$
$$0 = 2(z-b^+)F_p^+ + G_{p+1}^+ + G_{p+1},$$
$$0 = (z-b)^2 F_p + (z-b)G_{p+1} + a^2 F_p^+ - (a^-)^2 F_p^-, \qquad (0.45)$$

where $a^{(0)} = a^{(0)}(n)$, $b^{(0)} = b^{(0)}(n)$ in (0.43) has been replaced by $a = a(n, t_r)$, $b = b(n, t_r)$ in (0.45). Here

$$F_p(z) = \sum_{\ell=0}^{p} f_{p-\ell} z^\ell = \prod_{j=1}^{p}(z - \mu_j), \quad \widetilde{F}_r(z) = \sum_{s=0}^{r} \tilde{f}_{r-s} z^s,$$

$$G_{p+1}(z) = -z^{p+1} + \sum_{\ell=0}^{p} g_{p-\ell} z^{\ell} + f_{p+1},$$

$$\widetilde{G}_{r+1}(z) = -z^{r+1} + \sum_{s=0}^{r} \tilde{g}_{r-s} z^{s} + \tilde{f}_{r+1},$$

for fixed $p, r \in \mathbb{N}_0$. Introducing G_{p+1}, U, V_{p+1} and $\widetilde{G}_{r+1}$, $\widetilde{V}_{r+1}$ (replacing F_p by $\widetilde{F}_r$) as in (0.12)–(0.14), the basic equations (0.44), (0.45) are equivalent to the *Lax equations*

$$\frac{d}{dt_r} L - [\widetilde{P}_{2r+2}, L] = 0,$$

$$[P_{2p+2}, L] = 0,$$

and to the *zero-curvature equations*

$$U_{t_r} + U\widetilde{V}_{r+1} - \widetilde{V}_{r+1}^+ U = 0, \tag{0.46}$$

$$UV_{p+1} - V_{p+1}^+ U = 0. \tag{0.47}$$

Moreover, one computes in analogy to (0.22) and (0.23) that

$$\det(yI_2 - V_{p+1}(z, n+1, t_r)) - \det(yI_2 - V_{p+1}(z, n, t_r)) = 0,$$

$$\partial_{t_r} \det(yI_2 - V_{p+1}(z, n, t_r)) = 0,$$

and hence

$$\det(yI_2 - V_{p+1}(z, n, t_r)) = y^2 + \det(V_{p+1}(z, n, t_r)) \tag{0.48}$$

$$= y^2 - G_{p+1}^-(z, n)^2 + 4a^-(n)^2 F_p^-(z, n) F_p(z, n) = y^2 - R_{2p+2}(z),$$

is independent of $(n, t_r) \in \mathbb{Z} \times \mathbb{R}$. Thus,

$$G_{p+1}^2 - 4a^2 F_p F_p^+ = R_{2p+2},$$

$$(z-b)^4 F_p^2 - 2a^2(z-b)^2 F_p F_p^+ - 2(a^-)^2 (z-b)^2 F_p F_p^- + a^4 (F_p^+)^2$$

$$+ (a^-)^4 (F_p^-)^2 - 2a^2 (a^-)^2 F_p^+ F_p^- = (z-b)^2 R_{2p+2}(z),$$

$$(z-b)(z-b^+) G_{p+1}^2 - a^2 (G_{p+1}^- + G_{p+1})(G_{p+1} + G_{p+1}^+)$$

$$= (z-b)(z-b^+) R_{2p+2}(z),$$

hold as in the stationary context. The independence of (0.48) of t_r can be interpreted as follows: The rth Toda flow represents an isospectral deformation of the curve $\mathcal{K}_p$ defined in (0.27), in particular,[1] the branch points of $\mathcal{K}_p$ remain invariant under these

---

[1] Property (0.49) is weaker than the usually stated isospectral deformation of the Lax operator $L(t_r)$. However, the latter is a more delicate functional analytic problem since a, b need not be bounded and by the possibility of non-self-adjointness of $L(t_r)$. See, however, Theorem 1.62.

flows,
$$\partial_{t_r} E_m = 0, \quad m = 0, \ldots, 2p+1. \tag{0.49}$$

As in the stationary case, one can now introduce the basic meromorphic function ϕ on $\mathcal{K}_p$ by

$$\begin{aligned}\phi(P, n, t_r) &= \frac{y - G_{p+1}(z, n, t_r)}{2a(n, t_r) F_p(z, n, t_r)} \\ &= \frac{-2a(n, t_r) F_p(z, n+1, t_r)}{y + G_{p+1}(z, n, t_r)}, \quad P(z, y) \in \mathcal{K}_p,\end{aligned}$$

and a comparison with (0.46) and (0.47) then shows that ϕ satisfies the Riccati-type equations

$$a\phi(P) + a^-(\phi^-(P))^{-1} = z - b, \tag{0.50}$$

$$\begin{aligned}\phi_{t_r}(P) = {}&-2a\bigl(\widetilde{F}_r(z)\phi(P)^2 + \widetilde{F}_r^+(z)\bigr) + 2(z - b^+)\widetilde{F}_r^+(z)\phi(P) \\ &+ \bigl(\widetilde{G}_{r+1}^+(z) - \widetilde{G}_{r+1}(z)\bigr)\phi(P).\end{aligned} \tag{0.51}$$

Next, factorizing F_p as before,

$$F_p(z) = \prod_{j=1}^{p}(z - \mu_j),$$

one introduces points $\hat{\mu}_j(n, t_r)$, $\hat{\mu}_j^+(n, t_r)$ on $\mathcal{K}_p$ by

$$\hat{\mu}_j(n, t_r) = (\mu_j(n, t_r), -G_{p+1}(\mu_j(n, t_r), n, t_r)), \quad j = 1, \ldots, p,$$
$$\hat{\mu}_j^+(n, t_r) = (\mu_j^+(n, t_r), G_{p+1}(\mu_j^+(n, t_r), n, t_r)), \quad j = 1, \ldots, p,$$

and obtains for the divisor $(\phi(\,\cdot\,, n, t_r))$ of the meromorphic function ϕ,

$$(\phi(\,\cdot\,, n, t_r)) = \mathcal{D}_{P_{\infty_+}\hat{\underline{\mu}}^+(n, t_r)} - \mathcal{D}_{P_{\infty_-}\hat{\underline{\mu}}(n, t_r)},$$

as in the stationary context. Given $\phi(\,\cdot\,, n, t_r)$ one then defines the *time-dependent Baker–Akhiezer vector* $\psi(\,\cdot\,, n, n_0, t_r, t_{0,r})$ on $\mathcal{K}_p \setminus \{P_{\infty_\pm}\}$ by

$$\psi(P, n, n_0, t_r, t_{0,r})$$
$$= \exp\left(\int_{t_{0,r}}^{t_r} ds \bigl(2a(n_0, s)\widetilde{F}_r(z, n_0, s)\phi(P, n_0, s) + \widetilde{G}_{r+1}(z, n_0, s)\bigr)\right)$$
$$\times \begin{cases} \prod_{n'=n_0}^{n-1} \phi(P, n', t_r), & n \geq n_0 + 1, \\ 1, & n = n_0, \\ \prod_{n'=n}^{n_0-1} \phi(P, n', t_r)^{-1}, & n \leq n_0 - 1, \end{cases}$$

with

$$\phi(P, n, t_r) = \psi^+(P, n, n_0, t_r, t_{0,r})/\psi(P, n, n_0, t_r, t_{0,r}).$$

The Riccati-type equations (0.50), (0.51) satisfied by ϕ then show that

$$-V_{p+1,t_r} + [\widetilde{V}_{r+1}, V_{p+1}] = 0$$

in addition to (0.46), (0.47). Moreover, they yield again that the Baker–Akhiezer function ψ is the common formal eigenfunction of the commuting pair of Lax differential expressions $L(t_r)$ and $P_{2p+2}(t_r)$,

$$L\psi(P) = z\psi(P),$$
$$P_{2p+2}\psi(P) = y\psi(P), \quad P = (z, y) \in \mathcal{K}_p \setminus \{P_{\infty_\pm}\},$$
$$\psi_{t_r}(P) = \widetilde{P}_{2r+2}\psi(P)$$
$$= 2a\widetilde{F}_r(z)\psi^+(P) + \widetilde{G}_{r+1}(z)\psi(P),$$

and at the same time the Baker–Akhiezer vector Ψ (cf. (0.36)) satisfies the zero-curvature equations,

$$\Psi(P) = U(z)\Psi^-(P),$$
$$y\Psi^-(P) = V_{p+1}(z)\Psi^-(P), \quad P = (z, y) \in \mathcal{K}_p \setminus \{P_{\infty_\pm}\},$$
$$\Psi_{t_r}(P) = \widetilde{V}_{r+1}^+(z)\Psi(P).$$

The remaining time-dependent constructions closely follow our stationary outline. The time variation of the μ_j, $j = 1, \ldots, p$, is given by the *Dubrovin equations*[1]

$$\mu_{j,t_r} = -2\widetilde{F}_r(\mu_j) y(\hat{\mu}_j) \prod_{\substack{\ell=1 \\ \ell \neq j}}^{p} (\mu_j - \mu_\ell)^{-1}, \quad j = 1, \ldots, p. \qquad (0.52)$$

However, as in the stationary case, the formula (0.52) as well as (0.39) are not useful in the general complex-valued case where the μ_j, $j = 1, \ldots, p$ may be degenerate and may not remain bounded. Thus, a more elaborate procedure is required.

Let us first consider the case of real-valued and bounded sequences a, b, that is, the situation when the Lax operator L is self-adjoint. Given the curve $\mathcal{K}_p$ and an initial nonspecial Dirichlet divisor $\mathcal{D}_{\hat{\mu}(n_0, t_{0,r})} \in \text{Sym}^p(\mathcal{K}_p)$ at a point $(n_0, t_{0,r})$, one follows the stationary algorithm to construct a solution s-Tl$_p(a^{(0)}, b^{(0)}) = 0$. From each lattice point $n \in \mathbb{Z}$ one can use the time-dependent Dubrovin equations (0.52) to construct locally the solution $\mu_j(n, t_r)$ for t_r near $t_{0,r}$. Using the formulas (0.39) we find solutions of Tl$_r(a, b) = 0$ with $(a, b)|_{t_r=t_{0,r}} = (a^{(0)}, b^{(0)})$. However, this construction requires that the eigenvalues $\mu_j(n, t_r)$, $j = 1, \ldots, p$, remain distinct,

---

[1] To obtain a closed system of differential equations, one has to express $\widetilde{F}_r(\mu_j)$ solely in terms of $\mu_1, \ldots, \mu_p$ and $E_0, \ldots, E_{2p+1}$, see Lemma D.4.

which generally is only true in the self-adjoint case with real-valued initial data. In contrast, in the general complex-valued case where eigenvalues can be expected to collide, a considerably more refined approach is required. The Dubrovin equations (0.52) are replaced by a first-order autonomous system of $2p$ differential equations in the variables f_j, $j = 1, \ldots, p$, g_j, $j = 1, \ldots, p-1$, and $g_p + f_{p+1}$ which can be solved locally in a neighborhood $(t_{0,r} - T_0, t_{0,r} + T_0)$ of $t_{0,r}$. Next, one uses the general stationary algorithm to extend this solution from $\{n_0\} \times (t_{0,r} - T_0, t_{0,r} + T_0)$ to $\mathbb{Z} \times (t_{0,r} - T_0, t_{0,r} + T_0)$. For a carefully selected set $\mathcal{M}_1$ of full measure of initial divisors, the solution can even be extended to a global solution on $\mathbb{Z} \times \mathbb{R}$. Summarizing, we solve the following inverse problem: Given $\mathcal{K}_p$ and appropriate initial data

$$\underline{\hat{\mu}}(n_0, t_{0,r}) = \{\hat{\mu}_1(n_0), \ldots, \hat{\mu}_p(n_0, t_{0,r})\} \in \mathcal{M}_1,$$
$$\hat{\mu}_j(n_0, t_{0,r}) = \big(\mu_j(n_0, t_{0,r}), -G_{p+1}(\mu_j(n_0, t_{0,r}), n_0, t_{0,r})\big), \quad j = 1, \ldots, p,$$

where $\mathcal{M}_1 \subset \mathrm{Sym}^p(\mathcal{K}_p)$ is an appropriate set of nonspecial Dirichlet divisors, we develop an algorithm that defines finite nonspecial divisors $\underline{\hat{\mu}}(n, t_r)$ for all $(n, t_r) \in \mathbb{Z} \times \mathbb{R}$.

Having constructed $\mu_j(n, t_r)$, $j = 1, \ldots, p$, $(n, t_r) \in \mathbb{Z} \times \mathbb{R}$, one then shows that the analog of (0.40) remains valid and then leads to a solution a, b of (0.42), (0.43).

The corresponding representations of a, b, ϕ, and ψ in terms of the *Riemann theta function* associated with $\mathcal{K}_p$ are then obtained in close analogy to the stationary case. In particular, in the case of a, b, one obtains the *Its–Matveev formula*

$$a(n, t_r)^2 = \tilde{a}^2 \frac{\theta(\underline{A} - \underline{B} + \underline{B}n + \underline{C}_r t_r)\theta(\underline{A} + \underline{B} + \underline{B}n + \underline{C}_r t_r)}{\theta(\underline{A} + \underline{B}n + \underline{C}_r t_r)^2},$$
$$b(n, t_r) = \frac{1}{2}\sum_{m=0}^{2p+1} E_m - \sum_{j=1}^{p} \lambda_j \qquad (0.53)$$
$$- \sum_{j=1}^{p} c_j(p) \frac{\partial}{\partial w_j} \ln\left(\frac{\theta(\underline{A} + \underline{B}n + \underline{C}_r t_r + \underline{w})}{\theta(\underline{A} - \underline{B} + \underline{B}n + \underline{C}_r t_r + \underline{w})}\right)\bigg|_{\underline{w}=0}.$$

Here the constants $\tilde{a}, \lambda_j, c_j(p) \in \mathbb{C}$, $j = 1, \ldots, p$, and the constant vectors $\underline{B}, \underline{C}_r \in \mathbb{C}^p$ are uniquely determined by $\mathcal{K}_p$ (and its homology basis) and r, and the constant vector $\underline{A} \in \mathbb{C}^p$ is in one-to-one correspondence with the Dirichlet data $\underline{\hat{\mu}}(n_0, t_{0,r}) = (\hat{\mu}_1(n_0, t_{0,r}), \ldots, \hat{\mu}_p(n_0, t_{0,r})) \in \mathrm{Sym}^p(\mathcal{K}_p)$ at the initial point $(n_0, t_{0,r})$ as long as the divisor $\mathcal{D}_{\underline{\hat{\mu}}(n_0,t_{0,r})}$ is assumed to be nonspecial. Moreover, the theta function representation (0.53) remains valid as long as the divisor $\mathcal{D}_{\underline{\hat{\mu}}(n,t_r)}$ stays nonspecial. Again, one notes the remarkable fact that the argument of the theta functions in (0.53) is linear with respect to both n and t_r.

Introduction

The reader will have noticed that we used terms such as *completely integrable*, *soliton equations*, *isospectral deformations*, etc., without offering a precise definition for them. Arguably, an integrable system in connection with nonlinear evolution equations should possess several properties, including, for instance,

- infinitely many conservation laws
- isospectral deformations of a Lax operator
- action-angle variables, Hamiltonian formalism
- algebraic (spectral) curves
- infinitely many symmetries and transformation groups
- "explicit" solutions.

While many of these properties apply to particular systems of interest, there is simply no generally accepted definition to date of what constitutes an integrable system.[1] Thus, different schools have necessarily introduced different shades of integrability (Liouville integrability, analytic integrability, algebraically complete integrability, etc.); in this monograph we found it useful to focus on the existence of underlying algebraic curves and explicit representations of solutions in terms of corresponding Riemann theta functions and limiting situations thereof.

Finally, a brief discussion of the content of each chapter is in order (additional details are collected in the list of contents at the beginning of each chapter). Chapter 1 is devoted to the Toda hierarchy and its algebro-geometric solutions. In Chapter 2 we turn to the Kac–van Moerbeke equation. Rather than studying this equation independently, we exploit its intrinsic connection with the Toda lattice. Indeed, there exists a Miura-like transformation between the two integrable systems, allowing for a transfer of solutions between them. Next, in Chapter 3, we consider the Ablowitz–Ladik (AL) hierarchy (a complexified discrete nonlinear Schrödinger hierarchy) of differential-difference evolution equations and its algebro-geometric solutions.

Presentation: Each chapter, together with appropriate appendices compiled in the second part of this volume, is intended to be essentially self-contained and hence can be read independently from the remaining chapters. This attempt to organize chapters independently of one another comes at a price, of course: Similar arguments in the construction of algebro-geometric solutions for different hierarchies are repeated in different chapters. We believe this makes the results more easily accessible.

While we kept the style of presentation and the notation employed as close as possible to that used in Volume I, we emphasize that this volume is entirely self-contained and hence can be read independently of Volume I.

---

[1] See, also, Lakshmanan and Rajasekar (2003, Chs. 10, 14, App. I) and several contributions to Zakharov (1991) for an extensive discussion of various aspects of integrability.

References are deferred to detailed notes for each section at the end of every chapter. In addition to a comprehensive bibliographical documentation of the material dealt with in the main text, these notes also contain numerous additional comments and results (and occasionally hints to the literature of topics not covered in this monograph).

Succinctly written appendices, some of which summarize subjects of interest on their own, such as compact (and, in particular, hyperelliptic) Riemann surfaces, guarantee a fairly self-contained presentation, accessible at the advanced graduate level.

An extensive bibliography is included at the end of this volume. Its size reflects the enormous interest this subject generated over the past four decades. It underscores the wide variety of techniques employed to study completely integrable systems. Even though we undertook every effort to provide an exhaustive list of references, the result in the end must necessarily be considered incomplete. We regret any omissions that have occurred. Publications with three or more authors are abbreviated "First author et al. (year)" in the text. If more than one publication yields the same abbreviation, latin letters a,b,c, etc., are added after the year. In the bibliography, publications are alphabetically ordered using all authors' names and year of publication.

1
The Toda Hierarchy

> ...that I can offer you no better gift than the means of mastering in a very brief time, all that in the course of so many years, and at the cost of so many hardships and dangers, I have learned, and know.
>
> N. Machiavelli (1469–1527)[1]

1.1 Contents

The Toda lattice (Tl) equations,

$$a_t - a(b^+ - b) = 0,$$
$$b_t - 2(a^2 - (a^-)^2) = 0$$

for sequences $a = a(n, t)$, $b = b(n, t)$ (with $a^\pm(n, t) = a(n \pm 1, t)$, $b^\pm(n, t) = b(n \pm 1, t)$, $(n, t) \in \mathbb{Z} \times \mathbb{R}$), were originally derived in 1966 as a model for waves in lattices composed of particles interacting by nonlinear (exponential) forces between nearest neighbors.[2] This chapter focuses on the construction of algebro-geometric solutions of the Toda hierarchy as developed since the mid-1970s. Below we briefly summarize the principal content of each section. A more detailed discussion of the contents has been provided in the introduction to this volume.

Section 1.2.
- polynomial recursion formalism, Lax pairs (L, P_{2p+1})
- stationary and time-dependent Toda hierarchy
- Burchnall–Chaundy polynomial, hyperelliptic curve $\mathcal{K}_p$

Sections 1.3 and 1.4. (stationary)
- properties of ϕ and the Baker–Akhiezer function ψ
- Dubrovin equations for Dirichlet, Neumann, and other auxiliary divisors

---

[1] *The Prince*, Dover, New York, 1992, p. vii.
[2] A guide to the literature can be found in the detailed notes at the end of this chapter.

- trace formulas for a, b
- theta function representations for ϕ, ψ, and a, b
- examples
- the algebro-geometric initial value problem

Sections 1.5 and 1.6. (time-dependent)

- properties of ϕ and the Baker–Akhiezer function ψ
- Dubrovin equations for Dirichlet, Neumann, and other auxiliary divisors
- trace formulas for a, b
- theta function representations for ϕ, ψ, and a, b
- examples
- the algebro-geometric initial value problem

Section 1.7. (Hamiltonian formalism)

- asymptotic spectral parameter expansions of Riccati-type solutions
- local conservation laws
- variational derivatives
- Poisson brackets

This chapter relies on terminology and notions developed in connection with compact Riemann surfaces. A brief summary of key results as well as definitions of some of the main quantities can be found in Appendices A and B.

1.2 The Toda Hierarchy, Recursion Relations, Lax Pairs, and Hyperelliptic Curves

> I guess I should warn you, if I turn out to be particularly clear,
> you've probably misunderstood what I've said.
> *Alan Greenspan*[1]

In this section we provide the construction of the Toda hierarchy using a polynomial recursion formalism and derive the associated sequence of Toda Lax pairs (we also hint at zero-curvature pairs). Moreover, we discuss the Burchnall–Chaundy polynomial in connection with the stationary Toda hierarchy and the underlying hyperelliptic curve.

We denote by $\mathbb{C}^{\mathbb{Z}}$ the set of all complex-valued sequences indexed by $\mathbb{Z}$.

Throughout this section we suppose the following hypothesis.

---

[1] US economist. Quoted in *The New York Times*, October 28, 2005.

1.2 Fundamentals of the Toda Hierarchy

Hypothesis 1.1 *In the stationary case we assume that a, b satisfy*

$$a, b \in \mathbb{C}^{\mathbb{Z}}, \quad a(n) \neq 0, \; n \in \mathbb{Z}. \tag{1.1}$$

In the time-dependent case we assume that a, b satisfy

$$a(\cdot, t), b(\cdot, t) \in \mathbb{C}^{\mathbb{Z}}, \; t \in \mathbb{R}, \quad a(n, \cdot), b(n, \cdot) \in C^1(\mathbb{R}), \; n \in \mathbb{Z},$$
$$a(n, t) \neq 0, \; (n, t) \in \mathbb{Z} \times \mathbb{R}.$$

Actually, up to Remark 1.8 our analysis will be time-independent, and hence only the lattice variations of a and b will matter.

We denote by $S^{\pm}$ the shift operators acting on complex-valued sequences $f = \{f(n)\}_{n \in \mathbb{Z}} \in \mathbb{C}^{\mathbb{Z}}$ according to

$$(S^{\pm} f)(n) = f(n \pm 1), \; n \in \mathbb{Z}.$$

Moreover, we will frequently use the notation

$$f^{\pm} = S^{\pm} f, \quad f \in \mathbb{C}^{\mathbb{Z}}.$$

Consider the one-dimensional second-order difference expression

$$L = a S^+ + a^- S^- + b \tag{1.2}$$

of Jacobi-type, soon to be identified with the Lax differential expression of the Toda hierarchy. To construct the Toda hierarchy one needs a second difference expression of order $2p + 2$, denoted by P_{2p+2}, $p \in \mathbb{N}_0$, defined recursively in the following. We take the quickest route to the construction of P_{2p+2} and hence to that of the Toda hierarchy by starting from the recursion relation (1.3)–(1.5) below. Subsequently we will offer the motivation behind this approach (cf. Remark 1.7).

Define sequences $\{f_\ell(n)\}_{\ell \in \mathbb{N}_0}$ and $\{g_\ell(n)\}_{\ell \in \mathbb{N}_0}$ recursively by

$$f_0 = 1, \quad g_0 = -c_1, \tag{1.3}$$

$$2 f_{\ell+1} + g_\ell + g_\ell^- - 2 b f_\ell = 0, \quad \ell \in \mathbb{N}_0, \tag{1.4}$$

$$g_{\ell+1} - g_{\ell+1}^- + 2\big(a^2 f_\ell^+ - (a^-)^2 f_\ell^-\big) - b(g_\ell - g_\ell^-) = 0, \quad \ell \in \mathbb{N}_0. \tag{1.5}$$

We note that a enters only quadratically in f_ℓ and g_ℓ. Explicitly, one obtains

$$\begin{aligned}
f_0 &= 1, \\
f_1 &= b + c_1, \\
f_2 &= a^2 + (a^-)^2 + b^2 + c_1 b + c_2, \\
f_3 &= (a^-)^2(b^- + 2b) + a^2(b^+ + 2b) + b^3 \\
&\quad + c_1\big(a^2 + (a^-)^2 + b^2\big) + c_2 b + c_3, \text{ etc.,} \\
g_0 &= -c_1,
\end{aligned} \tag{1.6}$$

$$g_1 = -2a^2 - c_2,$$
$$g_2 = -2a^2(b + b^+) + c_1(-2a^2) - c_3, \quad \text{etc.}$$

Here $\{c_\ell\}_{\ell \in \mathbb{N}}$ denote summation constants which naturally arise when solving (1.5). Subsequently, it will also be useful to work with the corresponding homogeneous coefficients $\hat{f}_\ell$ and $\hat{g}_\ell$, defined by the vanishing of all summation constants c_k for $k = 1, \ldots, \ell + 1$,

$$\hat{f}_0 = 1, \quad \hat{f}_\ell = f_\ell \big|_{c_k=0,\, k=1,\ldots,\ell}, \quad \ell \in \mathbb{N}, \tag{1.7}$$

$$\hat{g}_0 = 0, \quad \hat{g}_\ell = g_\ell \big|_{c_k=0,\, k=1,\ldots,\ell+1}, \quad \ell \in \mathbb{N}. \tag{1.8}$$

By induction one infers that

$$f_\ell = \sum_{k=0}^{\ell} c_{\ell-k} \hat{f}_k, \quad \ell \in \mathbb{N}_0,$$

$$g_\ell = \sum_{k=1}^{\ell} c_{\ell-k} \hat{g}_k - c_{\ell+1}, \quad \ell \in \mathbb{N}, \tag{1.9}$$

introducing

$$c_0 = 1. \tag{1.10}$$

In a slight abuse of notation we will occasionally stress the dependence of f_ℓ and g_ℓ on a, b by writing $f_\ell(a, b)$, $g_\ell(a, b)$.

Remark 1.2 Using the nonlinear recursion relations (C.20), (C.21) in Theorem C.1, one infers inductively that all homogeneous elements $\hat{f}_\ell$ and $\hat{g}_\ell$, $\ell \in \mathbb{N}_0$, are polynomials in a, b, and some of their shifts. (Alternatively, one can prove directly by induction that the nonlinear recursion relations (C.20), (C.21) are equivalent to those in (1.3)–(1.5) with all summation constants put equal to zero, $c_\ell = 0$, $\ell \in \mathbb{N}$.)

Remark 1.3 As an efficient tool to distinguish between homogeneous and nonhomogeneous quantities $\hat{f}_\ell$, $\hat{g}_\ell$ and f_ℓ, g_ℓ, respectively, we now introduce the notation of degree as follows. Denote

$$f^{(r)} = S^{(r)} f, \quad f = \{f(n)\}_{n \in \mathbb{Z}} \in \mathbb{C}^{\mathbb{Z}}, \quad S^{(r)} = \begin{cases} (S^+)^r, & r \geq 0, \\ (S^-)^{-r}, & r < 0, \end{cases} \quad r \in \mathbb{Z},$$

and define

$$\deg\left(a^{(r)}\right) = \deg\left(b^{(r)}\right) = 1, \quad r \in \mathbb{Z}. \tag{1.11}$$

1.2 Fundamentals of the Toda Hierarchy

This then results in
$$\deg\left(\hat{f}_\ell^{(r)}\right) = \ell, \quad \ell \in \mathbb{N}_0, \ r \in \mathbb{Z},$$
$$\deg\left(\hat{g}_\ell^{(r)}\right) = \ell + 1, \quad \ell \in \mathbb{N}, \ r \in \mathbb{Z}, \tag{1.12}$$
using induction in the linear recursion relations (1.3)–(1.5).

Next we relate the homogeneous quantities $\hat{f}_\ell, \hat{g}_\ell$ to certain matrix elements of L^ℓ. For this purpose it is useful to introduce the standard basis $\{\delta_m\}_{m\in\mathbb{Z}}$ in $\ell^2(\mathbb{Z})$ by

$$\delta_m = \{\delta_{m,n}\}_{n\in\mathbb{Z}}, \ m \in \mathbb{Z}, \quad \delta_{m,n} = \begin{cases} 1, & m = n, \\ 0, & m \neq n. \end{cases} \tag{1.13}$$

The scalar product in $\ell^2(\mathbb{Z})$, denoted by $(\cdot, \cdot)$, is defined by
$$(f, g) = \sum_{n\in\mathbb{Z}} \overline{f(n)} g(n), \quad f, g \in \ell^2(\mathbb{Z}).$$

In the basis just introduced, the Jacobi difference expression L in (1.2) takes on the form

$$L = \begin{pmatrix} \ddots & \ddots & \ddots & \ddots & \ddots & & & & 0 \\ & 0 & a(-2) & b(-1) & a(-1) & 0 & & & \\ & & 0 & a(-1) & b(0) & a(0) & 0 & & \\ & & & 0 & a(0) & b(1) & a(1) & 0 & \\ & & & & 0 & a(1) & b(2) & a(2) & 0 \\ 0 & & & & & \ddots & \ddots & \ddots & \ddots \end{pmatrix}. \tag{1.14}$$

Here terms of the form $b(n)$ represent the diagonal (n, n)-entries, $n \in \mathbb{Z}$, in the infinite matrix (1.14).

Lemma 1.4 *Assume* (1.1) *and let* $n \in \mathbb{Z}, \ell \in \mathbb{N}_0$. *Then the homogeneous coefficients* $\{\hat{f}_\ell\}_{\ell\in\mathbb{N}_0}$ *and* $\{\hat{g}_\ell\}_{\ell\in\mathbb{N}_0}$ *satisfy*

$$\hat{f}_\ell(n) = (\delta_n, L^\ell \delta_n), \tag{1.15}$$
$$\hat{g}_\ell(n) = -2a(n)(\delta_{n+1}, L^\ell \delta_n). \tag{1.16}$$

Proof We abbreviate
$$\tilde{f}_\ell(n) = (\delta_n, L^\ell \delta_n), \quad \tilde{g}_\ell(n) = -2a(n)(\delta_{n+1}, L^\ell \delta_n).$$
Then
$$\tilde{f}_{\ell+1}(n) = (L\delta_n, L^\ell \delta_n) = -\frac{1}{2}(\tilde{g}_\ell(n) + \tilde{g}_\ell^-(n)) + b\tilde{f}_\ell(n), \tag{1.17}$$

and similarly,

$$\tilde{g}_{\ell+1} = b\tilde{g}_\ell - 2a^2 \tilde{f}_\ell^+ + \tilde{h}_\ell = b^+ \tilde{g}_\ell - 2a^2 \tilde{f}_\ell + \tilde{h}_\ell^+, \quad (1.18)$$

where

$$\tilde{h}_\ell(n) = -2a(n)a(n-1)(\delta_{n+1}, L^\ell \delta_{n-1}).$$

Eliminating $\tilde{h}_\ell$ in (1.18) results in

$$\tilde{g}_{\ell+1} - \tilde{g}_{\ell+1}^- = -2(a^2 \tilde{f}_\ell^+ - (a^-)^2 \tilde{f}_\ell^-) + b(\tilde{g}_\ell - \tilde{g}_\ell^-). \quad (1.19)$$

By inspection, (1.17) and (1.19) are equivalent to (1.3)–(1.5). In order to determine which solution of (1.3)–(1.5) has been found (i.e., determine the summation constants $c_1, \ldots, c_p$) we apply the notion of degree as introduced in Remark 1.3. Then (1.11) implies that $\hat{f}_\ell$ and $\hat{g}_{\ell-1}$ have degree ℓ and hence

$$c_0 = 1, \ c_\ell = 0, \quad \ell = 1, \ldots, p,$$

completing the proof. □

As a byproduct, (1.15) and (1.16) yield an alternative proof (by induction) that f_ℓ, g_ℓ, $\ell \in \mathbb{N}_0$, are polynomials in a, b, and some of their shifts.

Next we define difference expressions P_{2p+2} of order $2p+2$ by

$$P_{2p+2} = -L^{p+1} + \sum_{\ell=0}^{p}(g_{p-\ell} + 2af_{p-\ell}S^+)L^\ell + f_{p+1}, \quad p \in \mathbb{N}_0. \quad (1.20)$$

We record the first few P_{2p+2},

$$P_2 = aS^+ - a^- S^-,$$
$$P_4 = aa^+(S^+)^2 + a(b+b^+)S^+ - a^-(a^-)^2(S^-)^2$$
$$\quad - a^-(b+b^-)S^- + c_1(aS^+ - a^- S^-), \text{ etc.}$$

Introducing the corresponding homogeneous difference expressions $\widehat{P}_{2p+2}$ defined by

$$\widehat{P}_{2\ell+2} = P_{2\ell+2}\big|_{c_k=0,\, k=1,\ldots,\ell}, \quad \ell \in \mathbb{N}_0, \quad (1.21)$$

one finds

$$P_{2p+2} = \sum_{\ell=0}^{p} c_{p-\ell} \widehat{P}_{2\ell+2}. \quad (1.22)$$

1.2 Fundamentals of the Toda Hierarchy

Using the recursion (1.3)–(1.5), the commutator of P_{2p+2} and L can be computed explicitly and one obtains[1]

$$[P_{2p+2}, L] = -a\big(g_p^+ + g_p + f_{p+1}^+ + f_{p+1} - 2b^+ f_p^+\big)S^+ \\ + 2\big(-b(g_p + f_{p+1}) + a^2 f_p^+ - (a^-)^2 f_p^- + b^2 f_p\big) \\ - a^-\big(g_p + g_p^- + f_{p+1} + f_{p+1}^- - 2bf_p\big)S^-, \quad p \in \mathbb{N}_0. \quad (1.23)$$

In particular, (L, P_{2p+2}) represents the celebrated *Lax pair* of the Toda hierarchy. Varying $p \in \mathbb{N}_0$, the stationary Toda hierarchy is then defined in terms of the vanishing of the commutator of P_{2p+2} and L in (1.23) by

$$[P_{2p+2}, L] = 0, \quad p \in \mathbb{N}_0, \quad (1.24)$$

or equivalently, by

$$g_p + g_p^- + f_{p+1} + f_{p+1}^- - 2bf_p = 0, \quad (1.25)$$
$$-b(g_p + f_{p+1}) + a^2 f_p^+ - (a^-)^2 f_p^- + b^2 f_p = 0. \quad (1.26)$$

Using (1.4) with $\ell = p$ one concludes that (1.25) reduces to

$$f_{p+1} - f_{p+1}^- = 0, \quad (1.27)$$

that is, f_{p+1} is a lattice constant. Similarly, subtracting b times (1.25) from twice (1.26), and using (1.27) and (1.5) with $\ell = p$, one infers that g_{p+1} is a lattice constant as well,

$$g_{p+1} - g_{p+1}^- = 0. \quad (1.28)$$

Thus, varying $p \in \mathbb{N}_0$, equations (1.27) and (1.28) give rise to the stationary Toda hierarchy, which we introduce as follows

$$\text{s-Tl}_p(a, b) = \begin{pmatrix} f_{p+1}^+ - f_{p+1} \\ g_{p+1} - g_{p+1}^- \end{pmatrix} = 0, \quad p \in \mathbb{N}_0. \quad (1.29)$$

Explicitly,

$$\text{s-Tl}_0(a, b) = \begin{pmatrix} b^+ - b \\ 2((a^-)^2 - a^2) \end{pmatrix} = 0,$$

$$\text{s-Tl}_1(a, b) = \begin{pmatrix} (a^+)^2 - (a^-)^2 + (b^+)^2 - b^2 \\ 2(a^-)^2(b + b^-) - 2a^2(b^+ + b) \end{pmatrix}$$
$$+ c_1 \begin{pmatrix} b^+ - b \\ 2((a^-)^2 - a^2) \end{pmatrix} = 0,$$

---

[1] The recursion relations (1.3)–(1.5) are constructed in such a manner that the commutator of P_{2p+2} and L ceases to be a higher-order difference expression and reduces to second order only.

$$\text{s-Tl}_2(a,b) = \begin{pmatrix} (b^+)^3 - b^3 + 2(a^+)^2 b^+ - 2(a^-)^2 b \\ +a^2(b^+ - b) + (a^+)^2 b^{++} + (a^-)^2 b^{-} \\ 2(a^-)^2(b^2 + bb^- + (b^-)^2 + (a^-)^2 + (a^{--})^2) \\ -2a^2(b^2 + bb^+ + (b^+)^2 + a^2 + (a^+)^2) \end{pmatrix}$$

$$+ c_1 \begin{pmatrix} (a^+)^2 - (a^-)^2 + (b^+)^2 - b^2 \\ 2(a^-)^2(b + b^-) - 2a^2(b^+ + b) \end{pmatrix}$$

$$+ c_2 \begin{pmatrix} b^+ - b \\ 2((a^-)^2 - a^2) \end{pmatrix} = 0, \text{ etc.,}$$

represent the first few equations of the stationary Toda hierarchy. By definition, the set of solutions of (1.29), with p ranging in $\mathbb{N}_0$ and $c_\ell \in \mathbb{C}$, $\ell \in \mathbb{N}$, defines the class of algebro-geometric Toda solutions.

In the following we will frequently assume that a, b satisfy the pth stationary Toda system. By this we mean it satisfies one of the pth stationary Toda equations after a particular choice of summation constants $c_\ell \in \mathbb{C}$, $\ell = 1, \ldots, p$, $p \in \mathbb{N}$, has been made.

In accordance with our notation introduced in (1.7), (1.8), and (1.21), the corresponding homogeneous stationary Toda equations are defined by

$$\text{s-}\widehat{\text{Tl}}_p(a,b) = \text{s-Tl}_p(a,b)\big|_{c_\ell=0,\, \ell=1,\ldots,p} = 0, \quad p \in \mathbb{N}_0.$$

Now we are in a position to describe the connections with the usual approach to the Toda hierarchy equations. For this purpose it suffices to consider the homogeneous case only. Let T be a bounded operator in $\ell^2(\mathbb{Z})$. Given the standard basis (1.13) in $\ell^2(\mathbb{Z})$, we represent T by

$$T = \{T(m,n)\}_{(m,n)\in\mathbb{Z}^2}, \quad T(m,n) = (\delta_m, T\delta_n), \quad (m,n) \in \mathbb{Z}^2.$$

Moreover, we introduce the upper and lower triangular parts $T_\pm$ of T as

$$T_\pm = \{T_\pm(m,n)\}_{(m,n)\in\mathbb{Z}^2}, \quad T_\pm(m,n) = \begin{cases} T(m,n), & \pm(n-m) > 0, \\ 0, & \text{otherwise.} \end{cases}$$

Lemma 1.5 *The homogeneous Lax differential expression $\widehat{P}_{2p+2}$ satisfies*

$$\widehat{P}_{2p+2} = (L^{p+1})_+ - (L^{p+1})_-.$$

Proof We use induction on p. The case $p = 0$ is trivial. By (1.20) we need to show

$$\widehat{P}_{2p+2} = \widehat{P}_{2p} L + (\hat{g}_p + 2a\hat{f}_p S^+) - \hat{f}_p L + \hat{f}_{p+1}.$$

This can be done upon considering $(\delta_m, \widehat{P}_{2p+2} \delta_n)$ and making the case distinctions $m < n-1, m = n-1, m = n, m = n+1, m > n+1$ (for $m = n$ one can use (1.3)–(1.5)). $\square$

1.2 Fundamentals of the Toda Hierarchy

Next, we introduce polynomials F_p and G_{p+1} with respect to the spectral parameter $z \in \mathbb{C}$ of degree p and $p+1$, respectively, by

$$F_p(z) = \sum_{\ell=0}^{p} f_{p-\ell} z^\ell = \sum_{\ell=0}^{p} c_{p-\ell} \widehat{F}_\ell(z), \tag{1.30}$$

$$G_{p+1}(z) = -z^{p+1} + \sum_{\ell=0}^{p} g_{p-\ell} z^\ell + f_{p+1} = \sum_{\ell=1}^{p+1} c_{p+1-\ell} \widehat{G}_\ell(z), \tag{1.31}$$

where $\widehat{F}_\ell$ and $\widehat{G}_\ell$ denote the corresponding homogeneous polynomials defined by

$$\widehat{F}_0(z) = F_0(z) = 1,$$

$$\widehat{F}_\ell(z) = F_\ell(z)\big|_{c_k=0,\, k=1,\ldots,\ell} = \sum_{k=0}^{\ell} \hat{f}_{\ell-k} z^k, \quad \ell \in \mathbb{N}_0, \tag{1.32}$$

$$\widehat{G}_0(z) = G_0(z)\big|_{c_1=0} = 0, \quad \widehat{G}_1(z) = G_1(z) = -z + b, \tag{1.33}$$

$$\widehat{G}_{\ell+1}(z) = G_{\ell+1}(z)\big|_{c_k=0,\, k=1,\ldots,\ell} = -z^{\ell+1} + \sum_{k=0}^{\ell} \hat{g}_{\ell-k} z^k + \hat{f}_{\ell+1}, \quad \ell \in \mathbb{N}.$$

Explicitly, one obtains

$$\begin{aligned}
F_0 &= 1, \\
F_1 &= z + b + c_1, \\
F_2 &= z^2 + bz + a^2 + (a^-)^2 + b^2 + c_1(z+b) + c_2,\ \text{etc.}, \\
G_0 &= -c_1, \\
G_1 &= -z + b, \\
G_2 &= -z^2 + (a^-)^2 - a^2 + b^2 + c_1(-z+b), \\
G_3 &= -z^3 - 2a^2 z - a^2 b^+ + (a^-)^2 b^- - 2(a^-)^2 b + b^3 \\
&\quad + c_1\big(-z^2 + (a^-)^2 - a^2 + b^2\big) + c_2(-z+b),\ \text{etc.}
\end{aligned} \tag{1.34}$$

Considering the kernel of $L - z$ for $z \in \mathbb{C}$ (in the algebraic sense rather than in the functional analytic one),

$$\ker(L - z) = \{\psi : \mathbb{Z} \to \mathbb{C} \cup \{\infty\} \,|\, (L - z)\psi = 0\}, \quad z \in \mathbb{C},$$

one then computes for the restriction of P_{2p+2} to this space,

$$P_{2p+2}\big|_{\ker(L-z)} = \big(2aF_p(z)S^+ + G_{p+1}(z)\big)\big|_{\ker(L-z)}. \tag{1.35}$$

We emphasize that the result (1.35) is valid independently of whether or not P_{2p+2} and L commute. However, if one makes the additional assumption that P_{2p+2} and L

commute, we will prove in Theorem 1.6 that this implies an algebraic relationship between P_{2p+2} and L.

Given the result (1.35), the Lax relation (1.24) becomes

$$\begin{aligned} 0 = [P_{2p+2}, L]\big|_{\ker(L-z)} &= -(L-z)P_{2p+2}\big|_{\ker(L-z)} \\ &= \big(a\big(2(z-b^+)F_p^+ - 2(z-b)F_p + G_{p+1}^+ - G_{p+1}^-\big)S^+ \\ &\quad + \big(2(a^-)^2 F_p^- - 2a^2 F_p^+ + (z-b)(G_{p+1}^- - G_{p+1})\big)\big)\big|_{\ker(L-z)}, \end{aligned}$$

or equivalently,

$$2(z-b^+)F_p^+ - 2(z-b)F_p + G_{p+1}^+ - G_{p+1}^- = 0, \quad (1.36)$$

$$2a^2 F_p^+ - 2(a^-)^2 F_p^- + (z-b)(G_{p+1} - G_{p+1}^-) = 0. \quad (1.37)$$

Further manipulations of (1.36) and (1.37) then yield

$$2(z-b)F_p + G_{p+1} + G_{p+1}^- = 0, \quad (1.38)$$

$$(z-b)^2 F_p + (z-b)G_{p+1} + a^2 F_p^+ - (a^-)^2 F_p^- = 0, \quad p \in \mathbb{N}_0. \quad (1.39)$$

Indeed, adding $G_{p+1} - G_{p+1}$ to the left-hand side of (1.36) (neglecting a trivial summation constant) yields (1.38). Insertion of (1.38) into (1.37) implies (1.39). Equations (1.38) and (1.39) provide an alternative description of the stationary Toda hierarchy.

Combining equations (1.37) and (1.38) one infers that the expression R_{2p+2}, defined as

$$R_{2p+2}(z) = G_{p+1}(z,n)^2 - 4a(n)^2 F_p(z,n) F_p^+(z,n), \quad (1.40)$$

is a lattice constant, that is, $R_{2p+2} - R_{2p+2}^- = 0$, and hence depends on z only. Indeed,

$$\begin{aligned} (z-b)&(R_{2p+2} - R_{2p+2}^-) \\ &= (z-b)\big((G_{p+1} + G_{p+1}^-)(G_{p+1} - G_{p+1}^-) - 4F_p\big(a^2 F_p^+ - (a^-)^2 F_p^-\big)\big) \\ &= -(G_{p+1} + G_{p+1}^- + 2(z-b)F_p)2\big(a^2 F_p^+ - (a^-)^2 F_p^-\big) = 0, \end{aligned}$$

using (1.37) and (1.38). Thus, R_{2p+2} is a monic polynomial of degree $2p+2$. We denote its zeros[1] by $\{E_m\}_{m=0,\dots,2p+1}$ and hence write

$$R_{2p+2}(z) = \prod_{m=0}^{2p+1} (z - E_m), \quad \{E_m\}_{m=0,\dots,2p+1} \subset \mathbb{C}. \quad (1.41)$$

---

[1] The roots of R_{2p+2} are related to the spectrum of a bounded operator realization $\check{L}$ of L in $\ell^2(\mathbb{Z})$, assuming $a, b \in \ell^\infty(\mathbb{Z})$.

One can decouple (1.38) and (1.39) to obtain separate equations for F_p and G_{p+1}. For instance, computing G_{p+1} from (1.39) and inserting the result into (1.38) yields the following linear difference equation for F_p

$$(z-b)^2(z-b^-)F_p - (z-b^-)^2(z-b)F_p^- + \big((a^-)^2 F_p^- - a^2 F_p^+\big)(z-b^-)$$
$$+ \big((a^{--})^2 F_p^{--} - (a^-)^2 F_p\big)(z-b) = 0.$$

Similarly, insertion of (1.39) into (1.40) permits one to eliminate G_{p+1} and results in the following nonlinear difference equation for F_p,

$$(z-b)^4 F_p^2 - 2a^2(z-b)^2 F_p F_p^+ - 2(a^-)^2(z-b)^2 F_p F_p^- + a^4 (F_p^+)^2$$
$$+ (a^-)^4 (F_p^-)^2 - 2a^2(a^-)^2 F_p^+ F_p^- = (z-b)^2 R_{2p+2}(z). \tag{1.42}$$

On the other hand, computing F_p in terms of G_{p+1} and G_{p+1}^+ using (1.38) and inserting the result into (1.39) yields the following linear difference equation for G_{p+1}

$$a^2(z-b^-)(G_{p+1}^+ + G_{p+1}) - (a^-)^2(z-b^+)(G_{p+1}^- + G_{p+1}^{--})$$
$$+ (z-b^-)(z-b)(z-b^+)(G_{p+1}^- - G_{p+1}) = 0.$$

Finally, inserting the result for F_p into (1.40) yields the following nonlinear difference equation for G_{p+1}

$$(z-b)(z-b^+)G_{p+1}^2 - a^2(G_{p+1}^- + G_{p+1})(G_{p+1} + G_{p+1}^+)$$
$$= (z-b)(z-b^+)R_{2p+2}. \tag{1.43}$$

Equations analogous to (1.42) and (1.43) can be used to derive nonlinear recursion relations for the homogeneous coefficients $\hat{f}_\ell$ and $\hat{g}_\ell$ (i.e., the ones satisfying (1.7) and (1.8) in the case of vanishing integration constants) as proved in Theorem C.1 in Appendix C. This has interesting applications to the asymptotic expansion of the Green's function of L with respect to the spectral parameter and also yields a proof that $\hat{f}_\ell$ and $\hat{g}_\ell$ are polynomials in a, b, and some of their shifts (cf. Remark 1.2). In addition, as proven in Theorem C.2, (1.42) leads to an explicit determination of the integration constants $c_1, \ldots, c_p$ in (1.29) in terms of the zeros $E_0, \ldots, E_{2p+1}$ of the associated polynomial R_{2p+2} in (1.41). In fact, one can prove (cf. (C.23))

$$c_\ell = c_\ell(\underline{E}), \quad \ell = 0, \ldots, p, \tag{1.44}$$

where

$c_0(\underline{E}) = 1$,

$$c_k(\underline{E}) \tag{1.45}$$

$$= -\sum_{\substack{j_0,\ldots,j_{2p+1}=0 \\ j_0+\cdots+j_{2p+1}=k}}^{k} \frac{(2j_0)!\cdots(2j_{2p+1})!}{2^{2k}(j_0!)^2\cdots(j_{2p+1}!)^2(2j_0-1)\cdots(2j_{2p+1}-1)} E_0^{j_0}\cdots E_{2p+1}^{j_{2p+1}},$$

$$k=1,\ldots,p,$$

are symmetric functions of $\underline{E} = (E_0,\ldots,E_{2p+1})$.

The fact that the two difference expressions P_{2p+2} and L commute implies the existence of a polynomial relationship between them as detailed in the next result.

Theorem 1.6 *Assume Hypothesis* 1.1 *and suppose that P_{2p+2} and L commute, $[P_{2p+2}, L] = 0$, or equivalently, that* s-$Tl_p(a,b) = 0$ *for some* $p \in \mathbb{N}_0$. *Then L and P_{2p+2} satisfy an algebraic relationship of the type (cf.* (1.41))

$$\mathcal{F}_p(L, P_{2p+2}) = P_{2p+2}^2 - R_{2p+2}(L) = 0,$$
$$R_{2p+2}(z) = \prod_{m=0}^{2p+1}(z - E_m), \quad z \in \mathbb{C}. \tag{1.46}$$

Proof Using relations (1.20) and (1.35) one computes

$$\begin{aligned}
P_{2p+2}^2\big|_{\ker(L-z)} &= \left(P_{2p+2}\big|_{\ker(L-z)}\right)^2 \\
&= \left((2aF_pS^+ + G_{p+1})\big|_{\ker(L-z)}\right)^2 \\
&= \left(2aF_p(G_{p+1}^+ + G_{p+1} + 2(z-b^+)F_p^+)S^+ \right. \\
&\quad \left. + G_{p+1}^2 - 4a^2F_pF_p^+\right)\big|_{\ker(L-z)} \\
&= (G_{p+1}^2 - 4a^2F_pF_p^+)\big|_{\ker(L-z)} = R_{2p+2}(L)\big|_{\ker(L-z)}.
\end{aligned}$$

Thus one concludes that the finite-order difference expressions P_{2p+2}^2 and $R_{2p+2}(L)$ coincide on the nullspace of $L-z$. Since $z \in \mathbb{C}$ is arbitrary, and solutions $\psi(z)$ of $L\psi = z\psi$ for different values of z are linearly independent, one infers that (1.46) holds. □

The expression $\mathcal{F}_p(L, P_{2p+2})$ in (1.46) represents the Burchnall–Chaundy polynomial of the pair (L, P_{2p+2}). Equation (1.46) naturally leads to the hyperelliptic curve $\mathcal{K}_p$ of (arithmetic) genus $p \in \mathbb{N}_0$ (possibly with a singular affine part), where

$$\mathcal{K}_p: \mathcal{F}_p(z, y) = y^2 - R_{2p+2}(z) = 0,$$
$$R_{2p+2}(z) = \prod_{m=0}^{2p+1}(z - E_m), \quad \{E_m\}_{m=0,\ldots,2p+1} \subset \mathbb{C}. \tag{1.47}$$

Remark 1.7 At this point it is easy to motivate the recursion relation (1.3)–(1.5) used as our starting point for constructing the Toda hierarchy. If one is interested in

1.2 Fundamentals of the Toda Hierarchy

determining difference expressions P commuting with L (other than simply polynomials of L or the case where P and L are polynomials of a third difference expression), one can proceed as follows. Restricting P to the two-dimensional null space, $\ker(L-z)$, of $(L-z)$, one can systematically replace second-order shifts S^{++} by $(a^+)^{-1}(-a-(z-b^+)S^+)$ and hence effectively reduce P on $\ker(L-z)$ to a first-order difference expression of the type $P|_{\ker(L-z)} = (2aF(z)S^+ + G(z))|_{\ker(L-z)}$, where F and G are polynomials. Imposing commutativity of P and L on $\ker(L-z)$ then yields the relations (1.36) and (1.37) and hence (1.38) and (1.39) for F and G. Making the polynomial ansatz $F(z) = \sum_{\ell=0}^{p} f_{p-\ell} z^\ell$, $G(z) = z^{p+1} + \sum_{\ell=0}^{p} g_{p-\ell} z^\ell + f_{p+1}$ (by (1.38) the degree of G exceeds that of F by one) and inserting it into (1.37) and (1.38) then readily yields the recursion relation (1.3)–(1.5) for $f_0, \ldots, f_p, g_0, \ldots, g_p$. In other words, one obtains the beginning of the recursion relation (1.3)–(1.5) as well as relations (1.38), (1.39) defining the pth stationary Toda equations.

Remark 1.8 If a, b satisfy one of the stationary Toda equations in (1.29) for a particular value of p, s-$\mathrm{Tl}_p(a, b) = 0$, then they satisfy infinitely many such equations of order higher than p for certain choices of summation constants c_ℓ. In fact, they satisfy certain stationary Toda equations s-$\mathrm{Tl}_q(a, b) = 0$ for every $q \geq p+1$. This can be shown as in Remark 1.5 of Volume I.

Next we turn to the time-dependent Toda hierarchy. For that purpose the coefficients a and b are now considered as functions of both the lattice point and time. For each equation in the hierarchy, that is, for each $p \in \mathbb{N}_0$, we introduce a deformation (time) parameter $t_p \in \mathbb{R}$ in a, b, replacing $a(n), b(n)$ by $a(n, t_p), b(n, t_p)$. The second-order difference expression L (cf. (1.2)) now reads

$$L(t_p) = a(\,\cdot\,, t_p)S^+ + a^-(\,\cdot\,, t_p)S^- + b(\,\cdot\,, t_p).$$

The quantities $\{f_\ell\}_{\ell \in \mathbb{N}_0}$, $\{g_\ell\}_{\ell \in \mathbb{N}_0}$, and P_{2p+2}, $p \in \mathbb{N}_0$, are still defined by (1.3)–(1.5) and (1.20), respectively. The time-dependent Toda equations are then obtained by imposing the Lax commutator equations

$$L_{t_p}(t_p) - [P_{2p+2}(t_p), L(t_p)] = 0, \quad t_p \in \mathbb{R}. \tag{1.48}$$

Relation (1.48) implies

$$\begin{aligned}&\left(a_{t_p} + a(g_p^+ + g_p + f_{p+1}^+ + f_{p+1} - 2b^+ f_p^+)\right)S^+ \\ &- \left(-b_{t_p} + 2(-b(g_p + f_{p+1}) + a^2 f_p^+ - (a^-)^2 f_p^- + b^2 f_p)\right) \\ &+ \left(a_{t_p} + a(g_p^+ + g_p + f_{p+1}^+ + f_{p+1} - 2b^+ f_p^+)\right)^- S^- = 0.\end{aligned} \tag{1.49}$$

Inserting (1.4) and (1.5) with $\ell = p$ into (1.49) then yields
$$0 = L_{t_p} - [P_{2p+2}, L]$$
$$= \left(a_{t_p} - a(f^+_{p+1} - f_{p+1})\right) S^+ + \left(b_{t_p} + g_{p+1} - g^-_{p+1}\right)$$
$$+ \left(a_{t_p} - a(f^+_{p+1} - f_{p+1})\right)^- S^-.$$

Varying $p \in \mathbb{N}_0$, the collection of evolution equations
$$\text{Tl}_p(a,b) = \begin{pmatrix} a_{t_p} - a(f^+_{p+1} - f_{p+1}) \\ b_{t_p} + g_{p+1} - g^-_{p+1} \end{pmatrix} = 0, \quad (n, t_p) \in \mathbb{Z} \times \mathbb{R}, \ p \in \mathbb{N}_0, \quad (1.50)$$

then defines the time-dependent Toda hierarchy. Explicitly,
$$\text{Tl}_0(a,b) = \begin{pmatrix} a_{t_0} - a(b^+ - b) \\ b_{t_0} - 2(a^2 - (a^-)^2) \end{pmatrix} = 0,$$

$$\text{Tl}_1(a,b) = \begin{pmatrix} a_{t_1} - a((a^+)^2 - (a^-)^2 + (b^+)^2 - b^2) \\ b_{t_1} + 2(a^-)^2(b + b^-) - 2a^2(b^+ + b) \end{pmatrix}$$
$$+ c_1 \begin{pmatrix} -a(b^+ - b) \\ -2(a^2 - (a^-)^2) \end{pmatrix} = 0,$$

$$\text{Tl}_2(a,b) = \begin{pmatrix} a_{t_2} - a((b^+)^3 - b^3 + 2(a^+)^2 b^+ - 2(a^-)^2 b \\ + a^2(b^+ - b) + (a^+)^2 b^{++} + (a^-)^2 b^-) \\ b_{t_2} - 2a^2(b^2 + bb^+ + (b^+)^2 + a^2 + (a^+)^2) \\ +2(a^-)^2(b^2 + bb^- + (b^-)^2 + (a^-)^2 + (a^{--})^2) \end{pmatrix}$$
$$+ c_1 \begin{pmatrix} -a((a^+)^2 - (a^-)^2 + (b^+)^2 - b^2) \\ 2(a^-)^2(b + b^-) - 2a^2(b^+ + b) \end{pmatrix}$$
$$+ c_2 \begin{pmatrix} -a(b^+ - b) \\ -2(a^2 - (a^-)^2) \end{pmatrix} = 0, \text{ etc.,}$$

represent the first few equations of the time-dependent Toda hierarchy. The system of equations, $\text{Tl}_0(a,b) = 0$, is of course *the* Toda system.

The corresponding homogeneous Toda equations are then defined by
$$\widehat{\text{Tl}}_p(a,b) = \text{Tl}_p(a,b)\big|_{c_\ell = 0, \ \ell = 1, \dots, p}. \tag{1.51}$$

Restricting the Lax relation (1.48) to the kernel $\ker(L-z)$ one finds that
$$0 = \left(L_{t_p} - [P_{2p+2}, L]\right)\big|_{\ker(L-z)} = \left(L_{t_p} + (L-z) P_{2p+2}\right)\big|_{\ker(L-z)}$$
$$= \left(a\left(\frac{a_{t_p}}{a} - \frac{a^-_{t_p}}{a^-}\right) + 2(z - b^+) F^+_p - 2(z - b) F_p + G^+_{p+1} - G^-_{p+1}\right) S^+$$
$$+ \left(b_{t_p} + (z-b)\frac{a^-_{t_p}}{a^-} + 2(a^-)^2 F^-_p - 2a^2 F^+_p\right.$$
$$\left. + (z-b)(G^-_{p+1} - G_{p+1})\right)\bigg|_{\ker(L-z)}.$$

1.2 Fundamentals of the Toda Hierarchy

Hence one obtains

$$\frac{a_{t_p}}{a} - \frac{a_{t_p}^-}{a^-} = -2(z - b^+)F_p^+ + 2(z - b)F_p + G_{p+1}^- - G_{p+1}^+, \tag{1.52}$$

$$b_{t_p} = -(z - b)\frac{a_{t_p}^-}{a^-} + 2a^2 F_p^+ - 2(a^-)^2 F_p^- + (z - b)(G_{p+1} - G_{p+1}^-). \tag{1.53}$$

Further manipulations then lead to

$$a_{t_p} = -a\big(2(z - b^+)F_p^+(z) + G_{p+1}^+(z) + G_{p+1}(z)\big), \quad p \in \mathbb{N}_0, \tag{1.54}$$

$$b_{t_p} = 2\big((z - b)^2 F_p(z) + (z - b)G_{p+1}(z) + a^2 F_p^+(z) - (a^-)^2 F_p^-(z)\big), \tag{1.55}$$

$$p \in \mathbb{N}_0.$$

Indeed, adding $G_{p+1} - G_{p+1}$ to (1.52) (neglecting a trivial summation constant) then implies (1.54), and insertion of (1.54) into (1.53) then yields (1.55). Equations (1.54) and (1.55) give an alternative description of the time-dependent Toda hierarchy.

Remark 1.9 From (1.3)–(1.5) and (1.30), (1.31) one concludes that the coefficient a enters quadratically in F_p and G_{p+1}, and hence the Toda hierarchy (1.50) (respectively (1.29)) is invariant under the substitution

$$a \to a_\varepsilon = \{\varepsilon(n)a(n)\}_{n \in \mathbb{Z}}, \quad \varepsilon(n) \in \{1, -1\}, \ n \in \mathbb{Z}.$$

This result should be compared with the following lemma.

Lemma 1.10 *Suppose* $a, b \in \ell^\infty(\mathbb{Z})$ *with* $a(n) \neq 0$, $n \in \mathbb{Z}$, *and introduce* $a_\varepsilon \in \ell^\infty(\mathbb{Z})$ *by*

$$a_\varepsilon = \{\varepsilon(n)a(n)\}_{n \in \mathbb{Z}}, \quad \varepsilon(n) \in \{1, -1\}, \ n \in \mathbb{Z}.$$

Denote by $\check{L}$ *the bounded* $\ell^2(\mathbb{Z})$*-realization of the difference expression* L *in* (1.2) *and define* $\check{L}_\varepsilon$ *in* $\ell^2(\mathbb{Z})$ *with* L *replaced by* $L_\varepsilon = a_\varepsilon S^+ + a_\varepsilon^- S^- + b$. *Then* $\check{L}$ *and* $\check{L}_\varepsilon$ *are unitarily equivalent, that is, there exists a unitary operator* $U_{\tilde{\varepsilon}}$ *in* $\ell^2(\mathbb{Z})$ *such that*

$$\check{L}_\varepsilon = U_{\tilde{\varepsilon}} \check{L} U_{\tilde{\varepsilon}}^{-1}.$$

Proof $U_{\tilde{\varepsilon}}$ is explicitly represented by the infinite diagonal matrix

$$U_{\tilde{\varepsilon}} = \big(\tilde{\varepsilon}(n)\delta_{m,n}\big)_{m,n \in \mathbb{Z}}, \quad \tilde{\varepsilon}(n) \in \{+1, -1\}, \ n \in \mathbb{Z},$$

in the standard basis (1.13) of $\ell^2(\mathbb{Z})$ with

$$\varepsilon(n) = \tilde{\varepsilon}(n)\tilde{\varepsilon}(n + 1), \quad n \in \mathbb{Z}.$$

$\square$

We conclude this section by pointing out an alternative construction of the Toda hierarchy using a zero-curvature approach instead of Lax pairs (L, P_{2p+2}).

Remark 1.11 The zero-curvature formalism for the Toda hierarchy can be set up as follows. One defines the 2×2 matrix-valued polynomials with respect to $z \in \mathbb{C}$,

$$U(z) = \begin{pmatrix} 0 & 1 \\ -a^-/a & (z-b)/a \end{pmatrix}, \qquad (1.56)$$

$$V_{p+1}(z) = \begin{pmatrix} G^-_{p+1}(z) & 2a^- F^-_p(z) \\ -2a^- F_p(z) & 2(z-b)F_p + G_{p+1}(z) \end{pmatrix}, \quad p \in \mathbb{N}_0. \qquad (1.57)$$

Then the stationary part of this section can equivalently be based on the zero-curvature equation

$$0 = UV_{p+1} - V^+_{p+1}U \qquad (1.58)$$

$$= \frac{2}{a} \begin{pmatrix} 0 & 0 \\ a^-\big((z-b^+)F^+_p - (z-b)F_p\big) & a^2 F^+_p - (a^-)^2 F^-_p \\ +\tfrac{1}{2}(G^+_{p+1} - G^-_{p+1}) & +\tfrac{1}{2}(z-b)(G_{p+1} - G^+_{p+1}) \\ & +(z-b)^2 F_p - (z-b^+)(z-b)F^+_p \end{pmatrix}.$$

Thus, one obtains (1.36) from the (2, 1)-entry in (1.58). Insertion of (1.36) into the (2, 2)-entry of (1.58) then yields (1.37). Thus, one also obtains (1.38) and hence the (2, 2)-entry of V_{p+1} in (1.57) simplifies to

$$V_{p+1,2,2}(z) = -G^-_{p+1}(z) \qquad (1.59)$$

in the stationary case. Since $\det(U(z,n)) = a^-(n)/a(n) \neq 0$, $n \in \mathbb{Z}$, the zero-curvature equation (1.58) yields that $\det(V_{p+1}(z,n))$ is a lattice constant (i.e., independent of $n \in \mathbb{Z}$). Hence, the hyperelliptic curve $\mathcal{K}_p$ in (1.47) is then obtained from the characteristic equation of $V_{p+1}(z)$ by[1]

$$\det(yI_2 - V_{p+1}(z,n)) = y^2 + \det(V_{p+1}(z,n))$$
$$= y^2 - G^-_{p-1}(z,n)^2 + 4a^-(n)^2 F^-_p(z,n)F_p(z,n) = y^2 - R_{2p+2}(z) = 0,$$

using (1.59). Similarly, the time-dependent part (1.48)–(1.55) can equivalently be developed from the zero-curvature equation

$$0 = U_{t_p} + UV_{p+1} - V^+_{p+1}U. \qquad (1.60)$$

The (1, 1)- and (1, 2)-entry of (1.60) equals zero, the (2, 1)-entry yields (1.52), and inserting (1.52) into the (2, 2)-entry of (1.60) yields (1.53) and hence also the basic equations defining the time-dependent Toda hierarchy in (1.54), (1.55).

---

[1] I_2 denotes the identity matrix in $\mathbb{C}^2$.

1.3 The Stationary Toda Formalism

> Wenn ich nur erst die Sätze habe! Die Beweise werde ich schon finden.
>
> Bernhard Riemann[1]

As shown in Section 1.2, the stationary Toda hierarchy is intimately connected with pairs of commuting differential expressions P_{2p+2} and L of orders $2p + 2$ and 2, respectively, and a hyperelliptic curve $\mathcal{K}_p$. In this section we study this relationship more closely and present a detailed study of the stationary Toda hierarchy and its algebro-geometric solutions a, b. Our principal tools are derived from combining the polynomial recursion formalism introduced in Section 1.2 and a fundamental meromorphic function ϕ on $\mathcal{K}_p$, the analog of the Weyl–Titchmarsh function of L. With the help of ϕ we study the Baker–Akhiezer function ψ, the common eigenfunction of P_{2p+2} and L, trace formulas, and theta function representations of ϕ, ψ, a, and b.

Unless explicitly stated otherwise, we suppose throughout this section that

$$a, b \in \mathbb{C}^{\mathbb{Z}}, \quad a(n) \neq 0, \; n \in \mathbb{Z}, \tag{1.61}$$

and assume (1.29) (respectively (1.38), (1.39)) and (1.30), (1.31), and freely employ the formalism developed in (1.3)–(1.47), keeping $p \in \mathbb{N}_0$ fixed.

We recall the Burchnall–Chaundy curve

$$\begin{aligned}\mathcal{K}_p : \mathcal{F}_p(z, y) &= y^2 - R_{2p+2}(z) = 0, \\ R_{2p+2}(z) &= \prod_{m=0}^{2p+1}(z - E_m), \quad \{E_m\}_{m=0,\ldots,2p+1} \subset \mathbb{C},\end{aligned} \tag{1.62}$$

as introduced in (1.47). Throughout this section we assume $\mathcal{K}_p$ to be nonsingular, that is, we suppose that

$$E_m \neq E_{m'} \text{ for } m \neq m', \; m, m' = 0, 1, \ldots, 2p + 1. \tag{1.63}$$

$\mathcal{K}_p$ is compactified by joining two points $P_{\infty_\pm}$, $P_{\infty_+} \neq P_{\infty_-}$, but for notational simplicity the compactification is also denoted by $\mathcal{K}_p$. Points P on $\mathcal{K}_p \setminus \{P_{\infty_+}, P_{\infty_-}\}$ are represented as pairs $P = (z, y)$, where $y(\cdot)$ is the meromorphic function on $\mathcal{K}_p$ satisfying $\mathcal{F}_p(z, y) = 0$. The complex structure on $\mathcal{K}_p$ is then defined in the usual way, see Appendix B. Hence, $\mathcal{K}_p$ becomes a hyperelliptic Riemann surface of genus $p \in \mathbb{N}_0$ in a standard manner.

We also emphasize that by fixing the curve $\mathcal{K}_p$ (i.e., by fixing $E_0, \ldots, E_{2p+1}$), the summation constants $c_1, \ldots, c_p$ in $f_{p+1}^+ - f_{p+1}$ and $g_{p+1} - g_{p+1}^-$ (and hence in the corresponding stationary Tl_p equations) are uniquely determined as is clear from

---

[1] Quoted in O. Hölder, *Die Mathematische Methode*, Springer, Berlin, 1924, p. 487. ("If I only had the theorems first. The proofs I would surely find.")

(1.44), (1.45), which establish the summation constants c_ℓ as symmetric functions of $E_0, \ldots, E_{2p+1}$.

For notational simplicity we will usually tacitly assume that $p \in \mathbb{N}$. (The trivial case $p = 0$ is explicitly treated in Example 1.23.)

In the following, the zeros of the polynomial $F_p(\,\cdot\,, n)$ (cf. (1.30)) will play a special role. We denote them by $\{\mu_j(n)\}_{j=1,\ldots,p}$ and hence write

$$F_p(z) = \prod_{j=1}^{p}(z - \mu_j). \tag{1.64}$$

Similarly, we write

$$F_p^\pm(z) = \prod_{j=1}^{p}(z - \mu_j^\pm), \quad \mu_j^\pm(n) = \mu_j(n \pm 1), \; j = 1, \ldots, p, \; n \in \mathbb{Z}, \tag{1.65}$$

and recall that (cf. (1.40))

$$R_{2p+2} - G_{p+1}^2 = -4a^2 F_p F_p^+. \tag{1.66}$$

The next step is crucial; it permits us to "lift" the zeros μ_j and μ_j^+ of F_p and F_p^+ from the complex plane $\mathbb{C}$ to the curve $\mathcal{K}_p$. From (1.66) one infers that

$$R_{2p+2}(z) - G_{p+1}(z)^2 = 0, \quad z \in \{\mu_j, \mu_k^+\}_{j,k=1,\ldots,p}.$$

We now introduce $\{\hat{\mu}_j\}_{j=1,\ldots,p} \subset \mathcal{K}_p$ and $\{\hat{\mu}_j^+\}_{j=1,\ldots,p} \subset \mathcal{K}_p$ by

$$\hat{\mu}_j(n) = (\mu_j(n), -G_{p+1}(\mu_j(n), n)), \quad j = 1, \ldots, p, \; n \in \mathbb{Z}, \tag{1.67}$$

and

$$\hat{\mu}_j^+(n) = (\mu_j^+(n), G_{p+1}(\mu_j^+(n), n)), \quad j = 1, \ldots, p, \; n \in \mathbb{Z}. \tag{1.68}$$

Next, we introduce the fundamental meromorphic function $\phi(\,\cdot\,, n)$ on $\mathcal{K}_p$,

$$\phi(P, n) = \frac{y - G_{p+1}(z, n)}{2a(n) F_p(z, n)} \tag{1.69}$$

$$= \frac{-2a(n) F_p^+(z, n)}{y + G_{p+1}(z, n)}, \tag{1.70}$$

$$P = (z, y) \in \mathcal{K}_p, \; n \in \mathbb{Z},$$

with divisor $(\phi(\,\cdot\,, n))$ of $\phi(\,\cdot\,, n)$ given by

$$(\phi(\,\cdot\,, n)) = \mathcal{D}_{P_{\infty_+} \underline{\hat{\mu}}^+(n)} - \mathcal{D}_{P_{\infty_-} \underline{\hat{\mu}}(n)}, \tag{1.71}$$

using (1.64) and (1.65). Here we abbreviated

$$\underline{\hat{\mu}} = \{\hat{\mu}_1, \ldots, \hat{\mu}_p\}, \; \underline{\hat{\mu}}^+ = \{\hat{\mu}_1^+, \ldots, \hat{\mu}_p^+\} \in \mathrm{Sym}^p(\mathcal{K}_p).$$

1.3 The Stationary Toda Formalism

Given the function $\phi(\,\cdot\,,n)$, the meromorphic stationary Baker–Akhiezer function $\psi(\,\cdot\,,n,n_0)$ on $\mathcal{K}_p$ is then defined by

$$\psi(P,n,n_0) = \begin{cases} \prod_{n'=n_0}^{n-1} \phi(P,n'), & n \geq n_0 + 1, \\ 1, & n = n_0, \\ \prod_{n'=n}^{n_0-1} \phi(P,n')^{-1}, & n \leq n_0 - 1, \end{cases} \quad (1.72)$$

$$P \in \mathcal{K}_p \setminus \{P_{\infty_\pm}\}, \ (n,n_0) \in \mathbb{Z}^2,$$

with divisor

$$(\psi(\,\cdot\,,n,n_0)) = \mathcal{D}_{\underline{\hat{\mu}}(n)} - \mathcal{D}_{\underline{\hat{\mu}}(n_0)} + (n - n_0)(\mathcal{D}_{P_{\infty_+}} - \mathcal{D}_{P_{\infty_-}}). \quad (1.73)$$

In addition to ψ in (1.72) we also introduce the Baker–Akhiezer vector Ψ defined by

$$\Psi(P,n,n_0) = \begin{pmatrix} \psi(P,n,n_0) \\ \psi(P,n+1,n_0) \end{pmatrix}, \quad P \in \mathcal{K}_p \setminus \{P_{\infty_\pm}\}, \ (n,n_0) \in \mathbb{Z}^2. \quad (1.74)$$

Basic properties of ϕ, ψ, and Ψ are summarized in the following result. (We denote by $W(f,g) = a(fg^+ - f^+g)$ the (discrete) Wronskian of f and g, $f, g \in \mathbb{C}^{\mathbb{Z}}$.)

Lemma 1.12 *Suppose that a, b satisfy (1.61) and the pth stationary Toda system (1.29). Moreover, assume (1.62) and (1.63) and let $P = (z,y) \in \mathcal{K}_p \setminus \{P_{\infty_+}, P_{\infty_-}\}$, $(n,n_0) \in \mathbb{Z}^2$. Then ϕ satisfies the Riccati-type equation*

$$a\phi(P) + a^-\phi^-(P)^{-1} = z - b, \quad (1.75)$$

as well as

$$\phi(P)\phi(P^*) = \frac{F_p^+(z)}{F_p(z)}, \quad (1.76)$$

$$\phi(P) + \phi(P^*) = -\frac{G_{p+1}(z)}{aF_p(z)}, \quad (1.77)$$

$$\phi(P) - \phi(P^*) = \frac{y}{aF_p(z)}. \quad (1.78)$$

Moreover, ψ and Ψ satisfy

$$(L - z(P))\psi(P) = 0, \quad (P_{2p+2} - y(P))\psi(P) = 0, \quad (1.79)$$

$$U(z)\Psi^-(P) = \Psi(P), \quad (1.80)$$

$$V_{p+1}(z)\Psi^-(P) = y\Psi^-(P), \quad (1.81)$$

$$\psi(P,n,n_0)\psi(P^*,n,n_0) = \frac{F_p(z,n)}{F_p(z,n_0)}, \quad (1.82)$$

$$a(n)(\psi(P,n,n_0)\psi(P^*,n+1,n_0) + \psi(P,n+1,n_0)\psi(P^*,n,n_0))$$
$$= -G_{p+1}(z,n)/F_p(z,n_0), \tag{1.83}$$
$$W(\psi(P,\cdot,n_0),\psi(P^*,\cdot,n_0)) = -\frac{y}{F_p(z,n_0)}. \tag{1.84}$$

Proof Equation (1.37) implies

$$a\phi(P) + a^-\phi^-(P)^{-1}$$
$$= \frac{1}{2}(-G_{p+1}(z) + y)F_p(z)^{-1} - \frac{1}{2}(G^-_{p+1}(z) + y)F_p(z)^{-1}$$
$$= z - b(n) \tag{1.85}$$

which proves (1.75). Equations (1.76)–(1.78) then follow from (1.40) and (1.70). Clearly $\psi(\cdot,n,n_0)$ is meromorphic on $\mathcal{K}_p$ by (1.72) since $\phi(\cdot,n)$ is. $(L-z)\psi = 0$ follows from (1.85) and

$$\phi(P,n) = \psi(P,n+1,n_0)/\psi(P,n,n_0). \tag{1.86}$$

$(L-z)\psi = 0$, (1.86), and (1.35) then imply $P_{2p+2}\psi = 2aF_p\psi^+ + G_{p+1}\psi = (2aF_p\phi + G_{p+1})\psi = y\psi$. (1.82)–(1.84) are an immediate consequence of (1.76)–(1.78).

Equation (1.80) is an immediate consequence of the definition (1.56) of U and of $(L-z)\psi = 0$. Similarly, (1.81) follows from the definition (1.57) of V_{p+1} and the Riccati equation (1.75). □

The normalization chosen for the Baker–Akhiezer function ψ in (1.72) (basically, $\psi(P,n,n_0)$ equals $\tilde\psi(P,n)/\tilde\psi(P,n_0)$ for a certain (not necessarily normalized) solution $\tilde\psi$ of $(L-z)\psi = 0$) has some interesting consequences and is not quite as innocent as it may appear at first glance. In fact, by (1.73), one infers that its divisor of zeros and poles on $\mathcal{K}_p$ is precisely given by $\mathcal{D}_{\hat{\underline{\mu}}(n)} + n\mathcal{D}_{P_{\infty+}}$ and $\mathcal{D}_{\hat{\underline{\mu}}(n_0)} + n_0\mathcal{D}_{P_{\infty-}}$, respectively.

Equations (1.82)–(1.84) show that the basic identity (1.40), that is, $G^2_{p+1} - 4a^2F_pF^+_p = R_{2p+2}$, is equivalent to the elementary fact

$$a^2(\psi_{1,+}\psi_{2,-} + \psi_{1,-}\psi_{2,+})^2 - 4a^2\psi_{1,+}\psi_{1,-}\psi_{2,+}\psi_{2,-}$$
$$= a^2(\psi_{1,+}\psi_{2,-} - \psi_{1,-}\psi_{2,+})^2,$$

identifying $\psi(P) = \psi_{1,+}$, $\psi(P^*) = \psi_{1,-}$, $\psi^+(P) = \psi_{2,+}$, $\psi^+(P^*) = \psi_{2,-}$. This provides the intimate link between our approach and the squared function systems frequently employed in the literature in connection with algebro-geometric solutions of the Toda hierarchy.

1.3 The Stationary Toda Formalism

If $a, b \in \ell^\infty(\mathbb{Z})$, the zeros $\mu_j(n)$ of $F_p(\cdot, n)$ and the zeros $\mu_\ell^+(n)$ of $F_p^+(\cdot, n)$ are naturally associated with Dirichlet and Neumann boundary conditions of L at the point $n \in \mathbb{Z}$ and $(n+1) \in \mathbb{Z}$, respectively. In other words, the Dirichlet eigenvalues $\mu_j(n)$ are associated with the boundary condition $g(n) = 0$ for an element g in the domain of an appropriate $\ell^2(\mathbb{Z})$-operator realization of L, whereas the Neumann eigenvalues $\mu_\ell^+(n)$ correspond to the boundary condition $g(n+1) = 0$. Next, we "interpolate" between these two boundary conditions and consider the general case

$$g(n+1) + \beta g(n) = 0, \quad \beta \in \mathbb{R}.$$

The values $\beta = \infty$ (formally) and $\beta = 0$ then represent the Dirichlet and Neumann cases, respectively.

To this end we introduce the additional polynomial $K_{p+1}^\beta(z)$, $\beta \in \mathbb{R}$, of degree $p+1$ for $\beta \in \mathbb{R} \setminus \{0\}$ and degree p for $\beta = 0$ by

$$K_{p+1}^\beta(z) = F_p^+(z) - \beta a^{-1} G_{p+1}(z) + \beta^2 F_p(z) \tag{1.87}$$

$$= \begin{cases} \beta a^{-1} \prod_{\ell=0}^p (z - \lambda_\ell^\beta), & \beta \in \mathbb{R} \setminus \{0\} \\ \prod_{j=1}^p (z - \mu_j^+), & \beta = 0 \end{cases} \tag{1.88}$$

$$= \sum_{\ell=0}^p c_{p-\ell} \widehat{K}_{\ell+1}^\beta(z), \quad \beta \in \mathbb{R} \setminus \{0\}. \tag{1.89}$$

Here $\widehat{K}_{\ell+1}^\beta$ denote the corresponding homogeneous polynomials, defined by the vanishing of the integration constants c_k for $k = 1, \dots, \ell$,

$$\widehat{K}_1^\beta(z) = K_1^\beta(z) = \beta a^{-1} z - \beta a^{-1} b + 1 + \beta^2,$$

$$\widehat{K}_{\ell+1}^\beta(z) = K_{\ell+1}^\beta(z)\big|_{c_k = 0,\ k=1,\dots,\ell}, \quad \ell = 0, \dots, p.$$

In particular,

$$K_{p+1}^0(z) = F_p^+(z).$$

Explicitly, one computes

$$K_1^\beta = \beta a^{-1} z - \beta a^{-1} b + 1 + \beta^2,$$
$$K_2^\beta = \beta a^{-1} z^2 + (1+\beta^2) z + \beta^2 b - \beta a^{-1}\big((a^-)^2 - a^2 + b^2\big) + b^+$$
$$\quad + c_1(\beta a^{-1} z - \beta a^{-1} b + 1 + \beta^2), \quad \text{etc.}$$

Next, combining (1.69), (1.70), and (1.87) yields

$$\phi(P) + \beta = \frac{y - G_{p+1}(z) + 2\beta a F_p(z)}{2a F_p(z)}$$

$$= \frac{-2a K_{p+1}^\beta(z)}{y + G_{p+1}(z) - 2\beta a F_p(z)}.$$

One verifies as before (cf. Lemma 1.12) that

$$R_{2n+1}(z) - \left(G_{p+1}(z) - 2\beta a F_p(z)\right)^2 = -4a^2 F_p(z) K^\beta_{p+1}(z),$$

$$(\phi(P) + \beta)(\phi(P^*) + \beta) = \frac{K^\beta_{p+1}(z)}{F_p(z)},$$

$$(\psi(P, n+1, n_0) + \beta\psi(P, n, n_0))(\psi(P^*, n+1, n_0) + \beta\psi(P^*, n, n_0))$$
$$= \frac{K^\beta_{p+1}(z, n)}{F_p(z, n_0)},$$

where the Baker–Akhiezer function $\psi(\,\cdot\,, n, n_0)$ is defined in (1.72). The divisor $(\phi(\,\cdot\,, n) + \beta)$ of $\phi(\,\cdot\,, n) + \beta$, $\beta \in \mathbb{R} \setminus \{0\}$, is then given by

$$(\phi(\,\cdot\,, n) + \beta) = \mathcal{D}_{\hat{\lambda}^\beta_0(n)\hat{\underline{\lambda}}^\beta(n)} - \mathcal{D}_{P_{\infty_-}\hat{\underline{\mu}}(n)}, \quad \beta \in \mathbb{R} \setminus \{0\},$$

with

$$\hat{\lambda}^\beta_\ell(n) = (\lambda^\beta_\ell(n), G_{p+1}(\lambda^\beta_\ell(n), n) - 2\beta a F_p(\lambda^\beta_\ell(n), n)),$$
$$\ell = 0, \ldots, p, \ \beta \in \mathbb{R} \setminus \{0\}. \tag{1.90}$$

Remark 1.13 Our notation $\mathcal{D}_{\hat{\lambda}^\beta_0 \hat{\underline{\lambda}}^\beta}$, $\hat{\underline{\lambda}}^\beta = \{\hat{\lambda}^\beta_1, \ldots, \hat{\lambda}^\beta_p\}$, in the general case, where a, b are complex-valued, is somewhat misleading as

$$\mathcal{D}_{\hat{\lambda}^\beta_0 \hat{\underline{\lambda}}^\beta} = \sum_{\ell=0}^p \mathcal{D}_{\hat{\lambda}^\beta_\ell} \in \mathrm{Sym}^{p+1}(\mathcal{K}_p)$$

is symmetric in $\hat{\lambda}^\beta_0, \ldots, \hat{\lambda}^\beta_p$ and there is no natural way to distinguish $\hat{\lambda}^\beta_0$ from $\hat{\lambda}^\beta_\ell$, $\ell = 1, \ldots, p$. In particular,

$$\mathcal{D}_{\hat{\lambda}^\beta_0 \hat{\underline{\lambda}}^\beta} = \mathcal{D}_{\hat{\lambda}^\beta_\ell \hat{\underline{\lambda}}^{\beta,\ell}}$$

where

$$\hat{\underline{\lambda}}^{\beta,1} = \{\hat{\lambda}^\beta_0, \hat{\lambda}^\beta_2, \ldots, \hat{\lambda}^\beta_p\},$$
$$\hat{\underline{\lambda}}^{\beta,\ell} = \{\hat{\lambda}^\beta_0, \hat{\lambda}^\beta_1, \ldots, \hat{\lambda}^\beta_{\ell-1}, \hat{\lambda}^\beta_{\ell+1}, \hat{\lambda}^\beta_p\}, \quad \ell = 2, \ldots, p-1,$$
$$\hat{\underline{\lambda}}^{\beta,p} = \{\hat{\lambda}^\beta_0, \hat{\lambda}^\beta_1, \ldots, \hat{\lambda}^\beta_{p-1}\}.$$

In the special case, where a, b are real-valued, a distinction between $\hat{\lambda}^\beta_0$ and $\hat{\lambda}^\beta_\ell$, $\ell = 1, \ldots, p$, can be made naturally by supposing

$$\lambda^\beta_0 \in (-\infty, E_0] \cup [E_{2p+1}, \infty), \quad \lambda^\beta_\ell \in [E_{2\ell-1}, E_{2\ell}], \quad \ell = 1, \ldots, p.$$

For notational convenience in connection with positive divisors of degree p on $\mathcal{K}_p$ and their subsequent use in the associated p-dimensional theta function, we will

1.3 The Stationary Toda Formalism

keep the abbreviation $\mathcal{D}_{\hat{\lambda}_0^\beta; \hat{\lambda}}^{\beta}$ for general complex-valued a, b, but occasionally will caution the reader about this convention.

In the special case where $\{E_m\}_{m=0,\ldots,2p+1} \subset \mathbb{R}$, we will from now on always assume the ordering

$$E_m < E_{m+1}, \quad m = 0, 1, \ldots, 2p. \tag{1.91}$$

In particular, if $a, b \in \ell^\infty(\mathbb{Z})$ are assumed to be real-valued, then necessarily $\{\mu_j(n)\}_{j=1,\ldots,p} \subset \mathbb{R}$ and $\{\lambda_\ell^\beta(n)\}_{\ell=0,\ldots,p} \subset \mathbb{R}$, $\beta \in \mathbb{R} \setminus \{0\}$, for all $n \in \mathbb{Z}$, since one is then dealing with self-adjoint boundary value problems in $\ell^2(\mathbb{Z})$; hence, we will also always assume the ordering

$$\mu_j(n) < \mu_{j+1}(n), \quad j = 1, \ldots, p-1, \, n \in \mathbb{Z}, \tag{1.92}$$

$$\lambda_\ell^\beta(n) < \lambda_{\ell+1}^\beta(n), \quad \ell = 1, \ldots, p-1, \, n \in \mathbb{Z} \tag{1.93}$$

in this case.

There is apparently no simple discrete analog of the Dubrovin equations in the case of the Toda lattice, describing the variations of $\mu_j(n)$, $\lambda_\ell^\beta(n)$, $\beta \in \mathbb{R} \setminus \{0\}$, with respect to n by a first-order system of nonlinear difference equations. As a substitute we offer a continuous first-order system of nonlinear differential equations whose solution $\hat{\chi}(x)$ provides a continuous interpolation for $\hat{\mu}(n)$ in the notes to this section.

Next we analyze the behavior of $\lambda_\ell^\beta(n)$ as a function of the boundary condition parameter $\beta \in \mathbb{R}$. By (1.87) one concludes

$$\partial_\beta K_{p+1}^\beta(z) = -a^{-1} G_{p+1}(z) + 2\beta F_p(z) \tag{1.94}$$

and hence

$$\partial_\beta K_{p+1}^\beta(z)\big|_{z=\lambda_\ell^\beta} = -\beta a^{-1} \left(\partial_\beta \lambda_\ell^\beta\right) \prod_{\substack{m=0 \\ m \neq \ell}}^{n} (\lambda_\ell^\beta - \lambda_m^\beta)$$

$$= -a^{-1} G_{p+1}(\lambda_\ell^\beta) + 2\beta F_p(\lambda_\ell^\beta)$$

$$= -a^{-1} y(\hat{\lambda}_\ell^\beta) \tag{1.95}$$

by (1.90). This implies the following result for the β-variation of the eigenvalues $\lambda_\ell^\beta(n)$.

Lemma 1.14 *Suppose that a, b satisfy (1.61) and the pth stationary Toda system (1.29). Moreover, assume (1.62) and (1.63) and let $\beta \in \mathcal{U}$, where $\mathcal{U} \subset \mathbb{R} \setminus \{0\}$ is an open interval, and assume that the zeros $\lambda_\ell^\beta(n)$, $\ell = 0, \ldots, p$, of $K_{p+1}^\beta(\cdot, n)$*

remain distinct for $(n, \beta) \in \mathbb{Z} \times \mathcal{U}$. *Then* $\{\hat{\lambda}_\ell^\beta\}_{\ell=0,\ldots,p}$, *defined by* (1.90), *satisfies the following first-order system of differential equations*

$$\partial_\beta \lambda_\ell^\beta = \beta^{-1} y(\hat{\lambda}_\ell^\beta) \prod_{\substack{m=0 \\ m \neq \ell}}^{p} (\lambda_\ell^\beta - \lambda_m^\beta)^{-1}, \quad \ell = 0, \ldots, p.$$

Proof This follows from (1.95). □

Combining the polynomial recursion approach of Section 1.2 with (1.64) readily yields trace formulas for f_ℓ in terms of symmetric functions of the zeros μ_j of F_p. Similarly, a Lagrange interpolation formula involving G_{p+1} yields a trace formula for a^2. We focus on the simplest trace formulas only. For this purpose we find it convenient to introduce the abbreviation,

$$b^{(k)} = \frac{1}{2} \sum_{m=0}^{2p+1} E_m^k - \sum_{j=1}^{p} \mu_j^k, \quad k \in \mathbb{N}. \tag{1.96}$$

Lemma 1.15 *Suppose that a, b satisfy (1.61) and the pth stationary Toda system (1.29). Then,*

$$b = \frac{1}{2} \sum_{m=0}^{2p+1} E_m - \sum_{j=1}^{p} \mu_j. \tag{1.97}$$

In addition, if for all $n \in \mathbb{Z}$, $\mu_j(n) \neq \mu_k(n)$ for $j \neq k$, $j, k = 1, \ldots, p$, then,

$$a^2 = \frac{1}{2} \sum_{j=1}^{p} y(\hat{\mu}_j) \prod_{\substack{k=1 \\ k \neq j}}^{p} (\mu_j - \mu_k)^{-1} + \frac{1}{4}(b^{(2)} - b^2). \tag{1.98}$$

Proof The trace relation (1.97) follows by a comparison of powers of z^{p-1} in (1.30) and (1.64) for F_p, taking into account (1.6) for f_1. In order to prove (1.98) one can argue as follows. A simple computation reveals

$$G_{p+1}(z) + (z-b)F_p(z) \underset{|z| \to \infty}{=} (g_0 + f_1 - b)z^p + O(z^{p-1}) \underset{|z| \to \infty}{=} O(z^{p-1}), \tag{1.99}$$

since $g_0 + f_1 - b = 0$ by (1.6). Hence, applying Lagrange's interpolation formula (D.6) to $G_{p+1} + (z-b)F_p$ with interpolator F_p yields

$$G_{p+1}(z) + (z-b)F_p(z) = -\sum_{j=1}^{p} y(\hat{\mu}_j) \prod_{\substack{k=1 \\ k \neq j}}^{p} \frac{z - \mu_k}{\mu_j - \mu_k}, \quad z \in \mathbb{C}.$$

1.3 The Stationary Toda Formalism

Assuming $p \geq 2$ for simplicity (the case $p = 1$ can easily be handled separately), one then computes for the leading asymptotic terms in (1.99) as $|z| \to \infty$,

$$-z^{p+1} + g_0 z^p + g_1 z^{p-1} + z^{p+1} + f_1 z^p - b z^p + f_2 z^{p-1} - b f_1 z^{p-1} + O(z^{p-2})$$

$$\underset{|z|\to\infty}{=} -\sum_{j=1}^{p} y(\hat{\mu}_j) \prod_{\substack{k=1 \\ k \neq j}}^{p} (\mu_j - \mu_k)^{-1} z^{p-1} + O(z^{p-2}).$$

Comparing powers of z^{p-1} then yields

$$g_1 + f_2 - b f_1 = -\sum_{j=1}^{p} y(\hat{\mu}_j) \prod_{\substack{k=1 \\ k \neq j}}^{p} (\mu_j - \mu_k)^{-1}. \tag{1.100}$$

By (1.6), (1.100) is equivalent to

$$a^2 - (a^-)^2 = \sum_{j=1}^{p} y(\hat{\mu}_j) \prod_{\substack{k=1 \\ k \neq j}}^{p} (\mu_j - \mu_k)^{-1}. \tag{1.101}$$

On the other hand, using

$$f_1 = -\sum_{j=1}^{p} \mu_j, \quad f_2 = \sum_{\substack{j,k=1 \\ j<k}}^{p} \mu_j \mu_k,$$

one can rewrite (1.100) in the form

$$-2a^2 - c_2 + \sum_{\substack{j,k=1 \\ j<k}}^{p} \mu_j \mu_k + \left(\frac{1}{2} \sum_{m=0}^{2p+1} E_m - \sum_{j=1}^{p} \mu_j\right) \sum_{k=1}^{p} \mu_k$$

$$= -\sum_{j=1}^{p} y(\hat{\mu}_j) \prod_{\substack{k=1 \\ k \neq j}}^{p} (\mu_j - \mu_k)^{-1}. \tag{1.102}$$

Inserting c_1 and c_2 into (1.102), using (1.45), and taking into account (1.96) then yields (1.98). □

Combining (1.98) and (1.101) yields

$$(a^-)^2 = -\frac{1}{2} \sum_{j=1}^{p} y(\hat{\mu}_j) \prod_{\substack{k=1 \\ k \neq j}}^{p} (\mu_j - \mu_k)^{-1} + (1/4)(b^{(2)} - b^2).$$

The case where some of the μ_j coincide in (1.98) requires a more elaborate argument that will be presented in the next Section 1.4.

The situation for the general β-boundary conditions (1.94) is a bit different as described below.

Lemma 1.16 *Suppose that a, b satisfy (1.61) and the pth stationary Toda system (1.29). Then,*

$$a = (\beta + \beta^{-1})^{-1} \left(\frac{1}{2} \sum_{m=0}^{2p+1} E_m - \sum_{\ell=0}^{p} \lambda_\ell^\beta \right). \tag{1.103}$$

Proof This is an immediate consequence of comparing powers of z^p in (1.87) and (1.88). □

There appears to be no simple trace formula for b in terms of elementary symmetric functions of $\lambda_0^\beta, \ldots, \lambda_p^\beta$, though. Comparing powers of z^{p-1} in (1.87) and (1.88) only yields

$$b^+ + \beta^2 b = -(1 + \beta^2)c_1 - \beta(2a + c_2 a^{-1}) + \beta a \sum_{\substack{\ell_1, \ell_2 = 0 \\ \ell_1 < \ell_2}}^{p} \lambda_{\ell_1}^\beta \lambda_{\ell_2}^\beta. \tag{1.104}$$

Equations (1.97) and (1.98) are trace formulas for the algebro-geometric coefficients a, b. Equations (1.97), (1.98), (1.103), (1.104) (as well as the method of proof) indicate that higher-order trace formulas associated with the Toda hierarchy can be obtained from (1.64) and (1.88) comparing powers of z.

Next we turn to asymptotic properties of ϕ and ψ in a neighborhood of $P_{\infty\pm}$.

Lemma 1.17 *Suppose that a, b satisfy (1.61) and the pth stationary Toda system (1.29). Moreover, let $P = (z, y) \in \mathcal{K}_p \setminus \{P_{\infty_+}, P_{\infty_-}\}$, $(n, n_0) \in \mathbb{Z}^2$. Then,*

$$\phi(P) \underset{\zeta \to 0}{=} \begin{cases} a\zeta + ab^+ \zeta^2 + O(\zeta^3) & \text{as } P \to P_{\infty_+}, \\ a^{-1}\zeta^{-1} - a^{-1}b - a^{-1}(a^-)^2 \zeta + O(\zeta^2) & \text{as } P \to P_{\infty_-}, \end{cases} \quad \zeta = 1/z,$$

$$\tag{1.105}$$

$$\psi(P, n, n_0) \underset{\zeta \to 0}{=} \left(A(n, n_0) \zeta^{(n-n_0)} \right)^{\pm 1} (1 + O(\zeta)), \quad P \to P_{\infty_\pm}, \quad \zeta = 1/z, \tag{1.106}$$

where we used the abbreviation

$$A(n, n_0) = \begin{cases} \prod_{n'=n_0}^{n-1} a(n')^{-1}, & n \geq n_0 + 1, \\ 1, & n = n_0, \\ \prod_{n'=n}^{n_0-1} a(n'), & n \leq n_0 - 1. \end{cases}$$

1.3 The Stationary Toda Formalism

Proof The existence of the asymptotic expansion of ϕ in terms of the local coordinate $\zeta = 1/z$ near $P_{\infty\pm}$ (cf. (B.7)–(B.11)) is clear from the explicit form of ϕ in (1.69). Insertion of the polynomial F_p into (1.69) then yields the explicit expansion coefficients in (1.105). Alternatively, and more efficiently, one can insert the ansatz

$$\phi \underset{z\to\infty}{=} \phi_1 z^{-1} + \phi_2 z^{-2} + O(z^{-3})$$

into the Riccati-type equation (1.75). A comparison of powers of z^{-1} then proves the first line in (1.105). Similarly, inserting the ansatz

$$\phi \underset{z\to\infty}{=} \phi_{-1}z + \phi_0 + \phi_1 z^{-1} + O(z^{-2})$$

into the Riccati-type equation (1.75), a comparison of powers of z^{-1} then proves the second line in (1.105).

The existence of the corresponding asymptotic expansion of ψ in terms of the local coordinate $\zeta = 1/z$ near $P_{\infty\pm}$ is clear from the representation (1.72). Inserting the ansatz

$$\psi \underset{\zeta\to 0}{=} (\psi_{0,\pm} + O(\zeta))\zeta^{\mp(n-n_0)}$$

into the equation $a\psi^+ + a^-\psi^- = (z-b)\psi$ and comparing the coefficients of ζ^{-1} then yields the recursion relations

$$a\psi_{0,+}^+ = \psi_{0,+}, \quad a^-\psi_{0,-}^- = \psi_{0,-}$$

and hence proves (1.106) taking into account the normalization $\psi(P, n_0, n_0) = 1$.
$\square$

In addition to (1.105) one can use the Riccati-type equation (1.75) to derive a convergent expansion of ϕ around $P_{\infty\pm}$ and recursively determine the coefficients as in Lemma 1.56. Since this is not used later in this section, we omit further details at this point.

Since nonspecial divisors play a fundamental role in this section and the next, we now take a closer look at them.

Lemma 1.18 *Suppose that a, b satisfy (1.61) and the pth stationary Toda system (1.29). Moreover, assume (1.62) and (1.63) and let $n \in \mathbb{Z}$. Denote by $\mathcal{D}_{\hat{\underline{\mu}}}$, $\hat{\underline{\mu}} = (\hat{\mu}_1, \ldots, \hat{\mu}_p) \in \mathrm{Sym}^p(\mathcal{K}_p)$, the Dirichlet divisor of degree p associated with a, b, and ϕ defined according to (1.67), that is,*

$$\hat{\mu}_j(n) = (\mu_j(n), -G_{p+1}(\mu_j(n), n)) \in \mathcal{K}_p, \quad j = 1, \ldots, p.$$

Then $\mathcal{D}_{\hat{\underline{\mu}}(n)}$ is nonspecial for all $n \in \mathbb{Z}$.

Proof By Theorem A.32, $\mathcal{D}_{\underline{\hat{\mu}}(n)}$ is special if and only if $\{\hat{\mu}_1(n), \ldots, \hat{\mu}_p(n)\}$ contains at least one pair of the type $\{\hat{\mu}(n), \hat{\mu}^*(n)\}$. Hence $\mathcal{D}_{\underline{\hat{\mu}}(n)}$ is certainly nonspecial as long as the projections $\mu_j(n)$ of $\hat{\mu}_j(n)$ are mutually distinct, $\mu_j(n) \neq \mu_k(n)$ for $j \neq k$. On the other hand, if two or more projections coincide for some $n_0 \in \mathbb{Z}$, for instance,

$$\mu_{j_1}(n_0) = \cdots = \mu_{j_N}(n_0) = \mu_0, \quad N \in \{2, \ldots, p\},$$

then $G_{p+1}(\mu_0, n_0) \neq 0$ as long as $\mu_0 \notin \{E_0, \ldots, E_{2p+1}\}$. This fact immediately follows from (1.40) since $F_p(\mu_0, n_0) = 0$ but $R_{2p+2}(\mu_0) \neq 0$ by hypothesis. In particular, $\hat{\mu}_{j_1}(n_0), \ldots, \hat{\mu}_{j_N}(n_0)$ all meet on the same sheet since

$$\hat{\mu}_{j_r}(n_0) = (\mu_0, -G_{p+1}(\mu_0, n_0)), \quad r = 1, \ldots, N,$$

and hence no special divisor can arise in this manner. It remains to study the case where two or more projections collide at a branch point, say at $(E_{m_0}, 0)$ for some $n_0 \in \mathbb{Z}$. In this case one concludes

$$F_p(z, n_0) \underset{z \to E_{m_0}}{=} O\big((z - E_{m_0})^2\big)$$

and

$$G_{p+1}(E_{m_0}, n_0) = 0 \tag{1.107}$$

using again (1.40) and $F_p(E_{m_0}, n_0) = R_{2p+2}(E_{m_0}) = 0$. Since $G_{p+1}(\,\cdot\,, n_0)$ is a polynomial (of degree $p+1$), (1.107) implies

$$G_{p+1}(z, n_0) \underset{z \to E_{m_0}}{=} O((z - E_{m_0})).$$

Thus, using (1.40) once more, one obtains the contradiction,

$$O\big((z - E_{m_0})^2\big) \underset{z \to E_{m_0}}{=} R_{2p+2}(z)$$

$$\underset{z \to E_{m_0}}{=} (z - E_{m_0}) \Bigg(\prod_{\substack{m=1 \\ m \neq m_0}}^{2p+1} (E_{m_0} - E_m) + O(z - E_{m_0}) \Bigg).$$

Consequently, at most one $\hat{\mu}_j(n)$ can hit a branch point at a time and again no special divisor arises. Finally, by our hypotheses on a, b, $\hat{\mu}_j(n)$ stay finite for fixed $n \in \mathbb{Z}$ and hence never reach the points $P_{\infty_\pm}$. (Alternatively, by (1.105), $\hat{\mu}_j$ never reach the point P_{∞_-}. Hence, if some $\hat{\mu}_j$ tend to infinity, they all necessarily converge to P_{∞_+}.) Again no special divisor can arise in this manner, completing the proof. $\square$

If $a, b \in \ell^\infty(\mathbb{Z})$, the Dirichlet Jacobi operator $\check{L}_n^D$, the bounded operator realization of L with a Dirichlet boundary condition at the point $n \in \mathbb{Z}$ in $\ell^2((-\infty, n-1] \cap$

1.3 The Stationary Toda Formalism

$\mathbb{Z}) \oplus \ell^2([n+1, \infty) \cap \mathbb{Z})$ is self-adjoint, and one infers

$$|\mu_j(n)| \leq 2\|a\|_{\ell^\infty(\mathbb{Z})} + \|b\|_{\ell^\infty(\mathbb{Z})}, \quad j = 1, \ldots, p, \; n \in \mathbb{Z}.$$

Next, we shall provide an explicit representation of ϕ, Ψ, a, and b in terms of the Riemann theta function associated with $\mathcal{K}_p$. We freely employ the notation established in Appendices A and B. In order to avoid the trivial case $p = 0$ (considered in Examples 1.23 and 1.26) we assume $p \in \mathbb{N}$ for the remainder of this argument.

Let θ denote the Riemann theta function associated with $\mathcal{K}_p$ and introduce a fixed homology basis $\{a_j, b_j\}_{j=1,\ldots,p}$ on $\mathcal{K}_p$. Choosing as a convenient fixed base point the branch point $P_0 = (E_0, 0)$, the Abel maps $\underline{A}_{P_0}$ and $\underline{\alpha}_{P_0}$ are defined by (A.29) and (A.30) and the Riemann vector $\underline{\Xi}_{P_0}$ is given by (A.41). Let $\omega^{(3)}_{P_{\infty_+}, P_{\infty_-}}$ be the normal differential of the third kind holomorphic on $\mathcal{K}_p \setminus \{P_{\infty_+}, P_{\infty_-}\}$ with simple poles at P_{∞_+} and P_{∞_-} and residues $+1$ and -1, respectively (cf. (A.20)–(A.23), (B.40), (B.43)),

$$\omega^{(3)}_{P_{\infty_+}, P_{\infty_-}} = \frac{1}{y} \prod_{j=1}^{p}(z - \lambda_j)dz \underset{\zeta \to 0}{=} \pm(\zeta^{-1} + O(\zeta))d\zeta \text{ as } P \to P_{\infty_\pm}. \quad (1.108)$$

Here the constants $\{\lambda_j\}_{j=1}^{p} \subset \mathbb{C}$ are uniquely determined by employing the normalization

$$\int_{a_j} \omega^{(3)}_{P_{\infty_+}, P_{\infty_-}} = 0, \quad j = 1, \ldots, p \quad (1.109)$$

and ζ in (1.108) denotes the local coordinate

$$\zeta = 1/z \text{ for } P \text{ near } P_{\infty_\pm}.$$

Moreover,

$$\exp\left(\int_{P_0}^{P} \omega^{(3)}_{P_{\infty_+}, P_{\infty_-}}\right) \underset{\zeta \to 0}{=} (\tilde{a}\zeta)^{\pm 1}\left(\sum_{\ell=0}^{\infty} \tilde{b}_\ell \zeta^\ell\right)^{\pm 1} \quad (1.110)$$

$$\text{as } P \to P_{\infty_\pm}, \quad \zeta = 1/z,$$

where $\tilde{a}$, $\{\tilde{b}_\ell\}_{\ell \in \mathbb{N}_0}$ only depend on $\mathcal{K}_p$ (i.e., on $\{E_m\}_{m=0,\ldots,2p+1}$), $\tilde{a}$ is an integration constant, and

$$\tilde{b}_0 = 1, \quad \tilde{b}_1 = \frac{1}{2}\sum_{m=0}^{2p+1} E_m - \sum_{j=1}^{p} \lambda_j, \text{ etc.} \quad (1.111)$$

In order to prove (1.110) and (1.111) one integrates the expansion (B.43) term by term. The remaining contribution to the integral in (1.110) is then absorbed into the integration constant $\tilde{a}$. The vector of b-periods of the differential $\omega^{(3)}_{P_{\infty_+}, P_{\infty_-}}/(2\pi i)$

is denoted by

$$\underline{U}^{(3)} = (U_1^{(3)}, \ldots, U_p^{(3)}), \quad U_j^{(3)} = \frac{1}{2\pi i} \int_{b_j} \omega_{P_{\infty_+}, P_{\infty_-}}^{(3)}, \quad j = 1, \ldots, p. \quad (1.112)$$

By (A.23) one concludes

$$U_j^{(3)} = A_{P_{\infty_-}}(P_{\infty_+}) = 2A_{P_0}(P_{\infty_+}), \quad j = 1, \ldots, p. \quad (1.113)$$

Assuming $\mathcal{D}_{\underline{Q}}$ to be nonspecial, that is, $i(\mathcal{D}_{\underline{Q}}) = 0$, with $\underline{Q} = (Q_1, \ldots, Q_p)$, a special case of Riemann's vanishing theorem (cf. Theorem A.28) yields

$$\theta(\underline{\Xi}_{P_0} - \underline{A}_{P_0}(P) + \underline{\alpha}_{P_0}(\mathcal{D}_{\underline{Q}})) = 0 \text{ if and only if } P \in \{Q_1, \ldots, Q_p\}.$$

Hence the divisors (1.71) and (1.73) of $\phi(\,\cdot\,, n)$ and $\psi(\,\cdot\,, n, n_0)$ suggest considering expressions of the type

$$C(n) \frac{\theta(\underline{\Xi}_{P_0} - \underline{A}_{P_0}(P) + \underline{\alpha}_{P_0}(\mathcal{D}_{\underline{\hat{\mu}}^+(n)}))}{\theta(\underline{\Xi}_{P_0} - \underline{A}_{P_0}(P) + \underline{\alpha}_{P_0}(\mathcal{D}_{\underline{\hat{\mu}}(n)}))} \exp\left(\int_{P_0}^P \omega_{P_{\infty_+}, P_{\infty_-}}^{(3)} \right), \quad (1.114)$$

and

$$C(n, n_0) \frac{\theta(\underline{\Xi}_{P_0} - \underline{A}_{P_0}(P) + \underline{\alpha}_{P_0}(\mathcal{D}_{\underline{\hat{\mu}}(n)}))}{\theta(\underline{\Xi}_{P_0} - \underline{A}_{P_0}(P) + \underline{\alpha}_{P_0}(\mathcal{D}_{\underline{\hat{\mu}}(n_0)}))} \exp\left((n - n_0) \int_{P_0}^P \omega_{P_{\infty_+}, P_{\infty_-}}^{(3)} \right),$$

(1.115)

for ϕ and ψ, respectively, where $C(n)$ and $C(n, n_0)$ are independent of $P \in \mathcal{K}_p$.

In the following it will be convenient to use the abbreviation

$$\underline{z}(P, \underline{Q}) = \underline{\Xi}_{P_0} - \underline{A}_{P_0}(P) + \underline{\alpha}_{P_0}(\mathcal{D}_{\underline{Q}}),$$
$$P \in \mathcal{K}_p, \ \underline{Q} = \{Q_1, \ldots, Q_p\} \in \mathrm{Sym}^p(\mathcal{K}_p).$$

(1.116)

We note that by (A.48) and (A.49), $\underline{z}(\,\cdot\,, \underline{Q})$ is independent of the choice of base point P_0.

A comparison of (1.105), (1.110), and (1.114) at $P = P_{\infty_+}$ then yields

$$a(n) = C(n)\tilde{a} \frac{\theta(\underline{z}(P_{\infty_+}, \underline{\hat{\mu}}^+(n)))}{\theta(\underline{z}(P_{\infty_+}, \underline{\hat{\mu}}(n)))}, \quad n \in \mathbb{Z}. \quad (1.117)$$

By Abel's theorem (cf. Theorem A.16), (1.73) yields

$$\underline{\alpha}_{P_0}(\mathcal{D}_{\underline{\hat{\mu}}(n)}) = \underline{\alpha}_{P_0}(\mathcal{D}_{\underline{\hat{\mu}}(n_0)}) - \underline{A}_{P_{\infty_-}}(P_{\infty_+})(n - n_0)$$
$$= \underline{\alpha}_{P_0}(\mathcal{D}_{\underline{\hat{\mu}}(n_0)}) - 2\underline{A}_{P_0}(P_{\infty_+})(n - n_0), \quad (1.118)$$

and hence one infers

$$\underline{z}(P_{\infty_-}, \underline{\hat{\mu}}^+) = \underline{z}(P_{\infty_+}, \underline{\hat{\mu}}) \pmod{L_p}. \quad (1.119)$$

Given these preparations, the theta function representations for ϕ, ψ, a, and b then read as follows.

1.3 The Stationary Toda Formalism

Theorem 1.19 *Suppose that a, b satisfy (1.61) and the pth stationary Toda system (1.29). Moreover, assume (1.62), (1.63) and let $P \in \mathcal{K}_p \setminus \{P_{\infty_+}, P_{\infty_-}\}$ and $(n, n_0) \in \mathbb{Z}^2$. Then for each $n \in \mathbb{Z}$, $\mathcal{D}_{\underline{\hat{\mu}}(n)}$ is nonspecial. Moreover,*[1]

$$\phi(P, n) = C(n) \frac{\theta(\underline{z}(P, \underline{\hat{\mu}}^+(n)))}{\theta(\underline{z}(P, \underline{\hat{\mu}}(n)))} \exp\left(\int_{P_0}^{P} \omega_{P_{\infty_+}, P_{\infty_-}}^{(3)}\right), \tag{1.120}$$

and

$$\psi(P, n, n_0) = C(n, n_0) \frac{\theta(\underline{z}(P, \underline{\hat{\mu}}(n)))}{\theta(\underline{z}(P, \underline{\hat{\mu}}(n_0)))} \exp\left((n - n_0) \int_{P_0}^{P} \omega_{P_{\infty_+}, P_{\infty_-}}^{(3)}\right), \tag{1.121}$$

where $C(n)$ and $C(n, n_0)$ are given by

$$C(n) = C(n+1, n) = \left(\frac{\theta(\underline{z}(P_{\infty_+}, \underline{\hat{\mu}}^-(n)))}{\theta(\underline{z}(P_{\infty_+}, \underline{\hat{\mu}}^+(n)))}\right)^{1/2}, \tag{1.122}$$

$$C(n, n_0) = \begin{cases} \prod_{n'=n_0}^{n-1} C(n'), & n \geq n_0 + 1, \\ 1, & n = n_0, \\ \prod_{n'=n}^{n_0-1} C(n')^{-1}, & n \leq n_0 - 1, \end{cases} \tag{1.123}$$

$$= \left(\frac{\theta(\underline{z}(P_{\infty_+}, \underline{\hat{\mu}}(n_0)))\theta(\underline{z}(P_{\infty_+}, \underline{\hat{\mu}}^-(n_0)))}{\theta(\underline{z}(P_{\infty_+}, \underline{\hat{\mu}}(n)))\theta(\underline{z}(P_{\infty_+}, \underline{\hat{\mu}}^-(n)))}\right)^{1/2}. \tag{1.124}$$

Here the square root branch of $C(n)$ in (1.122) has to be chosen according to (1.117) and the square root branch of $C(n, n_0)$ in (1.124) is determined by that in (1.122) and by formula (1.123).

The Abel map linearizes the auxiliary divisor $\mathcal{D}_{\underline{\hat{\mu}}(n)}$ in the sense that

$$\underline{\alpha}_{P_0}(\mathcal{D}_{\underline{\hat{\mu}}(n)}) = \underline{\alpha}_{P_0}(\mathcal{D}_{\underline{\hat{\mu}}(n_0)}) - \underline{A}_{P_{\infty_-}}(P_{\infty_+})(n - n_0). \tag{1.125}$$

Finally, a, b are of the form

$$a(n)^2 = \tilde{a}^2 \frac{\theta(\underline{z}(P_{\infty_+}, \underline{\hat{\mu}}^-(n)))\theta(\underline{z}(P_{\infty_+}, \underline{\hat{\mu}}^+(n)))}{\theta(\underline{z}(P_{\infty_+}, \underline{\hat{\mu}}(n)))^2}, \tag{1.126}$$

$$b(n) = \frac{1}{2} \sum_{m=0}^{2p+1} E_m - \sum_{j=1}^{p} \lambda_j$$

$$- \sum_{j=1}^{p} c_j(p) \frac{\partial}{\partial w_j} \ln\left(\frac{\theta(\underline{z}(P_{\infty_+}, \underline{\hat{\mu}}(n)) + \underline{w})}{\theta(\underline{z}(P_{\infty_+}, \underline{\hat{\mu}}^-(n)) + \underline{w})}\right)\bigg|_{\underline{w}=0} \tag{1.127}$$

with $\tilde{a}$ introduced in (1.110).

---

[1] To avoid multi-valued expressions in formulas such as (1.120), (1.121), etc., we agree always to choose the same path of integration connecting P_0 and P and refer to Remark A.30 for additional tacitly assumed conventions.

Proof By (1.114) and (1.115) in order to prove (1.120) and (1.121) it only remains to determine the constants $C(n)$ and $C(n, n_0)$. By (1.82) one infers

$$\psi(P_{\infty_+}, n, n_0)\psi(P_{\infty_-}, n, n_0) = 1$$

and hence (1.114) and (1.116) yield

$$C(n, n_0)^2 = \frac{\theta(\underline{z}(P_{\infty_+}, \underline{\hat{\mu}}(n_0)))\theta(\underline{z}(P_{\infty_+}, \underline{\hat{\mu}}^-(n_0)))}{\theta(\underline{z}(P_{\infty_+}, \underline{\hat{\mu}}(n)))\theta(\underline{z}(P_{\infty_+}, \underline{\hat{\mu}}^-(n)))}.$$

Because of

$$\phi(P, n) = \psi(P, n+1, n),$$

one gets

$$C(n) = \left(\frac{\theta(\underline{z}(P_{\infty_+}, \underline{\hat{\mu}}^-(n)))}{\theta(\underline{z}(P_{\infty_+}, \underline{\hat{\mu}}^+(n)))}\right)^{1/2}.$$

The linearization property (1.125) has already been noted in (1.118).
Formula (1.117) then proves the representation (1.126) for a.
In order to determine b one can argue as follows: Introducing the short-hand notation

$$\delta_+ = 1, \quad \delta_- = 0,$$

one finds

$$\frac{\theta(\underline{z}(P, \underline{\hat{\mu}}^+(n)))}{\theta(\underline{z}(P, \underline{\hat{\mu}}(n)))} \underset{\zeta \to 0}{=} \frac{\theta(\underline{z}(P_{\infty_+}, \underline{\hat{\mu}}(n+\delta_\pm)))}{\theta(\underline{z}(P_{\infty_+}, \underline{\hat{\mu}}(n-1+\delta_\pm)))} \sum_{\ell=0}^{\infty} \tilde{\theta}_{\pm,\ell}(n)\zeta^\ell \text{ as } P \to P_{\infty_\pm},$$

(1.128)

where

$$\tilde{\theta}_{\pm,0}(n) = 1,$$

(1.129)

$$\tilde{\theta}_{\pm,1}(n) = \mp \sum_{j=1}^{p} c_j(p)\frac{\partial}{\partial w_j} \ln\left(\frac{\theta(\underline{z}(P_{\infty_+}, \underline{\hat{\mu}}(n+\delta_\pm) + \underline{w}))}{\theta(\underline{z}(P_{\infty_+}, \underline{\hat{\mu}}(n-1+\delta_\pm) + \underline{w}))}\right)\bigg|_{\underline{w}=0}, \text{ etc.,}$$

1.3 The Stationary Toda Formalism

and $c_j(k)$ are defined in (B.31). Equation (1.128) follows from (1.116), (1.119), and (B.33). Given (1.128), (1.129), one invokes (1.110) and (1.120) to obtain

$$\phi(P,n) \underset{\zeta\to 0}{=} \tilde{a}C(n)\frac{\theta(\underline{z}(P_{\infty_+},\underline{\hat{\mu}}^+(n)))}{\theta(\underline{z}(P_{\infty_+},\underline{\hat{\mu}}(n)))}\zeta$$

$$+ \tilde{a}C(n)\frac{\theta(\underline{z}(P_{\infty_+},\underline{\hat{\mu}}^+(n)))}{\theta(\underline{z}(P_{\infty_+},\underline{\hat{\mu}}(n)))}$$

$$\times \left(\tilde{b}_1 - \sum_{j=1}^p c_j(p)\frac{\partial}{\partial w_j}\ln\left(\frac{\theta(\underline{z}(P_{\infty_+},\underline{\hat{\mu}}^+(n))+\underline{w})}{\theta(\underline{z}(P_{\infty_+},\underline{\hat{\mu}}(n))+\underline{w})}\right)\bigg|_{\underline{w}=0}\right)\zeta^2$$

$$+ O(\zeta^3) \text{ as } P \to P_{\infty_+}. \tag{1.130}$$

A comparison of (1.105) at P_{∞_+} and (1.130), taking into account (1.117), then identifies b^+ and hence yields the representation (1.127) for b. □

Remark 1.20 Alternatively, one could have derived the expression (1.127) for b by evaluating the integral

$$I = \frac{1}{2\pi i}\int_{\partial\widehat{\mathcal{K}}_p} \tilde{\pi}(\,\cdot\,)\,d\ln(\theta(\underline{z}(\,\cdot\,,\underline{\hat{\mu}})))$$

$$= \sum_{j=1}^p \mu_j + \sum_{P\in\{P_{\infty_\pm}\}} \operatorname*{res}_P\left(\tilde{\pi}(\,\cdot\,)\,d\ln(\theta(\underline{z}(\,\cdot\,,\underline{\hat{\mu}})))\right),$$

using the residue theorem. A direct calculation shows that

$$I = \sum_{j=1}^p \int_{a_j} \tilde{\pi}\omega_j$$

and the trace relation (1.97) for b yields

$$b(n) = \frac{1}{2}\sum_{m=0}^{2p+1} E_m - \sum_{j=1}^p \int_{a_j}\tilde{\pi}\omega_j$$

$$- \sum_{j=1}^p c_j(p)\frac{\partial}{\partial w_j}\ln\left(\frac{\theta(\underline{z}(P_{\infty_+},\underline{\hat{\mu}}(n))+\underline{w})}{\theta(\underline{z}(P_{\infty_+},\underline{\hat{\mu}}^-(n))+\underline{w})}\right)\bigg|_{\underline{w}=0}, \quad n\in\mathbb{Z}. \tag{1.131}$$

A comparison of (1.127) and (1.131) then reveals that

$$\sum_{j=1}^p \int_{a_j}\tilde{\pi}\omega_j = \sum_{j=1}^p \lambda_j.$$

Next we derive an alternative theta function representation of b.

Corollary 1.21 *Under the hypotheses of Theorem 1.19, b admits the representation*

$$b(n) = E_0 - \tilde{a}\frac{\theta(\underline{z}(P_{\infty_+}, \underline{\hat{\mu}}^-(n)))\theta(\underline{z}(P_0, \underline{\hat{\mu}}^+(n)))}{\theta(\underline{z}(P_{\infty_+}, \underline{\hat{\mu}}(n)))\theta(\underline{z}(P_0, \underline{\hat{\mu}}(n)))}$$
$$- \tilde{a}\frac{\theta(\underline{z}(P_{\infty_+}, \underline{\hat{\mu}}(n)))\theta(\underline{z}(P_0, \underline{\hat{\mu}}^-(n)))}{\theta(\underline{z}(P_{\infty_+}, \underline{\hat{\mu}}^-(n)))\theta(\underline{z}(P_0, \underline{\hat{\mu}}(n)))}, \quad n \in \mathbb{Z}.$$

Proof It suffices to combine (1.75), (1.120), (1.122) (all at $P = P_0$), and (1.126). □

One can use Lemma B.1 in the stationary case to obtain an alternative proof of the fact that ϕ and ψ given by (1.120)–(1.124) coincide with the expressions (1.70) and (1.72) and satisfy the Riccati and Jacobi equations (1.75) and (1.79), respectively. We shall use precisely this strategy in the time-dependent context to be discussed in Section 1.5.

Combining (1.125) and (1.126), (1.127) shows the remarkable linearity of the theta function representations for a and b with respect to $n \in \mathbb{Z}$. In fact, one can rewrite (1.126), (1.127) as

$$a(n)^2 = \tilde{a}^2 \frac{\theta(\underline{A} - \underline{B} + \underline{B}n)\theta(\underline{A} + \underline{B} + \underline{B}n)}{\theta(\underline{A} + \underline{B}n)^2},$$

$$b(n) = \frac{1}{2}\sum_{m=0}^{2p+1} E_m - \sum_{j=1}^{p} \lambda_j \qquad (1.132)$$
$$- \sum_{j=1}^{p} c_j(p) \frac{\partial}{\partial w_j} \ln\left(\frac{\theta(\underline{A} + \underline{B}n + \underline{w})}{\theta(\underline{A} - \underline{B} + \underline{B}n + \underline{w})}\right)\bigg|_{\underline{w}=0},$$

where

$$\underline{A} = \Xi_{P_0} - \underline{A}_{P_0}(P_{\infty_+}) + \underline{A}_{P_{\infty_-}}(P_{\infty_+})n_0 + \underline{\alpha}_{P_0}(\mathcal{D}_{\underline{\hat{\mu}}(n_0)}), \qquad (1.133)$$

$$\underline{B} = -\underline{A}_{P_{\infty_-}}(P_{\infty_+}), \qquad (1.134)$$

$$\Lambda_0 = \frac{1}{2}\sum_{m=0}^{2p+1} E_m - \sum_{j=1}^{p} \lambda_j. \qquad (1.135)$$

Here the constants $\tilde{a}, \lambda_j, c_j(p) \in \mathbb{C}, j = 1, \ldots, p$, and the constant vector $\underline{B} \in \mathbb{C}^p$ are uniquely determined by $\mathcal{K}_p$ (and its homology basis), and the constant vector $\underline{A} \in \mathbb{C}^p$ is in one-to-one correspondence with the Dirichlet data $\underline{\hat{\mu}}(n_0) = (\hat{\mu}_1(n_0), \ldots, \hat{\mu}_p(n_0)) \in \text{Sym}^p(\mathcal{K}_p)$ at the initial point n_0 as long as the divisor $\mathcal{D}_{\underline{\hat{\mu}}(n_0)}$ is assumed to be nonspecial.

1.3 The Stationary Toda Formalism

Remark 1.22 The algebro-geometric coefficients a, b in (1.126), (1.127), respectively, (1.132), are complex-valued in general. To obtain real-valued coefficients, one needs to impose certain symmetry constraints on $\mathcal{K}_p$ and additional constraints on $\underline{A}$ in (1.133), which we will briefly indicate next. In particular, the formal self-adjointness of the Lax difference expression $L = aS^+ + a^- S^- + b$, with real-valued coefficients a, b, leads to the reality constraints

$$E_0 < E_1 < \cdots < E_{2p+1} \tag{1.136}$$

on the zeros of R_{2p+2}, that is, all branch points of $\mathcal{K}_p$ are assumed to be in real position.

We choose the homology basis $\{a_j, b_j\}_{j=1}^p$ according to Theorem A.38 (i) (cf. Figure B.2, implementing the additional constraint (1.136)). Moreover, we introduce the antiholomorphic involution $\rho_+ \colon (z, y) \mapsto (\bar{z}, \bar{y})$ as in Example A.37 (i). By Example A.37 (i), Theorem A.38 (cf. (A.61), (A.65)–(A.67)), (B.29), (B.30), (B.31), (B.34), and (B.35)–(B.37)) one infers that $(\mathcal{K}_p, \rho_+)$ is of dividing type and hence

$$r = p + 1, \quad \bar{\tau} = -\tau, \quad R = 0, \quad \overline{\theta(z)} = \theta(\bar{z}), \quad \underline{z} \in \mathbb{C}^p,$$
$$\rho_+(a_j) = a_j, \quad \rho_+(b_j) = -b_j, \quad j = 1, \ldots, p,$$
$$\underline{c}(k) \in \mathbb{R}^p, \quad k = 1, \ldots, p,$$
$$\tilde{a} \in \mathbb{R}, \quad \lambda_j \in \mathbb{R}, \quad j = 1, \ldots, p, \quad \Lambda_0 \in \mathbb{R}.$$

In particular,

$$\underline{B} \in \mathbb{R}^p.$$

The connected component of bounded, real-valued algebro-geometric Jacobi coefficients a^2, b in the Lax difference expression L can then be described as follows: The initial position of $\hat{\mu}_j(n_0) \in \mathcal{K}_p$ must be chosen in real position with its projections lying in the (closure of) spectral gaps of $\check{L}$, that is,

$$\mu_j(n_0) \in [E_{2j-1}, E_{2j}], \quad j = 1, \ldots, p, \tag{1.137}$$

implying

$$\underline{A} \in \mathbb{R}^p.$$

(This immediately shows that a^2 and b are real-valued. An additional argument also shows that a itself is real-valued.) One can show that all real-valued and bounded algebro-geometric Jacobi coefficients a, b arise in this manner. In particular, as n varies in $\mathbb{Z}$, the motion of the projection $\mu_j(n)$ of $\hat{\mu}_j(n) \in \mathcal{K}_p$ remains confined to the interval $[E_{2j-1}, E_{2j}]$ (the closure of the spectral gap (E_{2j-1}, E_{2j})). Since the initial divisor data $\hat{\mu}_j(n_0)$, with the projections $\mu_j(n_0)$ constrained by (1.137) for $j = 1, \ldots, p$, are independent of each other, the set of all initial divisors $\mathcal{D}_{\underline{\hat{\mu}}(n_0)}$

corresponds topologically to a product of p circles. Thus, the corresponding isospectral set of all bounded algebro-geometric Jacobi coefficients a^2, b, corresponding to a fixed curve $\mathcal{K}_p$, constrained by (1.136), can be identified with the p-dimensional real torus $\mathbb{T}^p$. Effective coordinates on this torus uniquely characterizing a^2, b are then the Dirichlet data $\underline{\hat{\mu}}(n_0) = (\hat{\mu}_1(n_0), \ldots, \hat{\mu}_p(n_0))$ (cf. also the notes to Section 1.3), or equivalently, Dirichlet divisors $\mathcal{D}_{\underline{\hat{\mu}}(n_0)}$ in real position constrained by (1.137). Formulas (1.126), (1.127) for a, b then provide a concrete representation of the elements of this isospectral torus $\mathbb{T}^p$. The coefficients a, b in (1.126), (1.127), in general, will be quasi-periodic[1] with respect to $n \in \mathbb{Z}$.

Real-valued Jacobi coefficients a, b associated with $\mathcal{K}_p$ constrained by (1.136) can also be constructed by "misplacing" one or several initial values $\mu_j(n_0)$ in the "wrong" spectral gap closure $(-\infty, E_0]$. This then results in additional connected but noncompact components of isospectral and singular, respectively, unbounded Jacobi coefficients a^2, b.

If in addition one is interested in periodic Jacobi coefficients a, b with a real period $\Omega > 0$, the additional periodicity constraints

$$\Omega \underline{B} \in \mathbb{Z}^p \setminus \{0\}$$

must be imposed.

Next we briefly consider the trivial case $p = 0$ excluded in Theorem 1.19.

Example 1.23 Assume $p = 0$, $P = (z, y) \in \mathcal{K}_0 \setminus \{P_{\infty_+}, P_{\infty_-}\}$, and let $(n, n_0) \in \mathbb{Z}^2$. Then,

$$\mathcal{K}_0 \colon \mathcal{F}_0(z, y) = y^2 - R_2(z) = y^2 - (z - E_0)(z - E_1) = 0,$$
$$E_0, E_1 \in \mathbb{C}, \ E_0 \neq E_1,$$
$$a(n) = a, \ a^2 = (E_1 - E_0)^2/16, \quad b(n) = b = (E_0 + E_1)/2,$$
$$\text{s-}\widehat{\mathrm{Tl}}_k(a, b) = 0, \ k \in \mathbb{N}_0,$$
$$L = aS^+ + aS^- + b, \quad P_2 = aS^+ - aS^-,$$
$$F_0(z, n) = 1, \quad G_1(z, n) = -z + b,$$
$$\phi(P, n) = \frac{y + z - b}{2a},$$
$$\psi(P, n, n_0) = \left(\frac{y + z - b}{2a}\right)^{n - n_0}.$$

---

[1] A sequence $f = \{f(n)\}_{n \in \mathbb{Z}} \in \mathbb{C}^{\mathbb{Z}}$ is called quasi-periodic with fundamental periods $(\omega_1, \ldots, \omega_N) \in (0, \infty)^N$ if the frequencies $2\pi/\omega_1, \ldots, 2\pi/\omega_N$ are linearly independent over $\mathbb{Q}$ and if there exists a continuous function $F \in C(\mathbb{R}^N)$, periodic of period 1 in each of its arguments, such that $f(n) = F(\omega_1^{-1} n, \ldots, \omega_N^{-1} n)$, $n \in \mathbb{Z}$. In particular, f becomes periodic with period $\omega > 0$ if and only if $\omega = m_j \omega_j$ for some $m_j \in \mathbb{N}$, $j = 1, \ldots, N$.

1.3 The Stationary Toda Formalism

We will end this section by providing some additional examples, which we hope will aid in illustrating the general results of this section. We also consider the case of Toda hierarchy solitons involving singular curves even though the principal results of this section were formulated for curves with nonsingular affine parts.

We start with the special case of real-valued periodic stationary Toda solutions and hence summarize the highlights of periodic Jacobi matrices.

Example 1.24 The case of real-valued periodic stationary Toda solutions.
We assume

$$a, b \in \mathbb{R}^{\mathbb{Z}}, \quad a(n) \neq 0, \; n \in \mathbb{Z},$$

and the periodicity condition

$$a(n+N) = a(n), \quad b(n+N) = b(n), \; n \in \mathbb{Z},$$

for some $N \in \mathbb{N}$. (In most formulas below we assume $N \geq 2$ and tacitly avoid the trivial case $N = 1$. The latter situation is treated in Example 1.26.) We agree to abbreviate

$$A = \prod_{n=1}^{N} a(n) = \prod_{n=1}^{N} a(n_0 + n), \quad B = \sum_{n=1}^{N} b(n) = \sum_{n=1}^{N} b(n_0 + n), \quad n_0 \in \mathbb{Z}.$$

Given the fundamental system of solutions $c(z, \cdot, n_0)$ and $s(z, \cdot, n_0)$ of $L\psi = z\psi$, satisfying the initial conditions

$$s(z, n_0, n_0) = 0, \quad s(z, n_0 + 1, n_0) = 1,$$
$$c(z, n_0, n_0) = 1, \quad c(z, n_0 + 1, n_0) = 0,$$

one defines the fundamental matrix

$$\Phi(z, n, n_0) = \begin{pmatrix} c(z, n, n_0) & s(z, n, n_0) \\ c(z, n+1, n_0) & s(z, n+1, n_0) \end{pmatrix}$$

$$= \begin{cases} U_n(z) \cdots U_{n_0+1}(z), & n \geq n_0 + 1, \\ I_2, & n = n_0, \\ U_{n+1}^{-1}(z) \cdots U_{n_0}^{-1}(z), & n \leq n_0 - 1, \end{cases}$$

where

$$U_m(z) = \frac{1}{a(m)} \begin{pmatrix} 0 & a(m) \\ -a(m-1) & z - b(m) \end{pmatrix},$$

$$U_m(z)^{-1} = \frac{1}{a(m-1)} \begin{pmatrix} z - b(m) & -a(m) \\ a(m-1) & 0 \end{pmatrix}.$$

Since
$$W(c(z,\cdot,n_0), s(z,\cdot,n_0)) = a(n_0),$$
an arbitrary solution ψ of $L\psi = z\psi$ is of the type
$$\psi(z,n) = \psi(z,n_0)c(z,n,n_0) + \psi(z,n_0+1)s(z,n,n_0),$$
or equivalently,
$$\begin{pmatrix} \psi(z,n) \\ \psi(z,n+1) \end{pmatrix} = \Phi(z,n,n_0) \begin{pmatrix} \psi(z,n_0) \\ \psi(z,n_0+1) \end{pmatrix}. \tag{1.138}$$

Moreover, one infers
$$\det(\Phi(z,n,n_0)) = \frac{a(n_0)}{a(n)},$$
$$\Phi(z,n,n_0) = \Phi(z,n,n_1)\Phi(z,n_1,n_0),$$
$$\Phi(z,n,n_0)^{-1} = \Phi(z,n_0,n). \tag{1.139}$$

The monodromy matrix $M(z,n)$ is then defined by
$$M(z,n) = \Phi(z,n+N,n)$$
and hence
$$M(z,n) = \Phi(z,n,n_0)M(z,n_0)\Phi(z,n,n_0)^{-1} \tag{1.140}$$
and
$$\det(M(z,n)) = 1.$$

The Floquet discriminant $\Delta(z)$ defined by
$$\Delta(z) = \frac{1}{2}\operatorname{tr}(M(z,n))$$
is independent of n (cf. (1.140)) and the Floquet multipliers $\rho_\pm(z)$ (the eigenvalues of $M(z,n)$) then read
$$\rho_\pm(z) = \Delta(z) \pm (\Delta(z)^2 - 1)^{1/2}. \tag{1.141}$$

Again by (1.140) they are independent of n and satisfy
$$\rho_+(z)\rho_-(z) = 1, \quad \rho_+(z) + \rho_-(z) = 2\Delta(z).$$

Let $\{\widetilde{E}_\ell\}_{\ell=0,\ldots,2N-1} \subset \mathbb{R}$ be the zeros of $\Delta(z)^2 - 1$ and write
$$\Delta(z)^2 - 1 = \frac{1}{4A^2} \prod_{\ell=0}^{2N-1}(z - \widetilde{E}_\ell) \tag{1.142}$$

1.3 The Stationary Toda Formalism

and

$$\Delta(z) \mp 1 = \frac{1}{2A} \prod_{j=1}^{N} (z - E_j^{\pm}). \tag{1.143}$$

The zeros $\{E_j^{\pm}\}_{1 \leq j \leq N}$ turn out to be the eigenvalues of the following periodic, respectively antiperiodic, Jacobi matrices $\check{L}_{n_0}^{\pm}$ in $\mathbb{C}^N$. More generally, define $\check{L}_{n_0}^{\theta}$ in $\mathbb{C}^N$ associated with the boundary conditions

$$a(n_0+N)\psi(n_0+N) = e^{i\theta} a(n_0)\psi(n_0), \quad \psi(n_0+N+1) = e^{i\theta}\psi(n_0+1), \quad 0 \leq \theta < 2\pi$$

by

$$\check{L}_{n_0}^{\theta} = \begin{pmatrix} b(n_0+1) & a(n_0+1) & 0 & \cdots & 0 & e^{-i\theta}a(n_0+N) \\ a(n_0+1) & b(n_0+2) & \ddots & & & 0 \\ 0 & \ddots & \ddots & \ddots & \ddots & \vdots \\ \vdots & & \ddots & \ddots & \ddots & 0 \\ 0 & & \ddots & \ddots & b(n_0+N-1) & a(n_0+N-1) \\ e^{i\theta}a(n_0+N) & 0 & \cdots & 0 & a(n_0+N-1) & b(n_0+N) \end{pmatrix},$$

$$0 \leq \theta < 2\pi.$$

One infers that $\check{L}_{n_0}^{\theta}$ and $\check{L}_{n_0}^{2\pi-\theta}$ are antiunitarily equivalent. The periodic, respectively antiperiodic, Jacobi matrices $\check{L}_{n_0}^{\pm}$ alluded to above are then defined by

$$\check{L}_{n_0}^{+} = \check{L}_{n_0}^{0}, \quad \check{L}_{n_0}^{-} = \check{L}_{n_0}^{\pi}.$$

The eigenvalues of $\check{L}_{n_0}^{\theta}$ are then given by

$$(\rho_+(z) - e^{i\theta})(\rho_-(z) - e^{i\theta}) = 0, \text{ that is, by } \Delta(z) = \cos(\theta).$$

They are simple for $\theta \in (0, \pi) \cup (\pi, 2\pi)$ and at most twice degenerate for $\theta = 0$ or π. The periodic and antiperiodic eigenvalues $\{E_j^{\pm}\}_{1 \leq j \leq N}$ (cf. (1.143)) satisfy the inequalities

$$E_1^{\pm} < E_1^{\mp} \leq E_2^{\mp} < E_2^{\pm} \leq E_3^{\pm} < \cdots < E_{N-1}^{(-1)^{N-1}} \leq E_N^{(-1)^{N-1}} < E_N^{(-1)^N},$$

$$\text{sgn}(A) = \pm(-1)^N.$$

In addition (cf. (1.142)),

$$\{E_j^{\pm}\}_{1 \leq j \leq N} = \{\widetilde{E}_\ell\}_{\ell=0,\ldots,2N-1}.$$

Another way to express these facts is to invoke the theory of direct integral decompositions

$$\ell^2(\mathbb{Z}) \cong \int_{[0,2\pi)}^{\oplus} \frac{d\theta}{2\pi} \ell^2((n_0+1, n_0+N)), \quad \check{L} \cong \int_{[0,2\pi)}^{\oplus} \frac{d\theta}{2\pi} \check{L}_{n_0}^{\theta},$$

where $\cong$ denotes unitary equivalence. In particular, the spectrum $\text{spec}(\check{L})$ of $\check{L}$ is characterized by

$$\text{spec}(\check{L}) = \{\lambda \in \mathbb{R} \mid |\Delta(\lambda)| \leq 1\} = \bigcup_{j=0}^{N-1} [\widetilde{E}_{2j}, \widetilde{E}_{2j+1}].$$

Returning to the square root $(\Delta(z)^2 - 1)^{1/2}$ in (1.141), we shall define it as follows. First, we fix its branch near $+\infty$ by

$$(\Delta(\lambda)^2 - 1)^{1/2} = -\operatorname{sgn}(A)|(\Delta(\lambda)^2 - 1)^{1/2}|, \quad \lambda > \widetilde{E}_{2N-1}.$$

Requiring $(\Delta(z)^2 - 1)^{1/2}$ to be analytic in $\mathbb{C} \setminus \bigcup_{j=0}^{N-1}[\widetilde{E}_{2j}, \widetilde{E}_{2j+1}]$ one then defines

$$(\Delta(\lambda)^2 - 1)^{1/2} = \lim_{\varepsilon \downarrow 0}(\Delta(\lambda + i\varepsilon)^2 - 1)^{1/2}, \quad \lambda \in \mathbb{R},$$

and analytically continues with respect to $z \in \mathbb{C} \setminus \bigcup_{j=0}^{N-1}[\widetilde{E}_{2j}, \widetilde{E}_{2j+1}]$.

As a consequence one obtains

$$|\rho_+(z)| \leq 1, \quad |\rho_-(z)| \geq 1 \tag{1.144}$$

and the (normalized) Floquet functions $\psi_\pm(z, \cdot, n_0)$ defined by

$$\psi_\pm(z, n, n_0) = c(z, n, n_0) + \phi_\pm(z, n)s(z, n, n_0)$$

then satisfy

$$\psi_\pm(z, n+N, n_0) = \rho_\pm(z)\psi_\pm(z, n, n_0) \tag{1.145}$$

and

$$W(\psi_-(z, \cdot, n_0), \psi_+(z, \cdot, n_0)) = a(n_0)(\phi_+(z, n_0) - \phi_-(z, n_0)).$$

In addition, one infers

$$\phi_\pm(z, n_0) = \phi_\pm(z, n_0 + N) = \frac{\rho_\pm(z) - c(z, n_0 + N, n_0)}{s(z, n_0 + N, n_0)}$$

$$= \frac{c(z, n_0 + N + 1, n_0)}{\rho_\pm(z) - s(z, n_0 + N + 1, n_0)}, \tag{1.146}$$

$$W(\psi_-(z, \cdot, n_0), \psi_+(z, \cdot, n_0)) = \frac{2a(n_0)(\Delta(z)^2 - 1)^{1/2}}{s(z, n_0 + N, n_0)}, \tag{1.147}$$

$$G(z, n, n) = \frac{s(z, n+N, n)}{2a(n)(\Delta(z)^2 - 1)^{1/2}} = \frac{\prod_{j=1}^{N-1}(z - \mu_j(n))}{\left(\prod_{\ell=0}^{2N-1}(z - \widetilde{E}_\ell)\right)^{1/2}}, \tag{1.148}$$

$$\psi_+(z, n, n_0)\psi_-(z, n, n_0) = \frac{a(n_0)s(z, n+N, n)}{a(n)s(z, n_0 + N, n_0)} = \prod_{j=1}^{N-1}\left(\frac{z - \mu_j(n)}{z - \mu_j(n_0)}\right). \tag{1.149}$$

If all spectral gaps of $\check{L}$ are "open", that is, the spectra of $\check{L}_{n_0}^{\pm}$ are both simple, we have

$$p = N - 1, \quad \left(\prod_{\ell=0}^{2N-1}(z - \widetilde{E}_\ell)\right)^{1/2} = R_{2p+2}(z)^{1/2} = 2A(\Delta(z)^2 - 1)^{1/2},$$

see (B.17)–(B.18). In the case where some spectral gaps "close", we introduce the index sets

$$J' = \{j' \in \{1, \ldots, N-1\} \mid \widetilde{E}_{2j'-1} = \widetilde{E}_{2j'}\},$$
$$J = \{0, 1, \ldots, 2N-1\} \setminus \{j', j'+1 \mid j' \in J'\},$$

and define

$$Q_{|J'|}(z) = \frac{1}{2A} \prod_{j' \in J'} (z - \widetilde{E}_{2j'-1}), \quad R_{2p+2}(z) = \prod_{j \in J}(z - \widetilde{E}_j).$$

In order to establish the connection with the notation employed earlier in this section and in Appendix A we agree to identify

$$\{\widetilde{E}_j\}_{j \in J} \text{ and } \{E_m\}_{0 \leq m \leq 2p+1}$$

and

$$\{\widetilde{\lambda}_{j'}\}_{j' \in \{1,\ldots,N-1\} \setminus J'} \text{ and } \{\lambda_j\}_{1 \leq j \leq p}.$$

Then one infers

$$p = N - 1 - |J'| = N - 1 - \deg(Q_{|J'|}) = \frac{1}{2}(|J| - 2),$$
$$(\Delta(z)^2 - 1)^{1/2} = R_{2p+2}(z)^{1/2} Q_{|J'|}(z),$$
$$\mathcal{K}_p \colon \mathcal{F}_p(z, y) = y^2 - R_{2p+2}(z) = y^2 - \prod_{m=0}^{2p+1}(z - E_m) = 0,$$

where $|J|$ and $|J'|$ abbreviate the cardinality of J and J', respectively. Finally, the N-periodic sequences a, b satisfy

$$\text{s-Tl}_p(a, b) = 0$$

for an appropriate set of summation constants $\{c_\ell\}_{\ell=1,\ldots,p} \subset \mathbb{C}$ (cf. (1.44)).

Next, we indicate a systematic approach to high-energy expansions of the functions $c(z, n, n_0)$ and $s(z, n, n_0)$. This will then be used to explicitly compute $\tilde{a}, \tilde{b}_1$, and B

in Lemma 1.25. First we note that (1.139) yields

$$\begin{aligned}
s(z, n+1, n_0) &= a(n_0)a(n)^{-1}c(z, n_0, n), \\
s(z, n, n_0) &= -a(n_0)a(n)^{-1}s(z, n_0, n), \\
c(z, n, n_0) &= a(n_0)a(n)^{-1}s(z, n_0+1, n), \\
c(z, n+1, n_0) &= -a(n_0)a(n)^{-1}c(z, n_0+1, n)
\end{aligned} \quad (1.150)$$

and (1.138) implies

$$\begin{aligned}
s(z, n, n_0+1) &= -a(n_0+1)a(n_0)^{-1}c(z, n, n_0), \\
c(z, n, n_0-1) &= -a(n_0-1)a(n_0)^{-1}s(z, n, n_0), \\
s(z, n, n_0-1) &= c(z, n, n_0) + (z - b(n_0))a(n_0)^{-1}s(z, n, n_0), \\
c(z, n, n_0+1) &= s(z, n, n_0) + (z - b(n_0+1))a(n_0+1)^{-1}c(z, n, n_0).
\end{aligned} \quad (1.151)$$

Next we define the Jacobi matrix $J_{n_0}(k)$ in $\mathbb{C}^k$

$$J_{n_0}(k) = \begin{pmatrix} b(n_0+1) & a(n_0+1) & 0 & \cdots & & 0 \\ a(n_0+1) & b(n_0+2) & \ddots & & \ddots & \vdots \\ 0 & \ddots & \ddots & \ddots & & 0 \\ \vdots & & \ddots & \ddots & b(n_0+k-1) & a(n_0+k-1) \\ 0 & \cdots & & 0 & a(n_0+k-1) & b(n_0+k) \end{pmatrix}$$

and introduce

$$P_{n_0}(n, k) = \frac{1}{n}\left(\operatorname{tr}(J_{n_0}(k)^n) - \sum_{j=1}^{n-1} P_{n_0}(j, k)\operatorname{tr}(J_{n_0}(k)^{n-j})\right).$$

One then obtains

$$s(z, n_0+k+1, n_0) = \frac{\det(z - J_{n_0}(k))}{\prod_{n=1}^{k} a(n_0+n)} = \frac{z^k - \sum_{\ell=1}^{k} P_{n_0}(\ell, k)z^{k-\ell}}{\prod_{n=1}^{k} a(n_0+n)}, \quad k \in \mathbb{N}. \quad (1.152)$$

Explicitly, one computes

$$\operatorname{tr}(J_{n_0}(k)) = \sum_{n=n_0+1}^{n_0+k} b(n),$$

$$\operatorname{tr}(J_{n_0}(k)^2) = \sum_{n=n_0+1}^{n_0+k} b(n)^2 + 2\sum_{n=n_0+1}^{n_0+k-1} a(n)^2, \quad (1.153)$$

1.3 The Stationary Toda Formalism

$$\operatorname{tr}(J_{n_0}(k)^3) = \sum_{n=n_0+1}^{n_0+k} b(n)^3 + 3 \sum_{n=n_0+1}^{n_0+k-1} a(n)^2(b(n) + b(n+1)), \text{ etc.}$$

Using (1.150) and (1.151) one can extend (1.152) to $k \leq -1$ and to corresponding results for $c(z, n, n_0)$. A direct calculation yields

$$c(z, n_0 + k + 1, n_0) = -\frac{a(n_0)z^{k-1}}{\prod_{n=1}^{k} a(n_0 + n)} \left(1 - z^{-1} \sum_{n=2}^{k} b(n_0 + n) + O(z^{-2})\right),$$

$$c(z, n_0 - k, n_0) = \frac{z^k}{\prod_{n=1}^{k} a(n_0 - n)} \left(1 - z^{-1} \sum_{n=0}^{k-1} b(n_0 - n) + O(z^{-2})\right),$$

$$s(z, n_0 + k + 1, n_0) = \frac{z^k}{\prod_{n=1}^{k} a(n_0 + n)} \left(1 - z^{-1} \sum_{n=1}^{k} b(n_0 + n) + O(z^{-2})\right),$$

$$s(z, n_0 - k, n_0) = -\frac{a(n_0)z^{k-1}}{\prod_{n=1}^{k} a(n_0 - n)} \left(1 - z^{-1} \sum_{n=1}^{k-1} b(n_0 - n) + O(z^{-2})\right),$$

$$k \in \mathbb{N}. \qquad (1.154)$$

We emphasize that (1.150)–(1.154) hold for general (not necessarily periodic or algebro-geometric finite-band) Jacobi operators. In the following we shall apply (1.154) to the periodic case. Equations (1.141), (1.144) yield the expansion

$$\rho_{\pm}(z) \underset{|z| \to \infty}{=} (1 \mp 1)\Delta(z) \pm \frac{1}{2\Delta(z)} + O(\Delta(z)^{-3})$$

$$\underset{|z| \to \infty}{=} \left(\frac{z^N}{A}\right)^{\mp 1} (1 + O(z^{-1}))$$

and (1.146) and (1.154) then imply

$$\phi_{\pm}(z, n) \underset{|z| \to \infty}{=} \left(\frac{a(n)}{z}\right)^{\pm 1} \left(1 \pm z^{-1} b(n + \delta_{\pm}) + O(z^{-2})\right),$$

$$\delta_{\pm} = \frac{1}{2}(1 \pm 1) = \begin{cases} 1, \\ 0. \end{cases} \qquad (1.155)$$

The relation

$$\psi_{\pm}(z, n, n_0) = \begin{cases} \prod_{n'=n_0}^{n-1} \phi_{\pm}(z, n'), & n \geq n_0 + 1, \\ 1, & n = n_0, \\ \prod_{n'=n}^{n_0-1} \phi_{\pm}(z, n')^{-1}, & n \leq n_0 - 1, \end{cases}$$

then yields

$$\psi_\pm(z, n_0 + k, n_0) = \left(z^{-k} \prod_{n=0}^{k-1} a(n_0 + n)\right)^{\pm 1} \quad (1.156)$$

$$\times \left(1 \pm z^{-1} \sum_{n=\delta_\pm}^{k(k-1)} b(n_0 + n) + O(z^{-2})\right), \quad k \in \mathbb{N},$$

$$\psi_\pm(z, n_0 - k, n_0) = \left(z^{-k} \prod_{n=1}^{k} a(n_0 - n)\right)^{\mp 1} \quad (1.157)$$

$$\times \left(1 \pm z^{-1} \sum_{n=\delta_\pm}^{k(k-1)} b(n_0 - n) + O(z^{-2})\right), \quad k \in \mathbb{N}.$$

Expansions (1.155)–(1.156) also hold in the general case if $\psi_\pm$ are the solutions of $L\psi = z\psi$ which are in $\ell^2((0, \pm\infty))$.

These expansions can now be employed to explicitly compute $\tilde{a}, \tilde{b}_1$, and B (cf. (1.110)–(1.111)).

Lemma 1.25 *In the periodic case one obtains (cf.* (1.110), (1.111))

$$\tilde{a} = -|A|^{1/N}, \quad (1.158)$$

$$\tilde{b}_1 = \frac{B}{N} = \frac{1}{2} \sum_{\ell=0}^{2N-1} \tilde{E}_\ell - \sum_{j=1}^{N} \tilde{\lambda}_j = \frac{1}{2} \sum_{m=0}^{2p+1} E_m - \sum_{j=1}^{p} \lambda_j. \quad (1.159)$$

Proof Combining (1.145), (1.156), and (1.110) yields

$$\rho_\pm(z) = \psi_\pm(z, n_0 + N, n_0) = (A/z^N)^{\pm 1}(1 \pm z^{-1}B + O(z^{-2}))$$
$$= \text{sgn}(A)(-\tilde{a}/z)^{\pm N}(1 \pm z^{-1}N\tilde{b}_1 + O(z^{-2}))$$

and hence (1.158) (noting $\tilde{a} < 0$) and the first equality in (1.159). Combining the latter and (1.111) (accounting for the possibility of closing spectral gaps) then yields the last two equalities in (1.159). □

The special cases of real-valued periodic stationary Toda solutions corresponding to $p = 0, 1$ are isolated next (in the case $p = 0$ we complement Example 1.23 and record additional results under the current real-valuedness assumptions on a and b):

Example 1.26 The case $p = 0$.
Let $a \in \mathbb{R} \setminus \{0\}, b \in \mathbb{R}, N \in \mathbb{N}$, and consider

$$a(n) = a, \quad b(n) = b, \; n \in \mathbb{Z}.$$

1.3 The Stationary Toda Formalism

One then verifies the following explicit formulas:

$$\phi_\pm(z, n) = \frac{1}{2a}\left(z - b \pm \sqrt{(z-b)^2 - 4a^2}\right) = \phi_\pm(z),$$

$$\psi_\pm(z, n, n_0) = \phi_\pm(z)^{(n-n_0)},$$

$$s(z, n, n_0) = \frac{a}{\sqrt{(z-b)^2 - 4a^2}}\left(\phi_+(z)^{(n-n_0)} - \phi_-(z)^{(n-n_0)}\right),$$

$$c(z, n, n_0) = -s(z, n-1, n_0),$$

$$\Delta(z) = \frac{1}{2}\left(\phi_+(z)^N + \phi_-(z)^N\right),$$

$$\rho_\pm(z) = \phi_\pm(z)^N,$$

$$A = a^N, \quad B = Nb,$$

$$\widetilde{E}_0 = -2|a| + b, \quad \widetilde{E}_{2j+1} = \widetilde{E}_{2j+2} = \mu_j(n) = -2|a|\cos(j\pi/N) + b,$$

$$j = 0, \ldots, N-2, \quad n \in \mathbb{Z}, \quad \widetilde{E}_{2N-1} = 2|a| + b,$$

$$J' = \{1, 2, \ldots, N-1\}, \quad J = \{0, 2N-1\},$$

$$E_0 = -2|a| + b, \quad E_1 = 2|a| + b,$$

$$|a| = \frac{1}{4}(E_1 - E_0), \quad b = \frac{1}{2}(E_0 + E_1),$$

$$\check{L} = a(S^+ + S^-) + b, \quad \text{dom}(\check{L}) = \ell^2(\mathbb{Z}),$$

$$\text{spec}(\check{L}) = [E_0, E_1] = [-2|a| + b, 2|a| + b],$$

$$\mathcal{K}_0: \mathcal{F}_0(z, y) = y^2 - R_2(z) = y^2 - (z - E_0)(z - E_1) = 0,$$

$$\tilde{a} = -|a|, \quad \tilde{b}_1 = b,$$

$$\text{s-}\widehat{\text{Tl}}_k(a, b) = 0, \quad k \in \mathbb{N}_0.$$

Example 1.27 The case $p = 1$.
Assume

$$E_0 < E_1 < E_2 < E_3, \tag{1.160}$$

and introduce the following objects:

$$\mathcal{K}_1: \mathcal{F}_1(z, y) = y^2 - R_4(z) = y^2 - \prod_{m=0}^{3}(z - E_m) = 0,$$

$$k = \left(\frac{(E_2 - E_1)(E_3 - E_0)}{(E_3 - E_1)(E_2 - E_0)}\right)^{1/2} \in (0, 1),$$

$$k' = \left(\frac{(E_3 - E_2)(E_1 - E_0)}{(E_3 - E_1)(E_2 - E_0)}\right)^{1/2} \in (0, 1),$$

$$k^2 + k'^2 = 1,$$

$$\bar{u}(z) = \left(\frac{(E_3 - E_1)(E_0 - z)}{(E_3 - E_0)(E_1 - z)}\right)^{1/2},$$

$$C = \frac{2}{((E_3 - E_1)(E_2 - E_0))^{1/2}}, \tag{1.161}$$

$$F(z, k) = \int_0^z \frac{dx}{((1 - x^2)(1 - k^2 x^2))^{1/2}},$$

$$E(z, k) = \int_0^z dx \left(\frac{1 - x^2}{1 - k^2 x^2}\right)^{1/2},$$

$$\Pi(z, \alpha^2, k) = \int_0^z \frac{dx}{(1 - \alpha^2 x^2)((1 - x^2)(1 - k^2 x^2))^{1/2}}, \quad \alpha^2 \in \mathbb{R},$$

$$K(k) = F(1, k), \quad E(k) = E(1, k), \quad \Pi(\alpha^2, k) = \Pi(1, \alpha^2, k).$$

(We note that all square roots are assumed to be positive for $x \in (0, 1)$.) Here F, E, and Π denote the Jacobi integral of the first, second, and third kind, respectively. One observes that $E(\,\cdot\,, k)$ has a simple pole at ∞ while $\Pi(\,\cdot\,, \alpha^2, k)$ has simple poles at $z = \pm \alpha^{-1}$.

Given these concepts we can now compute the basic objects in connection with the elliptic curve $\mathcal{K}_1$ as follows:

$$\omega_1 = \frac{dz}{2CK(k)y},$$

$$\tau_{1,1} = \int_{b_1} \omega_1 = iK(k')/K(k),$$

$$A_{P_0}(P) = \pm(2K(k))^{-1} F(\bar{u}(z), k) \quad (\text{mod } L_1), \quad P = (z, y) \in \Pi_\pm,$$

$$A_{P_0}(P_{\infty_+}) = (2K(k))^{-1} F\big(((E_3 - E_1)/(E_3 - E_0))^{1/2}, k\big) \quad (\text{mod } L_1),$$

$$\Xi = \frac{1}{2}(1 - \tau_{1,1}) \quad (\text{mod } L_1), \tag{1.162}$$

$$\omega^{(3)}_{P_{\infty_+}, P_{\infty_-}} = \frac{(z - \lambda_1)\, dz}{y}, \quad \lambda_1 = E_0 + \frac{E_1 - E_0}{K(k)} \Pi((E_2 - E_1)/(E_2 - E_0), k),$$

$$\int_{b_1} \omega^{(3)}_{P_{\infty_+}, P_{\infty_-}} = 2\pi i \Big(K(k)^{-1} F\big(((E_3 - E_1)/(E_3 - E_0))^{1/2}, k\big) + 1\Big),$$

$$\int_{P_0}^{P} \omega^{(3)}_{P_{\infty_+}, P_{\infty_-}}$$

$$= \pm C(E_1 - E_0)\Big(\Big(1 - K(k)^{-1} \Pi((E_2 - E_1)/(E_2 - E_0), k)\Big) F(\bar{u}(z), k)$$

$$- \Pi(\bar{u}(z), (E_3 - E_0)/(E_3 - E_1), k)\Big), \quad P = (z, y) \in \Pi_\pm.$$

The relation

$$A_{P_0}(\hat{\mu}_1(n)) = A_{P_0}(\hat{\mu}_1(n_0)) - 2(n - n_0) A_{P_0}(P_{\infty_+})$$

1.3 The Stationary Toda Formalism

for $(n, n_0) \in \mathbb{Z}^2$ then yields

$$\mu_1(n) = E_1\Big(1 - ((E_2 - E_1)/(E_2 - E_0))(E_0/E_1)$$
$$\times \operatorname{sn}^2\big(2K(k)\delta_1 - 2(n - n_0)F\big(((E_3 - E_1)/(E_3 - E_0))^{1/2}, k\big)\big)\Big)$$
$$\times \Big(1 - ((E_2 - E_1)/(E_2 - E_0))$$
$$\times \operatorname{sn}^2\big(2K(k)\delta_1 - 2(n - n_0)F\big(((E_3 - E_1)/(E_3 - E_0))^{1/2}, k\big)\big)\Big)^{-1},$$

where we abbreviated

$$A_{P_0}(\hat{\mu}_1(n_0)) = (-\delta_1 + \tfrac{1}{2}\tau_{1,1}) \pmod{L_1}$$

and

$$\operatorname{sn}(w) = z, \quad w = \int_0^z \frac{dx}{((1-x^2)(1-k^2x^2))^{1/2}} = F(z, k). \tag{1.163}$$

Moreover,

$$\operatorname{s-Tl}_1(a, b) = 0$$

for $c_1 = -\tfrac{1}{2}(E_0 + E_1 + E_2 + E_3)$.

Recalling that

$$\theta(z) = \vartheta_3(z) = \sum_{n \in \mathbb{Z}} \exp(2\pi i n z + \pi i \tau_{1,1} n^2), \tag{1.164}$$

the results collected in Example 1.27 now enable one to express a, b in terms of the quantities in (1.160)–(1.163), and (1.164). We omit further details at this point.

Finally, we consider the case of (not necessarily real-valued) *p-solitons* for the stationary Toda lattice hierarchy.

Example 1.28 The case of stationary Toda p-soliton solutions.
Let $p \in \mathbb{N}, n \in \mathbb{Z}$, introduce

$$\tau_0(n) = 1,$$
$$\tau_p(n) = \det(I_p + C_p(n)), \quad C_p(n) = \left(\check{c}_j\check{c}_k \frac{(z_j z_k)^{n+1}}{1 - z_j z_k}\right)_{j,k=1,\ldots,p},$$
$$\check{c}_j \in \mathbb{C} \setminus \{0\}, \ z_j \in \mathbb{C}, \ 0 < |z_j| < 1, \ z_j \neq z_k \text{ for } j \neq k, \ j, k \in \{1, \ldots, p\},$$

and assume that $\tau_{p-1}(n) \neq 0$, $\tau_p(n) \neq 0$ for all $n \in \mathbb{Z}$. Then,

$$a_p(n) = \frac{\tau_p(n+2)^{1/2}\tau_p(n)^{1/2}}{2\tau_p(n+1)},$$

$$b_p(n) = -\frac{1}{2}\left(\frac{z_p\tau_{p-1}(n+2)\tau_p(n)}{\tau_{p-1}(n+1)\tau_p(n+1)} + \frac{\tau_{p-1}(n)\tau_p(n+1)}{z_p\tau_{p-1}(n+1)\tau_p(n)}\right) + \frac{1}{2}(z_p + z_p^{-1}),$$

and

$$\text{s-Tl}_p(a_p, b_p) = 0$$

for an appropriate set of summation constants $\{c_\ell\}_{\ell=1,\ldots,p} \subset \mathbb{C}$ (cf. (1.44)). The associated singular curve is given by

$$\mathcal{F}_p(z, y) = y^2 - (z+1)(z-1)\prod_{j=1}^{p}\left(z - \frac{1}{2}(z_j + z_j^{-1})\right)^2 = 0,$$

$$E_0 = -1, \quad E_{2p+1} = 1, \quad E_{2j-1} = E_{2j} = \frac{1}{2}(z_j + z_j^{-1}), \quad j = 1, \ldots, p.$$

The conditions in Example 1.28 are satisfied in the case where a, b are real-valued, assuming

$$\check{c}_j > 0, \quad j = 1, \ldots, p,$$

$$\frac{1}{2}(z_p + z_p^{-1}) > \frac{1}{2}(z_{p-1} + z_{p-1}^{-1}) > \cdots > \frac{1}{2}(z_1 + z_1^{-1}) > 1,$$

and for sufficiently small (complex) perturbations thereof.

This completes our treatment of stationary algebro-geometric solutions of the Toda hierarchy. Before we now turn to the corresponding time-dependent case we describe a solution of the stationary algebro-geometric Toda hierarchy initial value problem with complex-valued initial data in the following section. Equivalently, we will present an algorithm solving the inverse algebro-geometric spectral problem for non-self-adjoint Jacobi operators.

1.4 The Stationary Toda Algebro-Geometric Initial Value Problem

> My method to overcome a difficulty is to go around it.
> *George Pólya*[1]

The aim of this section is to derive an algorithm that enables one to construct algebro-geometric solutions for the stationary Toda hierarchy for complex-valued initial data. To this effect we will develop a new algorithm for constructing stationary complex-valued algebro-geometric solutions of the Toda hierarchy, which is of independent

---
[1] *How to Solve It*, p. 181.

1.4 The Stationary Toda Algebro-Geometric IVP

interest as it solves the inverse algebro-geometric spectral problem for generally non-self-adjoint Lax operators $\check{L}$ in $\ell^2(\mathbb{Z})$ (i.e., operator realizations of the difference expression L in (1.2)), starting from a suitably chosen set of initial divisors of full measure. The generally non-self-adjoint behavior of the underlying Lax operator associated with general coefficients for the Toda hierarchy poses a variety of difficulties that we will briefly indicate next (a more detailed discussion will follow in the time-dependent context in Section 1.6).

In the special case of a self-adjoint Lax (i.e., Jacobi) operator $\check{L}$, where a and b are real-valued and bounded, real-valued algebro-geometric solutions of the stationary Toda hierarchy, or equivalently, self-adjoint Jacobi operators with real-valued algebro-geometric coefficients a, b are constructed as follows: One develops an algorithm that constructs finite nonspecial divisors $\mathcal{D}_{\hat{\underline{\mu}}(n)} \in \text{Sym}^p(\mathcal{K}_p)$ in real position for all $n \in \mathbb{Z}$ starting from an initial Dirichlet divisor $\mathcal{D}_{\hat{\underline{\mu}}(n_0)} \in \text{Sym}^p(\mathcal{K}_p)$ in an appropriate real position (i.e., with Dirichlet eigenvalues in appropriate spectral gaps of $\check{L}$). "Trace formulas" of the type (1.97) and (1.98) then construct the stationary real-valued solutions $a^{(0)}, b^{(0)}$ of s-Tl$_p(a, b) = 0$.

This approach works perfectly in the special self-adjoint case where the Dirichlet divisors $\hat{\underline{\mu}}(n) = (\hat{\mu}_1(n), \ldots, \hat{\mu}_p(n)) \in \text{Sym}^p(\mathcal{K}_p)$, $n \in \mathbb{Z}$, yield Dirichlet eigenvalues $\mu_1, \ldots, \mu_p$ of the Lax operator $\check{L}$ situated in p different spectral gaps of $\check{L}$ on the real axis. In particular, for fixed $n \in \mathbb{Z}$, the Dirichlet eigenvalues $\mu_j(n)$, $j = 1, \ldots, p$, are pairwise distinct and hence formula (1.98) for a is well-defined.

This situation drastically changes if complex-valued initial data $a^{(0)}, b^{(0)}$ or $\mathcal{D}_{\hat{\underline{\mu}}(n_0)}$ are permitted. In this case the Dirichlet eigenvalues μ_j, $j = 1, \ldots, p$, are no longer confined to separated spectral gaps of $\check{L}$ on the real axis and, in particular, they are in general no longer pairwise distinct and "collisions" between them can occur at certain values of $n \in \mathbb{Z}$. Thus, the stationary algorithm breaks down at such values of n. A priori, one has no control over such collisions, especially, it is not possible to identify initial conditions $\mathcal{D}_{\hat{\underline{\mu}}(n_0)}$ at some $n_0 \in \mathbb{Z}$ which avoid collisions for all $n \in \mathbb{Z}$. We solve this problem directly by explicitly permitting collisions from the outset. In particular, we properly modify the algorithm described above in the self-adjoint case by referring to a more general interpolation formalism (cf. the end of Appendix D) for polynomials, going beyond the usual Lagrange interpolation formulas. In this manner it will be shown that collisions of Dirichlet eigenvalues no longer pose a problem.

In addition, there is a second nontrivial complication in the non-self-adjoint case: Since the Dirichlet eigenvalues $\mu_j(n)$, $j = 1, \ldots, p$, are no longer confined to spectral gaps of $\check{L}$ on the real axis as n varies in $\mathbb{Z}$, it can no longer be guaranteed that $\mu_j(n)$, $j = 1, \ldots, p$, stay finite for all $n \in \mathbb{Z}$. As will be shown in this section, this phenomenon is related to certain deformations of the algebraic curve $\mathcal{K}_p$ under which for some $n_0 \in \mathbb{Z}$, $a(n_0) \to 0$ and $\mu_j(n_0 + 1) \to \infty$ for some $j \in \{1, \ldots, p\}$. We solve this particular problem by properly restricting the initial Dirichlet divisors $\mathcal{D}_{\underline{\hat{\mu}}(n_0)} \in \mathrm{Sym}^p(\mathcal{K}_p)$ to a dense set of full measure.

Summing up, we offer a new algorithm to solve the inverse algebro-geometric spectral problem for general (non-self-adjoint) Toda Lax operators, starting from a properly chosen dense set of initial divisors of full measure.

After this lengthy introduction we now embark on the corresponding inverse problem consisting of constructing a solution of (1.29) given certain initial data. More precisely, we seek to construct solutions $a, b \in \mathbb{C}^{\mathbb{Z}}$ satisfying the pth stationary Toda system (1.29) starting from a properly restricted set $\mathcal{M}_0$ of finite nonspecial Dirichlet divisor initial data $\mathcal{D}_{\underline{\hat{\mu}}(n_0)}$ at some fixed $n_0 \in \mathbb{Z}$,

$$\begin{aligned}
\underline{\hat{\mu}}(n_0) &= \{\hat{\mu}_1(n_0), \ldots, \hat{\mu}_p(n_0)\} \in \mathcal{M}_0, \quad \mathcal{M}_0 \subset \mathrm{Sym}^p(\mathcal{K}_p), \\
\hat{\mu}_j(n_0) &= \left(\mu_j(n_0), -G_{p+1}(\mu_j(n_0), n_0)\right), \quad j = 1, \ldots, p.
\end{aligned} \quad (1.165)$$

For notational convenience we will use the phrase that a, b *blow up* whenever the divisor $\mathcal{D}_{\underline{\hat{\mu}}}$ hits one of the points P_{∞_+} or P_{∞_-}.

Of course, we would like to ensure that the sequences obtained via our algorithm do not blow up. To investigate when this happens, we study the image of our divisors under the Abel map. A key ingredient in our analysis will be (1.125) which yields a linear discrete dynamical system on the Jacobi variety $J(\mathcal{K}_p)$. In particular, we will be led to investigate solutions $\mathcal{D}_{\underline{\hat{\mu}}}$ of the discrete initial value problem

$$\begin{aligned}
\underline{\alpha}_{Q_0}(\mathcal{D}_{\underline{\hat{\mu}}(n)}) &= \underline{\alpha}_{Q_0}(\mathcal{D}_{\underline{\hat{\mu}}(n_0)}) - (n - n_0)\underline{A}_{P_{\infty_-}}(P_{\infty_+}), \\
\underline{\hat{\mu}}(n_0) &= \{\hat{\mu}_1(n_0), \ldots, \hat{\mu}_p(n_0)\} \in \mathrm{Sym}^p(\mathcal{K}_p),
\end{aligned} \quad (1.166)$$

where $Q_0 \in \mathcal{K}_p$ is a given base point. Eventually, we will be interested in solutions $\mathcal{D}_{\underline{\hat{\mu}}}$ of (1.166) with initial data $\mathcal{D}_{\underline{\hat{\mu}}(n_0)}$ satisfying (1.165) and $\mathcal{M}_0$ to be specified as in (the proof of) Lemma 1.30.

Before proceeding to develop the stationary Toda algorithm, we briefly analyze the dynamics of (1.166).

Lemma 1.29 *Let $n \in \mathbb{Z}$ and suppose that $\mathcal{D}_{\underline{\hat{\mu}}(n)}$ is defined via (1.166) for some divisor $\mathcal{D}_{\underline{\hat{\mu}}(n_0)} \in \mathrm{Sym}^p(\mathcal{K}_p)$.*
(i) If $\mathcal{D}_{\underline{\hat{\mu}}(n)}$ is finite and nonspecial and $\mathcal{D}_{\underline{\hat{\mu}}(n+1)}$ is infinite, then $\mathcal{D}_{\underline{\hat{\mu}}(n+1)}$ contains P_{∞_+} but not P_{∞_-}.
(ii) If $\mathcal{D}_{\underline{\hat{\mu}}(n)}$ is nonspecial and $\mathcal{D}_{\underline{\hat{\mu}}(n+1)}$ is special, then $\mathcal{D}_{\underline{\hat{\mu}}(n)}$ contains P_{∞_+} at least twice.
(iii) Items (i) and (ii) hold if $n + 1$ is replaced by $n - 1$ and P_{∞_+} by P_{∞_-}.

1.4 The Stationary Toda Algebro-Geometric IVP

Proof (i) Suppose one point in $\mathcal{D}_{\underline{\hat{\mu}}(n+1)}$ equals P_{∞_-} and denote the remaining ones by $\mathcal{D}_{\underline{\tilde{\mu}}(n+1)}$. Then (1.166) implies that $\underline{\alpha}_{Q_0}(\mathcal{D}_{\underline{\tilde{\mu}}(n+1)}) + \underline{A}_{Q_0}(P_{\infty_+}) = \underline{\alpha}_{Q_0}(\mathcal{D}_{\underline{\hat{\mu}}(n)})$. Since we assumed $\mathcal{D}_{\underline{\hat{\mu}}(n)}$ to be nonspecial, we have $\mathcal{D}_{\underline{\hat{\mu}}(n)} = \mathcal{D}_{\underline{\tilde{\mu}}(n+1)} + \mathcal{D}_{P_{\infty_+}}$, contradicting finiteness of $\mathcal{D}_{\underline{\hat{\mu}}(n)}$.

(ii) Next, we choose Q_0 to be a branch point of $\mathcal{K}_p$ such that $\underline{A}_{Q_0}(P^*) = -\underline{A}_{Q_0}(P)$. If $\mathcal{D}_{\underline{\hat{\mu}}(n+1)}$ is special, then it contains a pair of points (Q, Q^*) whose contribution will cancel under the Abel map, that is, $\underline{\alpha}_{Q_0}(\mathcal{D}_{\underline{\hat{\mu}}(n+1)}) = \underline{\alpha}_{Q_0}(\mathcal{D}_{\underline{\hat{\nu}}(n+1)})$ for some $\mathcal{D}_{\underline{\hat{\nu}}(n+1)} \in \text{Sym}^{p-2}(\mathcal{K}_p)$. Invoking (1.166) then shows that $\underline{\alpha}_{Q_0}(\mathcal{D}_{\underline{\hat{\mu}}(n)}) = \underline{\alpha}_{Q_0}(\mathcal{D}_{\underline{\hat{\nu}}(n+1)}) + 2\underline{A}_{Q_0}(P_{\infty_+})$. As $\mathcal{D}_{\underline{\hat{\mu}}(n)}$ was assumed to be nonspecial, this shows that $\mathcal{D}_{\underline{\hat{\mu}}(n)} = \mathcal{D}_{\underline{\hat{\nu}}(n+1)} + 2\mathcal{D}_{P_{\infty_+}}$, as claimed.

(iii) This is proved analogously to item (i). □

This yields the following behavior of $\mathcal{D}_{\underline{\hat{\mu}}(n)}$ if we start with some nonspecial finite initial divisor $\mathcal{D}_{\underline{\hat{\mu}}(n_0)}$: As n increases, $\mathcal{D}_{\underline{\hat{\mu}}(n)}$ stays nonspecial as long as it remains finite. If it becomes infinite, then it is still nonspecial and contains P_{∞_+} at least once (but not P_{∞_-}). Further increasing n, all instances of P_{∞_+} will be rendered into P_{∞_-} step by step, until we have again a nonspecial divisor that has the same number of P_{∞_-} as the first infinite one had P_{∞_+}. Generically, one expects the subsequent divisor to be finite and nonspecial again.

Next we show that most initial divisors are well-behaved in the sense that their iterates stay away from $P_{\infty_\pm}$. Since we want to show that this set is of full measure, it will be convenient to identify $\text{Sym}^p(\mathcal{K}_p)$ with the Jacobi variety $J(\mathcal{K}_p)$ via the Abel map and take the Haar measure on $J(\mathcal{K}_p)$. Of course, the Abel map is only injective when restricted to the set of nonspecial divisors, but these are the only ones we are interested in.

Lemma 1.30 *The set $\mathcal{M}_0 \subset \text{Sym}^p(\mathcal{K}_p)$ of initial divisors $\mathcal{D}_{\underline{\hat{\mu}}(n_0)}$ for which $\mathcal{D}_{\underline{\hat{\mu}}(n)}$, defined via (1.166), is finite and hence nonspecial for all $n \in \mathbb{Z}$, forms a dense set of full measure in the set $\text{Sym}^p(\mathcal{K}_p)$ of positive divisors of degree p.*

Proof Let $\mathcal{M}_{\infty_\pm}$ be the set of divisors in $\text{Sym}^p(\mathcal{K}_p)$ for which (at least) one point is equal to P_{∞_+} or P_{∞_-}. The image $\underline{\alpha}_{Q_0}(\mathcal{M}_{\infty_\pm})$ of $\mathcal{M}_{\infty_\pm}$ is given by

$$\underline{\alpha}_{Q_0}(\mathcal{M}_{\infty_\pm}) \subseteq \bigcup_{P \in \{P_{\infty_+}, P_{\infty_-}\}} \underline{A}_{Q_0}(P) + \underline{\alpha}_{Q_0}(\text{Sym}^{p-1}(\mathcal{K}_p)) \subset J(\mathcal{K}_p).$$

Since the (complex) dimension of $\text{Sym}^{p-1}(\mathcal{K}_p)$ is $p - 1$, its image must be of measure zero by Sard's theorem. Similarly, let $\mathcal{M}_{\text{sp}}$ be the set of special divisors, then its image is given by

$$\underline{\alpha}_{Q_0}(\mathcal{M}_{\text{sp}}) = \underline{\alpha}_{Q_0}(\text{Sym}^{p-2}(\mathcal{K}_p)),$$

assuming Q_0 to be a branch point of $\mathcal{K}_p$. In particular, we conclude that $\underline{\alpha}_{Q_0}(\mathcal{M}_{\text{sp}}) \subset \underline{\alpha}_{Q_0}(\mathcal{M}_{\infty\pm})$ and thus $\underline{\alpha}_{Q_0}(\mathcal{M}_{\text{sing}}) = \underline{\alpha}_{Q_0}(\mathcal{M}_{\infty\pm})$ has measure zero, where

$$\mathcal{M}_{\text{sing}} = \mathcal{M}_{\infty\pm} \cup \mathcal{M}_{\text{sp}}.$$

Hence,

$$\bigcup_{n\in\mathbb{Z}} \left(\underline{\alpha}_{Q_0}(\mathcal{M}_{\text{sing}}) + n\underline{A}_{P_{\infty-}}(P_{\infty+})\right) \tag{1.167}$$

is of measure zero as well. But this last set contains all initial divisors which will hit $P_{\infty+}$ or $P_{\infty-}$ or become special at some $n \in \mathbb{Z}$. We denote by $\mathcal{M}_0$ the inverse image of the complement of the set (1.167) under the Abel map,

$$\mathcal{M}_0 = \underline{\alpha}_{Q_0}^{-1}\left(\text{Sym}^p(\mathcal{K}_p) \setminus \bigcup_{n\in\mathbb{Z}} \left(\underline{\alpha}_{Q_0}(\mathcal{M}_{\text{sing}}) + n\underline{A}_{P_{\infty-}}(P_{\infty+})\right)\right).$$

Since $\mathcal{M}_0$ is of full measure, it is automatically dense in $\text{Sym}^p(\mathcal{K}_p)$. □

We briefly illustrate some aspects of this analysis in the special case $p = 1$ (i.e., the case where (1.62) represents an elliptic Riemann surface) in more detail.

Example 1.31 The case $p = 1$.
In this case one has

$$F_1(z, n) = z - \mu_1(n),$$
$$G_2(z, n) = R_4(\hat{\mu}_1(n))^{1/2} + (z - b(n))F_1(z, n),$$
$$R_4(z) = \prod_{m=0}^{3}(z - E_m),$$

and hence a straightforward calculation shows that

$$G_2(z, n)^2 - R_4(z) = 4a(n)^2(z - \mu_1(n))(z - \mu_1(n+1))$$
$$= (z - \mu_1(n))(4a(n)^2 z - 4a(n)^2 b(n) + \widetilde{E}),$$

where

$$\widetilde{E} = \frac{1}{8}(E_0 + E_1 - E_2 - E_3)(E_0 - E_1 + E_2 - E_3)(E_0 - E_1 - E_2 + E_3).$$

Solving for $\mu_1(n + 1)$, one obtains

$$\mu_1(n+1) = b(n) - \frac{\widetilde{E}}{4a(n)^2}.$$

This shows that $\mu_1(n_0 + 1) \to \infty$, in fact, $\mu_1(n_0 + 1) = O(a(n_0)^{-2})$ as $a(n_0) \to 0$

1.4 The Stationary Toda Algebro-Geometric IVP

during an appropriate deformation of the parameters $E_m, m = 0, \ldots, 3$. In particular, as $a(n_0) \to 0$, one thus infers $b(n_0 + 1) \to \infty$ during such a deformation since

$$b(n) = \frac{1}{2} \sum_{m=0}^{3} E_m - \mu_1(n), \quad n \in \mathbb{Z},$$

specializing to $p = 1$ in the trace formula (1.97). Next, we illustrate the set $\mathcal{M}_\infty$ in the case $p = 1$. (We recall that $\mathcal{M}_{sp} = \emptyset$ and hence $\mathcal{M}_{sing} = \mathcal{M}_\infty$ if $p = 1$.) By (1.166) one infers

$$A_{P_{\infty_+}}(\hat{\mu}_1(n)) = A_{P_{\infty_+}}(\hat{\mu}_1(n_0)) + (n - n_0) A_{P_{\infty_+}}(P_{\infty_-}), \quad n, n_0 \in \mathbb{Z}. \quad (1.168)$$

We note that $\hat{\mu}_1 \in \mathcal{M}_\infty$ is equivalent to

there is an $n \in \mathbb{Z}$ such that $\hat{\mu}_1(n) = P_{\infty_+}$ (or P_{∞_-}). (1.169)

By (1.168), relation (1.169) is equivalent to

$$A_{P_{\infty_+}}(\hat{\mu}_1(n_0)) + A_{P_{\infty_+}}(P_{\infty_-})\mathbb{Z} = 0 \pmod{L_1}.$$

Thus, $\mathcal{D}_{\hat{\mu}_1(n_0)} \in \mathcal{M}_0 \subset \mathcal{K}_1$ if and only if

$$A_{P_{\infty_+}}(\hat{\mu}_1(n_0)) + A_{P_{\infty_+}}(P_{\infty_-})\mathbb{Z} \neq 0 \pmod{L_1}$$

or equivalently, if and only if

$$A_{P_{\infty_-}}(\hat{\mu}_1(n_0)) + A_{P_{\infty_-}}(P_{\infty_+})\mathbb{Z} \neq 0 \pmod{L_1}.$$

Next, we describe the stationary Toda algorithm. Since this is a somewhat lengthy affair, we will break it up into several steps.

The Stationary (Complex) Toda Algorithm:

We prescribe the following data:

(i) The set

$$\{E_m\}_{m=0}^{2p+1} \subset \mathbb{C}, \quad E_m \neq E_{m'} \text{ for } m \neq m', \ m, m' = 0, \ldots, 2p+1, \quad (1.170)$$

for some fixed $p \in \mathbb{N}$. Given $\{E_m\}_{m=0}^{2p+1}$, we introduce the function R_{2p+2} and the hyperelliptic curve $\mathcal{K}_p$ (with nonsingular affine part) according to (1.62).

(ii) The nonspecial divisor

$$\mathcal{D}_{\underline{\hat{\mu}}(n_0)} \in \operatorname{Sym}^p(\mathcal{K}_p),$$

where $\underline{\hat{\mu}}(n_0)$ is of the form

$$\underline{\hat{\mu}}(n_0) = \{\underbrace{\hat{\mu}_1(n_0), \ldots, \hat{\mu}_1(n_0)}_{p_1(n_0) \text{ times}}, \ldots, \underbrace{\hat{\mu}_{q(n_0)}, \ldots, \hat{\mu}_{q(n_0)}}_{p_{q(n_0)}(n_0) \text{ times}}\}$$

with
$$\hat{\mu}_k(n_0) = (\mu_k(n_0), y(\hat{\mu}_k(n_0))),$$
$$\mu_k(n_0) \neq \mu_{k'}(n_0) \text{ for } k \neq k', \ k, k' = 1, \ldots, q(n_0), \quad (1.171)$$

and
$$p_k(n_0) \in \mathbb{N}, \ k = 1, \ldots, q(n_0), \quad \sum_{k=1}^{q(n_0)} p_k(n_0) = p.$$

With $\{E_m\}_{m=0}^{2p+1}$ and $\mathcal{D}_{\underline{\hat{\mu}}(n_0)}$ prescribed, we next introduce the following quantities (for $z \in \mathbb{C}$):

$$F_p(z, n_0) = \prod_{k=1}^{q(n_0)} (z - \mu_k(n_0))^{p_k(n_0)}, \quad (1.172)$$

$$T_{p-1}(z, n_0) = -F_p(z, n_0) \sum_{k=1}^{q(n_0)} \sum_{\ell=0}^{p_k(n_0)-1} \frac{(d^\ell y(P)/d\zeta^\ell)\big|_{P=(\zeta,\eta)=\hat{\mu}_k(n_0)}}{\ell!(p_k(n_0) - \ell - 1)!} \quad (1.173)$$
$$\times \left(\frac{d^{p_k(n_0)-\ell-1}}{d\zeta^{p_k(n_0)-\ell-1}} \left((z-\zeta)^{-1} \prod_{k'=1, k'\neq k}^{q(n_0)} (\zeta - \mu_{k'}(n_0))^{-p_{k'}(n_0)} \right) \right)\bigg|_{\zeta=\mu_k(n_0)},$$

$$b(n_0) = \frac{1}{2} \sum_{m=0}^{2p+1} E_m - \sum_{k=1}^{q(n_0)} p_k(n_0)\mu_k(n_0), \quad (1.174)$$

$$G_{p+1}(z, n_0) = -(z - b(n_0))F_p(z, n_0) + T_{p-1}(z, n_0). \quad (1.175)$$

Here the sign of y in (1.173) is chosen according to (1.171).

Next we record a series of facts:

(I) By construction (cf. Lemma D.5),

$$T_{p-1}^{(\ell)}(\mu_k(n_0), n_0) = -\frac{d^\ell y(P)}{d\zeta^\ell}\bigg|_{P=(\zeta,\eta)=\hat{\mu}_k(n_0)} = G_{p+1}^{(\ell)}(\mu_k(n_0), n_0), \quad (1.176)$$
$$\ell = 0, \ldots, p_k(n_0) - 1, \ k = 1, \ldots, q(n_0),$$

(here the superscript (ℓ) denotes ℓ derivatives w.r.t. z) and hence

$$\hat{\mu}_k(n_0) = (\mu_k(n_0), -G_{p+1}(\mu_k(n_0), n_0)), \quad k = 1, \ldots, q(n_0).$$

(II) Since $\mathcal{D}_{\underline{\hat{\mu}}(n_0)}$ is nonspecial by hypothesis, one concludes that

$$p_k(n_0) \geq 2 \text{ implies } R_{2p+2}(\mu_k(n_0)) \neq 0, \quad k = 1, \ldots, q(n_0).$$

1.4 The Stationary Toda Algebro-Geometric IVP

(III) By **(I)** and **(II)** one computes

$$\frac{d^\ell \big(G_{p+1}(z, n_0)^2\big)}{dz^\ell}\bigg|_{z=\mu_k(n_0)} = \frac{d^\ell R_{2p+2}(z)}{dz^\ell}\bigg|_{z=\mu_k(n_0)}, \quad (1.177)$$

$z \in \mathbb{C}, \quad \ell = 0, \ldots, p_k(n_0) - 1, \quad k = 1, \ldots, q(n_0).$

(IV) By (1.175) and (1.177) one infers that F_p divides $R_{2p+2} - G_{p+1}^2$.

(V) By (1.174) and (1.175) one verifies that

$$R_{2p+2}(z) - G_{p+1}(z, n_0)^2 \underset{z \to \infty}{=} O(z^{2p}). \quad (1.178)$$

By **(IV)** and (1.178) we may write

$$R_{2p+2}(z) - G_{p+1}(z, n_0)^2 = F_p(z, n_0) \check{F}_{p-r}(z, n_0 + 1), \quad z \in \mathbb{C}, \quad (1.179)$$

for some $r \in \{0, \ldots, p\}$, where the polynomial $\check{F}_{p-r}$ has degree $p - r$. If, in fact, $\check{F}_0 = 0$, then $R_{2p+2}(z) = G_{p+1}(z, n_0)^2$ would yield double zeros of R_{2p+2}, contradicting our basic hypothesis (1.170). Thus we conclude that in the case $r = p$, $\check{F}_0$ cannot vanish identically and hence we may break up (1.179) in the following manner

$$\check{\phi}(P, n_0) = \frac{y - G_{p+1}(z, n_0)}{F_p(z, n_0)} = \frac{\check{F}_{p-r}(z, n_0 + 1)}{y + G_{p+1}(z, n_0)}, \quad P = (z, y) \in \mathcal{K}_p.$$

Next we decompose

$$\check{F}_{p-r}(z, n_0 + 1) = \check{C} \prod_{j=1}^{p-r} (z - \mu_j(n_0 + 1)), \quad z \in \mathbb{C}, \quad (1.180)$$

where $\check{C} \in \mathbb{C} \backslash \{0\}$ and $\{\mu_j(n_0+1)\}_{j=1}^{p-r} \subset \mathbb{C}$ (if $r = p$ we follow the usual convention and replace the product in (1.180) by 1). By inspection of the local zeros and poles as well as the behavior near $P_{\infty_\pm}$ of the function $\check{\phi}(\,\cdot\,, n_0)$, its divisor, $\big(\check{\phi}(\,\cdot\,, n_0)\big)$, is given by

$$\big(\check{\phi}(\,\cdot\,, n_0)\big) = \mathcal{D}_{P_{\infty_+}\hat{\underline{\mu}}(n_0+1)} - \mathcal{D}_{P_{\infty_-}\hat{\underline{\mu}}(n_0)},$$

where

$$\hat{\underline{\mu}}(n_0 + 1) = \{\hat{\mu}_1(n_0 + 1), \ldots, \hat{\mu}_{p-r}(n_0 + 1), \underbrace{P_{\infty_+}, \ldots, P_{\infty_+}}_{r \text{ times}}\}.$$

In particular,

$$\mathcal{D}_{\hat{\underline{\mu}}(n_0+1)} \text{ is a finite divisor if and only if } r = 0. \quad (1.181)$$

We note that

$$\underline{\alpha}_{Q_0}(\mathcal{D}_{\hat{\underline{\mu}}(n_0+1)}) = \underline{\alpha}_{Q_0}(\mathcal{D}_{\hat{\underline{\mu}}(n_0)}) - \underline{A}_{P_{\infty_-}}(P_{\infty_+}),$$

in accordance with (1.166).

(VI) Assuming that (1.178) is precisely of order z^{2p} as $z \to \infty$, that is, assuming $r = 0$ in (1.179), we rewrite (1.179) in the more appropriate manner

$$R_{2p+2}(z) - G_{p+1}(z, n_0)^2 = -4a(n_0)^2 F_p(z, n_0) F_p(z, n_0 + 1), \quad z \in \mathbb{C}, \quad (1.182)$$

where we introduced the coefficient $a(n_0)^2$ to make $F_p(\cdot, n_0 + 1)$ a monic polynomial of degree p. (We will later discuss conditions which indeed guarantee that $r = 0$, cf. (1.181) and the discussion in step **(XI)** below.) By construction, $F_p(\cdot, n_0 + 1)$ is then of the type

$$F_p(z, n_0 + 1) = \prod_{k=1}^{q(n_0+1)} (z - \mu_k(n_0 + 1))^{p_k(n_0+1)}, \quad \sum_{k=1}^{q(n_0+1)} p_k(n_0 + 1) = p,$$

$$\mu_k(n_0 + 1) \neq \mu_{k'}(n_0 + 1) \text{ for } k \neq k', \ k, k' = 1, \ldots, q(n_0 + 1), \quad z \in \mathbb{C},$$

and we define

$$\hat{\mu}_k(n_0+1) = (\mu_k(n_0+1), G_{p+1}(\mu_k(n_0+1), n_0)), \quad k = 1, \ldots, q(n_0+1). \quad (1.183)$$

Moreover, we introduce the divisor

$$\mathcal{D}_{\underline{\hat{\mu}}(n_0+1)} \in \text{Sym}^p(\mathcal{K}_p)$$

by

$$\underline{\hat{\mu}}(n_0 + 1) = \{\underbrace{\hat{\mu}_1(n_0+1), \ldots, \hat{\mu}_1(n_0+1)}_{p_1(n_0+1) \text{ times}}, \ldots, \underbrace{\hat{\mu}_{q(n_0+1)}, \ldots, \hat{\mu}_{q(n_0+1)}}_{p_{q(n_0+1)}(n_0+1) \text{ times}}\}.$$

In particular, because of definition (1.183), $\mathcal{D}_{\underline{\hat{\mu}}(n_0+1)}$ is nonspecial and hence

$$p_k(n_0 + 1) \geq 2 \text{ implies } R_{2p+2}(\mu_k(n_0+1)) \neq 0, \quad k = 1, \ldots, q(n_0+1).$$

Again we note that

$$\underline{\alpha}_{Q_0}(\mathcal{D}_{\underline{\hat{\mu}}(n_0+1)}) = \underline{\alpha}_{Q_0}(\mathcal{D}_{\underline{\hat{\mu}}(n_0)}) - \underline{A}_{P_{\infty_-}}(P_{\infty_+}),$$

in accordance with (1.166).

(VII) Introducing

$$b(n_0 + 1) = \frac{1}{2} \sum_{m=0}^{2p+1} E_m - \sum_{k=1}^{q(n_0+1)} p_k(n_0 + 1) \mu_k(n_0 + 1),$$

and interpolating $G_{p+1}(\cdot, n_0)$ with $F_p(\cdot, n_0 + 1)$ rather than $F_p(\cdot, n_0)$ yields

$$G_{p+1}(z, n_0) = -(z - b(n_0 + 1)) F_p(z, n_0 + 1) - T_{p-1}(z, n_0 + 1), \quad z \in \mathbb{C},$$

where

$$T_{p-1}(z, n_0 + 1) = F_p(z, n_0 + 1)$$

$$\times \sum_{k=1}^{q(n_0+1)} \sum_{\ell=0}^{p_k(n_0+1)-1} \frac{\left(d^\ell y(P)/d\zeta^\ell\right)\big|_{P=(\zeta,\eta)=\hat{\mu}_k(n_0+1)}}{\ell!(p_k(n_0+1)-\ell-1)!}$$

$$\times \left(\frac{d^{p_k(n_0+1)-\ell-1}}{d\zeta^{p_k(n_0+1)-\ell-1}}\left((z-\zeta)^{-1}\right.\right.$$

$$\left.\left.\times \prod_{k'=1, k'\neq k}^{q(n_0+1)} (\zeta - \mu_{k'}(n_0+1))^{-p_{k'}(n_0+1)}\right)\right)\bigg|_{\zeta=\mu_k(n_0+1)}. \quad (1.184)$$

Here the sign of y in (1.184) is chosen in accordance with (1.183), that is,

$$\hat{\mu}_k(n_0 + 1) = (\mu_k(n_0 + 1), G_{p+1}(\mu_k(n_0 + 1), n_0)), \quad k = 1, \ldots, q(n_0 + 1).$$

(VIII) An explicit computation of $a(n_0)^2$ then yields

$$a(n_0)^2 = \frac{1}{2} \sum_{k=1}^{q(n_0)} \frac{\left(d^{p_k(n_0)-1}y(P)/d\zeta^{p_k(n_0)-1}\right)\big|_{P=(\zeta,\eta)=\hat{\mu}_k(n_0)}}{(p_k(n_0)-1)!} \quad (1.185)$$

$$\times \prod_{k'=1, k'\neq k}^{q(n_0)} (\mu_k(n_0) - \mu_{k'}(n_0))^{-p_k(n_0)} + \frac{1}{4}(b^{(2)}(n_0) - b(n_0)^2).$$

Here and in the following we abbreviate

$$b^{(2)}(n) = \frac{1}{2} \sum_{m=0}^{2p+1} E_m^2 - \sum_{k=1}^{q(n)} p_k(n)\mu_k(n)^2 \quad (1.186)$$

for an appropriate range of $n \in \mathbb{N}$.

The result (1.185) is obtained as follows: Starting from the identity (1.182), one inserts the expressions (1.172) and (1.175) for $F_p(\cdot, n_0)$ and $G_{p+1}(\cdot, n_0)$, respectively, then inserts the explicit form (1.173) of $T_{p-1}(\cdot, n_0)$, and finally collects all terms of order z^{2p} as $z \to \infty$. An entirely elementary but fairly tedious calculation then produces (1.185).

In the special case $q(n_0) = p$, $p_k(n_0) = 1$, $k = 1, \ldots, p$, (1.185) and (1.186) reduce to (1.98) and (1.96) (for $k = 2$).

(IX) Introducing

$$G_{p+1}(z, n_0 + 1) = -(z - b(n_0 + 1))F_p(z, n_0 + 1) + T_{p-1}(z, n_0 + 1)$$

one then obtains

$$G_{p+1}(z, n_0 + 1) = -G_{p+1}(z, n_0) - 2(z - b(n_0 + 1))F_p(z, n_0 + 1). \quad (1.187)$$

(X) At this point one can iterate the procedure step by step to construct $F_p(\,\cdot\,,n)$, $G_{p+1}(\,\cdot\,,n), T_{p-1}(\,\cdot\,,n), a(n), b(n), \mu_k(n), k = 1,\ldots,q(n)$, etc., for $n \in [n_0,\infty) \cap \mathbb{Z}$, subject to the following assumption (cf. (1.181)) at each step:

$$\mathcal{D}_{\underline{\hat{\mu}}(n+1)} \text{ is a finite divisor (and hence } a(n) \neq 0) \text{ for all } n \in [n_0,\infty) \cap \mathbb{Z}. \quad (1.188)$$

The formalism is symmetric with respect to n_0 and can equally well be developed for $n \in (-\infty, n_0] \cap \mathbb{Z}$ subject to the analogous assumption

$$\mathcal{D}_{\underline{\hat{\mu}}(n-1)} \text{ is a finite divisor (and hence } a(n) \neq 0) \text{ for all } n \in (-\infty, n_0] \cap \mathbb{Z}. \quad (1.189)$$

Indeed, one first interpolates $G_{p+1}(\,\cdot\,,n_0-1)$ with the help of $F_p(\,\cdot\,,n_0)$, then with $F_p(\,\cdot\,,n_0-1)$, etc.

Moreover, we once again remark for consistency reasons that

$$\underline{\alpha}_{Q_0}(\mathcal{D}_{\underline{\hat{\mu}}(n)}) = \underline{\alpha}_{Q_0}(\mathcal{D}_{\underline{\hat{\mu}}(n_0)}) - (n-n_0)\underline{A}_{P_\infty_-}(P_{\infty_+}), \quad n \in \mathbb{Z},$$

in agreement with our starting point (1.166).

(XI) Choosing the initial data $\mathcal{D}_{\underline{\hat{\mu}}(n_0)}$ such that

$$\mathcal{D}_{\underline{\hat{\mu}}(n_0)} \in \mathcal{M}_0,$$

where $\mathcal{M}_0 \subset \mathrm{Sym}^p(\mathcal{K}_p)$ is the set of finite initial divisors introduced in Lemma 1.30, then guarantees that assumptions (1.188) and (1.189) are satisfied for all $n \in \mathbb{Z}$.

(XII) Performing these iterations for all $n \in \mathbb{Z}$, one then arrives at the following set of equations for F_p and G_{p+1} after the following elementary manipulations: Utilizing

$$G_{p+1}^2 - 4a^2 F_p F_p^+ = R_{2p+2} = (G_{p+1}^-)^2 - 4(a^-)^2 F_p^- F_p, \quad (1.190)$$

and inserting

$$G_{p+1}^+ = -G_{p+1} - 2(z - b^+) F_p^+ \quad (1.191)$$

into

$$G_{p+1}^2 - (G_{p+1}^-)^2 - 4a^2 F_p F_p^+ + 4(a^-)^2 F_p^- F_p = 0$$

then yields

$$2a^2 F_p^+ - 2(a^-)^2 F_p^- + (z - b)(G_{p+1} - G_{p+1}^-) = 0. \quad (1.192)$$

Subtracting (1.191) from its shifted version $G_{p+1} = -G_{p+1}^- - 2(z - b)F_p$ then also yields

$$2(z - b^+) F_p^+ - 2(z - b) F_p + G_{p+1}^+ - G_{p+1}^- = 0. \quad (1.193)$$

As discussed in Section 1.2, (1.192) and (1.193) are equivalent to the stationary Lax and zero-curvature equations (1.24) and (1.58), and hence to (1.29). At this stage we have verified the basic hypotheses of Section 1.3 (i.e., (1.61) and the assumption that

1.4 The Stationary Toda Algebro-Geometric IVP

a, b satisfy the pth stationary Toda system (1.29)) and hence all results of Section 1.3 apply.

Finally, we briefly summarize these considerations:

Theorem 1.32 *Let $n \in \mathbb{Z}$, suppose the set $\{E_m\}_{m=0}^{2p+1} \subset \mathbb{C}$ satisfies $E_m \neq E_{m'}$ for $m \neq m'$, $m, m' = 0, \ldots, 2p + 1$, and introduce the function R_{2p+2} and the hyperelliptic curve $\mathcal{K}_p$ as in (1.62). Choose a nonspecial divisor $\mathcal{D}_{\hat{\underline{\mu}}(n_0)} \in \mathcal{M}_0$, where $\mathcal{M}_0 \subset \mathrm{Sym}^p(\mathcal{K}_p)$ is the set of finite initial divisors introduced in Lemma 1.30. Then the stationary (complex) Toda algorithm as outlined in steps* **(I)–(XII)** *produces solutions a, b of the pth stationary Toda system,*

$$\text{s-Tl}_p(a, b) = \begin{pmatrix} f_{p+1}^+ - f_{p+1}^- \\ g_{p+1} - g_{p+1}^- \end{pmatrix} = 0, \quad p \in \mathbb{N}_0,$$

satisfying (1.61) *and*

$$a(n)^2 = \frac{1}{2} \sum_{k=1}^{q(n)} \frac{\left(d^{p_k(n)-1} y(P) / d\zeta^{p_k(n)-1} \right) \big|_{P=(\zeta,\eta)=\hat{\mu}_k(n)}}{(p_k(n) - 1)!}$$

$$\times \prod_{k'=1, k' \neq k}^{q(n)} (\mu_k(n) - \mu_{k'}(n))^{-p_k(n)} + \frac{1}{4}\left(b^{(2)}(n) - b(n)^2\right),$$

$$b(n) = \frac{1}{2} \sum_{m=0}^{2p+1} E_m - \sum_{k=1}^{q(n)} p_k(n)\mu_k(n), \quad n \in \mathbb{Z}.$$

Moreover, Lemmas 1.12 and 1.14–1.17 apply.

Remark 1.33 Suppose that the hypotheses of the previous theorem are satisfied and that $a(n_0), b(n_0), b(n_0+1), F_p(z, n_0), F_p(z, n_0+1), G_{p+1}(z, n_0)$, and $G_{p+1}(z, n_0+1)$ have already been computed using steps **(I)–(IX)**. Then, alternatively, one can use

$$(a^-)^2 F_p^- = a^2 F_p^+ + \frac{1}{2}(z - b)(G_{p+1} - G_{p+1}^+) + (z - b)^2 F_p$$
$$- (z - b^+)(z - b)F_p^+,$$
$$G_{p+1}^- = 2((z - b^+)F_p^+ - (z - b)F_p) + G_{p+1}^+$$

(derived from (1.58)) to compute $a(n), b(n), F_p(z, n), G_{p+1}(z, n)$ for $n < n_0$ and

$$a^+ F_p^{++} = aF_p - \frac{1}{2}(z - b)(G_{p+1}^+ - G_{p+1}),$$
$$G_{p+1}^{++} = G_{p+1} - 2((z - b^{++})F_p^{++} - (z - b^+)F_p^+)$$

to compute $a(n-1), b(n), F_p(z, n), G_{p+1}(z, n)$ for $n > n_0 + 1$.

Theta function representations of a and b can now be derived as in Theorem 1.19.

The stationary (complex) Toda algorithm as outlined in steps **(I)–(XII)**, starting from a nonspecial divisor $\mathcal{D}_{\hat{\underline{\mu}}(n_0)} \in \mathcal{M}_0$, represents a solution of the inverse algebro-geometric spectral problem for generally non-self-adjoint Jacobi operators. While we do not assume periodicity (or quasi-periodicity), let alone real-valuedness of the coefficients of the underlying Jacobi operator, once can view this algorithm as a continuation of the inverse periodic spectral problem started around 1975 (in the self-adjoint context).

We note that in general (i.e., unless one is, e.g., in the special periodic or self-adjoint case), $\mathcal{D}_{\hat{\underline{\mu}}(n)}$ will get arbitrarily close to $P_{\infty_\pm}$ since straight motions on the torus are generically dense. Thus, no uniform bound on the sequences $a(n), b(n)$ (and no uniform lower bound on $|a(n)|$) exists as n varies in $\mathbb{Z}$. In particular, these complex-valued algebro-geometric solutions of some of the equations of the stationary Toda hierarchy, generally, will not be quasi-periodic with respect to n.

1.5 The Time-Dependent Toda Formalism

> In theory there is no difference between theory and practice. In practice there is.
>
> *Yogi Berra*[1]

In this section we extend the algebro-geometric analysis of Section 1.3 to the time-dependent Toda hierarchy.

For most of this section we assume the following hypothesis.

Hypothesis 1.34 (*i*) *Assume that a, b satisfy*

$$a(\,\cdot\,, t), b(\,\cdot\,, t) \in \mathbb{C}^{\mathbb{Z}}, \ t \in \mathbb{R}, \quad a(n, \,\cdot\,), b(n, \,\cdot\,) \in C^1(\mathbb{R}), \ n \in \mathbb{Z}, \qquad (1.194)$$
$$a(n, t) \neq 0, \ (n, t) \in \mathbb{Z} \times \mathbb{R}.$$

(*ii*) *Suppose that the hyperelliptic curve $\mathcal{K}_p$, $p \in \mathbb{N}_0$, satisfies* (1.62) *and* (1.63).

The basic problem in the analysis of algebro-geometric solutions of the Toda hierarchy consists of solving the time-dependent rth Toda flow with initial data a stationary solution of the pth equation in the hierarchy. More precisely, given $p \in \mathbb{N}_0$, consider a solution $a^{(0)}, b^{(0)}$ of the pth stationary Toda system s-Tl$_p(a, b) = 0$, associated with the hyperelliptic curve $\mathcal{K}_p$ and a corresponding set of summation constants $\{c_\ell\}_{\ell=1,\dots,p} \subset \mathbb{C}$. Next, let $r \in \mathbb{N}_0$; we intend to construct solutions a, b of the rth Toda flow Tl$_r(a, b) = 0$ with $a(t_{0,r}) = a^{(0)}, b(t_{0,r}) = b^{(0)}$ for some $t_{0,r} \in \mathbb{R}$. To emphasize that the summation constants in the definitions of the stationary and

---

[1] American baseball player, 1925–.

1.5 The Time-Dependent Toda Formalism

the time-dependent Toda equations are independent of each other, we indicate this by adding a tilde on all the time-dependent quantities. Hence we shall employ the notation $\widetilde{P}_{2r+2}$, $\widetilde{V}_{r+1}$, $\widetilde{F}_r$, $\widetilde{G}_{r+1}$, $\tilde{f}_s$, $\tilde{g}_s$, $\tilde{c}_s$, in order to distinguish them from P_{2p+2}, V_{p+1}, F_p, G_{p+1}, f_ℓ, g_ℓ, c_ℓ, in the following. In addition, we will follow a more elaborate notation inspired by Hirota's τ-function approach and indicate the individual rth Toda flow by a separate time variable $t_r \in \mathbb{R}$.

Summing up, we are seeking solutions a, b of the time-dependent algebro-geometric initial value problem

$$\widetilde{\mathrm{Tl}}_r(a, b) = \begin{pmatrix} a_{t_r} - a(\tilde{f}^+_{p+1}(a,b) - \tilde{f}_{p+1}(a,b)) \\ b_{t_r} + \tilde{g}_{p+1}(a,b) - \tilde{g}^-_{p+1}(a,b) \end{pmatrix} = 0, \qquad (1.195)$$

$$(a, b)\big|_{t_r = t_{0,r}} = \big(a^{(0)}, b^{(0)}\big),$$

$$\text{s-Tl}_p\big(a^{(0)}, b^{(0)}\big) = \begin{pmatrix} -a\big(f^+_{p+1}(a^{(0)}, b^{(0)}) - f_{p+1}(a^{(0)}, b^{(0)})\big) \\ g_{p+1}(a^{(0)}, b^{(0)}) - g^-_{p+1}(a^{(0)}, b^{(0)}) \end{pmatrix} = 0 \qquad (1.196)$$

for some $t_{0,r} \in \mathbb{R}$, $p, r \in \mathbb{N}_0$, where $a = a(n, t_r)$, $b = b(n, t_r)$ satisfy (1.194) and a fixed curve $\mathcal{K}_p$ is associated with the stationary solutions $a^{(0)}, b^{(0)}$ in (1.196). In terms of Lax pairs this amounts to solving

$$\frac{d}{dt_r} L(t_r) - \big[\widetilde{P}_{2r+2}(t_r), L(t_r)\big] = 0, \quad t_r \in \mathbb{R}, \qquad (1.197)$$

$$[P_{2p+2}(t_{0,r}), L(t_{0,r})] = 0. \qquad (1.198)$$

In anticipating that the Toda flows are isospectral deformations of $L(t_{0,r})$, we are going a step further replacing (1.198) by

$$[P_{2p+2}(t_r), L(t_r)] = 0, \quad t_r \in \mathbb{R}. \qquad (1.199)$$

This then implies

$$P_{2p+2}(t_r)^2 = R_{2p+2}(L(t_r)) = \prod_{m=0}^{2p+1} (L(t_r) - E_m), \quad t_r \in \mathbb{R}.$$

Actually, instead of working with (1.197), (1.198), and (1.199), one can equivalently take the zero-curvature equations (1.60) as one's point of departure, that is, one can also start from

$$U_{t_r} + U\widetilde{V}_{r+1} - \widetilde{V}^+_{r+1} U = 0, \qquad (1.200)$$

$$U V_{p+1} - V^+_{p+1} U = 0. \qquad (1.201)$$

For further reference, we recall the relevant quantities here (cf. (1.30), (1.31), (1.56), (1.57)):

$$U(z) = \begin{pmatrix} 0 & 1 \\ -a^-/a & (z-b)/a \end{pmatrix},$$

$$V_{p+1}(z) = \begin{pmatrix} G^-_{p+1}(z) & 2a^- F^-_p(z) \\ -2a^- F_p(z) & 2(z-b)F_p + G_{p+1}(z) \end{pmatrix}, \quad (1.202)$$

$$\widetilde{V}_{r+1}(z) = \begin{pmatrix} \widetilde{G}^-_{r+1}(z) & 2a^- \widetilde{F}^-_r(z) \\ -2a^- \widetilde{F}_r(z) & 2(z-b)\widetilde{F}_r(z) + \widetilde{G}_{r+1}(z) \end{pmatrix},$$

and

$$F_p(z) = \sum_{\ell=0}^{p} f_{p-\ell} z^\ell = \prod_{j=1}^{p}(z-\mu_j), \quad f_0 = 1, \quad (1.203)$$

$$G_{p+1}(z) = -z^{p+1} + \sum_{\ell=0}^{p} g_{p-\ell} z^\ell + f_{p+1}, \quad g_0 = -c_1, \quad (1.204)$$

$$\widetilde{F}_r(z) = \sum_{s=0}^{r} \tilde{f}_{r-s} z^s, \quad \tilde{f}_0 = 1, \quad (1.205)$$

$$\widetilde{G}_{r+1}(z) = -z^{r+1} + \sum_{s=0}^{r} \tilde{g}_{r-s} z^s + \tilde{f}_{r+1}, \quad \tilde{g}_0 = -\tilde{c}_1, \quad (1.206)$$

for fixed $p, r \in \mathbb{N}_0$. Here $f_\ell, \tilde{f}_s, g_\ell$, and $\tilde{g}_s$, $\ell = 0, \ldots, p$, $s = 0, \ldots, r$, are defined as in (1.3)–(1.5) with appropriate sets of summation constants c_ℓ, $\ell \in \mathbb{N}$, and $\tilde{c}_k$, $k \in \mathbb{N}$. Explicitly, (1.200) and (1.201) are equivalent to (cf. (1.38), (1.39), (1.54), (1.55)),

$$a_{t_r} = -a\big(2(z-b^+)\widetilde{F}^+_r + \widetilde{G}^+_{r+1} + \widetilde{G}_{r+1}\big), \quad (1.207)$$

$$b_{t_r} = 2\big((z-b)^2 \widetilde{F}_r + (z-b)\widetilde{G}_{r+1} + a^2 \widetilde{F}^+_r - (a^-)^2 \widetilde{F}^-_r\big), \quad (1.208)$$

$$0 = 2(z-b^+)F^+_p + G^+_{p+1} + G_{p+1}, \quad (1.209)$$

$$0 = (z-b)^2 F_p + (z-b)G_{p+1} + a^2 F^+_p - (a^-)^2 F^-_p, \quad (1.210)$$

respectively. In particular, (1.40) holds in the present t_r-dependent setting, that is,

$$G^2_{p+1} - 4a^2 F_p F^+_p = R_{2p+2}.$$

As in the stationary context (1.67), (1.68) we introduce

$$\hat{\mu}_j(n, t_r) = (\mu_j(n, t_r), -G_{p+1}(\mu_j(n, t_r), n, t_r)) \in \mathcal{K}_p,$$
$$j = 1, \ldots, p, \ (n, t_r) \in \mathbb{Z} \times \mathbb{R}, \quad (1.211)$$

1.5 The Time-Dependent Toda Formalism

and

$$\hat{\mu}_j^+(n, t_r) = (\mu_j^+(n, t_r), G_{p+1}(\mu_j^+(n, t_r), n, t_r)) \in \mathcal{K}_p, \tag{1.212}$$
$$j = 1, \ldots, p, \ (n, t_r) \in \mathbb{Z} \times \mathbb{R},$$

and note that the regularity assumptions (1.194) on a, b imply continuity of μ_j with respect to $t_r \in \mathbb{R}$ (away from collisions of these zeros, μ_j are of course C^∞).

In analogy to (1.69), (1.70), one defines the following meromorphic function $\phi(\,\cdot\,, n, t_r)$ on $\mathcal{K}_p$,

$$\phi(P, n, t_r) = \frac{y - G_{p+1}(z, n, t_r)}{2a(n, t_r) F_p(z, n, t_r)} \tag{1.213}$$

$$= \frac{-2a(n, t_r) F_p(z, n+1, t_r)}{y + G_{p+1}(z, n, t_r)}, \tag{1.214}$$

$$P(z, y) \in \mathcal{K}_p, \ (n, t_r) \in \mathbb{Z} \times \mathbb{R},$$

with divisor $(\phi(\,\cdot\,, n, t_r))$ of $\phi(\,\cdot\,, n, t_r)$ given by

$$(\phi(\,\cdot\,, n, t_r)) = \mathcal{D}_{P_{\infty_+} \hat{\underline{\mu}}(n+1, t_r)} - \mathcal{D}_{P_{\infty_-} \hat{\underline{\mu}}(n, t_r)}, \tag{1.215}$$

using (1.203) and (1.211). Here we abbreviated

$$\hat{\underline{\mu}} = \{\hat{\mu}_1, \ldots, \hat{\mu}_p\}, \ \hat{\underline{\mu}}^+ = \{\hat{\mu}_1^+, \ldots, \hat{\mu}_p^+\} \in \mathrm{Sym}^p(\mathcal{K}_p).$$

The time-dependent Baker–Akhiezer function $\psi(\,\cdot\,, n, n_0, t_r, t_{0,r})$ is then defined in terms of ϕ by

$$\psi(P, n, n_0, t_r, t_{0,r})$$
$$= \exp\left(\int_{t_{0,r}}^{t_r} ds \big(2a(n_0, s) \widetilde{F}_r(z, n_0, s) \phi(P, n_0, s) + \widetilde{G}_{r+1}(z, n_0, s)\big)\right)$$
$$\times \begin{cases} \prod_{n'=n_0}^{n-1} \phi(P, n', t_r), & n \geq n_0 + 1, \\ 1, & n = n_0, \\ \prod_{n'=n}^{n_0-1} \phi(P, n', t_r)^{-1}, & n \leq n_0 - 1, \end{cases} \tag{1.216}$$
$$P = (z, y) \in \mathcal{K}_p \setminus \{P_{\infty_+}, P_{\infty_-}\}, \ (n, n_0, t_r, t_{0,r}) \in \mathbb{Z}^2 \times \mathbb{R}^2.$$

One observes that

$$\psi(P, n, n_0, t_r, t_{0,r}) = \psi(P, n, n_0, t_{0,r}, t_{0,r}) \psi(P, n_0, n_0, t_r, t_{0,r}), \tag{1.217}$$
$$P = (z, y) \in \mathcal{K}_p \setminus \{P_{\infty_+}, P_{\infty_-}\}, \ (n, n_0, t_r, t_{0,r}) \in \mathbb{Z}^2 \times \mathbb{R}^2.$$

As in the stationary context we also introduce the Baker–Akhiezer vector Ψ defined by

$$\Psi(P, n, n_0, t_r, t_{0,r}) = \begin{pmatrix} \psi(P, n, n_0, t_r, t_{0,r}) \\ \psi(P, n+1, n_0, t_r, t_{0,r}) \end{pmatrix},$$
$$P = (z, y) \in \mathcal{K}_p \setminus \{P_{\infty_+}, P_{\infty_-}\}, \ (n, n_0, t_r, t_{0,r}) \in \mathbb{Z}^2 \times \mathbb{R}^2.$$

The following lemma records basic properties of ϕ, ψ, and Ψ in analogy to the stationary case discussed in Lemma 1.12.

Lemma 1.35 *Assume Hypothesis 1.34 and suppose that* (1.207)–(1.210) *hold. In addition, let* $P = (z, y) \in \mathcal{K}_p \setminus \{P_{\infty_+}, P_{\infty_-}\}$, $(n, n_0, t_r, t_{0,r}) \in \mathbb{Z}^2 \times \mathbb{R}^2$, *and* $r \in \mathbb{N}_0$. *Then* ϕ *satisfies*

$$a\phi(P) + a^-(\phi^-(P))^{-1} = z - b, \tag{1.218}$$

$$\phi_{t_r}(P) = -2a\big(\widetilde{F}_r(z)\phi(P)^2 + \widetilde{F}_r^+(z)\big) + 2(z - b^+)\widetilde{F}_r^+(z)\phi(P)$$
$$+ \big(\widetilde{G}_{r+1}^+(z) - \widetilde{G}_{r+1}(z)\big)\phi(P), \tag{1.219}$$

$$\phi(P)\phi(P^*) = \frac{F_p^+(z)}{F_p(z)}, \tag{1.220}$$

$$\phi(P) - \phi(P^*) = \frac{y(P)}{aF_p(z)}, \tag{1.221}$$

$$\phi(P) + \phi(P^*) = -\frac{G_{p+1}(z)}{aF_p(z)}. \tag{1.222}$$

Moreover, ψ *and* Ψ *satisfy*

$$(L - z(P))\psi(P) = 0, \quad (P_{2p+2} - y(P))\psi(P) = 0, \tag{1.223}$$

$$\psi_{t_r}(P) = P_{2r+2}\,\psi(P) \tag{1.224}$$

$$= 2a\widetilde{F}_r(z)\psi^+(P) + \widetilde{G}_{r+1}(z)\psi(P), \tag{1.225}$$

$$U(z)\Psi^-(P) = \Psi(P), \tag{1.226}$$

$$V_{p+1}(z)\Psi^-(P) = y\Psi^-(P), \tag{1.227}$$

$$\Psi_{t_r}(P) = \widetilde{V}_{r+1}^+(z)\Psi(P), \tag{1.228}$$

$$\psi(P, n, n_0, t_r, t_{0,r})\psi(P^*, n, n_0, t_r, t_{0,r}) = \frac{F_p(z, n, t_r)}{F_p(z, n_0, t_{0,r})}, \tag{1.229}$$

$$a(n, t_r)\big(\psi(P, n, n_0, t_r, t_{0,r})\psi(P^*, n+1, n_0, t_r, t_{0,r})$$
$$+ \psi(P^*, n, n_0, t_r, t_{0,r})\psi(P, n+1, n_0, t_r, t_{0,r})\big) = -\frac{G_{p+1}(z, n, t_r)}{F_p(z, n_0, t_{0,r})}, \tag{1.230}$$

$$W(\psi(P, \cdot, n_0, t_r, t_{0,r}), \psi(P^*, \cdot, n_0, t_r, t_{0,r})) = -\frac{y(P)}{F_p(z, n_0, t_{0,r})}. \tag{1.231}$$

In addition, as long as the zeros $\mu_j(n_0, s)$ *of* $F_p(\,\cdot\,, n_0, s)$ *are all simple for* $s \in \mathcal{I}_\mu$, $\mathcal{I}_\mu \subseteq \mathbb{R}$ *an open interval,* ψ *is meromorphic on* $\mathcal{K}_p \setminus \{P_{\infty_+}, P_{\infty_-}\}$ *for* $(n, t_r, t_{0,r}) \in \mathbb{Z} \times \mathcal{I}_\mu^2$.

Proof The proof of this lemma, except for the time derivatives of ϕ and ψ, is essentially identical to that of Lemma 1.12.

1.5 The Time-Dependent Toda Formalism

We first study the time derivative of ϕ. The derivative of (1.218) reads

$$a_{t_r}\phi + a\phi_{t_r} + \frac{1}{(\phi^-)^2}(a_{t_r}^-\phi^- - a^-\phi_{t_r}^-) = -b_{t_r}. \tag{1.232}$$

A lengthy, but straightforward calculation, using (1.232), (1.38), and (1.39), shows that

$$\bigl(a(\phi^-)^2 - a^-S^-\bigr)\bigl(\phi_{t_r} + 2a(\widetilde{F}_r\phi^2 + \widetilde{F}_r^+)\bigr) - 2(z - b^+)\widetilde{F}_r^+\phi - \bigl(\widetilde{G}_{r+1}^+ - \widetilde{G}_{r+1}\bigr)\phi\bigr) = 0.$$

Thus

$$\phi_{t_r}(z, n, t_r) + 2a(n, t_r)\bigl(\widetilde{F}_r(z, n, t_r)\phi^2(z, n, t_r) + \widetilde{F}_r^+(z, n, t_r)\bigr)$$
$$- 2(z - b^+(n, t_r))\widetilde{F}_r^+(z, n, t_r)\phi(z, n, t_r)$$
$$- \bigl(\widetilde{G}_{r+1}^+(z, n, t_r) - \widetilde{G}_{r+1}(z, n, t_r)\bigr)\phi(z, n, t_r)$$
$$= C \begin{cases} \prod_{n'=1}^{n} B(z, n', t_r), & n \in \mathbb{N}, \\ 1, & n = 0, \\ \prod_{n'=0}^{n+1} B(z, n', t_r)^{-1}, & -n \in \mathbb{N}, \end{cases}$$

where

$$B(z, n, t_r) = \frac{a^-(n, t_r)}{a(n, t_r)\phi^-(z, n, t_r)^2}, \quad (n, t_r) \in \mathbb{Z} \times \mathbb{R}.$$

However, by studying the high-energy behavior of the left-hand side using (1.213) and that of the right-hand side, we infer that in fact $C = 0$, thereby proving (1.219). That ψ is meromorphic on $\mathcal{K}_p \setminus \{P_{\infty\pm}\}$ if $F_p(\cdot, n_0, t_r)$ has only simple zeros is a consequence of (1.213), (1.215), (1.216), and of

$$2a(n_0, s)\widetilde{F}_r(z, n_0, s)\phi(P, n_0, s) \underset{P \to \hat{\mu}_j(n_0, s)}{=} \frac{d}{ds}\ln\bigl(F_p(z, n_0, s)\bigr) + O(1)$$

$$\text{as } z \to \mu_j(n_0, s),$$

which follows from (1.233) below. (The proof of (1.233) in Lemma 1.36 below only requires (1.219) which has already been proven.)
The time derivative of ψ follows from (1.219) and the definition of ψ.
Equations (1.226) and (1.227) are proved as in Lemma 1.12 in the stationary context as t_r can be viewed as just an additional parameter. Equation (1.228) is an immediate consequence of (1.225) and the definition of $\widetilde{V}_{r+1}$ in (1.202). □

Next we consider the t_r-dependence of F_p and G_{p+1}.

Lemma 1.36 *Assume Hypothesis 1.34 and suppose that (1.207)–(1.210) hold. In addition, let $(z, n, t_r) \in \mathbb{C} \times \mathbb{Z} \times \mathbb{R}$. Then,*

$$F_{p,t_r} = 2\bigl(F_p\widetilde{G}_{r+1} - G_{p+1}\widetilde{F}_r\bigr), \tag{1.233}$$
$$G_{p+1,t_r} = 4a^2\bigl(F_p\widetilde{F}_r^+ - F_p^+\widetilde{F}_r\bigr). \tag{1.234}$$

In particular, (1.233) *and* (1.234) *are equivalent to*

$$V_{p+1,t_r} = [\widetilde{V}_{r+1}, V_{p+1}]. \tag{1.235}$$

Proof By taking the time derivative of (1.78) we find

$$\phi_{t_r}(P) - \phi_{t_r}(P^*) = -\frac{y(P)}{(aF_p)^2}\big(a_{t_r}F_p - aF_{p,t_r}\big).$$

Using (1.219) we can compute the left-hand side, and by inserting (1.52) for a_{t_r}, the result (1.233) follows.

In order to prove (1.234) we take the time derivative of (1.222) which gives

$$\phi_{t_r}(P) + \phi_{t_r}(P^*) = -\frac{G_{p+1,t_r}aF_p - G_{p+1}(a_{t_r}F_p - aF_{p,t_r})}{(aF_p)^2}.$$

Again using (1.219) on the left-hand side and (1.233) and (1.52) on the right-hand side yields the result. □

Next we turn to the Dubrovin equations for the time variation of the Dirichlet eigenvalues governed by the $\widetilde{\mathrm{Tl}}_r$ flow.

Lemma 1.37 (i) *Assume Hypothesis* 1.34 *and suppose that* (1.207)–(1.210) *hold on* $\mathbb{Z} \times \mathcal{I}_\mu$ *with* $\mathcal{I}_\mu \subseteq \mathbb{R}$ *an open interval. In addition, assume that the zeros* μ_j, $j = 1, \ldots, p$, *of* $F_p(\cdot)$ *remain distinct on* $\mathbb{Z} \times \mathcal{I}_\mu$. *Then* $\{\hat{\mu}_j\}_{j=1,\ldots,p}$, *defined in* (1.211), *satisfies the following first-order system of differential equations on* $\mathbb{Z} \times \mathcal{I}_\mu$,

$$\mu_{j,t_r} = -2\widetilde{F}_r(\mu_j)y(\hat{\mu}_j)\prod_{\substack{\ell=1\\\ell\neq j}}^{p}(\mu_j - \mu_\ell)^{-1}, \quad j = 1, \ldots, p, \tag{1.236}$$

with

$$\hat{\mu}_j(n,\cdot) \in C^\infty(\mathcal{I}_\mu, \mathcal{K}_p), \quad j = 1, \ldots, p, \, n \in \mathbb{Z}. \tag{1.237}$$

(ii) *Suppose in addition that a and b are real-valued and bounded. Moreover, assume the eigenvalue ordering*

$$E_m < E_{m+1}, \quad m = 0, 1, \ldots, 2p,$$

$$\mu_j(n, t_r) < \mu_{j+1}(n, t_r), \quad j = 1, \ldots, p-1, \, (n, t_r) \in \mathbb{Z} \times \mathbb{R}.$$

Then the Dirichlet eigenvalues $\{\mu_j(n, t_r)\}_{j=1,\ldots,p}$ *satisfy* (1.236) *on* $\mathbb{Z} \times \mathbb{R}$. *Furthermore, suppose that for all* $n \in \mathbb{Z}$, $\mu_j^{(0)}(n, t_{0,r}) \in [E_{2j-1}, E_{2j}]$, $j = 1, \ldots, p$ (*cf. the discussion in Remark* 1.77 *which introduces a sufficient condition for this entrapment to hold*), *then*

$$\mu_j(n, t_r) \in [E_{2j-1}, E_{2j}], \quad j = 1, \ldots, p, \, (n, t_r) \in \mathbb{Z} \times \mathbb{R}.$$

In particular, $\hat{\mu}_j(n, t_r)$ changes sheets whenever it hits E_{2j-1} or E_{2j}, and its projection $\mu_j(n, t_r)$ remains trapped in $[E_{2j-1}, E_{2j}]$, $j = 1, \ldots, p$, for all $(n, t_r) \in \mathbb{Z} \times \mathbb{R}$.

Proof Choosing $z = \mu_j(n, t_r)$ in (1.233) and combining the result with (1.64) and (1.67) yields (1.236). To prove (1.237) one can argue as follows: First, one notices that (1.236) is autonomous and the denominator on the right-hand side of (1.236) remains bounded away from zero by hypothesis. Thus, it remains to check what happens if the numerator on the right-hand side of (1.236) vanishes, that is, if μ_j hits one of the branch points $(E_m, 0)$ of $\mathcal{K}_p$. Hence we suppose

$$\mu_{j_0}(n_0, t_r) \to E_{m_0} \text{ as } t_r \to t_{0,r} \in \mathcal{I}_\mu,$$

for some $j_0 \in \{1, \ldots, p\}$, $m_0 \in \{0, \ldots, 2p+1\}$, $n_0 \in \mathbb{Z}$. Introducing

$$\zeta_{j_0}(n_0, t_r) = \sigma(\mu_{j_0}(n_0, t_r) - E_{m_0})^{1/2}, \quad \sigma = \pm 1,$$
$$\mu_{j_0}(n_0, t_r) = E_{m_0} + \zeta_{j_0}(n_0, t_r)^2$$

for t_r in a sufficiently small neighborhood of $t_{0,r}$, the Dubrovin equations (1.236) become

$$\zeta_{j_0,t_r}(n_0, t_r) \underset{t_r \to t_{0,r}}{=} c(\sigma)\tilde{F}_r(E_{m_0}, n_0, t_{0,r}) \left(\prod_{\substack{m=0 \\ m \neq m_0}}^{2p+1} (E_{m_0} - E_m) \right)^{1/2}$$

$$\times \left(\prod_{\substack{k=1 \\ k \neq j_0}}^{p} (E_{m_0} - \mu_k(n_0, t_r))^{-1} \right) \left(1 + O(\zeta_{j_0}(n_0, t_r)^2)\right)$$

for some $|c(\sigma)| = 1$ and one concludes (1.237).
Part (ii) follows as in Remark 1.77 since $t_r \in \mathbb{R}$ is now just an additional parameter. $\square$

When attempting to solve the Dubrovin system (1.236) it must be augmented with appropriate divisors $\mathcal{D}_{\underline{\hat{\mu}}(n,t_{0,r})} \in \text{Sym}^p \mathcal{K}_p$, $t_{0,r} \in \mathcal{I}_\mu$, as initial conditions.

Since the stationary trace formulas for Toda invariants in terms of symmetric functions of μ_j in Lemma 1.15 extend line by line to the corresponding time-dependent setting, we next record their t_r-dependent analogs without proof. For simplicity we again confine ourselves to the simplest cases only. We recall the abbreviation $b^{(k)}$ in (1.96).

Lemma 1.38 *Assume Hypothesis 1.34 and suppose that (1.207)–(1.210) hold. Then,*

$$b = \frac{1}{2} \sum_{m=0}^{2p+1} E_m - \sum_{j=1}^{p} \mu_j. \qquad (1.238)$$

In addition, if $\mu_j(n, t_r) \neq \mu_k(n, t_r)$ for $j \neq k$, $j, k = 1, \ldots, p$, for all $(n, t_r) \in \mathbb{Z} \times I_\mu$, $I_\mu \subseteq \mathbb{R}$ an open interval, then

$$a^2 = \frac{1}{2} \sum_{j=1}^{p} y(\hat{\mu}_j) \prod_{\substack{k=1 \\ k \neq j}}^{p} (\mu_j - \mu_k)^{-1} + \frac{1}{4}(b^{(2)} - b^2) \tag{1.239}$$

holds on $\mathbb{Z} \times I_\mu$.

Proof Equations (1.239) and (1.238) are obtained in precisely the same way as (1.97) and (1.98) taking into account (1.213), (1.218), (1.219), and (1.40) (for $n_0 \in \mathbb{Z}$). □

Next, we turn to the asymptotic expansions of ϕ and Ψ in a neighborhood of $P_{\infty_\pm}$.

Lemma 1.39 *Assume Hypothesis* 1.34 *and suppose that* (1.207)–(1.210) *hold. Moreover, let* $P = (z, y) \in \mathcal{K}_p \setminus \{P_{\infty_+}, P_{\infty_-}\}$, $(n, n_0, t_r, t_{0,r}) \in \mathbb{Z}^2 \times \mathbb{R}^2$. *Then* ϕ *has the asymptotic behavior*

$$\phi(P) \underset{\zeta \to 0}{=} \begin{cases} a\zeta + ab^+\zeta^2 + O(\zeta^3) & \text{as } P \to P_{\infty_+}, \\ a^{-1}\zeta^{-1} - a^{-1}b - a^{-1}(a^-)^2\zeta + O(\zeta^2) & \text{as } P \to P_{\infty_-}, \end{cases} \quad \zeta = 1/z.$$

The Baker–Akhiezer function ψ *has the asymptotic behavior*

$$\psi(P, n, n_0, t_r, t_{0,r}) \underset{\zeta \to 0}{=} \left(A(n, n_0, t_r)\zeta^{(n-n_0)}\right)^{\pm 1}(1 + O(\zeta)) \tag{1.240}$$

$$\times \exp\left(\mp (t_r - t_{0,r}) \sum_{s=0}^{r} \tilde{c}_{r-s}\zeta^{-s-1}\right), \quad P \to P_{\infty_\pm}, \quad \zeta = 1/z,$$

where we used the abbreviation

$$A(n, n_0, t_r) = \begin{cases} \prod_{n'=n_0}^{n-1} a(n', t_r)^{-1}, & n \geq n_0 + 1, \\ 1, & n = n_0, \\ \prod_{n'=n}^{n_0-1} a(n', t_r), & n \leq n_0 - 1. \end{cases}$$

Proof Since by the definition of ϕ in (1.213) the time parameter t_r can be viewed as an additional but fixed parameter, the asymptotic behavior of ϕ remains the same as in Lemma 1.17. Similarly, also the asymptotic behavior of $\psi_1(P, n, n_0, t_r, t_r)$ is derived in an identical fashion to that in Lemma 1.17. This proves (1.240) for $t_{0,r} = t_r$, that is,

$$\psi(P, n, n_0, t_r, t_r) \underset{\zeta \to 0}{=} \left(A(n, n_0, t_r)\zeta^{(n-n_0)}\right)^{\pm 1}(1 + O(\zeta)), \quad P \to P_{\infty_\pm}, \quad \zeta = 1/z. \tag{1.241}$$

1.5 The Time-Dependent Toda Formalism

By (1.217) it remains to investigate $\psi(P, n_0, n_0, t_r, t_{0,r})$. We note that (1.213), (1.216), and (1.233) imply

$$\psi(P, n_0, n_0, t_r, t_{0,r})$$
$$= \exp\left(\int_{t_{0,r}}^{t} ds \left(2a(n_0, s)\widetilde{F}_r(z, n_0, s)\phi(P, n_0, s) + \widetilde{G}_{r+1}(z, n_0, s)\right)\right)$$
$$= \exp\left(\int_{t_{0,r}}^{t} ds \left(\widetilde{F}_r(z, n_0, s)\frac{y(P) - G_{p+1}(z, n_0, s)}{F_p(z, n_0, s)} + \widetilde{G}_{r+1}(z, n_0, s)\right)\right). \tag{1.242}$$

Next we rewrite (1.242) in the form

$$\psi(P, n_0, n_0, t_r, t_{0,r}) = \exp\left(\int_{t_{0,r}}^{t} ds \left(\frac{1}{2}\frac{\frac{d}{ds}F_p(z, n_0, s)}{F_p(z, n_0, s)} + y(P)\frac{\widetilde{F}_r(z, n_0, s)}{F_p(z, n_0, s)}\right)\right)$$
$$= \left(\frac{F_p(z, n_0, t_r)}{F_p(z, n_0, t_{0,r})}\right)^{1/2} \exp\left(y(P)\int_{t_{0,r}}^{t} ds \frac{\widetilde{F}_r(z, n_0, s)}{F_p(z, n_0, s)}\right). \tag{1.243}$$

We claim that

$$y(P)\frac{\widetilde{F}_r(z, n, t_r)}{F_p(z, n, t_r)} \underset{\zeta \to 0}{=} \mp \sum_{q=0}^{r} \tilde{c}_{r-q}\zeta^{-q-1} + O(1) \text{ as } P \to P_{\infty_\pm}. \tag{1.244}$$

By (1.22), in order to prove (1.244), it suffices to prove the homogeneous case with $\tilde{c}_0 = 1$ and $\tilde{c}_q = 0, q = 1, \ldots, r$. Using the expression

$$G(z, n, n, t_r) = F_p(z, n, t_r)/y(P)$$

for the diagonal Green's function of $\check{L}(t_r)$, we may rewrite (1.244) in the form

$$\widetilde{F}_r(z, n, t_r)/z^{r+1} = z^{-1}\sum_{q=0}^{r} \hat{f}_{r-q}(n, t_r)z^{q-r}$$
$$\underset{z \to \infty}{=} -G(z, n, n, t_r) + O(z^{-r-1}). \tag{1.245}$$

Since

$$G(z, n, n, t_r) = \left(\delta_n, \left(\check{L}(t_r) - z\right)^{-1}\delta_n\right),$$

the Neumann expansion for $\left(\check{L}(t_r) - z\right)^{-1}$ then shows that (1.245) is equivalent to

$$z^{-1}\sum_{q=0}^{r} \hat{f}_{r-q}(n, t_r)z^{q-r} \underset{z \to \infty}{=} z^{-1}\sum_{q=0}^{r}\left(\delta_n, \check{L}(t_r)^q\delta_n\right)z^{-q} + O(z^{-r-2}). \tag{1.246}$$

But (1.246) is proven in (1.15) of Lemma 1.4. Combining (1.217), (1.241), (1.243), and (1.244) then proves (1.240). □

Finally, we note that Lemma 1.18 on nonspecial divisors in the stationary context extends to the present time-dependent situation without a change. Indeed, since $t_r \in \mathbb{R}$ just plays the role of a parameter, the proof of Lemma 1.18 extends line by line and is hence omitted.

Lemma 1.40 *Assume Hypothesis 1.34 and suppose that (1.207)–(1.210) hold. Moreover, let $(n, t_r) \in \mathbb{Z} \times \mathbb{R}$. Denote by $\mathcal{D}_{\underline{\hat{\mu}}}$, $\underline{\hat{\mu}} = (\hat{\mu}_1, \ldots, \hat{\mu}_p) \in \mathrm{Sym}^p(\mathcal{K}_p)$, the Dirichlet divisor of degree p associated with a, b, and ϕ defined according to (1.67), that is,*

$$\hat{\mu}_j(n, t_r) = (\mu_j(n, t_r), -G_{p+1}(\mu_j(n, t_r), n, t_r)) \in \mathcal{K}_p, \quad j = 1, \ldots, p.$$

Then $\mathcal{D}_{\underline{\hat{\mu}}(n,t_r)}$ is nonspecial for all $(n, t_r) \in \mathbb{Z} \times \mathbb{R}$.

In order to express $\phi(P, n, t_r)$ and $\psi(P, n, n_0, t_r, t_{0,r})$ in terms of the theta function of $\mathcal{K}_p$ we need some additional notation. Let $\omega^{(2)}_{P_{\infty\pm}, q}$ be the normalized abelian differential of the second kind (i.e., with vanishing a-periods) with a single pole at $P_{\infty\pm}$ of the form

$$\omega^{(2)}_{P_{\infty\pm}, q} \underset{\zeta \to 0}{=} \left(\zeta^{-2-q} + O(1)\right) d\zeta, \quad P \to P_{\infty\pm}, \; \zeta = 1/z, \; q \in \mathbb{N}_0.$$

Given the summation constants $\tilde{c}_1, \ldots, \tilde{c}_r$ in $\tilde{F}_r$, see (1.6), (1.51), (1.22), and (1.34), we then define

$$\widetilde{\Omega}^{(2)}_r = \sum_{q=0}^{r} (q+1) \tilde{c}_{r-q} \left(\omega^{(2)}_{P_{\infty+}, q} - \omega^{(2)}_{P_{\infty-}, q}\right), \quad \tilde{c}_0 = 1. \tag{1.247}$$

Since $\omega^{(2)}_{P_{\infty\pm}, q}$ were supposed to be normalized one has

$$\int_{a_j} \widetilde{\Omega}^{(2)}_r = 0, \quad j = 1, \ldots, p. \tag{1.248}$$

Moreover, writing

$$\omega_j = \left(\sum_{q=0}^{\infty} d_{j,q}(P_{\infty\pm}) \zeta^q\right) d\zeta = \pm \left(\sum_{q=0}^{\infty} d_{j,q}(P_{\infty+}) \zeta^q\right) d\zeta \text{ near } P_{\infty\pm}, \tag{1.249}$$

relations (A.19) and (B.33) yield

$$\underline{\widetilde{U}}^{(2)}_r = \left(\widetilde{U}^{(2)}_{r,1}, \ldots, \widetilde{U}^{(2)}_{r,p}\right),$$

$$\widetilde{U}^{(2)}_{r,j} = \frac{1}{2\pi i} \int_{b_j} \widetilde{\Omega}^{(2)}_r = 2 \sum_{q=0}^{r} \tilde{c}_{r-q} d_{j,q}(P_{\infty+}), \tag{1.250}$$

$$= 2 \sum_{q=0}^{r} \tilde{c}_{r-q} \sum_{k=1}^{p} c_j(k) \hat{c}_{k-p+q}(\underline{E}), \quad j = 1, \ldots, p.$$

1.5 The Time-Dependent Toda Formalism

We also recall the properties of the normal differential of the third kind $\omega_{P_{\infty_+},P_{\infty_-}}^{(3)}$ discussed in (1.108)–(1.111) and its vector of b-periods, $\underline{U}^{(3)}$, in (1.112), (1.113).

As in the stationary context (cf. (1.114) and (1.117)) one infers that ϕ is now of the form

$$\phi(P,n,t_r) = C(n,t_r)\frac{\theta(\underline{z}(P,\underline{\hat{\mu}}^+(n,t_r)))}{\theta(\underline{z}(P,\underline{\hat{\mu}}(n,t_r)))}\exp\left(\int_{P_0}^P \omega_{P_{\infty_+},P_{\infty_-}}^{(3)}\right)$$

for some $C(n,t_r)$ independent of P, and by comparison with Lemma 1.39 that

$$a(n,t_r) = C(n,t_r)\tilde{a}\frac{\theta(\underline{z}(P_{\infty_+},\underline{\hat{\mu}}^+(n,t_r)))}{\theta(\underline{z}(P_{\infty_+},\underline{\hat{\mu}}(n,t_r)))}, \quad (n,t_r) \in \mathbb{Z}\times\mathbb{R}. \tag{1.251}$$

Given these preparations, the theta function representations for ϕ, ψ, a, and b then read as follows.

Theorem 1.41 *Assume Hypothesis* 1.34 *and suppose that* (1.207)–(1.210) *hold. In addition, let* $P \in \mathcal{K}_p \setminus \{P_{\infty_+}, P_{\infty_-}\}$ *and* $(n,n_0,t_r,t_{0,r}) \in \mathbb{Z}^2 \times \mathbb{R}^2$. *Then for each* $(n,t_r) \in \mathbb{Z}\times\mathbb{R}$, $\mathcal{D}_{\underline{\hat{\mu}}(n,t_r)}$ *is nonspecial. Moreover*,[1]

$$\phi(P,n,t_r) = C(n,t_r)\frac{\theta(\underline{z}(P,\underline{\hat{\mu}}^+(n,t_r)))}{\theta(\underline{z}(P,\underline{\hat{\mu}}(n,t_r)))}\exp\left(\int_{P_0}^P \omega_{P_{\infty_+},P_{\infty_-}}^{(3)}\right), \tag{1.252}$$

$$\psi(P,n,n_0,t_r,t_{0,r}) = C(n,n_0,t_r,t_{0,r})\frac{\theta(\underline{z}(P,\underline{\hat{\mu}}(n,t_r)))}{\theta(\underline{z}(P,\underline{\hat{\mu}}(n_0,t_{0,r})))} \tag{1.253}$$

$$\times \exp\left((n-n_0)\int_{P_0}^P \omega_{P_{\infty_+},P_{\infty_-}}^{(3)} + (t_r - t_{0,r})\int_{P_0}^P \tilde{\Omega}_r^{(2)}\right),$$

where $C(n,t_r)$ *and* $C(n,n_0,t_r,t_{0,r})$ *are given by*

$$C(n,t_r) = C(n+1,n,t_r,t_r) = \left(\frac{\theta(\underline{z}(P_{\infty_+},\underline{\hat{\mu}}^-(n,t_r)))}{\theta(\underline{z}(P_{\infty_+},\underline{\hat{\mu}}^+(n,t_r)))}\right)^{1/2}, \tag{1.254}$$

$$C(n,n_0,t_r,t_{0,r}) = C(n,n_0,t_r,t_r)C(n_0,n_0,t_r,t_{0,r}) \tag{1.255}$$

$$= \left(\frac{\theta(\underline{z}(P_{\infty_+},\underline{\hat{\mu}}(n_0,t_{0,r})))\theta(\underline{z}(P_{\infty_+},\underline{\hat{\mu}}^-(n_0,t_{0,r})))}{\theta(\underline{z}(P_{\infty_+},\underline{\hat{\mu}}(n,t_r)))\theta(\underline{z}(P_{\infty_+},\underline{\hat{\mu}}^-(n,t_r)))}\right)^{1/2}, \tag{1.256}$$

---

[1] To avoid multi-valued expressions in formulas such as (1.120), (1.121), etc., we agree always to choose the same path of integration connecting P_0 and P and refer to Remark A.30 for additional tacitly assumed conventions.

with

$$C(n, n_0, t_r, t_r) = \begin{cases} \prod_{n'=n_0}^{n-1} C(n', t_r), & n \geq n_0 + 1, \\ 1, & n = n_0, \\ \prod_{n'=n}^{n_0-1} C(n', t_r)^{-1}, & n \leq n_0 - 1, \end{cases} \quad (1.257)$$

and

$$C(n_0, n_0, t_r, t_{0,r}) = \left(\frac{\theta(\underline{z}(P_{\infty_+}, \underline{\hat{\mu}}(n_0, t_{0,r}))) \theta(\underline{z}(P_{\infty_+}, \underline{\hat{\mu}}^-(n_0, t_{0,r})))}{\theta(\underline{z}(P_{\infty_+}, \underline{\hat{\mu}}(n_0, t_r))) \theta(\underline{z}(P_{\infty_+}, \underline{\hat{\mu}}^-(n_0, t_r)))} \right)^{1/2}.$$
(1.258)

Here the square root branch of $C(n, t_r)$ in (1.254) has to be chosen according to (1.251) and the square root branch of $C(n, n_0, t_r, t_{0,r})$ in (1.256) is determined with the help of (1.255), by the square root branch in (1.254), and by formula (1.257).

The Abel map linearizes the auxiliary divisor $\mathcal{D}_{\underline{\hat{\mu}}(n, t_r)}$ in the sense that

$$\underline{\alpha}_{P_0}(\mathcal{D}_{\underline{\hat{\mu}}(n, t_r)}) = \underline{\alpha}_{P_0}(\mathcal{D}_{\underline{\hat{\mu}}(n_0, t_{0,r})}) - \underline{U}^{(3)}(n - n_0) - \underline{\widetilde{U}}_r^{(2)}(t_r - t_{0,r}) \quad (\text{mod } L_p).$$
(1.259)

Finally, a, b are of the form

$$a(n, t_r)^2 = \tilde{a}^2 \frac{\theta(\underline{z}(P_{\infty_+}, \underline{\hat{\mu}}^-(n, t_r))) \theta(\underline{z}(P_{\infty_+}, \underline{\hat{\mu}}^+(n, t_r)))}{\theta(\underline{z}(P_{\infty_+}, \underline{\hat{\mu}}(n, t_r)))^2}, \quad (1.260)$$

$$b(n, t_r) = \frac{1}{2} \sum_{m=0}^{2p+1} E_m - \sum_{j=1}^{p} \lambda_j$$
$$- \sum_{j=1}^{p} c_j(p) \frac{\partial}{\partial w_j} \ln \left(\frac{\theta(\underline{z}(P_{\infty_+}, \underline{\hat{\mu}}(n, t_r) + \underline{w}))}{\theta(\underline{z}(P_{\infty_+}, \underline{\hat{\mu}}^-(n, t_r) + \underline{w}))} \right) \bigg|_{\underline{w}=0} \quad (1.261)$$

with $\tilde{a}$ introduced in (1.110).

Proof First of all we note that (1.252) and (1.253) are well-defined due to (1.109), (1.113), (1.248), (1.250), (A.34), and (A.35). Moreover, (1.252) for ϕ and (1.253) for ψ in the special case $t_{0,r} = t_r$ are clear from the stationary context that led to (1.114) and (1.115).

Denoting the right-hand side of (1.253) by $\Psi(P, n, n_0, t_r, t_{0,r})$, our goal is to prove $\psi = \Psi$. By inspection, one verifies

$$\Psi(P, n, n_0, t_r, t_{0,r}) = \Psi(P, n, n_0, t_r, t_r) \Psi(P, n_0, n_0, t_r, t_{0,r}). \quad (1.262)$$

1.5 The Time-Dependent Toda Formalism

Comparison of (1.72), (1.71), (1.120)–(1.122) and (1.213), (1.252)–(1.256) then yields

$$\psi(P, n+1, n, t_r, t_r) = \phi(P, n, t_r) = \Psi(P, n+1, n, t_r, t_r).$$

Moreover,

$$\psi(P, n, n_0, t_r, t_r) = \begin{cases} \prod_{n'=n_0}^{n-1} \phi(P, n', t_r), & n \geq n_0 + 1, \\ 1, & n = n_0, \\ \prod_{n'=n}^{n_0-1} \phi(P, n', t_r)^{-1}, & n \leq n_0 - 1, \end{cases}$$
$$= \Psi(P, n, n_0, t_r, t_r).$$

By (1.262) it remains to identify

$$\psi(P, n_0, n_0, t_r, t_{0,r}) = \Psi(P, n_0, n_0, t_r, t_{0,r}). \tag{1.263}$$

We start by recalling (cf. (1.242))

$$\psi(P, n_0, n_0, t_r, t_{0,r}) \tag{1.264}$$
$$= \exp\left(\int_{t_{0,r}}^{t} ds \left(\widetilde{F}_r(z, n_0, s) \frac{y(P) - G_{p+1}(z, n_0, s)}{F_p(z, n_0, s)} + \widetilde{G}_{r+1}(z, n_0, s) \right) \right).$$

In order to spot the zeros and poles of ψ on $\mathcal{K}_p \setminus \{P_{\infty_+}, P_{\infty_-}\}$ we need to expand the integrand in (1.264) near its singularities (the zeros $\mu_j(n_0, s)$ of $F_p(z, n_0, s)$). Using (1.236) one obtains

$$\psi(P, n_0, n_0, t_r, t_{0,r}) = \exp\left(\int_{t_{0,r}}^{t_r} ds \left(\frac{\frac{d}{ds}\mu_j(n_0, s)}{\mu_j(n_0, s) - z} + O(1) \right) \right)$$
$$= \begin{cases} (\mu_j(n_0, t_r) - z)O(1), & P \text{ near } \hat{\mu}_j(n_0, t_r) \neq \hat{\mu}_j(n_0, t_{0,r}), \\ O(1), & P \text{ near } \hat{\mu}_j(n_0, t_r) = \hat{\mu}_j(n_0, t_{0,r}), \\ (\mu_j(n_0, t_{0,r}) - z)^{-1}O(1), & P \text{ near } \hat{\mu}_j(n_0, t_{0,r}) \neq \hat{\mu}_j(n_0, t_r) \end{cases}$$

with $O(1) \neq 0$. Hence all zeros and poles of $\psi(P, n_0, n_0, t_r, t_{0,r})$ on $\mathcal{K}_p \setminus \{P_{\infty_+}, P_{\infty_-}\}$ are simple. Moreover, the poles of $\psi(P, n_0, n_0, t_r, t_{0,r})$ coincide with those of $\Psi(P, n_0, n_0, t_r, t_{0,r})$. Next we need to identify the essential singularities of $\psi(P, n_0, n_0, t_r, t_{0,r})$ at $P_{\infty_\pm}$. But these essential singularities have been identified in (1.240) and hence they coincide with those of $\Psi(P, n_0, n_0, t_r, t_{0,r})$ at $P_{\infty_\pm}$. Thus, we can apply Lemma B.1 to conclude (1.263) since $\mathcal{D}_{\underline{\hat{\mu}}(n, t_r)}$ is nonspecial for all $(n, t_p) \in \mathbb{Z} \times \mathbb{R}$. This yields (1.252) and (1.253).

The computation (cf. (D.13)–(D.18))

$$\partial_{t_r} \sum_{j=1}^{p} \underline{A}_{P_0,k}(\hat{\mu}_j(n,t_r)) = \partial_{t_r} \sum_{j=1}^{p} \int_{P_0}^{\hat{\mu}_j(n,t_r)} \omega_k$$

$$= \partial_{t_r} \sum_{j=1}^{p} \sum_{\ell=1}^{p} c_k(\ell) \int_{P_0}^{\hat{\mu}_j(n,t_r)} \frac{z^{\ell-1} dz}{y(P)}$$

$$= \sum_{j=1}^{p} \sum_{\ell=1}^{p} c_k(\ell) \frac{\mu_j(n,t_r)^{\ell-1}}{y(\hat{\mu}_j(n,t_r))} \partial_{t_r} \mu_j(n,t_r)$$

$$= -2i \sum_{j=1}^{p} \sum_{\ell=1}^{p} c_k(\ell) \frac{\mu_j(n,t_r)^{\ell-1}}{y(\hat{\mu}_j(n,t_r))} \frac{y(\hat{\mu}_j(n,t_r))}{\prod_{m \neq j}^{p} (\mu_j(n,t_r) - \mu_m(n,t_r))} \widetilde{F}_r(\mu_j(n,t_r))$$

$$= -2i \sum_{j=1}^{p} \sum_{\ell=1}^{p} c_k(\ell) U_p(\underline{\mu}(n,t_r))_{\ell,j} \widetilde{F}_r(\mu_j(n,t_r))$$

$$= -2i \sum_{\ell=1 \vee (p-r)}^{p} c_k(\ell) \tilde{d}_{r,p-\ell}(\underline{E}), \tag{1.265}$$

using Lemma D.3 and (D.18) in the final step (with $U_p(\underline{\mu})$ defined in (D.10)), then shows that the flows (1.236) are linearized by the Abel map. Equations (B.34) and (D.15) then yield

$$\frac{d}{dt_r} \underline{\alpha}_{P_0}(\mathcal{D}_{\underline{\hat{\mu}}(n,t_r)}) = -\underline{\widetilde{U}}_r^{(2)},$$

which proves (1.259).

Equations (1.254)–(1.256) and expressions (1.260) and (1.261) are then proved as in Theorem 1.19. □

Remark 1.42 (i) Since in the special case $r = 0$, that is, for the original Toda lattice equations, $\underline{U}_0^{(2)}$ simplifies to

$$\underline{U}_0^{(2)} = 2\underline{c}(p),$$

due to (1.249), (1.250), and (B.33), the linearization property (1.259) shows that the expression for $b(n,t_r)$ in (1.261) can be rewritten in the familiar form

$$b(n,t_0) = \frac{1}{2} \sum_{m=0}^{2p+1} E_m - \sum_{j=1}^{p} \lambda_j + \frac{1}{2} \frac{d}{dt_0} \ln\left(\frac{\theta(\underline{z}(P_{\infty_+}, \underline{\hat{\mu}}(n,t_0)))}{\theta(\underline{z}(P_{\infty_+}, \underline{\hat{\mu}}^-(n,t_0)))}\right),$$

$$(n,t_0) \in \mathbb{Z} \times \mathbb{R}.$$

1.5 The Time-Dependent Toda Formalism

(ii) Furthermore, expanding equation (1.225) around $P_{\infty_\pm}$ (still for $r=0$) shows that

$$\int_{P_0}^{P} \Omega_0^{(2)} \underset{\zeta \to 0}{=} \mp\big(\zeta^{-1} - \tilde{b}_1 + O(\zeta)\big) \text{ as } P \to P_{\infty_\pm}, \quad r = 0, \qquad (1.266)$$

where $\tilde{b}_1$ is defined in (1.111). Conversely, proving (1.225) as in the KdV case in Volume I (by expanding both sides in (1.225) around $P_{\infty_\pm}$ and using Lemma B.1) turns out to be equivalent to proving (1.266). However, since we are not aware of an independent proof of (1.266), we chose a different strategy in the proof of Theorem 1.41.

Moreover, in analogy to Corollary 1.21, $b(n, t_r)$ admits the following alternative theta function representation.

Corollary 1.43 $b(n, t_r)$ *admits the representation*

$$b(n, t_r) = E_0 - \tilde{a}\frac{\theta(\underline{z}(P_{\infty_+}, \underline{\hat{\mu}}^-(n, t_r)))\theta(\underline{z}(P_0, \underline{\hat{\mu}}^+(n, t_r)))}{\theta(\underline{z}(P_{\infty_+}, \underline{\hat{\mu}}(n, t_r)))\theta(\underline{z}(P_0, \underline{\hat{\mu}}(n, t_r)))}$$

$$- \tilde{a}\frac{\theta(\underline{z}(P_{\infty_+}, \underline{\hat{\mu}}(n, t_r)))\theta(\underline{z}(P_0, \underline{\hat{\mu}}^-(n, t_r)))}{\theta(\underline{z}(P_{\infty_+}, \underline{\hat{\mu}}^-(n, t_r)))\theta(\underline{z}(P_0, \underline{\hat{\mu}}(n, t_r)))}, \quad (n, t_r) \in \mathbb{Z} \times \mathbb{R}.$$

Since the proof of Corollary 1.43 is identical to that of 1.21 we omit further details.

Combining (1.259) and (1.260), (1.261) shows the remarkable linearity of the theta function representations for a and b with respect to $(n, t_r) \in \mathbb{Z} \times \mathbb{R}$. In fact, one can rewrite (1.260), (1.261) as

$$a(n)^2 = \tilde{a}^2 \frac{\theta(\underline{A} - \underline{B} + \underline{B}n + \underline{C}_r t_r)\theta(\underline{A} + \underline{B} + \underline{B}n + \underline{C}_r t_r)}{\theta(\underline{A} + \underline{B}n + \underline{C}_r t_r)^2},$$

$$b(n) = \frac{1}{2}\sum_{m=0}^{2p+1} E_m - \sum_{j=1}^{p} \lambda_j$$

$$- \sum_{j=1}^{p} c_j(p)\frac{\partial}{\partial w_j} \ln\left(\frac{\theta(\underline{A} + \underline{B}n + \underline{C}_r t_r + \underline{w})}{\theta(\underline{A} - \underline{B} + \underline{B}n + \underline{C}_r t_r + \underline{w})}\right)\bigg|_{\underline{w}=0},$$

where

$$\underline{A} = \Xi_{P_0} - \underline{A}_{P_0}(P_{\infty_+}) + \underline{U}^{(3)}n_0 + \underline{\tilde{U}}_r^{(2)}t_r + \alpha_{P_0}(\mathcal{D}_{\underline{\hat{\mu}}(n_0, t_{0,r})}),$$

$$\underline{B} = -\underline{U}^{(3)}, \quad \underline{C}_r = -\underline{\tilde{U}}_r^{(2)},$$

$$\Lambda_0 = \frac{1}{2}\sum_{m=0}^{2p+1} E_m - \sum_{j=1}^{p} \lambda_j.$$

Here the constants $\tilde{a}, \lambda_j, c_j(p) \in \mathbb{C}, j = 1, \ldots, p$, and the constant vectors $\underline{B}, \underline{C}_r \in \mathbb{C}^p$ are uniquely determined by $\mathcal{K}_p$ (and its homology basis) and r, and the constant vector $\underline{A} \in \mathbb{C}^p$ is in one-to-one correspondence with the Dirichlet data $\underline{\hat{\mu}}(n_0, t_{0,r}) = (\hat{\mu}_1(n_0, t_{0,r}), \ldots, \hat{\mu}_p(n_0, t_{0,r})) \in \operatorname{Sym}^p(\mathcal{K}_p)$ at the initial point $(n_0, t_{0,r})$ as long as the divisor $\mathcal{D}_{\underline{\hat{\mu}}(n_0, t_{0,r})}$ is assumed to be nonspecial.

Remark 1.44 The explicit representation (1.253) for ψ complements Lemma 1.35 and shows that ψ stays meromorphic on $\mathcal{K}_p \setminus \{P_{\infty_+}, P_{\infty_-}\}$ as long as $\mathcal{D}_{\underline{\hat{\mu}}}$ is nonspecial.

As in our stationary Section 1.3 we will end this section with some examples illustrating the general results. Again we also consider an example involving singular curves. We start with the elementary genus $p = 0$ example excluded thus far in our considerations of this section.

Example 1.45 Assume $p = 0, r \in \mathbb{N}_0, P = (z, y) \in \mathcal{K}_0 \setminus \{P_{\infty_+}, P_{\infty_-}\}$, and let $(n, n_0, t_r, t_{0,r}) \in \mathbb{Z}^2 \times \mathbb{R}^2$. Then,

$$\mathcal{K}_0 \colon \mathcal{F}_0(z, y) = y^2 - R_2(z) = y^2 - (z - E_0)(z - E_1) = 0,$$
$$E_0, E_1 \in \mathbb{C}, \ E_0 \neq E_1,$$
$$a(n) = a, \ a^2 = (E_1 - E_0)^2/16, \quad b(n) = b = \frac{1}{2}(E_0 + E_1),$$
$$\text{s-}\widehat{\operatorname{Tl}}_k(a, b) = 0, \ k \in \mathbb{N}_0, \quad \widetilde{\operatorname{Tl}}_r(a, b) = 0,$$
$$\phi(P, n) = \frac{y + z - b}{2a},$$
$$\psi(P, n, n_0, t_r, t_{0,r}) = \left(\frac{y + z - b}{2a}\right)^{(n - n_0)} \exp\left(y\widetilde{F}_r(z)(t_r - t_{0,r})\right).$$

Example 1.46 The case $p = 1, r = 0$.
We suppose

$$E_0 < E_1 < E_2 < E_3, \quad R_4(z) = \prod_{m=0}^{3}(z - E_m),$$

and record in addition to the objects introduced in (1.161) and (1.162) in connection

1.5 The Time-Dependent Toda Formalism

with the stationary $p = 1$ Example 1.27 the following formulas:

$$\Omega_0^{(2)} = \omega_{P_{\infty_+},0}^{(2)} - \omega_{P_{\infty_-},0}^{(2)},$$

$$2\pi i U_{0,1}^{(2)} = \int_{b_1} \Omega_0^{(2)} = 4\pi i c_1(1) = \frac{2\pi i}{CK(k)},$$

$$\int_{P_0}^{P} \Omega_0^{(2)} = \frac{1}{2}C(E_2 - E_0)\Big((E_3 - E_1)K(k)^{-1}E(k)F(\bar{u}(z), k)$$
$$- (E_3 - E_1)E(\bar{u}(z), k) - (E_3 - E_0)\big(1 - ((E_3 - E_0)/(E_3 - E_1))\bar{u}(z)^2\big)^{-1}$$
$$\times \bar{u}(z)\big(1 - \bar{u}(z)^2\big)^{1/2}\big(1 - k^2\bar{u}(z)^2\big)^{-1/2}\Big), \quad P = (z, y) \in \mathcal{K}_1.$$

In addition, one computes for $(n, n_0, t_0, t_{0,0}) \in \mathbb{Z}^2 \times \mathbb{R}^2$ (cf. (1.163))

$$A_{P_0}(\hat{\mu}_1(n, t_0)) = A_{P_0}(\hat{\mu}_1(n_0, t_{0,0})) - 2(n - n_0)A_{P_0}(P_{\infty_+}) - 2(t_0 - t_{0,0})c_1(1),$$

$$\mu_1(n, t_0) = E_1\Big(1 - ((E_2 - E_1)/(E_2 - E_0))(E_0/E_1)$$
$$\times \operatorname{sn}^2\big(2K(k)\delta_1 - 2(n - n_0)F\big(((E_3 - E_1)/(E_3 - E_0))^{1/2}, k\big)$$
$$+ 2C^{-1}(t_0 - t_{0,0})\big)\Big)$$
$$\times \Big(1 - ((E_2 - E_1)/(E_2 - E_0))$$
$$\times \operatorname{sn}^2\big(2K(k)\delta_1 - 2(n - n_0)F\big(((E_3 - E_1)/(E_3 - E_0))^{1/2}, k\big)$$
$$+ 2C^{-1}(t_0 - t_{0,0})\big)\Big)^{-1},$$

where we abbreviated

$$A_{P_0}(\hat{\mu}_1(n_0, t_{0,0})) = (-\delta_1 + \frac{1}{2}\tau_{1,1}) \pmod{L_1}.$$

Moreover,

$$\text{s-Tl}_1(a, b) = 0, \quad \widetilde{\text{Tl}}_0(a, b) = 0$$

for $c_1 = -\frac{1}{2}(E_0 + E_1 + E_2 + E_3)$.

Finally, we consider the case of (not necessarily real-valued) p-soliton solutions for the Toda lattice hierarchy.

Example 1.47 The case of Toda p-soliton solutions.
Let $p, r \in \mathbb{N}$, $(n, t_r) \in \mathbb{Z} \times I$, $I \subseteq \mathbb{R}$ an open interval, introduce

$$\tau_0(n, t_r) = 1,$$
$$\tau_p(n, t_r) = \det(I_p + C_p(n, t_r)),$$

$$C_p(n, t_r) = \left(\check{c}_j \check{c}_k \frac{(z_j z_k)^{n+1}}{1 - z_j z_k} \exp\left(\frac{1}{2}(z_j - z_j^{-1})\widetilde{F}_{0,r}(E_{2j})t_r \right.\right.$$
$$\left.\left. + \frac{1}{2}(z_k - z_k^{-1})\widetilde{F}_{0,r}(E_{2k})t_r \right) \right)_{j,k=1,\ldots,p},$$

$\check{c}_j \in \mathbb{C} \setminus \{0\}$, $z_j \in \mathbb{C}$, $0 < |z_j| < 1$, $z_j \neq z_k$ for $j \neq k$, $j, k \in \{1, \ldots, p\}$,

and assume $\tau_{p-1}(n, t_r) \neq 0$, $\tau_p(n, t_r) \neq 0$ for all $(n, t_r) \in \mathbb{N} \times I$. Here $\widetilde{F}_{0,r}(z)$ denotes the polynomial $\widetilde{F}_r(z)$ in (1.205) in the special case $N = 0$ with $a = \frac{1}{2}$, $b = 0$. Explicitly,

$$\widetilde{F}_{0,r}(z) = \sum_{s=0}^{r} \tilde{c}_{r-s} \widehat{F}_{0,s}(z), \tag{1.267}$$

$$\widehat{F}_{0,2k}(z) = \sum_{\ell=0}^{k} 2^{-2\ell} \binom{2\ell}{\ell} z^{2(k-\ell)}, \quad \widehat{F}_{0,2k+1}(z) = z\widehat{F}_{0,2k}(z), \quad k \in \mathbb{N}_0, \, z \in \mathbb{C},$$

for an appropriate set of summation constants $\{\tilde{c}_s\}_{s=1,\ldots,r} \subset \mathbb{C}$. Then,

$$a_p(n, t_r) = \frac{\tau_p(n+2, t_r)^{1/2} \tau_p(n, t_r)^{1/2}}{2\tau_p(n+1, t_r)},$$

$$b_p(n, t_r) = -\frac{1}{2}\left(\frac{z_p \tau_{p-1}(n+2, t_r)\tau_p(n, t_r)}{\tau_{p-1}(n+1, t_r)\tau_p(n+1, t_r)} + \frac{\tau_{p-1}(n, t_r)\tau_p(n+1, t_r)}{z_p \tau_{p-1}(n+1, t_r)\tau_p(n, t_r)}\right)$$
$$+ \frac{1}{2}(z_p + z_p^{-1}),$$

and

$$\text{s-Tl}_p(a_p, b_p) = 0, \quad \widetilde{\text{Tl}}_r(a_p, b_p) = 0$$

for an appropriate set of summation constants $\{c_\ell\}_{\ell=1,\ldots,p}$ (cf. (1.44)) and with $\{\tilde{c}_s\}_{s=1,\ldots,r}$ chosen as in (1.267). The associated singular curve is given by

$$\mathcal{F}_p(z, y) = y^2 - (z+1)(z-1) \prod_{j=1}^{p} \left(z - \frac{1}{2}(z_j + z_j^{-1})\right)^2 = 0,$$

$$E_0 = -1, \quad E_{2p+1} = 1, \quad E_{2j-1} = E_{2j} = \frac{1}{2}(z_j + z_j^{-1}), \quad j = 1, \ldots, p.$$

The conditions in Example 1.47 are satisfied in the case where a, b are real-valued for $I = \mathbb{R}$, assuming

$$\check{c}_j > 0, \quad j = 1, \ldots, p,$$
$$\frac{1}{2}(z_p + z_p^{-1}) > \frac{1}{2}(z_{p-1} + z_{p-1}^{-1}) > \cdots > \frac{1}{2}(z_1 + z_1^{-1}) > 1,$$

and for sufficiently small (complex) perturbations thereof (if I is small enough).

1.6 The Time-Dependent Toda Algebro-Geometric IVP

In exactly the same manner as in (1.176)–(1.177) one then infers that $F_p(\,\cdot\,, n_0, t_r)$ divides $R_{2p+2} - G_{p+1}^2$ (since t_r is just a fixed parameter). Moreover, arguing as in (1.178)–(1.181) we now assume that the polynomial

$$R_{2p+2}(z) - G_{p+1}(z, n_0, t_r)^2 \underset{z \to \infty}{=} O(z^{2p})$$

is precisely of maximal order $2p$ for all $t_r \in (t_{0,r} - T_0, t_{0,r} + T_0)$. One then obtains

$$R_{2p+2}(z) - G_{p+1}(z, n_0, t_r)^2 = -4a(n_0, t_r)^2 F_p(z, n_0, t_r) F_p(z, n_0 + 1, t_r),$$
$$(z, t_r) \in \mathbb{C} \times (t_{0,r} - T_0, t_{0,r} + T_0), \quad (1.296)$$

where we introduced the coefficient $a(n_0, t_r)^2$ to make $F_p(\,\cdot\,, n_0 + 1, t_r)$ a monic polynomial of degree p. As in Section 1.4, the assumption that the polynomial $F_p(\,\cdot\,, n_0 + 1, t_r)$ is precisely of order p is implied by the hypothesis that

$$\mathcal{D}_{\hat{\underline{\mu}}(n_0, t_r)} \in \mathcal{M}_0 \text{ for all } t_r \in (t_{0,r} - T_0, t_{0,r} + T_0), \quad (1.297)$$

a point we will revisit later (cf. Lemma 1.52). Given (1.296), we obtain consistency with (1.273) for $n = n_0$ and $t_r \in (t_{0,r} - T_0, t_{0,r} + T_0)$.

The explicit formula for $a(n_0, t_r)^2$ then reads (for $t_r \in (t_{0,r} - T_0, t_{0,r} + T_0)$)

$$a(n_0, t_r)^2 = \frac{1}{2} \sum_{k=1}^{q(n_0, t_r)} \frac{\left(d^{p_k(n_0,t_r)-1} y(P)/d\zeta^{p_k(n_0,t_r)-1}\right)\big|_{P=(\zeta,\eta)=\hat{\mu}_k(n_0,t_r)}}{(p_k(n_0, t_r) - 1)!}$$

$$\times \prod_{k'=1,\, k' \neq k}^{q(n_0, t_r)} (\mu_k(n_0, t_r) - \mu_{k'}(n_0, t_r))^{-p_k(n_0, t_r)}$$

$$+ \frac{1}{4}\left(b^{(2)}(n_0, t_r) - b(n_0, t_r)^2\right).$$

Here and in the following we use the abbreviation

$$b^{(2)}(n, t_r) = \frac{1}{2} \sum_{m=0}^{2p+1} E_m^2 - \sum_{k=1}^{q(n,t_r)} p_k(n, t_r) \mu_k(n, t_r)^2 \quad (1.298)$$

for appropriate ranges of $(n, t_r) \in \mathbb{N} \times \mathbb{R}$.

With (1.291)–(1.298) in place, we can now apply the stationary formalism as summarized in Theorem 1.32, subject to the additional hypothesis (1.297), for each fixed $t_r \in (t_{0,r} - T_0, t_{0,r} + T_0)$. This yields, in particular, the quantities

$$F_p, G_{p+1}, a, b, \text{ and } \hat{\underline{\mu}} \text{ for } (n, t_r) \in \mathbb{Z} \times (t_{0,r} - T_0, t_{0,r} + T_0), \quad (1.299)$$

which are of the form (1.291)–(1.298), replacing the fixed $n_0 \in \mathbb{Z}$ by an arbitrary $n \in \mathbb{Z}$. In addition, one has the following fundamental identities (cf. (1.187), (1.190), (1.192), and (1.193)), which we summarize in the following result.

Lemma 1.49 *Assume Hypothesis 1.48 and condition* (1.297). *Then the following relations are valid on* $\mathbb{C} \times \mathbb{Z} \times (t_{0,r} - T_0, t_{0,r} + T_0)$,

$$R_{2p+2} - G_{p+1}^2 + 4a^2 F_p F_p^+ = 0, \tag{1.300}$$

$$2(z - b^+)F_p^+ + G_{p+1}^+ + G_{p+1} = 0, \tag{1.301}$$

$$2a^2 F_p^+ - 2(a^-)^2 F_p^- + (z - b)(G_{p+1} - G_{p+1}^-) = 0, \tag{1.302}$$

$$2(z - b^+)F_p^+ - 2(z - b)F_p + G_{p+1}^+ - G_{p+1}^- = 0, \tag{1.303}$$

and hence the stationary part, (1.201), *of the algebro-geometric initial value problem holds,*

$$UV_{p+1} - V_{p+1}^+ U = 0 \ on \ \mathbb{C} \times \mathbb{Z} \times (t_{0,r} - T_0, t_{0,r} + T_0).$$

In particular, Lemmas 1.12 *and* 1.14–1.17 *apply.*

Lemma 1.49 now raises the following important consistency issue: On the one hand, one can solve the initial value problem (1.288), (1.289) at $n = n_0$ in some interval $t_r \in (t_{0,r} - T_0, t_{0,r} + T_0)$, and then extend the quantities F_p, G_{p+1} to all $\mathbb{C} \times \mathbb{Z} \times (t_{0,r} - T_0, t_{0,r} + T_0)$ using the stationary algorithm summarized in Theorem 1.32 as just recorded in Lemma 1.49. On the other hand, one can solve the initial value problem (1.288), (1.289) at $n = n_1$, $n_1 \neq n_0$, in some interval $t_r \in (t_{0,r} - T_1, t_{0,r} + T_1)$ with the initial condition obtained by applying the discrete algorithm to the quantities F_p, G_{p+1} starting at $(n_0, t_{0,r})$ and ending at $(n_1, t_{0,r})$. Consistency then requires that the two approaches yield the same result at $n = n_1$ for t_r in some open neighborhood of $t_{0,r}$.

Equivalently, and pictorially speaking, envisage a vertical t_r-axis and a horizontal n-axis. Then, consistency demands that first solving the initial value problem (1.288), (1.289) at $n = n_0$ in some t_r-interval around $t_{0,r}$ and using the stationary algorithm to extend F_p, G_{p+1} horizontally to $n = n_1$ and the same t_r-interval around $t_{0,r}$, or first applying the stationary algorithm starting at $(n_0, t_{0,r})$ to extend F_p, G_{p+1} horizontally to $(n_1, t_{0,r})$ and then solving the initial value problem (1.288), (1.289) at $n = n_1$ in some t_r-interval around $t_{0,r}$ should produce the same result at $n = n_1$ in a sufficiently small open t_r interval around $t_{0,r}$.

To settle this consistency issue, we will prove the following result. To this end we find it convenient to replace the initial value problem (1.288), (1.289) by the original t_r-dependent zero-curvature equation (1.200), $U_{t_r} + U\widetilde{V}_{r+1} - \widetilde{V}_{r+1}^+ U = 0$ on $\mathbb{C} \times \mathbb{Z} \times (t_{0,r} - T_0, t_{0,r} + T_0)$.

Lemma 1.50 *Assume Hypothesis 1.48 and condition* (1.297). *Moreover, suppose that* (1.285)–(1.287) *hold on* $\mathbb{C} \times \{n_0\} \times (t_{0,r} - T_0, t_{0,r} + T_0)$. *Then* (1.285)–(1.287)

1.6 The Time-Dependent Toda Algebro-Geometric IVP

hold on $\mathbb{C} \times \mathbb{Z} \times (t_{0,r} - T_0, t_{0,r} + T_0)$, that is,

$$F_{p,t_r}(z, n, t_r) = 2\big(F_p(z, n, t_r)\widetilde{G}_{r+1}(z, n, t_r) \\ - G_{p+1}(z, n, t_r)\widetilde{F}_r(z, n, t_r)\big), \tag{1.304}$$

$$G_{p+1,t_r}(z, n, t_r) = 4a(n, t_r)^2\big(F_p(z, n, t_r)\widetilde{F}_r^+(z, n, t_r) \\ - F_p^+(z, n, t_r)\widetilde{F}_r(z, n, t_r)\big), \tag{1.305}$$

$$R_{2p+2}(z) = G_{p+1}(z, n, t_r)^2 - 4a(n, t_r)^2 F_p(z, n, t_r) F_p^+(z, n, t_r). \tag{1.306}$$

Moreover,

$$\phi_{t_r}(P, n, t_r) = -2a(n, t_r)\big(\widetilde{F}_r(z, n, t_r)\phi(P, n, t_r)^2 + \widetilde{F}_r^+(z, n, t_r)\big) \\ + 2(z - b^+(n, t_r))\widetilde{F}_r^+(z, n, t_r)\phi(P, n, t_r) \tag{1.307} \\ + \big(\widetilde{G}_{r+1}^+(z, n, t_r) - \widetilde{G}_{r+1}(z, n, t_r)\big)\phi(P, n, t_r),$$

$$a_{t_r}(n, t_r) = -a(n, t_r)\big(2(z - b^+(n, t_r))\widetilde{F}_r^+(z, n, t_r) \\ + \widetilde{G}_{r+1}^+(z, n, t_r) + \widetilde{G}_{r+1}(z, n, t_r)\big), \tag{1.308}$$

$$b_{t_r}(n, t_r) = 2\big((z - b(n, t_r))^2 \widetilde{F}_r(z, n, t_r) + (z - b(n, t_r))\widetilde{G}_{r+1}(z, n, t_r) \\ + a(n, t_r)^2 \widetilde{F}_r^+(z, n, t_r) - (a^-(n, t_r))^2 \widetilde{F}_r^-(z, n, t_r)\big), \tag{1.309}$$

$$(z, n, t_r) \in \mathbb{C} \times \mathbb{Z} \times (t_{0,r} - T_0, t_{0,r} + T_0).$$

Proof By Lemma 1.49 we have (1.213), (1.214), (1.218), (1.220)–(1.222), and (1.300)–(1.303) for $(n, t_r) \in \mathbb{Z} \times (t_{0,r} - T_0, t_{0,r} + T_0)$ at our disposal. Differentiating (1.306) at $n = n_0$ with respect to t_r, inserting (1.304) and (1.305) at $n = n_0$, then yields

$$2F_p^+ a_{t_r} + aF_{p,t_r}^+ = 2a\big(G_{p+1}\widetilde{F}_r^+ - F_p^+ \widetilde{G}_{r+1}\big) \tag{1.310} \\ = 2F_p^+ a\big(-2(z - b^+)\widetilde{F}_r^+ - \widetilde{G}_{r+1}^+ - \widetilde{G}_{r+1}\big) + 2a\big(F_p^+ \widetilde{G}_{r+1}^+ - G_{p+1}^+ \widetilde{F}_r^+\big)$$

at $n = n_0$. By inspection,

$$F_p^+(z)\widetilde{G}_{r+1}^+(z) - G_{p+1}^+(z)\widetilde{F}_r^+(z) \underset{|z| \to \infty}{=} O(z^{p-1}). \tag{1.311}$$

This can be shown directly using formulas such as (1.30)–(1.33), (1.269), (1.270), (1.271), and (1.272). It also follows from (1.233) and the fact that F_p is a monic polynomial of degree p. Thus one concludes that

$$2F_p^+ a_{t_r} = 2F_p^+ a\big(-2(z - b^+)\widetilde{F}_r^+ - \widetilde{G}_{r+1}^+ - \widetilde{G}_{r+1}\big)$$

at $n = n_0$, and upon cancelling $2F_p^+$ that (1.308) holds at $n = n_0$. This and (1.310) then also prove that (1.304) holds at $n = n_0 + 1$.

Next, differentiating $2aF_p\phi = y - G_{p+1}$ at $n = n_0$ with respect to t_r, inserting (1.304), (1.305), and (1.308) at $n = n_0$, and using (1.214) to replace $2aF_p^+$ by

$-(y + G_{p+1})\phi$ and (1.213) to replace $(G_{p+1} - y)$ by $-2aF_p\phi$, yields (1.307) at $n = n_0$ upon cancelling the factor $2aF_p$.

Differentiating (1.301) with respect to t_r (fixing $n = n_0$), inserting (1.301) (to replace G_{p+1}^+), (1.305) at $n = n_0$, and (1.304) at $n = n_0 + 1$ yields

$$\begin{aligned}
0 &= -2F_p^+\big(b_{t_r}^+ - 2(z - b^+)^2 \widetilde{F}_r^+ + 2a^2 \widetilde{F}_r - 2(z - b^+)\widetilde{G}_{r+1}^+\big) \\
&\quad + 4(z - b^+)^2 F_p^+ \widetilde{F}_r^+ + 4(z - b^+) G_{p+1} \widetilde{F}_r^+ + 4(a)^2 F_p \widetilde{F}_r^+ + G_{p+1,t_r}^+ \\
&= -2F_p^+\big(b_{t_r}^+ - 2(z - b^+)^2 \widetilde{F}_r^+ - 2(z - b^+)\widetilde{G}_{r+1}^+ + 2a^2 \widetilde{F}_r - 2(a^+)^2 \widetilde{F}_r^{++}\big) \\
&\quad - 4(a^+)^2 F_p^+ \widetilde{F}_r^{++} + 4(z - b^+)^2 F_p^+ \widetilde{F}_r^+ + 4(z - b^+) G_{p+1} \widetilde{F}_r^+ \\
&\quad + 4a^2 F_p \widetilde{F}_r^+ + G_{p+1,t_r}^+ \\
&= -2F_p^+\big(b_{t_r}^+ - 2(z - b^+)^2 \widetilde{F}_r^+ - 2(z - b^+)\widetilde{G}_{r+1}^+ + 2a^2 \widetilde{F}_r - 2(a^+)^2 \widetilde{F}_r^{++}\big) \\
&\quad + G_{p+1,t_r}^+ - 4(a^+)^2 F_p^+ \widetilde{F}_r^{++} \\
&\quad + \big(4a^2 F_p + 4(z - b^+)^2 F_p^+ + 4(z - b^+)G_{p+1}\big)\widetilde{F}_r^+
\end{aligned} \quad (1.312)$$

at $n = n_0$. Combining (1.301) and (1.302) at $n = n_0$ one computes

$$4(a^+)^2 F_p^{++} = 4a^2 F_p + 4(z - b^+)^2 F_p^+ + 4(z - b^+) G_{p+1} \quad (1.313)$$

at $n = n_0$. Insertion of (1.313) into (1.312) then yields

$$\begin{aligned}
0 &= -2F_p^+\big(b_{t_r}^+ - 2(z - b^+)^2 \widetilde{F}_r^+ - 2(z - b^+)\widetilde{G}_{r+1}^+ + 2a^2 \widetilde{F}_r - 2(a^+)^2 \widetilde{F}_r^{++}\big) \\
&\quad + G_{p+1,t_r}^+ - 4(a^+)^2 F_p^+ \widetilde{F}_r^{++} + 4(a^+)^2 F_p^{++} \widetilde{F}_r^+
\end{aligned} \quad (1.314)$$

at $n = n_0$. In close analogy to (1.311) one observes that

$$F_p^+(z)\widetilde{F}_r^{++}(z) - F_p^{++}(z)\widetilde{F}_r^+(z) \underset{|z| \to \infty}{=} O(z^{p-1}) \text{ for } p \in \mathbb{N}.$$

Thus, since F_p^+ is a monic polynomial of degree p, (1.314) proves that

$$b_{t_r}^+ - 2(z - b^+)^2 \widetilde{F}_r^+ - 2(z - b^+)\widetilde{G}_{r+1}^+ + 2a^2 \widetilde{F}_r - 2(a^+)^2 \widetilde{F}_r^{++} = 0$$

at $n = n_0$, upon cancelling F_p^+. Thus, (1.309) holds at $n = n_0 + 1$. Simultaneously, this proves (1.305) at $n = n_0 + 1$.

Iterating the arguments just presented (and performing the analogous considerations for $n < n_0$) then extends these results to all lattice points $n \in \mathbb{Z}$ and hence proves (1.304)–(1.309) for $(z, n, t_r) \in \mathbb{C} \times \mathbb{Z} \times (t_{0,r} - T_0, t_{0,r} + T_0)$. □

We summarize Lemmas 1.49 and 1.50 next.

1.6 The Time-Dependent Toda Algebro-Geometric IVP

Theorem 1.51 *Assume Hypothesis* 1.48 *and condition* (1.297). *Moreover, suppose that*

$$f_j = f_j(n_0, t_r), \quad j = 1, \ldots, p,$$
$$g_j = g_j(n_0, t_r), \quad j = 1, \ldots, p-1,$$
$$g_p + f_{p+1} = g_p(n_0, t_r) + f_{p+1}(t_r)$$
for all $t_r \in (t_{0,r} - T_0, t_{0,r} + T_0)$,

satisfies the autonomous first-order system of ordinary differential equations (1.288) *(for fixed* $n = n_0$*),*

$$f_{j,t_r} = \mathcal{F}_j(f_1, \ldots, f_p, g_1, \ldots, g_{p-1}, g_p + f_{p+1}), \quad j = 1, \ldots, p,$$
$$g_{j,t_r} = \mathcal{G}_j(f_1, \ldots, f_p, g_1, \ldots, g_{p-1}, g_p + f_{p+1}), \quad j = 1, \ldots, p-1,$$
$$(g_p + f_{p+1})_{t_r} = \mathcal{G}_p(f_1, \ldots, f_p, g_1, \ldots, g_{p-1}, g_p + f_{p+1})$$

with initial condition

$$f_j(n_0, t_{0,r}), \quad j = 1, \ldots, p,$$
$$g_j(n_0, t_{0,r}), \quad j = 1, \ldots, p-1,$$
$$g_p(n_0, t_{0,r}) + f_{p+1}(t_{0,r}).$$

Then F_p *and* G_{p+1} *as constructed in* (1.269)–(1.299) *on* $\mathbb{C} \times \mathbb{Z} \times (t_{0,r} - T_0, t_{0,r} + T_0)$ *satisfy the zero-curvature equations* (1.200), (1.201), *and* (1.235) *on* $\mathbb{C} \times \mathbb{Z} \times (t_{0,r} - T_0, t_{0,r} + T_0)$,

$$U_{t_r} + U\widetilde{V}_{r+1} - \widetilde{V}_{r+1}^+ U = 0,$$
$$UV_{p+1} - V_{p+1}^+ U = 0,$$
$$V_{p+1,t_r} - [\widetilde{V}_{r+1}, V_{p+1}] = 0,$$

with U, V_{p+1}, *and* $\widetilde{V}_{r+1}$ *given by* (1.202). *In particular,* a, b *satisfy the algebro-geometric initial value problem* (1.195), (1.196) *on* $\mathbb{Z} \times (t_{0,r} - T_0, t_{0,r} + T_0)$,

$$\widetilde{\mathrm{Tl}}_r(a, b) = \begin{pmatrix} a_{t_r} - a(\tilde{f}_{p+1}^+(a,b) - \tilde{f}_{p+1}(a,b)) \\ b_{t_r} + \tilde{g}_{p+1}(a,b) - \tilde{g}_{p+1}^-(a,b) \end{pmatrix} = 0, \quad (1.315)$$

$$(a, b)\big|_{t_r = t_{0,r}} = (a^{(0)}, b^{(0)}),$$

$$\text{s-Tl}_p\left(a^{(0)}, b^{(0)}\right) = \begin{pmatrix} -a(f_{p+1}^+(a^{(0)}, b^{(0)}) - f_{p+1}(a^{(0)}, b^{(0)})) \\ g_{p+1}(a^{(0)}, b^{(0)}) - g_{p+1}^-(a^{(0)}, b^{(0)}) \end{pmatrix} = 0 \quad (1.316)$$

and are given by

$$a(n,t_r)^2 = \frac{1}{2}\sum_{k=1}^{q(n,t_r)} \frac{\left(d^{p_k(n,t_r)-1}y(P)/d\zeta^{p_k(n,t_r)-1}\right)\big|_{P=(\zeta,\eta)=\hat{\mu}_k(n,t_r)}}{(p_k(n,t_r)-1)!}$$

$$\times \prod_{k'=1, k'\neq k}^{q(n,t_r)} (\mu_k(n,t_r) - \mu_{k'}(n,t_r))^{-p_k(n,t_r)}$$

$$+ \frac{1}{4}\bigl(b^{(2)}(n,t_r) - b(n,t_r)^2\bigr),$$

$$b(n,t_r) = \frac{1}{2}\sum_{m=0}^{2p+1} E_m - \sum_{k=1}^{q(n,t_r)} p_k(n,t_r)\mu_k(n,t_r),$$

$$(z,n,t_r) \in \mathbb{Z} \times (t_{0,r} - T_0, t_{0,r} + T_0).$$

Moreover, Lemmas 1.12, 1.15, 1.18 *and* 1.35, 1.36, 1.37, 1.38, 1.40 *apply.*

In analogy to Lemma 1.30 we next show that also in the time-dependent case, most initial divisors are well-behaved in the sense that the corresponding divisor trajectory stays away from $P_{\infty\pm}$ for all $(n,t_r) \in \mathbb{Z} \times \mathbb{R}$.

Lemma 1.52 *The set $\mathcal{M}_1$ of initial divisors $\mathcal{D}_{\hat{\underline{\mu}}(n_0,t_{0,r})}$ for which $\mathcal{D}_{\hat{\underline{\mu}}(n,t_r)}$, defined via* (1.259), *is nonspecial and finite for all $(n,t_r) \in \mathbb{Z} \times \mathbb{R}$, forms a dense set of full measure in the set $\mathrm{Sym}^p(\mathcal{K}_p)$ of positive divisors of degree p.*

Proof Let $\mathcal{M}_{\text{sing}}$ be as introduced in the proof of Lemma 1.30. Then

$$\bigcup_{t_r \in \mathbb{R}} \left(\underline{\alpha}_{Q_0}(\mathcal{M}_{\text{sing}}) + t_r \underline{\widetilde{U}}_r^{(2)}\right)$$

$$\subseteq \bigcup_{P \in \{P_{\infty_+}, P_{\infty_-}\}} \bigcup_{t_r \in \mathbb{R}} \left(\underline{A}_{Q_0}(P) + \underline{\alpha}_{Q_0}(\mathrm{Sym}^{p-1}(\mathcal{K}_p)) + t_r \underline{\widetilde{U}}_r^{(2)}\right)$$

is of measure zero as well, since it is contained in the image of $\mathbb{R} \times \mathrm{Sym}^{p-1}(\mathcal{K}_p)$ which misses one real dimension in comparison to the $2p$ real dimensions of $J(\mathcal{K}_p)$. But then

$$\bigcup_{(n,t_r) \in \mathbb{Z} \times \mathbb{R}} \left(\underline{\alpha}_{Q_0}(\mathcal{M}_{\text{sing}}) + n\underline{A}_{P_{\infty_-}}(P_{\infty_+}) + t_r \underline{\widetilde{U}}_r^{(2)}\right) \qquad (1.317)$$

is also of measure zero. Applying $\underline{\alpha}_{Q_0}^{-1}$ to the complement of the set in (1.317) then yields a set $\mathcal{M}_1$ of full measure in $\mathrm{Sym}^p(\mathcal{K}_p)$. In particular, $\mathcal{M}_1$ is necessarily dense in $\mathrm{Sym}^p(\mathcal{K}_p)$. □

1.7 Toda Conservation Laws and the Hamiltonian Formalism 117

Theorem 1.53 *Let $\mathcal{D}_{\underline{\hat{\mu}}(n_0,t_{0,r})} \in \mathcal{M}_1$ be an initial divisor as in Lemma 1.52. Then the sequences a, b constructed from $\underline{\hat{\mu}}(n_0, t_{0,r})$ as described in Theorem 1.51 satisfy Hypothesis 1.34. In particular, the solution a, b of the algebro-geometric initial value problem* (1.315), (1.316) *is global in $(n, t_r) \in \mathbb{Z} \times \mathbb{R}$.*

Proof Starting with $\mathcal{D}_{\underline{\hat{\mu}}(n_0,t_{0,r})} \in \mathcal{M}_1$, the procedure outlined in this section and summarized in Theorem 1.51 leads to $\mathcal{D}_{\underline{\hat{\mu}}(n,t_r)}$ for all $(n, t_r) \in \mathbb{Z} \times (t_{0,r} - T_0, t_{0,r} + T_0)$ such that (1.259) holds. But if a, b should blow up, then $\mathcal{D}_{\underline{\hat{\mu}}(n,t_r)}$ must hit P_{∞_+} or P_{∞_-}, which is excluded by our choice of initial condition. □

Note, however, that in general (i.e., unless one is, e.g., in the special periodic or self-adjoint case), $\mathcal{D}_{\underline{\hat{\mu}}(n,t_r)}$ will get arbitrarily close to $P_{\infty_\pm}$ since straight motions on the torus are generically dense and hence no uniform bound on the sequences $a(n, t_r), b(n, t_r)$ (and no uniform lower bound on $|a(n, t_r)|$) exists as (n, t_r) varies in $\mathbb{Z} \times \mathbb{R}$. In particular, these complex-valued algebro-geometric solutions of the Toda hierarchy initial value problem, in general, will not be quasi-periodic with respect to n or t_r.

1.7 Toda Conservation Laws and the Hamiltonian Formalism

What is a Hilbert space?
Asked by David Hilbert[1]

In this section we deviate from the principal theme of this book and discuss the Green's function of an $\ell^2(\mathbb{Z})$-realization of the difference expression L and systematically derive high-energy expansions of solutions of an associated Riccati-type equation in connection with spatially sufficiently decaying sequences a and b, not necessarily associated with algebro-geometric coefficients. In addition, we derive local conservation laws and develop the Hamiltonian formalism for the Toda hierarchy including variational derivatives and Poisson brackets. At the end of this section we then hint at the necessary extensions to treat almost periodic (and hence quasi-periodic and periodic) coefficients a and b.

In connection with the asymptotic expansions of various quantities we now make the following strengthened assumptions on the coefficients a and b.

Hypothesis 1.54 *Suppose*

$$a, b \in \ell^\infty(\mathbb{Z}), \quad a(n) \neq 0 \text{ for all } n \in \mathbb{Z}.$$

---

[1] Quoted in S. G. Krantz, *Mathematical Apocrypha*, Mathematical Association of America, 2002, p. 89.

Given Hypothesis 1.54 we introduce the $\ell^2(\mathbb{Z})$-realization $\check{L}$ of the Jacobi difference expression L in (1.2) by

$$\check{L}f = Lf, \quad f \in \text{dom}(\check{L}) = \ell^2(\mathbb{Z}). \tag{1.318}$$

In addition, we introduce the half-line Dirichlet operators $\check{L}^D_{+,n_0}$ on $\ell^2([n_0,\infty) \cap \mathbb{Z})$ and $\check{L}^D_{-,n_0}$ on $\ell^2((-\infty, n_0] \cap \mathbb{Z})$, $n_0 \in \mathbb{Z}$, by

$$(\check{L}^D_{+,n_0} u)(n) = \begin{cases} a(n_0)u(n_0+1) + b(n_0)u(n_0), & n = n_0, \\ (Lu)(n), & n \in [n_0+1, \infty) \cap \mathbb{Z}, \end{cases}$$
$$u \in \text{dom}(\check{L}^D_{+,n_0}) = \ell^2([n_0,\infty) \cap \mathbb{Z}), \tag{1.319}$$
$$(\check{L}^D_{-,n_0} v)(n) = \begin{cases} a(n_0)v(n_0+1) + b(n_0)v(n_0), & n = n_0, \\ (Lv)(n), & n \in (-\infty, n_0-1] \cap \mathbb{Z}, \end{cases}$$
$$v \in \text{dom}(\check{L}^D_{-,n_0}) = \ell^2((-\infty, n_0] \cap \mathbb{Z}). \tag{1.320}$$

(At least formally, it is customary to join the Dirichlet-type boundary condition $u(n_0 - 1) = 0$ in the case of $\check{L}^D_{+,n_0}$ and $v(n_0 + 1) = 0$ in the case of $\check{L}^D_{-,n_0}$ in order to avoid the case distinction in (1.319) and (1.320).)

For future purposes we recall the notion of weak solutions associated with finite difference expressions: If R denotes a finite difference expression, then ψ is called a *weak solution* of $R\psi = z\psi$, for some $z \in \mathbb{C}$, if the relation holds pointwise for each lattice point, that is, if $((R - z)\psi)(n) = 0$ for all $n \in \mathbb{Z}$.

The following elementary result about a spectral inclusion of $\check{L}$ and $\check{L}^D_{\pm,n_0}$ will be useful later.

Theorem 1.55 *Suppose a, b satisfy Hypothesis 1.54. Then the numerical range of $\check{L}$, and hence in particular, the spectrum of $\check{L}$ and $\check{L}^D_{\pm,n_0}$ is contained in the closed ball centered at the origin of radius $2\|a\|_\infty + \|b\|_\infty$, that is,*

$$\text{spec}(\check{L}), \text{ spec}(\check{L}^D_{\pm,n_0}) \subseteq \overline{B(0; 2\|a\|_\infty + \|b\|_\infty)}, \quad n_0 \in \mathbb{Z}. \tag{1.321}$$

Proof We denote by $W(T)$ the numerical range of a bounded linear operator $T \in \mathcal{B}(\mathcal{H})$ in the complex, separable Hilbert space $\mathcal{H}$,

$$W(T) = \{(g, Tg)_{\mathcal{H}} \in \mathbb{C} \mid g \in \mathcal{H}, \|g\|_{\mathcal{H}} = 1\}.$$

It is well-known that $W(T)$ is convex and that its closure contains the spectrum of T,

$$\text{spec}(T) \subseteq \overline{W(T)}.$$

1.7 Toda Conservation Laws and the Hamiltonian Formalism

Elementary arguments then prove

$$|(f, \check{L}f)| \leq (2\|a\|_\infty + \|b\|_\infty)\|f\|^2, \quad f \in \ell^2(\mathbb{Z}),$$

$$\left|(w, \check{L}^D_{\pm,n_0} w)_{\ell^2([n_0, \pm\infty)\cap\mathbb{Z})}\right| \leq (2\|a\|_\infty + \|b\|_\infty)\|w\|^2_{\ell^2([n_0, \pm\infty)\cap\mathbb{Z})},$$

$$w \in \ell^2([n_0, \pm\infty) \cap \mathbb{Z}),$$

and hence (1.321). □

Since $\check{L}$ is a bounded second-order difference operator in $\ell^2(\mathbb{Z})$ with

$$\|\check{L}\| \leq 2\|a\|_\infty + \|b\|_\infty,$$

the resolvent $(\check{L} - zI)^{-1}$, $z \in \mathbb{C} \setminus \text{spec}(\check{L})$, of $\check{L}$ is thus a Carleman integral operator (the corresponding measures involved are of course discrete measures). Introducing

$$\psi_+(z, n) = ((\check{L} - zI)^{-1}\delta_0)(n), \quad n \geq 1, \quad z \in \mathbb{C} \setminus \text{spec}(\check{L}), \quad (1.322)$$

$$\psi_-(z, n) = ((\check{L} - zI)^{-1}\delta_0)(n), \quad n \leq -1, \quad z \in \mathbb{C} \setminus \text{spec}(\check{L}), \quad (1.323)$$

where

$$\delta_k(n) = \begin{cases} 1, & n = k, \\ 0, & n \in \mathbb{Z} \setminus \{k\}, \end{cases} \quad k \in \mathbb{Z},$$

one can use the Jacobi equation $L\psi_\pm(z) = z\psi_\pm(z)$ to extend $\psi_\pm(z, n)$ uniquely to all $n \in [0, \mp\infty) \cap \mathbb{Z}$ (this is possible since by Hypothesis 1.54, $a(n) \neq 0$ for all $n \in \mathbb{Z}$). Thus, one obtains,

$$L\psi_\pm(z) = z\psi_\pm(z), \quad \psi_\pm(z, \cdot) \in \ell^2([n_0, \pm\infty) \cap \mathbb{Z}), \quad z \in \mathbb{C} \setminus \text{spec}(\check{L}), \quad n_0 \in \mathbb{Z},$$

in the weak sense. We note that the Weyl–Titchmarsh-type solutions $\psi_\pm(z, \cdot)$ are unique up to normalization. Moreover, by the second inclusion in (1.321),

$$\psi_\pm(z, \cdot) \text{ is zero free for } |z| > (2\|a\|_\infty + \|b\|_\infty)$$

since $\psi_\pm(z, n_0) = 0$ for some $n_0 \in \mathbb{Z}$ yields an eigenvalue of $\check{L}^D_{\pm,n_0\pm1}$.

The Green's function $G(z, \cdot, \cdot)$ of $\check{L}$, that is, the integral kernel of the resolvent $(\check{L} - zI)^{-1}$ (with respect to discrete measures), is then given by

$$G(z, n, n') = \frac{1}{W(\psi_-(z), \psi_+(z))} \begin{cases} \psi_-(z, n)\psi_+(z, n'), & n \leq n', \\ \psi_-(z, n')\psi_+(z, n), & n \geq n', \end{cases} \quad (1.324)$$

$$z \in \mathbb{C} \setminus \text{spec}(\check{L}), \quad n, n' \in \mathbb{Z},$$

where $W(f, g)(n)$ denotes the Wronskian

$$W(f, g)(n) = a(n)(f(n)g(n+1) - f(n+1)g(n)), \quad n \in \mathbb{Z}, \tag{1.325}$$

of complex-valued sequences $f = \{f(n)\}_{n \in \mathbb{Z}}$ and $g = \{g(n)\}_{n \in \mathbb{Z}}$.

The corresponding Green's functions $G_{\pm,n_0}^D(z, \cdot, \cdot)$ of $\check{L}_{\pm,n_0}^D$, that is, the integral kernels of the resolvents $(\check{L}_{\pm,n_0}^D - zI)^{-1}$ (again with respect to discrete measures), are then given by

$$G_{+,n_0}^D(z, n, n')$$
$$= \frac{-1}{a(n_0 - 1)\psi_+(z, n_0 - 1)} \begin{cases} \phi_0(z, n, n_0 - 1)\psi_+(z, n'), & n \leq n', \\ \phi_0(z, n', n_0 - 1)\psi_+(z, n), & n \geq n', \end{cases} \tag{1.326}$$

$$z \in \mathbb{C} \setminus \operatorname{spec}(\check{L}_{+,n_0}^D), \ n, n' \in \mathbb{Z},$$

$$G_{-,n_0}^D(z, n, n')$$
$$= \frac{-1}{a(n_0 + 1)\psi_-(z, n_0 + 1)} \begin{cases} \psi_-(z, n)\phi_0(z, n', n_0 + 1), & n \leq n', \\ \psi_-(z, n')\phi_0(z, n, n_0 + 1), & n \geq n', \end{cases} \tag{1.327}$$

$$z \in \mathbb{C} \setminus \operatorname{spec}(\check{L}_{-,n_0}^D), \ n, n' \in \mathbb{Z},$$

where $\phi_0(z, \cdot, n_0)$ is a weak solution of

$$L\psi = z\psi, \tag{1.328}$$

satisfying the initial conditions

$$\phi_0(z, n_0, n_0) = 0, \quad \phi_0(z, n_0 + 1, n_0) = 1, \quad z \in \mathbb{C}.$$

Next, assuming that ψ satisfies (1.328), the function $\phi = \phi(z, n)$, introduced by

$$\phi = \frac{\psi^+}{\psi}, \tag{1.329}$$

satisfies the Riccati-type equation

$$a\phi + a^-(\phi^-)^{-1} + (b - z) = 0. \tag{1.330}$$

Defining

$$\phi_\pm(z, n) = \frac{\psi_\pm(z, n+1)}{\psi_\pm(z, n)}, \quad z \in \mathbb{C} \setminus \operatorname{spec}(\check{L}_{\pm,n\pm1}^D), \ n \in \mathbb{Z},$$

and using the Green's function representations (1.326), (1.327), one computes

$$\phi_+(z, n) = -a(n)G_{+,n+1}^D(z, n+1, n+1)$$
$$= -a(n)\bigl(\delta_{n+1}, (\check{L}_{+,n+1}^D - zI)^{-1}\delta_{n+1}\bigr)_{\ell^2([n+1,\infty) \cap \mathbb{Z})}, \tag{1.331}$$

$$z \in \mathbb{C} \setminus \operatorname{spec}(\check{L}_{+,n+1}^D), \ n \in \mathbb{Z},$$

1.7 Toda Conservation Laws and the Hamiltonian Formalism

and

$$\phi_-(z,n) = a(n)^{-1}\big(z - b(n) + a(n-1)^2 G^D_{-,n+1}(z, n-1, n-1)\big)$$
$$= a(n)^{-1}\big(z - b(n) \qquad (1.332)$$
$$+ a(n-1)^2 \big(\delta_{n-1}, \big(\check{L}^D_{+,n-1} - zI\big)^{-1}\delta_{n-1}\big)_{\ell^2((-\infty,n-1]\cap\mathbb{Z})}\big),$$
$$z \in \mathbb{C} \setminus \mathrm{spec}\big(\check{L}^D_{-,n-1}\big), \; n \in \mathbb{Z}.$$

We provide the following description of the asymptotic behavior of $\phi_\pm$ as $z \to \infty$.

Lemma 1.56 *Suppose a, b satisfy Hypothesis 1.54. Then $\phi_\pm$ has the following convergent expansion with respect to $1/z$ around $1/z = 0$,*

$$\phi_\pm(z) = \begin{cases} a \sum_{j=1}^{\infty} \phi_{+,j} z^{-j}, \\ \frac{1}{a} \sum_{j=-1}^{\infty} \phi_{-,j} z^{-j}, \end{cases} \qquad (1.333)$$

where

$$\phi_{+,1} = 1, \quad \phi_{+,2} = b^+,$$
$$\phi_{+,j+1} = b^+ \phi_{+,j} + (a^+)^2 \sum_{\ell=1}^{j-1} \phi^+_{+,j-\ell} \phi_{+,\ell}, \quad j \geq 2, \qquad (1.334)$$
$$\phi_{-,-1} = 1, \quad \phi_{-,0} = -b, \quad \phi_{-,1} = -(a^-)^2,$$
$$\phi_{-,j+1} = -b \phi^-_{-,j} - \sum_{\ell=0}^{j} \phi_{-,j-\ell} \phi^-_{-,\ell}, \quad j \geq 1. \qquad (1.335)$$

Proof Since $\check{L}^D_{\pm,n\pm 1}$ are bounded operators, (1.331) and (1.332) prove the existence of an analytic expansion of $\phi_\pm(z)$ with respect to $1/z$ around $1/z = 0$ for $|z|$ sufficiently large. Moreover, (1.331) and (1.332) yield the leading asymptotic behavior,

$$\phi_+(z) \underset{z\to\infty}{=} -az^{-1} + O(z^{-2})$$

and

$$\phi_-(z) \underset{z\to\infty}{=} \frac{z}{a} + O(1).$$

Thus, by making the ansatz (1.333) for $\phi_\pm$ and inserting it into (1.330) one finds (1.334) and (1.335). □

For the record we note the following explicit expressions,

$$\phi_{+,1} = 1,$$
$$\phi_{+,2} = b^+,$$
$$\phi_{+,3} = (a^+)^2 + (b^+)^2,$$
$$\phi_{+,4} = (a^+)^2(2b^+ + b^{++}) + (b^+)^3, \text{ etc.,}$$
$$\phi_{-,-1} = 1,$$
$$\phi_{-,0} = -b,$$
$$\phi_{-,1} = -(a^-)^2,$$
$$\phi_{-,2} = -(a^-)^2 b^-,$$
$$\phi_{-,3} = -(a^-)^2((a^{--})^2 + (b^-)^2), \text{ etc.}$$

Later on we will also need the convergent expansion of $\ln(\phi_+)$ with respect to $1/z$ for $1/|z|$ sufficiently small and hence we note that

$$\ln(\phi_+(z)) = \ln\left(a \sum_{j=1}^{\infty} \phi_{+,j} z^{-j}\right)$$
$$= \ln\left(\frac{a}{z}\right) + \ln\left(1 + \sum_{j=1}^{\infty} \phi_{+,j+1} z^{-j}\right)$$
$$= \ln\left(\frac{a}{z}\right) + \sum_{j=1}^{\infty} \rho_{+,j} z^{-j}, \quad (1.336)$$

where

$$\rho_{+,1} = \phi_{+,2}, \quad \rho_{+,j} = \phi_{+,j+1} - \sum_{\ell=1}^{j-1} \frac{\ell}{j} \phi_{+,j+1-\ell} \rho_{+,\ell}, \quad j \geq 2. \quad (1.337)$$

The first few explicitly read

$$\rho_{+,1} = b^+,$$
$$\rho_{+,2} = (a^+)^2 + \frac{1}{2}(b^+)^2,$$
$$\rho_{+,3} = \frac{1}{3}(b^+)^3 + (a^+)^2(b^+ + b^{++}), \text{ etc.}$$

1.7 Toda Conservation Laws and the Hamiltonian Formalism

Similarly, one finds for $1/|z|$ sufficiently small,

$$\ln(\phi_-(z)) = \ln\left(a^{-1}\sum_{j=-1}^{\infty}\phi_{-,j}z^{-j}\right)$$

$$= \ln\left(\frac{z}{a}\right) + \ln\left(1 + \sum_{j=1}^{\infty}\phi_{-,j-1}z^{-j}\right)$$

$$= \ln\left(\frac{z}{a}\right) + \sum_{j=1}^{\infty}\rho_{-,j}z^{-j},$$

where

$$\rho_{-,1} = \phi_{-,0}, \quad \rho_{-,j} = \phi_{-,j-1} - \sum_{\ell=1}^{j-1}\frac{\ell}{j}\phi_{-,j-1-\ell}\rho_{-,\ell}, \quad j \geq 2. \tag{1.338}$$

The first few explicitly read

$$\rho_{-,1} = -b,$$

$$\rho_{-,2} = -(a^-)^2 - \frac{1}{2}b^2,$$

$$\rho_{-,3} = -\frac{1}{3}b^3 - (a^-)^2(b^- + b), \text{ etc.}$$

Equations (1.337) and (1.338) may be written as

$$\rho_{\pm,1} = \phi_{\pm,1\pm 1}, \quad \rho_{\pm,j} = \phi_{\pm,j\pm 1} - \sum_{\ell=1}^{j-1}\frac{\ell}{j}\phi_{\pm,j\pm 1-\ell}\rho_{\pm,\ell}, \quad j \geq 2.$$

The next result shows that $\hat{f}_j$ and $\pm j\rho_{\pm,j}$ are equal up to terms that are total differences, that is, are of the form $(S^+ - I)d_{\pm,j}$ for some sequence $d_{\pm,j}$. The exact form of $d_{\pm,j}$ will not be needed later.

Lemma 1.57 *Suppose a, b satisfy Hypothesis 1.54. Then,*

$$\hat{f}_j = \pm j\rho_{\pm,j} + (S^+ - I)d_{\pm,j}, \quad j \in \mathbb{N}, \tag{1.339}$$

for some polynomials $d_{\pm,j}$, $j \in \mathbb{N}$, in a and b and certain shifts thereof.

Proof To shorten the notation we introduce the abbreviations

$$g(z,n) = G(z,n,n), \quad h(z,n) = G(z,n,n+1), \quad (z,n) \in \mathbb{C} \times \mathbb{Z}. \tag{1.340}$$

Thus,

$$\phi_+ = \frac{h}{g}.$$

124   *1 The Toda Hierarchy*

To increase readability we suppress the display of variables in the subsequent computations. First, one observes that

$$agg^+ - h(ah-1) = a\frac{\psi_+\psi_- - \psi_+^+\psi_-^+}{W^2} - \frac{\psi_+^+\psi_-}{W}\left(a\frac{\psi_+^+\psi_-}{W} - 1\right)$$

$$= \frac{\psi_+^+\psi_-}{W^2}\left(a\psi_+\psi_-^+ - a\psi_+^+\psi_- + W\right) = 0 \quad (1.341)$$

and thus,

$$\phi_+^2 = \frac{h^2}{g^2} = \frac{ah^2}{agg^+}\frac{g^+}{g} = \frac{ah}{ah-1}\frac{g^+}{g}.$$

This yields

$$\frac{d}{dz}\ln(\phi_+) = \frac{1}{2}\frac{d}{dz}\ln\left(\frac{ah}{ah-1}\right) + \frac{1}{2}\frac{d}{dz}\ln\left(\frac{g^+}{g}\right). \quad (1.342)$$

Next we study the first term on the right-hand side of (1.342), namely,

$$\frac{d}{dz}\ln\left(\frac{ah}{ah-1}\right) = \frac{h_z}{h} - \frac{ah_z}{ah-1}. \quad (1.343)$$

We claim that

$$\frac{h_z}{h} - \frac{ah_z}{ah-1} = (1-2ah)\left(\frac{g_z}{g} + \frac{g_z^+}{g^+}\right) + 4ah_z. \quad (1.344)$$

To this end we first take the logarithmic derivative of (1.341), that is, of $agg^+ = h(ah-1)$, to obtain

$$\frac{g_z}{g} + \frac{g_z^+}{g^+} = \frac{h_z}{h} + \frac{ah_z}{ah-1}.$$

Thus,

$$\frac{h_z}{h} - \frac{ah_z}{ah-1} - \left((1-2ah)\left(\frac{g_z}{g} + \frac{g_z^+}{g^+}\right) + 4ah_z\right)$$

$$= \frac{h_z}{h} - \frac{ah_z}{ah-1} - (1-2ah)\left(\frac{h_z}{h} + \frac{ah_z}{ah-1}\right) - 4ah_z$$

$$= -\frac{2ah_z}{ah-1}(1 + (ah-1) - ah) = 0.$$

Combining (1.342), (1.343), and (1.344) one finds

$$\frac{d}{dz}\ln(\phi_+) = \frac{1}{2}\left((1-2ah)\left(\frac{g_z}{g} + \frac{g_z^+}{g^+}\right) + 4ah_z\right) + \frac{1}{2}\frac{d}{dz}\ln\left(\frac{g^+}{g}\right)$$

$$= \frac{1}{2}(1-2ah)\left(\frac{g_z}{g} + \frac{g_z^+}{g^+}\right) + 2ah_z + \frac{1}{2}\frac{d}{dz}\ln\left(\frac{g^+}{g}\right). \quad (1.345)$$

1.7 Toda Conservation Laws and the Hamiltonian Formalism

Considering the first term on the right-hand side of (1.345) and subtracting g then yields

$$\frac{1}{2}(1-2ah)\left(\frac{g_z}{g}+\frac{g_z^+}{g^+}\right)+2ah_z-g$$

$$=\frac{1}{2}(1-2ah)\left(\frac{d}{dz}\ln(ah+a^-h^--1)+\frac{d}{dz}\ln(a^+h^++ah-1)\right.$$
$$\left.-\frac{d}{dz}\ln\left((z-b^+)(z-b)\right)\right)+2ah_z-\frac{ah+a^-h^--1}{z-b}$$

$$=\frac{1}{2}(1-2ah)\left(\frac{ah_z+a^-h_z^-}{ah+a^-h^--1}+\frac{a^+h_z^++ah_z}{a^+h^++ah-1}-\left(\frac{1}{z-b^+}+\frac{1}{z-b}\right)\right)$$
$$+2ah_z-\frac{ah+a^-h^-}{z-b}+\frac{1}{z-b}$$

$$=\frac{1}{2}\left(\frac{1}{z-b}-\frac{1}{z-b^+}\right)+\frac{ah}{z-b^+}-\frac{a^-h^-}{z-b}$$
$$+2ah_z+\frac{1}{2}(1-2ah)\left(\frac{ah_z+a^-h_z^-}{ah+a^-h^--1}+\frac{a^+h_z^++ah_z}{a^+h^++ah-1}\right),$$

using

$$(z-b)g-ah-a^-h^-+1=0.$$

The latter follows from the definition (1.340) of g and h (cf. (1.324)) and from $L\psi_-=z\psi_-$. By purely algebraic manipulations one obtains that

$$2ah_z+\frac{1}{2}(1-2ah)\left(\frac{ah_z+a^-h_z^-}{ah+a^-h^--1}+\frac{a^+h_z^++ah_z}{a^+h^++ah-1}\right)$$
$$=\left(\frac{a^-h^-ah_z-a^-h_z^-ah}{ah+a^-h^--1}-\frac{aha^+h_z^+-ah_za^+h^+}{a^+h^++ah-1}\right)$$
$$+\frac{1}{2}\left(\frac{a^+h_z^+-ah_z}{a^+h^++ah-1}-\frac{ah_z-a^-h_z^-}{ah+a^-h^--1}\right).$$

Summarizing the computations thus far, one finds

$$\frac{d}{dz}\ln(\phi_+)=g+\frac{1}{2}\frac{d}{dz}\ln\left(\frac{g^+}{g}\right)+\frac{1}{2}\left(\frac{1}{z-b}-\frac{1}{z-b^+}\right)+\frac{ah}{z-b^+}-\frac{a^-h^-}{z-b}$$
$$+\left(\frac{a^-h^-ah_z-a^-h_z^-ah}{ah+a^-h^--1}-\frac{aha^+h_z^+-ah_za^+h^+}{a^+h^++ah-1}\right)$$
$$+\frac{1}{2}\left(\frac{a^+h_z^+-ah_z}{a^+h^++ah-1}-\frac{ah_z-a^-h_z^-}{ah+a^-h^--1}\right)$$
$$=g+(S^+-I)\Phi \tag{1.346}$$

for some function $\Phi = \Phi(z,n)$. By (C.17) and (C.19) one obtains the convergent expansion

$$g(z,n) = G(z,n,n) \underset{|z|\to\infty}{=} -\sum_{j=0}^{\infty} \hat{f}_j(n) z^{-j-1}, \quad n \in \mathbb{Z}. \tag{1.347}$$

From (1.336) and (1.347) one then concludes

$$-z^{-1} - \sum_{j=1}^{\infty} j\rho_{+,j} z^{-j-1} = -\sum_{j=0}^{\infty} \hat{f}_j z^{-j-1} + (S^+ - I) \sum_{j=0}^{\infty} d_{+,j} z^{-j-1}$$

for some sequences $\{d_{+,j}(n)\}_{n\in\mathbb{Z}}$, $j \in \mathbb{N}_0$.
Noting that

$$\phi_- = \frac{g^+}{h},$$

one finds the analogous result regarding $\rho_{-,j}$ and $d_{-,j}$, $j \in \mathbb{N}$. $\square$

Remark 1.58 (i) Alternatively, one can derive (1.339) in a similar way to the corresponding result for the Ablowitz–Ladik model, see (3.357) and (3.358).
(ii) Closely related to (1.346) is the identity

$$\frac{d}{dz} \ln\left(1 - \frac{1}{ah}\right) = -2g + (S^+ - I)\left(-2\Phi + \frac{d}{dz} \ln(g)\right), \tag{1.348}$$

which follows from (1.342) and (1.346).

Remark 1.59 For later use in this section we recall the notion of degree of various quantities as introduced in Remark 1.3. One has

$$\deg\left(a^{(r)}\right) = \deg\left(b^{(r)}\right) = 1, \quad r \in \mathbb{Z},$$
$$\deg\left(\hat{f}_\ell\right) = \ell, \quad \deg\left(\hat{g}_\ell\right) = \ell + 1, \quad \ell \in \mathbb{N}.$$

Similarly, the recursion relations (1.334) and (1.335) yield inductively that

$$\deg\left(\phi_{\pm,j\pm1}\right) = j, \quad j \in \mathbb{N}_0.$$

Next, we turn to local conservation laws.
For this purpose we introduce the following assumption:

Hypothesis 1.60 *Suppose that $a, b \colon \mathbb{Z} \times \mathbb{R} \to \mathbb{C}$ satisfy*

$$\sup_{(n,t_p)\in\mathbb{Z}\times\mathbb{R}} \left(|a(n,t_p)| + |b(n,t_p)|\right) < \infty,$$

$a(n, \cdot), b(n, \cdot) \in C^1(\mathbb{R})$, $n \in \mathbb{Z}$, $a(n, t_p) \neq 0$ for all $(n, t_p) \in \mathbb{Z} \times \mathbb{R}$.

1.7 Toda Conservation Laws and the Hamiltonian Formalism

In accordance with the notation introduced in (1.318) we denote the bounded difference operator defined on $\ell^2(\mathbb{Z})$, generated by the finite difference expression P_{2p+2} in (1.20), by the symbol $\check{P}_{2p+2}$.

We start with the following existence result.

Theorem 1.61 *Assume Hypothesis* 1.60 *and suppose* a, b *satisfy* $\mathrm{Tl}_p(a, b) = 0$ *for some* $p \in \mathbb{N}_0$. *In addition, let* $t_p \in \mathbb{R}$ *and* $z \in \mathbb{C} \setminus \mathrm{spec}\,(\check{L})$. *Then there exist Weyl–Titchmarsh-type solutions* $\psi_\pm = \psi_\pm(z, n, t_p)$ *such that*

$$\psi_\pm(z, \cdot, t_p) \in \ell^2([n_0, \pm\infty) \cap \mathbb{Z}),\; n_0 \in \mathbb{Z}, \quad \psi_\pm(z, n, \cdot) \in C^1(\mathbb{R}), \quad (1.349)$$

and $\psi_\pm$ *simultaneously satisfy the following two equations in the weak sense*

$$\check{L}(t_p)\psi_\pm(z, \cdot, t_p) = z\psi_\pm(z, \cdot, t_p),$$
$$\psi_{\pm, t_p}(z, \cdot, t_p) = \check{P}_{2p+2}(t_p)\psi_\pm(z, \cdot, t_p), \quad (1.350)$$
$$= 2a(t_p)F_p(z, \cdot, t_p)\psi_\pm^+(z, \cdot, t_p) + G_{p+1}(z, \cdot, t_p)\psi_\pm(z, \cdot, t_p). \quad (1.351)$$

Moreover, the Wronskian

$$W(\psi_-(z, n, t_p), \psi_+(z, n, t_p)) \text{ is independent of } (n, t_p) \in \mathbb{Z} \times \mathbb{R}. \quad (1.352)$$

Proof Applying $(\check{L}(t) - zI)^{-1}$ to δ_0 (cf. (1.322) and (1.323)) yields the existence of Weyl–Titchmarsh-type solutions $\Psi_\pm$ of $L\psi = z\psi$ satisfying (1.349). Next, using the Lax commutator equation (1.48) one computes

$$z\Psi_{\pm, t_p} = (L\Psi_\pm)_{t_p} = L_{t_p}\Psi_\pm + L\Psi_{\pm, t_p} = [P_{2p+2}, L]\Psi_\pm + L\Psi_{\pm, t_p}$$
$$= zP_{2p+2}\Psi_\pm - LP_{2p+2}\Psi_\pm + L\Psi_{\pm, t_p}$$

and hence

$$(L - zI)(\Psi_{\pm, t_p} - P_{2p+2}\Psi_\pm) = 0.$$

Thus, $\Psi_\pm$ satisfy

$$\Psi_{\pm, t_p} - P_{2p+2}\Psi_\pm = C_\pm \Psi_\pm + D_\pm \Psi_\mp.$$

Introducing $\Psi_\pm = c_\pm \psi_\pm$, and choosing $c_\pm$ such that $c_{\pm, t_p} = C_\pm c_\pm$, one obtains

$$\psi_{\pm, t_p} - P_{2p+2}\psi_\pm = D_\pm \psi_\mp. \quad (1.353)$$

Since $\psi_\pm \in \ell^2([n_0, \pm\infty) \cap \mathbb{Z})$, $n_0 \in \mathbb{Z}$, and a, b satisfy Hypothesis 1.60, (1.35) shows that $P_{2p+2}\psi_\pm = (2aF_pS^+\psi_\pm + G_{p+1}\psi_\pm) \in \ell^2([n_0, \pm\infty) \cap \mathbb{Z})$. (Incidentally, this argument yields of course (1.351).) Moreover, since $\psi_\pm(z, n, t_p) = d_\pm(t_p)(\check{L}(t_p) - zI)^{-1}\delta_0)(n)$ for $n \in [\pm 1, \infty) \cap \mathbb{Z}$ and some $d_\pm \in C^1(\mathbb{R})$, the calculation

$$\psi_{\pm, t_p} = d_{\pm, t_p}(\check{L} - zI)^{-1}\delta_0 - d_\pm(\check{L} - zI)^{-1}\check{L}_{t_p}(\check{L} - zI)^{-1}\delta_0$$

also yields $\psi_{\pm,t_p} \in \ell^2([n_0, \pm\infty) \cap \mathbb{Z})$. But then $D_+ = 0$ in (1.353) since $\psi_\mp \notin \ell^2([n_0, \pm\infty) \cap \mathbb{Z})$. This proves (1.350).

Since $\psi_\pm(z, \cdot)$ satisfy $L\psi(z, \cdot) = z\psi(z, \cdot)$, the Wronskian (1.352) is independent of $n \in \mathbb{Z}$. To show also its t_p-independence, a computation reveals

$$\frac{d}{dt_p} W(\psi_-, \psi_+) = \left(\frac{a_{t_p}}{a} + 2(z - b^+)F_p^+ + G_{p+1}^+ + G_{p+1}\right) W(\psi_-, \psi_+) = 0$$

using (1.325), (1.351), and (1.54). □

For the rest of this section, $\psi_\pm$ will always refer to the Weyl–Titchmarsh solutions introduced in Theorem 1.61.

The next result recalls the existence of a propagator W_p associated with P_{2p+2}. (Below we denote by $\mathcal{B}(\mathcal{H})$ the Banach space of all bounded linear operators defined on the Hilbert space $\mathcal{H}$.)

Theorem 1.62 *Assume Hypothesis* 1.60 *and suppose* a, b *satisfy* $\mathrm{Tl}_p(a, b) = 0$ *for some* $p \in \mathbb{N}_0$. *Then there is a propagator* $W_p(s, t) \in \mathcal{B}(\ell^2(\mathbb{Z}))$, $(s, t) \in \mathbb{R}^2$, *satisfying*

$$(i) \quad W_p(t, t) = I, \quad t \in \mathbb{R}, \tag{1.354}$$

$$(ii) \quad W_p(r, s)W_p(s, t) = W_p(r, t), \quad (r, s, t) \in \mathbb{R}^3, \tag{1.355}$$

$$(iii) \quad W_p(s, t) \text{ is jointly strongly continuous in } (s, t) \in \mathbb{R}^2, \tag{1.356}$$

such that for fixed $t_0 \in \mathbb{R}$, $f_0 \in \ell^2(\mathbb{Z})$,

$$f(t) = W_p(t, t_0) f_0, \quad t \in \mathbb{R},$$

satisfies

$$\frac{d}{dt} f(t) = \check{P}_{2p+2}(t) f(t), \quad f(t_0) = f_0. \tag{1.357}$$

Moreover, $\check{L}(t)$ *is similar to* $\check{L}(s)$ *for all* $(s, t) \in \mathbb{R}^2$,

$$\check{L}(s) = W_p(s, t)\check{L}(t) W_p(s, t)^{-1}, \quad (s, t) \in \mathbb{R}^2. \tag{1.358}$$

This extends to appropriate functions of $\check{L}(t)$ *and so, in particular, to its resolvent* $(\check{L}(t) - zI)^{-1}$, $z \in \mathbb{C} \setminus \sigma(\check{L}(t))$, *and hence also yields*

$$\sigma(\check{L}(s)) = \sigma(\check{L}(t)), \quad (s, t) \in \mathbb{R}^2. \tag{1.359}$$

Consequently, the spectrum of $\check{L}(t)$ *is independent of* $t \in \mathbb{R}$.

1.7 Toda Conservation Laws and the Hamiltonian Formalism

Proof The existence of the propagator $W(\,\cdot\,,\,\cdot\,)$ satisfying (1.354)–(1.357) is a standard result (valid under even weaker hypotheses on a, b). In particular, the propagator W_p admits the norm convergent Dyson series

$$W_p(s,t) = I$$
$$+ \sum_{k \in \mathbb{N}} \int_s^t dt_1 \int_s^{t_1} dt_2 \cdots \int_s^{t_{k-1}} dt_k \, \check{P}_{2p+2}(t_1) \check{P}_{2p+2}(t_2) \cdots \check{P}_{2p+2}(t_k),$$

$$(s,t) \in \mathbb{R}^2.$$

Fixing $s \in \mathbb{R}$ and introducing the operator-valued function

$$\check{K}(t) = W_p(s,t)\check{L}(t)W_p(s,t)^{-1}, \quad t \in \mathbb{R}, \tag{1.360}$$

one computes

$$\check{K}'(t)f = W_p(s,t)\big(\check{L}'(t) - \big[\check{P}_{2p+2}(t), \check{L}(t)\big]\big)W_p(s,t)^{-1}f = 0,$$

$$t \in \mathbb{R}, \, f \in \ell^2(\mathbb{Z}),$$

using the Lax commutator equation (1.48). Thus, $\check{K}$ is independent of $t \in \mathbb{R}$ and taking $t = s$ in (1.360), then yields $\check{K} = \check{L}(s)$ and thus proves (1.358). □

In the special case where $\check{L}(t)$, $t \in \mathbb{R}$, is self-adjoint, the operator $\check{P}_{2p+2}(t)$ is skew-adjoint, $\check{P}_{2p+2}(t)^* = -\check{P}_{2p+2}(t)$, $t \in \mathbb{R}$, and hence $W_p(s,t)$ is unitary for all $(s,t) \in \mathbb{R}^2$.

Next we briefly recall the Toda initial value problem in a setting convenient for our purpose.

Theorem 1.63 *Let $t_{0,p} \in \mathbb{R}$ and suppose $a^{(0)}, b^{(0)} \in \ell^\infty(\mathbb{Z})$. Then the pth Toda lattice initial value problem*

$$\mathrm{Tl}_p(a,b) = 0, \quad (a,b)\big|_{t_p = t_{0,p}} = \big(a^{(0)}, b^{(0)}\big) \tag{1.361}$$

for some $p \in \mathbb{N}_0$ has a unique, local, and smooth solution in time, that is, there exists a $T_0 > 0$ such that

$$a(\,\cdot\,), b(\,\cdot\,) \in C^\infty((t_{0,p} - T_0, t_{0,p} + T_0), \ell^\infty(\mathbb{Z})).$$

Remark 1.64 (*i*) As discussed in the notes to this section, Theorem 1.63 extends to the case where $a^{(0)}, b^{(0)} \in \ell^\infty(\mathbb{Z})$ is replaced, for instance, by

$$\big\{a^{(0)}(n)^2 - \tfrac{1}{4}\big\}_{n \in \mathbb{Z}}, \, \big\{b^{(0)}(n)\big\}_{n \in \mathbb{Z}} \in \ell^1(\mathbb{Z}).$$

This observation will be used in the proof of Theorem 1.74.
(*ii*) In the special case where $\check{L}(t)$, $t \in \mathbb{R}$, is self-adjoint, one obtains

$$\sup_{(n,t_p) \in \mathbb{N} \times (t_{0,p} - T_0, t_{0,p} + T_0)} \big(|a(n,t_p)| + |b(n,t_p)|\big) \leq 2\|\check{L}(t_p)\| = 2\|\check{L}(t_{0,p})\|,$$

using (the local version of) the isospectral property (1.359) of Toda flows. This then yields a unique, global, and smooth solution of the pth Toda lattice initial value problem (1.361). Moreover, along similar lines one can show that if a, b satisfy Hypothesis 1.60 and the pth Toda equation $\mathrm{Tl}_p(a, b) = 0$, then a, b are actually smooth with respect to $t_p \in \mathbb{R}$, that is,

$$a(n, \cdot), b(n, \cdot) \in C^\infty(\mathbb{R}), \quad n \in \mathbb{Z}.$$

Theorem 1.65 *Assume Hypothesis* 1.60 *and suppose* a, b *satisfy* $\mathrm{Tl}_p(a, b) = 0$ *for some* $p \in \mathbb{N}_0$. *Then the following infinite sequence of local conservation laws holds,*

$$\partial_{t_p} \ln(a) - (S^+ - I)f_{p+1} = 0, \qquad (1.362)$$

$$\partial_{t_p} \rho_{\pm,j} + (S^+ - I)J_{\pm,p,j} = 0, \quad j \in \mathbb{N}, \qquad (1.363)$$

where

$$J_{\pm,p,j} = -2\left(a^2 \sum_{\ell=0}^{p} f_\ell \phi_{\pm,p+j-\ell}\right), \qquad (1.364)$$

and $\phi_{\pm,j}$, $j \in \mathbb{N}$, *are given by* (1.334) *and* (1.335), *and* $\rho_{\pm,j}$, $j \in \mathbb{N}$, *are given by* (1.337) *and* (1.338).

Proof Using (1.351) and (1.329) one computes

$$\partial_{t_p} \ln\left(\frac{\psi_+^+}{\psi_+}\right) = \frac{\psi_{+,t_p}^+}{\psi_+^+} - \frac{\psi_{+,t_p}}{\psi_+}$$
$$= 2a^+ F_p^+ \phi_+^+ - 2aF_p\phi_+ + G_{p+1}^+ - G_{p+1}$$
$$= (S^+ - I)(2aF_p\phi_+ + G_{p+1}).$$

Utilizing the expansion (1.333) of ϕ_+ as $z \to \infty$ one finds

$$2aF_p\phi_+ + G_{p+1} = 2a^2\left(\sum_{\ell=0}^{p} f_{p-\ell}z^\ell\right)\left(\sum_{j=1}^{\infty} \phi_{+,j}z^{-j}\right) - z^{p+1} + \sum_{\ell=0}^{p} g_{p-\ell}z^\ell$$
$$+ f_{p+1}$$
$$= 2a^2 \sum_{\ell=1}^{p} f_{p-\ell}\phi_{+,\ell} + g_p + f_{p+1} + \sum_{k=1}^{p}\left(2a^2 \sum_{\ell=1}^{p} f_{p-\ell}\phi_{+,\ell} + g_{p-k}\right)z^k$$
$$- z^{p+1} + 2a^2 \sum_{j=1}^{\infty}\left(\sum_{\ell=0}^{p} f_\ell \phi_{+,p+j-\ell}\right)z^{-j}.$$

1.7 Toda Conservation Laws and the Hamiltonian Formalism

On the other hand, using (1.329) and the expansions (1.333) and (1.336), one concludes

$$\partial_{t_p} \ln\left(\frac{\psi_+^+}{\psi_+}\right) = \partial_{t_p} \ln(\phi_+) = \frac{a_{t_p}}{a} + \sum_{j=1}^{\infty} (\partial_{t_p} \rho_{+,j}) z^{-j}.$$

Combining these equations one infers

$$\frac{a_{t_p}}{a} + \sum_{j=1}^{\infty} (\partial_{t_p} \rho_{+,j}) z^{-j} = (S^+ - 1)\left(2a^2 \sum_{\ell=1}^{p} f_{p-\ell} \phi_{+,\ell} + g_p + f_{p+1}\right.$$

$$+ \sum_{k=1}^{p} \left(2a^2 \sum_{\ell=k+1}^{p} f_{p-\ell} \phi_{+,\ell-k} + g_{p-k}\right) z^k - z^{p+1}$$

$$\left. + 2a^2 \sum_{j=1}^{\infty} \sum_{\ell=0}^{p} f_\ell \phi_{+,p+j-\ell} z^{-j}\right)$$

$$= (S^+ - 1)\left(2a^2 \sum_{\ell=1}^{p} f_{p-\ell} \phi_{+,\ell} + g_p + f_{p+1}\right)$$

$$+ (S^+ - 1)\left(2a^2 \sum_{j=1}^{\infty} \sum_{\ell=0}^{p} f_\ell \phi_{+,p+j-\ell} z^{-j}\right)$$

$$= (S^+ - 1) f_{p+1} + (S^+ - 1)\left(2a^2 \sum_{j=1}^{\infty} \sum_{\ell=0}^{p} f_\ell \phi_{+,p+j-\ell} z^{-j}\right) \quad (1.365)$$

and hence (1.363), (1.364) in the case of $\rho_{+,j}$. Here we used the first of the equations in $\text{Tl}_p(a,b) = 0$ as well as the fact that the left-hand side of (1.365) contains no positive powers of z.

Similarly, one can start with ϕ_- and finds

$$\partial_{t_p} \ln\left(\frac{\psi_-^+}{\psi_-}\right) = (S^+ - 1)(2aF_p \phi_- + G_{p+1})$$

$$= (S^+ - 1)\left(2\phi_{-,0} f_p + 2\sum_{\ell=0}^{p-1} f_\ell \phi_{-,p-\ell} + g_p + f_{p+1}\right)$$

$$+ (S^+ - 1) \sum_{j=1}^{p-1} \left(2 \sum_{\ell=0}^{p-1-j} f_\ell \phi_{-,p-j-\ell} + 2\phi_{-,-1} f_{p-j+1}\right.$$

$$\left. + 2\phi_{-,0} f_{p-j} + g_{p-j}\right) z^j$$

$$+ (S^+ - 1)((2\phi_{-,-1} f_{p-1} + 2\phi_{-,0} f_0 + g_0) z^p + (2\phi_{-,-1} f_0 - 1) z^{p+1})$$

$$+ (S^+ - 1) \sum_{j=1}^{\infty} \left(2 \sum_{\ell=0}^{p} f_\ell \phi_{-,p+j-\ell}\right) z^{-j}.$$

Thus,

$$-\frac{a_{t_p}}{a} + \sum_{j=1}^{\infty}(\partial_{t_p}\rho_{-,j})z^{-j} = \partial_{t_p}\ln(\phi_+) = \partial_{t_p}\ln\left(\frac{\psi_+^+}{\psi_+}\right)$$

$$= (S^+ - I)\left(2\phi_{-,0}f_p + 2\sum_{\ell=0}^{p-1}f_\ell\phi_{-,p-\ell} + g_p + f_{p+1}\right)$$

$$+ (S^+ - I)\sum_{j=1}^{\infty}\left(2\sum_{\ell=0}^{p}f_\ell\phi_{-,p+j-\ell}\right)z^{-j}, \qquad (1.366)$$

implying

$$\partial_{t_p}\rho_{-,j} = (S^+ - I)\left(2\sum_{\ell=0}^{p}f_\ell\phi_{-,p+j-\ell}\right), \quad j\in\mathbb{N} \qquad (1.367)$$

and hence (1.363), (1.364) in the case of $\rho_{-,j}$. □

Remark 1.66 We emphasize that the sequence (1.367) yields no new conservation laws, and so the latter is equivalent to that in (1.365).

The first local conservation law, (1.362), is of course nothing but the first equation in $\mathrm{Tl}_p(a,b) = 0$, namely $a_{t_p} = a(f_{p+1}^+ - f_{p+1})$. The second equation in $\mathrm{Tl}_p(a,b) = 0$ is (1.363) for $j = 1$, namely, $b_{t_p}^+ = -(S^+ - I)g_{p+1}$. Indeed, using the second equation in $\mathrm{Tl}_p(a,b) = 0$ one infers that $2a^2\sum_{\ell=0}^{p}\hat{f}_\ell\phi_{+,p+1-\ell} = -\hat{g}_{p+1}$ up to a constant, that is, an element in the kernel of $S^+ - I$. Using the notion of a degree (cf. Remark 1.59), one concludes that the constant equals zero.

The first few local conservation laws explicitly read as follows:
(i) $p = 0$:

$$\partial_{t_0}\rho_{+,j} = 2(S^+ - I)\phi_{+,j},$$

in particular,

$$j = 1: \quad \partial_{t_0}b^+ = 2(S^+ - I)a^2,$$

$$j = 2: \quad \partial_{t_0}\left((a^+)^2 + \frac{1}{2}(b^+)^2\right) = 2(S^+ - I)a^2 b^+.$$

(ii) $p = 1$:

$$\partial_{t_1}\rho_{+,j} = 2(S^+ - I)(\phi_{+,j+1} + (b + c_1)\phi_{+,j}),$$

in particular,

$$j = 1: \quad \partial_{t_1}b^+ = 2(S^+ - I)\left(a^2(b^+ + b) + c_1\right),$$

$$j = 2: \quad \partial_{t_1}\left((a^+)^2 + \frac{1}{2}(b^+)^2\right) = 2(S^+ - I)\left(a^2((a^+)^2 + b^+(b + b^+)) + c_1 b^+\right).$$

1.7 Toda Conservation Laws and the Hamiltonian Formalism 133

Using Lemma 1.57, one observes that one can replace $\rho_{\pm,j}$ in (1.363) by $\hat{f}_j$ by suitably adjusting the right-hand side.

An obvious consequence of the local conservation laws is that, assuming sufficient decay of the sequences $a^2 - \frac{1}{4}$ and b, one obtains

$$\frac{d}{dt_p} \sum_{n \in \mathbb{Z}} \rho_{\pm,j}(n, t_p) = \frac{d}{dt_p} \sum_{n \in \mathbb{Z}} \hat{f}_j(n, t_p) = 0.$$

Remark 1.67 As a byproduct, (1.365) also yields the following relations:

$$(S^+ - I)\left(2a^2 \sum_{\ell=k+1}^{p} f_{p-\ell}\phi_{+,\ell-k} + g_p\right) = 0,$$

$$(S^+ - I)\left(2a^2 \sum_{\ell=1}^{p} f_{p-\ell}\phi_{+,\ell} + g_{p-k}\right) = 0, \quad k = 1, \ldots, p.$$

Similarly, (1.366) yields

$$(S^+ - I)\left(2\phi_{-,0} f_p + 2 \sum_{\ell=0}^{p-1} f_\ell \phi_{-,p-\ell} + g_p + 2 f_{p+1}\right) = 0,$$

$$(S^+ - I)\left(2 \sum_{\ell=0}^{p-1-j} f_\ell \phi_{-,p-j-\ell} + 2\phi_{-,-1} f_{p-j+1} + 2\phi_{-,0} f_{p-j} + g_{p-j}\right) = 0,$$

$$j = 1, \ldots, p - 1.$$

We now turn to the Hamiltonian formalism and start with a short review of variational derivatives for discrete systems. Consider the functional

$$\mathcal{F} \colon \ell^1(\mathbb{Z})^\kappa \to \mathbb{C},$$

$$\mathcal{F}(u) = \sum_{n \in \mathbb{Z}} F\big(u(n), u^{(+1)}(n), u^{(-1)}(n), \ldots, u^{(k)}(n), u^{(-k)}(n)\big) \quad (1.368)$$

for some $\kappa \in \mathbb{N}$ and $k \in \mathbb{N}_0$, where $F \colon \mathbb{C}^{(2k+1)\kappa} \to \mathbb{C}$ is C^1 with respect to the $(2k+1)\kappa$ complex-valued entries and where

$$u^{(s)} = S^{(s)} u, \quad S^{(s)} = \begin{cases} (S^+)^s u & \text{if } s \geq 0, \\ (S^-)^{-s} u & \text{if } s < 0, \end{cases} \quad u \in \ell^\infty(\mathbb{Z})^\kappa.$$

For brevity we write

$$F(u) = F\big(u(n), u^{(+1)}(n), u^{(-1)}(n), \ldots, u^{(k)}(n), u^{(-k)}(n)\big), \quad (1.369)$$

and it is assumed that $\{F(u)\}_{n \in \mathbb{Z}} \in \ell^1(\mathbb{Z})$ and that F is a polynomial in u and some of its shifts.

The functional $\mathcal{F}$ is Frechet differentiable and one computes for any $v \in \ell^1(\mathbb{Z})^\kappa$ for the differential $d\mathcal{F}$

$$(d\mathcal{F})_u(v) = \frac{d}{d\epsilon}\mathcal{F}(u+\epsilon v)\Big|_{\epsilon=0}$$

$$= \sum_{n \in \mathbb{Z}} \left(\frac{\partial F(u)}{\partial u} v(n) + \frac{\partial F(u)}{\partial u^{(+1)}} v^{(+1)}(n) + \frac{\partial F(u)}{\partial u^{(-1)}} v^{(-1)}(n) \right.$$

$$\left. + \cdots + \frac{\partial F(u)}{\partial u^{(k)}} v^{(k)}(n) + \frac{\partial F(u)}{\partial u^{(-k)}} v^{(-k)}(n) \right)$$

$$= \sum_{n \in \mathbb{Z}} \left(\frac{\partial F(u)}{\partial u} + S^{(-1)} \frac{\partial F(u)}{\partial u^{(+1)}} + S^{(+1)} \frac{\partial F(n,u)}{\partial u^{(-1)}} \right.$$

$$\left. + \cdots + S^{(-k)} \frac{\partial F(u)}{\partial u^{(k)}} + S^{(k)} \frac{\partial F(u)}{\partial u^{(-k)}} \right) v(n), \quad (1.370)$$

assuming

$$\{F(u)\}_{n \in \mathbb{Z}}, \ \left\{ \frac{\partial F(u)}{\partial u^{(\pm j)}} \right\}_{n \in \mathbb{Z}} \in \ell^1(\mathbb{Z}), \quad j=1,\ldots,k. \quad (1.371)$$

Because of the result (1.370), we thus introduce the gradient and the variational derivative of $\mathcal{F}$ by

$$(\nabla \mathcal{F})_u = \frac{\delta F}{\delta u}$$
$$= \frac{\partial F}{\partial u} + S^{(-1)} \frac{\partial F}{\partial u^{(+1)}} + S^{(+1)} \frac{\partial F}{\partial u^{(-1)}} + \cdots + S^{(-k)} \frac{\partial F}{\partial u^{(k)}} + S^{(k)} \frac{\partial F}{\partial u^{(-k)}},$$

assuming (1.371). Thus,

$$(d\mathcal{F})_u(v) = \sum_{n \in \mathbb{Z}} (\nabla \mathcal{F})_u(n) v(n) = \sum_{n \in \mathbb{Z}} \frac{\delta F}{\delta u}(n) v(n). \quad (1.372)$$

To establish the connection with the Toda hierarchy we make the following assumption for the remainder of this section (it will be strengthened later on, though).

Hypothesis 1.68 *Suppose*

$$a, b \in \ell^1(\mathbb{Z}), \quad a^{-1} \in \ell^\infty(\mathbb{Z}).$$

We introduce the difference expressions

$$\mathcal{D} = \begin{pmatrix} 0 & D_1 \\ D_2 & 0 \end{pmatrix}, \quad D_1 = aS^+ - a, \ D_2 = a - a^- S^-,$$

$$D^{-1} = \begin{pmatrix} 0 & D_2^{-1} \\ D_1^{-1} & 0 \end{pmatrix}, \quad (D^{-1})^\dagger = \begin{pmatrix} 0 & (D_1^{-1})^\dagger \\ (D_2^{-1})^\dagger & 0 \end{pmatrix},$$

1.7 Toda Conservation Laws and the Hamiltonian Formalism

where

$$(D_1^{-1}u)(n) = \sum_{m=-\infty}^{n-1} \frac{u(m)}{a(m)}, \quad (D_2^{-1}u)(n) = \frac{1}{a(n)} \sum_{m=-\infty}^{n} u(m),$$

$$((D_1^{-1})^\dagger u)(n) = \frac{1}{a(n)} \sum_{m=n+1}^{\infty} u(m), \quad ((D_2^{-1})^\dagger u)(n) = \sum_{m=n}^{\infty} \frac{u(m)}{a(m)}, \quad u \in \ell^1(\mathbb{Z}).$$

Viewing $\mathcal{D}, \mathcal{D}^{-1}$ and $D_1, D_1^{-1}, D_2, D_2^{-1}$ as operators on $\ell^1(\mathbb{Z})^2$ and $\ell^1(\mathbb{Z})$, respectively, one concludes that

$$\mathcal{D}\mathcal{D}^{-1} = I_{\ell^1(\mathbb{Z})^2}, \quad D_1 D_1^{-1} = I_{\ell^1(\mathbb{Z})}, \quad D_2 D_2^{-1} = I_{\ell^1(\mathbb{Z})}.$$

Next, let $\mathcal{F}$ be a functional of the type

$$\mathcal{F} \colon \ell^1(\mathbb{Z})^2 \to \mathbb{C}, \tag{1.373}$$

$$\mathcal{F}(a, b) = \sum_{n \in \mathbb{Z}} F\big(a, b, a^{(+1)}, b^{(+1)}, a^{(-1)}, b^{(-1)}, \ldots, a^{(+k)}, b^{(+k)}, a^{(-k)}, b^{(-k)}\big),$$

assuming

$$\{F(a,b)\}_{n \in \mathbb{Z}}, \; \left\{\frac{\partial F(a,b)}{\partial a^{(\pm j)}}\right\}_{n \in \mathbb{Z}}, \; \left\{\frac{\partial F(a,b)}{\partial b^{(\pm j)}}\right\}_{n \in \mathbb{Z}} \in \ell^1(\mathbb{Z}), \quad j = 1, \ldots, k.$$

For simplicity of notation we again abbreviate (1.373) by

$$\mathcal{F}(a,b) = \sum_{n \in \mathbb{Z}} F(a,b)$$

in the following. The gradient $\nabla \mathcal{F}$ and symplectic gradient $\nabla_s \mathcal{F}$ of $\mathcal{F}$ are then defined by

$$(\nabla \mathcal{F})_{a,b} = \begin{pmatrix} (\nabla \mathcal{F})_a \\ (\nabla \mathcal{F})_b \end{pmatrix} = \begin{pmatrix} \frac{\delta \mathcal{F}}{\delta a} \\ \frac{\delta \mathcal{F}}{\delta b} \end{pmatrix}$$

and

$$(\nabla_s \mathcal{F})_{a,b} = \mathcal{D}(\nabla \mathcal{F})_{a,b} = \mathcal{D}\begin{pmatrix} (\nabla \mathcal{F})_a \\ (\nabla \mathcal{F})_b \end{pmatrix},$$

respectively. In addition, we introduce the weakly nondegenerate closed 2-form

$$\Omega \colon \ell^1(\mathbb{Z})^2 \times \ell^1(\mathbb{Z})^2 \to \mathbb{C},$$

$$\Omega(u,v) = \frac{1}{2} \sum_{n \in \mathbb{Z}} \Big((\mathcal{D}^{-1}u)(n) \cdot v(n) + u(n) \cdot ((\mathcal{D}^{-1})^\dagger v)(n)\Big).$$

One then concludes that

$$\Omega(\mathcal{D}u, v) = \sum_{n\in\mathbb{Z}} u(n) \cdot v(n) = \sum_{n\in\mathbb{Z}} \big(u_1(n)v_1(n) + u_2(n)v_2(n)\big)$$
$$= \langle u, v\rangle_{\ell^2(\mathbb{Z})^2}, \quad u, v \in \ell^1(\mathbb{Z})^2,$$

where $\langle \cdot, \cdot \rangle_{\ell^2(\mathbb{Z})^2}$ denotes the "real" inner product in $\ell^2(\mathbb{Z})^2$, that is,

$$\langle \cdot, \cdot \rangle_{\ell^2(\mathbb{Z})^2} : \ell^2(\mathbb{Z})^2 \times \ell^2(\mathbb{Z})^2 \to \mathbb{C},$$
$$\langle u, v\rangle_{\ell^2(\mathbb{Z})^2} = \sum_{n\in\mathbb{Z}} u(n) \cdot v(n) = \sum_{n\in\mathbb{Z}} \big(u_1(n)v_1(n) + u_2(n)v_2(n)\big).$$

In addition, one obtains

$$(d\mathcal{F})_{a,b}(v) = \langle (\nabla\mathcal{F})_{a,b}, v\rangle_{\ell^2(\mathbb{Z})^2} = \Omega(\mathcal{D}(\nabla\mathcal{F})_{a,b}, v) = \Omega((\nabla_s\mathcal{F})_{a,b}, v).$$

Given two functionals $\mathcal{F}_1, \mathcal{F}_2$ we define their Poisson bracket by

$$\{\mathcal{F}_1, \mathcal{F}_2\} = d\mathcal{F}_1(\nabla_s\mathcal{F}_2) = \Omega(\nabla_s\mathcal{F}_1, \nabla_s\mathcal{F}_2)$$
$$= \Omega(\mathcal{D}\nabla\mathcal{F}_1, \mathcal{D}\nabla\mathcal{F}_2) = \langle \nabla\mathcal{F}_1, \mathcal{D}\nabla\mathcal{F}_2\rangle_{\ell^2(\mathbb{Z})^2}$$
$$= \sum_{n\in\mathbb{Z}} \begin{pmatrix} \frac{\delta F_1}{\delta a}(n) \\ \frac{\delta F_1}{\delta b}(n) \end{pmatrix} \cdot \mathcal{D} \begin{pmatrix} \frac{\delta F_2}{\delta a}(n) \\ \frac{\delta F_2}{\delta b}(n) \end{pmatrix}. \tag{1.374}$$

One then verifies that both the Jacobi identity

$$\{\{\mathcal{F}_1, \mathcal{F}_2\}, \mathcal{F}_3\} + \{\{\mathcal{F}_2, \mathcal{F}_3\}, \mathcal{F}_1\} + \{\{\mathcal{F}_3, \mathcal{F}_1\}, \mathcal{F}_2\} = 0,$$

as well as the Leibniz rule

$$\{\mathcal{F}_1, \mathcal{F}_2\mathcal{F}_3\} = \{\mathcal{F}_1, \mathcal{F}_2\}\mathcal{F}_3 + \mathcal{F}_2\{\mathcal{F}_1, \mathcal{F}_3\},$$

hold.

If $\mathcal{F}$ is a smooth functional and (a, b) develops according to a Hamiltonian flow with Hamiltonian $\mathcal{H}$, that is,

$$\begin{pmatrix} a \\ b \end{pmatrix}_t = (\nabla_s\mathcal{H})_{a,b} = \mathcal{D}(\nabla\mathcal{H})_{a,b} = \mathcal{D}\begin{pmatrix} \frac{\delta H}{\delta a} \\ \frac{\delta H}{\delta b} \end{pmatrix}, \tag{1.375}$$

then

$$\frac{d\mathcal{F}}{dt} = \frac{d}{dt}\sum_{n\in\mathbb{Z}} F(a(n), b(n))$$
$$= \sum_{n\in\mathbb{Z}} \begin{pmatrix} \frac{\delta F}{\delta a}(n) \\ \frac{\delta F}{\delta b}(n) \end{pmatrix} \cdot \begin{pmatrix} a(n) \\ b(n) \end{pmatrix}_t = \sum_{n\in\mathbb{Z}} \begin{pmatrix} \frac{\delta F}{\delta a}(n) \\ \frac{\delta F}{\delta b}(n) \end{pmatrix} \cdot \mathcal{D}\begin{pmatrix} \frac{\delta H}{\delta a}(n) \\ \frac{\delta H}{\delta b}(n) \end{pmatrix}$$
$$= \{\mathcal{F}, \mathcal{H}\}. \tag{1.376}$$

1.7 Toda Conservation Laws and the Hamiltonian Formalism 137

Here, and in the remainder of this section and the next, time-dependent equations such as (1.376) are viewed locally in time, that is, assumed to hold on some open t-interval $\mathbb{I} \subseteq \mathbb{R}$.

If a functional $\mathcal{G}$ is in involution with the Hamiltonian $\mathcal{H}$, that is,

$$\{\mathcal{G}, \mathcal{H}\} = 0,$$

then it is conserved, that is,

$$\frac{d\mathcal{G}}{dt} = 0.$$

Next, we turn to the specifics of the Toda hierarchy.

Lemma 1.69 *Assume Hypothesis* 1.54. *Then,*

$$\frac{\delta \hat{f}_\ell}{\delta a} = -\frac{\ell}{a} \hat{g}_{\ell-1}, \quad \ell \in \mathbb{N}, \tag{1.377}$$

$$\frac{\delta \hat{f}_\ell}{\delta b} = \ell \hat{f}_{\ell-1}, \quad \ell \in \mathbb{N}. \tag{1.378}$$

Proof With our assumptions on (a, b) we only know that $\hat{f}_\ell \in \ell^\infty(\mathbb{Z})$. We start by deriving (1.378). To that end we introduce the functional

$$\widehat{\mathcal{F}}_{\ell,N}(a, b) = \sum_{n \in \mathbb{Z}} \hat{f}_\ell(n) \chi_N(n),$$

where χ_N is the characteristic function of the set $[-N, N] \cap \mathbb{Z}$. Then one finds

$$(d\widehat{\mathcal{F}}_{\ell,N}(a, b))_b(v) = \sum_{n \in \mathbb{Z}} \left(\chi_N(n) \frac{\partial \hat{f}_\ell}{\partial b}(n) + \chi_N^-(n) \left(S^{(-1)} \frac{\partial \hat{f}_\ell}{\partial b^{(+1)}} \right)(n) \right.$$

$$\left. + \chi_N^+(n) \left(S^{(+1)} \frac{\partial \hat{f}_\ell}{\partial b^{(-1)}} \right)(n) + \cdots \right) v(n)$$

$$\underset{N \to \infty}{\to} \sum_{n \in \mathbb{Z}} \left(\frac{\partial \hat{f}_\ell}{\partial b}(n) + \left(S^{(-1)} \frac{\partial \hat{f}_\ell}{\partial b^{(+1)}} \right)(n) + \left(S^{(+1)} \frac{\partial \hat{f}_\ell}{\partial b^{(-1)}} \right)(n) + \cdots \right) v(n)$$

$$= \sum_{n \in \mathbb{Z}} \frac{\delta \hat{f}_\ell}{\delta b}(n) v(n), \quad v \in \ell^1(\mathbb{Z}).$$

On the other hand, recalling that $(\cdot, \cdot)$ denotes the usual scalar product in $\ell^2(\mathbb{Z})$,

$$(d\widehat{\mathcal{F}}_{\ell,N}(a, b))_b(v)$$

$$= \frac{d}{d\epsilon} \widehat{\mathcal{F}}_{\ell,N}(b + \epsilon v)\Big|_{\epsilon=0} = \sum_{n \in \mathbb{Z}} \left(\delta_n, \sum_{k=0}^{\ell-1} L^k v L^{\ell-1-k} \delta_n \right) \chi_N(n)$$

$$= \sum_{n \in \mathbb{Z}} \left(\delta_n, \sum_{k=0}^{\ell-1} L^{\ell-1} v \delta_n \right) \chi_N(n)$$

$$+ \sum_{n \in \mathbb{Z}} \left(\delta_n, \sum_{k=0}^{\ell-1} [L^k v, L^{\ell-1-k}] \delta_n \right) \chi_N(n)$$

$$\underset{N \to \infty}{\to} \sum_{k=0}^{\ell-1} \sum_{n \in \mathbb{Z}} (\delta_n, L^{\ell-1} \delta_n) v(n) + \sum_{k=0}^{\ell-1} \sum_{n \in \mathbb{Z}} (\delta_n, [L^k v, L^{\ell-1-k}] \delta_n)$$

$$= \ell \sum_{n \in \mathbb{Z}} (\delta_n, L^{\ell-1} \delta_n) v(n) = \ell \sum_{n \in \mathbb{Z}} \hat{f}_{\ell-1}(n) v(n), \quad v \in \ell^1(\mathbb{Z}),$$

using (1.15), and the general result that for bounded operators $A, B \in \mathcal{B}(\mathcal{H})$ on a separable, complex Hilbert space $\mathcal{H}$ with AB and BA trace class operators, their commutator is traceless, that is, $\text{tr}([A, B]) = 0$. Combining the two expressions, one concludes that (1.378) holds.

Next we turn to the proof of (1.377). To this end one first observes that, as before,

$$(d\widehat{\mathcal{F}}_{\ell,N}(a,b))_a(v) \underset{N \to \infty}{\to} \sum_{n \in \mathbb{Z}} \left(\frac{\partial \hat{f}_\ell}{\partial a}(n) + S^{(-1)} \left(\frac{\partial \hat{f}_\ell}{\partial a^{(+1)}} \right)(n) \right.$$

$$\left. + S^{(+1)} \left(\frac{\partial \hat{f}_\ell}{\partial a^{(-1)}} \right)(n) + \cdots \right) v(n)$$

$$= \sum_{n \in \mathbb{Z}} \frac{\delta \hat{f}_\ell}{\delta a}(n) v(n), \quad v \in \ell^1(\mathbb{Z}).$$

Furthermore, one computes

$$(d\widehat{\mathcal{F}}_{\ell,N}(a,b))_a(v) = \frac{d}{d\epsilon} \widehat{\mathcal{F}}_{\ell,N}(a+\epsilon v) \Big|_{\epsilon=0}$$

$$= \frac{d}{d\epsilon} \sum_{n \in \mathbb{Z}} \left(\delta_n, \left((a+\epsilon v) S^+ + (a+\epsilon v)^- S^- + b \right)^\ell \delta_n \right) \chi_N(n) \Big|_{\epsilon=0}$$

$$= \sum_{n \in \mathbb{Z}} \left(\delta_n, \sum_{k=0}^{\ell-1} (L^k (vS^+ + v^- S^-) L^{\ell-1-k}) \delta_n \right) \chi_N(n)$$

$$= \sum_{k=0}^{\ell-1} \sum_{n \in \mathbb{Z}} (\delta_n, (vS^+ + v^- S^-) L^{\ell-1} \delta_n) v(n) \chi_N(n)$$

$$+ \sum_{k=0}^{\ell-1} \sum_{n \in \mathbb{Z}} (\delta_n, [L^k (vS^+ + v^- S^-), L^{\ell-1-k}] \delta_n) \chi_N(n)$$

1.7 Toda Conservation Laws and the Hamiltonian Formalism

$$\underset{N\to\infty}{\to} \sum_{k=0}^{\ell-1} \sum_{n\in\mathbb{Z}} \left(\delta_n, (vS^+ + v^-S^-)L^{\ell-1}\delta_n\right)v(n)$$

$$+ \sum_{k=0}^{\ell-1} \operatorname{tr}\left([L^k(vS^+ + v^-S^-), L^{\ell-1-k}]\right)$$

$$= \ell \sum_{n\in\mathbb{Z}} \left(\left(\delta_n, vS^+L^{\ell-1}\delta_n\right) + \left(\delta_n, v^-S^-L^{\ell-1}\delta_n\right)\right)$$

$$= \ell \sum_{n\in\mathbb{Z}} \left(v(n)\left(S^-\delta_n, L^{\ell-1}\delta_n\right) + v(n-1)\left(S^+\delta_n, L^{\ell-1}\delta_n\right)\right)$$

$$= \ell \sum_{n\in\mathbb{Z}} \left(v(n)\left(\delta_{n+1}, L^{\ell-1}\delta_n\right) + v(n-1)\left(\delta_{n-1}, L^{\ell-1}\delta_n\right)\right)$$

$$= \ell \sum_{n\in\mathbb{Z}} \left(v(n)\left(\delta_{n+1}, L^{\ell-1}\delta_n\right) + v(n)\left(\delta_n, L^{\ell-1}\delta_{n+1}\right)\right)$$

$$= 2\ell \sum_{n\in\mathbb{Z}} \left(\delta_{n+1}, L^{\ell-1}\delta_n\right)v(n)$$

$$= -\ell \sum_{n\in\mathbb{Z}} \frac{1}{a(n)} \hat{g}_{\ell-1}(n) v(n), \quad v \in \ell^1(\mathbb{Z}).$$

Thus (1.377) is established. □

For the remainder of this section we now introduce the following assumption.

Hypothesis 1.70 *In addition to Hypothesis* 1.54 *suppose that*

$$\{a(n)^2 - \tfrac{1}{4}\}_{n\in\mathbb{Z}}, \{b(n)\}_{n\in\mathbb{Z}} \in \ell^1(\mathbb{Z}).$$

We fix $\ell \in \mathbb{N}$ and want to show that a suitably renormalized functional $\widehat{\mathcal{F}}_\ell$ is well-defined. We define it by subtracting the limit of each $\hat{f}_\ell(n)$ as $n \to \infty$. Each $\hat{f}_\ell$ is a polynomial in a, b and certain shifts thereof. Terms with a and shifts enter homogeneously and only in even powers. Assume that $\hat{f}_\ell(n) \to \lambda_\ell$ as $|n| \to \infty$. For ℓ odd, we know that $\lambda_\ell = 0$ because each term contains at least one b or certain shifts thereof. Then $\{\hat{f}_\ell(n) - \lambda_\ell\}_{n\in\mathbb{Z}} \in \ell^1(\mathbb{Z})$. We can see this as follows. Each $\hat{f}_\ell$ is a finite sum of terms containing a and b with shifts. Terms with b are already summable. Only terms exclusively with a have a nonzero limit and hence are nonsummable. Their general form will be $\alpha_1(n)\cdots\alpha_k(n)$ with $\alpha_j(n) = (a^{(m_j)}(n))^{2p_j} \to \tilde{\lambda}_j$ as $|n| \to \infty$, where (m_j) denotes the shifts. Then

$$\alpha_1(n)\cdots\alpha_k(n) - \tilde{\lambda}_1\cdots\tilde{\lambda}_k = (\alpha_1(n) - \tilde{\lambda}_1)\alpha_2(n)\cdots\alpha_k(n)$$
$$+ \tilde{\lambda}_1(\alpha_2(n) - \tilde{\lambda}_2)\alpha_2(n)\cdots\alpha_k(n)$$
$$+ \cdots + \tilde{\lambda}_1\cdots\tilde{\lambda}_{k-1}(\alpha_k(n) - \tilde{\lambda}_k).$$

Thus we see that

$$\{\alpha_1(n)\cdots\alpha_k(n) - \tilde{\lambda}_1\cdots\tilde{\lambda}_k\}_{n\in\mathbb{Z}} \in \ell^1(\mathbb{Z}),$$

and hence

$$\{\hat{f}_\ell(n) - \lambda_\ell\}_{n\in\mathbb{Z}} \in \ell^1(\mathbb{Z}).$$

Thus, the functional

$$\widehat{\mathcal{F}}_\ell(a,b) = \sum_{n\in\mathbb{Z}} (\hat{f}_\ell(n) - \lambda_\ell)$$

is well-defined.

In the following it is convenient to introduce the abbreviation $\mathbb{N}_{-1} = \mathbb{N}\cup\{-1, 0\}$.

Theorem 1.71 *Assume Hypothesis 1.70 and let $p \in \mathbb{N}_{-1}$. In addition, define*

$$\widehat{\mathcal{H}}_p = \frac{1}{p+2}\sum_{n\in\mathbb{Z}}(\hat{f}_{p+2}(n) - \lambda_{p+2}), \tag{1.379}$$

where

$$\lambda_{p+2} = \lim_{|n|\to\infty} \hat{f}_{p+2}(n).$$

Then,

$$(\nabla\widehat{\mathcal{H}}_p)_a = \frac{1}{p+2}\frac{\delta \hat{f}_{p+2}}{\delta a} = -\frac{1}{a}\hat{g}_{p+1}, \quad (\nabla\widehat{\mathcal{H}}_p)_b = \frac{1}{p+2}\frac{\delta \hat{f}_{p+2}}{\delta b} = \hat{f}_{p+1}.$$

Moreover, the homogeneous pth Toda equations then take on the form

$$\widehat{\mathrm{Tl}}_p(a,b) = \begin{pmatrix} a_{t_p} \\ b_{t_p} \end{pmatrix} - \mathcal{D}\begin{pmatrix} (\nabla\widehat{\mathcal{H}}_p)_a \\ (\nabla\widehat{\mathcal{I}}_p)_b \end{pmatrix} = 0, \quad p \in \mathbb{N}_{-1}. \tag{1.380}$$

Proof This follows directly from Lemma 1.69 and (1.50). □

Remark 1.72 In (1.380) we also introduced the trivial linear flow $\widehat{\mathrm{Tl}}_{-1}$ given by

$$\widehat{\mathrm{Tl}}_{-1}(a,b) = \begin{pmatrix} a_{t_{-1}} \\ b_{t_{-1}} \end{pmatrix} - \mathcal{D}\begin{pmatrix} (\nabla\widehat{\mathcal{H}}_{-1})_a \\ (\nabla\widehat{\mathcal{I}}_{-1})_b \end{pmatrix} = \begin{pmatrix} a_{t_{-1}} \\ b_{t_{-1}} \end{pmatrix} = 0.$$

Next, we consider the general case, where

$$\mathrm{Tl}_p(a,b) = \begin{pmatrix} a_{t_p} - a(f_{p+1}^+ - f_{p+1}) \\ b_{t_p} + g_{p+1} - g_{p+1}^- \end{pmatrix} = 0, \quad p \in \mathbb{N}_{-1},$$

1.7 Toda Conservation Laws and the Hamiltonian Formalism

and

$$f_p = \sum_{k=0}^{p} c_{p-k} \hat{f}_k, \quad g_p = \sum_{k=1}^{p} c_{p-k} \hat{g}_k - c_{p+1}, \quad p \in \mathbb{N}_0.$$

Introducing

$$\mathcal{H}_{-1} = \widehat{\mathcal{H}}_{-1}, \quad \mathcal{H}_p = \sum_{k=0}^{p} c_{p-k} \widehat{\mathcal{H}}_k, \quad p \in \mathbb{N}_0, \tag{1.381}$$

one then obtains

$$\mathrm{Tl}_p(a,b) = \begin{pmatrix} a_{t_p} \\ b_{t_p} \end{pmatrix} - \mathcal{D} \begin{pmatrix} (\nabla \mathcal{H}_p)_a \\ (\nabla \mathcal{H}_p)_b \end{pmatrix} = 0, \quad p \in \mathbb{N}_{-1}. \tag{1.382}$$

Theorem 1.73 *Assume Hypothesis 1.70 and suppose that a, b satisfy the system $\mathrm{Tl}_p(a,b) = 0$ for some $p \in \mathbb{N}_{-1}$. Then,*

$$\frac{d\mathcal{H}_r}{dt_p} = 0, \quad r \in \mathbb{N}_{-1}. \tag{1.383}$$

Proof From (1.363) and (1.339) (cf. Remark 1.66) one obtains

$$\frac{d\hat{f}_{r+2}}{dt_p} = (S^+ - I)\beta_{r+2}, \quad r \in \mathbb{N}_{-1},$$

for some $\beta_{r+2}, r \in \mathbb{N}_{-1}$, which are polynomials in a and b and certain shifts thereof. Using definition (1.379) of $\widehat{\mathcal{H}}_r$, the result (1.383) follows in the homogeneous case and then by linearity in the general case. □

Theorem 1.74 *Assume Hypothesis 1.70 and let $p, r \in \mathbb{N}_{-1}$. Then,*

$$\{\mathcal{H}_p, \mathcal{H}_r\} = 0, \tag{1.384}$$

that is, $\mathcal{H}_p$ and $\mathcal{H}_r$ are in involution for all $p, r \in \mathbb{N}_{-1}$.

Proof By Remark 1.64 (i), there exists $T_0 > 0$ such that the initial value problem

$$\mathrm{Tl}_p(a,b) = 0, \quad (a,b)\big|_{t=0} = \left(a^{(0)}, b^{(0)}\right),$$

where $a^{(0)}, b^{(0)}$ satisfy Hypothesis 1.70, has a unique and continuous solution $a(t), b(t)$ satisfying Hypothesis 1.70 for each $t \in [0, T_0)$. For this solution we know that

$$\frac{d}{dt}\mathcal{H}_p(t) = \{\mathcal{H}_r(t), \mathcal{H}_p(t)\} = 0.$$

Next, let $t \downarrow 0$. Then

$$0 = \{\mathcal{H}_r(t), \mathcal{H}_p(t)\} \xrightarrow[t \downarrow 0]{} \{\mathcal{H}_r(0), \mathcal{H}_p(0)\} = \{\mathcal{H}_r, \mathcal{H}_p\}\big|_{(a,b)=(a^{(0)}, b^{(0)})}.$$

Since $a^{(0)}, b^{(0)}$ are arbitrary coefficients satisfying Hypothesis 1.70 one concludes (1.384). □

There is also a second Hamiltonian structure for the Toda hierarchy. Rewriting the linear recursion (1.3)–(1.5) in the form

$$f_{\ell+1}^+ - f_{\ell+1}^- = -\frac{1}{2}(g_\ell^+ - g_\ell^-) + (b^+ f_\ell^+ - b f_\ell),$$
$$g_{\ell+1}^- - g_{\ell+1}^- = -2(a^2 f_\ell^+ - (a^-)^2 f_\ell^-) + b(g_\ell - g_\ell^-), \quad \ell \in \mathbb{N}_0,$$
(1.385)

one finds

$$\begin{pmatrix} \frac{1}{2}(f_{n+1}^+ - f_{n+1}) \\ g_{n+1}^- - g_{n+1} \end{pmatrix} = \widetilde{\mathcal{D}} \begin{pmatrix} -\frac{1}{a} g_n \\ f_n \end{pmatrix},$$

where we abbreviated

$$\widetilde{\mathcal{D}} = \begin{pmatrix} \frac{1}{2} a(S^+ - S^-) a & a(S^+ - I) b \\ b(I - S^-) a & 2(S^+ - S^-) a^2 \end{pmatrix}.$$
(1.386)

One can introduce a second Poisson bracket $\{\{\cdot, \cdot\}\}$, defined by

$$\{\{\mathcal{F}_1, \mathcal{F}_2\}\} = \sum_{n \in \mathbb{Z}} \begin{pmatrix} \frac{\delta F_1}{\delta a}(n) \\ \frac{\delta F_1}{\delta b}(n) \end{pmatrix} \cdot \widetilde{\mathcal{D}} \begin{pmatrix} \frac{\delta F_2}{\delta a}(n) \\ \frac{\delta F_2}{\delta b}(n) \end{pmatrix}.$$
(1.387)

This second Poisson bracket is also skew-symmetric and satisfies the Jacobi identity and the Leibniz rule. As in Theorem 1.74, one verifies that all $\mathcal{H}_p$, $p \in \mathbb{N}_{-1}$, are in involution also with respect to this second Poisson bracket (1.387), that is,

$$\{\{\mathcal{H}_r, \mathcal{H}_p\}\} = 0, \quad p, r \in \mathbb{N}_{-1}.$$

Combining (1.379), (1.381), (1.382), and (1.385), (1.386) permits one to write the pth Toda equation in another Hamiltonian form

$$\mathrm{Tl}_p(a, b) = \begin{pmatrix} a_{t_p} \\ b_{t_p} \end{pmatrix} - \widetilde{\mathcal{D}} \begin{pmatrix} (\nabla \mathcal{H}_{p-1})_a \\ (\nabla \mathcal{H}_{p-1})_b \end{pmatrix} = 0, \quad p \in \mathbb{N}_0.$$

Finally, we now very briefly sketch an extension of the Hamiltonian formalism to the case of almost periodic coefficients a and b.

We start by recalling the notions of quasi-periodic and almost periodic sequences: First, a bounded continuous function $f \colon \mathbb{R} \to \mathbb{C}$ is called *Bohr almost periodic* (or *uniformly almost periodic*) if it is the limit of a sequence of trigonometric polynomials on $\mathbb{R}$ in the uniform (sup norm) topology. Then a sequence $f = \{f(n)\}_{n \in \mathbb{Z}}$ is called *almost periodic* if there exists a Bohr almost periodic function g on $\mathbb{R}$ such that $f(n) = g(n)$ for all $n \in \mathbb{Z}$. For completeness we also recall that a sequence f is called *quasiperiodic* with fundamental periods $(\omega_1, \ldots, \omega_N) \in (0, \infty)^N$ if the

1.7 Toda Conservation Laws and the Hamiltonian Formalism

frequencies $2\pi/\omega_1, \ldots, 2\pi/\omega_N$ are linearly independent over $\mathbb{Q}$ and if there exists a continuous function $F \in C(\mathbb{R}^N)$, periodic of period 1 in each of its arguments,

$$F(x_1, \ldots, x_j + 1, \ldots, x_N) = F(x_1, \ldots, x_N), \quad x_j \in \mathbb{R}, \ j = 1, \ldots, N,$$

such that

$$f(n) = F(\omega_1^{-1} n, \ldots, \omega_N^{-1} n), \quad n \in \mathbb{Z}.$$

Any quasiperiodic sequence is also almost periodic.

For any almost periodic sequence $f = \{f(n)\}_{n \in \mathbb{Z}}$, the mean value $\langle f \rangle$ of f, defined by

$$\langle f \rangle = \lim_{N \uparrow \infty} \frac{1}{2N+1} \sum_{n=n_0-N}^{n_0+N} f(n),$$

exists and is independent of $n_0 \in \mathbb{Z}$.

We assume that u has the frequency module $\mathcal{M}(u)$ and given a density F as in (1.369), equation (1.368) then becomes

$$\mathcal{F}(u) = \lim_{N \uparrow \infty} \frac{1}{2N+1} \sum_{n=-N}^{N} F\left(u(n), u^{(+1)}(n), u^{(-1)}(n), \ldots, u^{(k)}(n), u^{(-k)}(n)\right)$$

$$= \langle F(u) \rangle.$$

Supposing that the frequency module $\mathcal{M}(v)$ of v satisfies $\mathcal{M}(v) \subseteq \mathcal{M}(u)$, the analog of (1.370) then reads

$$(d\mathcal{F})_u(v) = \frac{d}{d\epsilon} \mathcal{F}(u + \epsilon v)\Big|_{\epsilon=0}$$

$$= \lim_{N \uparrow \infty} \sum_{n=-N}^{N} \left(\frac{\partial F(n,u)}{\partial u} + S^{(-1)} \frac{\partial F(n,u)}{\partial u^{(+1)}} + S^{(+1)} \frac{\partial F(n,u)}{\partial u^{(-1)}} \right.$$

$$\left. + \cdots + S^{(-k)} \frac{\partial F(n,u)}{\partial u^{(k)}} + S^{(k)} \frac{\partial F(n,u)}{\partial u^{(-k)}} \right) v(n)$$

$$= \langle (\nabla \mathcal{F})_u v \rangle = \left\langle \frac{\delta F}{\delta u} v \right\rangle.$$

Assuming a and b are almost periodic with frequency module $\mathcal{M}$, we again consider functionals as in (1.373),

$$\mathcal{F}(a,b) = \left\langle F(a, b, a^{(+1)}, b^{(+1)}, a^{(-1)}, b^{(-1)}, \ldots, a^{(+k)}, b^{(+k)}, a^{(-k)}, b^{(-k)}) \right\rangle$$

$$= \langle F(a,b) \rangle,$$

and in analogy to (1.374), the Poisson brackets of two functionals $\mathcal{F}_1, \mathcal{F}_2$ are then given by

$$\{\mathcal{F}_1, \mathcal{F}_2\} = \left\langle \begin{pmatrix} \frac{\delta F_1}{\delta a}(n) \\ \frac{\delta F_1}{\delta b}(n) \end{pmatrix} \cdot \mathcal{D} \begin{pmatrix} \frac{\delta F_2}{\delta a}(n) \\ \frac{\delta F_2}{\delta b}(n) \end{pmatrix} \right\rangle.$$

Again one verifies that both the Jacobi identity as well as the Leibniz rule hold in this case. Moreover, as in (1.375) and (1.376), if $\mathcal{F}$ is a smooth functional and (a, b) develops according to a Hamiltonian flow with Hamiltonian $\mathcal{H}$, that is,

$$\begin{pmatrix} a \\ b \end{pmatrix}_t = (\nabla_s \mathcal{H})_{a,b} = \mathcal{D}(\nabla \mathcal{H})_{a,b} = \mathcal{D} \begin{pmatrix} \frac{\delta H}{\delta a} \\ \frac{\delta H}{\delta b} \end{pmatrix},$$

then

$$\frac{d\mathcal{F}}{dt} = \frac{d}{dt} \langle F(a, b) \rangle = \{\mathcal{F}, \mathcal{H}\}.$$

Next, assuming in addition that $1/a \in \ell^\infty(\mathbb{N})$ and that $\langle 1/a \rangle$ stays finite for t varying in some open time interval $\mathbb{I} \subseteq \mathbb{R}$, we now introduce the fundamental function w by

$$w(z) = \langle \phi_+(z, \cdot) \rangle \qquad (1.388)$$

for $|z|$ sufficiently large. Since

$$w'(z) = -\langle g(z, \cdot) \rangle, \quad z \in \mathbb{C} \setminus \operatorname{spec}(\check{L}),$$

w in (1.388) extends analytically to $z \in \mathbb{C} \setminus \operatorname{spec}(\check{L})$.

One observes the asymptotic expansion

$$w(z) = \langle \ln(\phi_+(z, \cdot)) \rangle \underset{z \to \infty}{=} \langle \ln(a/z) \rangle + \sum_{j \in \mathbb{N}} \langle \rho_{+,j} \rangle z^{-j}$$
$$\underset{z \to \infty}{=} \langle \ln(a/z) \rangle + \sum_{j \in \mathbb{N}} j^{-1} \langle \hat{f}_j \rangle z^{-j}$$

and computes

$$\frac{\delta w(z)}{\delta a}(n) = \frac{1}{a(n)} - 2G(z, n, n+1), \quad z \in \mathbb{C} \setminus \operatorname{spec}(\check{L}), \, n \in \mathbb{Z},$$
$$\frac{\delta w(z)}{\delta b}(n) = -g(z, n), \quad z \in \mathbb{C} \setminus \operatorname{spec}(\check{L}), \, n \in \mathbb{Z}.$$

Introducing

$$\widehat{\mathcal{H}}_{-1} = \mathcal{H}_{-1} = \langle \hat{f}_1 \rangle, \quad \widehat{\mathcal{H}}_p = \frac{1}{p+2} \langle \hat{f}_{p+2} \rangle, \quad \mathcal{H}_p = \sum_{k=0}^{p} c_{p-k} \widehat{\mathcal{H}}_k, \quad p \in \mathbb{N}_0,$$

the Toda equations again take on the form

$$\mathrm{Tl}_p(a,b) = \begin{pmatrix} a_{t_p} \\ b_{t_p} \end{pmatrix} - \mathcal{D} \begin{pmatrix} (\nabla \mathcal{H}_p)_a \\ (\nabla \mathcal{H}_p)_b \end{pmatrix} = 0, \quad p \in \mathbb{N}_{-1}.$$

Finally, one can show that $w(z_1)$ and $w(z_2)$ are in involution for arbitrary $z_1, z_2 \in \mathbb{C} \setminus \mathrm{spec}\,(\check{L})$, and hence obtains

$$\{w(z_1), w(z_2)\} = 0, \quad z_1, z_2 \in \mathbb{C} \setminus \mathrm{spec}\,(\check{L}),$$

$$\{w(z), \mathcal{H}_p\} = 0, \quad \{\mathcal{H}_p, \mathcal{H}_r\} = 0, \quad z \in \mathbb{C} \setminus \mathrm{spec}\,(\check{L}), \quad p, r \in \mathbb{N}_{-1}.$$

Naturally, these considerations apply to the special periodic case in which $\langle c \rangle$ for a periodic sequence c on $\mathbb{Z}$ is to be interpreted as the periodic mean value.

1.8 Notes

Pereant qui ante nos nostra dixerunt.
Aelius Donatus (4th century)[1]

Section 1.1. For historical facts and key references leading up to the exponential lattice introduced by Toda, we refer to our discussion at the beginning of the introduction. In the remainder of these notes we shall primarily focus on the Toda lattice and its associated hierarchy.

The equations of motion for a chain of particles (of equal mass $m = 1$) with nearest neighbor interactions, are of the type

$$x_{tt}(n,t) = V'(x(n+1,t) - x(n,t)) - V'(x(n,t) - x(n-1,t)), \quad (n,t) \in \mathbb{Z} \times \mathbb{R}, \tag{1.389}$$

where $x(n,t)$ denotes the displacement of the nth particle from its equilibrium position at time t, and $V(\cdot)$ the interaction potential (with $-V'(y) = -\frac{dV(y)}{dy}$ the corresponding force). The special case $V(y) = \frac{1}{2}y^2$, in accordance with Hooke's law, represents the case of a linear lattice,

$$x_{tt}(n,t) = x(n+1,t) - 2x(n,t) + x(n-1,t)), \quad (n,t) \in \mathbb{Z} \times \mathbb{R}.$$

Motivated by the numerical investigations by Fermi, Pasta, and Ulam (cf. also the very recent historical account by Dauxois (2008), highlighting the contributions by M. T. Menzel to this first ever numerical experiment), and especially, by those of Ford and Waters, Toda published a particular nonlinear lattice in 1967, an exponential lattice, that supported a periodic solution. He chose (cf. Toda (1967a,b))

$$V(y) = e^{-y} + y - 1, \quad y \in \mathbb{R},$$

[1] Quoted by his pupil St. Jerome ("To the devil with those who published before us.").

such that $V(y) = \frac{1}{2}y^2 + O(y^3)$ as $y \to 0$. Introducing Flaschka's variables (cf. Flaschka (1974a,b))

$$a(n,t) = \frac{1}{2}\exp(\tfrac{1}{2}(x(n,t)-x(n+1,t))), \quad b(n,t) = -\frac{1}{2}x_t(n,t), \quad (n,t) \in \mathbb{Z}\times\mathbb{R},$$

equation (1.389) becomes the first-order Toda lattice system

$$a_t(n,t) - a(n,t)\big(b(n+1,t) - b(n,t)\big) = 0, \tag{1.390}$$

$$b_t(n,t) - 2\big(a(n,t)^2 - a(n-1,t)^2\big) = 0, \quad (n,t) \in \mathbb{Z} \times \mathbb{R}. \tag{1.391}$$

For very recent work on FPU chains we refer, for instance, to Henrici and Kappeler (2008).

As mentioned in the introduction, the integrability in the finite-dimensional periodic case was first established by Hénon (1974) and soon thereafter by Flaschka (1974b) (see also Flaschka (1975), Flaschka and McLaughlin (1976a), Kac and van Moerbeke (1975a), van Moerbeke (1976)). Flaschka proved the integrability of the doubly infinite Toda lattice (1.391) on $\mathbb{Z}$ in 1974 by establishing a Lax pair for it with Lax operator the tridiagonal Jacobi operator on $\mathbb{Z}$ (a discrete Sturm–Liouville-type operator, cf. Flaschka (1974a)). Nearly simultaneously, this was independently observed by Manakov (1975). Soon after, integrability of the finite nonperiodic Toda lattice was established by Moser (1975a).

For books containing material on the Toda lattice, its soliton solutions, etc., we refer to Eilenberger (1983), Faddeev and Takhtajan (1987), Kupershmidt (1985), Novikov et al. (1984, Sect. I.7), Teschl (2000), Toda (1989a,b). For general reviews on the Toda lattice and its solutions, the interested reader can consult, for instance, Dubrovin et al. (1990), McLaughlin (1989), Suris (2006), Toda (1970; 1975; 1976), Teschl (2001), and Krüger and Teschl (2007).

Integrable discretizations of the Toda lattice (and other integrable systems) are discussed in Suris (2003).

Since the paper Bulla et al. (1998) and the monograph Teschl (2000) focused on real-valued Toda hierarchy solutions, we put the emphasis on the complex-valued case (and hence on the case of a non-self-adjoint Lax operator) in this chapter.

Finally, for various Lie-algebraic extensions of the (one and two-dimensional, periodic and nonperiodic) Toda lattice, and applications to various algorithms, which are not discussed in this volume, we refer, for instance, to Adler (1979), Adler et al. (1993), Adler and van Moerbeke (1980a,b; 1982; 1991; 1995), Bloch et al. (1990; 1992), Bloch and Gekhtman (1998; 2007), Casian and Kodama (2006; 2007), Chu (1985; 1994), Chu and Norris (1988), Damianou (2004), Deift and Li (1989; 1991), Deift et al. (1986; 1989), Driessel (1986), Ercolani et al. (1993), Faybusovich (1992; 1994), Faybusovich and Gekhtman (2000; 2001), Flaschka (1994), Gekhtman (1998), Gekhtman and Shapiro (1999), Gel'fand and Zakharevich (2000), Guest

1.8 Notes

(1997), Kodama and Ye (1996a), Koelling et al. (2004), Kostant (1979), Leznov and Saveliev (1992, Sect. 4.2), Li (1997), McDaniel and Smolinsky (1992; 1997; 1998), Mikhailov et al. (1981), Olshanetsky and Perelomov (1979; 1981; 1994), Perelomov (1990, Ch. 4), Razumov and Saveliev (1997), Reyman and Semenov-Tian-Shansky (1979; 1981; 1994), Symes (1980a,b), Takasaki (1984), Ueno and Takasaki (1984), Watkins (1984; 1993), and the extensive list of references contained therein.

Section 1.2. The approach presented in this section closely follows Bulla et al. (1998).

The construction of the Toda hierarchy using a recursive approach is patterned after work by Alber (1991a) (see also Alber (1989; 1991b)).

Burchnall–Chaundy theory, the formalism underlying commuting difference expressions, is discussed in Glazman (1965, Sect. 67), Krichever (1978), van Moerbeke and Mumford (1979), Mumford (1978), Naĭman (1962; 1964). The special case involving a second-order difference expression L in Theorem 1.6 then leads to hyperelliptic curves not branched at infinity.

Lemma 1.10 is well-known, it can be found, for instance, in Eilenberger (1983, p. 141).

Interesting connections between biorthogonal Laurent polynomials, Töplitz determinants, τ-functions, and generalized Toda lattices are studied in Bertola and Gekhtman (2007). More on τ-functions can be found in Kajiwara et al. (2007). The connection between Jacobi operators, orthogonal polynomials, and continued fractions is detailed in Deift (1999).

For connections between Willmore surfaces and the Baker–Akhiezer vector of the Toda lattice we refer to Babich (1996). For connections between the Toda lattice and the construction of minimal tori in $\mathbb{C}^3$, see Sharipov (1991). A connection between the Toda lattice and discrete curves in $\mathbb{CP}^1$ is discussed in Hoffmann and Kutz (2004), Kutz (2003).

Intriguing connections between peakons, strings, and the finite Toda lattice are discussed in Alber et al. (2000), Beals et al. (2001), and Ragnisco and Bruschi (1996).

Connections between heat kernel expansions and the Toda lattice hierarchy were discussed in Iliev (2007).

Section 1.3. Most of the material presented in this section has been taken from Bulla et al. (1998).

As in previous chapters, the fundamental meromorphic function $\phi(\,\cdot\,, n_0)$ on $\mathcal{K}_p$ defined in (1.69), is in many respects the key object of our algebro-geometric formalism. Again, in the special self-adjoint case, where $a, b \in \ell^\infty(\mathbb{Z})$ and a, b, and E_m, $m = 0, \ldots, 2p + 1$, are real-valued, its two branches are intimately related to the two Dirichlet half-line Weyl m-functions $m_{\pm,0}(\,\cdot\,, n_0)$ associated with proper closed realizations $\check{L}$ of the difference expression $L = aS^+ + a^- S^- + b$ in $\ell^2((n_0, \pm\infty))$. In particular, the spectral properties of the self-adjoint realization of $\check{L}$ in $\ell^2(\mathbb{Z})$ (as well as those of the self-adjoint Dirichlet-type operators in $\ell^2((n_0, \pm\infty))$) can be inferred directly from $\phi(\,\cdot\,, n_0)$. (For a detailed spectral theoretic treatment of self-adjoint Jacobi operators, we refer, for instance, to Berezanskii (1968, Ch. VII), Carmona and Lacroix (1990), Teschl (2000, Part 1).)

A look at (1.69)–(1.71) shows that $\phi(\,\cdot\,, n)$ links the Dirichlet divisor $\mathcal{D}_{\hat{\underline{\mu}}(n)}$ and its shift $\mathcal{D}_{\hat{\underline{\mu}}(n+1)}$, the Neumann divisor. This is of course a direct consequence of the identity (1.61) together with the factorizations of $F_p(\,\cdot\,, n)$ in (1.64). This construction of positive divisors of degree p (respectively, $p + 1$, since the points $P_{\infty\pm}$ are also involved) on hyperelliptic curves $\mathcal{K}_p$ of genus p is analogous to that of Jacobi (1846) with applications to the KdV case by Mumford (1984, Sect. III a).1) and McKean (1985).

Trace formulas of the type (1.97), (1.96) in Lemma 1.15 can be found, for instance, in Date and Tanaka (1976a,b), Dubrovin et al. (1976), van Moerbeke (1976). (For systematic generalizations, involving Krein's spectral shift function, cf. Gesztesy and Simon (1996; 1997), Teschl (1998; 1999a), and the monograph Teschl (2000, Chs. 6, 8).)

Expression (1.127) for b in Theorem 1.19, in terms of the Riemann theta function associated with $\mathcal{K}_p$, apparently, was found by different groups around 1976. It appeared in papers by Dubrovin et al. (1976) and Date and Tanaka (1976a,b). An explicit formula for a in terms of theta functions was not derived in these papers, partly, since a (as well as the original Toda variables P and Q, cf. Toda (1989b, Sects. 3.1, 4.6)) in principle, follows from b. The explicit theta function formula (1.126) for a was derived a bit later by Krichever (1978), Kričever (1982), Krichever (1982; 1983) (cf. also the appendix written by Krichever in Dubrovin (1981)). While Dubrovin, Matveev, and Novikov as well as Date and Tanaka consider the special periodic case, Krichever treats both the periodic and quasi-periodic cases.

There was also an interesting parallel to the periodic KdV case (which was solved by Marčenko (1974a,b) shortly before the theta function representations by Its and Matveev (1975a,b)): The inverse periodic Toda problem had been solved slightly earlier by Kac and van Moerbeke (1975a), with a very detailed account to be found in the seminal paper van Moerbeke (1976). Additional results on isospectral deformations can be found in McKean (1979), van Moerbeke (1979), and van Moerbeke and Mumford (1979).

Since this initial period, many authors presented reviews and slightly varying approaches to algebro-geometric (respectively, periodic) solutions of (stationary and time-dependent) equations of the Toda hierarchy. We mention, for instance, the extensive treatments in Dubrovin et al. (1990), Flaschka (1975), Iguchi (1992a,b), McKean (1979), McKean and van Moerbeke (1980), van Moerbeke (1979),

1.8 Notes

van Moerbeke and Mumford (1979), Mumford (1978), and in the monographs Novikov et al. (1984, Sect. I.7, App., Sect. 9), Teschl (2000, Chs. 9, 13), Toda (1989b, Ch. 4), Toda (1989a, Chs. 26–30). We note that Aptekarev (1986) derived the theta function representation for a using the theory of orthogonal polynomials associated with measures supported on a system of contours. In this context we also refer to Lukashov (2004), Pastur (2006), Peherstorfer (1995), and Zhedanov (1990), and the references cited therein. For various additional material we refer to Deift and Trubowitz (1981) concerning the continuum limit of the periodic Jacobi matrix inverse problem, to Deift and McLaughlin (1998), Kuijlaars (2000), Kuijlaars and McLaughlin (2001) and the references therein for discussions of the continuum limit of the Toda lattice, to Gieseker (1996) for limiting connections between the periodic Toda and KdV hierarchies, to Iguchi (1992b), Smirnov (1989) in connection with expressions for a, b in terms of elliptic functions, to Iguchi (1992a), Sodin and Yuditskiĭ (1994; 1997), in the context of infinite genus situations, to Deift et al. (1995a,b; 1996) for dealing with forced lattice vibrations, to Boley and Golub (1984; 1987), Ferguson (1980), Zhernakov (1986) for additional discussions of the periodic inverse spectral problem, and to Antony and Krishna (1992; 1994), Carmona and Kotani (1987), Carmona and Lacroix (1990, Sects. VII.2.3, VII.4.2), Knill (1993b), Kotani and Simon (1988), in connection with almost-periodic and random inverse spectral problems.

The solution of certain discrete Peierls models for quasi-one-dimensional conducting polymers in connection with finite-band Toda solutions has been studied in Brazovskii et al. (1982), Dzyaloshinskii and Krichever (1983), Krichever (1982; 1983) (see also the references therein).

With the exception of the references Krichever (1978; 1982), Kričever (1982), van Moerbeke (1979), van Moerbeke and Mumford (1979), and Mumford (1978), the references discussed thus far focus on real-valued algebro-geometric sequences a and b. Assuming all $a(n) > 0$, $n \in \mathbb{Z}$, and b real-valued, the isospectral manifold is a torus $\mathbb{T}^p$ (equivalently, the real part of the Jacobi variety of the underlying hyperelliptic curve $\mathcal{K}_p$), as shown by van Moerbeke (1976). Here p, the genus of the curve $\mathcal{K}_p$, equals the number of nondegenerate gaps in the spectrum of the $\ell^2(\mathbb{Z})$-realization $\check{L}$ of the finite-difference expression L in (1.2), where we only count the (bounded) spectral gaps in the interval $[\inf \mathrm{spec}\, (\check{L}), \sup \mathrm{spec}\, (\check{L})]$, not the unbounded one near $\pm \infty$. Equivalently, the spectrum of $\check{L}$ consists of p compact intervals and hence reads

$$\mathrm{spec}\,(\check{L}) = \bigcup_{\ell=1}^{p+1} [E_{2\ell-2}, E_{2\ell-1}], \quad E_0 < E_1 < \cdots < E_{2p+1} \qquad (1.392)$$

for some $p \in \mathbb{N}$. In particular, if a, b are N-periodic real-valued sequences (i.e., for some $N \in \mathbb{N}$, $a(n + N) = a(n)$, $b(n + N) = b(n)$ for all $n \in \mathbb{Z}$) then $p \leq N - 1$.

Some delicate questions in connection with the boundary of isospectral manifolds, when flows hit the theta divisor, are discussed in Adler et al. (1993), Kodama (2002), and van Moerbeke (1993).

Apart from the realization of the p-dimensional isospectral torus $I_p((a_0, b_0))$ of a given base pair (a_0, b_0) of algebro-geometric coefficients, with fixed band spectrum as in (1.392) and fixed sign of all the coefficients a, in terms of the Riemann theta functions associated with $\mathcal{K}_p$ and positive Dirichlet-type divisors $\mathcal{D}_{\underline{\hat{\mu}}(n_0)}$ of degree p in Date and Tanaka (1976a,b), Dubrovin et al. (1976), there also exist explicit realizations of $I_p((a_0, b_0))$ in terms of $2p$ Darboux transformations, representable as a $2p \times 2p$ Wronski determinant of certain Baker–Akhiezer functions. This torus has been explicitly described by Gesztesy and Teschl (1996). The method can be considered a finite-difference analog of that in Buys and Finkel (1984) and Finkel et al. (1987), Gesztesy et al. (1996b) (see also Teschl (1997) for generalizations to arbitrary, not necessarily (quasi-)periodic, base pairs (a_0, b_0) with gaps in their essential spectrum).

Isospectral manifolds in connection with Toda flows in various settings (including non-abelian generalizations) have attracted a lot of interest and we refer, for instance, to Bättig et al. (1993), Bloch et al. (1990), Davis (1987), Faybusovich (1992), Fried (1986), Gibson (2002), Leite et al. (2008), van Moerbeke (1976; 1979), van Moerbeke and Mumford (1979), Tomei (1984), and the references therein. If one relaxes the sign restriction on the coefficients $a(n)$ and assumes N-periodicity of the real-valued sequences a and b, with $\prod_{n=1}^N a(n) \neq 0$, the corresponding isospectral set of periodic $N \times N$ Jacobi matrices (i.e., the associated tridiagonal $N \times N$ matrices with elements $a(N)$ in the lower left and upper right corners) breaks up into 2^N connected components of the type $\mathbb{T}^p$, according to the various sign combinations of the $a(n)$ (cf. van Moerbeke (1976)). Further relaxing the condition $\prod_{n=1}^N a(n) \neq 0$ and simply studying the isospectral set of periodic self-adjoint tridiagonal matrices then leads to intriguing topological and cohomological questions. The interested reader is invited to consult Bloch et al. (1990), Davis (1987), Deift et al. (1993), Fried (1986), Kodama and Ye (1996b; 1998), Tomei (1984), and van Moerbeke (1976).

Corollary 1.21 is taken from Bulla et al. (1998).

There is an extensive literature on the theory of periodic Jacobi matrices, see, for instance, Adler et al. (1993), Babelon et al. (2003, Ch. 6), Bättig et al. (1993), Date and Tanaka (1976a), Deift and Li (1991), Kac and van Moerbeke (1975a), Kato (1983), McKean (1979), Teschl (2000, Ch. 7), Toda (1989a,b), van Moerbeke (1976; 1979), and van Moerbeke and Mumford (1979). Our presentation of periodic Jacobi operators is taken from Bulla et al. (1998).

Examples 1.26 and 1.27 are taken from Bulla et al. (1998). We refer, for instance, to Byrd and Friedman (1971) for details on Jacobi elliptic integrals used in Example 1.27.

1.8 Notes

The possibility of obtaining stationary soliton solutions from degenerating hyperelliptic curves (i.e., pinching handles Fay (1973, Ch. III), etc.) is analogous to our discussion in the notes to Section 1.3 in Volume I. The alternative approach using Darboux transformations seems to be a bit simpler to implement in practice and we refer the interested reader to Bulla et al. (1998), Gesztesy et al. (1993), Gesztesy and Teschl (1996), Matveev and Salle (1991, Sects. 2.5, 5.3), Teschl (1995; 1997; 1999b), and the monograph Teschl (2000, Chs. 11, 14). The limit $N \to \infty$ in the stationary N-soliton solutions has been studied in Gesztesy and Renger (1997) and relative to general backgrounds in Renger (1999).

Constraints on scattering data to guarantee short-range solitons relative to algebro-geometric finite-band solutions were found in Teschl (2007). In this context we also refer to Egorova et al. (2006) and Michor (2005), Michor and Teschl (2007b) in connection with scattering theory relative to algebro-geometric finite-band backgrounds, and to Egorova et al. (2007b; 2008) in the context of scattering theory relative to steplike algebro-geometric finite-band backgrounds.

In analogy to the KdV context, real-valued algebro-geometric coefficients a and b are reflectionless (cf., e.g., Gesztesy and Simon (1996) for the corresponding definition). The interested reader may want to consult Clark et al. (2005), Gesztesy et al. (1996a), Gesztesy and Simon (1996), Gesztesy and Yuditskii (2006), Gesztesy and Zinchenko (to appear), Sodin and Yuditskiĭ (1994; 1997), Teschl (1998), and the monograph Teschl (2000, Ch. 8), for more details on reflectionless Jacobi operators.

In the case of periodic complex-valued sequences a, b, the $\ell^2(\mathbb{Z})$-spectrum associated with L consists of $p + 1$ regular analytic arcs in the complex plane. Compared to its real-valued counterpart, this case appears to have been much less studied in the literature, see, however, the detailed discussion in Batchenko and Gesztesy (2005a).

The Toda hierarchy differs from the soliton hierarchies for the continuous models studied in Volume I in the sense that it does not seem to have simple Dubrovin equations that govern the n-dependence of $\underline{\hat{\mu}}(n, t_r)$. We now show how to obtain a first-order system of nonlinear differential equations, whose solution $\underline{\hat{\chi}}(x, t_r)$ coincides with $\underline{\hat{\mu}}(n, t_r)$ at the integer points $x = n \in \mathbb{Z}$. Since the t_r-dependence of $\underline{\hat{\chi}}$ and $\underline{\hat{\mu}}$ plays no role in this argument, we ignore this dependence in the following result.

Lemma 1.75 (Dubrovin-type equations) *Assume* (1.62) *and* (1.63) *and let* $x \in \widetilde{\mathcal{I}}_\chi$, *where* $\widetilde{\mathcal{I}}_\chi \subseteq \mathbb{R}$ *is an open interval. Abbreviate*[1] $\underline{\Delta} = (\Delta_1, \ldots, \Delta_p) =$

---
[1] Here $\underline{A}_{Q_0}$ denotes the Abel map as defined in (A.29).

$-\underline{A}_{P_{\infty_-}}(P_{\infty_+})$ and consider the Dubrovin-type system on $\widetilde{\mathcal{I}}_\chi$,

$$\chi_{j,x}(x) = \frac{y(\hat{\chi}_j(x))}{\prod_{\substack{\ell=1\\ \ell\neq j}}^p (\chi_j(x) - \chi_\ell(x))} \sum_{m=1}^p \Phi_{p-m}^{(j)}(\underline{\chi}(x)) \sum_{n=1}^p C_{m,n} \Delta_n, \quad (1.393)$$

$\{\hat{\chi}_j(x_0)\}_{j=1,\dots,p} \subset \mathcal{K}_p$

for some $x_0 \in \widetilde{\mathcal{I}}_\chi$, *where* $\chi_1(x_0), \dots, \chi_p(x_0)$ *are assumed to be distinct and* $C_{m,n}$ *is defined in* (B.30). *Then there exists an open interval* $\mathcal{I}_\chi \subseteq \widetilde{\mathcal{I}}_\chi$, *with* $x_0 \in \mathcal{I}_\chi$, *such that the initial value problem* (1.393) *has a unique solution* $\{\hat{\chi}_j\}_{j=1,\dots,p} \subset \mathcal{K}_p$ *satisfying*

$$\hat{\chi}_j \in C^\infty(\mathcal{I}_\chi, \mathcal{K}_p), \quad j = 1, \dots, p,$$

and χ_j, $j = 1, \dots, p$, *remain distinct on* $\mathcal{I}_\chi$. *Moreover, suppose* $x_0 = n_0 \in \mathcal{I}_\chi \cap \mathbb{Z}$ *and*

$$\hat{\chi}_j(n_0) = \hat{\mu}_j(n_0) \in \mathcal{K}_p, \quad j = 1, \dots, p,$$

with $\mu_1(n_0), \dots, \mu_p(n_0)$ *assumed to be distinct and* $\hat{\mu}_j(n_0)$ *satisfying*

$$\hat{\mu}_j(n_0) = (\mu_j(n_0), -G_{p+1}(\mu_j(n_0), n_0)), \quad j = 1, \dots, p,$$
$$F_p(\mu_j(n_0), n_0) = 0, \quad j = 1, \dots, p.$$

Then the solution $\underline{\hat{\chi}} = \underline{\hat{\chi}}(x)$ *of the initial value problem* (1.393) *coincides with* $\underline{\hat{\mu}}(n)$ *defined in* (1.67) *at all integer values* $x = n \in \mathcal{I}_\chi \cap \mathbb{Z}$, *that is,*

$$\hat{\chi}_j(x)\big|_{x=n} = (\mu_j(n), -G_{p+1}(\mu_j(n), n)),$$
$$F_p(\chi_j(x), n)\big|_{x=n} = 0, \quad j = 1, \dots, p, \ n \in \mathcal{I}_\chi \cap \mathbb{Z}.$$

Proof First we recall (cf. (1.67) and (1.118))

$$\underline{\alpha}_{P_0}(\mathcal{D}_{\underline{\hat{\mu}}(n)}) - \underline{\alpha}_{P_0}(\mathcal{D}_{\underline{\hat{\mu}}(n_0)}) = -\underline{A}_{P_{\infty_-}}(P_{\infty_+})(n - n_0) = \underline{\Delta}(n - n_0).$$

Since $\mathcal{D}_{\underline{\hat{\chi}}(x)}$ is nonspecial, we only need to establish that $\underline{\hat{\chi}}$ satisfies

$$\underline{\alpha}_{P_0}(\mathcal{D}_{\underline{\hat{\chi}}(x)}) - \underline{\alpha}_{P_0}(\mathcal{D}_{\underline{\hat{\chi}}(x_0)}) = -\underline{A}_{P_{\infty_-}}(P_{\infty_+})(x - x_0) = \underline{\Delta}(x - x_0),$$

that is, we need to show that

$$\partial_x \underline{\alpha}_{P_0}(\mathcal{D}_{\underline{\hat{\chi}}(x)}) = \underline{\Delta}. \quad (1.394)$$

1.8 Notes

However, equation (1.394) follows along the lines of (1.265): Using (1.393), (D.10), (D.11), (D.12), and (B.31), (1.394) is a consequence of the following computation:

$$\begin{aligned}
\frac{d}{dx}\sum_{j=1}^{p}\underline{A}_{P_0,k}(\hat{\chi}_j(x)) &= \frac{d}{dx}\sum_{j=1}^{p}\int_{P_0}^{\hat{\chi}_j(x)}\omega_k \\
&= \frac{d}{dx}\sum_{j=1}^{p}\sum_{\ell=1}^{p}c_k(\ell)\int_{P_0}^{\hat{\chi}_j(x)}\frac{z^{\ell-1}\,dz}{y(P)} \\
&= \sum_{j=1}^{p}\sum_{\ell=1}^{p}c_k(\ell)\frac{\chi_j(x)^{\ell-1}}{y(\hat{\chi}_j(x))}\frac{d}{dx}\chi_j(x) \\
&= \sum_{j=1}^{p}\sum_{\ell=1}^{p}c_k(\ell)\frac{\chi_j(x)^{\ell-1}}{y(\hat{\chi}_j(x))}\frac{y(\hat{\chi}_j(x))}{\prod_{m\neq j}^{p}(\chi_j(x)-\chi_m(x))} \\
&\quad\times\sum_{m=1}^{p}\Phi_{p-m}^{(j)}(\underline{\chi}(x))\sum_{n=1}^{p}C_{m,n}\Delta_n \\
&= \sum_{j=1}^{p}\sum_{\ell=1}^{p}c_k(\ell)U_p(\underline{\chi}(x))_{\ell,j}\sum_{m=1}^{p}\Phi_{p-m}^{(j)}(\underline{\chi}(x))\sum_{n=1}^{p}C_{m,n}\Delta_n \\
&= \sum_{j=1}^{p}\sum_{\ell=1}^{p}\sum_{m=1}^{p}c_k(\ell)U_p(\underline{\chi}(x))_{\ell,j}U_p(\underline{\chi}(x))_{j,m}^{-1}(\underline{\chi}(x))\sum_{n=1}^{p}C_{m,n}\Delta_n \\
&= \sum_{\ell=1}^{p}\sum_{n=1}^{p}c_k(\ell)C_{\ell,n}\Delta_n \\
&= \sum_{\ell=1}^{p}\sum_{n=1}^{p}C_{k,\ell}^{-1}C_{\ell,n}\Delta_n \\
&= \Delta_k,\quad k=1,\ldots,p.
\end{aligned}$$

□

Thus, the solution $\hat{\underline{\chi}}(x)$ of (1.393) provides a continuous interpolation for $\hat{\underline{\mu}}(n)$. In principle, it might happen that $\mathcal{I}_\chi \cap \mathbb{Z} = \emptyset$ due to collisions of μ_j. In this case one has to resort to appropriate symmetric functions of the functions μ_j as indicated in the next remark.

The Dubrovin equations (1.393) for continuous interpolations of Dirichlet-type eigenvalues in Lemma 1.75 appeared in Gesztesy and Holden (2002). The possibility of a finite difference analog of Dubrovin-type equations is mentioned in Dubrovin et al. (1976), but without providing details. This finite difference equation idea was also discussed by Nijhoff (2000). In this context, relations between discrete and continuous integrable systems are also treated in Alber (1991a,b).

Remark 1.76 If two or more of the $\hat{\chi}_j(x)$ coincide at a point $x = x_0$, the Dubrovin-type equations (1.393) are ill-defined. However, in new and symmetric variables, the solution remains smooth. We illustrate this in the case when two $\hat{\chi}_j$ collide at a point x_0. Specifically, assume that $\hat{\chi}_1(x_0) = \hat{\chi}_2(x_0)$ but $\hat{\chi}_1(x) \neq \hat{\chi}_2(x)$ for $x \neq x_0$. All the remaining $\hat{\chi}_j$, $j > 2$, are supposed to be distinct from one another, and also from $\hat{\chi}_1$ and $\hat{\chi}_2$, near x_0. Consider new variables $\sigma_1 = \chi_1 + \chi_2$ and $\sigma_2 = \chi_1 \chi_2$. In these variables we record for x near x_0 that

$$\sigma_{1,x}(x) = \chi_{1,x}(x) + \chi_{2,x}(x)$$
$$= \frac{1}{\chi_1 - \chi_2} \bigg(\frac{y(\hat{\chi}_1)}{\prod_{\ell=3}^p (\chi_1 - \chi_\ell)} \sum_{m=1}^p \Phi_{p-m}^{(1)}(\underline{\chi}) \sum_{n=1}^p C_{m,n} \Delta_n$$
$$- \frac{y(\hat{\chi}_2)}{\prod_{\ell=3}^p (\chi_2 - \chi_\ell)} \sum_{m=1}^p \Phi_{p-m}^{(2)}(\underline{\chi}) \sum_{n=1}^p C_{m,n} \Delta_n \bigg)$$
$$= \frac{1}{\chi_1 - \chi_2} \big(A(\chi_1)\chi_2 + B(\chi_1) - A(\chi_2)\chi_1 - B(\chi_2) \big)$$
$$= A(\chi_1(x)) + A'(\chi_1(x))\chi_1(x) + B'(\chi_1(x)) + O(x - x_0)$$

by Taylor expanding the auxiliary functions $A(\chi_j)$ and $B(\chi_j)$ (that depend smoothly on χ_ℓ for $\ell > 2$ as well as χ_j) around χ_1. In the same way one finds

$$\sigma_{2,x}(x) = \chi_{1,x}(x)\chi_2(x) + \chi_1(x)\chi_{2,x}(x)$$
$$= \frac{1}{\chi_1(x) - \chi_2(x)} \big((A(\chi_1)\chi_2 + B(\chi_1))\chi_2 - (A(\chi_2)\chi_1 - B(\chi_2))\chi_1 \big)$$
$$= 2A(\chi_1(x))\chi_1(x) + A'(\chi_1(x))\chi_1(x)^2 + B'(\chi_1(x)) + O(x - x_0).$$

Clearly this analysis extends to the models treated in Volume I and to the AL hierarchy treated in Chapter 3.

Remark 1.77 Suppose in addition that a and b are real-valued and bounded. Moreover, assume the eigenvalue ordering (1.91) and (1.92). Furthermore, assume that the initial data in (1.393) are constrained to lie in spectral gaps, that is, $\chi_j(x_0) \in [E_{2j-1}, E_{2j}]$, $j = 1, \ldots, p$. Then,

$$\chi_j(x) \in [E_{2j-1}, E_{2j}], \quad j = 1, \ldots, p, \ x \in \mathbb{R}.$$

In particular, $\hat{\chi}_j(x)$ changes sheets whenever it hits E_{2j-1} or E_{2j}, and its projection $\chi_j(x)$ remains trapped in $[E_{2j-1}, E_{2j}]$, $j = 1, \ldots, p$, for all $x \times \mathbb{R}$.

1.8 Notes

Indeed, this follows from (1.393), $\underline{\Delta} \in \mathbb{R}^p$, $y(\hat{\chi}_j) \in \mathbb{R}$ for $\chi_j \in [E_{2j-1}, E_{2j}]$, $j = 1, \ldots, p$, since $R_{2p+2}(\lambda)^{1/2}$ is real-valued for $\lambda \in [E_{2j-1}, E_{2j}]$, $j = 1, \ldots, p$, by (B.18), and the fact that $C_{j,k} \in \mathbb{R}$, $j, k = 1, \ldots, p$, by (B.36).

Section 1.4. The results of this section are taken from the paper Gesztesy et al. (2008b).

For the construction of self-adjoint Jacobi operators with real-valued algebro-geometric coefficients a, b alluded to in the beginning of this section, we refer, for instance, to Bulla et al. (1998) and Teschl (2000, Sect. 8.3).

For Sard's theorem in connection with Lemma 1.30, we refer, for instance, to Abraham et al. (1988, Sect. 3.6).

The stationary (complex) Toda algorithm as outlined in steps **(I)–(XII)** represents a new solution of the inverse algebro-geometric spectral problem for generally non-self-adjoint Jacobi operators. In particular, one can view this algorithm as a continuation of the inverse periodic spectral problem started in 1975 (in the self-adjoint context) by Kac and van Moerbeke (1975a,c) and Flaschka (1975), continued in the seminal papers by van Moerbeke (1976), Date and Tanaka (1976a), and Dubrovin et al. (1976), and further developed by Krichever (1978), McKean (1979), van Moerbeke and Mumford (1979), Mumford (1978), and others, in part in the more general quasi-periodic algebro-geometric case.

That straight motions on the torus are generically dense is of course a well-known fact (see e.g. Arnold (1989, Sect. 51) or Katok and Hasselblatt (1995, Sects. 1.4, 1.5)). For quasi-periodicity of sequences indexed by $n \in \mathbb{Z}$ one can consult, for instance, Pastur and Figotin (1992, p. 31). For the special case of complex-valued and quasi-periodic Jacobi matrices where all quasi-periods are real-valued, we refer to Batchenko and Gesztesy (2005a,b).

We emphasize that the approach described in this section is not limited to the stationary Toda hierarchy, respectively, to the inverse algebro-geometric spectral problem of non-self-adjoint Jacobi operators, but applies universally to the construction of stationary algebro-geometric solutions of integrable lattice hierarchies of soliton equations. In particular, it is also applied to the Ablowitz–Ladik hierarchy as discussed in Gesztesy et al. (2007b) and in Section 3.5 of this monograph.

We also note that while the periodic case with complex-valued a, b is of course included in our analysis, we throughout consider the more general algebro-geometric case (in which a, b need not be quasi-periodic).

Section 1.5. The approach presented in this section closely follows Bulla et al. (1998).

Since many of the references provided in connection with Section 1.3 treat the time-dependent Toda equations and not just stationary Toda solutions, we will now mainly focus on issues markedly different from stationary ones and topics not yet covered. Connections between the Toda lattice and associated continuous integrable systems were discussed in Alber (1991a) and Gesztesy and Holden (2002).

In analogy to its stationary analog in Section 1.3, the role of $\phi(P, n, t_r)$ defined in (1.213) is still central to Section 1.5 and the corresponding facts recorded in the notes to Section 1.3 still apply.

The Dubrovin equations (1.236) in Lemma 1.37 were found simultaneously with the theta function representations discussed in the notes to Section 1.3. As in the corresponding KdV context, they are usually discussed in connection with the simplest cases $r = 0, 1$ only.

Since the proof of Lemma 1.38 is identical to that in the corresponding stationary case, the remarks in connection with the trace formulas in Lemma 1.15 in the notes to Section 1.3 apply again.

The linearization property (1.259) of the Abel map and formulas (1.260), (1.261) for the pair $(a(n, t_r), b(n, t_r))$ in terms of the Riemann theta function associated with $\mathcal{K}_p$ were again found simultaneously with their stationary counterparts and thus the historical development sketched in this connection in the notes to Section 1.3 remains valid in the context of Theorem 1.41.

Corollary 1.43 is taken from Bulla et al. (1998).

Although we mentioned p-soliton solutions of the Toda hierarchy at the end of Sections 1.3 and 1.5, we did not explicitly study degenerations of quasi-periodic algebro-geometric solutions as certain periods tend to infinity, or alternatively, systematically apply Darboux-type transformations (i.e., various commutation methods) that can lead to Bäcklund transformations between the Toda and Kac–van Moerbeke hierarchies as well as auto-Bäcklund transformations for the Toda hierarchy. The connections with the Kac–van Moerbeke hierarchy will be dealt with in detail in our next chapter. Here we just recall that this singularization procedure creates soliton-type solutions relative to a remaining algebro-geometric background solution (whose associated Riemann surface has appropriately diminished genus) and leads to a singular curve for the combined soliton and background Toda solution, etc. The possibility of obtaining soliton solutions from degenerating hyperelliptic curves (i.e., pinching handles, Fay (1973, Ch. III), etc.) is analogous to our discussion in the notes to Section 1.3. The alternative approach using Darboux transformations seems to be a bit simpler to implement in practice and we refer the interested reader to Bulla et al. (1998), Gesztesy et al. (1993), Gesztesy and Teschl (1996), Matveev and Salle (1991, Sects. 2.5, 5.3), Sun et al. (2005), Teschl (1997; 1999b), Teschl (2000, Chs. 11, 14). Infinite soliton solutions obtained as limits of N-soliton solutions as $N \to \infty$, have been discussed in Gesztesy and Renger (1997) and relative to general backgrounds in Renger (1999). For solutions expressed in terms of Fredholm determinants we also refer to Widom (1997). Different approaches to infinite soliton

solutions were undertaken in Orlov (2006) and Schiebold (1998; 2005). Soliton solutions on algebro-geometric finite-band backgrounds are treated in Egorova et al. (to appear). The inverse scattering transform for the Toda hierarchy relative to algebro-geometric quasi-periodic backgrounds is studied in Egorova et al. (2007a).
For Toda flows in the Hilbert space $\ell^2(\mathbb{N})$ we refer to Deift et al. (1985) (see also Li (1987)).

Section 1.6. The results of this section are taken from the paper Gesztesy et al. (2008b).
In the special case of a self-adjoint Lax (i.e., Jacobi) operator $\breve{L}$, where a and b are real-valued and bounded, the two-step procedure to construct the solution of this algebro-geometric initial value problem alluded to at the beginning of this section is discussed in detail in Bulla et al. (1998) and Teschl (2000, Sect. 13.2).
Unique solvability of the autonomous system (1.288), (1.289) with polynomial right-hand sides is a standard result (see, e.g., Walter (1998, Sect. III.10)).
Again we refer, for instance, to Arnold (1989, Sect. 51) or Katok and Hasselblatt (1995, Sects. 1.4, 1.5) for the well-known fact that straight motions on the torus are generically dense and as in the notes to Section 1.4 one can consult Pastur and Figotin (1992, p. 31) for quasi-periodicity with respect to sequences indexed by $n \in \mathbb{N}$.
We emphasize that the approach described in this section is not limited to the Toda hierarchy but applies universally to constructing algebro-geometric solutions of $(1 + 1)$-dimensional integrable soliton equations of differential-difference (i.e., lattice) type. Moreover, the principal idea of replacing the Dubrovin-type equations by a first-order system of the type (1.288) is also relevant in the context of general non-self-adjoint Lax operators for the continuous models in $(1 + 1)$-dimensions. In particular, the models studied in detail in Gesztesy and Holden (2003b) can be revisited from this point of view. However, the fact that the set in (1.317) is of measure zero relies on the fact that n varies in the countable set $\mathbb{Z}$ and hence is not applicable to continuous models in $(1 + 1)$-dimensions.
We also note that while the periodic case with complex-valued a, b is of course included in our analysis, we throughout consider the more general algebro-geometric case (in which a, b need not be quasi-periodic).
Alternative approaches to the integration of the Toda initial value problem have been studied by Gekhtman (1991a,b), Kudryavtsev (2002), Vinnikov and Yuditskii (2002), and Zhernakov (1987).
The following material is not treated in this volume but illustrates alternative ways to integrate the Toda lattice in various situations.
In connection with the Toda lattice half-line initial value problem, using a moment problem approach, we refer, for instance to Berezanskiĭ (1985), Berezanski (1986), Peherstorfer (2001), Sakhnovich (1989), Shmoish (1989), Yurko (1995). The

corresponding nonabelian problem is studied in Berezanskii and Gekhtman (1990), Berezanskii et al. (1986), Gekhtman (1990). Rapidly decreasing solutions of the half-line Toda lattice initial value problem, using scattering theoretic techniques, were studied in Khanmamedov (2005b).

Further generalizations including non-isospectral Toda flows can be found in Berezansky (1990; 1996), Berezanskii and Smoish (1990; 1994).

The asymptotic time behavior of solutions of the Toda lattice with rapidly decreasing initial data is discussed in Kamvissis (1993). Stability of the periodic (and algebro-geometric quasi-periodic) Toda lattice under short-range perturbations is treated in Kamvissis and Teschl (2007a,b). The asymptotic time behavior of the Toda lattice initial value problem with initial data tending to different asymptotic constants as $n \to \pm\infty$ has been studied in Guseinov and Khanmamedov (1999).

Finally, singularities of the finite Toda lattice in the complex domain were studied in Casian and Kodama (2002a–c), Flaschka (1988), Flaschka and Haine (1991), Gekhtman and Shapiro (1997), Kodama and Ye (1996a,b; 1998). In the case of the periodic Toda lattice we refer to Adler (1981), Adler et al. (1993), Kodama and Ye (1998), and Kodama (2002).

Section 1.7. The material in this section is primarily taken from Gesztesy and Holden (2006).

The Toda hierarchy and its interpretation as a Hamiltonian system has received enormous attention in the past. Without being able to provide a comprehensive review of the vast literature we mention that Lie algebraic approaches to Hamiltonian and gradient structures in the asymmetric nonperiodic Toda flows, Poisson maps, etc., are studied in many of the references provided at the end of the Notes to Section 1.1. The symplectic structure and action-angle variables for the periodic Toda lattice are discussed, for instance, in Bättig et al. (1993; 1995), Guillot (1994), Li (1997). Master symmetries for the finite nonperiodic Toda lattice are treated in Damianou (1993), the second Hamiltonian structure for the periodic Toda lattice is discussed in Beffa (1997), and multiple Hamiltonian structures for various Bogoyavlensky–Toda systems, Poisson manifolds, master symmetries, recursion operators, etc., can be found in Damianou (2000), Damianou and Fernandes (2002), Faybusovich and Gekhtman (2000), and Fernandes (1993). Moreover, Poisson brackets are studied in Faddeev and Takhtajan (1987, Sect. III.4), Flaschka and McLaughlin (1976a,b), Manakov (1975), McLaughlin (1989), Novikov et al. (1984, Sect. II.4), Tsiganov (2007) and infinitely many constants of motion appeared for instance, in Eilenberger (1983, Sect. 7.3), Faddeev and Takhtajan (1987, Sect. III.2.4), Flaschka (1974a,b; 1975), Flaschka and McLaughlin (1976a), Gesztesy and Holden (1994), Hénon (1974), Manakov (1975), McLaughlin (1975), Takebe (1990), Teschl (2000, Sect. 13.4), Toda (1989b, Sect. 3.7), and Wadati (1976). An elementary approach to the

1.8 Notes

infinite sequence of Toda hierarchy conservation laws was presented in Zhang and Chen (2002a) (but it seems to lack an explicit and recursive structure). Moreover, a Hamiltonian formalism for the Toda hierarchy, based on recursion operators and hence on an algebraic formalism familiar from the one involving formal pseudo-differential expressions in connection with the Gelfand–Dickey hierarchy, was developed in Zhang and Chen (2002b). A connection between the Hamiltonian formalism of the Toda lattice and the Weyl–Titchmarsh function associated with $\breve{L}$ is discussed in Vaninsky (2003). For interesting connections with orthogonal polynomials in this context we also refer to Peherstorfer et al. (2007).

In spite of this large body of literature on the Toda Hamiltonian formalism, the recursive and most elementary approach to local conservation laws of the infinite Toda hierarchy as presented in Gesztesy and Holden (2006), apparently, had not been noted in the literature before. Moreover, the treatment of Poisson brackets and variational derivatives, and their connections with the diagonal Green's function of the underlying Lax operator $\breve{L}$ in Gesztesy and Holden (2006), put the Toda hierarchy on precisely the same level as the KdV hierarchy with respect to these particular aspects of the Hamiltonian formalism.

Next we mention some references supporting more technical points in this section: For a collection of results concerning the numerical range of operators used in the proof of Theorem 1.55 we refer, for instance, to Gustafson and Rao (1997, Ch. 1).

Carleman integral kernels mentioned in connection with (1.322), (1.323) are discussed in Weidmann (1980, Sect. 6.2).

Equation (1.348) (which has been observed in Batchenko and Gesztesy (2005b) in the algebro-geometric context) can be viewed as the analog of a well-known identity for the diagonal Green's function of Schrödinger operators as discussed in Lemma 1.61 (i) in Volume I (see also (1.299) in Volume I). Relevant references in the Schrödinger context are, for instance, Carmona and Lacroix (1990, p. 369), Gel'fand and Dikii (1975), Gesztesy and Holden (2003b, pp. 99 and 122), and Johnson and Moser (1982).

The existence of the propagator $W_p(\,\cdot\,,\,\cdot\,)$ satisfying (1.354)–(1.357) in Theorem 1.62 is a standard result and follows, for instance, from Reed and Simon (1975, Theorem X.69) (under even weaker hypotheses on a, b).

Theorem 1.63 follows from standard results in Abraham et al. (1988, Sect. 4.1) and has been exploited in the self-adjoint case in Teschl (2000, Sect. 12.2). More precisely, local existence and uniqueness as well as smoothness of the solution of the initial value problem (1.361) (cf. also (1.50)) follows from Abraham et al. (1988, Theorem 4.1.5) since f_{p+1} and g_{p+1} depend only polynomially on a, b and certain of their shifts, and the fact that the Toda flows are autonomous.

Remark 1.64 (i) follows from Abraham et al. (1988, Theorem 4.1.5). To apply the Banach space setting of Abraham et al. (1988, Theorem 4.1.5) in this situation, one introduces a new sequence

$$\{c^{(0)}(n)\}_{n\in\mathbb{Z}} = \{a^{(0)}(n)^2 - \tfrac{1}{4}\}_{n\in\mathbb{Z}},$$

and then demands

$$\{b^{(0)}(n)\}_{n\in\mathbb{Z}}, \{c^{(0)}(n)\}_{n\in\mathbb{Z}} \in \ell^1(\mathbb{Z}),$$

upon substituting $\{a^{(0)}(n)^2\}_{n\in\mathbb{Z}}$ by $\{c^{(0)}(n) + \tfrac{1}{4}\}_{n\in\mathbb{Z}}$ in the equations of the Toda hierarchy. The latter is possible by Remark 1.9 since only a^2 but not a itself enters the Toda hierarchy of evolution equations.

In the self-adjoint context, the results of Theorems 1.61 and 1.62 appeared in Teschl (2000, Theorem 12.4, Lemma 12.16) (see also Gesztesy et al. (1993) for some relevant results in this context).

The unique, global, and smooth solution of the pth Toda lattice initial value problem (1.361) alluded to in Remark 1.64 (ii) follows from a further application of Abraham et al. (1988, Proposition 4.1.22).

The general result that for bounded operators $A, B \in \mathcal{B}(\mathcal{H})$ on a separable, complex Hilbert space $\mathcal{H}$ with AB and BA trace class operators, their commutator is traceless, that is, $\operatorname{tr}([A, B]) = 0$, used in the proof of Lemma 1.69, can be found in Deift (1978) and Simon (2005d, Corollary 3.8).

For the brief sketch of the Hamiltonian formalism in the case of almost periodic coefficients a and b we followed Johnson and Moser (1982) in the context of one-dimensional almost periodic Schrödinger operators and Carmona and Kotani (1987) in the corresponding discrete case (see also the treatment in Carmona and Lacroix (1990, Sects. VII.1, VII.2)). In particular, we note that w in (1.388) is the appropriate discrete analog of the function $\langle m_+(z)\rangle = \langle -1/G(z, \cdot, \cdot)\rangle$ introduced by Johnson and Moser (1982) in the case of almost periodic Schrödinger operators. (Here m_+ denotes the analog of the right half-line Weyl–Titchmarsh coefficient and $G(z, \cdot, \cdot)$ the diagonal Green's function of the underlying almost periodic Schrödinger operator $H = -d^2/dx^2 + V$ in $L^2(\mathbb{R})$.)

For basic properties of quasi-periodic and almost periodic functions and sequences we refer, for instance, to Corduneanu (1989), Levitan and Zhikov (1982), and Pastur and Figotin (1992, p. 30–31).

2
The Kac–van Moerbeke Hierarchy

A formal manipulator in mathematics often experiences the
discomforting feeling that his pencil surpasses him in
intelligence.

Howard W. Eves[1]

2.1 Contents

The Kac–van Moerbeke (KM) lattice is a system of differential-difference equations
that is closely related to the Toda lattice by a Miura-type transformation familiar from
analogous connections between the KdV and the modified KdV hierarchy. In fact,
the relation is so close that we will depart from the strategy of the previous chapters
in Volumes I and II in the construction of the associated hierarchy and directly exploit
this intimate connection to transport the algebro-geometric solutions constructed in
Chapter 1 for the Toda hierarchy to that of the Kac–van Moerbeke hierarchy.

The equation for the Kac–van Moerbeke lattice,

$$\mathrm{KM}_0(\rho) = \rho_t - \rho\left((\rho^+)^2 - (\rho^-)^2\right) = 0 \tag{2.1}$$

for a sequence $\rho = \rho(n, t)$ (with $\rho^\pm(n, t) = \rho(n \pm 1, t)$, $(n, t) \in \mathbb{Z} \times \mathbb{R}$, etc.), was
originally derived in the mid-1970s.[2]

The transformation $c = 2\rho^2$ transforms the KM system (2.1) into the Langmuir
lattice

$$c_t - c\left(c^+ - c^-\right) = 0, \tag{2.2}$$

of relevance to the spectrum of Langmuir oscillations in plasmas. Moreover, when
written instead in the form

$$R_t - \frac{1}{2}\left(\exp(-R^+) - \exp(-R^-)\right) = 0, \tag{2.3}$$

---

[1] Quoted in *Return to Mathematical Circles*, Boston: Prindle, Weber & Schmidt, 1988.
[2] A guide to the literature can be found in the detailed notes at the end of this chapter.

2 The Kac–van Moerbeke Hierarchy

it is frequently called the Volterra system. By introducing $\rho = \frac{1}{2}\exp(R/2)$, one verifies that (2.3) reduces to (2.1).

This chapter focuses on the Miura-type transformation between the Toda and KM hierarchies and as an explicit example constructs algebro-geometric KM solutions from algebro-geometric Toda hierarchy solutions. Below we briefly summarize the principal content of each section. A more detailed discussion of the contents has been provided in the introduction to this volume.

Section 2.2.
- Lax pairs (M, Q_{2p+2}), Miura-type transformation
- stationary and time-dependent KM hierarchy
- Burchnall–Chaundy polynomials, hyperelliptic curves

Section 2.3. (stationary)
- properties of ϕ_k and the Baker–Akhiezer function ψ_k, $k = 1, 2$
- theta function representations for ϕ_1, ψ_1, $\phi_{2,\pm}$, $\psi_{2,\pm}$, a_1, b_1, $a_{2,\pm}$, $b_{2,\pm}$, and $\rho_\pm$

Section 2.4. (time-dependent)
- theta function representations for a_1, b_1, $a_{2,\pm}$, $b_{2,\pm}$, and $\rho_\pm$

Due to its close ties with the Toda hierarchy dealt with in Chapter 1, this chapter also relies on terminology and notions developed in connection with compact Riemann surfaces. A brief summary of key results as well as definitions of some of the main quantities can be found in Appendices A and B.

2.2 The KM Hierarchy and its Relation to the Toda Hierarchy

> We can't define anything precisely.
> R. Feynman[1]

In this section we introduce the basic notation for the Kac–van Moerbeke hierarchy and describe how it connects with the Toda hierarchy.

Hypothesis 2.1 *In the stationary case we assume that ρ satisfies*

$$\rho \in \mathbb{C}^{\mathbb{Z}}, \quad \rho(n) \neq 0, \ n \in \mathbb{Z}. \tag{2.4}$$

In the time-dependent case we assume that ρ satisfies

$$\rho(\,\cdot\,, t) \in \mathbb{C}^{\mathbb{Z}}, \quad t \in \mathbb{R}, \quad \rho(n, \,\cdot\,) \in C^1(\mathbb{R}), \quad n \in \mathbb{Z},$$
$$\rho(n, t) \neq 0, \ (n, t) \in \mathbb{Z} \times \mathbb{R}. \tag{2.5}$$

---

[1] *The Feynman Lectures on Physics* (1998), Addison Wesley, Vol. I, 8-2.

2.2 The KM Hierarchy and its Relation to the Toda Hierarchy

We introduce the "even" and "odd" parts of ρ by

$$\rho_e(n) = \rho(2n), \quad \rho_o(n) = \rho(2n+1), \quad n \in \mathbb{Z}, \tag{2.6}$$

and consider the first-order difference expressions[1] in $\ell^\infty(\mathbb{Z})$

$$A = \rho_o S^+ + \rho_e, \quad A^\dagger = \rho_o^- S^- + \rho_e, \tag{2.7}$$

which in turn enable one to define matrix-valued difference expressions M and Q_{2p+2} in $\ell^\infty(\mathbb{Z}) \otimes \mathbb{C}^2$ as follows,

$$M = \begin{pmatrix} 0 & A^\dagger \\ A & 0 \end{pmatrix}, \tag{2.8}$$

$$Q_{2p+2} = \begin{pmatrix} P_{1,2p+2} & 0 \\ 0 & P_{2,2p+2} \end{pmatrix} = P_{1,2p+2} \oplus P_{2,2p+2}, \quad p \in \mathbb{N}_0. \tag{2.9}$$

Here $P_{k,2p+2}$, $k = 1, 2$, are defined as in (1.20) respectively (1.35), that is,

$$P_{k,2p+2} = -L_k^{p+1} + \sum_{j=0}^{p} \left(g_{k,j} + 2a_k f_{k,j} S^+ \right) L_k^{p-j} + f_{k,p+1}, \tag{2.10}$$

$$P_{k,2p+2}\big|_{\ker(L_k - z)} = \left(2a_k F_{k,p}(z) S^+ + G_{k,p+1}(z) \right)\big|_{\ker(L_k - z)}, \tag{2.11}$$

$$L_k = a_k S^+ + a_k^- S^- + b_k, \quad k = 1, 2, \tag{2.12}$$

and $\{f_{k,p,j}\}_{j=0,\dots,p}$ and $\{g_{k,p+1,j}\}_{j=0,\dots,p+1}$, as well as $F_{k,p}(z)$ and $G_{k,p+1}(z)$ are defined as in (1.3)–(1.5) and (1.31) with

$$a_1 = \rho_e \rho_o, \quad b_1 = \rho_e^2 + (\rho_o^-)^2, \tag{2.13}$$

$$a_2 = \rho_e^+ \rho_o, \quad b_2 = \rho_e^2 + \rho_o^2, \tag{2.14}$$

and an equal set of summation constants $\{c_\ell\}_{\ell \in \mathbb{N}}$ employed in $F_{k,p}(z)$ and $G_{k,p+1}(z)$ for $k = 1, 2$.

One then verifies the factorizations

$$L_1 = A^\dagger A, \quad L_2 = AA^\dagger.$$

Moreover, one proves that

$$M^2 = \begin{pmatrix} A^\dagger A & 0 \\ 0 & AA^\dagger \end{pmatrix} = \begin{pmatrix} L_1 & 0 \\ 0 & L_2 \end{pmatrix} = L_1 \oplus L_2 \tag{2.15}$$

and that[2]

$$\ker(M - w) = \ker(M^2 - z) = \ker(L_1 - z) \oplus \ker(L_2 - z), \quad w^2 = z. \tag{2.16}$$

---

[1] When ρ is real-valued, $A^\dagger$ equals the formal adjoint of the difference expression A.
[2] We shall use (2.15) and (2.16) only in an algebraic sense as in (1.35). However, (2.15) and (2.16) are easily seen to be valid in a functional analytic sense as well.

Using

$$[Q_{2p+2}, M] = \begin{pmatrix} 0 & -P_{1,2p+2}A^\dagger + A^\dagger P_{2,2p+2} \\ -P_{2,2p+2}A + AP_{1,2p+2} & 0 \end{pmatrix},$$

and postulating the stationary Lax equation

$$[Q_{2p+2}, M] = 0, \tag{2.17}$$

one concludes after some computations employing (2.6)–(2.16) that

$$\begin{aligned} 2\rho_e \rho_o^2 \big(F_{1,p}^+ - F_{2,p} \big) - \rho_e \big(G_{1,p+1} - G_{2,p+1} \big) &= 0, \\ 2\rho_o (\rho_e^+)^2 \big(F_{1,p}^+ - F_{2,p}^+ \big) - \rho_o \big(G_{1,p+1}^+ - G_{2,p+1} \big) &= 0. \end{aligned} \tag{2.18}$$

Taking $z = 0$ in (2.18) then results in

$$\begin{aligned} 2\rho_o^2 (f_{1,p}^+ - f_{2,p}) - (g_{1,p} + f_{1,p+1} - g_{2,p} - f_{2,p+1}) &= 0, \\ 2(\rho_e^+)^2 (f_{1,p}^+ - f_{2,p}^+) - (g_{1,p}^+ + f_{1,p+1}^+ - g_{2,p} - f_{2,p+1}) &= 0. \end{aligned}$$

Moreover, using the relations

$$\begin{aligned} 2\rho_o^2 (f_{1,\ell}^+ - f_{2,\ell}) - (g_{1,\ell} - g_{2,\ell}) &= 0, \quad \ell \in \mathbb{N}_0, \\ 2\rho_e^2 (f_{1,\ell} - f_{2,\ell}) - (g_{1,\ell} - g_{2,\ell}^-) &= 0, \quad \ell \in \mathbb{N}_0, \end{aligned}$$

equation (2.17) finally reduces to the pair of equations

$$\begin{aligned} -\rho_e (f_{2,p+1} - f_{1,p+1}) &= 0, \\ \rho_o (f_{2,p+1} - f_{1,p+1}^+) &= 0 \end{aligned}$$

or equivalently, to

$$\underline{\text{s-KM}}_p(\rho) = (\text{s-KM}_p(\rho)_e, \text{s-KM}_p(\rho)_o)^\top$$
$$= \begin{pmatrix} -\rho_e (f_{2,p+1} - f_{1,p+1}) \\ \rho_o (f_{2,p+1} - f_{1,p+1}^+) \end{pmatrix} = 0, \quad p \in \mathbb{N}_0. \tag{2.19}$$

Equation (2.19) can be combined into one by introducing

$$\text{s-KM}_p(\rho)(n) = \begin{cases} \text{s-KM}_p(\rho)_e(\frac{n}{2}), & n \text{ even,} \\ \text{s-KM}_p(\rho)_o(\frac{n-1}{2}), & n \text{ odd,} \end{cases} \quad n \in \mathbb{Z}. \tag{2.20}$$

An alternative and recursive approach to arrive at (2.20) proceeds as follows. Introducing

$$\gamma_\ell(2n) = f_{1,\ell}(n), \quad \gamma_\ell(2n+1) = f_{2,\ell}(n), \quad \ell \in \mathbb{N}_0, \ n \in \mathbb{Z}, \tag{2.21}$$

$$\omega_\ell(2n) = g_{1,\ell}(n), \quad \omega_\ell(2n+1) = g_{2,\ell}(n), \quad \ell \in \mathbb{N}_0, \ n \in \mathbb{Z}, \tag{2.22}$$

2.2 The KM Hierarchy and its Relation to the Toda Hierarchy

one notices the recursion (implied by that in (1.3)–(1.5)),

$$\gamma_0 = 1, \quad \omega_0 = -c_1, \tag{2.23}$$

$$2\gamma_{\ell+1} + \omega_\ell + \omega_\ell^{--} - 2(\rho^2 + (\rho^-)^2)\gamma_\ell = 0, \quad \ell \in \mathbb{N}_0, \tag{2.24}$$

$$\omega_{\ell+1} - \omega_{\ell+1}^{--} + 2\big((\rho\rho^+)^2\gamma_\ell^{++} - (\rho^-\rho^{--})^2\gamma_\ell^{--}\big)$$
$$- (\rho^2 + (\rho^-)^2)(\omega_\ell - \omega_\ell^{--}) = 0, \quad \ell \in \mathbb{N}_0. \tag{2.25}$$

Explicitly, one finds

$$\gamma_0 = 1,$$
$$\gamma_1 = \rho^2 + (\rho^-)^2 + c_1, \quad \text{etc.},$$
$$\omega_0 = -c_1,$$
$$\omega_1 = -2(\rho\rho^+)^2 - c_2, \quad \text{etc.}$$

Equation (2.19) finally is then equivalent to

$$\text{s-KM}_p(\rho) = -\rho\big(\gamma_{p+1}^+ - \gamma_{p+1}\big), \quad p \in \mathbb{N}_0. \tag{2.26}$$

Varying $p \in \mathbb{N}_0$ in (2.26) then gives rise to the stationary Kac–van Moerbeke (KM) hierarchy. Thus,

$$\text{s-KM}_0(\rho) = -\rho\left((\rho^+)^2 - (\rho^-)^2\right) = 0,$$
$$\text{s-KM}_1(\rho) = -\rho\left((\rho^+)^4 - (\rho^-)^4 + (\rho^{++})^2(\rho^+)^2 + (\rho^+)^2\rho^2 - \rho^2(\rho^-)^2\right.$$
$$\left. -(\rho^-)^2(\rho^{--})^2\right) + c_1(-\rho)\left((\rho^+)^2 - (\rho^-)^2\right) = 0, \quad \text{etc.},$$

represent the first few equations of the stationary KM hierarchy. By definition, the set of solutions of (2.26), with p ranging in $\mathbb{N}_0$ and $c_\ell \in \mathbb{C}$, $\ell \in \mathbb{N}$, defines the class of algebro-geometric KM solutions.

We can now describe the Burchnall–Chaundy polynomial and the algebraic curve associated with the pair (M, Q_{2p+2}).

Lemma 2.2 *Assume that M and Q_{2p+2} commute. Then,*

$$Q_{2p+2}^2 = \begin{pmatrix} R_{1,2p+2}(L_1) & 0 \\ 0 & R_{2,2p+2}(L_2) \end{pmatrix} = R_{1,2p+2}(L_1) \oplus R_{2,2p+2}(L_2).$$

If in addition L_1 and L_2 are isospectral in the sense that their Burchnall–Chaundy polynomials coincide and then equal

$$\mathcal{F}_p(L_k, P_{k,2p+2}) = P_{k,2p+2}^2 - R_{2p+2}(L_k) = 0, \quad k = 1, 2,$$

$$R_{2p+2}(z) = \prod_{m=0}^{2p+1}(z - E_m), \quad \{E_m\}_{m=0,\ldots,2p+1} \subset \mathbb{C},$$

the Burchnall–Chaundy polynomial $\mathcal{G}_p(M, Q_{2p+2})$ of the pair (M, Q_{2p+2}) is given by

$$\mathcal{G}_p(M, Q_{2p+2}) = Q_{2p+2}^2 - R_{2p+2}(M^2) = 0. \tag{2.27}$$

Proof One infers from (2.9) that

$$Q_{2p+2}^2 = \begin{pmatrix} P_{1,2p+2}^2 & 0 \\ 0 & P_{2,2p+2}^2 \end{pmatrix} = \begin{pmatrix} R_{1,2p+2}(L_1) & 0 \\ 0 & R_{2,2p+2}(L_2) \end{pmatrix}$$

$$= \begin{pmatrix} R_{2p+2}(L_1) & 0 \\ 0 & R_{2p+2}(L_2) \end{pmatrix} = R_{2p+2}(M^2)$$

using (1.40). $\square$

We note that the affine part of the curve associated with the Burchnall–Chaundy polynomial (2.27) of the pair (M, Q_{2p+2}), that is,

$$y^2 = \prod_{m=0}^{2p+1} (w - E_m^{1/2})(w + E_m^{1/2}) = \prod_{m=0}^{2p+1} (w^2 - E_m), \tag{2.28}$$

is singular if and only if $E_m = 0$ for some $m = 0, \ldots, 2p + 1$.

Now we turn to the time-dependent KM hierarchy. For that purpose the coefficients ρ as well as corresponding coefficients a_k and b_k, $k = 1, 2$, are now considered as functions of both the lattice point and time. For each equation in the hierarchy, that is, for each $p \in \mathbb{N}_0$, we now introduce a deformation (time) parameter $t_p \in \mathbb{R}$ in a_k, b_k, ρ, replacing $a_k(n)$, $b_k(n)$, $\rho(n)$ by $a_k(n, t_p)$, $b_k(n, t_p)$, $\rho(n, t_p)$, $k = 1, 2$, etc.

The time-dependent KM equations are then obtained by imposing the Lax commutator equations

$$0 = M_{t_p} - [Q_{2p+2}, M]$$
$$= \begin{pmatrix} 0 & A_{t_p}^\dagger - P_{1,2p+2}A^\dagger + A^\dagger P_{2,2p+2} \\ A_{t_p} - P_{2,2p+2}A + AP_{1,2p+2} & 0 \end{pmatrix}.$$

In analogy to the stationary case one then obtains

$$\begin{aligned} \rho_{e,t_p} - 2\rho_e \rho_o^2 (F_{1,p}^+ - F_{2,p}) + \rho_e (G_{1,p+1} - G_{2,p+1}) &= 0, \\ \rho_{o,t_p} + 2\rho_o (\rho_e^+)^2 (F_{1,p}^+ - F_{2,p}^+) - \rho_o (G_{1,p+1}^+ - G_{2,p+1}) &= 0. \end{aligned} \tag{2.29}$$

Taking again $z = 0$ in (2.29) then results in

$$\underline{\mathrm{KM}}_p(\rho) = (\mathrm{KM}_p(\rho)_e, \mathrm{KM}_p(\rho)_o)^\top$$
$$= \begin{pmatrix} \rho_{e,t_p} - \rho_e(f_{2,p+1} - f_{1,p+1}) \\ \rho_{o,t_p} + \rho_o(f_{2,p+1} - f_{1,p+1}^+) \end{pmatrix} = 0, \quad p \in \mathbb{N}_0. \tag{2.30}$$

2.2 The KM Hierarchy and its Relation to the Toda Hierarchy

As in the stationary case, we may combine the even and odd part of the lattice to obtain

$$\mathrm{KM}_p(\rho)(n) = \begin{cases} \mathrm{KM}_p(\rho)_e(\frac{n}{2}), & n \text{ even,} \\ \mathrm{KM}_p(\rho)_o(\frac{n-1}{2}), & n \text{ odd,} \end{cases} \quad n \in \mathbb{Z}.$$

Moreover, using the recursive approach (2.21)–(2.25) employed in the stationary context, (2.30) become

$$\mathrm{KM}_p(\rho) = \rho_{t_p} - \rho\big(\gamma^+_{p+1} - \gamma_{p+1}\big), \quad p \in \mathbb{N}_0. \tag{2.31}$$

Varying $p \in \mathbb{N}_0$ in (2.26) then gives rise to the time-dependent Kac–van Moerbeke hierarchy. Explicitly,

$$\mathrm{KM}_0(\rho) = \rho_{t_0} - \rho\left((\rho^+)^2 - (\rho^-)^2\right) = 0,$$

$$\mathrm{KM}_1(\rho) = \rho_{t_1} - \rho\Big((\rho^+)^4 - (\rho^-)^4 + (\rho^{++})^2(\rho^+)^2 + (\rho^+)^2\rho^2 - \rho^2(\rho^-)^2$$

$$- (\rho^-)^2(\rho^{--})^2\Big) + c_1(-\rho)\left((\rho^+)^2 - (\rho^-)^2\right) = 0, \text{ etc.,}$$

represent the first few equations of the time-dependent KM hierarchy.

Remark 2.3 In analogy to Remark 1.9 one infers that the coefficients ρ_e and ρ_o enter F_p and G_{p+1} quadratically so that the KM hierarchy (2.31) (as well as the stationary hierarchy (2.26)) is invariant under the substitution

$$\rho \to \rho_\varepsilon = \{\varepsilon(n)\rho(n)\}_{n \in \mathbb{Z}}, \quad \varepsilon(n) \in \{1, -1\}, \; n \in \mathbb{Z}.$$

Next, we will detail the relation between the Toda and the Kac–van Moerbeke hierarchies. We have already seen the connection between $P_{k,2p+2}$ for $k = 1, 2$ and Q_{2p+2} in (2.15).

Introducing the notation familiar from Section 1.5 (cf. (1.195))

$$\widetilde{\mathrm{Tl}}_r(a, b) = (\widetilde{\mathrm{Tl}}_r(a, b)_1, \widetilde{\mathrm{Tl}}_r(a, b)_2)^T = \begin{pmatrix} a_{t_r} - a\big(\tilde{f}^+_{r+1}\big) - \tilde{f}_{r+1}\big) \\ b_{t_r} + \tilde{g}_{r+1} - \tilde{g}^-_{r+1} \end{pmatrix}, \quad r \in \mathbb{N}_0,$$

one verifies the Miura-type identity

$$\widetilde{\mathrm{Tl}}_r(a_k, b_k) = W_k \widetilde{\underline{\mathrm{KM}}}_r(\rho), \quad r \in \mathbb{N}_0, \; k = 1, 2, \tag{2.32}$$

where W_k denote the 2×2 matrix-valued difference expressions

$$W_1 = \begin{pmatrix} \rho_o & \rho_e \\ 2\rho_e & 2\rho_o^- S^- \end{pmatrix}, \quad W_2 = \begin{pmatrix} \rho_o S^+ & \rho_e^+ \\ 2\rho_e & 2\rho_o \end{pmatrix}.$$

Here we employed the same set of summation constants $\{\tilde{c}_s\}_{s \in \mathbb{N}}$ in $\widetilde{F}_r$ and $\widetilde{G}_{r+1}$ and hence in $\widetilde{\mathrm{Tl}}_r(a_k, b_k)$ and $\widetilde{\underline{\mathrm{KM}}}_r(\rho)$ in (2.32) for $k = 1, 2$.

2 The Kac–van Moerbeke Hierarchy

Relation (2.32) is the discrete analog of Miura's well-known identity that links the KdV and mKdV hierarchy (cf. Section 2.2 in Volume I). Of course, (2.32)–(2.33) equally apply to the stationary case, where $a_{k,t_r} = b_{k,t_r} = \rho_{t_r} = 0, k = 1, 2$.

Using that $\widetilde{\mathrm{KM}}_r(\rho) = 0$ if and only if $\underline{\widetilde{\mathrm{KM}}}_r(\rho) = 0$, identity (2.32) immediately yields the implication

$$\widetilde{\mathrm{KM}}_r(\rho) = 0 \implies \widetilde{\mathrm{Tl}}_r(a_k, b_k) = 0, \quad k = 1, 2, \tag{2.33}$$

that is, given a solution ρ of the $\widetilde{\mathrm{KM}}_r$ equation (2.31) (respectively (2.26)), one obtains two solutions, (a_1, b_1) and (a_2, b_2), of the $\widetilde{\mathrm{Tl}}_r$ equations (1.50) related to each other by the Miura-type transformation (2.13), (2.14).

Next we will describe a method to reverse the implication in (2.33), that is, starting from a solution, say (a_1, b_1) of $\widetilde{\mathrm{Tl}}_r(a_1, b_1) = 0$, we construct a solution ρ of the $\widetilde{\mathrm{KM}}_r$ equation (2.31) (respectively (2.26)) and another $\widetilde{\mathrm{Tl}}_r$ solution (a_2, b_2) of (1.50) related to each other by the Miura-type transformation (2.13), (2.14).

We consider the stationary case first.

Theorem 2.4 *Suppose that a, b satisfy (1.61) and the pth stationary Toda system (1.29),*

$$\text{s-Tl}_p(a_1, b_1) = 0$$

for some $p \in \mathbb{N}_0$. In addition, assume the existence of weak solutions $\psi_{1,\pm}$ (not necessarily distinct) of $L_1\psi = 0$ which are nonzero, $\psi_{1,\pm}(n) \neq 0, n \in \mathbb{Z}$. Moreover, introduce for some constant $\sigma \in \mathbb{C}$,

$$\psi_{1,\sigma} = \frac{1}{2}(1-\sigma)\psi_{1,-} + \frac{1}{2}(1+\sigma)\psi_{1,+}, \tag{2.34}$$

$$\rho_{e,\sigma} = (a_1\psi_{1,\sigma}^+/\psi_{1,\sigma})^{1/2}, \tag{2.35}$$

$$\rho_{o,\sigma} = (a_1\psi_{1,\sigma}/\psi_{1,\sigma}^+)^{1/2}, \tag{2.36}$$

$$\rho_\sigma(n) = \begin{cases} \rho_{e,\sigma}(m), & n = 2m, \\ \rho_{o,\sigma}(m), & n = 2m+1, \end{cases} \quad m \in \mathbb{Z}, \tag{2.37}$$

$$a_{2,\sigma} = \rho_{e,\sigma}^+ \rho_{o,\sigma}, \tag{2.38}$$

$$b_{2,\sigma} = \rho_{e,\sigma}^2 + \rho_{o,\sigma}^2, \tag{2.39}$$

and suppose that $\psi_{1,\sigma}(n) \neq 0, n \in \mathbb{Z}$. Then, ρ_σ and $a_{2,\sigma}, b_{2,\sigma}$ satisfy (2.4) and (1.61), respectively, and

$$\text{s-KM}_p(\rho_\sigma) = 0, \quad \text{s-Tl}_p(a_{2,\sigma}, b_{2,\sigma}) = 0.$$

In addition, L_1 and $L_{2,\sigma}$ satisfy (2.12)–(2.14).

Proof This follows from an explicit computation (cf. (2.50)–(2.52)). □

2.2 The KM Hierarchy and its Relation to the Toda Hierarchy

In the time-dependent case one first has to show that one can find a solution of the equation $L_1\psi = z\psi$ with the proper time-dependence. To prepare for this we denote by $c = c(z, n, t_r)$ and $s = s(z, n, t_r)$ a fundamental system of solutions of $L_1\psi = z\psi$ satisfying the initial conditions

$$c(z, 0, t_r) = s(z, 1, t_r) = 1, \quad c(z, 1, t_r) = s(z, 0, t_r) = 0,$$

and where the coefficients a_1 and b_1 of L_1 satisfy $\widetilde{\mathrm{Tl}}_r(a_1, b_1) = 0$ for some $r \in \mathbb{N}_0$. Then the monodromy matrix defined by

$$\Phi(z, n, t_r) = \begin{pmatrix} c(z, n, t_r) & s(z, n, t_r) \\ c(z, n+1, t_r) & s(z, n+1, t_r) \end{pmatrix}, \quad (z, n, t_r) \in \mathbb{C} \times \mathbb{Z} \times \mathbb{R},$$

satisfies

$$(L_1 - z)\left(\frac{d}{dt_r} - \widetilde{P}_{2r+2}\right)\Phi = 0,$$

implying

$$\left(\frac{d}{dt_r} - \widetilde{P}_{2r+2}\right)\Phi(z, n, t_r) = \Phi(z, n, t_r) C_r(z, t_r)$$

for some 2×2 matrix C_r. In particular,

$$\Phi_{t_r} = \widetilde{P}_{2r+2}\Phi + \Phi C_r.$$

Lemma 2.5 *Assume that a_1, b_1 satisfy (1.194) and $\widetilde{\mathrm{Tl}}_r(a_1, b_1) = 0$ for some $r \in \mathbb{N}_0$. Let $\psi_1^{(0)}$ be a weak solution of $L_1\psi_1^{(0)} = z\psi_1^{(0)}$ for $t_r = t_{0,r}$ for some $t_{0,r} \in \mathbb{R}$. Then there is a unique weak solution ψ_1 of*

$$L_1\psi = z\psi, \quad \psi_{t_r} = \widetilde{P}_{2r+2}\psi, \tag{2.40}$$

$\psi_1(n, \cdot) \in C^1(\mathbb{R})$, $n \in \mathbb{Z}$, *such that*

$$\psi_1(\cdot, t_{0,r}) = \psi_1^{(0)}. \tag{2.41}$$

Proof The general solution of $L_1\psi = z\psi$ can be written as

$$\psi(z, n, t_r) = \psi(z, 0, t_r)c(z, n, t_r) + \psi(z, 1, t_r)s(z, n, t_r), \quad (z, n, t_r) \in \mathbb{C} \times \mathbb{Z} \times \mathbb{R},$$

and using (2.40) one concludes that (2.40), (2.41) is equivalent to the linear first-order system

$$\begin{pmatrix} \psi(z, 0, t_r) \\ \psi(z, 1, t_r) \end{pmatrix}_{t_r} = -C_r(z, t_r)\begin{pmatrix} \psi(z, 0, t_r) \\ \psi(z, 1, t_r) \end{pmatrix} \tag{2.42}$$

with initial data

$$\begin{pmatrix} \psi(z, 0, t_{0,r}) \\ \psi(z, 1, t_{0,r}) \end{pmatrix} = \begin{pmatrix} \psi_1^{(0)}(z, 0) \\ \psi_1^{(0)}(z, 1) \end{pmatrix}. \tag{2.43}$$

170   2 The Kac–van Moerbeke Hierarchy

Unique solvability of the first-order system (2.42), (2.43) then completes the proof. □

At this point we can present the key result on how to transfer solutions of the Toda hierarchy to that of the Kac–van Moerbeke hierarchy.

Theorem 2.6 *Let* $r \in \mathbb{N}_0$ *and suppose that* a_1, b_1 *satisfy assumptions* (1.194) *and* $\widetilde{\mathrm{Tl}}_r(a_1, b_1) = 0$ *on* $\mathbb{Z} \times I$, $I \subseteq \mathbb{R}$ *an open interval. In addition, assume the existence of weak solutions* $\psi_{1,\pm}(n, \cdot) \in C^1(I)$, $n \in \mathbb{Z}$ *(not necessarily distinct), of*

$$L_1 \psi_{1,\pm} = 0, \quad \psi_{1,\pm,t_r} = \widetilde{P}_{1,2r+2} \psi_{1,\pm}, \quad t_r \in I,$$

which are nonzero, $\psi_{1,\pm}(n, t_r) \neq 0$, $(n, t_r) \in \mathbb{Z} \times I$. *Moreover, introduce for some* $\sigma \in C^1(I)$,

$$\psi_{1,\sigma} = \frac{1}{2}(1-\sigma)\psi_{1,-} + \frac{1}{2}(1+\sigma)\psi_{1,+}, \tag{2.44}$$

$$\rho_{e,\sigma} = (a_1 \psi_{1,\sigma}^+ / \psi_{1,\sigma})^{1/2}, \tag{2.45}$$

$$\rho_{o,\sigma} = (a_1 \psi_{1,\sigma} / \psi_{1,\sigma}^+)^{1/2}, \tag{2.46}$$

$$\rho_\sigma(n, t_r) = \begin{cases} \rho_{e,\sigma}(m, t_r), & n = 2m, \\ \rho_{o,\sigma}(m, t_r), & n = 2m+1, \end{cases} \quad m \in \mathbb{Z}, \ t_r \in I, \tag{2.47}$$

$$a_{2,\sigma} = \rho_{e,\sigma}^+ \rho_{o,\sigma}, \tag{2.48}$$

$$b_{2,\sigma} = \rho_{e,\sigma}^2 + \rho_{o,\sigma}^2, \tag{2.49}$$

and suppose that $\psi_{1,\sigma}(n, t_r) \neq 0$, $(n, t_r) \in \mathbb{Z} \times I$. *Then* ρ_σ *and* $a_{2,\sigma}, b_{2,\sigma}$ *satisfy* (2.5) *and* (1.194) *on* $\mathbb{Z} \times I$, *respectively, and*

$$\widetilde{\mathrm{KM}}_r(\rho_\sigma) = 0, \quad \widetilde{\mathrm{Tl}}_r(a_{2,\sigma}, b_{2,\sigma}) = 0 \text{ on } \mathbb{Z} \times I,$$

if and only if $\sigma_{t_r} = 0$ *or* $W(\psi_{1,-}, \psi_{1,+}) = 0$.

In addition, L_1 *and* $L_{2,\sigma}$ *satisfy* (2.12)–(2.14).

Proof The existence of the solutions $\psi_\pm$ follows from Lemma 2.5. Let $(n, t_r) \in \mathbb{Z} \times I$. An explicit computation then shows that

$$\mathrm{KM}_r(\rho_\sigma)(2n, t_r) = -\frac{1}{4}\sigma(t_r)_{t_r} \rho_{e,\sigma}(n, t_r)^{-1} \psi_{1,\sigma}(n, t_r)^{-2} W(\psi_{1,-}, \psi_{1,+}), \tag{2.50}$$

$$\mathrm{KM}_r(\rho_\sigma)(2n+1, t_r) = \frac{1}{4}\sigma(t_r)_{t_r} \rho_{o,\sigma}(n, t_r)^{-1} \psi_{1,\sigma}(n+1, t_r)^{-2} W(\psi_{1,-}, \psi_{1,+}), \tag{2.51}$$

2.2 The KM Hierarchy and its Relation to the Toda Hierarchy

and similarly,

$$\widetilde{\text{Tl}}_r(a_{2,\sigma}, b_{2,\sigma}) = \frac{1}{4}\sigma_{t_r} W(\psi_{1,-}, \psi_{1,+}) \begin{pmatrix} (\psi_{1,\sigma}^+)^{-2}\left(-\frac{\rho_{o,\sigma}}{\rho_{e,\sigma}^+} + \frac{\rho_{e,\sigma}^+}{\rho_{o,\sigma}}\right) \\ 2(\psi_{1,\sigma}^{-2} - (\psi_{1,\sigma}^+)^{-2}) \end{pmatrix}. \quad (2.52)$$

$\square$

We emphasize that Theorem 2.6 applies to the transfer of general solutions (not just algebro-geometric ones) from the Toda to the KM hierarchy.

Since the cases $\sigma = \pm 1$ with $\psi_{1,\pm}$ being the branches of the Baker–Akhiezer function associated with the algebro-geometric finite-band operator L_1 will be the most important ones for us in the following, we shall identify $\psi_{1,\pm 1} = \psi_{1,\pm}$, $\rho_{\pm 1} = \rho_\pm$, $a_{2,\pm 1} = a_{2,\pm}$, $b_{2,\pm 1} = b_{2,\pm}$, $L_{2,\pm 1} = L_{2,\pm}$, etc., for notational convenience throughout the remainder of this chapter.

We conclude this section with the following fact:

Remark 2.7 Let $r \in \mathbb{N}_0$, and consider the $(2r+1)$th equation in the Toda hierarchy. Assume that all the even-order constants used in the construction of the Tl_{2r+1} equation vanish, that is, $c_{2j} = 0$, $j = 1, \ldots, r$. Then one has the following identity

$$\text{Tl}_{2r+1}(a, 0) = \text{KM}_r(a), \quad a \in \mathbb{C}^{\mathbb{Z}},$$

where the odd constants for the Toda hierarchy coincide with the corresponding constants in the Kac–van Moerbeke hierarchy. In particular, this implies that *any* solution (algebro-geometric or not) of $\text{KM}_r(a) = 0$ also gives rise to a solution of the type $\text{Tl}_{2r+1}(a, 0) = 0$ for the special Toda hierarchy where all the odd constants vanish. Similarly, starting with *any* solution with b identically zero of the Toda hierarchy where all the odd constants vanish, $\text{Tl}_{2r+1}(a, 0) = 0$, provides a solution of the Kac–van Moerbeke hierarchy, $\text{KM}_r(a) = 0$.

To apply this relation for algebro-geometric solutions, starting with a solution of the special Toda hierarchy, one has to identify quasi-periodic algebro-geometric solutions where b vanishes identically. This happens if and only if both the spectrum and the Dirichlet divisors are symmetric with respect to the reflection $z \mapsto -z$. In particular, for an N-soliton solution (cf. Example 1.47), b vanishes identically if and only if the eigenvalues come in pairs, $\pm E$, and the norming constants associated with each pair are equal. Similarly, starting with an algebro-geometric solution $\text{KM}_r(a) = 0$, we conclude that $(a, 0)$ is an algebro-geometric solution, $\text{Tl}_{2r+1}(a, 0) = 0$.

2.3 The Stationary KM Formalism

There's no sense in being precise when
you don't even know what you're talking about.

John von Neumann[1]

The principal focus of this section is the construction of algebro-geometric finite-band solutions of the stationary Kac–van Moerbeke hierarchy.

We start with some notations. Let a_1, b_1 be the stationary algebro-geometric p-band solutions (1.126), (1.127) and denote the corresponding Dirichlet eigenvalues and divisor by $\{\mu_{1,j}\}_{j=1,...,p}$ and $\mathcal{D}_{\hat{\underline{\mu}}_1}$, etc. We then define L_1, ϕ_1, and ψ_1 as in (1.2), (1.69), (1.72) (resp., (1.120)–(1.124)). Next, introducing[2]

$$Q_0 = (0, y(Q_0)) \in \Pi_+, \tag{2.53}$$

and assuming

$$\psi_1(Q_0, n, n_0) \neq 0, \quad n \in \mathbb{Z},$$

we identify the branches $\psi_{1,\pm}(0, n, n_0)$ of $\psi_1(Q_0, n, n_0)$ (i.e., the restrictions of $\psi_1(Q_0, n, n_0)$ to the upper and lower sheets $\Pi_\pm$ of $\mathcal{K}_p$) with the two solutions $\psi_{1,\pm}(n)$ in Theorem 2.4, that is,

$$\psi_{1,\pm}(n) = \psi_{1,\pm}(0, n, n_0), \quad (n, n_0) \in \mathbb{Z}^2.$$

This then enables one to construct $a_{2,\pm}, b_{2,\pm}$, and $\rho_\pm$ as in Theorem 2.4 in the case $\sigma = \pm 1$ (cf. (2.35)–(2.39)). For convenience of the reader we now list these formulas here, viz.

$$\rho_{e,\pm} = (a_1 \psi_{1,\pm}^+ / \psi_{1,\pm})^{1/2}, \tag{2.54}$$

$$\rho_{o,\pm} = (a_1 \psi_{1,\pm} / \psi_{1,\pm}^+)^{1/2}, \tag{2.55}$$

$$\rho_\pm(n) = \begin{cases} \rho_{e,\pm}(m), & n = 2m, \\ \rho_{o,\pm}(m), & n = 2m+1, \end{cases} \quad m \in \mathbb{Z}, \tag{2.56}$$

$$a_{2,\pm} = \rho_{e,\pm}^+ \rho_{o,\pm}, \tag{2.57}$$

$$b_{2,\pm} = \rho_{e,\pm}^2 + \rho_{o,\pm}^2. \tag{2.58}$$

Given $a_{2,\pm}, b_{2,\pm}$, and $\rho_\pm$ one then defines $L_{2,\pm}, \phi_{2,\pm}, \psi_{2,\pm}, A_\pm, M_\pm$, and $Q_\pm$ as in (1.2), (1.69), (1.72), (2.7)–(2.9). In particular,

$$L_1 = A_\pm^\dagger A_\pm, \quad L_{2,\pm} = A_\pm A_\pm^\dagger.$$

---

[1] Quoted on http://en.wikiquote.org/wiki/John_von_Neumann.
[2] We chose $Q_0 \in \Pi_+$ for notational simplicity only and note that Q_0 is permitted to coincide with a branch point of $\mathcal{K}_p$.

2.3 The Stationary KM Formalism

Next we want to determine the branches of ϕ_1. To this end one defines

$$\phi_{1,\pm} = \rho_{e,\pm}/\rho_{o,\pm},$$

and hence verifies

$$a_1\phi_{1,\pm} + (a_1^-/\phi_{1,\pm}^-) = -b_1.$$

A comparison with the Riccati-type equation (1.75) then yields that

$$\phi_{1,\pm} = \phi_{1,\pm}(0,\cdot)$$

are indeed the branches of $\phi_1(Q_0,\cdot)$. In particular, (1.72) and (1.79) imply

$$\psi_{1,\pm}(n) = \begin{cases} \prod_{m=n_0}^{n-1} \phi_{1,\pm}(m), & n \geq n_0+1, \\ 1, & n = n_0, \\ \prod_{m=n}^{n_0-1} \phi_{1,\pm}(m)^{-1}, & n \leq n_0-1, \end{cases} \quad (n,n_0) \in \mathbb{Z}^2,$$

and

$$L_1\psi_{1,\pm} = 0, \quad \psi_{1,\pm}(n) \neq 0, \ n \in \mathbb{Z},$$

since $\psi_{1,\pm} = \psi_{1,\pm}(0,\cdot,n_0)$ are the branches of $\psi_1(Q_0,\cdot,n_0)$. Next, defining[1]

$$\phi_{2,\pm,\mp} = \rho_{o,\pm}/\rho_{e,\pm}^+, \tag{2.59}$$

and

$$\psi_{2,\pm,\mp}(n) = \begin{cases} \prod_{m=n_0}^{n-1} \phi_{2,\pm,\mp}(m), & n \geq n_0+1, \\ 1, & n = n_0, \\ \prod_{m=n}^{n_0-1} \phi_{2,\pm,\mp}(m)^{-1}, & n \leq n_0-1, \end{cases} \quad (n,n_0) \in \mathbb{Z}^2,$$

one verifies

$$a_{2,\pm}\phi_{2,\pm,\mp} + (a_{2,\pm}^-/\phi_{2,\pm,\mp}^-) = -b_{2,\pm}$$

and

$$L_{2,\pm}\psi_{2,\pm,\mp} = 0.$$

In order to derive the theta function representations for $a_{2,\pm}$, $b_{2,\pm}$, and $\rho_\pm$, we first recall the ones for a_1, b_1, $\phi_1(Q_0)$, and $\psi_1(Q_0)$ as proven in Theorem 1.19. In the context of this section and the following it will be useful to slightly change the

---

[1] A remark concerning our notation: The first $\pm$ relates to the difference expression $L_{2,\pm}$, the second $\pm$ to the upper or lower sheet. Thus $\phi_{2,+,-}$ equals, as we shall see in (2.75), the function $\phi_{2,+}$ associated with $L_{2,+}$ and the lower sheet Π_-. Only the combinations $\phi_{2,+,-}$ and $\phi_{2,-,+}$ occur in (2.59).

notation we employed thus far for $\underline{z}(P, Q)$ as defined in (1.116). We will now use the following notation instead

$$\underline{z}_1(P, n) = \underline{\Xi}_{P_0} - \underline{A}_{P_0}(P) + \underline{\alpha}_{P_0}(\mathcal{D}_{\underline{\hat{\mu}}_1(n)}),$$
$$\underline{z}_1(n) = \underline{z}_1(P_{\infty_+}, n), \quad P \in \mathcal{K}_p, \; n \in \mathbb{Z},$$
(2.60)

and a bit later append it by considering also the branches of $\underline{z}_1(\,\cdot\,, n)$, $n \in \mathbb{Z}$. Given the notation (2.60), the principal results of Theorem 1.19 may be recast in the following form (for $(n, n_0) \in \mathbb{Z}^2$),

$$a_1(n) = \tilde{a}\big(\theta(\underline{z}_1(n+1))\theta(\underline{z}_1(n-1))/\theta(\underline{z}_1(n))^2\big)^{1/2}, \tag{2.61}$$

$$b_1(n) = \frac{1}{2}\sum_{m=0}^{2p+1} E_m - \sum_{j=1}^{p} \lambda_j - \sum_{j=1}^{p} c_j(p)\frac{\partial}{\partial w_j} \ln\left(\frac{\theta(\underline{z}_1(n) + \underline{w})}{\theta(\underline{z}_1(n-1) + \underline{w})}\right)\bigg|_{\underline{w}=0}, \tag{2.62}$$

$$\phi_1(Q_0, n) = \left(\frac{\theta(\underline{z}_1(n-1))}{\theta(\underline{z}_1(n+1))}\right)^{1/2} \frac{\theta(\underline{z}_1(Q_0, n+1))}{\theta(\underline{z}_1(Q_0, n))} \exp\left(\int_{P_0}^{Q_0} \omega^{(3)}_{P_{\infty_+}, P_{\infty_-}}\right), \tag{2.63}$$

$$\psi_1(Q_0, n, n_0) = C_1(n, n_0)\frac{\theta(\underline{z}_1(Q_0, n))}{\theta(\underline{z}_1(Q_0, n_0))} \exp\left((n - n_0)\int_{P_0}^{Q_0} \omega^{(3)}_{P_{\infty_+}, P_{\infty_-}}\right). \tag{2.64}$$

Here $\tilde{a}$ is defined in (1.110) and $C_1(n)$ and $C_1(n, n_0)$ are defined as in (1.122) and (1.124) with $\underline{\hat{\mu}}$ replaced by $\underline{\hat{\mu}}_1$, that is,

$$C_1(n) = C_1(n+1, n) = \left(\frac{\theta(\underline{z}_1(n-1))}{\theta(\underline{z}_1(n+1))}\right)^{1/2}, \tag{2.65}$$

$$C_1(n, n_0) = \begin{cases} \prod_{m=n_0}^{n-1} C_1(m), & n \geq n_0 + 1, \\ 1, & n = n_0, \\ \prod_{m=n}^{n_0-1} C_1(m)^{-1}, & n \leq n_0 - 1, \end{cases}$$

$$= \left(\frac{\theta(\underline{z}_1(n_0))\theta(\underline{z}_1(n_0-1))}{\theta(\underline{z}_1(n))\theta(\underline{z}_1(n-1))}\right)^{1/2}, \quad (n, n_0) \in \mathbb{Z}^2. \tag{2.66}$$

Next, we explicitly need the branches $\phi_{1,\pm}(z, n)$ of $\phi_1(P, n)$, $P = (z, y) \in \mathcal{K}_p$, $n \in \mathbb{Z}$. For this purpose we first introduce the branches associated with $\underline{z}_1(\,\cdot\,, n)$ in (2.60) as follows,

$$\underline{z}_{1,\pm}(z, n) = \underline{\Xi}_{P_0} \mp \underline{A}_{P_0}(P) + \underline{\alpha}_{P_0}(\mathcal{D}_{\underline{\hat{\mu}}_1(n)}), \quad P = (z, y) \in \Pi_+, \tag{2.67}$$
$$\underline{z}_{1,\pm}(n) = \underline{\Xi}_{P_0} \mp \underline{A}_{P_0}(P_{\infty_+}) + \underline{\alpha}_{P_0}(\mathcal{D}_{\underline{\hat{\mu}}_1(n)}), \quad \underline{z}_{1,+}(n) = \underline{z}_1(n), \; n \in \mathbb{Z},$$

2.3 The Stationary KM Formalism

with all paths of integration starting from P_0 and ending at P or P_{∞_+} in (2.67) chosen to lie on Π_+. The branches $\phi_{1,\pm}(z, n)$ of $\phi_1(P, n)$ then read explicitly,

$$\phi_{1,\pm}(z, n) = \left(\frac{\theta(\underline{z}_1(n-1))}{\theta(\underline{z}_1(n+1))}\right)^{1/2} \frac{\theta(\underline{z}_{1,\pm}(z, n+1))}{\theta(\underline{z}_{1,\pm}(z, n))} \exp\left(\pm \int_{E_0}^z \omega_{P_{\infty_+}, P_{\infty_-}}^{(3)+}\right),$$

$$P = (z, y) \in \Pi_+, \ n \in \mathbb{Z}, \quad (2.68)$$

again with all paths of integration starting from P_0 and ending at P or P_{∞_+} in (2.68) chosen to lie on Π_+. (Here $\int_{E_0}^z \omega_{P_{\infty_+}, P_{\infty_-}}^{(3)+}$ is a short-hand for $\int_{P_0}^P \omega_{P_{\infty_+}, P_{\infty_-}}^{(3)}$, $P = (z, y) \in \Pi_+$, etc.)

Together with (2.54), (2.59), and

$$\phi_{1,\pm}(z, n) = \psi_{1,\pm}(z, n+1)/\psi_{1,\pm}(z, n), \quad (z, n) \in \mathbb{C} \times \mathbb{Z},$$

this yields the following theta function representations (for $n \in \mathbb{Z}$),

$$\rho_{e,\pm}(n) = (a_1(n)\phi_{1,\pm}(0, n))^{1/2} \quad (2.69)$$

$$= \left(\tilde{a}\frac{\theta(\underline{z}_1(n-1))\theta(\underline{z}_{1,\pm}(0, n+1))}{\theta(\underline{z}_1(n))\theta(\underline{z}_{1,\pm}(0, n))}\right)^{1/2} \exp\left(\pm \frac{1}{2} \int_{E_0}^0 \omega_{P_{\infty_+}, P_{\infty_-}}^{(3)+}\right),$$

$$\rho_{o,\pm}(n) = (a_1(n)\phi_{1,\pm}(0, n)^{-1})^{1/2} \quad (2.70)$$

$$= \left(\tilde{a}\frac{\theta(\underline{z}_1(n+1))\theta(\underline{z}_{1,\pm}(0, n))}{\theta(\underline{z}_1(n))\theta(\underline{z}_{1,\pm}(0, n+1))}\right)^{1/2} \exp\left(\mp \frac{1}{2} \int_{E_0}^0 \omega_{P_{\infty_+}, P_{\infty_-}}^{(3)+}\right),$$

$$\phi_{2,\pm,\mp}(n) = \rho_{o,\pm}(n)/\rho_{e,\pm}(n+1) \quad (2.71)$$

$$= \left(\frac{\theta(\underline{z}_{1,\pm}(0, n))}{\theta(\underline{z}_{1,\pm}(0, n+2))}\right)^{1/2} \frac{\theta(\underline{z}_1(n+1))}{\theta(\underline{z}_1(n))} \exp\left(\mp \int_{E_0}^0 \omega_{P_{\infty_+}, P_{\infty_-}}^{(3)+}\right).$$

We can also express these functions in terms of the function $\underline{z}_{2,\pm}$ defined by

$$\underline{z}_{2,\pm}(P, n) = \underline{\Xi}_{P_0} - \underline{A}_{P_0}(P) + \underline{\alpha}_{P_0}(\mathcal{D}_{\hat{\underline{\mu}}_{2,\pm}(n)}),$$

$$\underline{z}_{2,\pm}(n) = \underline{z}_{2,\pm}(P_{\infty_+}, n), \quad P \in \mathcal{K}_p, \ n \in \mathbb{Z}, \quad (2.72)$$

where[1]

$$\underline{\alpha}_{P_0}(\mathcal{D}_{\hat{\underline{\mu}}_{2,\pm}(n)}) = \underline{\alpha}_{P_0}(\mathcal{D}_{\hat{\underline{\mu}}_1(n)}) \mp \underline{A}_{P_0}(Q_0) - \underline{A}_{P_0}(P_{\infty_+}), \quad n \in \mathbb{Z}, \quad (2.73)$$

describes the connection[2] between the Dirichlet divisors $\mathcal{D}_{\hat{\underline{\mu}}_{2,\pm}}$ of $L_{2,\pm} = A_\pm A_\pm^\dagger$ and $\mathcal{D}_{\hat{\underline{\mu}}_1}$ of $L_1 = A_\pm^\dagger A_\pm$.

---

[1] Here the choice $Q_0 \in \Pi_+$ has been used.
[2] In the special case where $E_0 = 0$ and hence $P_0 = Q_0$, one obtains $\hat{\mu}_{2,+,j} = \hat{\mu}_{2,-,j}$ for $j = 1, \ldots, p$, and the sign ambiguity in (2.73) vanishes.

2 The Kac–van Moerbeke Hierarchy

The branches of $z_{2,\pm}(P, n)$ are then denoted by

$$\underline{z}_{2,\varepsilon,\varepsilon'}(z, n) = \underline{\Xi}_{P_0} - \varepsilon'\underline{A}_{P_0}(P) + \underline{\alpha}_{P_0}(\mathcal{D}_{\underline{\hat{\mu}}_{2,\varepsilon}(n)}), \quad P = (z, y) \in \Pi_+, \quad (2.74)$$

$$\underline{z}_{2,\varepsilon,\varepsilon'}(n) = \underline{\Xi}_{P_0} - \varepsilon'\underline{A}_{P_0}(P_{\infty_+}) + \underline{\alpha}_{P_0}(\mathcal{D}_{\underline{\hat{\mu}}_{2,\varepsilon}(n)}), \quad \underline{z}_{2,\varepsilon,+}(n) = \underline{z}_{2,\varepsilon}(n),$$

$$\varepsilon, \varepsilon' \in \{+, -\}, \, n \in \mathbb{Z},$$

with all paths of integration starting from P_0 and ending at P or P_{∞_+} in (2.74) chosen to lie on Π_+.

At this point $\phi_{2,+,-}, \phi_{2,-,+}$ in (2.71) are seen to be the branches of the function $\phi_{2,\pm}(Q_0, n)$, where the meromorphic function $\phi_{2,\pm}(\,\cdot\,, n)$ on $\mathcal{K}_p$ is defined by

$$\phi_{2,\pm}(P, n) = \left(\frac{\theta(\underline{z}_{2,\pm}(n-1))}{\theta(\underline{z}_{2,\pm}(n+1))}\right)^{1/2} \frac{\theta(\underline{z}_{2,\pm}(P, n+1))}{\theta(\underline{z}_{2,\pm}(P, n))} \exp\left(\int_{P_0}^{P} \omega_{P_{\infty_+}, P_{\infty_-}}^{(3)}\right),$$

$$P = (z, y) \in \mathcal{K}_p \, n \in \mathbb{Z}, \quad (2.75)$$

by noticing that

$$\underline{z}_{2,\pm,\mp}(0, n) = \underline{z}_{1,+}(n) = \underline{z}_1(n), \quad z_{2,\pm,+}(n-1) = \underline{z}_{2,\pm}(n-1) = \underline{z}_{1,\pm}(0, n). \quad (2.76)$$

The relations in (2.76) follow by combining (1.125), (2.53), (2.60), (2.67), (2.72), (2.73), and (2.74). In particular, we may rewrite and extend (2.71) in the form[1]

$$\phi_{2,\varepsilon,\varepsilon'}(n) = \left(\frac{\theta(\underline{z}_{2,\varepsilon}(n-1))}{\theta(\underline{z}_{2,\varepsilon}(n+1))}\right)^{1/2} \frac{\theta(\underline{z}_{2,\varepsilon,-\varepsilon'}(n+1))}{\theta(\underline{z}_{2,\varepsilon,-\varepsilon'}(n))} \exp\left(\varepsilon' \int_{E_0}^{0} \omega_{P_{\infty_+}, P_{\infty_-}}^{(3)+}\right),$$

$$\varepsilon, \varepsilon' \in \{+, -\}, \, n \in \mathbb{Z}. \quad (2.77)$$

The divisor of $\phi_{2,\pm}(\,\cdot\,, n)$ thus reads

$$(\phi_{2,\pm}(\,\cdot\,, n)) = \mathcal{D}_{\underline{\hat{\mu}}_{2,\pm}(n+1)} - \mathcal{D}_{\underline{\hat{\mu}}_{2,\pm}(n)} + \mathcal{D}_{P_{\infty_+}} - \mathcal{D}_{P_{\infty_-}}, \quad n \in \mathbb{Z},$$

in analogy to that of $\phi_1(\,\cdot\,, n)$ (cf. (1.71))

$$(\phi_1(\,\cdot\,, n)) = \mathcal{D}_{\underline{\hat{\mu}}_1(n+1)} - \mathcal{D}_{\underline{\hat{\mu}}_1(n)} + \mathcal{D}_{P_{\infty_+}} - \mathcal{D}_{P_{\infty_-}}, \quad n \in \mathbb{Z}. \quad (2.78)$$

Given (2.77) (resp., (2.69)–(2.78)) one can now express $a_{2,\pm}, b_{2,\pm}$, and $\rho_\pm$ in terms of theta functions as follows.

Theorem 2.8 *Let $(n, n_0) \in \mathbb{Z}^2$ and suppose that a_1, b_1 satisfy (1.61) and the pth stationary Toda system (1.29) associated with the hyperelliptic curve $\mathcal{K}_p$ such that (1.62) and (1.63) are satisfied. Then for each $n \in \mathbb{Z}$, $\mathcal{D}_{\underline{\hat{\mu}}_1(n)}$ is nonspecial and a_1 and b_1 are given by (2.61) and (2.62). Given the associated quantities L_1, ϕ_1, and ψ_1 according to (1.2), (1.69), (1.72) (resp., (1.120)–(1.124)) one constructs $a_{2,\pm}, b_{2,\pm}$,*

---

[1] Here any combination of $\varepsilon, \varepsilon' \in \{+, -\}$ is permitted.

2.3 The Stationary KM Formalism

and $\rho_\pm$ according to (2.54)–(2.58), and hence $L_{2,\pm}$, $\phi_{2,\pm}$, $\psi_{2,\pm}$, $A_\pm$, $M_\pm$, and $Q_\pm$ as in (1.2), (1.69), (1.31), (2.7)–(2.9) such that $L_1 = A_\pm^\dagger A_\pm$ and $L_{2,\pm} = A_\pm A_\pm^\dagger$. Then $\rho_\pm$ and $a_{2,\pm}$, $b_{2,\pm}$ are of the form

$$\rho_{e,\pm}(n) = \left(\tilde{a}\frac{\theta(\underline{z}_1(n-1))\theta(\underline{z}_{1,\pm}(0,n+1))}{\theta(\underline{z}_1(n))\theta(\underline{z}_{1,\pm}(0,n))}\right)^{1/2} \exp\left(\pm\frac{1}{2}\int_{E_0}^0 \omega_{P_{\infty+},P_{\infty-}}^{(3)+}\right)$$

$$= \left(\tilde{a}\frac{\theta(\underline{z}_{2,\pm}(n))\theta(\underline{z}_{2,\pm,\mp}(0,n-1))}{\theta(\underline{z}_{2,\pm}(n-1))\theta(\underline{z}_{2,\pm,\mp}(0,n))}\right)^{1/2} \exp\left(\pm\frac{1}{2}\int_{E_0}^0 \omega_{P_{\infty+},P_{\infty-}}^{(3)+}\right), \tag{2.79}$$

$$\rho_{o,\pm}(n) = \left(\tilde{a}\frac{\theta(\underline{z}_1(n+1))\theta(\underline{z}_{1,\pm}(0,n))}{\theta(\underline{z}_1(n))\theta(\underline{z}_{1,\pm}(0,n+1))}\right)^{1/2} \exp\left(\mp\frac{1}{2}\int_{E_0}^0 \omega_{P_{\infty+},P_{\infty-}}^{(3)+}\right)$$

$$= \left(\tilde{a}\frac{\theta(\underline{z}_{2,\pm}(n-1))\theta(\underline{z}_{2,\pm,\mp}(0,n+1))}{\theta(\underline{z}_{2,\pm}(n))\theta(\underline{z}_{2,\pm,\mp}(0,n))}\right)^{1/2} \exp\left(\mp\frac{1}{2}\int_{E_0}^0 \omega_{P_{\infty+},P_{\infty-}}^{(3)+}\right), \tag{2.80}$$

$$\rho_\pm(n) = \begin{cases} \rho_{e,\pm}(m), & n=2m, \\ \rho_{o,\pm}(m), & n=2m+1, \end{cases} \quad m \in \mathbb{Z}, \tag{2.81}$$

$$a_{2,\pm}(n) = \tilde{a}\big(\theta(\underline{z}_{2,\pm}(n+1))\theta(\underline{z}_{2,\pm}(n-1))/\theta(\underline{z}_{2,\pm}(n))^2\big)^{1/2}, \tag{2.82}$$

$$b_{2,\pm}(n) = \frac{1}{2}\sum_{m=0}^{2p+1} E_m - \sum_{j=1}^p \lambda_j - \sum_{j=1}^p c_j(p)\frac{\partial}{\partial w_j}\ln\left(\frac{\theta(\underline{z}_{2,\pm}(n)+\underline{w})}{\theta(\underline{z}_{2,\pm}(n-1)+\underline{w})}\right)\bigg|_{\underline{w}=0}. \tag{2.83}$$

Moreover, $L_{2,\pm}$ are isospectral to L_1 in the sense that they correspond to the same hyperelliptic curve $\mathcal{K}_p$ and the corresponding Dirichlet divisor $\mathcal{D}_{\hat{\underline{\mu}}_{2,\pm}(n)}$ is nonspecial for each $n \in \mathbb{Z}$ and satisfies

$$\underline{\alpha}_{P_0}(\mathcal{D}_{\hat{\underline{\mu}}_{2,\pm}(n)}) = \underline{\alpha}_{P_0}(\mathcal{D}_{\hat{\underline{\mu}}_1(n)}) \mp \underline{A}_{P_0}(Q_0) - \underline{A}_{P_0}(P_{\infty+}) \tag{2.84}$$

$$= \underline{\alpha}_{P_0}(\mathcal{D}_{\hat{\underline{\mu}}_1(n_0)}) - 2(n-n_0)\underline{A}_{P_0}(P_{\infty+}) \mp \underline{A}_{P_0}(Q_0) - \underline{A}_{P_0}(P_{\infty+}).$$

Finally, a_1, b_1 and $a_{2,\pm}$, $b_{2,\pm}$ are related via the Miura-type transformation

$$a_1 = \rho_{e,\pm}\rho_{o,\pm}, \quad b_1 = \rho_{e,\pm}^2 + (\rho_{o,\pm}^-)^2, \tag{2.85}$$

$$a_{2,\pm} = \rho_{e,\pm}^+\rho_{o,\pm}, \quad b_{2,\pm} = \rho_{e,\pm}^2 + \rho_{o,\pm}^2. \tag{2.86}$$

Proof To prove (2.79) and (2.80) it suffices to combine (2.56), (2.69), (2.70), and (2.76). Equation (2.82) is clear from (2.57), (2.69), (2.70), and (2.76). Equation (2.84) directly follows from (2.73) and (2.66) and (2.83) can be derived from an

expansion of $\phi_{2,\pm}(P, n)$ near $P = P_{\infty_+}$ exactly as in the proof of Theorem 1.19. The Miura transformation (2.85), (2.86) is clear from (2.54), (2.55), (2.57), and (2.58).
□

Remark 2.9 Equations (2.73), respectively (2.84), illustrate the effect of commutation, that is, the transformation $L_1 = A_\pm^\dagger A_\pm \to L_{2,\pm} = A_\pm A_\pm^\dagger$, as translations by $-\underline{A}_{P_0}(Q_0) - \underline{A}_{P_0}(P_{\infty_+})$ on the Jacobi variety $J(\mathcal{K}_p)$.

We conclude this section with the trivial case $p = 0$ excluded thus far.

Example 2.10 Assume $p = 0$, $n \in \mathbb{Z}$, and
$$a_1(n) = a \in \mathbb{C} \setminus \{0\}, \quad b_1(n) = b \in \mathbb{C}$$
(cf. Example 1.23). Then one computes,
$$\rho_{e,\pm}(n) = 2^{-1/2}\big(b \pm (b^2 - 4a^2)^{1/2}\big)^{1/2},$$
$$\rho_{o,\pm}(n) = \rho_{e,\mp}(n),$$
$$\rho_\pm(n) = \begin{cases} 2^{-1/2}\big(b \pm (b^2 - 4a^2)^{1/2}\big)^{1/2}, & n = 2m, \\ 2^{-1/2}\big(b \mp (b^2 - 4a^2)^{1/2}\big)^{1/2}, & n = 2m + 1, \end{cases} \quad m \in \mathbb{Z},$$
$$a_{2,\pm}(n) = a, \quad b_{2,\pm}(n) = b,$$
$$\text{s-KM}_k(\rho_\pm) = 0, \ k \in \mathbb{N}_0.$$

2.4 The Time-Dependent KM Formalism

Many a little makes a mickle.
Scottish proverb

Finally, we describe how to transfer solutions of the time-dependent algebro-geometric initial value problem for the Toda hierarchy, derived in detail in Section 1.5, to solutions of the time-dependent algebro-geometric initial value problem for the KM hierarchy using the principal transfer result, Theorem 2.6.

First we recall that the basic problem in the analysis of algebro-geometric solutions of the Toda hierarchy consists of solving the time-dependent rth Toda flow with initial data a stationary solution of the pth equation in the hierarchy. More precisely, one is seeking solutions a_1, b_1 of the time-dependent algebro-geometric initial value problem for the Toda hierarchy

$$\widetilde{\text{Tl}}_r(a_1, b_1) = 0, \quad (a_1, b_1)\big|_{t_r = t_{0,r}} = \big(a_1^{(0)}, b_1^{(0)}\big), \tag{2.87}$$
$$\text{s-Tl}_p\big(a_1^{(0)}, b_1^{(0)}\big) = 0 \tag{2.88}$$

2.4 The Time-Dependent KM Formalism

for some $t_{0,r} \in \mathbb{R}$, $p, r \in \mathbb{N}_0$, where $a_1 = a_1(n, t_r)$, $b_1 = b_1(n, t_r)$ satisfy (1.194) and a fixed curve $\mathcal{K}_p$ is associated with the stationary solutions $a_1^{(0)}, b_1^{(0)}$ in (2.88).

Given the solution a_1, b_1 of (2.87), (2.88) we intend to construct solutions $\rho_\pm$ of the time-dependent algebro-geometric initial value problem for the KM hierarchy satisfying (2.5) and solutions $a_{2,\pm}, b_{2,\pm}$ of the time-dependent algebro-geometric initial value problem for the Toda hierarchy satisfying (1.194), that is,

$$\widetilde{\mathrm{KM}}_r(\rho_\pm) = 0, \quad \rho_\pm\big|_{t_r=t_{0,r}} = \rho_\pm^{(0)}, \qquad (2.89)$$

$$\text{s-KM}_p\left(\rho_\pm^{(0)}\right) = 0,$$

and

$$\widetilde{\mathrm{Tl}}_r(a_{2,\pm}, b_{2,\pm}) = 0, \quad (a_{2,\pm}, b_{2,\pm})\big|_{t_r=t_{0,r}} = \left(a_{2,\pm}^{(0)}, b_{2,\pm}^{(0)}\right),$$

$$\text{s-Tl}_p\left(a_{2,\pm}^{(0)}, b_{2,\pm}^{(0)}\right) = 0, \qquad (2.90)$$

related to each other by the Miura-type transformation (2.13), (2.14). Moreover, the solutions $a_{2,\pm}, b_{2,\pm}$ are isospectral to a_1, b_1 in the sense that they are associated with the same underlying curve $\mathcal{K}_p$ as a_1, b_1.

Associated with a_1, b_1 satisfying (2.87), (2.88) we introduce L_1, ϕ_1, and ψ_1 as in (1.2), (1.213), and (1.216) (resp., (1.252)–(1.256)). Since the time-dependent Baker–Akhiezer function ψ_1 satisfies (1.223) and (1.224) one can now apply Theorem 2.6 by identifying $\psi_{1,\pm}$ in Theorem 2.6 with the branches of the Baker–Akhiezer function ψ_1 as follows. Introducing $Q_0 = (0, y(Q_0)) \in \Pi_+$ as in (2.53) and assuming

$$\psi_1(Q_0, n, n_0, t_r, t_{0,r}) \neq 0, \quad (n, t_r) \in \mathbb{Z} \times I_0,$$

where I_0 is an open interval with $t_{0,r} \in I_0$, one identifies the branches $\psi_{1,\pm}(0, n, n_0, t_r, t_{0,r})$ of $\psi_1(Q_0, n, n_0, t_r, t_{0,r})$ with the two solutions $\psi_{1,\pm}(n, t_r)$ in Theorem 2.4, that is,

$$\psi_{1,\pm}(n, t_r) = \psi_{1,\pm}(0, n, n_0, t_r, t_{0,r}), \quad (n, n_0, t_r, t_{0,r}) \in \mathbb{Z}^2 \times I_0^2.$$

This then enables one to construct $a_{2,\pm}, b_{2,\pm}$, and $\rho_\pm$ according to (2.29)–(2.49), choosing $\sigma = \pm 1$. In particular, one verifies (2.89)–(2.90) on $\mathbb{Z} \times I_0$ as a consequence of Theorems 2.4 (for the Miura-type identities satisfied at $t_r = t_{0,r}$) and 2.6.

It remains to construct the theta function representations of $a_1, b_1, a_{2,\pm}, b_{2,\pm}$, and $\rho_\pm$. Given the results in Theorem 1.41 one can now proceed as in the stationary section replacing $\underline{z}_{1,\varepsilon}(z, n)$, $\underline{z}_{2,\varepsilon,\varepsilon'}(z, n)$ by $\underline{z}_{1,\varepsilon}(z, n, t_r)$, $\underline{z}_{2,\varepsilon,\varepsilon'}(z, n, t_r)$, which in turn is accomplished by replacing $\mathcal{D}_{\underline{\hat{\mu}}_1(n)}$ by $\mathcal{D}_{\underline{\hat{\mu}}_1(n,t_r)}$ and $\mathcal{D}_{\underline{\hat{\mu}}_{2,\varepsilon}(n)}$ by $\mathcal{D}_{\underline{\hat{\mu}}_{2,\varepsilon}(n,t_r)}$, $(n, t_r) \in \mathbb{Z} \times I_0$. In particular, $\mathcal{D}_{\underline{\hat{\mu}}_1(n,t_r)}$ and $\mathcal{D}_{\underline{\hat{\mu}}_{2,\varepsilon}(n,t_r)}$ are nonspecial for all $(n, t_r) \in$

$\mathbb{Z} \times I_0$ and (cf. (1.113), (1.125), (1.250), (1.259), (1.259))

$$\underline{\alpha}_{P_0}(\mathcal{D}_{\underline{\hat{\mu}}_{2,\pm}(n,t)}) = \underline{\alpha}_{P_0}(\mathcal{D}_{\underline{\hat{\mu}}_1(n,t_r)}) \mp \underline{A}_{P_0}(Q_0) - \underline{A}_{P_0}(P_{\infty_+})$$
$$= \underline{\alpha}_{P_0}(\mathcal{D}_{\underline{\hat{\mu}}_1(n_0,t_{0,r})}) - (n - n_0)\underline{U}^{(3)} - (t_r - t_{0,r})\underline{\widetilde{U}}_r^{(2)} \mp \underline{A}_{P_0}(Q_0) - \underline{A}_{P_0}(P_{\infty_+}).$$

We now summarize these findings in the final result of this section.

Theorem 2.11 *Let $(n, t_r) \in \mathbb{Z} \times I_0$. Then the theta function representations of the solutions $a_1, b_1, \rho_\pm$, and $a_{2,\pm}, b_{2,\pm}$ of the algebro-geometric initial value problems (2.87)–(2.90) on $\mathbb{Z} \times I_0$ are given by*

$$a_1(n, t_r) = \tilde{a} \left(\frac{\theta(\underline{z}_1(n+1, t_r))\theta(\underline{z}_1(n-1, t_r))}{\theta(\underline{z}_1(n, t_r))^2} \right)^{1/2},$$

$$b_1(n, t_r) = \frac{1}{2} \sum_{m=0}^{2p+1} E_m - \sum_{j=1}^{p} \lambda_j$$
$$- \sum_{j=1}^{p} c_j(p) \frac{\partial}{\partial w_j} \ln \left(\frac{\theta(\underline{z}_1(n, t_r) + \underline{w})}{\theta(\underline{z}_1(n-1, t_r) + \underline{w})} \right) \bigg|_{\underline{w}=\underline{0}},$$

$$\rho_{e,\pm}(n, t_r) = \left(\tilde{a} \frac{\theta(\underline{z}_1(n-1, t_r))\theta(\underline{z}_{1,\pm}(0, n+1, t_r))}{\theta(\underline{z}_1(n, t_r))\theta(\underline{z}_{1,\pm}(0, n, t_r))} \right)^{1/2}$$
$$\times \exp\left(\pm \frac{1}{2} \int_{E_0}^{0} \omega_{P_{\infty_+}, P_{\infty_-}}^{(3)+} \right)$$
$$= \left(\tilde{a} \frac{\theta(\underline{z}_{2,\pm}(n, t_r))\theta(\underline{z}_{2,\pm,\mp}(0, n-1, t_r))}{\theta(\underline{z}_{2,\pm}(n-1, t_r))\theta(\underline{z}_{2,\pm,\mp}(0, n, t_r))} \right)^{1/2}$$
$$\times \exp\left(\pm \frac{1}{2} \int_{E_0}^{0} \omega_{P_{\infty_+}, P_{\infty_-}}^{(3)+} \right),$$

$$\rho_{o,\pm}(n, t_r) = \left(\tilde{a} \frac{\theta(\underline{z}_1(n+1, t_r))\theta(\underline{z}_{1,\pm}(0, n, t_r))}{\theta(\underline{z}_1(n, t_r))\theta(\underline{z}_{1,\pm}(0, n+1, t_r))} \right)^{1/2}$$
$$\times \exp\left(\mp \frac{1}{2} \int_{E_0}^{0} \omega_{P_{\infty_+}, P_{\infty_-}}^{(3)+} \right)$$
$$= \left(\tilde{a} \frac{\theta(\underline{z}_{2,\pm}(n-1, t_r))\theta(\underline{z}_{2,\pm,\mp}(0, n+1, t_r))}{\theta(\underline{z}_{2,\pm}(n, t_r))\theta(\underline{z}_{2,\pm,\mp}(0, n, t_r))} \right)^{1/2}$$
$$\times \exp\left(\mp \frac{1}{2} \int_{E_0}^{0} \omega_{P_{\infty_+}, P_{\infty_-}}^{(3)+} \right),$$

$$\rho_{\pm}(n, t_r) = \begin{cases} \rho_{e,\pm}(m, t_r), & n = 2m, \\ \rho_{o,\pm}(m, t_r), & n = 2m + 1, \end{cases} \quad m \in \mathbb{Z},$$

$$a_{2,\pm}(n, t_r) = \tilde{a} \left(\frac{\theta(\underline{z}_{2,\pm}(n+1, t_r))\theta(\underline{z}_{2,\pm}(n-1, t_r))}{\theta(\underline{z}_{2,\pm}(n, t_r))^2} \right)^{1/2},$$

$$b_{2,\pm}(n, t_r) = \frac{1}{2} \sum_{m=0}^{2p+1} E_m - \sum_{j=1}^{p} \lambda_j$$

$$- \sum_{j=1}^{p} c_j(p) \frac{\partial}{\partial w_j} \ln \left(\frac{\theta(\underline{z}_{2,\pm}(n, t_r) + \underline{w})}{\theta(\underline{z}_{2,\pm}(n-1, t_r) + \underline{w})} \right) \Bigg|_{\underline{w}=0}.$$

In addition, a_1, b_1 and $a_{2,\pm}$, $b_{2,\pm}$ are related via the Miura-type transformation on $\mathbb{Z} \times I_0$,

$$a_1 = \rho_{e,\pm}\rho_{o,\pm}, \quad b_1 = \rho_{e,\pm}^2 + (\rho_{o,\pm}^-)^2,$$
$$a_{2,\pm} = \rho_{e,\pm}^+\rho_{o,\pm}, \quad b_{2,\pm} = \rho_{e,\pm}^2 + \rho_{o,\pm}^2.$$

2.5 Notes

Who controls the past controls the future.
Who controls the present controls that past.

George Orwell[1]

The material of this chapter is primarily taken from Gesztesy et al. (1993) and Bulla et al. (1998).

Section 2.1. Complete integrability and a Lax pair for what we call the Kac–van Moerbeke (KM) system, seems to have been found independently and nearly at the same time by Kac and van Moerbeke (1975b–d), and Manakov (1975). More precisely, Kac and van Moerbeke considered the finite nonperiodic, the semi-infinite, and the periodic cases, while Manakov studied the doubly infinite system on $\mathbb{Z}$. Soon after, isospectral deformations of the finite nonperiodic system were also discussed in Moser (1975b), who also notes in a footnote that the Kac–van Moerbeke equation, apparently, was known to Hénon in 1973. The first equation in the KM hierarchy had been isolated by Zakharov et al. (1974) in their study of the spectrum of Langmuir waves, and hence it is also called the Langmuir lattice (cf. Novikov et al. (1984, p. 45)). However, the same equation arises in various different circumstances: For instance, it is called the Volterra lattice (in connection with a predator–prey model) and intimately related to a nonlinear network system as discussed in Hirota (1975),

[1] *Nineteen Eighty-four* (1949), London: Secker & Warburg, Book One, Chapter 3.

Hirota and Satsuma (1976a,b), Ladik and Chiu (1977), and Wadati (1976). Moreover, it is also called the KdV difference equation in Dubrovin et al. (1976) (or simply the discrete KdV equation in Sharipov (1990)).

The KM equation, originally, was introduced in the form

$$x_t(n,t) = \frac{1}{2}\left(e^{-x(n-1,t)} - e^{-x(n+1,t)}\right), \quad (n,t) \in \mathbb{Z} \times \mathbb{R}. \tag{2.91}$$

The transformation

$$\rho(n,t) = \frac{1}{2}e^{-\frac{1}{2}x(n,t)}, \quad (n,t) \in \mathbb{Z} \times \mathbb{R},$$

then puts (2.91) into the form used in this monograph

$$\rho_t(n,t) = \rho(n,t)\left(\rho(n+1,t)^2 - \rho(n-1,t)^2\right), \quad (n,t) \in \mathbb{Z} \times \mathbb{R}.$$

Analogously to our treatment of the Toda lattice, the finite nonperiodic KM lattice and its various extensions are not discussed in this volume. For extensions of the finite nonperiodic (one and two-dimensional) KM lattice connected with root systems of Lie algebras we refer, for instance, to Leznov and Saveliev (1992, Sect. 4.2) and the references therein.

Section 2.2. The approach presented in this section is essentially modeled after Gesztesy et al. (1993) (see also Bulla et al. (1998), Teschl (1999b), and Chapter 14 in the monograph Teschl (2000)).

The supersymmetric approach described in (2.8)–(2.15) demonstrates that the Kac–van Moerbeke (KM) hierarchy is a modified Toda hierarchy precisely in the manner the Drinfeld–Sokolov (DS) hierarchy is a modified version of the Gel'fand–Dickey (GD) hierarchy. The latter represents an extension of the well-known connection between the modified Korteweg–de Vries (mKdV) hierarchy and the Korteweg–de Vries (KdV) hierarchy. The connection between all these hierarchies and their modified counterparts is based on (suitable generalizations of) Miura-type transformations which in turn are based on factorizations of differential (respectively, difference) expressions. The literature on this subject is too extensive to be quoted here in full. The interested reader can consult, for instance, Gesztesy (1989; 1992), Gesztesy et al. (1991; 1994), Gesztesy and Svirsky (1995), Toda (1989b, Ch. 3), Wadati (1976), and the references cited therein.

In the present case of the Toda and modified Toda, respectively, Kac–van Moerbeke hierarchies, a connection between the KM and Toda systems was known to Hénon in 1973 according to a footnote in Moser (1975b). The discrete analog of the Miura transformation (2.13), (2.14) is due to Kac and van Moerbeke (1975b). For a detailed discussion see Toda (1989b, Sect. 3.8), Toda (1989a, Ch. 20). The connection between the discrete Miura transformation and factorization methods of the Lax

2.5 Notes

operator was first systematically employed by Adler (1981) and further developed in Gesztesy et al. (1993) (where further details on the Tl and KM systems can be found). Transformations between KM and Toda lattices were also investigated by Wadati (1976). In the half-line case a detailed discussion of this relationship can be found in Peherstorfer (2001).

For a functional analytic treatment in connection with (2.15) and (2.16) we refer to Gesztesy et al. (1991, Theorems 2.1 and 2.3).

The ω_ℓ-recursion relation in (2.23)–(2.25) can be found in Teschl (1999b) and in Section 14.1 of the monograph Teschl (2000).

Theorem 2.4 was first proved in the case $r = 0$ in Gesztesy et al. (1993) and extended to the general case $r \in \mathbb{N}_0$ by Teschl (1999b).

The connection between the Toda and KM systems alluded to in Remark 2.7 appears to be due to Kac and van Moerbeke (1975b,c), and Manakov (1975) who seem to have found this independently and nearly at the same time. This relation is also mentioned in Dubrovin et al. (1976, p. 70), Novikov et al. (1984, Sect. I.7), Perelomov (1990, p. 207), and in Toda (1989b, Sect. 3.8). The case $b = 0$ for algebrogeometric Toda lattice solutions is also discussed in Kac and van Moerbeke (1975c) and Dubrovin et al. (1976, pp. 109, 111), Dubrovin et al. (1990, p. 269) (see also Veselov (1991)). For details and additional facts in connection with Remark 2.7, especially, the extension of that connection to the whole KM hierarchy, we refer to Michor and Teschl (2007a).

Bäcklund transformations for the Toda hierarchy and connections with the KM system based on factorization techniques were also studied by Knill (1993a,c). Miura transformations for integrable extensions of the KM equation due to Bogoyavlensky (1988; 1990), Bogoyavlenskii (1991) are studied in Inoue and Wadati (1997).

The following references are included to provide the reader with a glimpse of the large amount of material available that could not be treated in this volume.

The integration of the KM equation on the half-line using a moment problem approach was undertaken by Kac and van Moerbeke (1975b), Khanmamedov (2005a), Peherstorfer (2001), and Yamazaki (1987; 1989; 1990; 1992). The integration of certain nonabelian KM lattices on the half-line was studied by Osipov (1997). Connections with the Hamburger moment problem and τ-functions are discussed in Nakamura and Kodama (1995).

The bi-Hamiltonian formalism and the continuum limit of the KM hierarchy are studied in Zeng and Rauch-Wojciechowski (1995a), the multi-Hamiltonian structure is discussed in Agrotis and Damianou (2007). Conservation laws of the KM and nonlinear self-dual network equation are derived in Wadati and Watanabe (1977). Bäcklund transformations between Toda and Volterra lattices on half-lines are discussed in Rolanía and Heredero (2006) and Vekslerchik (2005). A link between the KM and lattice sinh-Gordon model is established in Volkov (1988).

The integrability of a generalized KM lattice with periodic boundary conditions is studied in Inoue (2003; 2004).

For an alternative approach to the Volterra hierarchy in terms of τ-functions we refer to Vekslerchik (2005).

Section 2.3. The material in this section is taken from Bulla et al. (1998, Sects. 5, 9).

Various aspects of the periodic KM lattice (in the real-valued case) were discussed in the following references: The algebro-geometric approach can be found in Dai and Geng (2003), Dubrovin et al. (1976), Kac and van Moerbeke (1975c), and Sharipov (1990) (the latter reference, actually, considers integrable extensions of the KM equation due to Bogoyavlensky (1988; 1990), Bogoyavlenskii (1991)). Algebro-geometric Poisson brackets, etc., are studied in Fernandes and Vanhaecke (2001), Penskoi (1998a), and Veselov and Penskoi (1998; 1999). Isospectral sets of Jacobi matrices with zero diagonal entries and gradient flows for the KM lattice were considered by Penskoi (1998b; 2007; 2008).

In the important special case where $a > 0$ and b is real-valued, and all branch points of the underlying hyperelliptic Toda curve $\mathcal{K}_p$ are in real position and ordered according to

$$0 \leq E_0 < E_1 < \cdots < E_{2p+1},$$

all p-gap sequences (a, b) associated with the corresponding Jacobi operator $\check{L}$ are parametrized by the initial conditions

$$\{(\mu_j(n_0), \sigma_j(n_0))\}_{j=1,\ldots,p}, \ \sigma_j(n_0) = \pm \text{ for } \hat{\mu}_j(n_0) \in \Pi_\pm, j = 1, \ldots, p. \quad (2.92)$$

Here one omits $\sigma_j(n_0)$ in the special case where $\mu_j(n_0) \in \{E_{2j-1}, E_{2j}\}$. With this restriction in mind, (2.92) represents the product of p circles S^1 when varying $\mu_j(n_0)$ (independently of $\mu_\ell(n_0)$, $\ell \neq j$) in $[E_{2j-1}, E_{2j}]$, $j = 1, \ldots, p$. In other words, the isospectral set of all $(p+1)$-band sequences (a, b) associated with $\check{L}$ can be identified with the p-dimensional torus $T^p = \times_{j=1}^p S^1$. Theorem 2.8 then provides a concrete realization of all elements in T^p, that is, of all isospectral $(p+1)$-band sequences (a, b). Again by Theorem 2.8, the same applies to the set of all $(2p+1)$-gap, respectively, $2p$-gap sequences ρ associated with the Dirac-type operator M, depending on whether $E_0 > 0$ or $E_0 = 0$. More precisely, assuming $\inf\left(\text{spec}\left(\check{L}_1\right)\right) = E_0 > 0$ (and hence $\inf\left(\text{spec}\left(\check{L}_{2,\pm}\right)\right) = E_0 > 0$), the isospectral set of all $(2p+1)$-gap sequences ρ in connection with the nonsingular hyperelliptic curve $\mathcal{K}_{2p+1}$ of genus $2p+1$ (cf. (2.28)), is again parametrized bijectively by the Dirichlet divisor $\mathcal{D}_{\hat{\underline{\mu}}_1(n_0)}$ (respectively by the analog of (2.92)), as is demonstrated in (2.84). In particular, ρ_+ and ρ_- in (2.79)–(2.81) represent two independent (yet equivalent) concrete realizations of the isospectral manifold T^p of all $(2p+1)$-gap

2.5 Notes

sequences ρ associated with M. In the case where $\inf\bigl(\text{spec}\,(\check{L}_1)\bigr) = E_0 = 0$ (and hence $\inf\bigl(\text{spec}\,(\check{L}_{2,\pm})\bigr) = E_0 = 0$), the curve $\mathcal{K}_{2p+1}$ of arithmetic genus $2p+1$ is singular (cf. (2.28)), yet $\mathcal{D}_{\underline{\hat{\mu}}_1(n_0)}$ still parametrizes the corresponding isospectral set of $2p$-gap sequences $\rho = \rho_\pm$ in a one-to-one and onto fashion. In particular, ρ in (2.79)–(2.81) then represents a concrete realization of the isospectral torus T^p of all $2p$-gap sequences ρ associated with M.

For additional spectral theoretic results in the special case where a, b are real-valued we refer to Bulla et al. (1998) and Teschl (2000).

Section 2.4. The material in this section closely follows Bulla et al. (1998, Sects. 6, 9).

3
The Ablowitz–Ladik Hierarchy

Never express yourself more clearly than you think.
Niels Bohr[1]

3.1 Contents

The Ablowitz–Ladik (AL) equations

$$-i\alpha_t - (1 - \alpha\beta)(\alpha^- + \alpha^+) + 2\alpha = 0,$$
$$-i\beta_t + (1 - \alpha\beta)(\beta^- + \beta^+) - 2\beta = 0 \quad (3.1)$$

for sequences $\alpha = \alpha(n, t)$, $\beta = \beta(n, t)$ (with $\alpha^{\pm}(n, t) = \alpha(n \pm 1, t)$, $\beta^{\pm}(n, t) = \beta(n \pm 1, t)$, etc.), $(n, t) \in \mathbb{Z} \times \mathbb{R}$, were derived in the mid-1970s by Ablowitz and Ladik in a series of papers[2] in an attempt to use inverse scattering methods to analyze certain integrable differential-difference systems. In particular, the system (3.1) can be viewed as an integrable discretization of the AKNS-ZS system. This chapter focuses on the construction of algebro-geometric solutions of the AL hierarchy. Below we briefly summarize the principal content of each section.

Section 3.2.
- Laurent polynomial recursion formalism, zero-curvature pairs $U, V_{\underline{p}}$
- stationary and time-dependent Ablowitz–Ladik hierarchy
- Burchnall–Chaundy Laurent polynomial, hyperelliptic curve $\mathcal{K}_p$

Section 3.3.
- Lax pairs for the Ablowitz–Ladik hierarchy

Sections 3.4 and 3.5. (stationary)
- properties of ϕ and the Baker–Akhiezer vector Ψ

---
[1] As quoted in A. Pais, *Niels Bohr's Times, In Physics, Philosophy, and Polity*, Clarendon Press, Oxford, 1991, p. 178.
[2] A guide to the literature can be found in the detailed notes at the end of this chapter.

3.2 Fundamentals of the AL Hierarchy

- auxiliary divisors
- trace formulas for α, β
- theta function representations for ϕ, Ψ, and α, β
- the algebro-geometric initial value problem

Sections 3.6 and 3.7. (time-dependent)

- properties of ϕ and the Baker–Akhiezer vector Ψ
- auxiliary divisors
- trace formulas for α, β
- theta function representations for ϕ, Ψ, and α, β
- the algebro-geometric initial value problem

Section 3.8. (Hamiltonian formalism)

- asymptotic spectral parameter expansions of Riccati-type solutions
- local conservation laws
- variational derivatives
- Poisson brackets

This chapter relies on terminology and notions developed in connection with compact Riemann surfaces. A brief summary of key results as well as definitions of some of the main quantities can be found in Appendices A and B.

3.2 The Ablowitz–Ladik Hierarchy, Recursion Relations, Zero-Curvature Pairs, and Hyperelliptic Curves

Consistency is the last refuge of the unimaginative.
Oscar Wilde[1]

In this section we provide the construction of the Ablowitz–Ladik hierarchy employing a polynomial recursion formalism and derive the associated sequence of Ablowitz–Ladik zero-curvature pairs. Moreover, we discuss the hyperelliptic curve underlying the stationary Ablowitz–Ladik hierarchy.

We denote by $\mathbb{C}^\mathbb{Z}$ the set of complex-valued sequences indexed by $\mathbb{Z}$.

Throughout this section we suppose the following hypothesis.

Hypothesis 3.1 *In the stationary case we assume that α, β satisfy*

$$\alpha, \beta \in \mathbb{C}^\mathbb{Z}, \quad \alpha(n)\beta(n) \notin \{0, 1\}, \quad n \in \mathbb{Z}. \tag{3.2}$$

[1] *The Relation of Art to Dress* in (London) Sunday Telegraph (2/28/1885); reprinted in Aristotle at Afternoon Tea: The Rare Oscar Wilde (1991).

In the time-dependent case we assume that α, β satisfy

$$\alpha(\cdot, t), \beta(\cdot, t) \in \mathbb{C}^{\mathbb{Z}}, \quad t \in \mathbb{R}, \quad \alpha(n, \cdot), \beta(n, \cdot) \in C^1(\mathbb{R}), \quad n \in \mathbb{Z},$$
$$\alpha(n,t)\beta(n,t) \notin \{0,1\}, \quad (n,t) \in \mathbb{Z} \times \mathbb{R}. \tag{3.3}$$

For a discussion of assumptions (3.2) and (3.3) we refer to Remark 3.20.

Actually, up to Remark 3.11 our analysis will be time-independent and hence only the lattice variations of α and β will matter.

We denote by $S^{\pm}$ the shift operators acting on complex-valued sequences $f = \{f(n)\}_{n\in\mathbb{Z}} \in \mathbb{C}^{\mathbb{Z}}$ according to

$$(S^{\pm} f)(n) = f(n \pm 1), \quad n \in \mathbb{Z}.$$

Moreover, we will frequently use the notation

$$f^{\pm} = S^{\pm} f, \quad f \in \mathbb{C}^{\mathbb{Z}}.$$

To construct the Ablowitz–Ladik hierarchy we will try to generalize (3.1) by considering the 2×2 matrix

$$U(z) = \begin{pmatrix} z & \alpha \\ z\beta & 1 \end{pmatrix}, \quad z \in \mathbb{C}, \tag{3.4}$$

and making the ansatz

$$V_{\underline{p}}(z) = i \begin{pmatrix} G_{\underline{p}}^-(z) & -F_{\underline{p}}^-(z) \\ H_{\underline{p}}^-(z) & -K_{\underline{p}}^-(z) \end{pmatrix}, \quad \underline{p} = (p_-, p_+) \in \mathbb{N}_0^2, \tag{3.5}$$

where $G_{\underline{p}}, K_{\underline{p}}, F_{\underline{p}}$, and $H_{\underline{p}}$ are chosen as Laurent polynomials, namely[1]

$$G_{\underline{p}}(z) = \sum_{\ell=1}^{p_-} z^{-\ell} g_{p_--\ell,-} + \sum_{\ell=0}^{p_+} z^{\ell} g_{p_+-\ell,+},$$

$$F_{\underline{p}}(z) = \sum_{\ell=1}^{p_-} z^{-\ell} f_{p_--\ell,-} + \sum_{\ell=0}^{p_+} z^{\ell} f_{p_+-\ell,+},$$

$$H_{\underline{p}}(z) = \sum_{\ell=1}^{p_-} z^{-\ell} h_{p_--\ell,-} + \sum_{\ell=0}^{p_+} z^{\ell} h_{p_+-\ell,+}, \tag{3.6}$$

$$K_{\underline{p}}(z) = \sum_{\ell=1}^{p_-} z^{-\ell} k_{p_--\ell,-} + \sum_{\ell=0}^{p_+} z^{\ell} k_{p_+-\ell,+}.$$

---

[1] Throughout this monograph, a sum is interpreted as zero whenever the upper limit in the sum is strictly less than its lower limit.

3.2 Fundamentals of the AL Hierarchy

Without loss of generality we will only look at the time-independent case and add time later on. Then the stationary zero-curvature equation,

$$0 = UV_{\underline{p}} - V_{\underline{p}}^+ U, \qquad (3.7)$$

is equivalent to the following relationships between the Laurent polynomials

$$UV_{\underline{p}} - V_{\underline{p}}^+ U = i \begin{pmatrix} z(G_{\underline{p}}^- - G_{\underline{p}}) + z\beta F_{\underline{p}} + \alpha H_{\underline{p}}^- & F_{\underline{p}} - zF_{\underline{p}}^- - \alpha(G_{\underline{p}} + K_{\underline{p}}^-) \\ z\beta(G_{\underline{p}}^- + K_{\underline{p}}) - zH_{\underline{p}} + H_{\underline{p}}^- & -z\beta F_{\underline{p}}^- - \alpha H_{\underline{p}} + K_{\underline{p}} - K_{\underline{p}}^- \end{pmatrix},$$

respectively, to

$$z(G_{\underline{p}}^- - G_{\underline{p}}) + z\beta F_{\underline{p}} + \alpha H_{\underline{p}}^- = 0, \qquad (3.8)$$

$$z\beta F_{\underline{p}}^- + \alpha H_{\underline{p}} - K_{\underline{p}} + K_{\underline{p}}^- = 0, \qquad (3.9)$$

$$-F_{\underline{p}} + zF_{\underline{p}}^- + \alpha(G_{\underline{p}} + K_{\underline{p}}^-) = 0, \qquad (3.10)$$

$$z\beta(G_{\underline{p}}^- + K_{\underline{p}}) - zH_{\underline{p}} + H_{\underline{p}}^- = 0. \qquad (3.11)$$

Lemma 3.2 *Suppose the Laurent polynomials defined in* (3.6) *satisfy the zero-curvature equation* (3.7), *then*

$$f_{0,+} = 0, \quad h_{0,-} = 0, \quad g_{0,\pm} = g_{0,\pm}^-, \quad k_{0,\pm} = k_{0,\pm}^-,$$

$$k_{\ell,\pm} - k_{\ell,\pm}^- = g_{\ell,\pm} - g_{\ell,\pm}^-, \ \ell = 0, \ldots, p_\pm - 1, \quad g_{p_+,+} - g_{p_+,+}^- = k_{p_+,+} - k_{p_+,+}^-. \qquad (3.12)$$

Proof Comparing coefficients at the highest order of z in (3.9) and the lowest in (3.8) immediately yields $f_{0,+} = 0$, $h_{0,-} = 0$. Then $g_{0,+} = g_{0,+}^-$, $k_{0,-} = k_{0,-}^-$ are necessarily lattice constants by (3.8), (3.9). Since $\det(U(z)) \neq 0$ for $z \in \mathbb{C} \setminus \{0\}$ by (3.2), (3.7) yields $\mathrm{tr}(V_{\underline{p}}^+) = \mathrm{tr}(UV_{\underline{p}}U^{-1}) = \mathrm{tr}(V_{\underline{p}})$ and hence

$$G_{\underline{p}} - G_{\underline{p}}^- = K_{\underline{p}} - K_{\underline{p}}^-,$$

implying (3.12). Taking $\ell = 0$ in (3.12) then yields $g_{0,-} = g_{0,-}^-$ and $k_{0,+} = k_{0,+}^-$. □

In particular, this lemma shows that we can choose

$$k_{\ell,\pm} = g_{\ell,\pm}, \ 0 \leq \ell \leq p_\pm - 1, \quad k_{p_+,+} = g_{p_+,+}$$

without loss of generality (since this can always be achieved by adding a Laurent polynomial times the identity to $V_{\underline{p}}$, which does not affect the zero-curvature equation). Hence the ansatz (3.6) can be refined as follows (it is more convenient in the following to relabel $h_{p_+,+} = h_{p_--1,-}$ and $k_{p_+,+} = g_{p_-,-}$, and hence, $g_{p_-,-} = g_{p_+,+}$),

$$F_{\underline{p}}(z) = \sum_{\ell=1}^{p_-} f_{p_--\ell,-}z^{-\ell} + \sum_{\ell=0}^{p_+-1} f_{p_+-1-\ell,+}z^{\ell}, \tag{3.13}$$

$$G_{\underline{p}}(z) = \sum_{\ell=1}^{p_-} g_{p_--\ell,-}z^{-\ell} + \sum_{\ell=0}^{p_+} g_{p_+-\ell,+}z^{\ell}, \tag{3.14}$$

$$H_{\underline{p}}(z) = \sum_{\ell=0}^{p_--1} h_{p_--1-\ell,-}z^{-\ell} + \sum_{\ell=1}^{p_+} h_{p_+-\ell,+}z^{\ell}, \tag{3.15}$$

$$K_{\underline{p}}(z) = G_{\underline{p}}(z) \text{ since } g_{p_-,-} = g_{p_+,+}. \tag{3.16}$$

In particular, (3.16) renders $V_{\underline{p}}$ in (3.5) traceless in the stationary context. We emphasize, however, that equation (3.16) ceases to be valid in the time-dependent context: In the latter case (3.16) needs to be replaced by

$$K_{\underline{p}}(z) = \sum_{\ell=0}^{p_-} g_{p_--\ell,-}z^{-\ell} + \sum_{\ell=1}^{p_+} g_{p_+-\ell,+}z^{\ell} = G_{\underline{p}}(z) + g_{p_-,-} - g_{p_+,+}. \tag{3.17}$$

Plugging the refined ansatz (3.13)–(3.16) into the zero-curvature equation (3.7) and comparing coefficients then yields the following result.

Lemma 3.3 *Suppose that U and $V_{\underline{p}}$ satisfy the zero-curvature equation (3.7). Then the coefficients $\{f_{\ell,\pm}\}_{\ell=0,\ldots,p_\pm-1}$, $\{g_{\ell,\pm}\}_{\ell=0,\ldots,p_\pm}$, and $\{h_{\ell,\pm}\}_{\ell=0,\ldots,p_\pm-1}$ of $F_{\underline{p}}$, $G_{\underline{p}}$, $H_{\underline{p}}$, and $K_{\underline{p}}$ in (3.13)–(3.16) satisfy the following relations*

$$g_{0,+} = \tfrac{1}{2}c_{0,+}, \quad f_{0,+} = -c_{0,+}\alpha^+, \quad h_{0,+} = c_{0,+}\beta,$$
$$g_{\ell+1,+} - g_{\ell+1,+}^- = \alpha h_{\ell,+}^- + \beta f_{\ell,+}, \quad 0 \le \ell \le p_+ - 1,$$
$$f_{\ell+1,+}^- = f_{\ell,+} - \alpha(g_{\ell+1,+} + g_{\ell+1,+}^-), \quad 0 \le \ell \le p_+ - 2,$$
$$h_{\ell+1,+} = h_{\ell,+}^- + \beta(g_{\ell+1,+} + g_{\ell+1,+}^-), \quad 0 \le \ell \le p_+ - 2,$$

and

$$g_{0,-} = \tfrac{1}{2}c_{0,-}, \quad f_{0,-} = c_{0,-}\alpha, \quad h_{0,-} = -c_{0,-}\beta^+,$$
$$g_{\ell+1,-} - g_{\ell+1,-}^- = \alpha h_{\ell,-} + \beta f_{\ell,-}^-, \quad 0 \le \ell \le p_- - 1,$$
$$f_{\ell+1,-} = f_{\ell,-}^- + \alpha(g_{\ell+1,-} + g_{\ell+1,-}^-), \quad 0 \le \ell \le p_- - 2,$$
$$h_{\ell+1,-}^- = h_{\ell,-} - \beta(g_{\ell+1,-} + g_{\ell+1,-}^-), \quad 0 \le \ell \le p_- - 2.$$

3.2 Fundamentals of the AL Hierarchy

Here $c_{0,\pm} \in \mathbb{C}$ are given constants. In addition, (3.7) reads

$$0 = UV_{\underline{p}} - V_{\underline{p}}^+ U$$
$$= i \begin{pmatrix} 0 & -\alpha(g_{p_+,+} + g_{p_-,-}^-) \\ & +f_{p_+-1,+} - f_{p_--1,-}^- \\ z(\beta(g_{p_+,+}^- + g_{p_-,-}) & 0 \\ -h_{p_--1,-} + h_{p_+-1,+}^-) & \end{pmatrix}. \quad (3.18)$$

Given Lemma 3.3, we now introduce the sequences $\{f_{\ell,\pm}\}_{\ell \in \mathbb{N}_0}$, $\{g_{\ell,\pm}\}_{\ell \in \mathbb{N}_0}$, and $\{h_{\ell,\pm}\}_{\ell \in \mathbb{N}_0}$ recursively by

$$g_{0,+} = \tfrac{1}{2} c_{0,+}, \quad f_{0,+} = -c_{0,+}\alpha^+, \quad h_{0,+} = c_{0,+}\beta, \quad (3.19)$$
$$g_{\ell+1,+} - g_{\ell+1,+}^- = \alpha h_{\ell,+}^- + \beta f_{\ell,+}, \quad \ell \in \mathbb{N}_0, \quad (3.20)$$
$$f_{\ell+1,+}^- = f_{\ell,+} - \alpha(g_{\ell+1,+} + g_{\ell+1,+}^-), \quad \ell \in \mathbb{N}_0, \quad (3.21)$$
$$h_{\ell+1,+} = h_{\ell,+}^- + \beta(g_{\ell+1,+} + g_{\ell+1,+}^-), \quad \ell \in \mathbb{N}_0, \quad (3.22)$$

and

$$g_{0,-} = \tfrac{1}{2} c_{0,-}, \quad f_{0,-} = c_{0,-}\alpha, \quad h_{0,-} = -c_{0,-}\beta^+, \quad (3.23)$$
$$g_{\ell+1,-} - g_{\ell+1,-}^- = \alpha h_{\ell,-} + \beta f_{\ell,-}^-, \quad \ell \in \mathbb{N}_0, \quad (3.24)$$
$$f_{\ell+1,-} = f_{\ell,-}^- + \alpha(g_{\ell+1,-} + g_{\ell+1,-}^-), \quad \ell \in \mathbb{N}_0, \quad (3.25)$$
$$h_{\ell+1,-}^- = h_{\ell,-} - \beta(g_{\ell+1,-} + g_{\ell+1,-}^-), \quad \ell \in \mathbb{N}_0. \quad (3.26)$$

For later use we also introduce

$$f_{-1,\pm} = h_{-1,\pm} = 0. \quad (3.27)$$

Remark 3.4 The sequences $\{f_{\ell,+}\}_{\ell \in \mathbb{N}_0}$, $\{g_{\ell,+}\}_{\ell \in \mathbb{N}_0}$, and $\{h_{\ell,+}\}_{\ell \in \mathbb{N}_0}$ can be computed recursively as follows: Assume that $f_{\ell,+}$, $g_{\ell,+}$, and $h_{\ell,+}$ are known. Equation (3.20) is a first-order difference equation in $g_{\ell+1,+}$ that can be solved directly and yields a local lattice function that is determined up to a new constant denoted by $c_{\ell+1,+} \in \mathbb{C}$. Relations (3.21) and (3.22) then determine $f_{\ell+1,+}$ and $h_{\ell+1,+}$, etc. The sequences $\{f_{\ell,-}\}_{\ell \in \mathbb{N}_0}$, $\{g_{\ell,-}\}_{\ell \in \mathbb{N}_0}$, and $\{h_{\ell,-}\}_{\ell \in \mathbb{N}_0}$ are determined similarly.

Upon setting

$$\gamma = 1 - \alpha\beta,$$

one explicitly obtains

$$f_{0,+} = c_{0,+}(-\alpha^+),$$
$$f_{1,+} = c_{0,+}\big(-\gamma^+\alpha^{++} + (\alpha^+)^2\beta\big) + c_{1,+}(-\alpha^+),$$
$$g_{0,+} = \tfrac{1}{2}c_{0,+},$$
$$g_{1,+} = c_{0,+}(-\alpha^+\beta) + \tfrac{1}{2}c_{1,+},$$
$$g_{2,+} = c_{0,+}\big((\alpha^+\beta)^2 - \gamma^+\alpha^{++}\beta - \gamma\alpha^+\beta^-\big) + c_{1,+}(-\alpha^+\beta) + \tfrac{1}{2}c_{2,+},$$
$$h_{0,+} = c_{0,+}\beta,$$
$$h_{1,+} = c_{0,+}\big(\gamma\beta^- - \alpha^+\beta^2\big) + c_{1,+}\beta,$$
$$f_{0,-} = c_{0,-}\alpha,$$
$$f_{1,-} = c_{0,-}\big(\gamma\alpha^- - \alpha^2\beta^+\big) + c_{1,-}\alpha,$$
$$g_{0,-} = \tfrac{1}{2}c_{0,-},$$
$$g_{1,-} = c_{0,-}(-\alpha\beta^+) + \tfrac{1}{2}c_{1,-},$$
$$g_{2,-} = c_{0,-}\big((\alpha\beta^+)^2 - \gamma^+\alpha\beta^{++} - \gamma\alpha^-\beta^+\big) + c_{1,-}(-\alpha\beta^+) + \tfrac{1}{2}c_{2,-},$$
$$h_{0,-} = c_{0,-}(-\beta^+),$$
$$h_{1,-} = c_{0,-}\big(-\gamma^+\beta^{++} + \alpha(\beta^+)^2\big) + c_{1,-}(-\beta^+), \quad \text{etc.}$$

Here $\{c_{\ell,\pm}\}_{\ell\in\mathbb{N}}$ denote summation constants which naturally arise when solving the difference equations for $g_{\ell,\pm}$ in (3.20), (3.24).

In particular, by (3.18), the stationary zero-curvature relation (3.7), $0 = UV_{\underline{p}} - V_{\underline{p}}^+ U$, is equivalent to

$$-\alpha(g_{p_+,+} + g_{p_-,-}^-) + f_{p_+-1,+} - f_{p_--1,-}^- = 0, \tag{3.28}$$
$$\beta(g_{p_+,+}^- + g_{p_-,-}) + h_{p_+-1,+}^- - h_{p_--1,-} = 0. \tag{3.29}$$

Thus, varying $p_\pm \in \mathbb{N}_0$, equations (3.28) and (3.29) give rise to the stationary Ablowitz–Ladik (AL) hierarchy which we introduce as follows

$$\text{s-AL}_{\underline{p}}(\alpha,\beta) = \begin{pmatrix} -\alpha(g_{p_+,+} + g_{p_-,-}^-) + f_{p_+-1,+} - f_{p_--1,-}^- \\ \beta(g_{p_+,+}^- + g_{p_-,-}) + h_{p_+-1,+}^- - h_{p_--1,-} \end{pmatrix} = 0, \tag{3.30}$$
$$\underline{p} = (p_-, p_+) \in \mathbb{N}_0^2.$$

Explicitly (recalling $\gamma = 1 - \alpha\beta$ and taking $p_- = p_+$ for simplicity),

$$\text{s-AL}_{(0,0)}(\alpha,\beta) = \begin{pmatrix} -c_{(0,0)}\alpha \\ c_{(0,0)}\beta \end{pmatrix} = 0,$$
$$\text{s-AL}_{(1,1)}(\alpha,\beta) = \begin{pmatrix} -\gamma(c_{0,-}\alpha^- + c_{0,+}\alpha^+) - c_{(1,1)}\alpha \\ \gamma(c_{0,+}\beta^- + c_{0,-}\beta^+) + c_{(1,1)}\beta \end{pmatrix} = 0,$$

3.2 Fundamentals of the AL Hierarchy

$$\text{s-AL}_{(2,2)}(\alpha,\beta) = \begin{pmatrix} -\gamma(c_{0,+}\alpha^{++}\gamma^+ + c_{0,-}\alpha^{--}\gamma^- - \alpha(c_{0,+}\alpha^+\beta^- + c_{0,-}\alpha^-\beta^+)) \\ -\beta(c_{0,-}(\alpha^-)^2 + c_{0,+}(\alpha^+)^2)) \\ \gamma(c_{0,-}\beta^{++}\gamma^+ + c_{0,+}\beta^{--}\gamma^- - \beta(c_{0,+}\alpha^+\beta^- + c_{0,-}\alpha^-\beta^+) \\ -\alpha(c_{0,+}(\beta^-)^2 + c_{0,-}(\beta^+)^2)) \end{pmatrix}$$
$$+ \begin{pmatrix} -\gamma(c_{1,-}\alpha^- + c_{1,+}\alpha^+) - c_{(2,2)}\alpha \\ \gamma(c_{1,+}\beta^- + c_{1,-}\beta^+) + c_{(2,2)}\beta \end{pmatrix} = 0, \text{ etc.,}$$

represent the first few equations of the stationary Ablowitz–Ladik hierarchy. Here we introduced

$$c_{\underline{p}} = (c_{p,-} + c_{p,+})/2, \quad p_\pm \in \mathbb{N}_0. \tag{3.31}$$

By definition, the set of solutions of (3.30), with $p_\pm$ ranging in $\mathbb{N}_0$ and $c_{\ell,\pm} \in \mathbb{C}$, $\ell \in \mathbb{N}_0$, represents the class of algebro-geometric Ablowitz–Ladik solutions.

In the special case $\underline{p} = (1,1)$, $c_{0,\pm} = 1$, and $c_{(1,1)} = -2$, one obtains the stationary version of the Ablowitz–Ladik system (3.1)

$$\begin{pmatrix} -\gamma(\alpha^- + \alpha^+) + 2\alpha \\ \gamma(\beta^- + \beta^+) - 2\beta \end{pmatrix} = 0.$$

Subsequently, it will also be useful to work with the corresponding homogeneous coefficients $\hat{f}_{\ell,\pm}$, $\hat{g}_{\ell,\pm}$, and $\hat{h}_{\ell,\pm}$, defined by the vanishing of all summation constants $c_{k,\pm}$ for $k = 1, \dots, \ell$, and choosing $c_{0,\pm} = 1$,

$$\hat{f}_{0,+} = -\alpha^+, \quad \hat{f}_{0,-} = \alpha, \quad \hat{f}_{\ell,\pm} = f_{\ell,\pm}|_{c_{0,\pm}=1,\, c_{j,\pm}=0,\, j=1,\dots,\ell}, \quad \ell \in \mathbb{N}, \tag{3.32}$$
$$\hat{g}_{0,\pm} = \tfrac{1}{2}, \quad \hat{g}_{\ell,\pm} = g_{\ell,\pm}|_{c_{0,\pm}=1,\, c_{j,\pm}=0,\, j=1,\dots,\ell}, \quad \ell \in \mathbb{N}, \tag{3.33}$$
$$\hat{h}_{0,+} = \beta, \quad \hat{h}_{0,-} = -\beta^+, \quad \hat{h}_{\ell,\pm} = h_{\ell,\pm}|_{c_{0,\pm}=1,\, c_{j,\pm}=0,\, j=1,\dots,\ell}, \quad \ell \in \mathbb{N}. \tag{3.34}$$

By induction one infers that

$$f_{\ell,\pm} = \sum_{k=0}^{\ell} c_{\ell-k,\pm}\hat{f}_{k,\pm}, \quad g_{\ell,\pm} = \sum_{k=0}^{\ell} c_{\ell-k,\pm}\hat{g}_{k,\pm}, \quad h_{\ell,\pm} = \sum_{k=0}^{\ell} c_{\ell-k,\pm}\hat{h}_{k,\pm}. \tag{3.35}$$

In a slight abuse of notation we will occasionally stress the dependence of $f_{\ell,\pm}$, $g_{\ell,\pm}$, and $h_{\ell,\pm}$ on α, β by writing $f_{\ell,\pm}(\alpha,\beta)$, $g_{\ell,\pm}(\alpha,\beta)$, and $h_{\ell,\pm}(\alpha,\beta)$.

Remark 3.5 Using the nonlinear recursion relations (C.50)–(C.55) recorded in Theorem C.3, one infers inductively that all homogeneous elements $\hat{f}_{\ell,\pm}$, $\hat{g}_{\ell,\pm}$, and $\hat{h}_{\ell,\pm}$, $\ell \in \mathbb{N}_0$, are polynomials in α, β, and some of their shifts. (As an alternative, one can prove directly by induction that the nonlinear recursion relations (C.50)–(C.55) are equivalent to that in (3.19)–(3.26) with all summation constants put equal to zero, $c_{\ell,\pm} = 0, \ell \in \mathbb{N}$.)

Remark 3.6 As an efficient tool to later distinguish between nonhomogeneous and homogeneous quantities $f_{\ell,\pm}, g_{\ell,\pm}, h_{\ell,\pm}$, and $\hat{f}_{\ell,\pm}, \hat{g}_{\ell,\pm}, \hat{h}_{\ell,\pm}$, respectively, we now introduce the notion of degree as follows. Denote

$$f^{(r)} = S^{(r)}f, \quad f = \{f(n)\}_{n \in \mathbb{Z}} \in \mathbb{C}^{\mathbb{Z}}, \quad S^{(r)} = \begin{cases} (S^+)^r, & r \geq 0, \\ (S^-)^{-r}, & r < 0, \end{cases} \quad r \in \mathbb{Z},$$

and define

$$\deg\left(\alpha^{(r)}\right) = r, \quad \deg\left(\beta^{(r)}\right) = -r, \quad r \in \mathbb{Z}.$$

This then results in

$$\deg\left(\hat{f}_{\ell,+}^{(r)}\right) = \ell + 1 + r, \quad \deg\left(\hat{f}_{\ell,-}^{(r)}\right) = -\ell + r, \quad \deg\left(\hat{g}_{\ell,\pm}^{(r)}\right) = \pm \ell, \tag{3.36}$$
$$\deg\left(\hat{h}_{\ell,+}^{(r)}\right) = \ell - r, \quad \deg\left(\hat{h}_{\ell,-}^{(r)}\right) = -\ell - 1 - r, \quad \ell \in \mathbb{N}_0, \, r \in \mathbb{Z},$$

using induction in the linear recursion relations (3.19)–(3.26).

In accordance with our notation introduced in (3.32)–(3.34), the corresponding homogeneous stationary Ablowitz–Ladik equations are defined by

$$\text{s-}\widehat{\text{AL}}_{\underline{p}}(\alpha, \beta) = \text{s-AL}_{\underline{p}}(\alpha, \beta)\big|_{c_{0,\pm}=1, \, c_{\ell,\pm}=0, \, \ell=1,\ldots,p_{\pm}}, \quad \underline{p} = (p_-, p_+) \in \mathbb{N}_0^2.$$

We also note the following useful result.

Lemma 3.7 *The coefficients $f_{\ell,\pm}, g_{\ell,\pm}$, and $h_{\ell,\pm}$ satisfy the relations*

$$\begin{aligned} g_{\ell,+} - g_{\ell,+}^- &= \alpha h_{\ell,+}^- + \beta f_{\ell,+}^-, \quad \ell \in \mathbb{N}_0, \\ g_{\ell,-} - g_{\ell,-}^- &= \alpha h_{\ell,-}^- + \beta f_{\ell,-}, \quad \ell \in \mathbb{N}_0. \end{aligned} \tag{3.37}$$

Moreover, we record the following symmetries,

$$\hat{f}_{\ell,\pm}(c_{0,\pm}, \alpha, \beta) = \hat{h}_{\ell,\mp}(c_{0,\mp}, \beta, \alpha), \quad \hat{g}_{\ell,\pm}(c_{0,\pm}, \alpha, \beta) = \hat{g}_{\ell,\mp}(c_{0,\mp}, \beta, \alpha), \quad \ell \in \mathbb{N}_0. \tag{3.38}$$

Proof The relations (3.37) are derived as follows:

$$\begin{aligned} \alpha h_{\ell+1,+}^- + \beta f_{\ell+1,+}^- &= \alpha h_{\ell,+}^- + \alpha\beta(g_{\ell+1,+} + g_{\ell+1,+}^-) + \beta f_{\ell,+} \\ &\quad - \alpha\beta(g_{\ell+1,+} + g_{\ell+1,+}^-) \\ &= \alpha h_{\ell,+}^- + \beta f_{\ell,+} = g_{\ell+1,+} - g_{\ell+1,+}^-, \end{aligned}$$

3.2 Fundamentals of the AL Hierarchy

and

$$\alpha h^-_{\ell+1,-} + \beta f_{\ell+1,-} = \alpha h_{\ell,-} - \alpha\beta(g_{\ell+1,-} + g^-_{\ell+1,-}) + \beta f^-_{\ell,-}$$
$$+ \alpha\beta(g_{\ell+1,-} + g^-_{\ell+1,-})$$
$$= \alpha h_{\ell,-} + \beta f^-_{\ell,-} = g_{\ell+1,-} + g^-_{\ell+1,-}.$$

The statement (3.38) follows by showing that $\hat{h}_{\ell,\mp}(\beta, \alpha)$ and $\hat{g}_{\ell,\mp}(\beta, \alpha)$ satisfy the same recursion relations as those of $\hat{f}_{\ell,\pm}(\alpha, \beta)$ and $\hat{g}_{\ell,\pm}(\alpha, \beta)$, respectively. That the recursion constants are the same, follows from the observation that the corresponding coefficients have the proper degree. $\square$

Next we turn to the Laurent polynomials $F_{\underline{p}}$, $G_{\underline{p}}$, $H_{\underline{p}}$, and $K_{\underline{p}}$ defined in (3.13)–(3.15) and (3.17). Explicitly, one obtains

$$F_{(0,0)} = 0,$$
$$F_{(1,1)} = c_{0,-}\alpha z^{-1} + c_{0,+}(-\alpha^+),$$
$$F_{(2,2)} = c_{0,-}\alpha z^{-2} + \big(c_{0,-}(\gamma\alpha^- - \alpha^2\beta^+) + c_{1,-}\alpha\big)z^{-1}$$
$$+ c_{0,+}\big(-\gamma^+\alpha^{++} + (\alpha^+)^2\beta\big) + c_{1,+}(-\alpha^+) + c_{0,+}(-\alpha^+)z,$$
$$G_{(0,0)} = \tfrac{1}{2}c_{0,+},$$
$$G_{(1,1)} = \tfrac{1}{2}c_{0,-}z^{-1} + c_{0,+}(-\alpha^+\beta) + \tfrac{1}{2}c_{1,+} + \tfrac{1}{2}c_{0,+}z,$$
$$G_{(2,2)} = \tfrac{1}{2}c_{0,-}z^{-2} + \big(c_{0,-}(-\alpha\beta^+) + \tfrac{1}{2}c_{1,-}\big)z^{-1}$$
$$+ c_{0,+}\big((\alpha^+\beta)^2 - \gamma^+\alpha^{++}\beta - \gamma\alpha^+\beta^-\big) + c_{1,+}(-\alpha^+\beta) + \tfrac{1}{2}c_{2,+}$$
$$+ \big(c_{0,+}(-\alpha^+\beta) + \tfrac{1}{2}c_{1,+}\big)z + \tfrac{1}{2}c_{0,+}z^2,$$
$$H_{(0,0)} = 0,$$
$$H_{(1,1)} = c_{0,-}(-\beta^+) + c_{0,+}\beta z,$$
$$H_{(2,2)} = c_{0,-}(-\beta^+)z^{-1} + c_{0,-}\big(-\gamma^+\beta^{++} + \alpha(\beta^+)^2\big) + c_{1,-}(-\beta^+)$$
$$+ \big(c_{0,+}(\gamma\beta^- - \alpha^+\beta^2) + c_{1,+}\beta\big)z + c_{0,+}\beta z^2,$$
$$K_{(0,0)} = \tfrac{1}{2}c_{0,-},$$
$$K_{(1,1)} = \tfrac{1}{2}c_{0,-}z^{-1} + c_{0,-}(-\alpha\beta^+) + \tfrac{1}{2}c_{1,-} + \tfrac{1}{2}c_{0,+}z,$$
$$K_{(2,2)} = \tfrac{1}{2}c_{0,-}z^{-2} + \big(c_{0,-}(-\alpha\beta^+) + \tfrac{1}{2}c_{1,-}\big)z^{-1}$$
$$+ c_{0,-}\big((\alpha\beta^+)^2 - \gamma^+\alpha\beta^{++} - \gamma\alpha^-\beta^+\big) + c_{1,-}(-\alpha\beta^+) + \tfrac{1}{2}c_{2,-}$$
$$+ \big(c_{0,+}(-\alpha^+\beta) + \tfrac{1}{2}c_{1,+}\big)z + \tfrac{1}{2}c_{0,+}z^2, \text{ etc.}$$

The corresponding homogeneous quantities are defined by ($\ell \in \mathbb{N}_0$)

$$\widehat{F}_{0,\mp}(z) = 0, \quad \widehat{F}_{\ell,-}(z) = \sum_{k=1}^{\ell} \hat{f}_{\ell-k,-} z^{-k}, \quad \widehat{F}_{\ell,+}(z) = \sum_{k=0}^{\ell-1} \hat{f}_{\ell-1-k,+} z^k,$$

$$\widehat{G}_{0,-}(z) = 0, \quad \widehat{G}_{\ell,-}(z) = \sum_{k=1}^{\ell} \hat{g}_{\ell-k,-} z^{-k},$$

$$\widehat{G}_{0,+}(z) = \frac{1}{2}, \quad \widehat{G}_{\ell,+}(z) = \sum_{k=0}^{\ell} \hat{g}_{\ell-k,+} z^k,$$

$$\widehat{H}_{0,\mp}(z) = 0, \quad \widehat{H}_{\ell,-}(z) = \sum_{k=0}^{\ell-1} \hat{h}_{\ell-1-k,-} z^{-k}, \quad \widehat{H}_{\ell,+}(z) = \sum_{k=1}^{\ell} \hat{h}_{\ell-k,+} z^k,$$

$$\widehat{K}_{0,-}(z) = \frac{1}{2}, \quad \widehat{K}_{\ell,-}(z) = \sum_{k=0}^{\ell} \hat{g}_{\ell-k,-} z^{-k} = \widehat{G}_{\ell,-}(z) + \hat{g}_{\ell,-},$$

$$\widehat{K}_{0,+}(z) = 0, \quad \widehat{K}_{\ell,+}(z) = \sum_{k=1}^{\ell} \hat{g}_{\ell-k,+} z^k = \widehat{G}_{\ell,+}(z) - \hat{g}_{\ell,+}.$$

Similarly, with $F_{\ell_+,+}, G_{\ell_+,+}, H_{\ell_+,+}$, and $K_{\ell_+,+}$ denoting the polynomial parts of $F_{\underline{\ell}}, G_{\underline{\ell}}, H_{\underline{\ell}}$, and $K_{\underline{\ell}}$, respectively, and $F_{\ell_-,-}, G_{\ell_-,-}, H_{\ell_-,-}$, and $K_{\ell_-,-}$ denoting the Laurent parts of $F_{\underline{\ell}}, G_{\underline{\ell}}, H_{\underline{\ell}}$, and $K_{\underline{\ell}}, \underline{\ell} = (\ell_-, \ell_+) \in \mathbb{N}_0$, such that

$$F_{\underline{\ell}}(z) = F_{\ell_-,-}(z) + F_{\ell_+,+}(z), \quad G_{\underline{\ell}}(z) = G_{\ell_-,-}(z) + G_{\ell_+,+}(z),$$
$$H_{\underline{\ell}}(z) = H_{\ell_-,-}(z) + H_{\ell_+,+}(z), \quad K_{\underline{\ell}}(z) = K_{\ell_-,-}(z) + K_{\ell_+,+}(z),$$

one finds that

$$F_{\ell_\pm,\pm} = \sum_{k=1}^{\ell_\pm} c_{\ell_\pm-k,\pm} \widehat{F}_{k,\pm}, \quad H_{\ell_\pm,\pm} = \sum_{k=1}^{\ell_\pm} c_{\ell_\pm-k,\pm} \widehat{H}_{k,\pm},$$

$$G_{\ell_-,-} = \sum_{k=1}^{\ell_-} c_{\ell_--k,-} \widehat{G}_{k,-}, \quad G_{\ell_+,+} = \sum_{k=0}^{\ell_+} c_{\ell_+-k,+} \widehat{G}_{k,+},$$

$$K_{\ell_-,-} = \sum_{k=0}^{\ell_-} c_{\ell_--k,-} \widehat{K}_{k,-}, \quad K_{\ell_+,+} = \sum_{k=1}^{\ell_+} c_{\ell_+-k,+} \widehat{K}_{k,+}.$$

In addition, one immediately obtains the following relations from (3.38):

3.2 Fundamentals of the AL Hierarchy

Lemma 3.8 *Let $\ell \in \mathbb{N}_0$. Then,*

$$\widehat{F}_{\ell,\pm}(\alpha, \beta, z, n) = \widehat{H}_{\ell,\mp}(\beta, \alpha, z^{-1}, n),$$
$$\widehat{H}_{\ell,\pm}(\alpha, \beta, z, n) = \widehat{F}_{\ell,\mp}(\beta, \alpha, z^{-1}, n),$$
$$\widehat{G}_{\ell,\pm}(\alpha, \beta, z, n) = \widehat{G}_{\ell,\mp}(\beta, \alpha, z^{-1}, n),$$
$$\widehat{K}_{\ell,\pm}(\alpha, \beta, z, n) = \widehat{K}_{\ell,\mp}(\beta, \alpha, z^{-1}, n).$$

Returning to the stationary Ablowitz–Ladik hierarchy, we will frequently assume in the following that α, β satisfy the $\underline{p}$th stationary Ablowitz–Ladik system s-AL$_{\underline{p}}(\alpha, \beta) = 0$, supposing a particular choice of summation constants $c_{\ell,\pm} \in \mathbb{C}$, $\ell = 0, \ldots, p_\pm$, $p_\pm \in \mathbb{N}_0$, has been made.

Remark 3.9 (*i*) The particular choice $c_{0,+} = c_{0,-} = 1$ in (3.30) yields the stationary Ablowitz–Ladik equation. Scaling $c_{0,\pm}$ with the same constant then amounts to scaling $V_{\underline{p}}$ with this constant which drops out in the stationary zero-curvature equation (3.7).
(*ii*) Different ratios between $c_{0,+}$ and $c_{0,-}$ will lead to different stationary hierarchies. In particular, the choice $c_{0,+} = 2$, $c_{0,-} = \cdots = c_{p_--1,-} = 0$, $c_{p_-,-} \neq 0$, yields the stationary Baxter–Szegő hierarchy. However, in this case some parts from the recursion relation for the negative coefficients still remain. In fact, (3.26) reduces to $g_{p_-,-} - g_{p_-,-}^- = \alpha h_{p_--1,-}$, $h_{p_--1,-} = 0$ and thus requires $g_{p_-,-}$ to be a constant in (3.30) and (3.54). Moreover, $f_{p_--1,-} = 0$ in (3.30) in this case.
(*iii*) Finally, by Lemma 3.8, the choice $c_{0,+} = \cdots = c_{p_+-1,+} = 0$, $c_{p_+,+} \neq 0$, $c_{0,-} = 2$ again yields the Baxter–Szegő hierarchy, but with α and β interchanged.

Next, taking into account (3.16), one infers that the expression $R_{\underline{p}}$, defined as

$$R_{\underline{p}} = G_{\underline{p}}^2 - F_{\underline{p}} H_{\underline{p}}, \tag{3.39}$$

is a lattice constant, that is, $R_{\underline{p}} - R_{\underline{p}}^- = 0$, since taking determinants in the stationary zero-curvature equation (3.7) immediately yields

$$\gamma\bigl(-(G_{\underline{p}}^-)^2 + F_{\underline{p}}^- H_{\underline{p}}^- + G_{\underline{p}}^2 - F_{\underline{p}} H_{\underline{p}}\bigr)z = 0.$$

Hence, $R_{\underline{p}}(z)$ only depends on z, and assuming in addition to (3.2) that

$$c_{0,\pm} \in \mathbb{C} \setminus \{0\}, \quad \underline{p} = (p_-, p_+) \in \mathbb{N}_0^2 \setminus \{(0,0)\},$$

one may write $R_{\underline{p}}$ as

$$R_{\underline{p}}(z) = \left(\frac{c_{0,+}}{2z^{p_-}}\right)^2 \prod_{m=0}^{2p+1}(z - E_m), \quad \{E_m\}_{m=0}^{2p+1} \subset \mathbb{C} \setminus \{0\}, \quad p = p_- + p_+ - 1 \in \mathbb{N}_0.$$
$$\tag{3.40}$$

Moreover, (3.39) also implies

$$\lim_{z \to 0} 4z^{2p_-} R_{\underline{p}}(z) = c_{0,+}^2 \prod_{m=0}^{2p+1} (-E_m) = c_{0,-}^2,$$

and hence,

$$\prod_{m=0}^{2p+1} E_m = \frac{c_{0,-}^2}{c_{0,+}^2}. \tag{3.41}$$

Relation (3.39) allows one to introduce a hyperelliptic curve $\mathcal{K}_p$ of (arithmetic) genus $p = p_- + p_+ - 1$ (possibly with a singular affine part), where

$$\mathcal{K}_p: \mathcal{F}_p(z, y) = y^2 - 4c_{0,+}^{-2} z^{2p_-} R_{\underline{p}}(z) = y^2 - \prod_{m=0}^{2p+1} (z - E_m) = 0, \quad p = p_- + p_+ - 1. \tag{3.42}$$

Remark 3.10 In the special case $p_- = p_+$ and $c_{\ell,+} = c_{\ell,-}$, $\ell = 0, \ldots, p_-$, the symmetries of Lemma 3.8 also hold for $F_{\underline{p}}, G_{\underline{p}},$ and $H_{\underline{p}}$ and thus $R_{\underline{p}}(1/z) = R_{\underline{p}}(z)$ and hence the numbers $E_m, m = 0, \ldots, 2p+1$, come in pairs $(E_k, 1/E_k)$, $k = 1, \ldots, p+1$.

Equations (3.8)–(3.11) and (3.39) permit one to derive nonlinear difference equations for $F_{\underline{p}}, G_{\underline{p}},$ and $H_{\underline{p}}$ separately. One obtains

$$\left((\alpha^+ + z\alpha)^2 F_{\underline{p}} - z(\alpha^+)^2 \gamma F_{\underline{p}}^-\right)^2 - 2z\alpha^2 \gamma^+ \left((\alpha^+ + z\alpha)^2 F_{\underline{p}} + z(\alpha^+)^2 \gamma F_{\underline{p}}^-\right) F_{\underline{p}}^+$$
$$+ z^2 \alpha^4 (\gamma^+)^2 (F_{\underline{p}}^+)^2 = 4(\alpha\alpha^+)^2 (\alpha^+ + \alpha z)^2 R_{\underline{p}}, \tag{3.43}$$

$$(\alpha^+ + z\alpha)(\beta + z\beta^+)(z + \alpha^+\beta)(1 + z\alpha\beta^+) G_{\underline{p}}^2$$
$$+ z(\alpha^+ \gamma G_{\underline{p}}^- + z\alpha\gamma^+ G_{\underline{p}}^+)(z\beta^+ \gamma G_{\underline{p}}^- + \beta\gamma^+ G_{\underline{p}}^+)$$
$$- z\gamma\left((\alpha^+\beta + z^2\alpha\beta^+)(2 - \gamma^+) + 2z(1 - \gamma^+)(2 - \gamma)\right) G_{\underline{p}}^- G_{\underline{p}}$$
$$- z\gamma^+\left(2z(1 - \gamma)(2 - \gamma^+) + (\alpha^+\beta + z^2\alpha\beta^+)(2 - \gamma)\right) G_{\underline{p}}^+ G_{\underline{p}}$$
$$= (\alpha^+\beta - z^2\alpha\beta^+)^2 R_{\underline{p}}, \tag{3.44}$$

$$z^2\left((\beta^+)^2 \gamma H_{\underline{p}}^- - \beta^2 \gamma^+ H_{\underline{p}}^+\right)^2 - 2z(\beta + z\beta^+)^2\left((\beta^+)^2 \gamma H_{\underline{p}}^- + \beta^2 \gamma^+ H_{\underline{p}}^+\right) H_{\underline{p}}$$
$$+ (\beta + z\beta^+)^4 H_{\underline{p}}^2 = 4z^2(\beta\beta^+)^2(\beta + \beta^+ z)^2 R_{\underline{p}}. \tag{3.45}$$

Equations analogous to (3.43)–(3.45) can be used to derive nonlinear recursion relations for the homogeneous coefficients $\hat{f}_{\ell,\pm}, \hat{g}_{\ell,\pm}$, and $\hat{h}_{\ell,\pm}$ (i.e., the ones satisfying (3.32)–(3.34) in the case of vanishing summation constants) as proved in

3.2 Fundamentals of the AL Hierarchy

Theorem C.3. This then yields a proof that $\hat{f}_{\ell,\pm}$, $\hat{g}_{\ell,\pm}$, and $\hat{h}_{\ell,\pm}$ are polynomials in α, β, and some of their shifts (cf. Remark 3.5). In addition, as proven in Theorem C.4, (3.43) leads to an explicit determination of the summation constants $c_{1,\pm}, \ldots, c_{p_\pm,\pm}$ in (3.30) in terms of the zeros $E_0, \ldots, E_{2p+1}$ of the associated Laurent polynomial $R_{\underline{p}}$ in (3.40). In fact, one can prove (cf. (C.59))

$$c_{\ell,\pm} = c_{0,\pm} c_\ell(\underline{E}^{\pm 1}), \quad \ell = 0, \ldots, p_\pm, \tag{3.46}$$

where

$$c_0(\underline{E}^{\pm 1}) = 1,$$

$$c_k(\underline{E}^{\pm 1}) = - \sum_{\substack{j_0,\ldots,j_{2p+1}=0 \\ j_0+\cdots+j_{2p+1}=k}}^{k} \frac{(2j_0)! \cdots (2j_{2p+1})!}{2^{2k}(j_0!)^2 \cdots (j_{2p+1}!)^2 (2j_0 - 1) \cdots (2j_{2p+1} - 1)}$$

$$\times E_0^{\pm j_0} \cdots E_{2p+1}^{\pm j_{2p+1}}, \quad k \in \mathbb{N}, \tag{3.47}$$

are symmetric functions of $\underline{E}^{\pm 1} = (E_0^{\pm 1}, \ldots, E_{2p+1}^{\pm 1})$ introduced in (C.4) and (C.5).

Remark 3.11 If α, β satisfy one of the stationary Ablowitz–Ladik equations in (3.30) for a particular value of $\underline{p}$, s-AL$_{\underline{p}}(\alpha, \beta) = 0$, then they satisfy infinitely many such equations of order higher than $\underline{p}$ for certain choices of summation constants $c_{\ell,\pm}$.

Finally we turn to the time-dependent Ablowitz–Ladik hierarchy. For that purpose the coefficients α and β are now considered as functions of both the lattice point and time. For each system in the hierarchy, that is, for each $\underline{p}$, we introduce a deformation (time) parameter $t_{\underline{p}} \in \mathbb{R}$ in α, β, replacing $\alpha(n), \beta(n)$ by $\alpha(n, t_{\underline{p}}), \beta(n, t_{\underline{p}})$. Moreover, the definitions (3.4), (3.5), and (3.13)–(3.15) of $U, V_{\underline{p}}$, and $F_{\underline{p}}, G_{\underline{p}}, H_{\underline{p}}, K_{\underline{p}}$, respectively, still apply; however, equation (3.16) now needs to be replaced by (3.17) in the time-dependent context.

Imposing the zero-curvature relation

$$U_{t_{\underline{p}}} + U V_{\underline{p}} - V_{\underline{p}}^+ U = 0, \quad \underline{p} \in \mathbb{N}_0^2, \tag{3.48}$$

then results in the equations

$$0 = U_{t_{\underline{p}}} + U V_{\underline{p}} - V_{\underline{p}}^+ U$$

$$= i \begin{pmatrix} z(G_{\underline{p}}^- - G_{\underline{p}}) & -i\alpha_{t_{\underline{p}}} + F_{\underline{p}} - zF_{\underline{p}}^- \\ +z\beta F_{\underline{p}} + \alpha H_{\underline{p}}^- & -\alpha(G_{\underline{p}} + K_{\underline{p}}^-) \\ -iz\beta_{t_{\underline{p}}} + z\beta(G_{\underline{p}}^- & -z\beta F_{\underline{p}}^- - \alpha H_{\underline{p}} \\ +K_{\underline{p}}) - zH_{\underline{p}} + H_{\underline{p}}^- & +K_{\underline{p}} - K_{\underline{p}}^- \end{pmatrix}$$

$$= i \begin{pmatrix} 0 & -i\alpha_{t_{\underline{p}}} - \alpha(g_{p_+,+} + g_{p_-,-}^-) \\ & +f_{p_+-1,+} - f_{p_--1,-}^- \\ z(-i\beta_{t_{\underline{p}}} + \beta(g_{p_+,+}^- + g_{p_-,-}) & 0 \\ -h_{p_--1,-} + h_{p_+-1,+}^-) & \end{pmatrix}, \quad (3.49)$$

or equivalently,

$$\alpha_{t_{\underline{p}}} = i(zF_{\underline{p}}^- + \alpha(G_{\underline{p}} + K_{\underline{p}}^-) - F_{\underline{p}}), \tag{3.50}$$

$$\beta_{t_{\underline{p}}} = -i(\beta(G_{\underline{p}}^- + K_{\underline{p}}) - H_{\underline{p}} + z^{-1}H_{\underline{p}}^-), \tag{3.51}$$

$$0 = z(G_{\underline{p}}^- - G_{\underline{p}}) + z\beta F_{\underline{p}} + \alpha H_{\underline{p}}^-, \tag{3.52}$$

$$0 = z\beta F_{\underline{p}}^- + \alpha H_{\underline{p}} + K_{\underline{p}}^- - K_{\underline{p}}. \tag{3.53}$$

Varying $\underline{p} \in \mathbb{N}_0^2$, the collection of evolution equations

$$\mathrm{AL}_{\underline{p}}(\alpha,\beta) = \begin{pmatrix} -i\alpha_{t_{\underline{p}}} - \alpha(g_{p_+,+} + g_{p_-,-}^-) + f_{p_+-1,+} - f_{p_--1,-}^- \\ -i\beta_{t_{\underline{p}}} + \beta(g_{p_+,+}^- + g_{p_-,-}) - h_{p_--1,-} + h_{p_+-1,+}^- \end{pmatrix} = 0,$$

$$t_{\underline{p}} \in \mathbb{R}, \ \underline{p} = (p_-, p_+) \in \mathbb{N}_0^2,$$
(3.54)

then defines the time-dependent Ablowitz–Ladik hierarchy. Explicitly, taking $p_- = p_+$ for simplicity,

$$\mathrm{AL}_{(0,0)}(\alpha,\beta) = \begin{pmatrix} -i\alpha_{t_{(0,0)}} - c_{(0,0)}\alpha \\ -i\beta_{t_{(0,0)}} + c_{(0,0)}\beta \end{pmatrix} = 0,$$

$$\mathrm{AL}_{(1,1)}(\alpha,\beta) = \begin{pmatrix} -i\alpha_{t_{(1,1)}} - \gamma(c_{0,-}\alpha^- + c_{0,+}\alpha^+) - c_{(1,1)}\alpha \\ -i\beta_{t_{(1,1)}} + \gamma(c_{0,+}\beta^- + c_{0,-}\beta^+) + c_{(1,1)}\beta \end{pmatrix} = 0,$$

$$\mathrm{AL}_{(2,2)}(\alpha,\beta)$$
$$= \begin{pmatrix} -i\alpha_{t_{(2,2)}} - \gamma\big(c_{0,+}\alpha^{++}\gamma^+ + c_{0,-}\alpha^{--}\gamma^- - \alpha(c_{0,+}\alpha^+\beta^- + c_{0,-}\alpha^-\beta^+) \\ \qquad -\beta(c_{0,-}(\alpha^-)^2 + c_{0,+}(\alpha^+)^2)\big) \\ -i\beta_{t_{(2,2)}} + \gamma\big(c_{0,-}\beta^{++}\gamma^+ + c_{0,+}\beta^{--}\gamma^- - \beta(c_{0,+}\alpha^+\beta^- + c_{0,-}\alpha^-\beta^+) \\ \qquad -\alpha(c_{0,+}(\beta^-)^2 + c_{0,-}(\beta^+)^2)\big) \end{pmatrix}$$
$$+ \begin{pmatrix} -\gamma(c_{1,-}\alpha^- + c_{1,+}\alpha^+) - c_{(2,2)}\alpha \\ \gamma(c_{1,+}\beta^- + c_{1,-}\beta^+) + c_{(2,2)}\beta \end{pmatrix} = 0, \text{ etc.,}$$

represent the first few equations of the time-dependent Ablowitz–Ladik hierarchy. Here we recall the definition of $c_{\underline{p}}$ in (3.31).

The special case $\underline{p} = (1,1)$, $c_{0,\pm} = 1$, and $c_{(1,1)} = -2$, that is,

$$\begin{pmatrix} -i\alpha_{t_{(1,1)}} - \gamma(\alpha^- + \alpha^+) + 2\alpha \\ -i\beta_{t_{(1,1)}} + \gamma(\beta^- + \beta^+) - 2\beta \end{pmatrix} = 0,$$

represents *the* Ablowitz–Ladik system (3.1).

3.2 Fundamentals of the AL Hierarchy

The corresponding homogeneous equations are then defined by

$$\widehat{\mathrm{AL}}_{\underline{p}}(\alpha, \beta) = \mathrm{AL}_{\underline{p}}(\alpha, \beta)\big|_{c_{0,\pm}=1,\, c_{\ell,\pm}=0,\, \ell=1,\ldots,p_{\pm}} = 0, \quad \underline{p} = (p_-, p_+) \in \mathbb{N}_0^2.$$

By (3.54), (3.20), and (3.24), the time derivative of $\gamma = 1 - \alpha\beta$ is given by

$$\gamma_{t_{\underline{p}}} = i\gamma\big((g_{p_+,+} - g^-_{p_+,+}) - (g_{p_-,-} - g^-_{p_-,-})\big), \tag{3.55}$$

or equivalently, by

$$\gamma_{t_{\underline{p}}} = i\gamma\big(z^{-1}\alpha H^-_{\underline{p}} - \alpha H_{\underline{p}} + \beta F_{\underline{p}} - z\beta F^-_{\underline{p}}\big), \tag{3.56}$$

using (3.50)–(3.53). (Alternatively, this follows from computing the trace of $U_{t_{\underline{p}}}U^{-1} = V_{\underline{p}}^+ - UV_{\underline{p}}U^{-1}$.) For instance, if α, β satisfy $\mathrm{AL}_1(\alpha, \beta) = 0$, then

$$\gamma_{t_1} = i\gamma\big(\alpha(c_{0,-}\beta^+ + c_{0,+}\beta^-) - \beta(c_{0,+}\alpha^+ + c_{0,-}\alpha^-)\big).$$

Remark 3.12 From (3.8)–(3.11) and the explicit computations of the coefficients $f_{\ell,\pm}$, $g_{\ell,\pm}$, and $h_{\ell,\pm}$, one concludes that the zero-curvature equation (3.49) and hence the Ablowitz–Ladik hierarchy is invariant under the scaling transformation

$$\alpha \to \alpha_c = \{c\,\alpha(n)\}_{n\in\mathbb{Z}}, \quad \beta \to \beta_c = \{\beta(n)/c\}_{n\in\mathbb{Z}}, \quad c \in \mathbb{C} \setminus \{0\}.$$

Moreover, $R_{\underline{p}} = G^2_{\underline{p}} - H_{\underline{p}}F_{\underline{p}}$ and hence $\{E_m\}_{m=0}^{2p+1}$ are invariant under this transformation. Furthermore, choosing $c = e^{ic_{\underline{p}}t}$, one verifies that it is no restriction to assume $c_{\underline{p}} = 0$. This also shows that stationary solutions α, β can only be constructed up to a multiplicative constant.

Remark 3.13 (i) The special choices $\beta = \pm\overline{\alpha}$, $c_{0,\pm} = 1$ lead to the discrete nonlinear Schrödinger hierarchy. In particular, choosing $c_{(1,1)} = -2$ yields the discrete nonlinear Schrödinger equations in their usual form, with

$$-i\alpha_t - (1 \mp |\alpha|^2)(\alpha^- + \alpha^+) + 2\alpha = 0,$$

the first nonlinear element of the hierarchy. The choice $\beta = \overline{\alpha}$ is called the *defocusing* case, $\beta = -\overline{\alpha}$ represents the *focusing* case of the discrete nonlinear Schrödinger hierarchy.
(ii) The alternative choice $\beta = \overline{\alpha}$, $c_{0,\pm} = \mp i$, leads to the hierarchy of Schur flows. In particular, choosing $c_{(1,1)} = 0$, yields

$$\alpha_t - (1 - |\alpha|^2)(\alpha^+ - \alpha^-) = 0 \tag{3.57}$$

as the first nonlinear element of this hierarchy.

3.3 Lax Pairs for the Ablowitz–Ladik Hierarchy

> Such errors as do not depend upon wrong reasoning can be of no
> great consequence & may be corrected by the Reader.
>
> *Isaac Newton*[1]

In this section we introduce Lax pairs for the AL hierarchy and prove the equivalence of the zero-curvature and Lax representation.

Throughout this section we suppose Hypothesis 3.1. We start by relating the homogeneous coefficients $\hat{f}_{\ell,\pm}$, $\hat{g}_{\ell,\pm}$, and $\hat{h}_{\ell,\pm}$ to certain matrix elements of L, where L will later be identified as the Lax difference expression associated with the Ablowitz–Ladik hierarchy. For this purpose it is useful to introduce the standard basis $\{\delta_m\}_{m \in \mathbb{Z}}$ in $\ell^2(\mathbb{Z})$ by

$$\delta_m = \{\delta_{m,n}\}_{n \in \mathbb{Z}}, \ m \in \mathbb{Z}, \quad \delta_{m,n} = \begin{cases} 1, & m = n, \\ 0, & m \neq n. \end{cases} \tag{3.58}$$

The scalar product in $\ell^2(\mathbb{Z})$, denoted by $(\cdot, \cdot)$, is defined by

$$(f, g) = \sum_{n \in \mathbb{Z}} \overline{f(n)} g(n), \quad f, g \in \ell^2(\mathbb{Z}).$$

In the standard basis just defined, we introduce the difference expression L by

$$L = \begin{pmatrix} \ddots & \ddots & \ddots & \ddots & \ddots & & & \\ & 0 & -\alpha(0)\rho(-1) & -\beta(-1)\alpha(0) & -\alpha(1)\rho(0) & \rho(0)\rho(1) & & 0 \\ & \rho(-1)\rho(0) & \beta(-1)\rho(0) & -\beta(0)\alpha(1) & \beta(0)\rho(1) & 0 & & \\ & & 0 & -\alpha(2)\rho(1) & -\beta(1)\alpha(2) & -\alpha(3)\rho(2) & \rho(2)\rho(3) & \\ & & & \rho(1)\rho(2) & \beta(1)\rho(2) & -\beta(2)\alpha(3) & \beta(2)\rho(3) & 0 \\ & 0 & & & & \ddots & \ddots & \ddots \end{pmatrix}$$

$$\tag{3.59}$$

$$= \bigg(- \beta(n)\alpha(n+1)\delta_{m,n} + \big(\beta(n-1)\rho(n)\delta_{\text{odd}}(n)$$

$$- \alpha(n+1)\rho(n)\delta_{\text{even}}(n)\big)\delta_{m,n-1}$$

$$+ \big(\beta(n)\rho(n+1)\delta_{\text{odd}}(n) - \alpha(n+2)\rho(n+1)\delta_{\text{even}}(n)\big)\delta_{m,n+1} \tag{3.60}$$

$$+ \rho(n+1)\rho(n+2)\delta_{\text{even}}(n)\delta_{m,n+2}$$

$$+ \rho(n-1)\rho(n)\delta_{\text{odd}}(n)\delta_{m,n-2}\bigg)_{m,n \in \mathbb{Z}}$$

---

[1] Quoted in R. Westfall, *The Life of Isaac Newton*, Cambridge University Press, Cambridge, 1993, p. 275.

3.3 Lax Pairs for the AL Hierarchy

$$= \rho^- \rho \, \delta_{\text{even}} \, S^{--} + (\beta^- \rho \, \delta_{\text{even}} - \alpha^+ \rho \, \delta_{\text{odd}}) S^- - \beta \alpha^+$$

$$+ (\beta \rho^+ \delta_{\text{even}} - \alpha^{++} \rho^+ \delta_{\text{odd}}) S^+ + \rho^+ \rho^{++} \delta_{\text{odd}} \, S^{++}, \tag{3.61}$$

where δ_{even} and δ_{odd} denote the characteristic functions of the even and odd integers,

$$\delta_{\text{even}} = \chi_{2\mathbb{Z}}, \quad \delta_{\text{odd}} = 1 - \delta_{\text{even}} = \chi_{2\mathbb{Z}+1}.$$

In particular, terms of the form $-\beta(n)\alpha(n+1)$ represent the diagonal (n, n)-entries, $n \in \mathbb{Z}$, in the infinite matrix (3.59). In addition, we used the abbreviation

$$\rho = \gamma^{1/2} = (1 - \alpha\beta)^{1/2}. \tag{3.62}$$

Next, we introduce the unitary operator $U_{\tilde{\varepsilon}}$ in $\ell^2(\mathbb{Z})$ by

$$U_{\tilde{\varepsilon}} = \big(\tilde{\varepsilon}(n)\delta_{m,n}\big)_{(m,n)\in\mathbb{Z}^2}, \quad \tilde{\varepsilon}(n) \in \{1, -1\}, \; n \in \mathbb{Z},$$

and the sequence $\varepsilon = \{\varepsilon(n)\}_{n\in\mathbb{Z}} \in \mathbb{C}^{\mathbb{Z}}$ by

$$\varepsilon(n) = \tilde{\varepsilon}(n-1)\tilde{\varepsilon}(n), \; n \in \mathbb{Z}.$$

Assuming $\alpha, \beta \in \ell^\infty(\mathbb{Z})$, a straightforward computation then shows that

$$\check{L}_\varepsilon = U_{\tilde{\varepsilon}} \check{L} U_{\tilde{\varepsilon}}^{-1},$$

where L_ε is associated with the sequences $\alpha_\varepsilon = \alpha$, $\beta_\varepsilon = \beta$, and $\rho_\varepsilon = \varepsilon\rho$, and $\check{L}$ and $\check{L}_\varepsilon$ are the bounded operator realizations of L and L_ε in $\ell^2(\mathbb{Z})$, respectively. Moreover, the recursion formalism in (3.19)–(3.26) yields coefficients which are polynomials in α, β and some of their shifts and hence depends only quadratically on ρ. As a result, the choice of square root of $\rho(n)$, $n \in \mathbb{Z}$, in (3.62) is immaterial when introducing the AL hierarchy via the Lax equations (3.79).

The matrix representation of L^{-1} is then obtained from that of L in (3.59) by taking the formal adjoint of L and subsequently exchanging α and β

$$L^{-1} = \Big(-\alpha(n)\beta(n+1)\delta_{m,n} + \big(\alpha(n-1)\rho(n)\delta_{\text{even}}(n)$$
$$- \beta(n+1)\rho(n)\delta_{\text{odd}}(n)\big)\delta_{m,n-1} \tag{3.63}$$
$$+ \big(\alpha(n)\rho(n+1)\delta_{\text{even}}(n) - \beta(n+2)\rho(n+1)\delta_{\text{odd}}(n)\big)\delta_{m,n+1}$$
$$+ \rho(n+1)\rho(n+2)\delta_{\text{odd}}(n)\delta_{m,n+2} + \rho(n-1)\rho(n)\delta_{\text{even}}(n)\delta_{m,n-2} \Big)_{m,n\in\mathbb{Z}}$$
$$= \rho^- \rho \, \delta_{\text{odd}} \, S^{--} + (\alpha^- \rho \, \delta_{\text{odd}} - \beta^+ \rho \, \delta_{\text{even}}) S^- - \alpha\beta^+$$
$$+ (\alpha \rho^+ \delta_{\text{odd}} - \beta^{++}\rho^+ \delta_{\text{even}}) S^+ + \rho^+ \rho^{++} \delta_{\text{even}} \, S^{++}. \tag{3.64}$$

L and L^{-1} lead to bounded operators in $\ell^2(\mathbb{Z})$ if α and β are bounded sequences. However, this is of no importance in the context of Lemma 3.14 below as we only apply the five-diagonal matrices L and L^{-1} to basis vectors of the type δ_m.

Next, we discuss a useful factorization of L. For this purpose we introduce the sequence of 2×2 matrices $\theta(n)$, $n \in \mathbb{Z}$, by

$$\theta(n) = \begin{pmatrix} -\alpha(n) & \rho(n) \\ \rho(n) & \beta(n) \end{pmatrix}, \quad n \in \mathbb{Z}, \tag{3.65}$$

and two difference expressions D and E by their matrix representations in the standard basis (3.58) of $\ell^2(\mathbb{Z})$

$$D = \begin{pmatrix} \ddots & & 0 \\ & \theta(2n-2) & \\ & & \theta(2n) \\ 0 & & & \ddots \end{pmatrix}, \quad E = \begin{pmatrix} \ddots & & 0 \\ & \theta(2n-1) & \\ & & \theta(2n+1) \\ 0 & & & \ddots \end{pmatrix},$$

where

$$\begin{pmatrix} D(2n-1, 2n-1) & D(2n-1, 2n) \\ D(2n, 2n-1) & D(2n, 2n) \end{pmatrix} = \theta(2n),$$

$$\begin{pmatrix} E(2n, 2n) & E(2n, 2n+1) \\ E(2n+1, 2n) & E(2n+1, 2n+1) \end{pmatrix} = \theta(2n+1), \quad n \in \mathbb{Z}.$$

Then L can be factorized into

$$L = DE. \tag{3.66}$$

Explicitly, D and E are given by

$$D = \rho\, \delta_{\text{even}}\, S^- - \alpha^+\, \delta_{\text{odd}} + \beta\, \delta_{\text{even}} + \rho^+\, \delta_{\text{odd}}\, S^+, \tag{3.67}$$

$$E = \rho\, \delta_{\text{odd}}\, S^- + \beta\, \delta_{\text{odd}} - \alpha^+\, \delta_{\text{even}} + \rho^+\, \delta_{\text{even}}\, S^+, \tag{3.68}$$

and their inverses are of the form

$$D^{-1} = \rho\, \delta_{\text{even}}\, S^- - \beta^+\, \delta_{\text{odd}} + \alpha\, \delta_{\text{even}} + \rho^+\, \delta_{\text{odd}}\, S^+, \tag{3.69}$$

$$E^{-1} = \rho\, \delta_{\text{odd}}\, S^- + \alpha\, \delta_{\text{odd}} - \beta^+\, \delta_{\text{even}} + \rho^+\, \delta_{\text{even}}\, S^+. \tag{3.70}$$

The next result details the connections between L and the recursion coefficients $f_{\ell,\pm}$, $g_{\ell,\pm}$, and $h_{\ell,\pm}$.

Lemma 3.14 *Let $n \in \mathbb{Z}$. Then the homogeneous coefficients $\{\hat{f}_{\ell,\pm}\}_{\ell \in \mathbb{N}_0}$, $\{\hat{g}_{\ell,\pm}\}_{\ell \in \mathbb{N}_0}$, and $\{\hat{h}_{\ell,\pm}\}_{\ell \in \mathbb{N}_0}$ satisfy the following relations:*

$$\hat{f}_{\ell,+}(n) = (\delta_n, EL^\ell \delta_n)\delta_{\text{even}}(n) + (\delta_n, L^\ell D \delta_n)\delta_{\text{odd}}(n), \quad \ell \in \mathbb{N}_0,$$

$$\hat{f}_{\ell,-}(n) = (\delta_n, D^{-1}L^{-\ell}\delta_n)\delta_{\text{even}}(n) + (\delta_n, L^{-\ell}E^{-1}\delta_n)\delta_{\text{odd}}(n), \quad \ell \in \mathbb{N}_0,$$

3.3 Lax Pairs for the AL Hierarchy

$$\hat{g}_{0,\pm} = 1/2, \quad \hat{g}_{\ell,\pm}(n) = (\delta_n, L^{\pm\ell}\delta_n), \quad \ell \in \mathbb{N}, \tag{3.71}$$

$$\hat{h}_{\ell,+}(n) = (\delta_n, L^\ell D\delta_n)\delta_{\text{even}}(n) + (\delta_n, EL^\ell\delta_n)\delta_{\text{odd}}(n), \quad \ell \in \mathbb{N}_0,$$

$$\hat{h}_{\ell,-}(n) = (\delta_n, L^{-\ell}E^{-1}\delta_n)\delta_{\text{even}}(n) + (\delta_n, D^{-1}L^{-\ell}\delta_n)\delta_{\text{odd}}(n) \quad \ell \in \mathbb{N}_0.$$

Proof Using (3.66)–(3.70) we show that the sequences defined in (3.71) satisfy the recursion relations of Lemma C.5 respectively relation (3.20). For n even,

$$\hat{g}_{\ell,+}(n) - \hat{g}_{\ell,+}(n-1) = (\delta_n, DEL^{\ell-1}\delta_n) - (\delta_{n-1}, DEL^{\ell-1}\delta_{n-1})$$

$$= (D^*\delta_n, EL^{\ell-1}\delta_n) - (D^*\delta_{n-1}, EL^{\ell-1}\delta_{n-1})$$

$$= \beta(n)(\delta_n, EL^{\ell-1}\delta_n) + \rho(n)(\delta_{n-1}, EL^{\ell-1}\delta_n)$$

$$+ \alpha(n)(\delta_{n-1}, EL^{\ell-1}\delta_{n-1}) - \rho(n)(\delta_n, EL^{\ell-1}\delta_{n-1})$$

$$= \beta(n)\hat{f}_{\ell-1,+}(n) + \alpha(n)\hat{h}_{\ell-1,+}(n-1),$$

since $(EL^\ell)^\top = EL^\ell$ by (3.65), (3.66). Moreover,

$$\hat{f}_{\ell,+}(n) = (\delta_n, EL^\ell\delta_n) = (E^*\delta_n, L^\ell\delta_n)$$

$$= -\alpha(n+1)(\delta_n, L^\ell\delta_n) + \rho(n+1)(\delta_{n+1}, L^\ell\delta_n)$$

$$+ \alpha(n+1)(\delta_{n+1}, L^\ell\delta_{n+1}) - \alpha(n+1)(\delta_{n+1}, L^\ell\delta_{n+1})$$

$$= \hat{f}_{\ell-1,+}(n+1) - \alpha(n+1)\big(\hat{g}_{\ell,+}(n+1) + \hat{g}_{\ell,+}(n)\big),$$

$$\hat{h}_{\ell,+}(n) = (\delta_n, L^\ell D\delta_n) = \beta(n)(\delta_n, L^\ell\delta_n) + \rho(n)(\delta_n, L^\ell\delta_{n-1})$$

$$+ \beta(n)(\delta_{n-1}, L^\ell\delta_{n-1}) - \beta(n)(\delta_{n-1}, L^\ell\delta_{n-1})$$

$$= \hat{h}_{\ell-1,+}(n-1) + \beta(n)\big(\hat{g}_{\ell,+}(n) + \hat{g}_{\ell,+}(n-1)\big),$$

that is, the coefficients satisfy (C.77). The remaining cases follow analogously. □

Finally, we derive an explicit expression for the Lax pair for the Ablowitz–Ladik hierarchy, but first we need some notation. Let T be a bounded operator in $\ell^2(\mathbb{Z})$. Given the standard basis (3.58) in $\ell^2(\mathbb{Z})$, we represent T by

$$T = \big(T(m,n)\big)_{(m,n)\in\mathbb{Z}^2}, \quad T(m,n) = (\delta_m, T\delta_n), \quad (m,n) \in \mathbb{Z}^2.$$

Actually, for our purpose below, it is sufficient that T is an N-diagonal matrix for some $N \in \mathbb{N}$. Moreover, we introduce the upper and lower triangular parts $T_\pm$ of T by

$$T_\pm = \big(T_\pm(m,n)\big)_{(m,n)\in\mathbb{Z}^2}, \quad T_\pm(m,n) = \begin{cases} T(m,n), & \pm(n-m) > 0, \\ 0, & \text{otherwise.} \end{cases} \tag{3.72}$$

3 The Ablowitz–Ladik Hierarchy

Next, consider the finite difference expression $P_{\underline{p}}$ defined by

$$P_{\underline{p}} = \frac{i}{2}\sum_{\ell=1}^{p_+} c_{p_+-\ell,+}\big((L^\ell)_+ - (L^\ell)_-\big) - \frac{i}{2}\sum_{\ell=1}^{p_-} c_{p_--\ell,-}\big((L^{-\ell})_+ - (L^{-\ell})_-\big)$$
$$- \frac{i}{2} c_{\underline{p}} Q_d, \quad \underline{p} \in \mathbb{N}_0^2, \tag{3.73}$$

with L given by (3.59) and Q_d denoting the doubly infinite diagonal matrix

$$Q_d = \big((-1)^k \delta_{k,\ell}\big)_{k,\ell \in \mathbb{Z}}. \tag{3.74}$$

Before we prove that $(L, P_{\underline{p}})$ is indeed the Lax pair for the Ablowitz–Ladik hierarchy, we derive one more representation of $P_{\underline{p}}$ in terms of L.

We denote by $\ell_0(\mathbb{Z})$ the set of complex-valued sequences of compact support. If R denotes a finite difference expression, then ψ is called a weak solution of $R\psi = z\psi$, for some $z \in \mathbb{C}$, if the relation holds pointwise for each lattice point, that is, if $((R - z)\psi)(n) = 0$ for all $n \in \mathbb{Z}$.

Lemma 3.15 *Let $\psi \in \ell_0^\infty(\mathbb{Z})$. Then the difference expression $P_{\underline{p}}$ defined in (3.73) acts on ψ by*

$$(P_{\underline{p}}\psi)(n) = i\Bigg(-\sum_{\ell=1}^{p_-} f_{p_--\ell,-}(n)(EL^{-\ell}\psi)(n) - \sum_{\ell=0}^{p_+-1} f_{p_+-1-\ell,+}(n)(EL^\ell\psi)(n)$$
$$+ \sum_{\ell=1}^{p_-} g_{p_--\ell,-}(n)(L^{-\ell}\psi)(n) + \sum_{\ell=1}^{p_+} g_{p_+-\ell,+}(n)(L^\ell\psi)(n)$$
$$+ \frac{1}{2}\big(g_{p_-,-}(n) + g_{p_+,+}(n)\big)\psi(n)\Bigg)\delta_{\text{odd}}(n) \tag{3.75}$$
$$+ i\Bigg(\sum_{\ell=0}^{p_--1} h_{p_--1-\ell,-}(n)(D^{-1}L^{-\ell}\psi)(n)$$
$$+ \sum_{\ell=1}^{p_+} h_{p_+-\ell,+}(n)(D^{-1}L^\ell\psi)(n)$$
$$- \sum_{\ell=1}^{p_-} g_{p_--\ell,-}(n)(L^{-\ell}\psi)(n) - \sum_{\ell=1}^{p_+} g_{p_+-\ell,+}(n)(L^\ell\psi)(n)$$
$$- \frac{1}{2}\big(g_{p_-,-}(n) + g_{p_+,+}(n)\big)\psi(n)\Bigg)\delta_{\text{even}}(n), \quad n \in \mathbb{Z}.$$

3.3 Lax Pairs for the AL Hierarchy

In addition, if u is a weak solution of $Lu(z) = zu(z)$, then

$$\begin{aligned}(P_{\underline{p}}u(z))(n) \\ = \Big(&-iF_{\underline{p}}(z,n)(Eu(z))(n) + \frac{i}{2}(G_{\underline{p}}(z,n) + K_{\underline{p}}(z,n))u(z,n)\Big)\delta_{\text{odd}}(n) \\ + \Big(&iH_{\underline{p}}(z,n)(D^{-1}u(z))(n) - \frac{i}{2}(G_{\underline{p}}(z,n) + K_{\underline{p}}(z,n))u(z,n)\Big)\delta_{\text{even}}(n),\end{aligned}$$

$$n \in \mathbb{Z}, \qquad (3.76)$$

in the weak sense.

Proof We consider the case where n is even and use induction on $\underline{p} = (p_-, p_+)$. The case n odd is analogous. For $\underline{p} = (0,0)$, the formulas (3.75) and (3.73) match. Denote by $\widehat{P}_{\underline{p}}$ the corresponding homogeneous difference expression where all summation constants $c_{k,\pm}$, $k = 1, \ldots, p_\pm$, vanish. We have to show that

$$i\widehat{P}_{\underline{p}} = i\widehat{P}^+_{p_+-1}L - \hat{h}_{p_+-1,+}D^{-1}L + \frac{1}{2}(\hat{g}_{p_+-1,+}L + \hat{g}_{p_+,+})$$
$$+ i\widehat{P}^-_{p_--1}L^{-1} - \hat{h}_{p_--1,-}D^{-1} + \frac{1}{2}(\hat{g}_{p_--1,-}L^{-1} + \hat{g}_{p_-,-}),$$

where $\widehat{P}^\pm_j$ correspond to the powers of L in (3.73), $\widehat{P}^\pm_j = \frac{1}{2}((L^{\pm j})_\pm - (L^{\pm j})_\mp)$. This can be done upon considering $(\delta_m, \widehat{P}_{\underline{p}}\delta_n)$ and making appropriate case distinctions $m = n$, $m > n$, and $m < n$.
Using (3.60), (3.63), (3.66)–(3.70), (3.72), and Lemma 3.14, one verifies, for instance, in the case $m = n$,

$$\begin{aligned}&(\delta_n, i\widehat{P}^+_{p_+}\delta_n) \\ &= (\delta_n, i\widehat{P}^+_{p_+-1}L\delta_n) + \alpha(n+1)\hat{h}_{p_+-1,+}(n) \\ &\quad + \frac{1}{2}(\hat{g}_{p_+,+}(n) - \alpha(n+1)\beta(n)\hat{g}_{p_+-1,+}(n)) \\ &= \Big(\delta_n, \frac{1}{2}((L^{p_+-1})_+ - (L^{p_+-1})_-) \\ &\qquad \times (\alpha^{++}\rho^+\delta_{n-1} + \alpha^+\beta\delta_n + \alpha^+\rho\delta_{n+1} - \rho^-\rho\delta_{n+2})\Big) \\ &\quad + \alpha(n+1)\hat{h}_{p_+-1,+}(n) + \frac{1}{2}(\hat{g}_{p_+,+}(n) - \alpha(n+1)\beta(n)\hat{g}_{p_+-1,+}(n)) \\ &= -\frac{1}{2}\alpha(n+1)\rho(n)(\delta_n, L^{p_+-1}\delta_{n-1}) + \frac{1}{2}\alpha(n+2)\rho(n+1)(\delta_n, L^{p_+-1}\delta_{n+1}) \\ &\quad - \frac{1}{2}\rho(n+1)\rho(n+2)(\delta_n, L^{p_+-1}\delta_{n+2}) + \alpha(n+1)\hat{h}_{p_+-1,+}(n) \\ &\quad + \frac{1}{2}(\hat{g}_{p_+,+}(n) - \alpha(n+1)\beta(n)\hat{g}_{p_+-1,+}(n))\end{aligned}$$

$$= -\alpha(n+1)\rho(n)(\delta_n, L^{p_+-1}\delta_{n-1})$$
$$- \alpha(n+1)\beta(n)\hat{g}_{p_+-1,+}(n) + \alpha(n+1)\hat{h}_{p_+-1,+}(n)$$
$$= 0,$$

since by Lemma 3.14,

$$\hat{g}_{p_+,+}(n) = (\delta_n, L^{p_+-1}L\delta_n),$$
$$\hat{h}_{p_+-1,+}(n) = (\delta_n, L^{p_+-1}D\delta_n) = \beta(n)(\delta_n, L^{p_+-1}\delta_n) + \rho(n)(\delta_n, L^{p_+-1}\delta_{n-1}).$$

Similarly,

$$(\delta_n, i\widehat{P}_{p_-}^-\delta_n)$$
$$= (\delta_n, i\widehat{P}_{p_--1}^- L^{-1}\delta_n) - \alpha(n)\hat{h}_{p_--1,-}(n)$$
$$+ \frac{1}{2}(\hat{g}_{p_-,-}(n) - \alpha(n)\beta(n+1)\hat{g}_{p_--1,-}(n))$$
$$= \left(\delta_n, \frac{1}{2}((L^{1-p_-})_+ - (L^{1-p_-})_-) \right.$$
$$\left. \times (\rho^+\rho^{++}\delta_{n-2} + \alpha\rho^+\delta_{n-1} - \alpha\beta^+\delta_n + \alpha^-\rho\delta_{n+1})\right)$$
$$- \alpha(n)\hat{h}_{p_--1,-}(n) + \frac{1}{2}(\hat{g}_{p_-,-}(n) - \alpha(n)\beta(n+1)\hat{g}_{p_--1,-}(n))$$
$$= -\frac{1}{2}\rho(n-1)\rho(n)(\delta_n, L^{1-p_-}\delta_{n-2}) - \frac{1}{2}\alpha(n-1)\rho(n)(\delta_n, L^{1-p_-}\delta_{n-1})$$
$$+ \frac{1}{2}\alpha(n)\rho(n+1)(\delta_n, L^{1-p_-}\delta_{n+1}) - \alpha(n)\hat{h}_{p_--1,-}(n)$$
$$+ \frac{1}{2}(\hat{g}_{p_-,-}(n) - \alpha(n)\beta(n+1)\hat{g}_{p_--1,-}(n))$$
$$= \alpha(n)\rho(n+1)(\delta_n, L^{1-p_-}\delta_{n+1}) - \alpha(n)\beta(n+1)\hat{g}_{p_--1,-}(n)$$
$$- \alpha(n)\hat{h}_{p_--1,-}(n)$$
$$= 0,$$

where we used Lemma 3.14 and (3.26) at $\ell = p_- - 2$ for the last equality. This proves the case $m = n$. The remaining cases $m > n$ and $m < n$ are settled in a similar fashion.

Equality (3.76) then follows from $Lu(z) = zu(z)$ and (3.13)–(3.16). $\square$

3.3 Lax Pairs for the AL Hierarchy

Next, we introduce the difference expression $P_{\underline{p}}^\top$ by

$$P_{\underline{p}}^\top = -\frac{i}{2}\sum_{\ell=1}^{p_+} c_{p_+-\ell,+}\big(((L^\top)^\ell)_+ - ((L^\top)^\ell)_-\big) \tag{3.77}$$

$$+ \frac{i}{2}\sum_{\ell=1}^{p_-} c_{p_--\ell,-}\big(((L^\top)^{-\ell})_+ - ((L^\top)^{-\ell})_-\big) - \frac{i}{2}c_{\underline{p}}\, Q_d, \quad \underline{p} \in \mathbb{N}_0^2,$$

with $L^\top = ED$ the difference expression associated with the transpose of the infinite matrix (3.59) in the standard basis of $\ell^2(\mathbb{Z})$ and Q_d denoting the doubly infinite diagonal matrix in (3.74). Here we used

$$(M_+)^\top = (M^\top)_-, \quad (M_-)^\top = (M^\top)_+$$

for a finite difference expression M in the standard basis of $\ell^2(\mathbb{Z})$.

For later purpose in Section 3.8 we now mention the analog of Lemma 3.15 for the difference expression $P_{\underline{p}}^\top$:

Lemma 3.16 *Let $\chi \in \ell_0^\infty(\mathbb{Z})$. Then the difference expression $P_{\underline{p}}^\top$ defined in (3.77) acts on χ by*

$$(P_{\underline{p}}^\top \chi)(n) = i\bigg(-\sum_{\ell=0}^{p_--1} h_{p_--1-\ell,-}(n)(E^{-1}(L^\top)^{-\ell}\chi)(n)$$

$$-\sum_{\ell=1}^{p_+} h_{p_+-\ell,+}(n)(E^{-1}(L^\top)^\ell \chi)(n)$$

$$+ \sum_{\ell=1}^{p_-} g_{p_--\ell,-}(n)((L^\top)^{-\ell}\chi)(n) + \sum_{\ell=1}^{p_+} g_{p_+-\ell,+}(n)((L^\top)^\ell \chi)(n)$$

$$+ \frac{1}{2}\big(g_{p_-,-}(n) + g_{p_+,+}(n)\big)\chi(n)\bigg)\delta_{\text{odd}}(n)$$

$$+ i\bigg(\sum_{\ell=1}^{p_-} f_{p_--\ell,-}(n)(D(L^\top)^{-\ell}\chi)(n)$$

$$+ \sum_{\ell=0}^{p_+-1} f_{p_+-1-\ell,+}(n)(D(L^\top)^\ell \chi)(n)$$

$$- \sum_{\ell=1}^{p_-} g_{p_--\ell,-}(n)((L^\top)^{-\ell}\chi)(n)$$

$$- \sum_{\ell=1}^{p_+} g_{p_+-\ell,+}(n)((L^\top)^\ell \chi)(n)$$

$$- \frac{1}{2}\big(g_{p_-,-}(n) + g_{p_+,+}(n)\big)\chi(n)\bigg)\delta_{\text{even}}(n), \quad n \in \mathbb{Z}.$$

210   3 The Ablowitz–Ladik Hierarchy

In addition, if v is a weak solution of $L^\top v(z) = zv(z)$, then

$$\bigl(P_{\underline{p}}^\top v(z)\bigr)(n)$$
$$= -i\Bigl(H_{\underline{p}}(z,n)(E^{-1}v(z))(n) - \frac{1}{2}(G_{\underline{p}}(z,n) + K_{\underline{p}}(z,n))v(z,n)\Bigr)\delta_{\mathrm{odd}}(n)$$
$$+ i\Bigl(F_{\underline{p}}(z,n)(Dv(z))(n) - \frac{1}{2}(G_{\underline{p}}(z,n) + K_{\underline{p}}(z,n))v(z,n)\Bigr)\delta_{\mathrm{even}}(n),$$

$$n \in \mathbb{Z}, \quad (3.78)$$

in the weak sense.

Proof It suffices to consider the case where n is even. As in the proof of Lemma 3.15 we have to show that the corresponding homogeneous difference expression $\widehat{P}_{\underline{p}}^\top$ satisfies

$$i\widehat{P}_{\underline{p}}^\top = i(\widehat{P}_{p_+-1}^\top)^+ L^\top - \hat{f}_{p_+-1,+}D + \frac{1}{2}\bigl(\hat{g}_{p_+-1,+}L^\top + \hat{g}_{p_+,+}\bigr)$$
$$+ i(\widehat{P}_{p_--1}^\top)^- (L^\top)^{-1} - \hat{f}_{p_--1,-}D(L^\top)^{-1}$$
$$+ \frac{1}{2}\bigl(\hat{g}_{p_--1,-}(L^\top)^{-1} + \hat{g}_{p_-,-}\bigr),$$

where $(\widehat{P}_j^\top)^\pm = \frac{i}{2}\bigl(((L^\top)^{\pm j})_\mp - ((L^\top)^{\pm j})_\pm\bigr)$. Note that $L^\top$, $(L^\top)^{-1}$ are given by (3.60), (3.63) with δ_{even} and δ_{odd} interchanged. Using (3.67)–(3.70), (3.72), one verifies, for instance, in the case $m = n$,

$$\bigl(\delta_n, i(\widehat{P}_{p_+}^\top)^+ \delta_n\bigr)$$
$$= \bigl(\delta_n, i(\widehat{P}_{p_+-1}^\top)^+(\rho^+\rho^{++}\delta_{n-2} + \beta\rho^+\delta_{n-1} - \alpha^+\beta\delta_n + \beta^-\rho\delta_{n+1})\bigr)$$
$$- \beta(n)\hat{f}_{p_+-1,+}(n) + \frac{1}{2}\bigl(\hat{g}_{p_+,+}(n) - \alpha(n+1)\beta(n)\hat{g}_{p_+-1,+}(n)\bigr)$$
$$= -\frac{1}{2}\beta(n-1)\rho(n)\bigl(\delta_n, (L^\top)^{p_+-1}\delta_{n-1}\bigr)$$
$$+ \frac{1}{2}\beta(n)\rho(n+1)\bigl(\delta_n, (L^\top)^{p_+-1}\delta_{n+1}\bigr)$$
$$- \frac{1}{2}\rho(n-1)\rho(n)\bigl(\delta_n, (L^\top)^{p_+-1}\delta_{n-2}\bigr) - \beta(n)\hat{f}_{p_+-1,+}(n)$$
$$+ \frac{1}{2}\bigl(\hat{g}_{p_+,+}(n) - \alpha(n+1)\beta(n)\hat{g}_{p_+-1,+}(n)\bigr)$$
$$= \beta(n)\bigl(-\alpha(n+1)\hat{g}_{p_+-1,+}(n) + \rho(n+1)\bigl(\delta_n, (L^\top)^{p_+-1}\delta_{n+1}\bigr)\bigr)$$
$$- \beta(n)\hat{f}_{p_+-1,+}(n)$$
$$= 0,$$

since Lemma 3.14 reads in terms of $L^\top$ for n even

$$\hat{g}_{\ell,\pm}(n) = (\delta_n, (L^\top)^{\pm\ell}\delta_n), \quad \hat{f}_{\ell,+}(n) = (\delta_n, (L^\top)^\ell E \delta_n).$$

Similarly,

$$\begin{aligned}
&(\delta_n, i(\widehat{P}_{p_-}^\top)^- \delta_n)\\
&= (\delta_n, i(\widehat{P}_{p_--1}^\top)^-(-\beta^{++}\rho^+\delta_{n-1} - \alpha\beta^+\delta_n - \beta^+\rho\delta_{n+1} + \rho^-\rho\delta_{n+2}))\\
&\quad + \beta(n+1)\hat{f}_{p_--1,-}(n) + \frac{1}{2}\big(\hat{g}_{p_-,-}(n) - \alpha(n)\beta(n+1)\hat{g}_{p_--1,-}(n)\big)\\
&= -\frac{1}{2}\beta(n+1)\rho(n)(\delta_n, (L^\top)^{1-p_-}\delta_{n-1})\\
&\quad + \frac{1}{2}\beta(n+2)\rho(n+1)(\delta_n, (L^\top)^{1-p_-}\delta_{n+1})\\
&\quad - \frac{1}{2}\rho(n+1)\rho(n+2)(\delta_n, (L^\top)^{1-p_-}\delta_{n+2}) + \beta(n+1)\hat{f}_{p_--1,-}(n)\\
&\quad + \frac{1}{2}\big(\hat{g}_{p_-,-}(n) - \alpha(n)\beta(n+1)\hat{g}_{p_--1,-}(n)\big)\\
&= -\beta(n+1)\big(\alpha(n)\hat{g}_{p_--1,-}(n) + \rho(n)(\delta_n, (L^\top)^{1-p_-}\delta_{n-1})\big)\\
&\quad + \beta(n+1)\hat{f}_{p_--1,-}(n)\\
&= 0,
\end{aligned}$$

where we used $\hat{f}_{\ell,-}(n) = (\delta_n, (L^\top)^{-\ell}D^{-1}\delta_n)$. This proves the case $m = n$. The remaining cases $m > n$ and $m < n$ are settled in a similar fashion. Equality (3.78) then follows from $L^\top v(z) = zv(z)$ and (3.13)–(3.16). □

Given these preliminaries, one can now prove the following result, the proof of which is based on fairly tedious computations. We present them here in detail as these results have not appeared in print before.

Theorem 3.17 *Assume Hypothesis* 3.1. *Then, for each $\underline{p} \in \mathbb{N}_0^2$, the $\underline{p}$th stationary Ablowitz–Ladik equation* s-AL$_{\underline{p}}(\alpha, \beta) = 0$ *in* (3.30) *is equivalent to the vanishing of the commutator of $P_{\underline{p}}$ and L,*

$$[P_{\underline{p}}, L] = 0.$$

In addition, the $\underline{p}$th time-dependent Ablowitz–Ladik equation AL$_{\underline{p}}(\alpha, \beta) = 0$ *in* (3.54) *is equivalent to the Lax commutator equations*

$$L_{t_{\underline{p}}}(t_{\underline{p}}) - [P_{\underline{p}}(t_{\underline{p}}), L(t_{\underline{p}})] = 0, \quad t_{\underline{p}} \in \mathbb{R}. \tag{3.79}$$

In particular, the pair of difference expressions $(L, P_{\underline{p}})$ represents the Lax pair for the Ablowitz–Ladik hierarchy of nonlinear differential-difference evolution equations.

3 The Ablowitz–Ladik Hierarchy

Proof Let $f \in \ell_0(\mathbb{Z})$. We first choose n to be even and apply formulas (3.75) to compute the commutator $([P_{\underline{p}}, L]f)(n)$ by rewriting $D^{-1}L^\ell = EL^{\ell-1}$ and using (3.61), (3.67), and (3.68). This yields

$$
\begin{aligned}
i&([P_{\underline{p}}, L]f)(n) \\
&= \bigg(\sum_{\ell=1}^{p_+} \rho^- \rho(g^-_{p_+-\ell,+} - g^{--}_{p_+-\ell,+}) \\
&\quad - \sum_{\ell=0}^{p_+-1} \rho^- \rho(\alpha^- h^{--}_{p_+-1-\ell,+} + \beta^- f^-_{p_+-1-\ell,+})\bigg)(L^\ell f)(n-2) \\
&\quad + \bigg(\sum_{\ell=1}^{p_+} \rho(2\beta^- g^-_{p_+-\ell,+} - h^-_{p_+-\ell,+}) \\
&\quad + \sum_{\ell=0}^{p_+-1} \rho((\rho^-)^2 h^{--}_{p_+-1-\ell,+} - (\beta^-)^2 f^-_{p_+-1-\ell,+})\bigg)(L^\ell f)(n-1) \\
&\quad + \bigg(\sum_{\ell=1}^{p_+} \rho^+(\beta(g^+_{p_+-\ell,+} + g^-_{p_+-\ell,+}) - h_{p_+-\ell,+}) \\
&\quad + \sum_{\ell=0}^{p_+-1} \rho^+\big(h^-_{p_+-1-\ell,+} - \beta\beta^+ f^+_{p_+-1-\ell,+} \\
&\quad - \beta\alpha^+ h_{p_+-1-\ell,+}\big)\bigg)(L^\ell f)(n+1) \\
&\quad + \bigg(\sum_{\ell=2}^{p_++1} (g_{p_++1-\ell,+} - g^-_{p_++1-\ell,+}) \\
&\quad + \sum_{\ell=1}^{p_+} \Big(\alpha^+\big(\beta(g_{p_+-\ell,+} - g^-_{p_+-\ell,+}) + h_{p_+-\ell,+}\big) - \alpha h^-_{p_+-\ell,+}\Big) \\
&\quad + \sum_{\ell=0}^{p_+-1} \Big(\beta(\alpha^+)^2 h_{p_+-1-\ell,+} - \beta(\rho^+)^2 f^+_{p_+-1-\ell,+} \\
&\quad - \alpha^+ h^-_{p_+-1-\ell,+}\Big)\bigg)(L^\ell f)(n) \\
&\quad + \sum_{\ell=1}^{p_-} \rho^- \rho(g^{--}_{p_--\ell,-} - g^-_{p_--\ell,-} - \beta^- f^-_{p_--\ell,-} - \alpha^- h^{--}_{p_--\ell,-})(L^{-\ell} f)(n-2) \\
&\quad + \bigg(\sum_{\ell=1}^{p_-} \rho\big(\beta^-(2g^-_{p_--\ell,-} - \beta^- f^-_{p_--\ell,-}) + \rho(\rho^-)^2 h^{--}_{p_--\ell,-}\big) \\
&\quad - \sum_{\ell=0}^{p_--1} \rho h^-_{p_--1-\ell,-}\bigg)(L^{-\ell} f)(n-1)
\end{aligned}
$$

3.3 Lax Pairs for the AL Hierarchy

$$+ \left(\sum_{\ell=1}^{p_-} \left(\beta\rho^+(g^+_{p_--\ell,-} + g^-_{p_--\ell,-} - \alpha^+ h_{p_--\ell,-} - \beta^+ f^+_{p_--\ell,-})\right.\right.$$

$$\left. + \rho^+ h^-_{p_--\ell,-}\right) - \sum_{\ell=0}^{p_--1} \rho^+ h_{p_--1-\ell,-}\right)(L^{-\ell}f)(n+1)$$

$$+ \left(\sum_{\ell=1}^{p_-} \left(\beta\alpha^+(g_{p_--\ell,-} - g^-_{p_--\ell,-} + \alpha^+ h_{p_--\ell,-} + \beta^+ f^+_{p_--\ell,-}) - \beta f^+_{p_--\ell,-}\right.\right.$$

$$\left. - \alpha^+ h^-_{p_--\ell,-}\right) + \sum_{\ell=0}^{p_--1} \left(g_{p_--1-\ell,-} - g^-_{p_--1-\ell,-} - \alpha h^-_{p_--1-\ell,-}\right.$$

$$\left.\left. + \alpha^+ h_{p_--1-\ell,-}\right)\right)(L^{-\ell}f)(n)$$

$$+ \frac{1}{2}\Big(\beta\rho^+\big(g^+_{p_-,-} + g^+_{p_+,+} + g_{p_-,-} + g_{p_+,+}\big)f(n+1)$$

$$+ \beta^-\rho\big(g^-_{p_-,-} + g^-_{p_+,+} + g_{p_-,-} + g_{p_+,+}\big)f(n-1)$$

$$- \rho^-\rho\big(g^{--}_{p_-,-} + g^{--}_{p_+,+} - g_{p_-,-} - g_{p_+,+}\big)f(n-2)\Big),$$

where we added the terms

$$0 = -\sum_{\ell=1}^{p_+} g^-_{p_+-\ell,+}(L^{\ell+1}f)(n) + \sum_{\ell=1}^{p_+} g^-_{p_+-\ell,+}(L^{\ell+1}f)(n)$$

$$= -\sum_{\ell=2}^{p_++1} g^-_{p_++1-\ell,+}(L^\ell f)(n) + \sum_{\ell=1}^{p_+} g^-_{p_+-\ell,+}L(L^\ell f)(n),$$

$$0 = -\sum_{\ell=1}^{p_+} h^-_{p_+-\ell,+}(D^{-1}L^\ell f)(n) + \sum_{\ell=1}^{p_+} h^-_{p_+-\ell,+}(EL^{\ell-1}f)(n)$$

$$= -\sum_{\ell=1}^{p_+} h^-_{p_+-\ell,+}\Big(\alpha(L^\ell f)(n) + \rho(L^\ell f)(n-1)\Big)$$

$$+ \sum_{\ell=0}^{p_+-1} h^-_{p_+-1-\ell,+}\Big(-\alpha^+(L^\ell f)(n) + \rho^+(L^\ell f)(n+1)\Big), \qquad (3.80)$$

$$0 = -\sum_{\ell=1}^{p_-} g^-_{p_--\ell,-}(L^{-\ell+1}f)(n) + \sum_{\ell=1}^{p_-} g^-_{p_--\ell,-}(L^{-\ell+1}f)(n)$$

$$= -\sum_{\ell=0}^{p_--1} g^-_{p_--1-\ell,-}(L^{-\ell}f)(n) + \sum_{\ell=1}^{p_-} g^-_{p_--\ell,-}L(L^{-\ell}f)(n),$$

$$0 = -\sum_{\ell=1}^{p_-} h^-_{p_--\ell,-}(D^{-1}L^{-\ell+1}f)(n) + \sum_{\ell=1}^{p_-} h^-_{p_--\ell,-}(EL^{-\ell}f)(n)$$

$$= -\sum_{\ell=0}^{p_--1} h^-_{p_--1-\ell,-}\Big(\alpha(L^{-\ell}f)(n) + \rho(L^{-\ell}f)(n-1)\Big)$$

$$+ \sum_{\ell=1}^{p_-} h^-_{p_--\ell,-}\Big(-\alpha^+(L^{-\ell}f)(n) + \rho^+(L^{-\ell}f)(n+1)\Big).$$

Next we apply the recursion relations (3.19)–(3.26). In addition, we also use

$$\alpha^+ h_{p_--\ell,-} + \beta^+ f^+_{p_--\ell,-} = \alpha^+\big(h^+_{p_--1-\ell,-} - \beta^+(g^+_{p_--\ell,-} + g_{p_--\ell,-})\big)$$
$$+ \beta^+\big(f_{p_--1-\ell,-} + \alpha^+(g^+_{p_--\ell,-} + g_{p_--\ell,-})\big)$$
$$= g^+_{p_--\ell,-} - g_{p_--\ell,-}.$$

This implies,

$$i([P_{\underline{p}}, L]f)(n)$$

$$= \sum_{\ell=1}^{p_+-1} \rho^-\rho\Big(g^-_{p_+-\ell,+} - g^{--}_{p_+-\ell,+} - \alpha^- h^{--}_{p_+-1-\ell,+} - \beta^- f^-_{p_+-1-\ell,+}\Big)(L^\ell f)(n-2)$$

$$+ \sum_{\ell=1}^{p_+-1} \Big(\beta^-\rho(g^-_{p_+-\ell,+} - g^{--}_{p_+-\ell,+} - \alpha^- h^{--}_{p_+-1-\ell,+} - \beta^- f^-_{p_+-1-\ell,+})$$
$$+ \rho\big(\beta^-(g^-_{p_+-\ell,+} + g^{--}_{p_+-\ell,+}) + h^{--}_{p_+-1-\ell,+} - h^-_{p_+-\ell,+}\big)\Big)(L^\ell f)(n-1)$$

$$+ \sum_{\ell=1}^{p_+-1} \Big(\beta\rho^+(g^+_{p_+-\ell,+} - g_{p_+-\ell,+} - \alpha^+ h_{p_+-1-\ell,+} - \beta^+ f^+_{p_+-1-\ell,+})$$
$$+ \rho^+\big(\beta(g_{p_+-\ell,+} + g^-_{p_+-\ell,+}) + h^-_{p_+-1-\ell,+} - h_{p_+-\ell,+}\big)\Big)(L^\ell f)(n+1)$$

$$+ \bigg(\sum_{\ell=1}^{p_+-1} \big(g_{p_++1-\ell,+} - g^-_{p_++1-\ell,+} - \alpha h^-_{p_+-\ell,+} + \beta\alpha^+(g^+_{p_+-\ell,+} + g_{p_+-\ell,+})$$
$$- \beta f^+_{p_+-1-\ell,+}\big)$$

$$+ \sum_{\ell=1}^{p_+-1} \alpha^+\big(\beta(-g_{p_+-\ell,+} - g^-_{p_+-\ell,+}) + h_{p_+-\ell,+} - h^-_{p_+-1-\ell,+}\big)$$

$$+ \sum_{\ell=0}^{p_+-1} \beta\alpha^+\big(g_{p_+-\ell,+} - g^+_{p_+-\ell,+} + \alpha^+ h_{p_+-1-\ell,+}$$
$$+ \beta^+ f^+_{p_+-1-\ell,+}\big)\bigg)(L^\ell f)(n)$$

3.3 Lax Pairs for the AL Hierarchy

$$+ \sum_{\ell=1}^{p_-} \rho^- \rho (g^-_{p_--\ell,-} - g^{--}_{p_--\ell,-} - \beta^- f^-_{p_--\ell,-} - \alpha^- h^{--}_{p_--\ell,-})(L^{-\ell} f)(n-2)$$

$$+ \sum_{\ell=1}^{p_--1} \Big(\beta^- \rho (g^-_{p_--\ell,-} - g^{--}_{p_--\ell,-} - \beta^- f^-_{p_--\ell,-} - \alpha^- h^{--}_{p_--\ell,-})$$
$$\quad + \rho \big(\beta^- (g^-_{p_--\ell,-} + g^{--}_{p_--\ell,-}) + h^{--}_{p_--\ell,-}$$
$$\quad - h^-_{p_--1-\ell,-} \big) \Big) (L^{-\ell} f)(n-1)$$

$$+ \sum_{\ell=1}^{p_--1} \Big(\beta \rho^+ (g^+_{p_--\ell,-} - g_{p_--\ell,-} - \alpha^+ h_{p_--\ell,-} - \beta^+ f^+_{p_--\ell,-})$$
$$\quad + \rho^+ \big(\beta (g_{p_--\ell,-} + g^-_{p_--\ell,-}) + h^-_{p_--\ell,-}$$
$$\quad - h_{p_--1-\ell,-} \big) \Big) (L^{-\ell} f)(n+1)$$

$$+ \bigg(\sum_{\ell=1}^{p_--1} \Big(g_{p_--1-\ell,-} - g^-_{p_--1-\ell,-} - \alpha h^-_{p_--1-\ell,-} \Big)$$
$$\quad + \sum_{\ell=1}^{p_-} \beta \big(\alpha^+ (g^+_{p_--\ell,-} + g_{p_--\ell,-}) - f^+_{p_--\ell,-} \big)$$
$$\quad + \sum_{\ell=1}^{p_-} \beta \alpha^+ \big(g_{p_--\ell,-} - g^+_{p_--\ell,-} + \alpha^+ h_{p_--\ell,-} + \beta^+ f^+_{p_--\ell,-} \big)$$
$$\quad + \sum_{\ell=1}^{p_--1} \alpha^+ \big(\beta (-g_{p_--\ell,-} - g^-_{p_--\ell,-}) - h^-_{p_--\ell,-}$$
$$\quad + h_{p_--1-\ell,-} \big) \bigg) (L^{-\ell} f)(n)$$

$$+ \rho^- \rho (g^-_{0,+} - g^{--}_{0,+})(L^{p_+} f)(n-2)$$
$$- \rho^- \rho (\alpha^- h^{--}_{p_++-1,+} + \beta^- f^-_{p_++-1,+}) f(n-2)$$
$$+ \rho (2\beta^- g^-_{0,+} - h^-_{0,+})(L^{p_+} f)(n-1)$$
$$+ \rho \big((\rho^-)^2 h^{--}_{p_++-1,+} - (\beta^-)^2 f^-_{p_++-1,+} \big) f(n-1)$$
$$+ \rho^+ \big(\beta (g^+_{0,+} + g^-_{0,+}) - h_{0,+} \big) (L^{p_+} f)(n+1)$$
$$- \rho^+ \big(\beta (\alpha^+ h_{p_++-1,+} + \beta^+ f^+_{p_++-1,+}) - h^-_{p_++-1,+} \big) f(n+1)$$
$$+ (g_{0,+} - g^-_{0,+})(L^{p_++1} f)(n)$$
$$+ \big(g_{1,+} - g^-_{1,+} - \alpha h^-_{0,+} + \beta \alpha^+ (g_{0,+} - g^-_{0,+}) + \alpha^+ h_{0,+} \big) (L^{p_+} f)(n)$$
$$+ \big(g_{p_--1,-} - g^-_{p_--1,-} - \alpha h^-_{p_--1,-} - \beta f_{p_--1,-} \big) f(n)$$

$$
\begin{aligned}
&+ \beta\big(\alpha^+(g^+_{p_+,+} - g_{p_+,+}) + f_{p_--1,-} - f^+_{p_+-1,+}\big)f(n) \\
&+ \alpha^+\big(\beta(\beta f_{p_+-1,+} + \alpha h^-_{p_+-1,+}) + h_{p_--1,-} - h^-_{p_+-1,+}\big)f(n) \\
&- \big(\beta\alpha^+(g_{0,-} + g^-_{0,-}) - \beta f_{-1,-} + \alpha^+ h^-_{0,-}\big)(L^{-p_-}f)(n) \\
&- \big(\alpha h^-_{p_+-1,+} + \beta f_{p_+-1,+}\big)\big(\beta \rho^+ f(n+1) + \beta^- \rho f(n-1) + \rho^- \rho f(n-2)\big) \\
&+ \Big(\beta^- \rho(g^-_{0,-} + g^-_{0,-} - \beta^- f^-_{0,-}) + (\rho^-)^2 \rho h^{--}_{0,-}\Big)(L^{-p_-}f)(n-1) \\
&+ \Big(\beta \rho^+(g^+_{0,-} + g^-_{0,-} - \beta^+ f^+_{0,-} - \alpha^+ h_{0,-}) + \rho^+ h^-_{0,-}\Big)(L^{-p_-}f)(n+1) \\
&- \rho h^-_{p_--1,-} f(n-1) - \rho^+ h_{p_--1,-} f(n+1) \\
&+ \frac{1}{2}\Big(\beta \rho^+(g^+_{p_-,-} + g^+_{p_+,+} + g_{p_-,-} + g_{p_+,+})f(n+1) \\
&\qquad + \beta^- \rho(g^-_{p_-,-} + g^-_{p_+,+} + g_{p_-,-} + g_{p_+,+})f(n-1) \\
&\qquad - \rho^- \rho(g^{--}_{p_-,-} + g^{--}_{p_+,+} - g_{p_-,-} - g_{p_+,+})f(n-2)\Big) \\
&= \frac{\rho^- \rho}{2}(g^{--}_{p_+,+} - g^{--}_{p_-,-} + g_{p_-,-} - g_{p_+,+})f(n-2) \\
&+ \Big(\rho\big(\beta^-(g^{--}_{p_+,+} + g^-_{p_-,-}) - h^-_{p_--1,-} + h^{--}_{p_+-1,+}\big) \\
&\qquad + \frac{\beta^- \rho}{2}(g^-_{p_+,+} + g_{p_-,-} - g_{p_+,+} - g^-_{p_-,-})\Big)f(n-1) \\
&+ \Big(\beta\big(\alpha^+(g^+_{p_+,+} + g_{p_-,-}) + f_{p_--1,-} - f^+_{p_+-1,+}\big) \\
&\qquad - \alpha^+\big(\beta(g^-_{p_+,+} + g_{p_-,-}) - h_{p_--1,-} + h^-_{p_+-1,+}\big)\Big)f(n) \\
&+ \Big(\rho^+\big(\beta(g^-_{p_+,+} + g_{p_-,-}) - h_{p_--1,-} + h^-_{p_+-1,+}\big) \\
&\qquad + \frac{\beta \rho^+}{2}(g_{p_+,+} + g^+_{p_-,-} - g^+_{p_+,+} - g_{p_-,-})\Big)f(n+1),
\end{aligned}
$$

where we also used (3.37).

Performing the same calculation for n odd one obtains

$$
\begin{aligned}
&i([P_{\underline{p}}, L]f)(n) \\
&= \sum_{\ell=1}^{p_+-1} \rho^+ \rho^{++}\Big(g^{++}_{p_+-\ell,+} - g^+_{p_+-\ell,+} - \alpha^{++} h^+_{p_+-1-\ell,+} \\
&\qquad\qquad - \beta^{++} f^{++}_{p_+-1-\ell,+}\Big)(L^\ell f)(n+2) \\
&+ \sum_{\ell=1}^{p_+-1}\Big(\alpha^{++}\rho^+(g^+_{p_+-\ell,+} - g^{++}_{p_+-\ell,+} + \alpha^{++} h^+_{p_+-1-\ell,+} + \beta^{++} f^{++}_{p_+-1-\ell,+}\Big)
\end{aligned}
$$

3.3 Lax Pairs for the AL Hierarchy

$$+ \rho^+\big(\alpha^{++}(g^{++}_{p_+-\ell,+} + g^+_{p_+-\ell,+}) - f^{++}_{p_+-1-\ell,+}$$

$$+ f^+_{p_+-\ell,+}\big)\Big)(L^\ell f)(n+1)$$

$$+ \sum_{\ell=1}^{p_+-1}\Big(\alpha^+\rho(g^-_{p_+-\ell,+} - g_{p_+-\ell,+} + \alpha h^-_{p_+-1-\ell,+} + \beta f_{p_+-1-\ell,+})$$

$$+ \rho\big(\alpha^+(g^+_{p_+-\ell,+} + g_{p_+-\ell,+}) - f^+_{p_+-1-\ell,+} + f_{p_+-\ell,+}\big)\Big)(L^\ell f)(n-1)$$

$$+ \bigg(\sum_{\ell=2}^{p_+-1}\big(g^+_{p_++1-\ell,+} - g_{p_++1-\ell,+} - \alpha^+ h_{p_+-\ell,+} - \beta^+ f^+_{p_+-\ell,+}\big)$$

$$+ \sum_{\ell=1}^{p_+-1}\alpha^+\big(\beta(-g_{p_+-\ell,+} - g^-_{p_+-\ell,+}) + h_{p_+-\ell,+} - h^-_{p_+-1-\ell,+}\big)$$

$$+ \sum_{\ell=1}^{p_+-1}\beta\big(\alpha^+(g^+_{p_+-\ell,+} + g_{p_+-\ell,+}) - f^+_{p_+-1-\ell,+} + f_{p_+-\ell,+}\big)\bigg)(L^\ell f)(n)$$

$$+ \sum_{\ell=1}^{p_-}\rho^+\rho^{++}\big(g^{++}_{p_--\ell,-} - g^+_{p_--\ell,-} - \beta^{++} f^{++}_{p_--\ell,-}$$

$$- \alpha^{++} h^+_{p_--\ell,-}\big)(L^{-\ell} f)(n+2)$$

$$+ \bigg(\sum_{\ell=1}^{p_-}\alpha^{++}\rho^+\big(g^+_{p_--\ell,-} - g^{++}_{p_--\ell,-} + \beta^{++} f^{++}_{p_--\ell,-} + \alpha^{++} h^+_{p_--\ell,-}\big)$$

$$+ \sum_{\ell=1}^{p_--1}\rho^+\big(\alpha^{++}(g^{++}_{p_--\ell,-} + g^+_{p_--\ell,-}) - f^{++}_{p_--\ell,-}$$

$$+ f^+_{p_--1-\ell,-}\big)\bigg)(L^{-\ell} f)(n+1)$$

$$+ \bigg(\sum_{\ell=1}^{p_-}\alpha^+\rho\big(g^-_{p_--\ell,-} - g_{p_--\ell,-} + \alpha h^-_{p_--\ell,-} + \beta f_{p_--\ell,-}\big)$$

$$+ \sum_{\ell=1}^{p_--1}\rho\big(\alpha^+(g^+_{p_--\ell,-} + g_{p_--\ell,-}) - f^+_{p_--\ell,-}$$

$$+ f_{p_--1-\ell,-}\big)\bigg)(L^{-\ell} f)(n-1)$$

$$+ \bigg(\sum_{\ell=1}^{p_--1}\big(g^+_{p_--1-\ell,-} - g_{p_--1-\ell,-} - \alpha^+ h_{p_--1-\ell,-} - \beta^+ f^+_{p_--1-\ell,-}\big)$$

$$+ \sum_{\ell=1}^{p_--1} \beta\big(\alpha^+(g^+_{p_--\ell,-} + g_{p_--\ell,-}) - f^+_{p_--\ell,-} + f_{p_--1-\ell,-}\big)$$

$$+ \sum_{\ell=1}^{p_-} \alpha^+\big(\beta(-g_{p_--\ell,-} - g^-_{p_--\ell,-}) - h^-_{p_--\ell,-} + h_{p_--1-\ell,-}\big)\bigg)(L^{-\ell}f)(n)$$

$$+ \rho^+\rho^{++}(g^{++}_{0,+} - g^+_{0,+})(L^{p_+}f)(n+2)$$

$$+ \rho^+\rho^{++}\Big(\frac{1}{2}(g^{++}_{p_+,+} - g_{p_+,+}) - \alpha^{++}h^+_{p_+-1,+} - \beta^{++}f^{++}_{p_+-1,+}\Big)f(n+2)$$

$$+ \rho^+(2\alpha^{++}g^+_{0,+} + f^+_{0,+})(L^{p_+}f)(n+1)$$

$$+ \alpha^{++}\rho^+\Big(\frac{1}{2}(g^+_{p_+,+} + g_{p_+,+}) + \alpha^{++}h^+_{p_+-1,+} + \beta^{++}f^{++}_{p_+-1,+}\Big)f(n+1)$$

$$- \rho^+ f^{++}_{p_+-1,+}f(n+1)$$

$$+ \alpha^+\rho(g^-_{0,+} - g_{0,+})(L^{p_+}f)(n-1)$$

$$+ \rho\big(\alpha^+(g_{0,+} + g^+_{0,+}) + f_{0,+}\big)(L^{p_+}f)(n-1)$$

$$+ \alpha^+\rho\Big(\frac{1}{2}(g^-_{p_+,+} + g_{p_+,+}) + \alpha h^-_{p_+-1,+} + \beta f_{p_+-1,+}\Big)f(n-1)$$

$$- \rho f^+_{p_+-1,+}f(n-1)$$

$$+ (g^+_{0,+} - g_{0,+})(L^{p_++1}f)(n)$$

$$+ (g^+_{1,+} - g_{1,+} - \alpha^+ h_{0,+} - \beta^+ f^+_{0,+})(L^{p_+}f)(n)$$

$$+ \alpha^+\beta(g^+_{0,+} - g_{0,+})(L^{p_+}f)(n) + (\alpha^+ h_{0,+} + \beta f_{0,+})(L^{p_+}f)(n)$$

$$- (\beta^+ f^+_{p_+-1,+} + \alpha^+ h_{p_+-1,+})(Lf)(n)$$

$$+ \big(\alpha^+\beta(g_{p_+,+} - g^-_{p_+,+}) - \alpha^+ h^-_{p_+-1,+} - \beta f^+_{p_+-1,+}\big)f(n)$$

$$+ \frac{1}{2}\rho^+\rho^{++}(g^{++}_{p_-,-} - g_{p_-,-})f(n+2)$$

$$+ \rho^+\big(\alpha^{++}(g^+_{0,-} + g^{++}_{0,-}) - f^{++}_{0,-}\big)(L^{-p_-}f)(n+1)$$

$$+ \rho^+\Big(\frac{1}{2}\alpha^{++}(g_{p_-,-} + g^+_{p_+,+}) + f^+_{p_--1,-}\Big)f(n+1)$$

$$+ \rho\big(\alpha^+(g^+_{0,-} + g_{0,-}) - f^+_{0,-}\big)(L^{-p_-}f)(n-1)$$

$$+ \rho\Big(\frac{1}{2}\alpha^+(g^-_{p_-,-} + g_{p_+,+}) + f_{p_--1,-}\Big)f(n-1)$$

$$+ \beta\big(\alpha^+(g^+_{0,-} + g_{0,-}) - f^+_{0,-}\big)(L^{-p_-}f)(n)$$

$$+ (g^+_{p_--1,-} - g_{p_--1,-} - \beta^+ f^+_{p_--1,-} + \beta f_{p_--1,-})f(n)$$

$$- \alpha^+ h_{-1,-}f(n) \tag{3.81}$$

3.3 Lax Pairs for the AL Hierarchy   219

$$= \frac{\rho^+\rho^{++}}{2}\left(g^{++}_{p_-,-} - g^{++}_{p_+,+} + g_{p_+,+} - g_{p_-,-}\right) f(n+2)$$
$$+ \left(\rho^+\left(\alpha^{++}(g^{++}_{p_+,+} + g^+_{p_-,-}) - f^{++}_{p_+-1,+} + f^+_{p_--1,-}\right)\right.$$
$$+ \frac{\alpha^{++}\rho^+}{2}\left(g^+_{p_+,+} + g_{p_-,-} - g_{p_+,+} - g^+_{p_-,-}\right)\bigg) f(n+1)$$
$$+ \bigg(\beta\big(\alpha^+(g^+_{p_+,+} + g_{p_-,-}) + f_{p_--1,-} - f^+_{p_+-1,+}\big)$$
$$- \alpha^+\big(\beta(g^-_{p_+,+} + g_{p_-,-}) - h_{p_--1,-} + h^-_{p_+-1,+}\big)\bigg) f(n)$$
$$+ \bigg(\rho\big(\alpha^+(g^+_{p_+,+} + g_{p_-,-}) + f_{p_--1,-} - f^+_{p_+-1,+}\big)$$
$$+ \frac{\alpha^+\rho}{2}\left(g_{p_+,+} + g^-_{p_-,-} - g^-_{p_+,+} - g_{p_-,-}\right)\bigg) f(n-1).$$

Similarly to (3.80) we have added and subtracted the following terms in (3.81)

$$\sum_{\ell=1}^{p_+} g^+_{p_+-\ell,+}(L^{\ell+1}f)(n), \quad \sum_{\ell=1}^{p_+} f^+_{p_+-\ell,+}(D^{-1}L^\ell f)(n),$$

$$\sum_{\ell=1}^{p_-} g^+_{p_--\ell,-}(L^{-\ell+1}f)(n), \quad \sum_{\ell=1}^{p_-} f^+_{p_--\ell,-}(D^{-1}L^{-\ell+1}f)(n).$$

Comparing coefficients finally shows that (3.79) is equivalent to

$$(\rho^-\rho)_{t_{\underline{p}}} = \rho^-\rho(C^- + C), \tag{3.82}$$
$$(\alpha\rho^-)_{t_{\underline{p}}} = \rho^- A + \alpha\rho^- C^-, \tag{3.83}$$
$$(\beta\rho^+)_{t_{\underline{p}}} = \rho^+ B + \beta\rho^+ C^+, \tag{3.84}$$
$$(\alpha^+\beta)_{t_{\underline{p}}} = \beta A^+ + \alpha^+ B, \tag{3.85}$$

where

$$A = i\big(\alpha(g_{p_+,+} + g^-_{p_-,-}) - f_{p_+-1,+} + f^-_{p_--1,-}\big),$$
$$B = i\big(-\beta(g^-_{p_+,+} + g_{p_-,-}) + h_{p_--1,-} - h^-_{p_+-1,+}\big),$$
$$C = \frac{i}{2}\big(g_{p_+,+} + g^-_{p_-,-} - g^-_{p_+,+} - g_{p_-,-}\big).$$

In particular, (3.54) implies (3.79) since, by (3.55),

$$\rho_{t_{\underline{p}}} = \frac{i}{2}\rho\big(g_{p_+,+} + g^-_{p_-,-} - g^-_{p_+,+} - g_{p_-,-}\big).$$

To prove the converse assertion (i.e., that (3.79) implies (3.54)), we argue as follows:

Rewriting (3.83) and (3.84) using $\rho = \gamma^{1/2} = (1-\alpha\beta)^{1/2}$ and (3.82) yields

$$(1+\frac{\alpha\beta}{2\gamma})\alpha_{t_{\underline{p}}} + \frac{\alpha^2}{2\gamma}\beta_{t_{\underline{p}}} = A - \alpha C,$$

$$\frac{\beta^2}{2\gamma}\alpha_{t_{\underline{p}}} + (1+\frac{\alpha\beta}{2\gamma})\beta_{t_{\underline{p}}} = B - \beta C.$$

This linear system is uniquely solvable since its determinant equals γ^{-1} and the solution reads

$$\alpha_{t_{\underline{p}}} = A - \frac{\alpha}{2}(\beta A + \alpha B + 2\gamma C),$$

$$\beta_{t_{\underline{p}}} = B - \frac{\beta}{2}(\beta A + \alpha B + 2\gamma C).$$

Using (3.20) and (3.24) it is straightforward to check that $\beta A + \alpha B + 2\gamma C = 0$ which shows that the converse assertion also holds. $\square$

In the special stationary case, where $P_{\underline{p}}$ and L commute, $[P_{\underline{p}}, L] = 0$, one can prove that $P_{\underline{p}}$ and L satisfy an algebraic relationship of the type (cf. (3.341), (3.342))

$$P_{\underline{p}}^2 + R_{\underline{p}}(L) = P_{\underline{p}}^2 + \frac{1}{4}c_{0,+}^2 L^{-2p_-}\prod_{m=0}^{2p+1}(L - E_m) = 0, \qquad (3.86)$$

where (cf. (3.40))

$$\underline{p} = (p_-, p_+) \in \mathbb{N}_0^2 \setminus \{(0,0)\}, \quad \{E_m\}_{m=0}^{2p+1} \subset \mathbb{C}\setminus\{0\}, \quad p = p_- + p_+ - 1 \in \mathbb{N}_0.$$

Hence, the expression $P_{\underline{p}}^2 + R_{\underline{p}}(L)$ in (3.86) represents the Burchnall–Chaundy Laurent polynomial for the Lax pair $(L, P_{\underline{p}})$.

3.4 The Stationary Ablowitz–Ladik Formalism

> If I am given a formula, and I am ignorant of its meaning, it cannot teach me anything, but if I already know it what does the formula teach me?
>
> St. Augustine (354–430)

This section is devoted to a detailed study of the stationary Ablowitz–Ladik hierarchy and its algebro-geometric solutions. Our principal tools are derived from combining the Laurent polynomial recursion formalism introduced in Section 3.2 and a fundamental meromorphic function ϕ on a hyperelliptic curve $\mathcal{K}_p$. With the help of ϕ we study the Baker–Akhiezer vector Ψ, trace formulas, and theta function representations of ϕ, Ψ, α, and β.

3.4 The Stationary AL Formalism

Unless explicitly stated otherwise, we suppose in this section that

$$\alpha, \beta \in \mathbb{C}^{\mathbb{Z}}, \quad \alpha(n)\beta(n) \notin \{0, 1\}, \quad n \in \mathbb{Z}, \tag{3.87}$$

and assume (3.4), (3.5), (3.7), (3.13)–(3.16), (3.19)–(3.26), (3.27), (3.30), (3.39), (3.40), keeping $p \in \mathbb{N}_0$ fixed.

We recall the hyperelliptic curve

$$\mathcal{K}_p : \mathcal{F}_p(z, y) = y^2 - 4c_{0,+}^{-2} z^{2p_-} R_{\underline{p}}(z) = y^2 - \prod_{m=0}^{2p+1} (z - E_m) = 0, \tag{3.88}$$

$$R_{\underline{p}}(z) = \left(\frac{c_{0,+}}{2z^{p_-}}\right)^2 \prod_{m=0}^{2p+1} (z - E_m), \quad \{E_m\}_{m=0}^{2p+1} \subset \mathbb{C} \setminus \{0\}, \quad p = p_- + p_+ - 1,$$

as introduced in (3.42). Throughout this section we assume $\mathcal{K}_p$ to be nonsingular, that is, we suppose that

$$E_m \neq E_{m'} \text{ for } m \neq m', \quad m, m' = 0, 1, \ldots, 2p+1. \tag{3.89}$$

$\mathcal{K}_p$ is compactified by joining two points $P_{\infty_\pm}$, $P_{\infty_+} \neq P_{\infty_-}$, but for notational simplicity the compactification is also denoted by $\mathcal{K}_p$. Points P on $\mathcal{K}_p \setminus \{P_{\infty_+}, P_{\infty_-}\}$ are represented as pairs $P = (z, y)$, where $y(\cdot)$ is the meromorphic function on $\mathcal{K}_p$ satisfying $\mathcal{F}_p(z, y) = 0$. The complex structure on $\mathcal{K}_p$ is then defined in the usual manner. Hence, $\mathcal{K}_p$ becomes a hyperelliptic Riemann surface of genus p in a standard manner.

We also emphasize that by fixing the curve $\mathcal{K}_p$ (i.e., by fixing $E_0, \ldots, E_{2p+1}$), the summation constants $c_{1,\pm}, \ldots, c_{p_\pm,\pm}$ in $f_{p_\pm,\pm}$, $g_{p_\pm,\pm}$, and $h_{p_\pm,\pm}$ (and hence in the corresponding stationary s-AL$_{\underline{p}}$ equations) are uniquely determined as is clear from (3.46), (3.47), which establish the summation constants $c_{\ell,\pm}$ as symmetric functions of $E_0^{\pm 1}, \ldots, E_{2p+1}^{\pm 1}$.

For notational simplicity we will usually tacitly assume that $p \in \mathbb{N}$ and hence $\underline{p} \in \mathbb{N}_0^2 \setminus \{(0,0), (0,1), (1,0)\}$. (The trivial case $\underline{p} = 0$ is explicitly treated in Example 3.26.)

We denote by $\{\mu_j(n)\}_{j=1,\ldots,p}$ and $\{\nu_j(n)\}_{j=1,\ldots,p}$ the zeros of $(\cdot)^{p_-} F_{\underline{p}}(\cdot, n)$ and $(\cdot)^{p_- - 1} H_{\underline{p}}(\cdot, n)$, respectively. Thus, we may write

$$F_{\underline{p}}(z) = -c_{0,+} \alpha^+ z^{-p_-} \prod_{j=1}^{p} (z - \mu_j), \tag{3.90}$$

$$H_{\underline{p}}(z) = c_{0,+} \beta z^{-p_- + 1} \prod_{j=1}^{p} (z - \nu_j), \tag{3.91}$$

and we recall that (cf. (3.39))

$$R_{\underline{p}} - G_{\underline{p}}^2 = -F_{\underline{p}} H_{\underline{p}}. \tag{3.92}$$

The next step is crucial; it permits us to "lift" the zeros μ_j and ν_j from the complex plane $\mathbb{C}$ to the curve $\mathcal{K}_p$. From (3.92) one infers that

$$R_{\underline{p}}(z) - G_{\underline{p}}(z)^2 = 0, \quad z \in \{\mu_j, \nu_k\}_{j,k=1,\ldots,p}.$$

We now introduce $\{\hat{\mu}_j\}_{j=1,\ldots,p} \subset \mathcal{K}_p$ and $\{\hat{\nu}_j\}_{j=1,\ldots,p} \subset \mathcal{K}_p$ by

$$\hat{\mu}_j(n) = (\mu_j(n), (2/c_{0,+})\mu_j(n)^{p-}G_{\underline{p}}(\mu_j(n), n)), \quad j = 1, \ldots, p, \ n \in \mathbb{Z}, \tag{3.93}$$

and

$$\hat{\nu}_j(n) = (\nu_j(n), -(2/c_{0,+})\nu_j(n)^{p-}G_{\underline{p}}(\nu_j(n), n)), \quad j = 1, \ldots, p, \ n \in \mathbb{Z}. \tag{3.94}$$

We also introduce the points $P_{0,\pm}$ by

$$P_{0,\pm} = (0, \pm(c_{0,-}/c_{0,+})) \in \mathcal{K}_p, \quad \frac{c_{0,-}^2}{c_{0,+}^2} = \prod_{m=0}^{2p+1} E_m.$$

We emphasize that $P_{0,\pm}$ and $P_{\infty_\pm}$ are not necessarily on the same sheet of $\mathcal{K}_p$.

Next, we briefly recall our conventions used in connection with divisors on $\mathcal{K}_p$. A map, $\mathcal{D} \colon \mathcal{K}_p \to \mathbb{Z}$, is called a divisor on $\mathcal{K}_p$ if $\mathcal{D}(P) \neq 0$ for only finitely many $P \in \mathcal{K}_p$. The set of divisors on $\mathcal{K}_p$ is denoted by $\mathrm{Div}(\mathcal{K}_p)$. We shall employ the following (additive) notation for divisors,

$$\mathcal{D}_{Q_0 \underline{Q}} = \mathcal{D}_{Q_0} + \mathcal{D}_{\underline{Q}}, \quad \mathcal{D}_{\underline{Q}} = \mathcal{D}_{Q_1} + \cdots + \mathcal{D}_{Q_m},$$
$$\underline{Q} = \{Q_1, \ldots, Q_m\} \in \mathrm{Sym}^m \mathcal{K}_p, \quad Q_0 \in \mathcal{K}_p, \ m \in \mathbb{N},$$

where for any $Q \in \mathcal{K}_p$,

$$\mathcal{D}_Q \colon \mathcal{K}_p \to \mathbb{N}_0, \quad P \mapsto \mathcal{D}_Q(P) = \begin{cases} 1 & \text{for } P = Q, \\ 0 & \text{for } P \in \mathcal{K}_p \setminus \{Q\}, \end{cases}$$

and $\mathrm{Sym}^n \mathcal{K}_p$ denotes the nth symmetric product of $\mathcal{K}_p$. In particular, one can identify $\mathrm{Sym}^m \mathcal{K}_p$ with the set of nonnegative divisors $0 \leq \mathcal{D} \in \mathrm{Div}(\mathcal{K}_p)$ of degree m. Moreover, for a nonzero, meromorphic function f on $\mathcal{K}_p$, the divisor of f is denoted by (f). Two divisors $\mathcal{D}, \mathcal{E} \in \mathrm{Div}(\mathcal{K}_p)$ are called equivalent, denoted by $\mathcal{D} \sim \mathcal{E}$, if and only if $\mathcal{D} - \mathcal{E} = (f)$ for some $f \in \mathcal{M}(\mathcal{K}_p) \setminus \{0\}$. The divisor class $[\mathcal{D}]$ of $\mathcal{D}$ is then given by $[\mathcal{D}] = \{\mathcal{E} \in \mathrm{Div}(\mathcal{K}_p) \mid \mathcal{E} \sim \mathcal{D}\}$. We recall that

$$\deg((f)) = 0, \ f \in \mathcal{M}(\mathcal{K}_p) \setminus \{0\},$$

where the degree $\deg(\mathcal{D})$ of $\mathcal{D}$ is given by $\deg(\mathcal{D}) = \sum_{P \in \mathcal{K}_p} \mathcal{D}(P)$.

3.4 The Stationary AL Formalism

Next we introduce the fundamental meromorphic function on $\mathcal{K}_p$ by

$$\phi(P, n) = \frac{(c_{0,+}/2)z^{-p_-}y + G_{\underline{p}}(z, n)}{F_{\underline{p}}(z, n)} \tag{3.95}$$

$$= \frac{-H_{\underline{p}}(z, n)}{(c_{0,+}/2)z^{-p_-}y - G_{\underline{p}}(z, n)}, \tag{3.96}$$

$$P = (z, y) \in \mathcal{K}_p, \ n \in \mathbb{Z},$$

with divisor $(\phi(\,\cdot\,, n))$ of $\phi(\,\cdot\,, n)$ given by

$$(\phi(\,\cdot\,, n)) = \mathcal{D}_{P_{0,-}\hat{\underline{\nu}}(n)} - \mathcal{D}_{P_{\infty_-}\hat{\underline{\mu}}(n)}, \tag{3.97}$$

using (3.90) and (3.91). Here we abbreviated

$$\hat{\underline{\mu}} = \{\hat{\mu}_1, \ldots, \hat{\mu}_p\}, \ \hat{\underline{\nu}} = \{\hat{\nu}_1, \ldots, \hat{\nu}_p\} \in \mathrm{Sym}^p(\mathcal{K}_p).$$

Given $\phi(\,\cdot\,, n)$, the meromorphic stationary Baker–Akhiezer vector $\Psi(\,\cdot\,, n, n_0)$ on $\mathcal{K}_p$ is then defined by

$$\Psi(P, n, n_0) = \begin{pmatrix} \psi_1(P, n, n_0) \\ \psi_2(P, n, n_0) \end{pmatrix},$$

$$\psi_1(P, n, n_0) = \begin{cases} \prod_{n'=n_0+1}^{n} \left(z + \alpha(n')\phi^-(P, n')\right), & n \geq n_0 + 1, \\ 1, & n = n_0, \\ \prod_{n'=n+1}^{n_0} \left(z + \alpha(n')\phi^-(P, n')\right)^{-1}, & n \leq n_0 - 1, \end{cases} \tag{3.98}$$

$$\psi_2(P, n, n_0) = \phi(P, n_0) \begin{cases} \prod_{n'=n_0+1}^{n} \left(z\beta(n')\phi^-(P, n')^{-1} + 1\right), & n \geq n_0 + 1, \\ 1, & n = n_0, \\ \prod_{n'=n+1}^{n_0} \left(z\beta(n')\phi^-(P, n')^{-1} + 1\right)^{-1}, & n \leq n_0 - 1. \end{cases}$$

Basic properties of ϕ and Ψ are summarized in the following result.

Lemma 3.18 *Suppose α, β satisfy (3.87) and the $\underline{p}$th stationary Ablowitz–Ladik system (3.30). Moreover, assume (3.88) and (3.89) and let $P = (z, y) \in \mathcal{K}_p \setminus \{P_{\infty_+}, P_{\infty_-}, P_{0,+}, P_{0,-}\}$, $(n, n_0) \in \mathbb{Z}^2$. Then ϕ satisfies the Riccati-type equation*

$$\alpha\phi(P)\phi^-(P) - \phi^-(P) + z\phi(P) = z\beta, \tag{3.99}$$

as well as

$$\phi(P)\phi(P^*) = \frac{H_{\underline{p}}(z)}{F_{\underline{p}}(z)}, \tag{3.100}$$

$$\phi(P) + \phi(P^*) = 2\frac{G_{\underline{p}}(z)}{F_{\underline{p}}(z)}, \tag{3.101}$$

$$\phi(P) - \phi(P^*) = c_{0,+}z^{-p_-}\frac{y(P)}{F_{\underline{p}}(z)}. \tag{3.102}$$

The vector Ψ satisfies

$$U(z)\Psi^-(P) = \Psi(P), \tag{3.103}$$

$$V_{\underline{p}}(z)\Psi^-(P) = -\frac{i}{2}c_{0,+}z^{-p_-}y\Psi^-(P), \tag{3.104}$$

$$\psi_2(P, n, n_0) = \phi(P, n)\psi_1(P, n, n_0), \tag{3.105}$$

$$\psi_1(P, n, n_0)\psi_1(P^*, n, n_0) = z^{n-n_0}\frac{F_{\underline{p}}(z, n)}{F_{\underline{p}}(z, n_0)}\Gamma(n, n_0), \tag{3.106}$$

$$\psi_2(P, n, n_0)\psi_2(P^*, n, n_0) = z^{n-n_0}\frac{H_{\underline{p}}(z, n)}{F_{\underline{p}}(z, n_0)}\Gamma(n, n_0), \tag{3.107}$$

$$\psi_1(P, n, n_0)\psi_2(P^*, n, n_0) + \psi_1(P^*, n, n_0)\psi_2(P, n, n_0)$$
$$= 2z^{n-n_0}\frac{G_{\underline{p}}(z, n)}{F_{\underline{p}}(z, n_0)}\Gamma(n, n_0), \tag{3.108}$$

$$\psi_1(P, n, n_0)\psi_2(P^*, n, n_0) - \psi_1(P^*, n, n_0)\psi_2(P, n, n_0) \tag{3.109}$$
$$= -c_{0,+}z^{n-n_0-p_-}\frac{y}{F_{\underline{p}}(z, n_0)}\Gamma(n, n_0),$$

where we used the abbreviation

$$\Gamma(n, n_0) = \begin{cases} \prod_{n'=n_0+1}^{n} \gamma(n'), & n \geq n_0 + 1, \\ 1, & n = n_0, \\ \prod_{n'=n+1}^{n_0} \gamma(n')^{-1}, & n \leq n_0 - 1. \end{cases} \tag{3.110}$$

Proof To prove (3.99) one uses the definition (3.95) of ϕ and equations (3.8), (3.10), and (3.39) to obtain

$$\alpha\phi(P)\phi^-(P) - \phi(P)^- + z\phi(P) - z\beta$$
$$= \frac{1}{F_{\underline{p}}F_{\underline{p}}^-}\Big(\alpha G_{\underline{p}}G_{\underline{p}}^- + (c_{0,+}/2)z^{-p_-}y\alpha(G_{\underline{p}} + G_{\underline{p}}^-) + \alpha R_{\underline{p}}$$
$$- (G_{\underline{p}}^- + (c_{0,+}/2)z^{-p_-}y)F_{\underline{p}} + z(G_{\underline{p}} + (c_{0,+}/2)z^{-p_-}y)F_{\underline{p}}^- - z\beta F_{\underline{p}}F_{\underline{p}}^-\Big)$$
$$= \frac{1}{F_{\underline{p}}F_{\underline{p}}^-}\Big(\alpha G_{\underline{p}}(G_{\underline{p}} + G_{\underline{p}}^-) + F_{\underline{p}}(-\alpha H_{\underline{p}} - G_{\underline{p}}^- - z\beta F_{\underline{p}}^-) + zF_{\underline{p}}^- G_{\underline{p}}\Big) = 0.$$

Equations (3.100)–(3.102) are clear from the definitions of ϕ and y. By definition of ψ, (3.105) holds for $n = n_0$. By induction,

$$\frac{\psi_2(P, n, n_0)}{\psi_1(P, n, n_0)} = \frac{z\beta(n)\phi^-(P, n)^{-1} + 1}{z + \alpha(n)\phi^-(P, n)}\frac{\psi_2^-(P, n, n_0)}{\psi_1^-(P, n, n_0)} = \frac{z\beta(n) + \phi^-(P, n)}{z + \alpha(n)\phi^-(P, n)},$$

3.4 The Stationary AL Formalism

and hence ψ_2/ψ_1 satisfies the Riccati-type equation (3.99)

$$\alpha(n)\phi^-(P,n)\frac{\psi_2(P,n,n_0)}{\psi_1(P,n,n_0)} - \phi^-(P,n) + z\frac{\psi_2(P,n,n_0)}{\psi_1(P,n,n_0)} - z\beta(n) = 0.$$

This proves (3.105).

The definition of ψ implies

$$\psi_1(P,n,n_0) = (z + \alpha(n)\phi^-(P,n))\psi_1^-(P,n,n_0)$$
$$= z\psi_1^-(P,n,n_0) + \alpha(n)\psi_2^-(P,n,n_0),$$
$$\psi_2(P,n,n_0) = (z\beta(n)\phi^-(P,n)^{-1} + 1)\psi_2^-(P,n,n_0)$$
$$= z\beta(n)\psi_1^-(P,n,n_0) + \psi_2^-(P,n,n_0),$$

which proves (3.103). Property (3.104) follows from (3.105) and the definition of ϕ.
To prove (3.106) one can use (3.8) and (3.10)

$$\psi_1(P)\psi_1(P^*) = (z+\alpha\phi^-(P))(z+\alpha\phi^-(P^*))\psi_1^-(P)\psi_1^-(P^*)$$
$$= \frac{1}{F_{\underline{p}}^-}(z^2 F_{\underline{p}}^- + 2z\alpha G_{\underline{p}}^- + \alpha^2 H_{\underline{p}}^-)\psi_1^-(P)\psi_1^-(P^*)$$
$$= \frac{1}{F_{\underline{p}}^-}(z^2 F_{\underline{p}}^- - z\alpha\beta F_{\underline{p}} + z\alpha(G_{\underline{p}} + G_{\underline{p}}^-))\psi_1^-(P)\psi_1^-(P^*)$$
$$= z\gamma \frac{F_{\underline{p}}}{F_{\underline{p}}^-}\psi_1^-(P)\psi_1^-(P^*).$$

Equation (3.107) then follows from (3.100), (3.105), and (3.106). Finally, equation (3.108) (resp. (3.109)) is proved by combining (3.101) and (3.105) (resp. (3.102) and (3.105)). □

Combining the Laurent polynomial recursion approach of Section 3.2 with (3.90) and (3.91) readily yields trace formulas for $f_{\ell,\pm}$ and $h_{\ell,\pm}$ in terms of symmetric functions of the zeros μ_j and ν_k of $(\cdot)^{p-}F_{\underline{p}}$ and $(\cdot)^{p-{-}1}H_{\underline{p}}$, respectively. For simplicity we just record the simplest cases.

Lemma 3.19 *Suppose α, β satisfy (3.87) and the $\underline{p}$th stationary Ablowitz–Ladik system (3.30). Then,*

$$\frac{\alpha}{\alpha^+} = (-1)^{p+1}\frac{c_{0,+}}{c_{0,-}}\prod_{j=1}^{p}\mu_j, \qquad (3.111)$$

$$\frac{\beta^+}{\beta} = (-1)^{p+1}\frac{c_{0,+}}{c_{0,-}}\prod_{j=1}^{p}\nu_j, \qquad (3.112)$$

$$\sum_{j=1}^{p} \mu_j = \alpha^+ \beta - \gamma^+ \frac{\alpha^{++}}{\alpha^+} - \frac{c_{1,+}}{c_{0,+}}, \tag{3.113}$$

$$\sum_{j=1}^{p} \nu_j = \alpha^+ \beta - \gamma \frac{\beta^-}{\beta} - \frac{c_{1,+}}{c_{0,+}}. \tag{3.114}$$

Proof We compare coefficients in (3.13) and (3.90)

$$z^{p_-} F_{\underline{p}}(z) = f_{0,-} + \cdots + z^{p_- + p_+ - 2} f_{1,+} + z^{p_- + p_+ - 1} f_{0,+}$$

$$= c_{0,+} \alpha^+ \left((-1)^{p+1} \prod_{j=1}^{p} \mu_j + \cdots + z^{p_- + p_+ - 2} \sum_{j=1}^{p} \mu_j - z^{p_- + p_+ - 1} \right)$$

and use $f_{0,-} = c_{0,-} \alpha$ and $f_{1,+} = c_{0,+}((\alpha^+)^2 \beta - \gamma^+ \alpha^{++}) - \alpha^+ c_{1,+}$ which yields (3.111) and (3.113). Similarly, one employs $h_{0,-} = -c_{0,-} \beta^+$ and $h_{1,+} = c_{0,+}(\gamma \beta^- - \alpha^+ \beta^2) + \beta c_{1,+}$ for the remaining formulas (3.112) and (3.114). □

Remark 3.20 The trace formulas in Lemma 3.19 illustrate why we assumed the condition $\alpha(n)\beta(n) \neq 0$ for all $n \in \mathbb{N}$ throughout this chapter. Moreover, the following section shows that this condition is intimately connected with admissible divisors $\mathcal{D}_{\hat{\underline{\mu}}}, \mathcal{D}_{\hat{\underline{\nu}}}$ avoiding the exceptional points $P_{\infty_\pm}, P_{0,\pm}$. On the other hand, as is clear from the matrix representation (3.59) of the Lax difference expression L, if $\alpha(n_0)\beta(n_0) = 1$ for some $n_0 \in \mathbb{N}$, and hence $\rho(n_0) = 0$, the infinite matrix L splits into a direct sum of two half-line matrices $L_\pm(n_0)$ (in analogy to the familiar singular case of infinite Jacobi matrices $aS^+ + a^- S^- + b$ on $\mathbb{Z}$ with $a(n_0) = 0$). This explains why we assumed $\alpha(n)\beta(n) \neq 1$ for all $n \in \mathbb{N}$ throughout this chapter.

Next we turn to asymptotic properties of ϕ and Ψ in a neighborhood of $P_{\infty_\pm}$ and $P_{0,\pm}$.

Lemma 3.21 *Suppose α, β satisfy (3.87) and the pth stationary Ablowitz–Ladik system (3.30). Moreover, let $P = (z, y) \in \mathcal{K}_p \setminus \{P_{\infty_+}, P_{\infty_-}, P_{0,+}, P_{0,-}\}, (n, n_0) \in \mathbb{Z}^2$. Then ϕ has the asymptotic behavior*

$$\phi(P) \underset{\zeta \to 0}{=} \begin{cases} \beta + \beta^- \gamma \zeta + O(\zeta^2), & P \to P_{\infty_+}, \\ -(\alpha^+)^{-1} \zeta^{-1} + (\alpha^+)^{-2} \alpha^{++} \gamma^+ + O(\zeta), & P \to P_{\infty_-}, \end{cases} \quad \zeta = 1/z, \tag{3.115}$$

$$\phi(P) \underset{\zeta \to 0}{=} \begin{cases} \alpha^{-1} - \alpha^{-2} \alpha^- \gamma \zeta + O(\zeta^2), & P \to P_{0,+}, \\ -\beta^+ \zeta - \beta^{++} \gamma^+ \zeta^2 + O(\zeta^3), & P \to P_{0,-}, \end{cases} \quad \zeta = z. \tag{3.116}$$

3.4 The Stationary AL Formalism

The components of the Baker–Akhiezer vector Ψ have the asymptotic behavior

$$\psi_1(P, n, n_0) \underset{\zeta \to 0}{=} \begin{cases} \zeta^{n_0-n}(1 + O(\zeta)), & P \to P_{\infty_+}, \\ \frac{\alpha^+(n)}{\alpha^+(n_0)} \Gamma(n, n_0) + O(\zeta), & P \to P_{\infty_-}, \end{cases} \quad \zeta = 1/z, \quad (3.117)$$

$$\psi_1(P, n, n_0) \underset{\zeta \to 0}{=} \begin{cases} \frac{\alpha(n)}{\alpha(n_0)} + O(\zeta), & P \to P_{0,+}, \\ \zeta^{n-n_0} \Gamma(n, n_0)(1 + O(\zeta)), & P \to P_{0,-}, \end{cases} \quad \zeta = z, \quad (3.118)$$

$$\psi_2(P, n, n_0) \underset{\zeta \to 0}{=} \begin{cases} \beta(n)\zeta^{n_0-n}(1 + O(\zeta)), & P \to P_{\infty_+}, \\ -\frac{1}{\alpha^+(n_0)} \Gamma(n, n_0)\zeta^{-1}(1 + O(\zeta)), & P \to P_{\infty_-}, \end{cases} \quad \zeta = 1/z,$$

(3.119)

$$\psi_2(P, n, n_0) \underset{\zeta \to 0}{=} \begin{cases} \frac{1}{\alpha(n_0)} + O(\zeta), & P \to P_{0,+}, \\ -\beta^+(n) \Gamma(n, n_0)\zeta^{n+1-n_0}(1 + O(\zeta)), & P \to P_{0,-}, \end{cases} \quad \zeta = z.$$

(3.120)

The divisors (ψ_j) of ψ_j, $j = 1, 2$, are given by

$$(\psi_1(\cdot, n, n_0)) = \mathcal{D}_{\underline{\hat{\mu}}(n)} - \mathcal{D}_{\underline{\hat{\mu}}(n_0)} + (n - n_0)(\mathcal{D}_{P_{0,-}} - \mathcal{D}_{P_{\infty_+}}), \quad (3.121)$$

$$(\psi_2(\cdot, n, n_0)) = \mathcal{D}_{\underline{\hat{\nu}}(n)} - \mathcal{D}_{\underline{\hat{\mu}}(n_0)} + (n - n_0)(\mathcal{D}_{P_{0,-}} - \mathcal{D}_{P_{\infty_+}}) + \mathcal{D}_{P_{0,-}} - \mathcal{D}_{P_{\infty_-}}.$$

(3.122)

Proof The existence of the asymptotic expansion of ϕ in terms of the local coordinate $\zeta = 1/z$ near $P_{\infty_\pm}$, respectively, $\zeta = z$ near $P_{0,\pm}$ is clear from the explicit form of ϕ in (3.95) and (3.96). Insertion of the Laurent polynomials F_p into (3.95) and H_p into (3.96) then yields the explicit expansion coefficients in (3.115) and (3.116). Alternatively, and more efficiently, one can insert each of the following asymptotic expansions

$$\phi(P) \underset{z \to \infty}{=} \phi_{-1} z + \phi_0 + \phi_1 z^{-1} + O(z^{-2}),$$
$$\phi(P^*) \underset{z \to \infty}{=} \phi_0 + \phi_1 z^{-1} + O(z^{-2}),$$
$$\phi(P) \underset{z \to 0}{=} \phi_0 + \phi_1 z + O(z^2),$$
$$\phi(P^*) \underset{z \to 0}{=} \phi_1 z + \phi_2 z^2 + O(z^3)$$

(3.123)

into the Riccati-type equation (3.99) and, upon comparing coefficients of powers of z, which determines the expansion coefficients ϕ_k in (3.123), one concludes (3.115) and (3.116).

Next we compute the divisor of ψ_1. By (3.98) it suffices to compute the divisor of $z + \alpha \phi^-(P)$. First of all we note that

$$z + \alpha \phi^-(P) = \begin{cases} z + O(1), & P \to P_{\infty_+}, \\ \frac{\alpha^+}{\alpha}\gamma + O(z^{-1}), & P \to P_{\infty_-}, \\ \frac{\alpha}{\alpha^-} + O(z), & P \to P_{0,+}, \\ \gamma z + O(z^2), & P \to P_{0,-}, \end{cases}$$

which establishes (3.117) and (3.118). Moreover, the poles of the function $z + \alpha \phi^-(P)$ in $\mathcal{K}_p \setminus \{P_{0,\pm}, P_{\infty_\pm}\}$ coincide with the ones of $\phi^-(P)$, and so it remains to compute the missing p zeros in $\mathcal{K}_p \setminus \{P_{0,\pm}, P_{\infty_\pm}\}$. Using (3.10), (3.16), (3.39), and $y(\hat{\mu}_j) = (2/c_{0,+})\mu_j^{p-}G_{\underline{p}}(\mu_j)$ (cf. (3.93)) one computes

$$z + \alpha \phi^-(P) = z + \alpha \frac{(c_{0,+}/2)z^{-p_-}y + G_{\underline{p}}^-}{F_{\underline{p}}^-}$$

$$= \frac{F_{\underline{p}} + \alpha((c_{0,+}/2)z^{-p_-}y - G_{\underline{p}})}{F_{\underline{p}}^-}$$

$$= \frac{F_{\underline{p}}}{F_{\underline{p}}^-} + \alpha \frac{(c_{0,+}/2)^2 z^{-2p_-} y^2 - G_{\underline{p}}^2}{F_{\underline{p}}^-((c_{0,+}/2)z^{-p_-}y + G_{\underline{p}})}$$

$$= \frac{F_{\underline{p}}}{F_{\underline{p}}^-}\left(1 + \frac{\alpha H_{\underline{p}}}{(c_{0,+}/2)z^{-p_-}y + G_{\underline{p}}}\right) \underset{P \to \hat{\mu}_j}{=} \frac{F_{\underline{p}}(P)}{F_{\underline{p}}^-(P)} O(1).$$

Hence the sought after zeros are at $\hat{\mu}_j$, $j = 1, \ldots, p$ (with the possibility that a zero at $\hat{\mu}_j$ is cancelled by a pole at $\hat{\mu}_j^-$).

Finally, the behavior of ψ_2 follows immediately using $\psi_2 = \phi \psi_1$. □

In addition to (3.115), (3.116) one can use the Riccati-type equation (3.99) to derive a convergent expansion of ϕ around $P_{\infty_\pm}$ and $P_{0,\pm}$ and recursively determine the coefficients as in Lemma 3.21. Since this is not used later in this section, we omit further details at this point.

Since nonspecial divisors play a fundamental role in the derivation of theta function representations of algebro-geometric solutions of the AL hierarchy, we now take a closer look at them.

Lemma 3.22 *Suppose α, β satisfy (3.87) and the $\underline{p}$th stationary Ablowitz–Ladik system (3.30). Moreover, assume (3.88) and (3.89) and let $n \in \mathbb{Z}$. Let $\mathcal{D}_{\hat{\underline{\mu}}}$, $\hat{\underline{\mu}} = \{\hat{\mu}_1, \ldots, \hat{\mu}_p\}$, and $\mathcal{D}_{\hat{\underline{\nu}}}$, $\hat{\underline{\nu}} = \{\hat{\nu}_1, \ldots, \hat{\nu}_p\}$, be the pole and zero divisors of degree p,*

3.4 The Stationary AL Formalism

respectively, associated with α, β, and ϕ defined according to (3.93) and (3.94), that is,

$$\hat{\mu}_j(n) = (\mu_j(n), (2/c_{0,+})\mu_j(n)^{p^-} G_{\underline{p}}(\mu_j(n), n)), \quad j = 1, \ldots, p,$$

$$\hat{\nu}_j(n) = (\nu_j(n), -(2/c_{0,+})\nu_j(n)^{p^-} G_{\underline{p}}(\nu_j(n), n)), \quad j = 1, \ldots, p.$$

Then $\mathcal{D}_{\underline{\hat{\mu}}(n)}$ and $\mathcal{D}_{\underline{\hat{\nu}}(n)}$ are nonspecial for all $n \in \mathbb{Z}$.

Proof We provide a detailed proof in the case of $\mathcal{D}_{\underline{\hat{\mu}}(n)}$. By Theorem A.32, $\mathcal{D}_{\underline{\hat{\mu}}(n)}$ is special if and only if $\{\hat{\mu}_1(n), \ldots, \hat{\mu}_p(n)\}$ contains at least one pair of the type $\{\hat{\mu}(n), \hat{\mu}(n)^*\}$. Hence $\mathcal{D}_{\underline{\hat{\mu}}(n)}$ is certainly nonspecial as long as the projections $\mu_j(n)$ of $\hat{\mu}_j(n)$ are mutually distinct, $\mu_j(n) \neq \mu_k(n)$ for $j \neq k$. On the other hand, if two or more projections coincide for some $n_0 \in \mathbb{Z}$, for instance,

$$\mu_{j_1}(n_0) = \cdots = \mu_{j_N}(n_0) = \mu_0, \quad N \in \{2, \ldots, p\},$$

then $G_{\underline{p}}(\mu_0, n_0) \neq 0$ as long as $\mu_0 \notin \{E_0, \ldots, E_{2p+1}\}$. This fact immediately follows from (3.39) since $F_{\underline{p}}(\mu_0, n_0) = 0$ but $R_{\underline{p}}(\mu_0) \neq 0$ by hypothesis. In particular, $\hat{\mu}_{j_1}(n_0), \ldots, \hat{\mu}_{j_N}(n_0)$ all meet on the same sheet since

$$\hat{\mu}_{j_r}(n_0) = (\mu_0, (2/c_{0,+})\mu_0^{p^-} G_{\underline{p}}(\mu_0, n_0)), \quad r = 1, \ldots, N,$$

and hence no special divisor can arise in this manner. Remaining to be studied is the case where two or more projections collide at a branch point, say at $(E_{m_0}, 0)$ for some $n_0 \in \mathbb{Z}$. In this case one concludes

$$F_{\underline{p}}(z, n_0) \underset{z \to E_{m_0}}{=} O\big((z - E_{m_0})^2\big)$$

and

$$G_{\underline{p}}(E_{m_0}, n_0) = 0 \tag{3.124}$$

using again (3.39) and $F_{\underline{p}}(E_{m_0}, n_0) = R_{\underline{p}}(E_{m_0}) = 0$. Since $G_{\underline{p}}(\,\cdot\,, n_0)$ is a Laurent polynomial, (3.124) implies $G_{\underline{p}}(z, n_0) \underset{z \to E_{m_0}}{=} O((z - E_{m_0}))$. Thus, using (3.39) once more, one obtains the contradiction,

$$O\big((z - E_{m_0})^2\big) \underset{z \to E_{m_0}}{=} R_{\underline{p}}(z)$$

$$\underset{z \to E_{m_0}}{=} \left(\frac{c_{0,+}}{2 E_{m_0}^{p^-}}\right)^2 (z - E_{m_0}) \bigg(\prod_{\substack{m=0 \\ m \neq m_0}}^{2p+1} (E_{m_0} - E_m) + O(z - E_{m_0})\bigg).$$

Consequently, at most one $\hat{\mu}_j(n)$ can hit a branch point at a time and again no special divisor arises. Finally, by our hypotheses on α, β, $\hat{\mu}_j(n)$ stay finite for fixed $n \in \mathbb{Z}$ and hence never reach the points $P_{\infty_\pm}$. (Alternatively, by (3.115), $\hat{\mu}_j$ never reaches the point P_{∞_+}. Hence, if some $\hat{\mu}_j$ tend to infinity, they all necessarily converge to P_{∞_-}.) Again no special divisor can arise in this manner.

The proof for $\mathcal{D}_{\underline{\hat{\nu}}(n)}$ is analogous (replacing $F_{\underline{p}}$ by $H_{\underline{p}}$ and noticing that by (3.115), ϕ has no zeros near $P_{\infty_\pm}$), thereby completing the proof. □

Next, we shall provide an explicit representation of ϕ, Ψ, α, and β in terms of the Riemann theta function associated with $\mathcal{K}_p$. We freely employ the notation established in Appendices A and B. (We recall our tacit assumption $p \in \mathbb{N}$ to avoid the trivial case $p = 0$.)

Let θ denote the Riemann theta function associated with $\mathcal{K}_p$ and introduce a fixed homology basis $\{a_j, b_j\}_{j=1,\ldots,p}$ on $\mathcal{K}_p$. Choosing as a convenient fixed base point one of the branch points, $Q_0 = (E_{m_0}, 0)$, the Abel maps $\underline{A}_{Q_0}$ and $\underline{\alpha}_{Q_0}$ are defined by (A.29) and (A.30) and the Riemann vector $\underline{\Xi}_{Q_0}$ is given by (A.41). Let $\omega^{(3)}_{P_+, P_-}$ be the normal differential of the third kind holomorphic on $\mathcal{K}_p \setminus \{P_+, P_-\}$ with simple poles at $P_\pm$ and residues ± 1, respectively. In particular, one obtains for $\omega^{(3)}_{P_{0,-}, P_{\infty_\pm}}$,

$$\omega^{(3)}_{P_{0,-}, P_{\infty_\pm}} = \left(\frac{y + y_{0,-}}{z} \mp \prod_{j=1}^{p}(z - \lambda_{\pm, j})\right)\frac{dz}{2y}, \quad P_{0,-} = (0, y_{0,-}), \quad (3.125)$$

where the constants $\{\lambda_{\pm, j}\}_{j=1}^{p} \subset \mathbb{C}$ are uniquely determined by employing the normalization

$$\int_{a_j} \omega^{(3)}_{P_{0,-}, P_{\infty_\pm}} = 0, \quad j = 1, \ldots, p.$$

The explicit formula (3.125) then implies the following asymptotic expansions (using the local coordinate $\zeta = z$ near $P_{0,\pm}$ and $\zeta = 1/z$ near $P_{\infty_\pm}$),

$$\int_{Q_0}^{P} \omega^{(3)}_{P_{0,-}, P_{\infty_-}} \underset{\zeta \to 0}{=} \begin{Bmatrix} 0 \\ \ln(\zeta) \end{Bmatrix} + \omega_0^{0,\pm}(P_{0,-}, P_{\infty_-}) + O(\zeta) \text{ as } P \to P_{0,\pm}, \quad (3.126)$$

$$\int_{Q_0}^{P} \omega^{(3)}_{P_{0,-}, P_{\infty_-}} \underset{\zeta \to 0}{=} \begin{Bmatrix} 0 \\ -\ln(\zeta) \end{Bmatrix} + \omega_0^{\infty\pm}(P_{0,-}, P_{\infty_-}) + O(\zeta) \text{ as } P \to P_{\infty_\pm}, \quad (3.127)$$

$$\int_{Q_0}^{P} \omega^{(3)}_{P_{0,-}, P_{\infty_+}} \underset{\zeta \to 0}{=} \begin{Bmatrix} 0 \\ \ln(\zeta) \end{Bmatrix} + \omega_0^{0,\pm}(P_{0,-}, P_{\infty_+}) + O(\zeta) \text{ as } P \to P_{0,\pm}, \quad (3.128)$$

$$\int_{Q_0}^{P} \omega^{(3)}_{P_{0,-}, P_{\infty_+}} \underset{\zeta \to 0}{=} \begin{Bmatrix} -\ln(\zeta) \\ 0 \end{Bmatrix} + \omega_0^{\infty\pm}(P_{0,-}, P_{\infty_+}) + O(\zeta) \text{ as } P \to P_{\infty_\pm}. \quad (3.129)$$

3.4 The Stationary AL Formalism

Lemma 3.23 *With $\omega_0^{\infty\sigma}(P_{0,-}, P_{\infty\pm})$ and $\omega_0^{0,\sigma'}(P_{0,-}, P_{\infty\pm})$, $\sigma,\sigma' \in \{+,-\}$, defined as in (3.126)–(3.129) one has*

$$\exp\left(\omega_0^{0,-}(P_{0,-}, P_{\infty\pm}) - \omega_0^{\infty+}(P_{0,-}, P_{\infty\pm}) \right. \\ \left. - \omega_0^{\infty-}(P_{0,-}, P_{\infty\pm}) + \omega_0^{0,+}(P_{0,-}, P_{\infty\pm})\right) = 1. \tag{3.130}$$

Proof Pick $Q_{1,\pm} = (z_1, \pm y_1) \in \mathcal{K}_p \setminus \{P_{\infty\pm}\}$ in a neighborhood of $P_{\infty\pm}$ and $Q_{2,\pm} = (z_2, \pm y_2) \in \mathcal{K}_p \setminus \{P_{0,\pm}\}$ in a neighborhood of $P_{0,\pm}$. Without loss of generality one may assume that $P_{\infty+}$ and $P_{0,+}$ lie on the same sheet. Then by (3.125),

$$\int_{Q_0}^{Q_{2,-}} \omega_{P_{0,-},P_{\infty-}}^{(3)} - \int_{Q_0}^{Q_{1,+}} \omega_{P_{0,-},P_{\infty-}}^{(3)} - \int_{Q_0}^{Q_{1,-}} \omega_{P_{0,-},P_{\infty-}}^{(3)} + \int_{Q_0}^{Q_{2,+}} \omega_{P_{0,-},P_{\infty-}}^{(3)}$$
$$= \int_{Q_0}^{Q_{2,+}} \frac{dz}{z} - \int_{Q_0}^{Q_{1,+}} \frac{dz}{z} = \ln(z_2) - \ln(z_1) + 2\pi i k,$$

for some $k \in \mathbb{Z}$. On the other hand, by (3.126)–(3.129) one obtains

$$\int_{Q_0}^{Q_{2,-}} \omega_{P_{0,-},P_{\infty-}}^{(3)} - \int_{Q_0}^{Q_{1,+}} \omega_{P_{0,-},P_{\infty-}}^{(3)} - \int_{Q_0}^{Q_{1,-}} \omega_{P_{0,-},P_{\infty-}}^{(3)} + \int_{Q_0}^{Q_{2,+}} \omega_{P_{0,-},P_{\infty-}}^{(3)}$$
$$= \ln(z_2) + \ln(1/z_1) + \omega_0^{0,-}(P_{0,-}, P_{\infty-}) - \omega_0^{\infty+}(P_{0,-}, P_{\infty-})$$
$$- \omega_0^{\infty-}(P_{0,-}, P_{\infty-}) + \omega_0^{0,+}(P_{0,-}, P_{\infty-}) + O(z_2) + O(1/z_1),$$

and hence the part of (3.130) concerning $\omega_{P_{0,-},P_{\infty-}}^{(3)}$ follows. The corresponding result for $\omega_{P_{0,-},P_{\infty+}}^{(3)}$ is proved analogously. □

In the following it will be convenient to use the abbreviation

$$\underline{z}(P, \underline{Q}) = \underline{\Xi}_{Q_0} - \underline{A}_{Q_0}(P) + \underline{\alpha}_{Q_0}(\mathcal{D}_{\underline{Q}}),$$
$$P \in \mathcal{K}_p, \quad \underline{Q} = \{Q_1, \ldots, Q_p\} \in \mathrm{Sym}^p(\mathcal{K}_p).$$

We note that $\underline{z}(\,\cdot\,, \underline{Q})$ is independent of the choice of base point Q_0. For later use we state the following result.

Lemma 3.24 *The following relations hold:*

$$\underline{z}(P_{\infty+}, \underline{\hat{\mu}}^+) = \underline{z}(P_{\infty-}, \underline{\hat{\nu}}) = \underline{z}(P_{0,-}, \underline{\hat{\mu}}) = \underline{z}(P_{0,+}, \underline{\hat{\nu}}^+), \tag{3.131}$$
$$\underline{z}(P_{\infty+}, \underline{\hat{\nu}}^+) = \underline{z}(P_{0,-}, \underline{\hat{\nu}}), \quad \underline{z}(P_{0,+}, \underline{\hat{\mu}}^+) = \underline{z}(P_{\infty-}, \underline{\hat{\mu}}). \tag{3.132}$$

232   3 The Ablowitz–Ladik Hierarchy

Proof We indicate the proof of some of the relations to be used in (3.146) and (3.147). Let $\underline{\hat{\lambda}}$ denote either $\underline{\hat{\mu}}$ or $\underline{\hat{\nu}}$. Then,

$$\begin{aligned}
\underline{z}(P_{0,+}, \underline{\hat{\lambda}}^+) &= \underline{\Xi}_{Q_0} - \underline{A}_{Q_0}(P_{0,+}) + \underline{\alpha}_{Q_0}(\mathcal{D}_{\underline{\hat{\lambda}}^+}) \\
&= \underline{\Xi}_{Q_0} - \underline{A}_{Q_0}(P_{0,+}) + \underline{\alpha}_{Q_0}(\mathcal{D}_{\underline{\hat{\lambda}}}) + \underline{A}_{P_{0,-}}(P_{\infty_+}) \\
&= \underline{\Xi}_{Q_0} - \underline{A}_{Q_0}(P_{\infty_-}) + \underline{\alpha}_{Q_0}(\mathcal{D}_{\underline{\hat{\lambda}}}) \\
&= \underline{z}(P_{\infty_-}, \underline{\hat{\lambda}}), \\
\underline{z}(P_{\infty_+}, \underline{\hat{\lambda}}^+) &= \underline{\Xi}_{Q_0} - \underline{A}_{Q_0}(P_{\infty_+}) + \underline{\alpha}_{Q_0}(\mathcal{D}_{\underline{\hat{\lambda}}^+}) \\
&= \underline{\Xi}_{Q_0} - \underline{A}_{Q_0}(P_{\infty_+}) + \underline{\alpha}_{Q_0}(\mathcal{D}_{\underline{\hat{\lambda}}}) + \underline{A}_{P_{0,-}}(P_{\infty_+}) \\
&= \underline{\Xi}_{Q_0} - \underline{A}_{Q_0}(P_{0,-}) + \underline{\alpha}_{Q_0}(\mathcal{D}_{\underline{\hat{\lambda}}}) \\
&= \underline{z}(P_{0,-}, \underline{\hat{\lambda}}), \quad \text{etc.}
\end{aligned}$$

Here we used $\underline{A}_{Q_0}(P^*) = -\underline{A}_{Q_0}(P)$, $P \in \mathcal{K}_p$, since Q_0 is a branch point of $\mathcal{K}_p$, and $\underline{\alpha}_{Q_0}(\mathcal{D}_{\underline{\hat{\lambda}}^+}) = \underline{\alpha}_{Q_0}(\mathcal{D}_{\underline{\hat{\lambda}}}) + \underline{A}_{P_{0,-}}(P_{\infty_+})$. The latter equality immediately follows from (3.121) in the case $\underline{\hat{\lambda}} = \underline{\hat{\mu}}$ and from combining (3.97) and (3.122) in the case $\underline{\hat{\lambda}} = \underline{\hat{\nu}}$. $\square$

Given these preparations, the theta function representations of ϕ, ψ_1, ψ_2, α, and β then read as follows.

Theorem 3.25 *Suppose α, β satisfy (3.87) and the $\underline{p}$th stationary Ablowitz–Ladik system (3.30). Moreover, assume hypothesis (3.88) and (3.89), and let $P \in \mathcal{K}_p \setminus \{P_{\infty_+}, P_{\infty_-}, P_{0,+}, P_{0,-}\}$ and $(n, n_0) \in \mathbb{Z}^2$. Then for each $n \in \mathbb{Z}$, $\mathcal{D}_{\underline{\hat{\mu}}(n)}$ and $\mathcal{D}_{\underline{\hat{\nu}}(n)}$ are nonspecial. Moreover,*

$$\phi(P, n) = C(n) \frac{\theta(\underline{z}(P, \underline{\hat{\nu}}(n)))}{\theta(\underline{z}(P, \underline{\hat{\mu}}(n)))} \exp\left(\int_{Q_0}^{P} \omega^{(3)}_{P_{0,-}, P_{\infty_-}} \right), \qquad (3.133)$$

$$\psi_1(P, n, n_0) = C(n, n_0) \frac{\theta(\underline{z}(P, \underline{\hat{\mu}}(n)))}{\theta(\underline{z}(P, \underline{\hat{\mu}}(n_0)))} \exp\left((n - n_0) \int_{Q_0}^{P} \omega^{(3)}_{P_{0,-}, P_{\infty_+}} \right), \qquad (3.134)$$

$$\psi_2(P, n, n_0) = C(n) C(n, n_0) \frac{\theta(\underline{z}(P, \underline{\hat{\nu}}(n)))}{\theta(\underline{z}(P, \underline{\hat{\mu}}(n_0)))}$$

$$\times \exp\left(\int_{Q_0}^{P} \omega^{(3)}_{P_{0,-}, P_{\infty_-}} + (n - n_0) \int_{Q_0}^{P} \omega^{(3)}_{P_{0,-}, P_{\infty_+}} \right), \qquad (3.135)$$

3.4 The Stationary AL Formalism

where

$$C(n) = (-1)^{n-n_0} \exp\left((n-n_0)(\omega_0^{0,-}(P_{0,-}, P_{\infty-}) - \omega_0^{\infty+}(P_{0,-}, P_{\infty-}))\right)$$
$$\times \frac{1}{\alpha(n_0)} \exp\left(-\omega_0^{0,+}(P_{0,-}, P_{\infty-})\right) \frac{\theta(\underline{z}(P_{0,+}, \hat{\underline{\mu}}(n_0)))}{\theta(\underline{z}(P_{0,+}, \hat{\underline{\nu}}(n_0)))}, \quad (3.136)$$

$$C(n, n_0) = \exp\left(-(n-n_0)\omega_0^{\infty+}(P_{0,-}, P_{\infty+})\right) \frac{\theta(\underline{z}(P_{\infty+}, \hat{\underline{\mu}}(n_0)))}{\theta(\underline{z}(P_{\infty+}, \hat{\underline{\mu}}(n)))}. \quad (3.137)$$

The Abel map linearizes the auxiliary divisors $\mathcal{D}_{\hat{\underline{\mu}}(n)}$ *and* $\mathcal{D}_{\hat{\underline{\nu}}(n)}$ *in the sense that*

$$\underline{\alpha}_{Q_0}(\mathcal{D}_{\hat{\underline{\mu}}(n)}) = \underline{\alpha}_{Q_0}(\mathcal{D}_{\hat{\underline{\mu}}(n_0)}) + \underline{A}_{P_{0,-}}(P_{\infty+})(n-n_0), \quad (3.138)$$

$$\underline{\alpha}_{Q_0}(\mathcal{D}_{\hat{\underline{\nu}}(n)}) = \underline{\alpha}_{Q_0}(\mathcal{D}_{\hat{\underline{\nu}}(n_0)}) + \underline{A}_{P_{0,-}}(P_{\infty+})(n-n_0), \quad (3.139)$$

in addition,

$$\underline{\alpha}_{Q_0}(\mathcal{D}_{\hat{\underline{\nu}}(n)}) = \underline{\alpha}_{Q_0}(\mathcal{D}_{\hat{\underline{\mu}}(n)}) - \underline{A}_{Q_0}(P_{0,-}) + \underline{A}_{Q_0}(P_{\infty-})$$
$$= \underline{\alpha}_{Q_0}(\mathcal{D}_{\hat{\underline{\mu}}(n)}) + \underline{A}_{P_{0,-}}(P_{\infty-}). \quad (3.140)$$

Finally, α, β *are of the form*

$$\alpha(n) = \alpha(n_0)(-1)^{n-n_0}$$
$$\times \exp\left(-(n-n_0)(\omega_0^{0,-}(P_{0,-}, P_{\infty-}) - \omega_0^{\infty+}(P_{0,-}, P_{\infty-}))\right)$$
$$\times \frac{\theta(\underline{z}(P_{0,+}, \hat{\underline{\nu}}(n_0)))\theta(\underline{z}(P_{0,+}, \hat{\underline{\mu}}(n)))}{\theta(\underline{z}(P_{0,+}, \hat{\underline{\mu}}(n_0)))\theta(\underline{z}(P_{0,+}, \hat{\underline{\nu}}(n)))}, \quad (3.141)$$

$$\beta(n) = \beta(n_0)(-1)^{n-n_0}$$
$$\times \exp\left((n-n_0)(\omega_0^{0,-}(P_{0,-}, P_{\infty-}) - \omega_0^{\infty+}(P_{0,-}, P_{\infty-}))\right)$$
$$\times \frac{\theta(\underline{z}(P_{\infty+}, \hat{\underline{\mu}}(n_0)))\theta(\underline{z}(P_{\infty+}, \hat{\underline{\nu}}(n)))}{\theta(\underline{z}(P_{\infty+}, \hat{\underline{\nu}}(n_0)))\theta(\underline{z}(P_{\infty+}, \hat{\underline{\mu}}(n)))}, \quad (3.142)$$

$$\alpha(n)\beta(n) = \exp\left(\omega_0^{\infty+}(P_{0,-}, P_{\infty-}) - \omega_0^{0,+}(P_{0,-}, P_{\infty-})\right)$$
$$\times \frac{\theta(\underline{z}(P_{0,+}, \hat{\underline{\mu}}(n)))\theta(\underline{z}(P_{\infty+}, \hat{\underline{\nu}}(n)))}{\theta(\underline{z}(P_{0,+}, \hat{\underline{\nu}}(n)))\theta(\underline{z}(P_{\infty+}, \hat{\underline{\mu}}(n)))}, \quad (3.143)$$

$$\Gamma(n, n_0) = \exp\left((n-n_0)(\omega_0^{0,-}(P_{0,-}, P_{\infty+}) - \omega_0^{\infty+}(P_{0,-}, P_{\infty+}))\right)$$
$$\times \frac{\theta(\underline{z}(P_{0,-}, \hat{\underline{\mu}}(n)))\theta(\underline{z}(P_{\infty+}, \hat{\underline{\mu}}(n_0)))}{\theta(\underline{z}(P_{0,-}, \hat{\underline{\mu}}(n_0)))\theta(\underline{z}(P_{\infty+}, \hat{\underline{\mu}}(n)))}. \quad (3.144)$$

Proof Applying Abel's theorem (cf. Theorem A.16, (A.31)) to (3.97) proves (3.140), and applying it to (3.121), (3.122) results in (3.138) and (3.139). By Lemma 3.22,

$\mathcal{D}_{\hat{\underline{\mu}}}$ and $\mathcal{D}_{\hat{\underline{\nu}}}$ are nonspecial. By equation (3.97) and Theorem A.28, $\phi(P,n)\exp\bigl(-\int_{Q_0}^{P}\omega_{P_{0,-},P_{\infty_-}}^{(3)}\bigr)$ must be of the type

$$\phi(P,n)\exp\left(-\int_{Q_0}^{P}\omega_{P_{0,-},P_{\infty_-}}^{(3)}\right) = C(n)\frac{\theta(\underline{z}(P,\hat{\underline{\nu}}(n)))}{\theta(\underline{z}(P,\hat{\underline{\mu}}(n)))} \quad (3.145)$$

for some constant $C(n)$. A comparison of (3.145) and the asymptotic relations (3.115) then yields, with the help of (3.126), (3.127) and (3.131), (3.132), the following expressions for α and β:

$$\begin{aligned}(\alpha^+)^{-1} &= C^+ e^{\omega_0^{0,+}(P_{0,-},P_{\infty_-})}\frac{\theta(\underline{z}(P_{0,+},\hat{\underline{\nu}}^+))}{\theta(\underline{z}(P_{0,+},\hat{\underline{\mu}}^+))}\\ &= C^+ e^{\omega_0^{0,+}(P_{0,-},P_{\infty_-})}\frac{\theta(\underline{z}(P_{\infty_-},\hat{\underline{\nu}}))}{\theta(\underline{z}(P_{\infty_-},\hat{\underline{\mu}}))}\\ &= -C e^{\omega_0^{\infty_-}(P_{0,-},P_{\infty_-})}\frac{\theta(\underline{z}(P_{\infty_-},\hat{\underline{\nu}}))}{\theta(\underline{z}(P_{\infty_-},\hat{\underline{\mu}}))}.\end{aligned} \quad (3.146)$$

Similarly one obtains

$$\begin{aligned}\beta^+ &= C^+ e^{\omega_0^{\infty_+}(P_{0,-},P_{\infty_-})}\frac{\theta(\underline{z}(P_{\infty_+},\hat{\underline{\nu}}^+))}{\theta(\underline{z}(P_{\infty_+},\hat{\underline{\mu}}^+))}\\ &= C^+ e^{\omega_0^{\infty_+}(P_{0,-},P_{\infty_-})}\frac{\theta(\underline{z}(P_{0,-},\hat{\underline{\nu}}))}{\theta(\underline{z}(P_{0,-},\hat{\underline{\mu}}))}\\ &= -C e^{\omega_0^{0,-}(P_{0,-},P_{\infty_-})}\frac{\theta(\underline{z}(P_{0,-},\hat{\underline{\nu}}))}{\theta(\underline{z}(P_{0,-},\hat{\underline{\mu}}))}.\end{aligned} \quad (3.147)$$

Here we used (3.138) and (3.139), more precisely,

$$\underline{\alpha}_{Q_0}(\mathcal{D}_{\hat{\underline{\mu}}^+}) = \underline{\alpha}_{Q_0}(\mathcal{D}_{\hat{\underline{\mu}}}) + \underline{A}_{P_{0,-}}(P_{\infty_+}), \quad \underline{\alpha}_{Q_0}(\mathcal{D}_{\hat{\underline{\nu}}^+}) = \underline{\alpha}_{Q_0}(\mathcal{D}_{\hat{\underline{\nu}}}) + \underline{A}_{P_{0,-}}(P_{\infty_+}). \quad (3.148)$$

Thus, one concludes

$$C(n+1) = -\exp\bigl(\omega_0^{0,-}(P_{0,-},P_{\infty_-}) - \omega_0^{\infty_+}(P_{0,-},P_{\infty_-})\bigr)C(n), \quad n\in\mathbb{Z}, \quad (3.149)$$

and

$$C(n+1) = -\exp\bigl(\omega_0^{\infty_-}(P_{0,-},P_{\infty_-}) - \omega_0^{0,+}(P_{0,-},P_{\infty_-})\bigr)C(n), \quad n\in\mathbb{Z},$$

which is consistent with (3.130). The first-order difference equation (3.149) then implies

$$C(n) = (-1)^{(n-n_0)}\exp\bigl((n-n_0)(\omega_0^{0,-}(P_{0,-},P_{\infty_-}) - \omega_0^{\infty_+}(P_{0,-},P_{\infty_-}))\bigr)$$
$$\times C(n_0), \quad n,n_0\in\mathbb{Z}. \quad (3.150)$$

3.4 The Stationary AL Formalism

Thus one infers (3.141) and (3.142). Moreover, (3.150) and taking $n = n_0$ in the first line in (3.146) yield (3.136). Dividing the first line in (3.147) by the first line in (3.146) then proves (3.143).

By (3.121) and Theorem A.28, $\psi_1(P, n, n_0)$ must be of the type (3.134). A comparison of (3.98), (3.115), and (3.134) as $P \to P_{\infty_+}$ (with local coordinate $\zeta = 1/z$) then yields

$$\psi_1(P, n, n_0) \underset{\zeta \to 0}{=} \zeta^{n_0 - n}(1 + O(\zeta))$$

and

$$\psi_1(P, n, n_0) \underset{\zeta \to 0}{=} C(n, n_0) \frac{\theta(\underline{z}(P_{\infty_+}, \underline{\hat{\mu}}(n)))}{\theta(\underline{z}(P_{\infty_+}, \underline{\hat{\mu}}(n_0)))}$$
$$\times \exp\left[(n - n_0)\omega_0^{\infty_+}(P_{0,-}, P_{\infty_+})\right]\zeta^{n_0 - n}(1 + O(\zeta))$$

proving (3.137). Equation (3.135) is clear from (3.105), (3.133), and (3.134). Finally, a comparison of (3.118) and (3.134) as $P \to P_{0,-}$ (with local coordinate $\zeta = z$) yields

$$\psi_1(P, n, n_0) \underset{\zeta \to 0}{=} \Gamma(n, n_0)\zeta^{n - n_0}(1 + O(\zeta))$$
$$\underset{\zeta \to 0}{=} C(n, n_0) \frac{\theta(\underline{z}(P_{0,-}, \underline{\hat{\mu}}(n)))}{\theta(\underline{z}(P_{0,-}, \underline{\hat{\mu}}(n_0)))} \exp\left((n - n_0)\omega_0^{0,-}(P_{0,-}, P_{\infty_+})\right)$$
$$\times \zeta^{n - n_0}(1 + O(\zeta))$$

and hence

$$\Gamma(n, n_0) = C(n, n_0) \frac{\theta(\underline{z}(P_{0,-}, \underline{\hat{\mu}}(n)))}{\theta(\underline{z}(P_{0,-}, \underline{\hat{\mu}}(n_0)))} \exp\left((n - n_0)\omega_0^{0,-}(P_{0,-}, P_{\infty_+})\right)$$
$$= \exp\left((n - n_0)(\omega_0^{0,-}(P_{0,-}, P_{\infty_+}) - \omega_0^{\infty_+}(P_{0,-}, P_{\infty_+}))\right)$$
$$\times \frac{\theta(\underline{z}(P_{0,-}, \underline{\hat{\mu}}(n)))\theta(\underline{z}(P_{\infty_+}, \underline{\hat{\mu}}(n_0)))}{\theta(\underline{z}(P_{0,-}, \underline{\hat{\mu}}(n_0)))\theta(\underline{z}(P_{\infty_+}, \underline{\hat{\mu}}(n)))},$$

using (3.137). $\square$

Combining (3.138), (3.139) and (3.141), (3.142) shows the remarkable linearity of the theta function representations for α and β with respect to $n \in \mathbb{Z}$.

We note that the apparent n_0-dependence of $C(n)$ in the right-hand side of (3.136) actually drops out to ensure the n_0-independence of ϕ in (3.133).

The theta function representations (3.141), (3.142) for α, β and that for Γ in (3.144) also show that $\gamma(n) \notin \{0, 1\}$ for all $n \in \mathbb{Z}$, and hence condition (3.87) is satisfied for the stationary algebro-geometric AL solutions discussed in this section, provided the associated divisors $\mathcal{D}_{\underline{\hat{\mu}}}(n)$ and $\mathcal{D}_{\underline{\hat{\nu}}}(n)$ stay away from $P_{\infty_\pm}$, $P_{0,\pm}$ for all $n \in \mathbb{Z}$.

We conclude this section with the trivial case $\underline{p} = 0$ excluded thus far.

Example 3.26 Assume $\underline{p} = 0$ and $c_{0,+} = c_{0,-} = c_0 \neq 0$ (we recall that $g_{p_+,+} = g_{p_-,-}$). Then,

$$F_{(0,0)} = \widehat{F}_{(0,0)} = H_{(0,0)} = \widehat{H}_{(0,0)} = 0, \quad G_{(0,0)} = K_{(0,0)} = \frac{1}{2}c_0,$$

$$\widehat{G}_{(0,0)} = \widehat{K}_{(0,0)} = \frac{1}{2}, \quad R_{(0,0)} = \frac{1}{4}c_0^2,$$

$$\alpha = \beta = 0,$$

$$U = \begin{pmatrix} z & 0 \\ 0 & 1 \end{pmatrix}, \quad V_{(0,0)} = \frac{ic_0}{2}\begin{pmatrix} 1 & 0 \\ 0 & -1 \end{pmatrix}.$$

Introducing

$$\Psi_+(z, n, n_0) = \begin{pmatrix} z^{n-n_0} \\ 0 \end{pmatrix}, \quad \Psi_-(z, n, n_0) = \begin{pmatrix} 0 \\ 1 \end{pmatrix}, \quad n, n_0 \in \mathbb{Z},$$

one verifies the equations

$$U\Psi_\pm^- = \Psi_\pm, \quad V_{(0,0)}\Psi_\pm^- = \pm\frac{ic_0}{2}\Psi_\pm^-.$$

3.5 The Stationary Ablowitz–Ladik Algebro-Geometric Initial Value Problem

> I don't see how he can EVEN finish, if he doesn't begin.
> *Lewis Carroll*[1]

The aim of this section is to derive an algorithm that enables one to construct algebro-geometric solutions for the stationary Ablowitz–Ladik hierarchy for general complex-valued initial data. To this effect we will develop a new algorithm for constructing stationary complex-valued algebro-geometric solutions of the Ablowitz–Ladik hierarchy, which is of independent interest as it solves the inverse algebro-geometric spectral problem for general (i.e., non-normal) AL Lax operators $\breve{L}$ in (3.61), starting from a suitably chosen set of initial divisors of full measure.

The generally non-normal behavior of the underlying Lax operator $\breve{L}$ associated with general coefficients for the Ablowitz–Ladik hierarchy poses a variety of difficulties that we will briefly indicate next: First of all, given general initial data $\alpha^{(0)}, \beta^{(0)}$ or divisors $\mathcal{D}_{\underline{\hat{\mu}}(n_0)}$ in general complex position, the Dirichlet eigenvalues μ_j, $j = 1, \ldots, p$, are in general not pairwise distinct and "collisions" between them

---

[1] *Alice's Adventures in Wonderland* (1865), Project Gutenberg Etext, p. 70.

3.5 The Stationary AL Algebro-Geometric IVP

can occur at certain values of $n \in \mathbb{Z}$. A priori, one has no control over such collisions, especially, it is not possible to identify initial conditions $\mathcal{D}_{\hat{\underline{\mu}}(n_0)}$ at some $n_0 \in \mathbb{Z}$, which avoid collisions for all $n \in \mathbb{Z}$. We solve this problem by explicitly permitting collisions from the outset by referring to a general interpolation formalism (cf. Appendix D) for polynomials, going beyond the usual Lagrange interpolation formulas. In this manner it will be shown that collisions of Dirichlet eigenvalues no longer pose a problem. In addition, there is a second complication since it cannot be guaranteed that $\mu_j(n)$ and $\nu_j(n)$, $j = 1, \ldots, p$, stay finite and nonzero for all $n \in \mathbb{Z}$. We solve this particular problem by properly restricting the initial Dirichlet and Neumann divisors $\mathcal{D}_{\hat{\underline{\mu}}(n_0)}, \mathcal{D}_{\hat{\underline{\nu}}(n_0)} \in \mathrm{Sym}^p \mathcal{K}_p$ to a dense set of full measure.

Next we embark on the corresponding inverse problem that consists of constructing a solution of (3.30) given certain initial data. More precisely, we seek to construct solutions $\alpha, \beta \in \mathbb{C}^{\mathbb{Z}}$ satisfying the pth stationary Ablowitz–Ladik system (3.30) starting from a properly restricted set $\mathcal{M}_0$ of admissible nonspecial Dirichlet divisor initial data $\mathcal{D}_{\hat{\underline{\mu}}(n_0)}$ at some fixed $n_0 \in \mathbb{Z}$,

$$\hat{\underline{\mu}}(n_0) = \{\hat{\mu}_1(n_0), \ldots, \hat{\mu}_p(n_0)\} \in \mathcal{M}_0, \quad \mathcal{M}_0 \subset \mathrm{Sym}^p(\mathcal{K}_p),$$
$$\hat{\mu}_j(n_0) = (\mu_j(n_0), (2/c_{0,+})\mu_j(n_0)^{p-} G_{\underline{p}}(\mu_j(n_0), n_0)), \quad j = 1, \ldots, p. \quad (3.151)$$

For notational convenience we will frequently use the phrase that α, β *blow up* in this manuscript whenever one of the divisors $\mathcal{D}_{\hat{\underline{\mu}}}$ or $\mathcal{D}_{\hat{\underline{\nu}}}$ hits one of the points $P_{\infty_\pm}$, $P_{0,\pm}$.

Of course we would like to ensure that the sequences α, β obtained via our algorithm do not blow up. To investigate when this happens, we study the image of our divisors under the Abel map. A key ingredient in our analysis will be (3.138), which yields a linear discrete dynamical system on the Jacobi variety $J(\mathcal{K}_p)$. In particular, we will be led to investigate solutions $\mathcal{D}_{\hat{\underline{\mu}}}, \mathcal{D}_{\hat{\underline{\nu}}}$ of the discrete initial value problem

$$\underline{\alpha}_{Q_0}(\mathcal{D}_{\hat{\underline{\mu}}(n)}) = \underline{\alpha}_{Q_0}(\mathcal{D}_{\hat{\underline{\mu}}(n_0)}) + (n - n_0)\underline{A}_{P_{0,-}}(P_{\infty_+}),$$
$$\hat{\underline{\mu}}(n_0) = \{\hat{\mu}_1(n_0), \ldots, \hat{\mu}_p(n_0)\} \in \mathrm{Sym}^p(\mathcal{K}_p), \quad (3.152)$$

respectively

$$\underline{\alpha}_{Q_0}(\mathcal{D}_{\hat{\underline{\nu}}(n)}) = \underline{\alpha}_{Q_0}(\mathcal{D}_{\hat{\underline{\mu}}(n_0)}) + \underline{A}_{P_{0,-}}(P_{\infty_-}) + (n - n_0)\underline{A}_{P_{0,-}}(P_{\infty_+}),$$
$$\hat{\underline{\nu}}(n_0) = \{\hat{\nu}_1(n_0), \ldots, \hat{\nu}_p(n_0)\} \in \mathrm{Sym}^p(\mathcal{K}_p), \quad (3.153)$$

where $Q_0 \in \mathcal{K}_p$ is a given base point. Eventually, we will be interested in solutions $\mathcal{D}_{\hat{\underline{\mu}}}, \mathcal{D}_{\hat{\underline{\nu}}}$ of (3.152), (3.153) with initial data $\mathcal{D}_{\hat{\underline{\mu}}(n_0)}$ satisfying (3.151) and $\mathcal{M}_0$ to be specified as in (the proof of) Lemma 3.27.

Before proceeding to develop the stationary Ablowitz–Ladik algorithm, we briefly analyze the dynamics of (3.152).

Lemma 3.27 Let $n \in \mathbb{Z}$ and suppose that $\mathcal{D}_{\hat{\underline{\mu}}(n)}$ is defined via (3.152) for some divisor $\mathcal{D}_{\hat{\underline{\mu}}(n_0)} \in \mathrm{Sym}^p(\mathcal{K}_p)$.
(i) If $\mathcal{D}_{\hat{\underline{\mu}}(n)}$ is nonspecial and does not contain any of the points $P_{0,\pm}$, $P_{\infty\pm}$, and $\mathcal{D}_{\hat{\underline{\mu}}(n+1)}$ contains one of the points $P_{0,\pm}$, $P_{\infty\pm}$, then $\mathcal{D}_{\hat{\underline{\mu}}(n+1)}$ contains $P_{0,-}$ or $P_{\infty-}$ but not $P_{\infty+}$ or $P_{0,+}$.
(ii) If $\mathcal{D}_{\hat{\underline{\mu}}(n)}$ is nonspecial and $\mathcal{D}_{\hat{\underline{\mu}}(n+1)}$ is special, then $\mathcal{D}_{\hat{\underline{\mu}}(n)}$ contains at least one of the points $P_{\infty+}$, $P_{\infty-}$ and one of the points $P_{0,+}$, $P_{0,-}$.
(iii) Item (i) holds if $n+1$ is replaced by $n-1$, $P_{\infty+}$ by $P_{\infty-}$, and $P_{0,+}$ by $P_{0,-}$.
(iv) Items (i)–(iii) also hold for $\mathcal{D}_{\hat{\underline{\nu}}(n)}$.

Proof (i) Suppose one point in $\mathcal{D}_{\hat{\underline{\mu}}(n+1)}$ equals $P_{\infty+}$ and denote the remaining ones by $\mathcal{D}_{\tilde{\underline{\mu}}(n+1)}$. Then (3.152) implies that $\underline{\alpha}_{Q_0}(\mathcal{D}_{\tilde{\underline{\mu}}(n+1)}) + \underline{A}_{Q_0}(P_{\infty+}) = \underline{\alpha}_{Q_0}(\mathcal{D}_{\hat{\underline{\mu}}(n)}) + \underline{A}_{P_{0,-}}(P_{\infty+})$. Since $\mathcal{D}_{\hat{\underline{\mu}}(n)}$ is assumed to be nonspecial one concludes $\mathcal{D}_{\hat{\underline{\mu}}(n)} = \mathcal{D}_{\tilde{\underline{\mu}}(n+1)} + \mathcal{D}_{P_{0,-}}$, contradicting our assumption on $\mathcal{D}_{\hat{\underline{\mu}}(n)}$. The statement for $P_{0,+}$ follows similarly; here we choose Q_0 to be a branch point of $\mathcal{K}_p$ such that $\underline{A}_{Q_0}(P^*) = -\underline{A}_{Q_0}(P)$.
(ii) Next, we choose Q_0 to be a branch point of $\mathcal{K}_p$. If $\mathcal{D}_{\hat{\underline{\mu}}(n+1)}$ is special, then it contains a pair of points (Q, Q^*) whose contribution will cancel under the Abel map, that is, $\underline{\alpha}_{Q_0}(\mathcal{D}_{\hat{\underline{\mu}}(n+1)}) = \underline{\alpha}_{Q_0}(\mathcal{D}_{\hat{\underline{\eta}}(n+1)})$ for some $\mathcal{D}_{\hat{\underline{\eta}}(n+1)} \in \mathrm{Sym}^{p-2}(\mathcal{K}_p)$. Invoking (3.152) then shows that $\underline{\alpha}_{Q_0}(\mathcal{D}_{\hat{\underline{\mu}}(n)}) = \underline{\alpha}_{Q_0}(\mathcal{D}_{\hat{\underline{\eta}}(n+1)}) + \underline{A}_{Q_0}(P_{\infty-}) + \underline{A}_{Q_0}(P_{0,-})$. As $\mathcal{D}_{\hat{\underline{\mu}}(n)}$ was assumed to be nonspecial, this shows that $\mathcal{D}_{\hat{\underline{\mu}}(n)} = \mathcal{D}_{\hat{\underline{\eta}}(n+1)} + \mathcal{D}_{P_{\infty-}} + \mathcal{D}_{P_{0,-}}$, as claimed.
(iii) This is proved as in item (i).
(iv) Since $\mathcal{D}_{\hat{\underline{\nu}}(n)}$ satisfies the same equation as $\mathcal{D}_{\hat{\underline{\mu}}(n)}$ in (3.152) (cf. (3.139)), items (i)–(iii) also hold for $\mathcal{D}_{\hat{\underline{\nu}}(n)}$. □

We also note the following result:

Lemma 3.28 Let $n \in \mathbb{Z}$ and assume that $\mathcal{D}_{\hat{\underline{\mu}}(n)}$ and $\mathcal{D}_{\hat{\underline{\nu}}(n)}$ are nonspecial. Then $\mathcal{D}_{\hat{\underline{\mu}}(n)}$ contains $P_{0,-}$ if and only if $\mathcal{D}_{\hat{\underline{\nu}}(n)}$ contains $P_{\infty-}$. Moreover, $\mathcal{D}_{\hat{\underline{\mu}}(n)}$ contains $P_{\infty+}$ if and only if $\mathcal{D}_{\hat{\underline{\nu}}(n)}$ contains $P_{0,+}$.

Proof Suppose a point in $\mathcal{D}_{\hat{\underline{\mu}}(n)}$ equals $P_{0,-}$ and denote the remaining ones by $\mathcal{D}_{\tilde{\underline{\mu}}(n)}$. By (3.140),

$$\underline{\alpha}_{Q_0}(\mathcal{D}_{\hat{\underline{\nu}}(n)}) = \underline{\alpha}_{Q_0}(\mathcal{D}_{\tilde{\underline{\mu}}(n)}) + \underline{A}_{Q_0}(P_{0,-}) + \underline{A}_{P_{0,-}}(P_{\infty-})$$
$$= \underline{\alpha}_{Q_0}(\mathcal{D}_{\tilde{\underline{\mu}}(n)}) + \underline{A}_{Q_0}(P_{\infty-}).$$

Since $\mathcal{D}_{\hat{\underline{\nu}}(n)}$ is nonspecial, $\mathcal{D}_{\hat{\underline{\nu}}(n)}$ contains $P_{\infty-}$, and vice versa. The second statement follows similarly. □

3.5 The Stationary AL Algebro-Geometric IVP

In the following we will call the points P_{∞_+}, P_{∞_-}, $P_{0,+}$, and $P_{0,-}$ *exceptional points*. Then Lemma 3.27 yields the following behavior of $\mathcal{D}_{\hat{\underline{\mu}}(n)}$ assuming one starts with some nonspecial initial divisor $\mathcal{D}_{\hat{\underline{\mu}}(n_0)}$ without exceptional points: As n increases, $\mathcal{D}_{\hat{\underline{\mu}}(n)}$ stays nonspecial as long as it does not include exceptional points. If an exceptional point appears, $\mathcal{D}_{\hat{\underline{\mu}}(n)}$ is still nonspecial and contains $P_{0,-}$ or P_{∞_-} at least once (but not $P_{0,+}$ and P_{∞_+}). Further increasing n, all instances of $P_{0,-}$ and P_{∞_-} will be rendered into $P_{0,+}$ and P_{∞_+}, until we have again a nonspecial divisor that has the same number of $P_{0,+}$ and P_{∞_+} as the first one had of $P_{0,-}$ and P_{∞_-}. Generically, one expects the subsequent divisor to be nonspecial without exceptional points again.

Next we show that most initial divisors are well-behaved in the sense that their iterates stay away from $P_{\infty_\pm}$, $P_{0,\pm}$. Since we want to show that this set is of full measure, it will be convenient to identify $\mathrm{Sym}^p(\mathcal{K}_p)$ with the Jacobi variety $J(\mathcal{K}_p)$ via the Abel map and take the Haar measure on $J(\mathcal{K}_p)$. Of course, the Abel map is only injective when restricted to the set of nonspecial divisors, but these are the only ones we are interested in.

Lemma 3.29 *The set $\mathcal{M}_0 \subset \mathrm{Sym}^p(\mathcal{K}_p)$ of initial divisors $\mathcal{D}_{\hat{\underline{\mu}}(n_0)}$ for which $\mathcal{D}_{\hat{\underline{\mu}}(n)}$ and $\mathcal{D}_{\hat{\underline{\nu}}(n)}$, defined via (3.152) and (3.153), are admissible (i.e., do not contain the points $P_{\infty_\pm}$, $P_{0,\pm}$) and hence are nonspecial for all $n \in \mathbb{Z}$, forms a dense set of full measure in the set $\mathrm{Sym}^p(\mathcal{K}_p)$ of positive divisors of degree p.*

Proof Let $\mathcal{M}_{\infty,0}$ be the set of divisors in $\mathrm{Sym}^p(\mathcal{K}_p)$ for which (at least) one point is equal to $P_{\infty_\pm}$ or $P_{0,\pm}$. The image $\underline{\alpha}_{Q_0}(\mathcal{M}_{\infty,0})$ of $\mathcal{M}_{\infty,0}$ is then contained in the following set,

$$\underline{\alpha}_{Q_0}(\mathcal{M}_{\infty,0}) \subseteq \bigcup_{P \in \{P_{0,\pm}, P_{\infty_\pm}\}} \left(\underline{A}_{Q_0}(P) + \underline{\alpha}_{Q_0}(\mathrm{Sym}^{p-1}(\mathcal{K}_p))\right) \subset J(\mathcal{K}_p).$$

Since the (complex) dimension of $\mathrm{Sym}^{p-1}(\mathcal{K}_p)$ is $p-1$, its image must be of measure zero by Sard's theorem. Similarly, let $\mathcal{M}_{\mathrm{sp}}$ be the set of special divisors, then its image is given by

$$\underline{\alpha}_{Q_0}(\mathcal{M}_{\mathrm{sp}}) = \underline{\alpha}_{Q_0}(\mathrm{Sym}^{p-2}(\mathcal{K}_p)),$$

assuming Q_0 to be a branch point. Thus, $\underline{\alpha}_{Q_0}(\mathcal{M}_{\mathrm{sp}}) \subset \underline{\alpha}_{Q_0}(\mathcal{M}_{\infty,0})$ and hence $\underline{\alpha}_{Q_0}(\mathcal{M}_{\mathrm{sing}}) = \underline{\alpha}_{Q_0}(\mathcal{M}_{\infty,0})$ has measure zero, where

$$\mathcal{M}_{\mathrm{sing}} = \mathcal{M}_{\infty,0} \cup \mathcal{M}_{\mathrm{sp}}.$$

Consequently,

$$\mathcal{S}_\mu = \bigcup_{n \in \mathbb{Z}} \left(\underline{\alpha}_{Q_0}(\mathcal{M}_{\mathrm{sing}}) + n\underline{A}_{P_{0,-}}(P_{\infty_+})\right) \quad \text{and} \quad \mathcal{S}_\nu = \mathcal{S}_\mu + \underline{A}_{P_{0,-}}(P_{\infty_-}) \quad (3.154)$$

are of measure zero as well. But the set $\mathcal{S}_\mu \cup \mathcal{S}_\nu$ contains all initial divisors for which $\mathcal{D}_{\hat{\underline{\mu}}(n)}$ or $\mathcal{D}_{\hat{\underline{\nu}}(n)}$ will hit $P_{\infty_\pm}$ or $P_{0,\pm}$, or become special at some $n \in \mathbb{Z}$. We denote by $\mathcal{M}_0$ the inverse image of the complement of the set $\mathcal{S}_\mu \cup \mathcal{S}_\nu$ under the Abel map,

$$\mathcal{M}_0 = \underline{\alpha}_{Q_0}^{-1}(\mathrm{Sym}^p(\mathcal{K}_p) \setminus (\mathcal{S}_\mu \cup \mathcal{S}_\nu)).$$

Since $\mathcal{M}_0$ is of full measure, it is automatically dense in $\mathrm{Sym}^p(\mathcal{K}_p)$. □

Next, we describe the stationary Ablowitz–Ladik algorithm. Since this is a somewhat lengthy affair, we will break it up into several steps.

The Stationary Ablowitz–Ladik Algorithm:

We prescribe the following data

(i) The coefficient $\alpha(n_0) \in \mathbb{C} \setminus \{0\}$ and the constant $c_{0,+} \in \mathbb{C} \setminus \{0\}$.

(ii) The set

$$\{E_m\}_{m=0}^{2p+1} \subset \mathbb{C} \setminus \{0\}, \quad E_m \neq E_{m'} \text{ for } m \neq m', \quad m, m' = 0, \ldots, 2p+1, \quad (3.155)$$

for some fixed $p \in \mathbb{N}$. Given $\{E_m\}_{m=0}^{2p+1}$, we introduce the function

$$R_{\underline{p}}(z) = \left(\frac{c_{0,+}}{2z^{p_-}}\right)^2 \prod_{m=0}^{2p+1} (z - E_m) \quad (3.156)$$

and the hyperelliptic curve $\mathcal{K}_p$ (with nonsingular affine part) according to (3.88).

(iii) The nonspecial divisor

$$\mathcal{D}_{\hat{\underline{\mu}}(n_0)} \in \mathrm{Sym}^p(\mathcal{K}_p),$$

where $\hat{\underline{\mu}}(n_0)$ is of the form

$$\hat{\underline{\mu}}(n_0) = \{\underbrace{\hat{\mu}_1(n_0), \ldots, \hat{\mu}_1(n_0)}_{p_1(n_0) \text{ times}}, \ldots, \underbrace{\hat{\mu}_{q(n_0)}, \ldots, \hat{\mu}_{q(n_0)}}_{p_{q(n_0)}(n_0) \text{ times}}\}$$

with

$$\begin{aligned}\hat{\mu}_k(n_0) &= (\mu_k(n_0), y(\hat{\mu}_k(n_0))), \\ \mu_k(n_0) &\neq \mu_{k'}(n_0) \text{ for } k \neq k', \; k, k' = 1, \ldots, q(n_0),\end{aligned} \quad (3.157)$$

and

$$p_k(n_0) \in \mathbb{N}, \; k = 1, \ldots, q(n_0), \quad \sum_{k=1}^{q(n_0)} p_k(n_0) = p.$$

3.5 The Stationary AL Algebro-Geometric IVP

With $\{E_m\}_{m=0}^{2p+1}$, $\mathcal{D}_{\hat{\underline{\mu}}(n_0)}$, $\alpha(n_0)$, and $c_{0,+}$ prescribed, we next introduce the following quantities (for $z \in \mathbb{C} \setminus \{0\}$):

$$\alpha^+(n_0) = \alpha(n_0) \left(\prod_{m=0}^{2p+1} E_m \right)^{1/2} \prod_{k=1}^{q(n_0)} \mu_k(n_0)^{-p_k(n_0)}, \qquad (3.158)$$

$$c_{0,-}^2 = c_{0,+}^2 \prod_{m=0}^{2p+1} E_m, \qquad (3.159)$$

$$F_{\underline{p}}(z, n_0) = -c_{0,+} \alpha^+(n_0) z^{-p_-} \prod_{k=1}^{q(n_0)} (z - \mu_k(n_0))^{p_k(n_0)}, \qquad (3.160)$$

$$G_{\underline{p}}(z, n_0) = \frac{1}{2} \left(\frac{1}{\alpha(n_0)} - \frac{z}{\alpha^+(n_0)} \right) F_{\underline{p}}(z, n_0) \qquad (3.161)$$

$$- \frac{z}{2\alpha^+(n_0)} F_{\underline{p}}(z, n_0) \sum_{k=1}^{q(n_0)} \sum_{\ell=0}^{p_k(n_0)-1} \frac{\left(d^\ell (\zeta^{-1} y(P))/d\zeta^\ell \right)\big|_{P=(\zeta,\eta)=\hat{\mu}_k(n_0)}}{\ell!(p_k(n_0) - \ell - 1)!}$$

$$\times \left(\frac{d^{p_k(n_0)-\ell-1}}{d\zeta^{p_k(n_0)-\ell-1}} \left((z-\zeta)^{-1} \prod_{k'=1, k' \neq k}^{q(n_0)} (\zeta - \mu_{k'}(n_0))^{-p_{k'}(n_0)} \right) \right)\bigg|_{\zeta=\mu_k(n_0)}.$$

Here the sign of the square root is chosen according to (3.157).

Next we record a series of facts:
(I) By construction (cf. Lemma D.5),

$$\frac{d^\ell \left(G_{\underline{p}}(z, n_0)^2 \right)}{dz^\ell}\bigg|_{z=\mu_k(n_0)} = \frac{d^\ell R_{\underline{p}}(z)}{dz^\ell}\bigg|_{z=\mu_k(n_0)}, \qquad (3.162)$$

$z \in \mathbb{C} \setminus \{0\}$, $\ell = 0, \ldots, p_k(n_0) - 1$, $k = 1, \ldots, q(n_0)$.

(II) Since $\mathcal{D}_{\hat{\underline{\mu}}(n_0)}$ is nonspecial by hypothesis, one concludes that

$$p_k(n_0) \geq 2 \text{ implies } R_{\underline{p}}(\mu_k(n_0)) \neq 0, \quad k = 1, \ldots, q(n_0).$$

(III) By (3.161) and (3.162) one infers that $F_{\underline{p}}$ divides $G_{\underline{p}}^2 - R_{\underline{p}}$.
(IV) By (3.156) and (3.161) one verifies that

$$G_{\underline{p}}(z, n_0)^2 - R_{\underline{p}}(z) \underset{z \to \infty}{=} O(z^{2p_+-1}), \qquad (3.163)$$

$$G_{\underline{p}}(z, n_0)^2 - R_{\underline{p}}(z) \underset{z \to 0}{=} O(z^{-2p_-+1}). \qquad (3.164)$$

By **(III)** and **(IV)** we may write

$$G_{\underline{p}}(z, n_0)^2 - R_{\underline{p}}(z) = F_{\underline{p}}(z, n_0) \check{H}_{q,r}(z, n_0), \quad z \in \mathbb{C} \setminus \{0\}, \qquad (3.165)$$

for some $q \in \{0, \ldots, p_- - 1\}$, $r \in \{0, \ldots, p_+\}$, where $\check{H}_{q,r}(z, n_0)$ is a Laurent polynomial of the form $c_{-q}z^{-q} + \cdots + c_r z^r$. If, in fact, $\check{H}_{0,0} = 0$, then $R_{\underline{p}}(z) = G_{\underline{p}}(z, n_0)^2$ would yield double zeros of $R_{\underline{p}}$, contradicting our basic hypothesis (3.155). Thus we conclude that in the case $r = q = 0$, $\check{H}_{0,0}$ cannot vanish identically and hence we may break up (3.165) in the following manner

$$\check{\phi}(P, n_0) = \frac{G_{\underline{p}}(z, n_0) + (c_{0,+}/2)z^{-p_-}y}{F_{\underline{p}}(z, n_0)} = \frac{\check{H}_{q,r}(z, n_0)}{G_{\underline{p}}(z, n_0) - (c_{0,+}/2)z^{-p_-}y},$$

$$P = (z, y) \in \mathcal{K}_p.$$

Next we decompose

$$\check{H}_{q,r}(z, n_0) = C z^{-q} \prod_{j=1}^{r+q} (z - \nu_j(n_0)), \quad z \in \mathbb{C} \setminus \{0\}, \tag{3.166}$$

where $C \in \mathbb{C} \setminus \{0\}$ and $\{\nu_j(n_0)\}_{j=1}^{r+q} \subset \mathbb{C}$ (if $r = q = 0$ we replace the product in (3.166) by 1). By inspection of the local zeros and poles as well as the behavior near $P_{0,\pm}$, $P_{\infty_\pm}$ of the function $\check{\phi}(\,\cdot\,, n_0)$ using

$$y(P) \underset{\zeta \to 0}{=} \begin{cases} \mp \zeta^{-2p}(1 + O(\zeta)), & P \to P_{\infty_\pm}, \ \zeta = 1/z, \\ \pm(c_{0,-}/c_{0,+}) + O(\zeta), & P \to P_{0,\pm}, \ \zeta = z, \end{cases}$$

its divisor, $(\check{\phi}(\,\cdot\,, n_0))$, is given by

$$(\check{\phi}(\,\cdot\,, n_0)) = \mathcal{D}_{P_{0,-}\hat{\underline{\nu}}(n_0)} - \mathcal{D}_{P_{\infty_-}\hat{\underline{\mu}}(n_0)},$$

where

$$\hat{\underline{\nu}}(n_0) = \{\underbrace{P_{0,-}, \ldots, P_{0,-}}_{p_- - 1 - q \text{ times}}, \hat{\nu}_1(n_0), \ldots, \hat{\nu}_{r+q}(n_0), \underbrace{P_{\infty_+}, \ldots, P_{\infty_+}}_{p_+ - r \text{ times}}\}.$$

In the following we call a positive divisor of degree p admissible if it does not contain any of the points $P_{\infty_\pm}$, $P_{0,\pm}$.

Hence,

$$\mathcal{D}_{\hat{\underline{\nu}}(n_0)} \text{ is an admissible divisor if and only if } r = p_+ \text{ and } q = p_- - 1. \tag{3.167}$$

We note that

$$\underline{\alpha}_{Q_0}(\mathcal{D}_{\hat{\underline{\nu}}(n_0)}) = \underline{\alpha}_{Q_0}(\mathcal{D}_{\hat{\underline{\mu}}(n_0)}) + \underline{A}_{P_{0,-}}(P_{\infty_-}),$$

in accordance with (3.140).

3.5 The Stationary AL Algebro-Geometric IVP

(V) Assuming that (3.163), (3.164) are precisely of order $z^{\pm(2p_\pm - 1)}$, that is, assuming $r = p_+$ and $q = p_- - 1$ in (3.165), we rewrite (3.165) in the more appropriate manner

$$G_{\underline{p}}(z, n_0)^2 - R_{\underline{p}}(z) = F_{\underline{p}}(z, n_0) H_{\underline{p}}(z, n_0), \quad z \in \mathbb{C} \setminus \{0\}. \tag{3.168}$$

(We will later discuss conditions which indeed guarantee that $q = p_- - 1$ and $r = p_+$, cf. (3.167) and the discussion in step **(X)** below.) By construction, $H_{\underline{p}}(\cdot, n_0)$ is then of the type

$$H_{\underline{p}}(z, n_0) = c_{0,+} \beta(n_0) z^{-p_- + 1} \prod_{k=1}^{\ell(n_0)} (z - \nu_k(n_0))^{s_k(n_0)}, \quad \sum_{k=1}^{\ell(n_0)} s_k(n_0) = p,$$

$$\nu_k(n_0) \neq \nu_{k'}(n_0) \text{ for } k \neq k', \; k, k' = 1, \ldots, \ell(n_0), \; z \in \mathbb{C} \setminus \{0\}, \tag{3.169}$$

where we introduced the coefficient $\beta(n_0)$. We define

$$\hat{\nu}_k(n_0) = (\nu_k(n_0), -(2/c_{0,+}) \nu_k(n_0)^{p_-} G_{\underline{p}}(\nu_k(n_0), n_0)), \quad k = 1, \ldots, \ell(n_0).$$

An explicit computation of $\beta(n_0)$ then yields

$$\alpha^+(n_0) \beta(n_0) = -\frac{1}{2} \sum_{k=1}^{q(n_0)} \frac{\left(d^{p_k(n_0)-1}(\zeta^{-1} y(P))/d\zeta^{p_k(n_0)-1}\right)\big|_{P=(\zeta,\eta)=\hat{\mu}_k(n_0)}}{(p_k(n_0) - 1)!}$$

$$\times \prod_{k'=1, k' \neq k}^{q(n_0)} (\mu_k(n_0) - \mu_{k'}(n_0))^{-p_k(n_0)}$$

$$+ \frac{1}{2} \left(\frac{\alpha^+(n_0)}{\alpha(n_0)} + \sum_{k=1}^{q(n_0)} p_k(n_0) \mu_k(n_0) - \frac{1}{2} \sum_{m=0}^{2p+1} E_m \right). \tag{3.170}$$

The result (3.170) is obtained by inserting the expressions (3.160), (3.161), and (3.169) for $F_{\underline{p}}(\cdot, n_0)$, $G_{\underline{p}}(\cdot, n_0)$, and $H_{\underline{p}}(\cdot, n_0)$ into (3.168) and collecting all terms of order $z^{2p_+ - 1}$.

(VI) Introduce

$$\beta^+(n_0) = \beta(n_0) \prod_{k=1}^{\ell(n_0)} \nu_k(n_0)^{s_k(n_0)} \left(\prod_{m=0}^{2p+1} E_m \right)^{-1/2}.$$

3 The Ablowitz–Ladik Hierarchy

(VII) Using $G_{\underline{p}}(z, n_0)$, $H_{\underline{p}}(z, n_0)$, $F_{\underline{p}}(z, n_0)$, $\beta(n_0)$, $\alpha^+(n_0)$, and $\beta^+(n_0)$, we next construct the $n_0 \pm 1$ terms from the following equations:

$$F_{\underline{p}}^- = \frac{1}{z\gamma}(\alpha^2 H_{\underline{p}} - 2\alpha G_{\underline{p}} + F_{\underline{p}}), \tag{3.171}$$

$$H_{\underline{p}}^- = \frac{z}{\gamma}(\beta^2 F_{\underline{p}} - 2\beta G_{\underline{p}} + H_{\underline{p}}), \tag{3.172}$$

$$G_{\underline{p}}^- = \frac{1}{\gamma}((1 + \alpha\beta)G_{\underline{p}} - \alpha H_{\underline{p}} - \beta F_{\underline{p}}), \tag{3.173}$$

respectively,

$$F_{\underline{p}}^+ = \frac{1}{z\gamma^+}((\alpha^+)^2 H_{\underline{p}} + 2\alpha^+ z G_{\underline{p}} + z^2 F_{\underline{p}}), \tag{3.174}$$

$$H_{\underline{p}}^+ = \frac{1}{z\gamma^+}((\beta^+ z)^2 F_{\underline{p}} + 2\beta^+ z G_{\underline{p}} + H_{\underline{p}}), \tag{3.175}$$

$$G_{\underline{p}}^+ = \frac{1}{z\gamma^+}((1 + \alpha^+\beta^+)z G_{\underline{p}} + \alpha^+ H_{\underline{p}} + \beta^+ z^2 F_{\underline{p}}). \tag{3.176}$$

Moreover,

$$(G_{\underline{p}}^-)^2 - F_{\underline{p}}^- H_{\underline{p}}^- = R_{\underline{p}}, \quad (G_{\underline{p}}^+)^2 - F_{\underline{p}}^+ H_{\underline{p}}^+ = R_{\underline{p}}.$$

Inserting (3.160), (3.161), and (3.169) in (3.171)–(3.173) one verifies

$$F_{\underline{p}}^-(z, n_0) \underset{z \to \infty}{=} -c_{0,+}\alpha(n_0) z^{p_+ - 1} + O(z^{p_+ - 2}),$$

$$H_{\underline{p}}^-(z, n_0) \underset{z \to \infty}{=} O(z^{p_+}),$$

$$F_{\underline{p}}^-(z, n_0) \underset{z \to 0}{=} O(z^{-p_-}),$$

$$H_{\underline{p}}^-(z, n_0) \underset{z \to 0}{=} -c_{0,-}\beta(n_0) z^{-p_- + 1} + O(z^{-p_- + 2}),$$

$$G_{\underline{p}}^-(z, n_0) = \tfrac{1}{2} c_{0,-} z^{-p_-} + \cdots + \tfrac{1}{2} c_{0,+} z^{p_+}.$$

The last equation implies

$$G_{\underline{p}}(z, n_0 - 1)^2 - R_{\underline{p}}(z) \underset{z \to \infty}{=} O(z^{2p_+ - 1}), \tag{3.177}$$

$$G_{\underline{p}}(z, n_0 - 1)^2 - R_{\underline{p}}(z) \underset{z \to 0}{=} O(z^{-2p_- + 1}), \tag{3.178}$$

so we may write

$$G_{\underline{p}}(z, n_0 - 1)^2 - R_{\underline{p}}(z) = \check{F}_{s, p_+ - 1}(z, n_0 - 1) \check{H}_{p_- - 1, r}(z, n_0 - 1), \quad z \in \mathbb{C} \setminus \{0\}, \tag{3.179}$$

3.5 The Stationary AL Algebro-Geometric IVP

for some $s \in \{1, \ldots, p_-\}, r \in \{1, \ldots, p_+\}$, where

$$\check{F}_{s,p_+-1}(n_0 - 1) = c_{-s}z^{-s} + \cdots - c_{0,+}\alpha(n_0)z^{p_+-1},$$
$$\check{H}_{p_--1,r}(n_0 - 1) = -c_{0,-}\beta(n_0)z^{-p_-+1} + \cdots + c_r z^r.$$

The right-hand side of (3.179) cannot vanish identically (since otherwise $R_{\underline{p}}(z) = G_{\underline{p}}(z, n_0 - 1)^2$ would yield double zeros of $R_{\underline{p}}(z)$), and hence,

$$\check{\phi}(P, n_0 - 1) = \frac{G_{\underline{p}}(z, n_0 - 1) + (c_{0,+}/2)z^{-p_-}y}{\check{F}_{s,p_+-1}(z, n_0 - 1)}$$
$$= \frac{\check{H}_{p_--1,r}(z, n_0 - 1)}{G_{\underline{p}}(z, n_0 - 1) - (c_{0,+}/2)z^{-p_-}y}, \quad P = (z, y) \in \mathcal{K}_p. \quad (3.180)$$

Next, we decompose

$$\check{F}_{s,p_+-1}(z, n_0 - 1) = -c_{0,+}\alpha(n_0)z^{-s} \prod_{j=1}^{p_+-1+s}(z - \mu_j(n_0 - 1)),$$

$$\check{H}_{p_--1,r}(z, n_0 - 1) = Cz^{-p_-+1} \prod_{j=1}^{p_--1+r}(z - \nu_j(n_0 - 1)),$$

where $C \in \mathbb{C} \setminus \{0\}$ and $\{\mu_j(n_0 - 1)\}_{j=1}^{p_+-1+s} \subset \mathbb{C}$, $\{\nu_j(n_0 - 1)\}_{j=1}^{p_--1+r} \subset \mathbb{C}$. The divisor of $\check{\phi}(\,\cdot\,, n_0 - 1)$ is then given by

$$(\check{\phi}(\,\cdot\,, n_0 - 1)) = \mathcal{D}_{P_{0,-}\underline{\hat{\nu}}(n_0-1)} - \mathcal{D}_{P_{\infty_-}\underline{\hat{\mu}}(n_0-1)},$$

where

$$\underline{\hat{\mu}}(n_0 - 1) = \{\underbrace{P_{0,+}, \ldots, P_{0,+}}_{p_- - s \text{ times}}, \hat{\mu}_1(n_0 - 1), \ldots, \hat{\mu}_{p_+-1+s}(n_0 - 1)\},$$

$$\underline{\hat{\nu}}(n_0 - 1) = \{\hat{\nu}_1(n_0 - 1), \ldots, \hat{\nu}_{p_--1+r}(n_0 - 1), \underbrace{P_{\infty_+}, \ldots, P_{\infty_+}}_{p_+ - r \text{ times}}\}.$$

In particular,

$$\mathcal{D}_{\underline{\hat{\mu}}(n_0-1)} \text{ is an admissible divisor if and only if } s = p_-, \quad (3.181)$$

$$\mathcal{D}_{\underline{\hat{\nu}}(n_0-1)} \text{ is an admissible divisor if and only if } r = p_+. \quad (3.182)$$

(VIII) Assuming that (3.177), (3.178) are precisely of order $z^{\pm(2p_\pm-1)}$, that is, assuming $s = p_-$ and $r = p_+$ in (3.180), we rewrite (3.180) as

$$G_{\underline{p}}(z, n_0 - 1)^2 - R_{\underline{p}}(z) = F_{\underline{p}}(z, n_0 - 1)H_{\underline{p}}(z, n_0 - 1), \quad z \in \mathbb{C} \setminus \{0\}.$$

By construction, $F_{\underline{p}}(\,\cdot\,, n_0 - 1)$ and $H_{\underline{p}}(\,\cdot\,, n_0 - 1)$ are then of the type

$$F_{\underline{p}}(z, n_0 - 1) = -c_{0,+}\alpha(n_0)z^{-p_-} \prod_{k=1}^{q(n_0-1)} (z - \mu_j(n_0 - 1))^{p_k(n_0-1)},$$

$$\sum_{k=1}^{q(n_0-1)} p_k(n_0 - 1) = p,$$

$\mu_k(n_0 - 1) \neq \mu_{k'}(n_0 - 1)$ for $k \neq k'$, $k, k' = 1, \ldots, q(n_0 - 1)$, $z \in \mathbb{C} \setminus \{0\}$,

$$H_{\underline{p}}(z, n_0 - 1) = c_{0,+}\beta(n_0 - 1)z^{-p_-+1} \prod_{k=1}^{\ell(n_0-1)} (z - \nu_k(n_0 - 1))^{s_k(n_0-1)},$$

$$\sum_{k=1}^{\ell(n_0-1)} s_k(n_0 - 1) = p,$$

$\nu_k(n_0 - 1) \neq \nu_{k'}(n_0 - 1)$ for $k \neq k'$, $k, k' = 1, \ldots, \ell(n_0 - 1)$, $z \in \mathbb{C} \setminus \{0\}$,

where we introduced the coefficient $\beta(n_0 - 1)$. We define

$$\hat{\mu}_k(n_0 - 1) = (\mu_k(n_0 - 1), (2/c_{0,+})\mu_k(n_0 - 1)^{p_-}G_{\underline{p}}(\mu_k(n_0 - 1), n_0 - 1)),$$

$$k = 1, \ldots, q(n_0 - 1),$$

$$\hat{\nu}_k(n_0 - 1) = (\nu_k(n_0 - 1), -(2/c_{0,+})\nu_k(n_0 - 1)^{p_-}G_{\underline{p}}(\nu_k(n_0 - 1), n_0 - 1)),$$

$$k = 1, \ldots, \ell(n_0 - 1).$$

(IX) At this point one can iterate the procedure step by step to construct $F_{\underline{p}}(\,\cdot\,, n)$, $G_{\underline{p}}(\,\cdot\,, n)$, $H_{\underline{p}}(\,\cdot\,, n)$, $\alpha(n)$, $\beta(n)$, $\mu_j(n)$, $\nu_j(n)$, etc., for $n \in (-\infty, n_0] \cap \mathbb{Z}$, subject to the following assumption (cf. (3.181), (3.182)) at each step:

$\mathcal{D}_{\hat{\underline{\mu}}(n-1)}$ is an admissible divisor (and hence $\alpha(n - 1) \neq 0$)   (3.183)

for all $n \in (-\infty, n_0] \cap \mathbb{Z}$,

$\mathcal{D}_{\hat{\underline{\nu}}(n-1)}$ is an admissible divisor (and hence $\beta(n - 1) \neq 0$)   (3.184)

for all $n \in (-\infty, n_0] \cap \mathbb{Z}$.

The formalism is symmetric with respect to n_0 and can equally well be developed for $n \in (-\infty, n_0] \cap \mathbb{Z}$ subject to the analogous assumption

$\mathcal{D}_{\hat{\underline{\mu}}(n+1)}$ is an admissible divisor (and hence $\alpha(n + 2) \neq 0$)   (3.185)

for all $n \in [n_0, \infty) \cap \mathbb{Z}$,

$\mathcal{D}_{\hat{\underline{\nu}}(n+1)}$ is an admissible divisor (and hence $\beta(n + 2) \neq 0$)   (3.186)

for all $n \in [n_0, \infty) \cap \mathbb{Z}$.

3.5 The Stationary AL Algebro-Geometric IVP

(X) Choosing the initial data $\mathcal{D}_{\hat{\underline{\mu}}(n_0)}$ such that

$$\mathcal{D}_{\hat{\underline{\mu}}(n_0)} \in \mathcal{M}_0,$$

where $\mathcal{M}_0 \subset \mathrm{Sym}^p(\mathcal{K}_p)$ is the set of admissible initial divisors introduced in Lemma 3.29, then guarantees that assumptions (3.183)–(3.186) are satisfied for all $n \in \mathbb{Z}$.

Equations (3.171)–(3.176) (for arbitrary $n \in \mathbb{Z}$) are then equivalent to s-AL$_{\underline{p}}(\alpha, \beta) = 0$.

At this stage we have verified the basic hypotheses of Section 3.4 (i.e., (3.87) and the assumption that α, β satisfy the $\underline{p}$th stationary AL system (3.30)) and hence all results of Section 3.4 apply.

In summary, we proved the following result:

Theorem 3.30 *Let $n \in \mathbb{Z}$, suppose the set $\{E_m\}_{m=0}^{2p+1} \subset \mathbb{C}$ satisfies $E_m \neq E_{m'}$ for $m \neq m', m, m' = 0, \ldots, 2p+1$, and introduce the function $R_{\underline{p}}$ and the hyperelliptic curve $\mathcal{K}_p$ as in (3.88). Choose $\alpha(n_0) \in \mathbb{C} \setminus \{0\}$, $c_{0,+} \in \mathbb{C} \setminus \{0\}$, and a nonspecial divisor $\mathcal{D}_{\hat{\underline{\mu}}(n_0)} \in \mathcal{M}_0$, where $\mathcal{M}_0 \subset \mathrm{Sym}^p(\mathcal{K}_p)$ is the set of admissible initial divisors introduced in Lemma 3.29. Then the stationary (complex) Ablowitz–Ladik algorithm as outlined in steps $(\mathbf{I})$–$(\mathbf{X})$ produces solutions α, β of the $\underline{p}$th stationary Ablowitz–Ladik system,*

$$\text{s-AL}_{\underline{p}}(\alpha, \beta) = \begin{pmatrix} -\alpha(g_{p_+,+}^- + g_{p_-,-}^-) + f_{p_+-1,+}^- - f_{p_--1,-}^- \\ \beta(g_{p_+,+}^- + g_{p_-,-}^-) + h_{p_+-1,+}^- - h_{p_--1,-}^- \end{pmatrix} = 0,$$

$$\underline{p} = (p_-, p_+) \in \mathbb{N}_0^2,$$

satisfying (3.87) and

$$\alpha(n) = \left(\prod_{m=0}^{2p+1} E_m\right)^{(n-n_0)/2} \mathcal{A}(n, n_0)\, \alpha(n_0), \tag{3.187}$$

$$\beta(n) = \Bigg(-\frac{1}{2}\sum_{k=1}^{q(n)} \frac{\left(d^{p_k(n)-1}\!\left(\zeta^{-1}y(P)\right)/d\zeta^{p_k(n)-1}\right)\big|_{P=(\zeta,\eta)=\hat{\mu}_k(n)}}{(p_k(n)-1)!}$$

$$\times \prod_{k'=1,\,k'\neq k}^{q(n)} (\mu_k(n) - \mu_{k'}(n))^{-p_k(n)}$$

$$+ \frac{1}{2}\Bigg(\Bigg(\prod_{m=0}^{2p+1} E_m\Bigg)^{1/2} \prod_{k=1}^{q(n)} \mu_k(n)^{-p_k(n)} + \sum_{k=1}^{q(n)} p_k(n)\mu_k(n) - \frac{1}{2}\sum_{m=0}^{2p+1} E_m\Bigg)\Bigg)$$

$$\times \Bigg(\prod_{m=0}^{2p+1} E_m\Bigg)^{-(n+1-n_0)/2} \mathcal{A}(n+1, n_0)^{-1} \alpha(n_0)^{-1}, \tag{3.188}$$

where

$$\mathcal{A}(n, n_0) = \begin{cases} \prod_{n'=n_0}^{n-1} \prod_{k=1}^{q(n')} \mu_k(n')^{-p_k(n')}, & n \geq n_0 + 1, \\ 1, & n = n_0, \\ \prod_{n'=n}^{n_0-1} \prod_{k=1}^{q(n')} \mu_k(n')^{p_k(n')}, & n \leq n_0 - 1. \end{cases}$$

Moreover, Lemmas 3.18–3.22 *apply.*

Finally, we briefly illustrate some aspects of this analysis in the special case $\underline{p} = (1, 1)$ (i.e., the case where (3.88) represents an elliptic Riemann surface) in more detail.

Example 3.31 The case $\underline{p} = (1, 1)$.
In this case one has

$$F_{(1,1)}(z, n) = -c_{0,+}\alpha(n+1)z^{-1}(z - \mu_1(n)),$$

$$G_{(1,1)}(z, n) = \frac{1}{2}\left(\frac{1}{\alpha(n)} - \frac{z}{\alpha(n+1)}\right)F_{(1,1)}(z, n) + R_{(1,1)}(\hat{\mu}_1(n))^{1/2},$$

$$R_{(1,1)}(z) = \left(\frac{c_{0,+}\alpha^+}{z}\right)^2 \prod_{m=0}^{3}(z - E_m),$$

and hence a straightforward calculation shows that

$$G_{(1,1)}(z, n)^2 - R_{(1,1)}(z) = -c_{0,+}^2\alpha(n+1)\beta(n)z^{-1}(z - \mu_1(n))(z - \nu_1(n))$$

$$= -\frac{c_{0,+}^2}{2z}(z - \mu_1(n))\left(\left(-\frac{y(\hat{\mu}_1(n))}{\mu_1(n)} + \frac{\check{E}^{1/2}}{\mu_1(n)} + \mu_1(n) - \frac{\widehat{E}^+}{2}\right)z\right.$$

$$\left. - \frac{\check{E}}{\mu_1(n)}\left(-\frac{1}{\check{E}^{1/2}}\frac{y(\hat{\mu}_1(n))}{\mu_1(n)} + \frac{\mu_1(n)}{\check{E}^{1/2}} + \frac{1}{\mu_1(n)} - \frac{\widehat{E}^-}{2}\right)\right),$$

where

$$\widehat{E}^\pm = \sum_{m=0}^{3} E_m^{\pm 1}, \quad \widetilde{E} = \prod_{m=0}^{3} E_m.$$

Solving for $\nu_1(n)$ one then obtains

$$\nu_1(n) = \frac{\widetilde{E}}{\mu_1(n)} \frac{-\frac{y(\hat{\mu}_1(n))}{\mu_1(n)} + \frac{\check{E}^{1/2}}{\mu_1(n)} + \mu_1(n) - \frac{\widehat{E}^+}{2}}{-\frac{1}{\check{E}^{1/2}}\frac{y(\hat{\mu}_1(n))}{\mu_1(n)} + \frac{\mu_1(n)}{\check{E}^{1/2}} + \frac{1}{\mu_1(n)} - \frac{\widehat{E}^-}{2}}.$$

Thus, $\nu_1(n_0)$ could be 0 or ∞ even if $\mu_1(n_0) \neq 0, \infty$.

3.6 The Time-Dependent Ablowitz–Ladik Formalism

Man skal ej læse for at sluge,
men for at se, hvad man kan bruge.

Henrik Ibsen[1]

In this section we extend the algebro-geometric analysis of Section 3.4 to the time-dependent Ablowitz–Ladik hierarchy.
For most of this section we assume the following hypothesis.

Hypothesis 3.32 (i) *Suppose that* α, β *satisfy*

$$\alpha(\cdot, t), \beta(\cdot, t) \in \mathbb{C}^{\mathbb{Z}}, \ t \in \mathbb{R}, \quad \alpha(n, \cdot), \beta(n, \cdot) \in C^1(\mathbb{R}), \ n \in \mathbb{Z},$$
$$\alpha(n, t)\beta(n, t) \notin \{0, 1\}, \ (n, t) \in \mathbb{Z} \times \mathbb{R}. \tag{3.189}$$

(ii) *Assume that the hyperelliptic curve* $\mathcal{K}_p$ *satisfies* (3.88) *and* (3.89).

The basic problem in the analysis of algebro-geometric solutions of the Ablowitz–Ladik hierarchy consists of solving the time-dependent $\underline{r}$th Ablowitz–Ladik flow with initial data a stationary solution of the $\underline{p}$th system in the hierarchy. More precisely, given $\underline{p} \in \mathbb{N}_0^2 \setminus \{(0, 0)\}$ we consider a solution $\alpha^{(0)}, \beta^{(0)}$ of the $\underline{p}$th stationary Ablowitz–Ladik system s-AL$_{\underline{p}}(\alpha^{(0)}, \beta^{(0)}) = 0$, associated with the hyperelliptic curve $\mathcal{K}_p$ and a corresponding set of summation constants $\{c_{\ell,\pm}\}_{\ell=1,\dots,p_\pm} \subset \mathbb{C}$. Next, let $\underline{r} = (r_-, r_+) \in \mathbb{N}_0^2$; we intend to construct a solution α, β of the $\underline{r}$th Ablowitz–Ladik flow AL$_{\underline{r}}(\alpha, \beta) = 0$ with $\alpha(t_{0,\underline{r}}) = \alpha^{(0)}, \beta(t_{0,\underline{r}}) = \beta^{(0)}$ for some $t_{0,\underline{r}} \in \mathbb{R}$. To emphasize that the summation constants in the definitions of the stationary and the time-dependent Ablowitz–Ladik equations are independent of each other, we indicate this by adding a tilde on all the time-dependent quantities. Hence we shall employ the notation $\tilde{V}_{\underline{r}}, \tilde{F}_{\underline{r}}, \tilde{G}_{\underline{r}}, \tilde{H}_{\underline{r}}, \tilde{K}_{\underline{r}}, \tilde{f}_{s,\pm}, \tilde{g}_{s,\pm}, \tilde{h}_{s,\pm}, \tilde{c}_{s,\pm}$, in order to distinguish them from $V_{\underline{p}}, F_{\underline{p}}, G_{\underline{p}}, H_{\underline{p}}, K_{\underline{p}}, f_{\ell,\pm}, g_{\ell,\pm}, h_{\ell,\pm}, c_{\ell,\pm}$, in the following. In addition, we will follow a more elaborate notation inspired by Hirota's τ-function approach and indicate the individual $\underline{r}$th Ablowitz–Ladik flow by a separate time variable $t_{\underline{r}} \in \mathbb{R}$.

Summing up, we are interested in solutions α, β of the time-dependent algebro-geometric initial value problem

$$\widetilde{\mathrm{AL}}_{\underline{r}}(\alpha, \beta) = \begin{pmatrix} -i\alpha_{t_{\underline{r}}} - \alpha(\tilde{g}_{r_+,+}^- + \tilde{g}_{r_-,-}^-) + \tilde{f}_{r_+-1,+}^- - \tilde{f}_{r_--1,-}^- \\ -i\beta_{t_{\underline{r}}} + \beta(\tilde{g}_{r_+,+}^- + \tilde{g}_{r_-,-}^-) - \tilde{h}_{r_--1,-} + \tilde{h}_{r_+-1,+}^- \end{pmatrix} = 0,$$

$$(\alpha, \beta)\big|_{t=t_{0,\underline{r}}} = (\alpha^{(0)}, \beta^{(0)}), \tag{3.190}$$

---

[1] *Peer Gynt* (1867), fourth act. ("One should not read to swallow all, but rather see what one has use for.")

$$\text{s-AL}_{\underline{p}}\left(\alpha^{(0)}, \beta^{(0)}\right) = \begin{pmatrix} -\alpha^{(0)}(g_{p_+,+} + g^-_{p_-,-}) + f_{p_+-1,+} - f^-_{p_--1,-} \\ \beta^{(0)}(g^-_{p_+,+} + g_{p_-,-}) - h_{p_--1,-} + h^-_{p_+-1,+} \end{pmatrix} = 0$$
(3.191)

for some $t_{0,\underline{r}} \in \mathbb{R}$, where $\alpha = \alpha(n, t_{\underline{r}})$, $\beta = \beta(n, t_{\underline{r}})$ satisfy (3.189) and a fixed curve $\mathcal{K}_p$ is associated with the stationary solutions $\alpha^{(0)}, \beta^{(0)}$ in (3.191). Here,

$$\underline{p} = (p_-, p_+) \in \mathbb{N}_0^2 \setminus \{(0,0)\}, \quad \underline{r} = (r_-, r_+) \in \mathbb{N}_0^2, \quad p = p_- + p_+ - 1.$$

In terms of the zero-curvature formulation this amounts to solving

$$U_{t_{\underline{r}}}(z, t_{\underline{r}}) + U(z, t_{\underline{r}}) \widetilde{V}_{\underline{r}}(z, t_{\underline{r}}) - \widetilde{V}_{\underline{r}}^+(z, t_{\underline{r}}) U(z, t_{\underline{r}}) = 0, \qquad (3.192)$$

$$U(z, t_{0,\underline{r}}) V_{\underline{p}}(z, t_{0,\underline{r}}) - V_{\underline{p}}^+(z, t_{0,\underline{r}}) U(z, t_{0,\underline{r}}) = 0. \qquad (3.193)$$

One can show (cf. Theorem 3.43) that the stationary Ablowitz–Ladik system (3.193) is in fact satisfied for all times $t_{\underline{r}} \in \mathbb{R}$: Thus, we actually impose

$$U_{t_{\underline{r}}}(z, t_{\underline{r}}) + U(z, t_{\underline{r}}) \widetilde{V}_{\underline{r}}(z, t_{\underline{r}}) - \widetilde{V}_{\underline{r}}^+(z, t_{\underline{r}}) U(z, t_{\underline{r}}) = 0, \qquad (3.194)$$

$$U(z, t_{\underline{r}}) V_{\underline{p}}(z, t_{\underline{r}}) - V_{\underline{p}}^+(z, t_{\underline{r}}) U(z, t_{\underline{r}}) = 0, \qquad (3.195)$$

instead of (3.192) and (3.193). For further reference, we recall the relevant quantities here (cf. (3.4), (3.5), (3.13)–(3.17)):

$$U(z) = \begin{pmatrix} z & \alpha \\ z\beta & 1 \end{pmatrix},$$

$$V_{\underline{p}}(z) = i \begin{pmatrix} G^-_{\underline{p}}(z) & -F^-_{\underline{p}}(z) \\ H^-_{\underline{p}}(z) & -G^-_{\underline{p}}(z) \end{pmatrix}, \quad \widetilde{V}_{\underline{r}}(z) = i \begin{pmatrix} \widetilde{G}^-_{\underline{r}}(z) & -\widetilde{F}^-_{\underline{r}}(z) \\ \widetilde{H}^-_{\underline{r}}(z) & -\widetilde{K}^-_{\underline{r}}(z) \end{pmatrix}, \qquad (3.196)$$

and

$$F_{\underline{p}}(z) = \sum_{\ell=1}^{p_-} f_{p_--\ell,-} z^{-\ell} + \sum_{\ell=0}^{p_+-1} f_{p_+-1-\ell,+} z^\ell = -c_{0,+} \alpha^+ z^{-p_-} \prod_{j=1}^{p}(z - \mu_j),$$

$$G_{\underline{p}}(z) = \sum_{\ell=1}^{p_-} g_{p_--\ell,-} z^{-\ell} + \sum_{\ell=0}^{p_+} g_{p_+-\ell,+} z^\ell,$$

$$H_{\underline{p}}(z) = \sum_{\ell=0}^{p_--1} h_{p_--1-\ell,-} z^{-\ell} + \sum_{\ell=1}^{p_+} h_{p_+-\ell,+} z^\ell = c_{0,+} \beta z^{-p_-+1} \prod_{j=1}^{p}(z - \nu_j),$$

$$\widetilde{F}_{\underline{r}}(z) = \sum_{s=1}^{r_-} \widetilde{f}_{r_--s,-} z^{-s} + \sum_{s=0}^{r_+-1} \widetilde{f}_{r_+-1-s,+} z^s, \qquad (3.197)$$

$$\widetilde{G}_{\underline{r}}(z) = \sum_{s=1}^{r_-} \widetilde{g}_{r_--s,-} z^{-s} + \sum_{s=0}^{r_+} \widetilde{g}_{r_+-s,+} z^s,$$

3.6 The Time-Dependent AL Formalism

$$\widetilde{H}_{\underline{r}}(z) = \sum_{s=0}^{r_--1} \tilde{h}_{r_--1-s,-} z^{-s} + \sum_{s=1}^{r_+} \tilde{h}_{r_+-s,+} z^s,$$

$$\widetilde{K}_{\underline{r}}(z) = \sum_{s=0}^{r_-} \tilde{g}_{r_--s,-} z^{-s} + \sum_{s=1}^{r_+} \tilde{g}_{r_+-s,+} z^s = \widetilde{G}_{\underline{r}}(z) + \tilde{g}_{r_-,-} - \tilde{g}_{r_+,+}$$

for fixed $\underline{p} \in \mathbb{N}_0^2 \setminus \{(0,0)\}$, $\underline{r} \in \mathbb{N}_0^2$. Here $f_{\ell,\pm}$, $\tilde{f}_{s,\pm}$, $g_{\ell,\pm}$, $\tilde{g}_{s,\pm}$, $h_{\ell,\pm}$, and $\tilde{h}_{s,\pm}$ are defined as in (3.19)–(3.26) with appropriate sets of summation constants $c_{\ell,\pm}$, $\ell \in \mathbb{N}_0$, and $\tilde{c}_{k,\pm}$, $k \in \mathbb{N}_0$. Explicitly, (3.194) and (3.195) are equivalent to (cf. (3.8)–(3.11), (3.50)–(3.53)),

$$\alpha_{t_{\underline{r}}} = i\left(z\widetilde{F}_{\underline{r}}^- + \alpha(\widetilde{G}_{\underline{r}}^- + \widetilde{K}_{\underline{r}}^-) - \widetilde{F}_{\underline{r}}\right), \tag{3.198}$$

$$\beta_{t_{\underline{r}}} = -i\left(\beta(\widetilde{G}_{\underline{r}}^- + \widetilde{K}_{\underline{r}}) - \widetilde{H}_{\underline{r}} + z^{-1}\widetilde{H}_{\underline{r}}^-\right), \tag{3.199}$$

$$0 = z(\widetilde{G}_{\underline{r}}^- - \widetilde{G}_{\underline{r}}) + z\beta\widetilde{F}_{\underline{r}} + \alpha\widetilde{H}_{\underline{r}}^-, \tag{3.200}$$

$$0 = z\beta\widetilde{F}_{\underline{r}}^- + \alpha\widetilde{H}_{\underline{r}} + \widetilde{K}_{\underline{r}}^- - \widetilde{K}_{\underline{r}}, \tag{3.201}$$

$$0 = z(G_{\underline{p}}^- - G_{\underline{p}}) + z\beta F_{\underline{p}} + \alpha H_{\underline{p}}^-, \tag{3.202}$$

$$0 = z\beta F_{\underline{p}}^- + \alpha H_{\underline{p}} - G_{\underline{p}} + G_{\underline{p}}^-, \tag{3.203}$$

$$0 = -F_{\underline{p}} + zF_{\underline{p}}^- + \alpha(G_{\underline{p}} + G_{\underline{p}}^-), \tag{3.204}$$

$$0 = z\beta(G_{\underline{p}} + G_{\underline{p}}^-) - zH_{\underline{p}} + H_{\underline{p}}^-, \tag{3.205}$$

respectively. In particular, (3.39) holds in the present $t_{\underline{r}}$-dependent setting, that is,

$$G_{\underline{p}}^2 - F_{\underline{p}} H_{\underline{p}} = R_{\underline{p}}.$$

As in the stationary context (3.93), (3.94) we introduce

$$\hat{\mu}_j(n, t_{\underline{r}}) = (\mu_j(n, t_{\underline{r}}), (2/c_{0,+})\mu_j(n, t_{\underline{r}})^{p_-} G_{\underline{p}}(\mu_j(n, t_{\underline{r}}), n, t_{\underline{r}})) \in \mathcal{K}_p, \\ j = 1, \ldots, p, \ (n, t_{\underline{r}}) \in \mathbb{Z} \times \mathbb{R}, \tag{3.206}$$

and

$$\hat{\nu}_j(n, t_{\underline{r}}) = (\nu_j(n, t_{\underline{r}}), -(2/c_{0,+})\nu_j(n, t_{\underline{r}})^{p_-} G_{\underline{p}}(\nu_j(n, t_{\underline{r}}), n, t_{\underline{r}})) \in \mathcal{K}_p, \\ j = 1, \ldots, p, \ (n, t_{\underline{r}}) \in \mathbb{Z} \times \mathbb{R}, \tag{3.207}$$

and note that the regularity assumptions (3.189) on α, β imply continuity of μ_j and ν_k with respect to $t_{\underline{r}} \in \mathbb{R}$ (away from collisions of these zeros, μ_j and ν_k are of course C^∞).

In analogy to (3.95), (3.96), one defines the following meromorphic function $\phi(\,\cdot\,, n, t_{\underline{r}})$ on $\mathcal{K}_p$,

$$\phi(P, n, t_{\underline{r}}) = \frac{(c_{0,+}/2)z^{-p_-}y + G_{\underline{p}}(z, n, t_{\underline{r}})}{F_{\underline{p}}(z, n, t_{\underline{r}})} \qquad (3.208)$$

$$= \frac{-H_{\underline{p}}(z, n, t_{\underline{r}})}{(c_{0,+}/2)z^{-p_-}y - G_{\underline{p}}(z, n, t_{\underline{r}})}, \qquad (3.209)$$

$$P = (z, y) \in \mathcal{K}_p, \ (n, t_{\underline{r}}) \in \mathbb{Z} \times \mathbb{R},$$

with divisor $(\phi(\,\cdot\,, n, t_{\underline{r}}))$ of $\phi(\,\cdot\,, n, t_{\underline{r}})$ given by

$$(\phi(\,\cdot\,, n, t_{\underline{r}})) = \mathcal{D}_{P_{0,-}\hat{\underline{\nu}}(n, t_{\underline{r}})} - \mathcal{D}_{P_{\infty_-}\hat{\underline{\mu}}(n, t_{\underline{r}})}. \qquad (3.210)$$

The time-dependent Baker–Akhiezer vector is then defined in terms of ϕ by

$$\Psi(P, n, n_0, t_{\underline{r}}, t_{0,\underline{r}}) = \begin{pmatrix} \psi_1(P, n, n_0, t_{\underline{r}}, t_{0,\underline{r}}) \\ \psi_2(P, n, n_0, t_{\underline{r}}, t_{0,\underline{r}}) \end{pmatrix},$$

$$\psi_1(P, n, n_0, t_{\underline{r}}, t_{0,\underline{r}}) = \exp\left(i \int_{t_{0,\underline{r}}}^{t_{\underline{r}}} ds \big(\widetilde{G}_{\underline{r}}(z, n_0, s) - \widetilde{F}_{\underline{r}}(z, n_0, s)\phi(P, n_0, s)\big)\right)$$

$$\times \begin{cases} \prod_{n'=n_0+1}^{n} \big(z + \alpha(n', t_{\underline{r}})\phi^-(P, n', t_{\underline{r}})\big), & n \geq n_0 + 1, \\ 1, & n = n_0, \\ \prod_{n'=n+1}^{n_0} \big(z + \alpha(n', t_{\underline{r}})\phi^-(P, n', t_{\underline{r}})\big)^{-1}, & n \leq n_0 - 1, \end{cases} \qquad (3.211)$$

$$\psi_2(P, n, n_0, t_{\underline{r}}, t_{0,\underline{r}}) = \exp\left(i \int_{t_{0,\underline{r}}}^{t_{\underline{r}}} ds \big(\widetilde{G}_{\underline{r}}(z, n_0, s) - \widetilde{F}_{\underline{r}}(z, n_0, s)\phi(P, n_0, s)\big)\right)$$

$$\times \phi(P, n_0, t_{\underline{r}}) \begin{cases} \prod_{n'=n_0+1}^{n} \big(z\beta(n', t_{\underline{r}})\phi^-(P, n', t_{\underline{r}})^{-1} + 1\big), & n \geq n_0 + 1, \\ 1, & n = n_0, \\ \prod_{n'=n+1}^{n_0} \big(z\beta(n', t_{\underline{r}})\phi^-(P, n', t_{\underline{r}})^{-1} + 1\big)^{-1}, & n \leq n_0 - 1, \end{cases}$$

$$P = (z, y) \in \mathcal{K}_p \setminus \{P_{\infty_+}, P_{\infty_-}, P_{0,+}, P_{0,-}\}, \ (n, t_{\underline{r}}) \in \mathbb{Z} \times \mathbb{R}. \qquad (3.212)$$

One observes that

$$\psi_1(P, n, n_0, t_{\underline{r}}, \tilde{t}_{\underline{r}}) = \psi_1(P, n_0, n_0, t_{\underline{r}}, \tilde{t}_{\underline{r}})\psi_1(P, n, n_0, t_{\underline{r}}, t_{\underline{r}}), \\ P = (z, y) \in \mathcal{K}_p \setminus \{P_{\infty_+}, P_{\infty_-}, P_{0,+}, P_{0,-}\}, \ (n, n_0, t_{\underline{r}}, \tilde{t}_{\underline{r}}) \in \mathbb{Z}^2 \times \mathbb{R}^2. \qquad (3.213)$$

The following lemma records basic properties of ϕ and Ψ in analogy to the stationary case discussed in Lemma 3.18.

Lemma 3.33 *Assume Hypothesis* 3.32 *and suppose that* (3.194), (3.195) *hold. In addition, let* $P = (z, y) \in \mathcal{K}_p \setminus \{P_{\infty_+}, P_{\infty_-}\}$, $(n, n_0, t_{\underline{r}}, t_{0,\underline{r}}) \in \mathbb{Z}^2 \times \mathbb{R}^2$. *Then* ϕ

3.6 The Time-Dependent AL Formalism

satisfies

$$\alpha\phi(P)\phi^-(P) - \phi^-(P) + z\phi(P) = z\beta, \tag{3.214}$$

$$\phi_{t_{\underline{r}}}(P) = i\widetilde{F}_{\underline{r}}\phi^2(P) - i\big(\widetilde{G}_{\underline{r}}(z) + \widetilde{K}_{\underline{r}}(z)\big)\phi(P) + i\widetilde{H}_{\underline{r}}(z), \tag{3.215}$$

$$\phi(P)\phi(P^*) = \frac{H_{\underline{p}}(z)}{F_{\underline{p}}(z)}, \tag{3.216}$$

$$\phi(P) + \phi(P^*) = 2\frac{G_{\underline{p}}(z)}{F_{\underline{p}}(z)}, \tag{3.217}$$

$$\phi(P) - \phi(P^*) = c_{0,+} z^{-p_-} \frac{y(P)}{F_{\underline{p}}(z)}. \tag{3.218}$$

Moreover, assuming $P = (z, y) \in \mathcal{K}_p \setminus \{P_{\infty_+}, P_{\infty_-}, P_{0,+}, P_{0,-}\}$, *then* Ψ *satisfies*

$$\psi_2(P, n, n_0, t_{\underline{r}}, t_{0,\underline{r}}) = \phi(P, n, t_{\underline{r}})\psi_1(P, n, n_0, t_{\underline{r}}, t_{0,\underline{r}}), \tag{3.219}$$

$$U(z)\Psi^-(P) = \Psi(P), \tag{3.220}$$

$$V_{\underline{p}}(z)\Psi^-(P) = -(i/2)c_{0,+}z^{-p_-}y\Psi^-(P), \tag{3.221}$$

$$\Psi_{t_{\underline{r}}}(P) = \widetilde{V}_{\underline{r}}^+(z)\Psi(P), \tag{3.222}$$

$$\psi_1(P, n, n_0, t_{\underline{r}}, t_{0,\underline{r}})\psi_1(P^*, n, n_0, t_{\underline{r}}, t_{0,\underline{r}}) = z^{n-n_0}\frac{F_{\underline{p}}(z, n, t_{\underline{r}})}{F_{\underline{p}}(z, n_0, t_{0,\underline{r}})}\Gamma(n, n_0, t_{\underline{r}}), \tag{3.223}$$

$$\psi_2(P, n, n_0, t_{\underline{r}}, t_{0,\underline{r}})\psi_2(P^*, n, n_0, t_{\underline{r}}, t_{0,\underline{r}}) = z^{n-n_0}\frac{H_{\underline{p}}(z, n, t_{\underline{r}})}{F_{\underline{p}}(z, n_0, t_{0,\underline{r}})}\Gamma(n, n_0, t_{\underline{r}}),$$

$$\psi_1(P, n, n_0, t_{\underline{r}}, t_{0,\underline{r}})\psi_2(P^*, n, n_0, t_{\underline{r}}, t_{0,\underline{r}})$$
$$+ \psi_1(P^*, n, n_0, t_{\underline{r}}, t_{0,\underline{r}})\psi_2(P, n, n_0, t_{\underline{r}}, t_{0,\underline{r}})$$
$$= 2z^{n-n_0}\frac{G_{\underline{p}}(z, n, t_{\underline{r}})}{F_{\underline{p}}(z, n_0, t_{0,\underline{r}})}\Gamma(n, n_0, t_{\underline{r}}), \tag{3.224}$$

$$\psi_1(P, n, n_0, t_{\underline{r}}, t_{0,\underline{r}})\psi_2(P^*, n, n_0, t_{\underline{r}}, t_{0,\underline{r}})$$
$$- \psi_1(P^*, n, n_0, t_{\underline{r}}, t_{0,\underline{r}})\psi_2(P, n, n_0, t_{\underline{r}}, t_{0,\underline{r}})$$
$$= -c_{0,+}z^{n-n_0-p_-}\frac{y}{F_{\underline{p}}(z, n_0, t_{0,\underline{r}})}\Gamma(n, n_0, t_{\underline{r}}), \tag{3.225}$$

where

$$\Gamma(n, n_0, t_{\underline{r}}) = \begin{cases} \prod_{n'=n_0+1}^{n} \gamma(n', t_{\underline{r}}), & n \geq n_0 + 1, \\ 1, & n = n_0, \\ \prod_{n'=n+1}^{n_0} \gamma(n', t_{\underline{r}})^{-1}, & n \leq n_0 - 1. \end{cases}$$

In addition, as long as the zeros $\mu_j(n_0, s)$ of $(\cdot)^{p_-} F_{\underline{p}}(\cdot, n_0, s)$ are all simple and distinct from zero for $s \in \mathcal{I}_\mu$, $\mathcal{I}_\mu \subseteq \mathbb{R}$ an open interval, $\Psi(\cdot, n, n_0, t_{\underline{r}}, t_{0,\underline{r}})$ is meromorphic on $\mathcal{K}_p \setminus \{P_{\infty_+}, P_{\infty_-}, P_{0,+}, P_{0,-}\}$ for $(n, t_{\underline{r}}, t_{0,\underline{r}}) \in \mathbb{Z} \times \mathcal{I}_\mu^2$.

Proof Equations (3.214), (3.216)–(3.221), and (3.223)–(3.225) are proved as in the stationary case, see Lemma 3.18. Thus, we turn to the proof of (3.215) and (3.222): Differentiating the Riccati-type equation (3.214) yields

$$\begin{aligned}
0 &= \left(\alpha\phi\phi^- - \phi^- + z\phi - z\beta\right)_{t_{\underline{r}}} \\
&= \alpha_{t_{\underline{r}}}\phi\phi^- + (\alpha\phi^- + z)\phi_{t_{\underline{r}}} + (\alpha\phi - 1)\phi^-_{t_{\underline{r}}} - z\beta_{t_{\underline{r}}} \\
&= \left((\alpha\phi^- + z) + (\alpha\phi - 1)S^-\right)\phi_{t_{\underline{r}}} + i\phi\phi^-\left(\alpha(\widetilde{G}_{\underline{r}} + \widetilde{K}^-_{\underline{r}}) + z\widetilde{F}^-_{\underline{r}} - \widetilde{F}_{\underline{r}}\right) \\
&\quad + iz\beta(\widetilde{G}^-_{\underline{r}} + \widetilde{K}_{\underline{r}}) + i(z\widetilde{H}_{\underline{r}} - \widetilde{H}^-_{\underline{r}}),
\end{aligned}$$

using (3.198) and (3.199). Next, one employs (3.99) to rewrite

$$(\alpha\phi^- + z) + (\alpha\phi - 1)S^- = \frac{1}{\phi}(z\beta + \phi^-) + \frac{z}{\phi^-}(\beta - \phi)S^-.$$

This allows one to calculate the right-hand side of (3.215) using (3.200) and (3.201)

$$\begin{aligned}
&\left((\alpha\phi^- + z) + (\alpha\phi - 1)S^-\right)\left(\widetilde{H}_{\underline{r}} + \widetilde{F}_{\underline{r}}\phi^2 - (\widetilde{G}_{\underline{r}} + \widetilde{K}_{\underline{r}})\phi\right) \\
&= (\alpha\phi^- + z)\widetilde{H}_{\underline{r}} + (\alpha\phi - 1)\widetilde{H}^-_{\underline{r}} + \phi(z\beta + \phi^-)\widetilde{F}_{\underline{r}} + z\phi^-(\beta - \phi)\widetilde{F}^-_{\underline{r}} \\
&\quad - (z\beta + \phi^-)(\widetilde{G}_{\underline{r}} + \widetilde{K}_{\underline{r}}) - z(\beta - \phi)(\widetilde{G}^-_{\underline{r}} + \widetilde{K}^-_{\underline{r}}) \\
&= \phi\phi^-(\widetilde{F}_{\underline{r}} - z\widetilde{F}^-_{\underline{r}}) + z\widetilde{H}_{\underline{r}} - \widetilde{H}^-_{\underline{r}} + \phi^-(\alpha\widetilde{H}_{\underline{r}} + z\beta\widetilde{F}^-_{\underline{r}}) + \phi(\alpha\widetilde{H}^-_{\underline{r}} + z\beta\widetilde{F}_{\underline{r}}) \\
&\quad - z\beta(\widetilde{G}_{\underline{r}} + \widetilde{K}_{\underline{r}} + \widetilde{G}^-_{\underline{r}} + \widetilde{K}^-_{\underline{r}}) - z\phi(\widetilde{G}^-_{\underline{r}} + \widetilde{K}^-_{\underline{r}}) - \phi^-(\widetilde{G}_{\underline{r}} + \widetilde{K}_{\underline{r}}) \\
&= \phi\phi^-(\widetilde{F}_{\underline{r}} - z\widetilde{F}^-_{\underline{r}}) + z\widetilde{H}_{\underline{r}} - \widetilde{H}^-_{\underline{r}} - z\beta(\widetilde{G}^-_{\underline{r}} + \widetilde{K}_{\underline{r}}) \\
&\quad + (z\phi - \phi^- - z\beta)(\widetilde{G}_{\underline{r}} + \widetilde{K}^-_{\underline{r}}) \\
&= \phi\phi^-(\widetilde{F}_{\underline{r}} - z\widetilde{F}^-_{\underline{r}}) + z\widetilde{H}_{\underline{r}} - \widetilde{H}^-_{\underline{r}} - z\beta(\widetilde{G}^-_{\underline{r}} + \widetilde{K}_{\underline{r}}) - \alpha\phi\phi^-(\widetilde{G}_{\underline{r}} + \widetilde{K}^-_{\underline{r}}).
\end{aligned}$$

Hence,

$$\left(\frac{1}{\phi}(z\beta + \phi^-) + \frac{z}{\phi^-}(\beta - \phi)S^-\right)\left(\phi_{t_{\underline{r}}} - i\widetilde{H}_{\underline{r}} - i\widetilde{F}_{\underline{r}}\phi^2 + i(\widetilde{G}_{\underline{r}} + \widetilde{K}_{\underline{r}})\phi\right) = 0. \quad (3.226)$$

Solving the first-order difference equation (3.226) then yields

$$\begin{aligned}
&\phi_{t_{\underline{r}}}(P, n, t_{\underline{r}}) - i\widetilde{F}_{\underline{r}}(z, n, t_{\underline{r}})\phi(P, n, t_{\underline{r}})^2 \\
&\quad + i(\widetilde{G}_{\underline{r}}(z, n, t_{\underline{r}}) + \widetilde{K}_{\underline{r}}(z, n, t_{\underline{r}}))\phi(P, n, t_{\underline{r}}) - i\widetilde{H}_{\underline{r}}(z, n, t_{\underline{r}}) \\
&= C(P, t_{\underline{r}})\begin{cases} \prod_{n'=1}^{n} B(P, n', t_{\underline{r}})/A(P, n', t_{\underline{r}}), & n \geq 1, \\ 1, & n = 0, \\ \prod_{n'=n+1}^{0} A(P, n', t_{\underline{r}})/B(P, n', t_{\underline{r}}), & n \leq -1 \end{cases} \quad (3.227)
\end{aligned}$$

3.6 The Time-Dependent AL Formalism

for some n-independent function $C(\,\cdot\,, t_{\underline{r}})$ meromorphic on $\mathcal{K}_p$, where

$$A = \phi^{-1}(z\beta + \phi^-), \quad B = -z(\phi^-)^{-1}(\beta - \phi).$$

The asymptotic behavior of $\phi(P, n, t_{\underline{r}})$ in (3.115) then yields (for $t_{\underline{r}} \in \mathbb{R}$ fixed)

$$\frac{B(P)}{A(P)} \underset{P \to P_{\infty_+}}{=} -(1 - \alpha\beta)(\beta^-)^{-1}z^{-1} + O(z^{-2}). \tag{3.228}$$

Since the left-hand side of (3.227) is of order $O(z^{r_+})$ as $P \to P_{\infty_+}$, and C is meromorphic, insertion of (3.228) into (3.227), taking $n \geq 1$ sufficiently large, then yields a contradiction unless $C = 0$. This proves (3.215).

Proving (3.222) is equivalent to showing

$$\psi_{1,t_{\underline{r}}} = i(\widetilde{G}_{\underline{r}} - \phi\widetilde{F}_{\underline{r}})\psi_1, \tag{3.229}$$

$$\psi_1 \phi_{t_{\underline{r}}} + \phi\psi_{1,t_{\underline{r}}} = i(\widetilde{H}_{\underline{r}} - \phi\widetilde{K}_{\underline{r}})\psi_1, \tag{3.230}$$

using (3.219). Equation (3.230) follows directly from (3.229) and from (3.215),

$$\psi_1 \phi_{t_{\underline{r}}} + \phi\psi_{1,t_{\underline{r}}} = \psi_1\big(i\widetilde{H}_{\underline{r}} + i\widetilde{F}_{\underline{r}}\phi^2 - i(\widetilde{G}_{\underline{r}} + \widetilde{K}_{\underline{r}})\phi + i(\widetilde{G}_{\underline{r}} - \phi\widetilde{F}_{\underline{r}})\phi\big)$$
$$= i(\widetilde{H}_{\underline{r}} - \phi\widetilde{K}_{\underline{r}})\psi_1.$$

To prove (3.229) we start from

$$(z + \alpha\phi^-)_{t_{\underline{r}}} = \alpha_{t_{\underline{r}}}\phi^- + \alpha\phi^-_{t_{\underline{r}}}$$
$$= \phi^- i\big(z\widetilde{F}_{\underline{r}}^- + \alpha(\widetilde{G}_{\underline{r}} + \widetilde{K}_{\underline{r}}^-) - \widetilde{F}_{\underline{r}}\big) + \alpha i\big(\widetilde{H}_{\underline{r}}^- + \widetilde{F}_{\underline{r}}^-(\phi^-)^2 - (\widetilde{G}_{\underline{r}}^- + \widetilde{K}_{\underline{r}}^-)\phi^-\big)$$
$$= i\alpha\phi^-(\widetilde{G}_{\underline{r}} - \widetilde{G}_{\underline{r}}^-) + i(z + \alpha\phi^-)\phi^-\widetilde{F}_{\underline{r}} - i\phi^-\widetilde{F}_{\underline{r}} + i\alpha\widetilde{H}_{\underline{r}}^-$$
$$= i(z + \alpha\phi^-)\big(\widetilde{G}_{\underline{r}} - \phi\widetilde{F}_{\underline{r}} - (\widetilde{G}_{\underline{r}}^- - \phi^-\widetilde{F}_{\underline{r}}^-)\big),$$

where we used (3.200) and (3.99) to rewrite

$$i\alpha\widetilde{H}_{\underline{r}}^- - i\phi^-\widetilde{F}_{\underline{r}} = iz(\widetilde{G}_{\underline{r}} - \widetilde{G}_{\underline{r}}^-) - \alpha\phi\phi^-\widetilde{F}_{\underline{r}} - z\phi\widetilde{F}_{\underline{r}}.$$

Abbreviating

$$\sigma(P, n_0, t_{\underline{r}}) = i\int_0^{t_{\underline{r}}} ds\big(\widetilde{G}_{\underline{r}}(z, n_0, s) - \widetilde{F}_{\underline{r}}(z, n_0, s)\phi(P, n_0, s)\big),$$

one computes for $n \geq n_0 + 1$,

$$\psi_{1,t_{\underline{r}}} = \left(\exp(\sigma) \prod_{n'=n_0+1}^{n} (z + \alpha\phi^-)(n') \right)_{t_{\underline{r}}}$$

$$= \sigma_{t_{\underline{r}}}\psi_1 + \exp(\sigma) \sum_{n'=n_0+1}^{n} (z + \alpha\phi^-)_{t_{\underline{r}}}(n') \prod_{\substack{n''=1 \\ n'' \neq n'}}^{n} (z + \alpha\phi^-)(n'')$$

$$= \psi_1 \left(\sigma_{t_{\underline{r}}} + i \sum_{n'=n_0+1}^{n} \left((\widetilde{G}_{\underline{r}} - \widetilde{F}_{\underline{r}}\phi)(n') - (\widetilde{G}_{\underline{r}} - \widetilde{F}_{\underline{r}}\phi)(n'-1) \right) \right)$$

$$= i(\widetilde{G}_{\underline{r}} - \widetilde{F}_{\underline{r}}\phi)\psi_1.$$

The case $n \leq n_0$ is handled analogously establishing (3.229).

That $\Psi(\cdot, n, n_0, t_{\underline{r}}, t_{0,\underline{r}})$ is meromorphic on $\mathcal{K}_p \setminus \{P_{\infty_{\pm}}, P_{0,\pm}\}$ if $F_{\underline{p}}(\cdot, n_0, t_{\underline{r}})$ has only simple zeros distinct from zero is a consequence of (3.208), (3.210), (3.211), (3.212), and of

$$-i\widetilde{F}_{\underline{r}}(z, n_0, s)\phi(P, n_0, s) \underset{P \to \hat{\mu}_j(n_0, s)}{=} \partial_s \ln\left(F_{\underline{p}}(z, n_0, s)\right) + O(1),$$

using (3.231). (Equation (3.231) in Lemma 3.34 follows from (3.215), (3.217), and (3.218) which have already been proven.) □

Next we consider the $t_{\underline{r}}$-dependence of $F_{\underline{p}}$, $G_{\underline{p}}$, and $H_{\underline{p}}$.

Lemma 3.34 *Assume Hypothesis 3.32 and suppose that (3.194), (3.195) hold. In addition, let $(z, n, t_{\underline{r}}) \in \mathbb{C} \times \mathbb{Z} \times \mathbb{R}$. Then,*

$$F_{\underline{p},t_{\underline{r}}} = -2iG_{\underline{p}}\widetilde{F}_{\underline{r}} + i(\widetilde{G}_{\underline{r}} + \widetilde{K}_{\underline{r}})F_{\underline{p}}, \tag{3.231}$$

$$G_{\underline{p},t_{\underline{r}}} = iF_{\underline{p}}\widetilde{H}_{\underline{r}} - iH_{\underline{p}}\widetilde{F}_{\underline{r}}, \tag{3.232}$$

$$H_{\underline{p},t_{\underline{r}}} = 2iG_{\underline{p}}\widetilde{H}_{\underline{r}} - i(\widetilde{G}_{\underline{r}} + \widetilde{K}_{\underline{r}})H_{\underline{p}}. \tag{3.233}$$

In particular, (3.231)–(3.233) are equivalent to

$$V_{\underline{p},t_{\underline{r}}} = [\widetilde{V}_{\underline{r}}, V_{\underline{p}}]. \tag{3.234}$$

Proof To prove (3.231) one first differentiates equation (3.218)

$$\phi_{t_{\underline{r}}}(P) - \phi_{t_{\underline{r}}}(P^*) = -c_{0,+}z^{-p_-}yF_{\underline{p}}^{-2}F_{\underline{p},t_{\underline{r}}}.$$

3.6 The Time-Dependent AL Formalism

The time derivative of ϕ given in (3.215) and (3.217) yields

$$\begin{aligned}\phi_{t_{\underline{r}}}(P) - \phi_{t_{\underline{r}}}(P^*) &= i\big(\widetilde{H}_{\underline{r}} + \widetilde{F}_{\underline{r}}\phi(P)^2 - (\widetilde{G}_{\underline{r}} + \widetilde{K}_{\underline{r}})\phi(P)\big) \\ &\quad - i\big(\widetilde{H}_{\underline{r}} + \widetilde{F}_{\underline{r}}\phi(P^*)^2 - (\widetilde{G}_{\underline{r}} + \widetilde{K}_{\underline{r}})\phi(P^*)\big) \\ &= i\widetilde{F}_{\underline{r}}(\phi(P) + \phi(P^*))(\phi(P) - \phi(P^*)) \\ &\quad - i(\widetilde{G}_{\underline{r}} + \widetilde{K}_{\underline{r}})(\phi(P) - \phi(P^*)) \\ &= 2ic_{0,+}z^{-p-}\widetilde{F}_{\underline{r}}yG_{\underline{p}}F_{\underline{p}}^{-2} - ic_{0,+}z^{-p-}(\widetilde{G}_{\underline{r}} + \widetilde{K}_{\underline{r}})yF_{\underline{p}}^{-1},\end{aligned}$$

and hence

$$F_{\underline{p},t_{\underline{r}}} = -2iG_{\underline{p}}\widetilde{F}_{\underline{r}} + i(\widetilde{G}_{\underline{r}} + \widetilde{K}_{\underline{r}})F_{\underline{p}}.$$

Similarly, starting from (3.217)

$$\phi_{t_{\underline{r}}}(P) + \phi_{t_{\underline{r}}}(P^*) = 2F_{\underline{p}}^{-2}(F_{\underline{p}}G_{\underline{p},t_{\underline{r}}} - F_{\underline{p},t_{\underline{r}}}G_{\underline{p}})$$

yields (3.232) and

$$0 = R_{\underline{p},t_{\underline{r}}} = 2G_{\underline{p}}G_{\underline{p},t_{\underline{r}}} - F_{\underline{p},t_{\underline{r}}}H_{\underline{p}} - F_{\underline{p}}H_{\underline{p},t_{\underline{r}}}$$

proves (3.233). □

Next we turn to the Dubrovin equations for the time variation of the zeros μ_j of $(\,\cdot\,)^{p-}F_{\underline{p}}$ and ν_j of $(\,\cdot\,)^{p--1}H_{\underline{p}}$ governed by the $\widetilde{\text{AL}}_{\underline{r}}$ flow.

Lemma 3.35 *Assume Hypothesis 3.32 and suppose that* (3.194), (3.195) *hold on* $\mathbb{Z} \times \mathcal{I}_\mu$ *with* $\mathcal{I}_\mu \subseteq \mathbb{R}$ *an open interval. In addition, assume that the zeros* μ_j, $j = 1,\ldots,p$, *of* $(\,\cdot\,)^{p-}F_{\underline{p}}(\,\cdot\,)$ *remain distinct and nonzero on* $\mathbb{Z} \times \mathcal{I}_\mu$. *Then* $\{\hat{\mu}_j\}_{j=1,\ldots,p}$, *defined in* (3.206), *satisfies the following first-order system of differential equations on* $\mathbb{Z} \times \mathcal{I}_\mu$,

$$\mu_{j,t_{\underline{r}}} = -i\widetilde{F}_{\underline{r}}(\mu_j)y(\hat{\mu}_j)(\alpha^+)^{-1}\prod_{\substack{k=1\\k\neq j}}^{p}(\mu_j - \mu_k)^{-1}, \quad j=1,\ldots,p, \qquad (3.235)$$

with

$$\hat{\mu}_j(n,\cdot) \in C^\infty(\mathcal{I}_\mu, \mathcal{K}_p), \quad j=1,\ldots,p, \ n \in \mathbb{Z}.$$

For the zeros ν_j, $j = 1,\ldots,p$, *of* $(\,\cdot\,)^{p--1}H_{\underline{p}}(\,\cdot\,)$, *identical statements hold with* μ_j *and* $\mathcal{I}_\mu$ *replaced by* ν_j *and* $\mathcal{I}_\nu$, *etc. (with* $\mathcal{I}_\nu \subseteq \mathbb{R}$ *an open interval). In particular,* $\{\hat{\nu}_j\}_{j=1,\ldots,p}$, *defined in* (3.207), *satisfies the first-order system on* $\mathbb{Z} \times I_\nu$,

$$\nu_{j,t_{\underline{r}}} = i\widetilde{H}_{\underline{r}}(\nu_j)y(\hat{\nu}_j)(\beta\nu_j)^{-1}\prod_{\substack{k=1\\k\neq j}}^{p}(\nu_j - \nu_k)^{-1}, \quad j=1,\ldots,p, \qquad (3.236)$$

258   3 The Ablowitz–Ladik Hierarchy

with

$$\hat{v}_j(n,\cdot) \in C^\infty(\mathcal{I}_v, \mathcal{K}_p), \quad j=1,\ldots,p, \ n \in \mathbb{Z}.$$

Proof It suffices to consider (3.235) for $\mu_{j,t_{\underline{r}}}$. Using the product representation for $F_{\underline{p}}$ in (3.197) and employing (3.206) and (3.231), one computes

$$F_{\underline{p},t_{\underline{r}}}(\mu_j) = \left(c_{0,+}\alpha^+\mu_j^{-p_-}\prod_{\substack{k=1\\k\neq j}}^{p}(\mu_j - \mu_k)\right)\mu_{j,t_{\underline{r}}} = -2iG_{\underline{p}}(\mu_j)\widetilde{F}_{\underline{r}}(\mu_j)$$

$$= -ic_{0,+}\mu_j^{-p_-}y(\hat{\mu}_j)\widetilde{F}_{\underline{r}}(\mu_j), \quad j=1,\ldots,p,$$

proving (3.235). The case of (3.236) for $v_{j,t_{\underline{r}}}$ is of course analogous using the product representation for $H_{\underline{p}}$ in (3.197) and employing (3.207) and (3.233). □

When attempting to solve the Dubrovin systems (3.235) and (3.236), they must be augmented with appropriate divisors $\mathcal{D}_{\hat{\underline{\mu}}(n_0,t_{0,\underline{r}})} \in \mathrm{Sym}^p \mathcal{K}_p$, $t_{0,\underline{r}} \in \mathcal{I}_\mu$, and $\mathcal{D}_{\hat{\underline{v}}(n_0,t_{0,\underline{r}})} \in \mathrm{Sym}^p \mathcal{K}_p$, $t_{0,\underline{r}} \in \mathcal{I}_v$, as initial conditions.

Since the stationary trace formulas for $f_{\ell,\pm}$ and $h_{\ell,\pm}$ in terms of symmetric functions of the zeros μ_j and v_k of $(\cdot)^{p_-}F_{\underline{p}}$ and $(\cdot)^{p_--1}H_{\underline{p}}$ in Lemma 3.19 extend line by line to the corresponding time-dependent setting, we next record their $t_{\underline{r}}$-dependent analogs without proof. For simplicity we again confine ourselves to the simplest cases only.

Lemma 3.36 *Assume Hypothesis 3.32 and suppose that* (3.194), (3.195) *hold. Then,*

$$\frac{\alpha}{\alpha^+} = (-1)^{p+1}\frac{c_{0,+}}{c_{0,-}}\prod_{j=1}^{p}\mu_j,$$

$$\frac{\beta^+}{\beta} = (-1)^{p+1}\frac{c_{0,+}}{c_{0,-}}\prod_{j=1}^{p}v_j,$$

$$\sum_{j=1}^{p}\mu_j = \alpha^+\beta - \gamma^+\frac{\alpha^{++}}{\alpha^+} - \frac{c_{1,+}}{c_{0,+}},$$

$$\sum_{j=1}^{p}v_j = \alpha^+\beta - \gamma\frac{\beta^-}{\beta} - \frac{c_{1,+}}{c_{0,+}}.$$

Next, we turn to the asymptotic expansions of ϕ and Ψ in a neighborhood of $P_{\infty\pm}$ and $P_{0,\pm}$.

3.6 The Time-Dependent AL Formalism 259

Lemma 3.37 *Assume Hypothesis 3.32 and suppose that* (3.194), (3.195) *hold. Moreover, let* $P = (z, y) \in \mathcal{K}_p \setminus \{P_{\infty_+}, P_{\infty_-}, P_{0,+}, P_{0,-}\}$, $(n, n_0, t_{\underline{r}}, t_{0,\underline{r}}) \in \mathbb{Z}^2 \times \mathbb{R}^2$. *Then* ϕ *has the asymptotic behavior*

$$\phi(P) \underset{\zeta \to 0}{=} \begin{cases} \beta + \beta^- \gamma \zeta + O(\zeta^2), & P \to P_{\infty_+}, \\ -(\alpha^+)^{-1} \zeta^{-1} + (\alpha^+)^{-2} \alpha^{++} \gamma^+ + O(\zeta), & P \to P_{\infty_-}, \end{cases} \quad \zeta = 1/z, \tag{3.237}$$

$$\phi(P) \underset{\zeta \to 0}{=} \begin{cases} \alpha^{-1} - \alpha^{-2} \alpha^- \gamma \zeta + O(\zeta^2), & P \to P_{0,+}, \\ -\beta^+ \zeta - \beta^{++} \gamma^+ \zeta^2 + O(\zeta^3), & P \to P_{0,-}, \end{cases} \quad \zeta = z. \tag{3.238}$$

The component ψ_1 *of the Baker–Akhiezer vector* Ψ *has the asymptotic behavior*

$$\psi_1(P, n, n_0, t_{\underline{r}}, t_{0,\underline{r}}) \underset{\zeta \to 0}{=} \exp\left(\pm \frac{i}{2}(t_{\underline{r}} - t_{0,\underline{r}}) \sum_{s=0}^{r_+} \tilde{c}_{r_+-s,+} \zeta^{-s}\right)(1 + O(\zeta))$$

$$\times \begin{cases} \zeta^{n_0 - n}, & P \to P_{\infty_+}, \\ \Gamma(n, n_0, t_{\underline{r}}) \frac{\alpha^+(n, t_{\underline{r}})}{\alpha^+(n_0, t_{0,\underline{r}})} & \\ \times \exp\left(i \int_{t_{0,\underline{r}}}^{t_{\underline{r}}} ds \left(\tilde{g}_{r_+,+}(n_0, s) - \tilde{g}_{r_-,-}(n_0, s)\right)\right), & P \to P_{\infty_-}, \end{cases} \quad \zeta = 1/z,$$

$$\tag{3.239}$$

$$\psi_1(P, n, n_0, t_{\underline{r}}, t_{0,\underline{r}}) \underset{\zeta \to 0}{=} \exp\left(\pm \frac{i}{2}(t_{\underline{r}} - t_{0,\underline{r}}) \sum_{s=0}^{r_-} \tilde{c}_{r_--s,-} \zeta^{-s}\right)(1 + O(\zeta))$$

$$\times \begin{cases} \frac{\alpha(n, t_{\underline{r}})}{\alpha(n_0, t_{0,\underline{r}})}, & P \to P_{0,+}, \\ \Gamma(n, n_0, t_{\underline{r}}) \zeta^{n - n_0} & \\ \times \exp\left(i \int_{t_{0,\underline{r}}}^{t_{\underline{r}}} ds \left(\tilde{g}_{r_+,+}(n_0, s) - \tilde{g}_{r_-,-}(n_0, s)\right)\right), & P \to P_{0,-}, \end{cases} \quad \zeta = z.$$

$$\tag{3.240}$$

Proof Since by the definition of ϕ in (3.208) the time parameter $t_{\underline{r}}$ can be viewed as an additional but fixed parameter, the asymptotic behavior of ϕ remains the same as in Lemma 3.21. Similarly, also the asymptotic behavior of $\psi_1(P, n, n_0, t_{\underline{r}}, t_{\underline{r}})$ is derived in an identical fashion to that in Lemma 3.21. This proves (3.239) and (3.240) for $t_{0,\underline{r}} = t_{\underline{r}}$, that is,

$$\psi_1(P, n, n_0, t_{\underline{r}}, t_{\underline{r}}) \underset{\zeta \to 0}{=} \begin{cases} \zeta^{n_0 - n}(1 + O(\zeta)), & P \to P_{\infty_+}, \\ \Gamma(n, n_0, t_{\underline{r}}) \frac{\alpha^+(n, t_{\underline{r}})}{\alpha^+(n_0, t_{\underline{r}})} + O(\zeta), & P \to P_{\infty_-}, \end{cases} \quad \zeta = 1/z,$$

$$\psi_1(P, n, n_0, t_{\underline{r}}, t_{\underline{r}}) \underset{\zeta \to 0}{=} \begin{cases} \frac{\alpha(n, t_{\underline{r}})}{\alpha(n_0, t_{\underline{r}})} + O(\zeta), & P \to P_{0,+}, \\ \Gamma(n, n_0, t_{\underline{r}}) \zeta^{n - n_0}(1 + O(\zeta)), & P \to P_{0,-}, \end{cases} \quad \zeta = z,$$

It remains to investigate

$$\psi_1(P, n_0, n_0, t_{\underline{r}}, t_{0,\underline{r}}) = \exp\left(i \int_{t_{0,\underline{r}}}^{t_{\underline{r}}} dt \big(\widetilde{G}_{\underline{r}}(z, n_0, t) - \widetilde{F}_{\underline{r}}(z, n_0, t)\phi(P, n_0, t)\big)\right). \tag{3.241}$$

The asymptotic expansion of the integrand is derived using Theorem C.4. Focusing on the homogeneous coefficients first, one computes as $P \to P_{\infty\pm}$,

$$\widehat{G}_{s,+} - \widehat{F}_{s,+}\phi = \widehat{G}_{s,+} - \widehat{F}_{s,+} \frac{G_{\underline{p}} + (c_{0,+}/2)z^{-\underline{p}_-}y}{F_{\underline{p}}}$$

$$= \widehat{G}_{s,+} - \widehat{F}_{s,+}\left(\frac{2z^{p_-}}{c_{0,+}}\frac{G_{\underline{p}}}{y} + 1\right)\left(\frac{2z^{p_-}}{c_{0,+}}\frac{F_{\underline{p}}}{y}\right)^{-1}$$

$$\underset{\zeta \to 0}{=} \pm\frac{1}{2}\zeta^{-s} + \frac{\hat{g}_{0,+} \mp \frac{1}{2}}{\hat{f}_{0,+}}\hat{f}_{s,+} + O(\zeta), \quad P \to P_{\infty\pm}, \ \zeta = 1/z.$$

Since

$$\widetilde{F}_{\underline{r}} \underset{\zeta \to 0}{=} \sum_{s=0}^{r_+} \tilde{c}_{r_+-s,+}\widehat{F}_{s,+} + O(\zeta), \quad \widetilde{G}_{\underline{r}} \underset{\zeta \to 0}{=} \sum_{s=0}^{r_+} \tilde{c}_{r_+-s,+}\widehat{G}_{s,+} + O(\zeta),$$

one infers from (3.237)

$$\widetilde{G}_{\underline{r}} - \widetilde{F}_{\underline{r}}\phi \underset{\zeta \to 0}{=} \frac{1}{2}\sum_{s=0}^{r_+} \tilde{c}_{r_+-s,+}\zeta^{-s} + O(\zeta), \quad P \to P_{\infty+}, \ \zeta = 1/z. \tag{3.242}$$

Insertion of (3.242) into (3.241) then proves (3.239) as $P \to P_{\infty+}$.
As $P \to P_{\infty-}$, we need one additional term in the asymptotic expansion of $\widetilde{F}_{\underline{r}}$, that is, we will use

$$\widetilde{F}_{\underline{r}} \underset{\zeta \to 0}{=} \sum_{s=0}^{r_+} \tilde{c}_{r_+-s,+}\widehat{F}_{s,+} + \sum_{s=0}^{r_-} \tilde{c}_{r_--s,-}\hat{f}_{s-1,-}\zeta + O(\zeta^2).$$

This then yields

$$\widetilde{G}_{\underline{r}} - \widetilde{F}_{\underline{r}}\phi \underset{\zeta \to 0}{=} -\frac{1}{2}\sum_{s=0}^{r_+} \tilde{c}_{r_+-s,+}\zeta^{-s} - (\alpha^+)^{-1}(\tilde{f}_{r_+,+} - \tilde{f}_{r_--1,-}) + O(\zeta).$$

Invoking (3.21) and (3.190) one concludes that

$$\tilde{f}_{r_--1,-} - \tilde{f}_{r_+,+} = -i\alpha^+_{t_{\underline{r}}} + \alpha^+(\tilde{g}_{r_+,+} - \tilde{g}_{r_-,-})$$

and hence

$$\widetilde{G}_{\underline{r}} - \widetilde{F}_{\underline{r}}\phi \underset{\zeta \to 0}{=} -\frac{1}{2}\sum_{s=0}^{r_+} \tilde{c}_{r_+-s,+}\zeta^{-s} - \frac{i\alpha^+_{t_{\underline{r}}}}{\alpha^+} + \tilde{g}_{r_+,+} - \tilde{g}_{r_-,-} + O(\zeta). \tag{3.243}$$

3.6 The Time-Dependent AL Formalism

Insertion of (3.243) into (3.241) then proves (3.239) as $P \to P_{\infty,-}$.
Using Theorem C.4 again, one obtains in the same manner as $P \to P_{0,\pm}$,

$$\widehat{G}_{s,-} - \widehat{F}_{s,-}\phi \underset{\zeta \to 0}{=} \pm \frac{1}{2}\zeta^{-s} - \hat{g}_{s,-} + \frac{\hat{g}_{0,-} \pm \frac{1}{2}}{\hat{f}_{0,-}}\hat{f}_{s,-} + O(\zeta). \tag{3.244}$$

Since

$$\widetilde{F}_{\underline{r}} \underset{\zeta \to 0}{=} \sum_{s=0}^{r_-} \tilde{c}_{r_- - s, -}\widehat{F}_{s,-} + \tilde{f}_{r_+ - 1, +} + O(\zeta), \quad P \to P_{0,\pm}, \ \zeta = z, \tag{3.245}$$

$$\widetilde{G}_{\underline{r}} \underset{\zeta \to 0}{=} \sum_{s=0}^{r_-} \tilde{c}_{r_- - s, -}\widehat{G}_{s,-} + \tilde{g}_{r_+, +} + O(\zeta), \quad P \to P_{0,\pm}, \ \zeta = z, \tag{3.246}$$

(3.244)–(3.246) yield

$$\widetilde{G}_{\underline{r}} - \widetilde{F}_{\underline{r}}\phi \underset{\zeta \to 0}{=} \pm\frac{1}{2}\sum_{s=0}^{r_-} \tilde{c}_{r_- - s, -}\zeta^{-s} + \tilde{g}_{r_+, +} - \tilde{g}_{r_-, -}$$

$$- \frac{\hat{g}_{0,-} \pm \frac{1}{2}}{\hat{f}_{0,-}}(\tilde{f}_{r_+ - 1, +} - \tilde{f}_{r_-, -}) + O(\zeta),$$

where we again used (3.238), (3.35), and (3.190). As $P \to P_{0,-}$, one thus obtains

$$\widetilde{G}_{\underline{r}} - \widetilde{F}_{\underline{r}}\phi \underset{\zeta \to 0}{=} -\frac{1}{2}\sum_{s=0}^{r_-} \tilde{c}_{r_- - s, -}\zeta^{-s} + \tilde{g}_{r_+, +} - \tilde{g}_{r_-, -}, \quad P \to P_{0,-}, \ \zeta = z. \tag{3.247}$$

Insertion of (3.247) into (3.241) then proves (3.240) as $P \to P_{0,-}$.
As $P \to P_{0,+}$, one obtains

$$\widetilde{G}_{\underline{r}} - \widetilde{F}_{\underline{r}}\phi \underset{\zeta \to 0}{=} \frac{1}{2}\sum_{s=0}^{r_-} \tilde{c}_{r_- - s, -}\zeta^{-s} + \tilde{g}_{r_+, +} - \tilde{g}_{r_-, -}$$

$$- \frac{1}{\alpha}(\tilde{f}_{r_+ - 1, +} - \tilde{f}_{r_-, -}) + O(\zeta)$$

$$\underset{\zeta \to 0}{=} \frac{1}{2}\sum_{s=0}^{r_-} \tilde{c}_{r_- - s, -}\zeta^{-s} - \frac{i\alpha_{t_{\underline{r}}}}{\alpha} + O(\zeta), \quad P \to P_{0,+}, \ \zeta = z, \tag{3.248}$$

using $\tilde{f}_{r_-,-} = \tilde{f}^-_{r_- - 1,-} + \alpha(\tilde{g}_{r_-,-} - \tilde{g}^-_{r_-,-})$ (cf. (3.25)) and (3.190). Insertion of (3.248) into (3.241) then proves (3.240) as $P \to P_{0,+}$. □

Next, we note that Lemma 3.22 on nonspecial divisors in the stationary context extends to the present time-dependent situation without a change. Indeed, since $t_{\underline{r}} \in \mathbb{R}$ just plays the role of a parameter, the proof of Lemma 3.22 extends line by line and is hence omitted.

Lemma 3.38 *Assume Hypothesis* 3.32 *and suppose that* (3.194), (3.195) *hold. Moreover, let* $(n, t_{\underline{r}}) \in \mathbb{Z} \times \mathbb{R}$. *Denote by* $\mathcal{D}_{\underline{\hat{\mu}}}$, $\underline{\hat{\mu}} = \{\hat{\mu}_1, \ldots, \hat{\mu}_p\}$, *and* $\mathcal{D}_{\underline{\hat{\nu}}}$, $\underline{\hat{\nu}} = \{\hat{\nu}_1, \ldots, \hat{\nu}_p\}$, *the pole and zero divisors of degree* p, *respectively, associated with* α, β, *and* ϕ *defined according to* (3.206) *and* (3.207), *that is,*

$$\hat{\mu}_j(n, t_{\underline{r}}) = (\mu_j(n, t_{\underline{r}}), (2/c_{0,+})\mu_j(n, t_{\underline{r}})^{p-}G_{\underline{p}}(\mu_j(n, t_{\underline{r}}), n, t_{\underline{r}})), \quad j = 1, \ldots, p,$$

$$\hat{\nu}_j(n, t_{\underline{r}}) = (\nu_j(n, t_{\underline{r}}), -(2/c_{0,+})\nu_j(n, t_{\underline{r}})^{p-}G_{\underline{p}}(\nu_j(n, t_{\underline{r}}), n, t_{\underline{r}})), \quad j = 1, \ldots, p.$$

Then $\mathcal{D}_{\underline{\hat{\mu}}(n,t_{\underline{r}})}$ *and* $\mathcal{D}_{\underline{\hat{\nu}}(n,t_{\underline{r}})}$ *are nonspecial for all* $(n, t_{\underline{r}}) \in \mathbb{Z} \times \mathbb{R}$.

We also note that

$$\Gamma(n, n_0, t_{\underline{r}}) = \Gamma(n, n_0, t_{0,\underline{r}})$$
$$\times \exp\left(i \int_{t_{0,\underline{r}}}^{t_{\underline{r}}} ds \big(\tilde{g}_{r_+,+}(n, s) - \tilde{g}_{r_+,+}(n_0, s) - \tilde{g}_{r_-,-}(n, s) + \tilde{g}_{r_-,-}(n_0, s)\big)\right),$$

which follows from (3.55), (3.110), and from

$$\Gamma(n, n_0, t_{\underline{r}})_{t_{\underline{r}}} = \sum_{j=n_0+1}^{n} \gamma(j, t_{\underline{r}})_{t_{\underline{r}}} \prod_{\substack{k=n_0+1 \\ k \neq j}}^{n} \gamma(j, t_{\underline{r}})$$
$$= i\big(\tilde{g}_{r_+,+}(n, t_{\underline{r}}) - \tilde{g}_{r_+,+}(n_0, t_{\underline{r}}) - \tilde{g}_{r_-,-}(n, t_{\underline{r}}) + \tilde{g}_{r_-,-}(n_0, t_{\underline{r}})\big)\Gamma(n, n_0, t_{\underline{r}})$$

after integration with respect to $t_{\underline{r}}$.

Next, we turn to the principal result of this section, the representation of ϕ, Ψ, α, and β in terms of the Riemann theta function associated with $\mathcal{K}_p$, assuming $p \in \mathbb{N}$ for the remainder of this section.

In addition to (3.125)–(3.130), let $\omega^{(2)}_{P_{\infty\pm},q}$ and $\omega^{(2)}_{P_{0,\pm},q}$ be the normalized differentials of the second kind with a unique pole at $P_{\infty\pm}$ and $P_{0,\pm}$, respectively, and principal parts

$$\omega^{(2)}_{P_{\infty\pm},q} \underset{\zeta \to 0}{=} \big(\zeta^{-2-q} + O(1)\big)d\zeta, \quad P \to P_{\infty\pm}, \ \zeta = 1/z, \ q \in \mathbb{N}_0, \quad (3.249)$$

$$\omega^{(2)}_{P_{0,\pm},q} \underset{\zeta \to 0}{=} \big(\zeta^{-2-q} + O(1)\big)d\zeta, \quad P \to P_{0,\pm}, \ \zeta = z, \ q \in \mathbb{N}_0, \quad (3.250)$$

with vanishing a-periods,

$$\int_{a_j} \omega^{(2)}_{P_{\infty\pm},q} = \int_{a_j} \omega^{(2)}_{P_{0,\pm},q} = 0, \quad j = 1, \ldots, p.$$

3.6 The Time-Dependent AL Formalism

Moreover, we define

$$\widetilde{\Omega}_{\underline{r}}^{(2)} = \frac{i}{2}\Bigg(\sum_{s=1}^{r_-} s\tilde{c}_{r_- -s,-}\big(\omega_{P_{0,+},s-1}^{(2)} - \omega_{P_{0,-},s-1}^{(2)}\big)$$
$$+ \sum_{s=1}^{r_+} s\tilde{c}_{r_+ -s,+}\big(\omega_{P_{\infty_+},s-1}^{(2)} - \omega_{P_{\infty_-},s-1}^{(2)}\big)\Bigg), \quad (3.251)$$

where $\tilde{c}_{\ell,\pm}$ are the summation constants in $\widetilde{F}_{\underline{r}}$. The corresponding vector of b-periods of $\widetilde{\Omega}_{\underline{r}}^{(2)}/(2\pi i)$ is then denoted by

$$\underline{\widetilde{U}}_{\underline{r}}^{(2)} = \big(\widetilde{U}_{\underline{r},1}^{(2)}, \ldots, \widetilde{U}_{\underline{r},p}^{(2)}\big), \quad \widetilde{U}_{\underline{r},j}^{(2)} = \frac{1}{2\pi i}\int_{b_j}\widetilde{\Omega}_{\underline{r}}^{(2)}, \quad j=1,\ldots,p.$$

Finally, we abbreviate

$$\widetilde{\Omega}_{\underline{r}}^{\infty,\pm} = \lim_{P\to P_{\infty_\pm}}\Bigg(\int_{Q_0}^P \widetilde{\Omega}_{\underline{r}}^{(2)} \pm \frac{i}{2}\sum_{s=0}^{r_+}\tilde{c}_{r_+ -s,+}\zeta^{-s}\Bigg),$$

$$\widetilde{\Omega}_{\underline{r}}^{0,\pm} = \lim_{P\to P_{0,\pm}}\Bigg(\int_{Q_0}^P \widetilde{\Omega}_{\underline{r}}^{(2)} \pm \frac{i}{2}\sum_{s=0}^{r_-}\tilde{c}_{r_- -s,-}\zeta^{-s}\Bigg).$$

Theorem 3.39 *Assume Hypothesis 3.32 and suppose that* (3.194), (3.195) *hold. In addition, let* $P \in \mathcal{K}_p\setminus\{P_{\infty_+}, P_{\infty_-}, P_{0,+}, P_{0,-}\}$ *and* $(n, n_0, t_{\underline{r}}, t_{0,\underline{r}}) \in \mathbb{Z}^2\times\mathbb{R}^2$. *Then for each* $(n, t_{\underline{r}}) \in \mathbb{Z}\times\mathbb{R}$, $\mathcal{D}_{\hat{\underline{\mu}}(n,t_{\underline{r}})}$ *and* $\mathcal{D}_{\hat{\underline{\nu}}(n,t_{\underline{r}})}$ *are nonspecial. Moreover,*

$$\phi(P, n, t_{\underline{r}}) = C(n, t_{\underline{r}})\frac{\theta(\underline{z}(P,\hat{\underline{\nu}}(n,t_{\underline{r}})))}{\theta(\underline{z}(P,\hat{\underline{\mu}}(n,t_{\underline{r}})))}\exp\Bigg(\int_{Q_0}^P \omega_{P_{0,-},P_{\infty_-}}^{(3)}\Bigg), \quad (3.252)$$

$$\psi_1(P, n, n_0, t_{\underline{r}}, t_{0,\underline{r}}) = C(n, n_0, t_{\underline{r}}, t_{0,\underline{r}})\frac{\theta(\underline{z}(P,\hat{\underline{\mu}}(n,t_{\underline{r}})))}{\theta(\underline{z}(P,\hat{\underline{\mu}}(n_0,t_{0,\underline{r}})))} \quad (3.253)$$
$$\times\exp\Bigg((n-n_0)\int_{Q_0}^P \omega_{P_{0,-},P_{\infty_+}}^{(3)} - (t_{\underline{r}}-t_{0,\underline{r}})\int_{Q_0}^P \widetilde{\Omega}_{\underline{r}}^{(2)}\Bigg),$$

$$\psi_2(P, n, n_0, t_{\underline{r}}, t_{0,\underline{r}}) = C(n, t_{\underline{r}})C(n, n_0, t_{\underline{r}}, t_{0,\underline{r}})\frac{\theta(\underline{z}(P,\hat{\underline{\nu}}(n,t_{\underline{r}})))}{\theta(\underline{z}(P,\hat{\underline{\mu}}(n_0,t_{0,\underline{r}})))} \quad (3.254)$$
$$\times\exp\Bigg(\int_{Q_0}^P \omega_{P_{0,-},P_{\infty_-}}^{(3)} + (n-n_0)\int_{Q_0}^P \omega_{P_{0,-},P_{\infty_+}}^{(3)} - (t_{\underline{r}}-t_{0,\underline{r}})\int_{Q_0}^P \widetilde{\Omega}_{\underline{r}}^{(2)}\Bigg),$$

where

$$C(n, t_{\underline{r}}) = \frac{(-1)^{n-n_0}}{\alpha(n_0, t_{\underline{r}})}\exp\big((n-n_0)(\omega_0^{0,-}(P_{0,-}, P_{\infty_-}) - \omega_0^{\infty,+}(P_{0,-}, P_{\infty_-}))\big)$$
$$\times\exp\big(-\omega_0^{0,+}(P_{0,-}, P_{\infty_-})\big)\frac{\theta(\underline{z}(P_{0,+},\hat{\underline{\mu}}(n_0,t_{\underline{r}})))}{\theta(\underline{z}(P_{0,+},\hat{\underline{\nu}}(n_0,t_{\underline{r}})))}, \quad (3.255)$$

$$C(n, n_0, t_{\underline{r}}, t_{0,\underline{r}}) = \frac{\theta(\underline{z}(P_{\infty_+}, \hat{\underline{\mu}}(n_0, t_{0,\underline{r}})))}{\theta(\underline{z}(P_{\infty_+}, \hat{\underline{\mu}}(n, t_{\underline{r}})))} \quad (3.256)$$
$$\times \exp\left((t_{\underline{r}} - t_{0,\underline{r}})\widetilde{\Omega}_{\underline{r}}^{\infty_+} - (n - n_0)\omega_0^{\infty_+}(P_{0,-}, P_{\infty_+})\right).$$

The Abel map linearizes the auxiliary divisors $\mathcal{D}_{\hat{\underline{\mu}}(n,t_{\underline{r}})}$ and $\mathcal{D}_{\hat{\underline{\nu}}(n,t_{\underline{r}})}$ in the sense that

$$\underline{\alpha}_{Q_0}(\mathcal{D}_{\hat{\underline{\mu}}(n,t_{\underline{r}})}) = \underline{\alpha}_{Q_0}(\mathcal{D}_{\hat{\underline{\mu}}(n_0,t_{0,\underline{r}})}) + \underline{A}_{P_{0,-}}(P_{\infty_+})(n - n_0) + \widetilde{\underline{U}}_{\underline{r}}^{(2)}(t_{\underline{r}} - t_{0,\underline{r}}), \quad (3.257)$$

$$\underline{\alpha}_{Q_0}(\mathcal{D}_{\hat{\underline{\nu}}(n,t_{\underline{r}})}) = \underline{\alpha}_{Q_0}(\mathcal{D}_{\hat{\underline{\nu}}(n_0,t_{0,\underline{r}})}) + \underline{A}_{P_{0,-}}(P_{\infty_+})(n - n_0) + \widetilde{\underline{U}}_{\underline{r}}^{(2)}(t_{\underline{r}} - t_{0,\underline{r}}). \quad (3.258)$$

Finally, α, β are of the form

$$\alpha(n, t_{\underline{r}}) = \alpha(n_0, t_{0,\underline{r}}) \exp\left((n - n_0)(\omega_0^{0,+}(P_{0,-}, P_{\infty_+}) - \omega_0^{\infty_+}(P_{0,-}, P_{\infty_+}))\right)$$
$$\times \exp\left((t_{\underline{r}} - t_{0,\underline{r}})(\widetilde{\Omega}_{\underline{r}}^{\infty_+} - \widetilde{\Omega}_{\underline{r}}^{0,+})\right)$$
$$\times \frac{\theta(\underline{z}(P_{0,+}, \hat{\underline{\mu}}(n, t_{\underline{r}})))\theta(\underline{z}(P_{\infty_+}, \hat{\underline{\mu}}(n_0, t_{0,\underline{r}})))}{\theta(\underline{z}(P_{0,+}, \hat{\underline{\mu}}(n_0, t_{0,\underline{r}})))\theta(\underline{z}(P_{\infty_+}, \hat{\underline{\mu}}(n, t_{\underline{r}})))}, \quad (3.259)$$

$$\beta(n, t_{\underline{r}}) = \frac{1}{\alpha(n_0, t_{0,\underline{r}})} \exp\left((n - n_0)(\omega_0^{0,+}(P_{0,-}, P_{\infty_+}) - \omega_0^{\infty_+}(P_{0,-}, P_{\infty_+}))\right)$$
$$\times \exp\left(\omega_0^{\infty_+}(P_{0,-}, P_{\infty_-}) - \omega_0^{0,+}(P_{0,-}, P_{\infty_-})\right)$$
$$\times \exp\left(-(t_{\underline{r}} - t_{0,\underline{r}})(\widetilde{\Omega}_{\underline{r}}^{\infty_+} - \widetilde{\Omega}_{\underline{r}}^{0,+})\right)$$
$$\times \frac{\theta(\underline{z}(P_{0,+}, \hat{\underline{\mu}}(n_0, t_{0,\underline{r}})))\theta(\underline{z}(P_{\infty_+}, \hat{\underline{\nu}}(n, t_{\underline{r}})))}{\theta(\underline{z}(P_{0,+}, \hat{\underline{\nu}}(n_0, t_{0,\underline{r}})))\theta(\underline{z}(P_{\infty_+}, \hat{\underline{\mu}}(n, t_{\underline{r}})))}, \quad (3.260)$$

and

$$\alpha(n, t_{\underline{r}})\beta(n, t_{\underline{r}}) = \exp\left(\omega_0^{\infty_+}(P_{0,-}, P_{\infty_-}) - \omega_0^{0,+}(P_{0,-}, P_{\infty_-})\right)$$
$$\times \frac{\theta(\underline{z}(P_{0,+}, \hat{\underline{\mu}}(n, t_{\underline{r}})))\theta(\underline{z}(P_{\infty_+}, \hat{\underline{\nu}}(n, t_{\underline{r}})))}{\theta(\underline{z}(P_{0,+}, \hat{\underline{\nu}}(n, t_{\underline{r}})))\theta(\underline{z}(P_{\infty_+}, \hat{\underline{\mu}}(n, t_{\underline{r}})))}. \quad (3.261)$$

Proof As in Theorem 3.25 one concludes that $\phi(P, n, t_{\underline{r}})$ is of the form (3.252) and that for $t_{0,\underline{r}} = t_{\underline{r}}, \psi_1(P, n, n_0, t_{\underline{r}}, t_{\underline{r}})$ is of the form

$$\psi_1(P, n, n_0, t_{\underline{r}}, t_{\underline{r}}) = C(n, n_0, t_{\underline{r}}, t_{\underline{r}}) \frac{\theta(\underline{z}(P, \hat{\underline{\mu}}(n, t_{\underline{r}})))}{\theta(\underline{z}(P, \hat{\underline{\mu}}(n_0, t_{\underline{r}})))}$$
$$\times \exp\left((n - n_0) \int_{Q_0}^{P} \omega_{P_{0,-}, P_{\infty_+}}^{(3)}\right).$$

To discuss $\psi_1(P, n, n_0, t_{\underline{r}}, t_{0,\underline{r}})$ we recall (3.213), that is,

$$\psi_1(P, n, n_0, t_{\underline{r}}, t_{0,\underline{r}}) = \psi_1(P, n, n_0, t_{\underline{r}}, t_{\underline{r}})\psi_1(P, n_0, n_0, t_{\underline{r}}, t_{0,\underline{r}}), \quad (3.262)$$

3.6 The Time-Dependent AL Formalism

and hence remaining to be studied is

$$\psi_1(P, n_0, n_0, t_{\underline{r}}, t_{0,\underline{r}}) = \exp\left(i \int_{t_{0,\underline{r}}}^{t_{\underline{r}}} ds\big(\widetilde{G}_{\underline{r}}(z, n_0, s) - \widetilde{F}_{\underline{r}}(z, n_0, s)\phi(P, n_0, s)\big)\right). \tag{3.263}$$

Introducing $\hat{\psi}_1(P)$ on $\mathcal{K}_p \setminus \{P_{\infty_+}, P_{\infty_-}\}$ by

$$\hat{\psi}_1(P, n_0, t_{\underline{r}}, t_{0,\underline{r}}) = C(n_0, n_0, t_{\underline{r}}, t_{0,\underline{r}}) \frac{\theta(\underline{z}(P, \hat{\underline{\mu}}(n_0, t_{\underline{r}})))}{\theta(\underline{z}(P, \hat{\underline{\mu}}(n_0, t_{0,\underline{r}})))}$$
$$\times \exp\left(-(t_{\underline{r}} - t_{0,\underline{r}}) \int_{Q_0}^{P} \widetilde{\Omega}_{\underline{r}}^{(2)}\right), \tag{3.264}$$

we intend to prove that

$$\psi_1(P, n_0, n_0, t_{\underline{r}}, t_{0,\underline{r}}) = \hat{\psi}_1(P, n_0, t_{\underline{r}}, t_{0,\underline{r}}),$$
$$P \in \mathcal{K}_p \setminus \{P_{\infty_+}, P_{\infty_-}\},\ n_0 \in \mathbb{Z},\ t_{\underline{r}}, t_{0,\underline{r}} \in \mathbb{R}, \tag{3.265}$$

for an appropriate choice of the normalization constant $C(n_0, n_0, t_{\underline{r}}, t_{0,\underline{r}})$ in (3.264). We start by noting that a comparison of (3.239), (3.240) and (3.249), (3.250), (3.251), (3.253) shows that ψ_1 and $\hat{\psi}_1$ have the same essential singularities at $P_{\infty_{\pm}}$ and $P_{0,\pm}$. Thus, we turn to the local behavior of ψ_1 and $\hat{\psi}_1$. By (3.264), $\hat{\psi}_1$ has zeros and poles at $\hat{\underline{\mu}}(n_0, t_{\underline{r}})$ and $\hat{\underline{\mu}}(n_0, t_{0,\underline{r}})$, respectively. Similarly, by (3.263), ψ_1 can have zeros and poles only at poles of $\phi(P, n_0, s)$, $s \in [t_{0,\underline{r}}, t_{\underline{r}}]$ (resp., $s \in [t_{\underline{r}}, t_{0,\underline{r}}]$). In the following we temporarily restrict $t_{0,\underline{r}}$ and $t_{\underline{r}}$ to a sufficiently small nonempty interval $I \subseteq \mathbb{R}$ and pick $n_0 \in \mathbb{Z}$ such that for all $s \in I$, $\mu_j(n_0, s) \neq \mu_k(n_0, s)$ for all $j \neq k$, $j, k = 1, \ldots, p$. One computes

$$i\widetilde{G}_{\underline{r}}(z, n_0, s) - i\widetilde{F}_{\underline{r}}(z, n_0, s)\phi(P, n_0, s)$$
$$= i\widetilde{G}_{\underline{r}}(z, n_0, s) - i\widetilde{F}_{\underline{r}}(z, n_0, s)\frac{(c_{0,+}/2)z^{-p_-}y + G_{\underline{p}}(z, n_0, s)}{F_{\underline{p}}(z, n_0, s)}$$
$$\underset{P \to \hat{\mu}_j(n_0,s)}{=} \frac{i\widetilde{F}_{\underline{r}}(\mu_j(n_0, s), n_0, s)y(\hat{\mu}_j(n_0, s))}{\alpha^+(n_0, s)(z - \mu_j(n_0, s))\prod_{\substack{k=1 \\ k \neq j}}^{p}(\mu_j(n_0, s) - \mu_k(n_0, s))} + O(1)$$
$$\underset{P \to \hat{\mu}_j(n_0,s)}{=} \frac{\partial}{\partial s}\ln\big(\mu_j(n_0, s) - z\big) + O(1). \tag{3.266}$$

Restricting P to a sufficiently small neighborhood $\mathcal{U}_j(n_0)$ of $\{\hat{\mu}_j(n_0, s) \in \mathcal{K}_p \mid s \in [t_{0,\underline{r}}, t_{\underline{r}}] \subseteq I\}$ such that $\hat{\mu}_k(n_0, s) \notin \mathcal{U}_j(n_0)$ for all $s \in [t_{0,\underline{r}}, t_{\underline{r}}] \subseteq I$ and all $k \in$

$\{1, \ldots, p\} \setminus \{j\}$, (3.264) and (3.266) imply

$$\psi_1(P, n_0, n_0, t_{\underline{r}}, t_{0,\underline{r}})$$
$$= \begin{cases} (\mu_j(n_0, t_{\underline{r}}) - z)O(1) & \text{as } P \to \hat{\mu}_j(n_0, t_{\underline{r}}) \neq \hat{\mu}_j(n_0, t_{0,\underline{r}}), \\ O(1) & \text{as } P \to \hat{\mu}_j(n_0, t_{\underline{r}}) = \hat{\mu}_j(n_0, t_{0,\underline{r}}), \\ (\mu_j(n_0, t_{0,\underline{r}}) - z)^{-1} O(1) & \text{as } P \to \hat{\mu}_j(n_0, t_{0,\underline{r}}) \neq \hat{\mu}_j(n_0, t_{\underline{r}}), \end{cases}$$
$$P = (z, y) \in \mathcal{K}_p,$$

with $O(1) \neq 0$. Thus, ψ_1 and $\hat{\psi}_1$ have the same local behavior and identical essential singularities at $P_{\infty_\pm}$ and $P_{0,\pm}$. By Lemma B.1, ψ_1 and $\hat{\psi}_1$ coincide up to a multiple constant (which may depend on $n_0, t_{\underline{r}}, t_{0,\underline{r}}$). This proves (3.265) for $t_{0,\underline{r}}, t_{\underline{r}} \in I$ and for n_0 as restricted above. By continuity with respect to divisors this extends to all $n_0 \in \mathbb{Z}$ since by hypothesis $\mathcal{D}_{\hat{\mu}(n,s)}$ remain nonspecial for all $(n, s) \in \mathbb{Z} \times \mathbb{R}$. Moreover, since by (3.263), for fixed P and n_0, $\psi_1(P, n_0, n_0, ., t_{0,\underline{r}})$ is entire in $t_{\underline{r}}$ (and this argument is symmetric in $t_{\underline{r}}$ and $t_{0,\underline{r}}$), (3.265) holds for all $t_{\underline{r}}, t_{0,\underline{r}} \in \mathbb{R}$ (for an appropriate choice of $C(n_0, n_0, t_{\underline{r}}, t_{0,\underline{r}})$). Together with (3.262), this proves (3.253) for all $(n, t_{\underline{r}}), (n_0, t_{0,\underline{r}}) \in \mathbb{Z} \times \mathbb{R}$. The expression (3.254) for ψ_2 then immediately follows from (3.252) and (3.253).

To determine the constant $C(n, n_0, t_{\underline{r}}, t_{0,\underline{r}})$ one compares the asymptotic expansions of $\psi_1(P, n, n_0, t_{\underline{r}}, t_{0,\underline{r}})$ for $P \to P_{\infty_+}$ in (3.239) and (3.253)

$$C(n, n_0, t_{\underline{r}}, t_{0,\underline{r}}) = \exp\left((t_{\underline{r}} - t_{0,\underline{r}}) \widetilde{\Omega}_{\underline{r}}^{\infty_+} - (n - n_0) \omega_0^{\infty_+}(P_{0,-}, P_{\infty_+})\right)$$
$$\times \frac{\theta(\underline{z}(P_{\infty_+}, \hat{\underline{\mu}}(n_0, t_{0,\underline{r}})))}{\theta(\underline{z}(P_{\infty_+}, \hat{\underline{\mu}}(n, t_{\underline{r}})))}.$$

Remaining to be computed are the expressions for α and β. Comparing the asymptotic expansions of $\psi_1(P, n, n_0, t_{\underline{r}}, t_{0,\underline{r}})$ for $P \to P_{0,+}$ in (3.240) and (3.253) shows

$$\frac{\alpha(n, t_{\underline{r}})}{\alpha(n_0, t_{0,\underline{r}})} = C(n, n_0, t_{\underline{r}}, t_{0,\underline{r}})$$
$$\times \exp\left((n - n_0) \omega_0^{0,+}(P_{0,-}, P_{\infty_+}) - (t_{\underline{r}} - t_{0,\underline{r}}) \widetilde{\Omega}_{\underline{r}}^{0,+}\right)$$
$$\times \frac{\theta(\underline{z}(P_{0,+}, \hat{\underline{\mu}}(n, t_{\underline{r}})))}{\theta(\underline{z}(P_{0,+}, \hat{\underline{\mu}}(n_0, t_{0,\underline{r}})))}$$

and inserting $C(n, n_0, t_{\underline{r}}, t_{0,\underline{r}})$ proves (3.259). Equation (3.255) for $C(n, t_{\underline{r}})$ follows as in the stationary case since $t_{\underline{r}}$ can be viewed as an additional but fixed parameter. By the first line of (3.146),

$$\alpha(n, t_{\underline{r}}) = \frac{1}{C(n, t_{\underline{r}})} \exp\left(-\omega_0^{0,+}(P_{0,-}, P_{\infty_-})\right) \frac{\theta(\underline{z}(P_{0,+}, \hat{\underline{\mu}}(n, t_{\underline{r}})))}{\theta(\underline{z}(P_{0,+}, \hat{\underline{\nu}}(n, t_{\underline{r}})))}. \quad (3.267)$$

Inserting the result (3.259) for $\alpha(n, t_{\underline{r}})$ into (3.267) then yields (using Lemma 3.24)

$$C(n, t_{\underline{r}}) = \frac{1}{\alpha(n_0, t_{0,\underline{r}})} \frac{\theta(\underline{z}(P_{0,+}, \hat{\underline{\mu}}(n_0, t_{0,\underline{r}})))}{\theta(\underline{z}(P_{\infty_+}, \hat{\underline{\mu}}(n_0, t_{0,\underline{r}})))} \exp\left((t_{\underline{r}} - t_{0,\underline{r}})(\widetilde{\Omega}_{\underline{r}}^{0,+} - \widetilde{\Omega}_{\underline{r}}^{\infty_+})\right)$$

$$\times \exp\left((n - n_0)(\omega_0^{\infty_+}(P_{0,-}, P_{\infty_+}) - \omega_0^{0,+}(P_{0,-}, P_{\infty_+})) - \omega_0^{0,+}(P_{0,-}, P_{\infty_-})\right).$$
(3.268)

Also, since the first line of (3.147) holds,

$$\beta(n, t_{\underline{r}}) = C(n, t_{\underline{r}}) \exp\left(\omega_0^{\infty_+}(P_{0,-}, P_{\infty_-})\right) \frac{\theta(\underline{z}(P_{\infty_+}, \hat{\underline{\nu}}(n, t_{\underline{r}})))}{\theta(\underline{z}(P_{\infty_+}, \hat{\underline{\mu}}(n, t_{\underline{r}})))}, \quad (3.269)$$

an insertion of (3.268) into (3.269), observing Lemma 3.24, yields equation (3.260) for $\beta(n, t_{\underline{r}})$. Finally, multiplying (3.267) and (3.269) proves (3.261). Single-valuedness of $\psi_1(\cdot, n_0, n_0, t_{\underline{r}}, t_{0,\underline{r}})$ on $\mathcal{K}_p$ implies

$$\underline{\alpha}_{Q_0}(\mathcal{D}_{\hat{\underline{\mu}}(n_0, t_{\underline{r}})}) = \underline{\alpha}_{Q_0}(\mathcal{D}_{\hat{\underline{\mu}}(n_0, t_{0,\underline{r}})}) + i(t_{\underline{r}} - t_{0,\underline{r}})\underline{\widetilde{U}}_{\underline{r}}^{(2)}. \quad (3.270)$$

Inserting (3.270) into (3.138),

$$\underline{\alpha}_{Q_0}(\mathcal{D}_{\hat{\underline{\mu}}(n, t_{\underline{r}})}) = \underline{\alpha}_{Q_0}(\mathcal{D}_{\hat{\underline{\mu}}(n_0, t_{\underline{r}})}) + \underline{A}_{P_{0,-}}(P_{\infty_+})(n - n_0),$$

one obtains the result (3.257). □

Combining (3.257), (3.258) and (3.259), (3.260) shows the remarkable linearity of the theta function representations for α and β with respect to $(n, t_r) \in \mathbb{Z} \times \mathbb{R}$.

Again we note that the apparent n_0-dependence of $C(n, t_{\underline{r}})$ in the right-hand side of (3.255) actually drops out to ensure the n_0-independence of ϕ in (3.252).

The theta function representations (3.259) and (3.260) for α and β, and the one for $\Gamma(\cdot, \cdot, t_{\underline{r}})$ analogous to that in (3.261) also show that $\gamma(n, t_r) \notin \{0, 1\}$ for all $(n, t_{\underline{r}}) \in \mathbb{Z} \times \mathbb{R}$. Hence, condition (3.189) is satisfied for the time-dependent algebro-geometric AL solutions discussed in this section, provided the associated divisors $\mathcal{D}_{\hat{\underline{\mu}}}(n, t_{\underline{r}})$ and $\mathcal{D}_{\hat{\underline{\nu}}}(n, t_{\underline{r}})$ stay away from $P_{\infty_{\pm}}, P_{0,\pm}$ for all $(n, t_{\underline{r}}) \in \mathbb{Z} \times \mathbb{R}$.

3.7 The Time-Dependent Ablowitz–Ladik Algebro-Geometric Initial Value Problem

> Es ist nicht genug, zu wissen,
> man muß auch anwenden.
>
> Johann Wolfgang von Goethe[1]

In this section we discuss the algebro-geometric initial value problem (3.190), (3.191) for the Ablowitz–Ladik hierarchy with complex-valued initial data and prove

---

[1] *Wilhelm Meisters Wanderjahre oder Die Entsagenden* (1821). ("Knowing is not enough, one also must apply.")

unique solvability globally in time for a set of initial (Dirichlet divisor) data of full measure. More precisely, we intend to describe a solution of the following problem: Given $\underline{p} \in \mathbb{N}_0^2 \setminus \{(0,0)\}$, assume $\alpha^{(0)}, \beta^{(0)}$ to be complex-valued solutions of the $\underline{p}$th stationary Ablowitz–Ladik system s-AL$_{\underline{p}}(\alpha^{(0)}, \beta^{(0)}) = 0$ associated with a prescribed nonsingular hyperelliptic curve $\mathcal{K}_p$ of genus p and let $\underline{r} = (r_-, r_+) \in \mathbb{N}_0^2$; we want to construct unique global solutions $\alpha = \alpha(n, t_{\underline{r}}), \beta = \beta(n, t_{\underline{r}})$ of the $\underline{r}$th Ablowitz–Ladik flow AL$_{\underline{r}}(\alpha, \beta) = 0$ with $\alpha(t_{0,\underline{r}}) = \alpha^{(0)}, \beta(t_{0,\underline{r}}) = \beta^{(0)}$ for some $t_{0,\underline{r}} \in \mathbb{R}$. Thus, we seek a unique global solution of the initial value problem

$$\widetilde{\mathrm{AL}}_{\underline{r}}(\alpha, \beta) = 0,$$
$$(\alpha, \beta)\big|_{t=t_{0,\underline{r}}} = \left(\alpha^{(0)}, \beta^{(0)}\right),$$
$$\text{s-AL}_{\underline{p}}\left(\alpha^{(0)}, \beta^{(0)}\right) = 0$$

for some $t_{0,\underline{r}} \in \mathbb{R}$, where $\alpha = \alpha(n, t_{\underline{r}}), \beta = \beta(n, t_{\underline{r}})$ satisfy

$$\alpha(\cdot, t), \beta(\cdot, t) \in \mathbb{C}^{\mathbb{Z}}, \ t \in \mathbb{R}, \quad \alpha(n, \cdot), \beta(n, \cdot) \in C^1(\mathbb{R}), \ n \in \mathbb{Z},$$
$$\alpha(n, t)\beta(n, t) \notin \{0, 1\}, \ (n, t) \in \mathbb{Z} \times \mathbb{R}.$$

Here,

$$\underline{p} = (p_-, p_+) \in \mathbb{N}_0^2 \setminus \{(0,0)\}, \quad \underline{r} = (r_-, r_+) \in \mathbb{N}_0^2, \quad p = p_- + p_+ - 1.$$

As in the stationary context treated in Section 3.5, the fact that the underlying Lax operator $\check{L}$ is non-normal poses two kinds of difficulties when general initial data $\alpha^{(0)}, \beta^{(0)}$ or divisors $\mathcal{D}_{\underline{\hat{\mu}}(n_0, t_{0,\underline{r}})}$ in general complex position are considered. In this case the Dirichlet eigenvalues $\mu_j, j = 1, \ldots, p$, are in general not pairwise distinct and "collisions" between them can occur at certain values of $(n, t_{\underline{r}}) \in \mathbb{Z} \times \mathbb{R}$. A priori, one has no control over such collisions, especially, it is not possible to identify initial conditions $\mathcal{D}_{\underline{\hat{\mu}}(n_0, t_{0,\underline{r}})}$ at some $(n_0, t_{0,\underline{r}}) \in \mathbb{Z} \times \mathbb{R}$ which avoid collisions for all $(n, t_{\underline{r}}) \in \mathbb{Z} \times \mathbb{R}$. Again we solve this problem by referring to a more general interpolation formalism (cf. Appendix D) for polynomials, going beyond the usual Lagrange interpolation formulas. In the time-dependent context we replace the first-order system of Dubrovin-type equations (3.235), augmented with the initial divisor $\mathcal{D}_{\underline{\hat{\mu}}(n_0, t_{0,\underline{r}})}$, by a different first-order system of differential equations (3.281), (3.287), and (3.288) with initial conditions (3.289) which focuses on symmetric functions of $\mu_1(n, t_{\underline{r}}), \ldots, \mu_p(n, t_{\underline{r}})$ rather than individual Dirichlet eigenvalues $\mu_j(n, t_{\underline{r}}), j = 1, \ldots, p$. In this manner it will be shown that collisions of Dirichlet eigenvalues no longer pose a problem. In addition, there is again a second complication since it cannot be guaranteed that $\mu_j(n, t_{\underline{r}})$ and $\nu_j(n, t_{\underline{r}}), j = 1, \ldots, p$, stay finite and nonzero for all $(n, t_{\underline{r}}) \in \mathbb{Z} \times \mathbb{R}$. We solve this particular problem in the stationary as well as time-dependent case by properly restricting the initial Dirichlet and Neumann divisors $\mathcal{D}_{\underline{\hat{\mu}}(n_0, t_{0,\underline{r}})}, \mathcal{D}_{\underline{\hat{\nu}}(n_0, t_{0,\underline{r}})} \in \mathrm{Sym}^p \mathcal{K}_p$ to a dense set of full measure.

3.7 The Time-Dependent AL Algebro-Geometric IVP

In short, our strategy will consist of the following:

(α) Replace the first-order autonomous Dubrovin-type system (3.235) of differential equations in $t_{\underline{r}}$ for the Dirichlet eigenvalues $\mu_j(n, t_{\underline{r}})$, $j = 1, \ldots, p$, augmented by appropriate initial conditions, by the first-order autonomous system (3.313), (3.314) for the coefficients $f_{\ell,\pm}, h_{\ell,\pm}, \ell = 1, \ldots, p_{\pm} - 1$, and $g_{\ell,\pm}, \ell = 1, \ldots, p_{\pm}$, with respect to $t_{\underline{r}}$. Solve this first-order autonomous system in some time interval $(t_{0,\underline{r}} - T_0, t_{0,\underline{r}} + T_0)$ under appropriate initial conditions at $(n_0, t_{0,\underline{r}})$ derived from an initial (nonspecial) Dirichlet divisor $\mathcal{D}_{\hat{\underline{\mu}}(n_0, t_{0,\underline{r}})}$.

(β) Use the stationary algorithm derived in Section 3.5 to extend the solution of step (α) from $\{n_0\} \times (t_{0,\underline{r}} - T_0, t_{0,\underline{r}} + T_0)$ to $\mathbb{Z} \times (t_{0,\underline{r}} - T_0, t_{0,\underline{r}} + T_0)$ (cf. Lemma 3.41).

(γ) Prove consistency of this approach, that is, show that the discrete algorithm of Section 3.5 is compatible with the time-dependent Lax and zero-curvature equations in the sense that first solving the autonomous system (3.313), (3.314) and then applying the discrete algorithm, or first applying the discrete algorithm and then solving the autonomous system (3.313), (3.314) yields the same result whenever the same endpoint $(n, t_{\underline{r}})$ is reached (cf. Lemma 3.42 and Theorem 3.43).

(δ) Prove that there is a dense set of initial conditions of full measure for which this strategy yields global solutions of the algebro-geometric Ablowitz–Ladik hierarchy initial value problem.

To set up this formalism we need some preparations. From the outset we make the following assumption.

Hypothesis 3.40 *Suppose that*

$$\alpha, \beta \in \mathbb{C}^{\mathbb{Z}} \text{ and } \alpha(n)\beta(n) \notin \{0, 1\} \text{ for all } n \in \mathbb{Z},$$

and assume that α, β satisfy the $\underline{p}$th stationary Ablowitz–Ladik equation (3.30). In addition, suppose that the affine part of the hyperelliptic curve $\mathcal{K}_p$ (cf. (3.88)) is nonsingular.

We introduce a deformation (time) parameter $t_{\underline{r}} \in \mathbb{R}$ in $\alpha = \alpha(t_{\underline{r}})$ and $\beta = \beta(t_{\underline{r}})$ and hence obtain $t_{\underline{r}}$-dependent quantities $f_{\ell,\pm} = f_{\ell,\pm}(t_{\underline{r}})$, $g_{\ell,\pm} = g_{\ell,\pm}(t_{\underline{r}})$, $h_{\ell,\pm} = h_{\ell,\pm}(t_{\underline{r}})$, $F_{\underline{p}}(z) = F_{\underline{p}}(z, t_{\underline{r}})$, $G_{\underline{p}}(z) = G_{\underline{p}}(z, t_{\underline{r}})$, $H_{\underline{p}}(z) = H_{\underline{p}}(z, t_{\underline{r}})$, etc. At a fixed initial time $t_{0,\underline{r}} \in \mathbb{R}$ we require that

$$(\alpha, \beta)|_{t_{\underline{r}} = t_{0,\underline{r}}} = \left(\alpha^{(0)}, \beta^{(0)}\right), \tag{3.271}$$

where $\alpha^{(0)} = \alpha(\cdot, t_{0,\underline{r}})$, $\beta^{(0)} = \beta(\cdot, t_{0,\underline{r}})$ satisfy the $\underline{p}$th stationary Ablowitz–Ladik system (3.30). As discussed in Section 3.5, in order to guarantee that the stationary solutions (3.271) can be constructed for all $n \in \mathbb{Z}$ one starts from a particular divisor

$$\mathcal{D}_{\hat{\underline{\mu}}(n_0, t_{0,\underline{r}})} \in \mathcal{M}_0, \tag{3.272}$$

where $\underline{\hat{\mu}}(n_0, t_{0,\underline{r}})$ is of the form

$$\underline{\hat{\mu}}(n_0, t_{0,\underline{r}})$$
$$= \{\underbrace{\hat{\mu}_1(n_0, t_{0,\underline{r}}), \ldots, \hat{\mu}_1(n_0, t_{0,\underline{r}})}_{p_1(n_0, t_{0,\underline{r}}) \text{ times}}, \ldots$$
$$\ldots, \underbrace{\hat{\mu}_{q(n_0, t_{0,\underline{r}})}(n_0, t_{0,\underline{r}}), \ldots, \hat{\mu}_{q(n_0, t_{0,\underline{r}})}(n_0, t_{0,\underline{r}})}_{p_{q(n_0, t_{0,\underline{r}})}(n_0, t_{0,\underline{r}}) \text{ times}}\}.$$

Moreover, as in Section 3.5 we prescribe the data

$$\alpha(n_0, t_{0,\underline{r}}) \in \mathbb{C} \setminus \{0\} \text{ and } c_{0,+} \in \mathbb{C} \setminus \{0\}, \tag{3.273}$$

and of course the hyperelliptic curve $\mathcal{K}_p$ with nonsingular affine part (cf. (3.155)). In addition, we introduce

$$\alpha^+(n_0, t_{0,\underline{r}}) = \alpha(n_0, t_{0,\underline{r}}) \left(\prod_{m=0}^{2p+1} E_m\right)^{1/2} \prod_{k=1}^{q(n_0, t_{0,\underline{r}})} \mu_k(n_0, t_{0,\underline{r}})^{-p_k(n_0, t_{0,\underline{r}})},$$
$$\tag{3.274}$$

$$F_{\underline{p}}(z, n_0, t_{0,\underline{r}}) = \sum_{\ell=1}^{p_-} f_{p_- - \ell, -}(n_0, t_{0,\underline{r}}) z^{-\ell} + \sum_{\ell=0}^{p_+ - 1} f_{p_+ - 1 - \ell, +}(n_0, t_{0,\underline{r}}) z^\ell$$

$$= -c_{0,+} \alpha^+(n_0, t_{0,\underline{r}}) z^{-p_-} \prod_{k=1}^{q(n_0, t_{0,\underline{r}})} (z - \mu_k(n_0, t_{0,\underline{r}}))^{p_k(n_0, t_{0,\underline{r}})}, \tag{3.275}$$

$$G_{\underline{p}}(z, n_0, t_{0,\underline{r}}) = \frac{1}{2}\left(\frac{1}{\alpha(n_0, t_{0,\underline{r}})} - \frac{z}{\alpha^+(n_0, t_{0,\underline{r}})}\right) F_{\underline{p}}(z, n_0, t_{0,\underline{r}})$$
$$- \frac{z}{2\alpha^+(n_0, t_{0,\underline{r}})} F_{\underline{p}}(z, n_0, t_{0,\underline{r}})$$
$$\times \sum_{k=1}^{q(n_0, t_{0,\underline{r}})} \sum_{\ell=0}^{p_k(n_0, t_{0,\underline{r}})-1} \frac{(d^\ell(\zeta^{-1} y(P))/d\zeta^\ell)|_{P=(\zeta, \eta)=\hat{\mu}_k(n_0, t_{0,\underline{r}})}}{\ell!(p_k(n_0, t_{0,\underline{r}}) - \ell - 1)!} \tag{3.276}$$
$$\times \left(\frac{d^{p_k(n_0, t_{0,\underline{r}})-\ell-1}}{d\zeta^{p_k(n_0, t_{0,\underline{r}})-\ell-1}}\left((z-\zeta)^{-1}\right.\right.$$
$$\left.\left.\times \prod_{k'=1,\, k'\neq k}^{q(n_0, t_{0,\underline{r}})} (\zeta - \mu_{k'}(n_0, t_{0,\underline{r}}))^{-p_{k'}(n_0, t_{0,\underline{r}})}\right)\right)\bigg|_{\zeta=\mu_k(n_0, t_{0,\underline{r}})},$$

in analogy to (3.158)–(3.161).

Our aim is to find an autonomous first-order system of ordinary differential equations with respect to $t_{\underline{r}}$ for $f_{\ell,\pm}$, $g_{\ell,\pm}$, and $h_{\ell,\pm}$ rather than for μ_j. We divide the

3.7 The Time-Dependent AL Algebro-Geometric IVP

differential equation

$$F_{\underline{p},t_{\underline{r}}} = -2iG_{\underline{p}}\widetilde{F}_{\underline{r}} + i(\widetilde{G}_{\underline{r}} + \widetilde{K}_{\underline{r}})F_{\underline{p}}$$

by $c_{0,+}z^{-p_-}y$ and rewrite it using Theorem C.4 as

$$\sum_{\ell=0}^{\infty} \hat{f}_{\ell,+,t_{\underline{r}}}\zeta^{\ell+1} = -2i\left(\sum_{s=1}^{r_-} \tilde{f}_{r_--s,-}\zeta^s + \sum_{s=0}^{r_+-1} \tilde{f}_{r_+-1-s,+}\zeta^{-s}\right)\sum_{\ell=0}^{\infty} \hat{g}_{\ell,+}\zeta^{\ell}$$

$$+ i\left(2\sum_{s=0}^{r_-} \tilde{g}_{r_--s,-}\zeta^s + 2\sum_{s=1}^{r_+} \tilde{g}_{r_+-s,+}\zeta^{-s} - \tilde{g}_{r_-,-} + \tilde{g}_{r_+,+}\right)\sum_{\ell=0}^{\infty} \hat{f}_{\ell,+}\zeta^{\ell+1},$$

$$P \to P_{\infty_-}, \ \zeta = 1/z. \quad (3.277)$$

The coefficients of ζ^{-s}, $s = 0, \ldots, r_+ - 1$, cancel since

$$\sum_{k=0}^{\ell} \tilde{f}_{\ell-k,+}\hat{g}_{k,+} = \sum_{k=0}^{\ell} \tilde{g}_{\ell-k,+}\hat{f}_{k,+}, \quad \ell \in \mathbb{N}_0. \quad (3.278)$$

In (3.278) we used (3.35),

$$\tilde{f}_{\ell,+} = \sum_{k=0}^{\ell} \tilde{c}_{\ell-k,+}\hat{f}_{k,+}, \quad \tilde{g}_{\ell,+} = \sum_{k=0}^{\ell} \tilde{c}_{\ell-k,+}\hat{g}_{k,+}.$$

Comparing coefficients in (3.277) then yields [1]

$$\hat{f}_{\ell,+,t_{\underline{r}}} = i\hat{f}_{\ell,+}(\tilde{g}_{r_+,+} - \tilde{g}_{r_-,-}) + 2i\sum_{k=0}^{r_+-1}(\tilde{g}_{k,+}\hat{f}_{r_++\ell-k,+} - \tilde{f}_{k,+}\hat{g}_{r_++\ell-k,+})$$

$$- 2i\sum_{k=(\ell+1-r_-)\vee 0}^{\ell} \hat{g}_{k,+}\tilde{f}_{r_--1-\ell+k,-} + 2i\sum_{k=(\ell+2-r_-)\vee 0}^{\ell} \hat{f}_{k,+}\tilde{g}_{r_--\ell+k,-},$$

$$\ell \in \mathbb{N}_0. \quad (3.279)$$

By (3.278), the last sum in (3.279) can be rewritten as

$$\sum_{j=0}^{r_+-1}(\tilde{g}_{j,+}\hat{f}_{r_++\ell-j,+} - \tilde{f}_{j,+}\hat{g}_{r_++\ell-j,+})$$

$$= \left(\sum_{j=0}^{r_++\ell} - \sum_{j=r_+}^{r_++\ell}\right)(\tilde{g}_{j,+}\hat{f}_{r_++\ell-j,+} - \tilde{f}_{j,+}\hat{g}_{r_++\ell-j,+})$$

$$= -\sum_{j=r_+}^{r_++\ell}(\tilde{g}_{j,+}\hat{f}_{r_++\ell-j,+} - \tilde{f}_{j,+}\hat{g}_{r_++\ell-j,+})$$

$$= \sum_{j=0}^{\ell}(\hat{g}_{j,+}\tilde{f}_{r_++\ell-j,+} - \hat{f}_{j,+}\tilde{g}_{r_++\ell-j,+}). \quad (3.280)$$

[1] $m \vee n = \max\{m, n\}$.

One performs a similar computation for $\hat{f}_{\ell,-,t_{\underline{r}}}$ using Theorem C.4 at $P \to P_{0,+}$. In summary, since $f_{k,\pm} = \sum_{\ell=0}^{k} c_{k-\ell,\pm} \hat{f}_{\ell,\pm}$, (3.279) and (3.280) yield the following autonomous first-order system (for fixed $n = n_0$)

$$f_{\ell,\pm,t_{\underline{r}}} = \mathcal{F}_{\ell,\pm}(f_{j,-}, f_{j,+}, g_{j,-}, g_{j,+}), \quad \ell = 0, \ldots, p_{\pm} - 1, \qquad (3.281)$$

with initial conditions

$$\begin{aligned} f_{\ell,\pm}(n_0, t_{0,\underline{r}}), & \quad \ell = 0, \ldots, p_{\pm} - 1, \\ g_{\ell,\pm}(n_0, t_{0,\underline{r}}), & \quad \ell = 0, \ldots, p_{\pm}, \end{aligned} \qquad (3.282)$$

where $\mathcal{F}_{\ell,\pm}$, $\ell = 0, \ldots, p_{\pm} - 1$, are polynomials in $2p + 3$ variables,

$$\begin{aligned} \mathcal{F}_{\ell,\pm} = & \, i f_{\ell,\pm}(\tilde{g}_{r_{\pm},\pm} - \tilde{g}_{r_{\mp},\mp}) \\ & + 2i \sum_{k=0}^{\ell} \bigl(f_{k,\pm}(\tilde{g}_{r_{\mp}-\ell+k,\mp} - \tilde{g}_{r_{\pm}+\ell-k,\pm}) \\ & \quad + g_{k,\pm}(\tilde{f}_{r_{\pm}+\ell-k,\pm} - \tilde{f}_{r_{\mp}-1-\ell+k,\mp}) \bigr) \\ & + 2i \sum_{k=0}^{\ell} c_{\ell-k,\pm} \times \begin{cases} 0, & 0 \leq k < r_{\mp} - 1, \\ \sum_{j=0}^{k-r_{\mp}} \hat{g}_{j,\pm} \tilde{f}_{r_{\mp}-1-k+j,\mp} \\ \quad - \sum_{j=0}^{k+1-r_{\mp}} \hat{f}_{j,\pm} \tilde{g}_{r_{\mp}-k+j,\mp}, & k \geq r_{\mp} - 1. \end{cases} \end{aligned} \qquad (3.283)$$

Explicitly, one obtains (for simplicity, $r_{\pm} > 1$)

$$\begin{aligned} \mathcal{F}_{0,\pm} = & \, i f_{0,\pm}(\tilde{g}_{r_{\mp},\mp} - \tilde{g}_{r_{\pm},\pm}) + 2i g_{0,\pm}(\tilde{f}_{r_{\pm},\pm} - \tilde{f}_{r_{\mp}-1,\mp}), \\ \mathcal{F}_{1,\pm} = & \, 2i f_{0,\pm}(\tilde{g}_{r_{\mp}-1,\mp} - \tilde{g}_{r_{\pm}+1,\pm}) + i f_{1,\pm}(\tilde{g}_{r_{\mp},\mp} - \tilde{g}_{r_{\pm},\pm}) \\ & + 2i g_{0,\pm}(\tilde{f}_{r_{\pm}+1,\pm} - \tilde{f}_{r_{\mp}-2,\mp}) + 2i g_{1,\pm}(\tilde{f}_{r_{\pm},\pm} - \tilde{f}_{r_{\mp}-1,\mp}), \text{ etc.} \end{aligned} \qquad (3.284)$$

By (3.274)–(3.276), the initial conditions (3.282) are uniquely determined by the initial divisor $\mathcal{D}_{\hat{\underline{\mu}}(n_0,t_{0,\underline{r}})}$ in (3.272) and by the data in (3.273).

Similarly, one transforms

$$\begin{aligned} G_{\underline{p},t_{\underline{r}}} &= i F_{\underline{p}} \tilde{H}_{\underline{r}} - i H_{\underline{p}} \tilde{F}_{\underline{r}}, \\ H_{\underline{p},t_{\underline{r}}} &= 2i G_{\underline{p}} \tilde{H}_{\underline{r}} - i \bigl(\tilde{G}_{\underline{r}} + \tilde{K}_{\underline{r}} \bigr) H_{\underline{p}} \end{aligned}$$

into (for fixed $n = n_0$) [1]

$$\hat{g}_{0,\pm,t_{\underline{r}}} = 0,$$

$$\hat{g}_{\ell,\pm,t_{\underline{r}}} = i \sum_{k=0}^{r_{\pm}-1} \bigl(\tilde{h}_{k,\pm} \hat{f}_{r_{\pm}-1+\ell-k,\pm} - \tilde{f}_{k,\pm} \hat{h}_{r_{\pm}-1+\ell-k,\pm} \bigr)$$

$$+ i \sum_{k=(\ell-r_{\mp}) \vee 0}^{\ell-1} \bigl(\hat{f}_{k,\pm} \tilde{h}_{r_{\mp}-\ell+k,\mp} - \hat{h}_{k,\pm} \tilde{f}_{r_{\mp}-\ell+k,\mp} \bigr)$$

---

[1] $m \vee n = \max\{m, n\}$.

3.7 The Time-Dependent AL Algebro-Geometric IVP

$$= i \sum_{k=0}^{\ell-1} \left(\hat{h}_{k,\pm} \tilde{f}_{r_\pm - 1 + \ell - k, \pm} - \hat{f}_{k,\pm} \tilde{h}_{r_\pm - 1 + \ell - k, \pm} \right)$$

$$+ i \sum_{k=(\ell - r_\mp) \vee 0}^{\ell-1} \left(\hat{f}_{k,\pm} \tilde{h}_{r_\mp - \ell + k, \mp} - \hat{h}_{k,\pm} \tilde{f}_{r_\mp - \ell + k, \mp} \right), \quad \ell \in \mathbb{N}, \quad (3.285)$$

$$\hat{h}_{\ell,\pm,t_{\underline{r}}} = i\hat{h}_{\ell,\pm} \left(\tilde{g}_{r_\mp,\mp} - \tilde{g}_{r_\pm,\pm} \right) + 2i \sum_{k=0}^{r_\pm - 1} \left(\hat{h}_{k,\pm} \tilde{g}_{r_\pm + \ell - k, \pm} - \tilde{g}_{k,\pm} \hat{h}_{r_\pm + \ell - k, +} \right)$$

$$+ 2i \sum_{k=(\ell - r_\mp + 1) \vee 0}^{\ell} \hat{g}_{k,\pm} \tilde{h}_{r_\mp - 1 - \ell + k, \mp} - 2i \sum_{k=(\ell - r_\mp) \vee 0}^{\ell} \hat{h}_{k,\pm} \tilde{g}_{r_\mp - \ell + k, \mp}$$

$$= i\hat{h}_{\ell,\pm} \left(\tilde{g}_{r_\mp,\mp} - \tilde{g}_{r_\pm,\pm} \right) + 2i \sum_{k=0}^{\ell} \left(\hat{h}_{k,\pm} \tilde{g}_{r_\pm + \ell - k, \pm} - \hat{g}_{k,\pm} \tilde{h}_{r_\pm + \ell - k, \pm} \right)$$

$$+ 2i \sum_{k=(\ell - r_\mp + 1) \vee 0}^{\ell} \hat{g}_{k,\pm} \tilde{h}_{r_\mp - 1 - \ell + k, \mp} - 2i \sum_{k=(\ell - r_\mp) \vee 0}^{\ell} \hat{h}_{k,\pm} \tilde{g}_{r_\mp - \ell + k, \mp},$$

$$\ell \in \mathbb{N}_0. \quad (3.286)$$

Summing over ℓ in (3.285), (3.286) then yields the following first-order system

$$g_{\ell,\pm,t_{\underline{r}}} = \mathcal{G}_{\ell,\pm}(f_{k,-}, f_{k,+}, h_{k,-}, h_{k,+}), \quad \ell = 0, \ldots, p_\pm, \quad (3.287)$$

$$h_{\ell,\pm,t_{\underline{r}}} = \mathcal{H}_{\ell,\pm}(g_{k,-}, g_{k,+}, h_{k,-}, h_{k,+}), \quad \ell = 0, \ldots, p_\pm - 1, \quad (3.288)$$

with initial conditions

$$\begin{aligned} f_{\ell,\pm}(n_0, t_{0,\underline{r}}), & \quad \ell = 0, \ldots, p_\pm - 1, \\ g_{\ell,\pm}(n_0, t_{0,\underline{r}}), & \quad \ell = 0, \ldots, p_\pm, \\ h_{\ell,\pm}(n_0, t_{0,\underline{r}}), & \quad \ell = 0, \ldots, p_\pm - 1, \end{aligned} \quad (3.289)$$

where $\mathcal{G}_{\ell,\pm}, \mathcal{H}_{\ell,\pm}$ are polynomials in $2p+2, 2p+3$ variables

$$\mathcal{G}_{\ell,\pm} = i \sum_{k=0}^{\ell-1} \left(f_{k,\pm}(\tilde{h}_{r_\mp - \ell + k, \mp} - \tilde{h}_{r_\pm - 1 + \ell - k, \pm}) \right.$$
$$\left. + h_{k,\pm}(\tilde{f}_{r_\pm - 1 + \ell - k, \pm} - \tilde{f}_{r_\mp - \ell + k, \mp}) \right)$$
$$- i \sum_{k=0}^{\ell-1} c_{\ell - 1 - k, \pm} \quad (3.290)$$
$$\times \begin{cases} 0, & 0 \leq k \leq r_\mp, \\ \sum_{j=0}^{k - r_\mp - 1} (\hat{f}_{j,\pm} \tilde{h}_{r_\mp - k + j, \mp} - \hat{h}_{j,\pm} \tilde{f}_{r_\mp - k + j, \mp}), & k > r_\mp, \end{cases}$$

$$\mathcal{H}_{\ell,\pm} = ih_{\ell,\pm}(\tilde{g}_{r_{\mp},\mp} - \tilde{g}_{r_{\pm},\pm})$$
$$+ 2i \sum_{k=0}^{\ell} \left(g_{k,\pm}(\tilde{h}_{r_{\mp}-1-\ell+k,\mp} - \tilde{h}_{r_{\pm}+\ell-k,\pm}) \right.$$
$$\left. + h_{k,\pm}(\tilde{g}_{r_{\pm}+\ell-k,\pm} - \tilde{g}_{r_{\mp}-\ell+k,\mp}) \right) \quad (3.291)$$
$$+ 2i \sum_{k=0}^{\ell} c_{\ell-k,\pm} \times \begin{cases} 0, & 0 \le k < r_{\mp}, \\ -\sum_{j=0}^{k-r_{\mp}} \hat{g}_{j,\pm} \tilde{h}_{r_{\mp}-1-k+j,\mp} \\ + \sum_{j=0}^{k-r_{\mp}-1} \hat{h}_{j,\pm} \tilde{g}_{r_{\mp}-k+j,\mp}, & k \ge r_{\mp}. \end{cases}$$

Explicitly (assuming $r_{\pm} > 2$),

$$\mathcal{G}_{0,\pm} = 0,$$
$$\mathcal{G}_{1,\pm} = if_{0,\pm}(\tilde{h}_{r_{\mp}-1,\mp} - \tilde{h}_{r_{\pm},\pm}) + ih_{0,\pm}(\tilde{f}_{r_{\pm},\pm} - \tilde{f}_{r_{\mp}-1,\mp}),$$
$$\mathcal{G}_{2,\pm} = if_{0,\pm}(\tilde{h}_{r_{\mp}-2,\mp} - \tilde{h}_{r_{\pm}+1,\pm}) + if_{1,\pm}(\tilde{h}_{r_{\mp}-1,\mp} - \tilde{h}_{r_{\pm},\pm})$$
$$\qquad + ih_{0,\pm}(\tilde{f}_{r_{\pm}+1,\pm} - \tilde{f}_{r_{\mp}-2,\mp}) + ih_{1,\pm}(\tilde{f}_{r_{\pm},\pm} - \tilde{f}_{r_{\mp}-1,\mp}), \text{ etc.,}$$
$$\mathcal{H}_{0,\pm} = 2ig_{0,\pm}(\tilde{h}_{r_{\mp}-1,\mp} - \tilde{h}_{r_{\pm},\pm}) + ih_{0,\pm}(\tilde{g}_{r_{\pm},\pm} - \tilde{g}_{r_{\mp},\mp}), \quad (3.292)$$
$$\mathcal{H}_{1,\pm} = 2ig_{0,\pm}(\tilde{h}_{r_{\mp}-2,\mp} - \tilde{h}_{r_{\pm}+1,\pm}) + 2ig_{1,\pm}(\tilde{h}_{r_{\mp}-1,\mp} - \tilde{h}_{r_{\pm},\pm})$$
$$\qquad + 2ih_{0,\pm}(\tilde{g}_{r_{\pm}+1,\pm} - \tilde{g}_{r_{\mp}-1,\mp}) + ih_{1,\pm}(\tilde{g}_{r_{\pm},\pm} - \tilde{g}_{r_{\mp},\mp}), \text{ etc.}$$

Again by (3.274)–(3.276), the initial conditions (3.289) are uniquely determined by the initial divisor $\mathcal{D}_{\hat{\underline{\mu}}(n_0,t_{0,\underline{r}})}$ in (3.272) and by the data in (3.273).

Since the system (3.281), (3.287), (3.288) is autonomous with polynomial right-hand sides, there exists a $T_0 > 0$, such that the first-order initial value problem (3.281), (3.287), (3.288) with initial conditions (3.289) has a unique solution

$$\begin{aligned} f_{\ell,\pm} &= f_{\ell,\pm}(n_0, t_{\underline{r}}), & \ell &= 0, \ldots, p_{\pm} - 1, \\ g_{\ell,\pm} &= g_{\ell,\pm}(n_0, t_{\underline{r}}), & \ell &= 0, \ldots, p_{\pm}, \\ h_{\ell,\pm} &= h_{\ell,\pm}(n_0, t_{\underline{r}}), & \ell &= 0, \ldots, p_{\pm} - 1, \end{aligned} \quad (3.293)$$

for all $t_{\underline{r}} \in (t_{0,\underline{r}} - T_0, t_{0,\underline{r}} + T_0)$.

Given the solution (3.293), we proceed as in Section 3.5 and introduce the following quantities (where $t_{\underline{r}} \in (t_{0,\underline{r}} - T_0, t_{0,\underline{r}} + T_0)$):

$$\alpha^+(n_0, t_{\underline{r}}) = \alpha(n_0, t_{\underline{r}}) \left(\prod_{m=0}^{2p+1} E_m \right)^{1/2} \prod_{k=1}^{q(n_0,t_{\underline{r}})} \mu_k(n_0, t_{\underline{r}})^{-p_k(n_0,t_{\underline{r}})}, \quad (3.294)$$

3.7 The Time-Dependent AL Algebro-Geometric IVP

$$F_{\underline{p}}(z, n_0, t_{\underline{r}}) = \sum_{\ell=1}^{p_-} f_{p_- -\ell, -}(n_0, t_{\underline{r}}) z^{-\ell} + \sum_{\ell=0}^{p_+ -1} f_{p_+ -1-\ell, +}(n_0, t_{\underline{r}}) z^{\ell}$$

$$= -c_{0,+} \alpha^+(n_0, t_{\underline{r}}) z^{-p_-} \prod_{k=1}^{q(n_0, t_{\underline{r}})} (z - \mu_k(n_0, t_{\underline{r}}))^{p_k(n_0, t_{\underline{r}})}, \quad (3.295)$$

$$G_{\underline{p}}(z, n_0, t_{\underline{r}}) = \frac{1}{2}\left(\frac{1}{\alpha(n_0, t_{\underline{r}})} - \frac{z}{\alpha^+(n_0, t_{\underline{r}})}\right) F_{\underline{p}}(z, n_0, t_{\underline{r}})$$

$$- \frac{z}{2\alpha^+(n_0, t_{\underline{r}})} F_{\underline{p}}(z, n_0, t_{\underline{r}})$$

$$\times \sum_{k=1}^{q(n_0, t_{\underline{r}})} \sum_{\ell=0}^{p_k(n_0, t_{\underline{r}})-1} \frac{\left(d^{\ell}(\zeta^{-1} y(P))/d\zeta^{\ell}\right)\big|_{P=(\zeta,\eta)=\hat{\mu}_k(n_0, t_{\underline{r}})}}{\ell!(p_k(n_0, t_{\underline{r}}) - \ell - 1)!} \quad (3.296)$$

$$\times \left(\frac{d^{p_k(n_0, t_{\underline{r}})-\ell-1}}{d\zeta^{p_k(n_0, t_{\underline{r}})-\ell-1}}\left((z-\zeta)^{-1}\right.\right.$$

$$\left.\left.\times \prod_{k'=1, k'\neq k}^{q(n_0, t_{\underline{r}})} (\zeta - \mu_{k'}(n_0, t_{\underline{r}}))^{-p_{k'}(n_0, t_{\underline{r}})}\right)\right)\Bigg|_{\zeta=\mu_k(n_0, t_{\underline{r}})}.$$

In particular, this leads to the divisor

$$\mathcal{D}_{\hat{\underline{\mu}}(n_0, t_{\underline{r}})} \in \text{Sym}^p(\mathcal{K}_p)$$

and the sign of y in (3.295) is chosen as usual by

$$\hat{\mu}_k(n_0, t_{\underline{r}}) = (\mu_k(n_0, t_{\underline{r}}), (2/c_{0,+})\mu_j(n_0, t_{\underline{r}})^{p_-} G_{\underline{p}}(\mu_k(n_0, t_{\underline{r}}), n_0, t_{\underline{r}})),$$
$$k = 1, \ldots, q(n_0, t_{\underline{r}}), \quad (3.297)$$

and

$$\hat{\underline{\mu}}(n_0, t_{\underline{r}}) = \{\underbrace{\mu_1(n_0, t_{\underline{r}}), \ldots, \mu_1(n_0, t_{\underline{r}})}_{p_1(n_0, t_{\underline{r}}) \text{ times}}, \ldots$$

$$\ldots, \underbrace{\mu_{q(n_0, t_{\underline{r}})}(n_0, t_{\underline{r}}), \ldots, \mu_{q(n_0, t_{\underline{r}})}(n_0, t_{\underline{r}})}_{p_{q(n_0, t_{\underline{r}})}(n_0, t_{\underline{r}}) \text{ times}}\}$$

with

$$\mu_k(n_0, t_{\underline{r}}) \neq \mu_{k'}(n_0, t_{\underline{r}}) \text{ for } k \neq k', \; k, k' = 1, \ldots, q(n_0, t_{\underline{r}}),$$

and

$$p_k(n_0, t_{\underline{r}}) \in \mathbb{N}, \; k = 1, \ldots, q(n_0, t_{\underline{r}}), \; \sum_{k=1}^{q(n_0, t_{\underline{r}})} p_k(n_0, t_{\underline{r}}) = p.$$

By construction (cf. (3.297)), the divisor $\mathcal{D}_{\hat{\underline{\mu}}(n_0, t_{\underline{r}})}$ is nonspecial for all $t_{\underline{r}} \in (t_{0,\underline{r}} - T_0, t_{0,\underline{r}} + T_0)$.

In exactly the same manner as in (3.162)–(3.165) one then infers that $F_{\underline{p}}(\,\cdot\,, n_0, t_{\underline{r}})$ divides $R_{\underline{p}} - G_{\underline{p}}^2$ (since $t_{\underline{r}}$ is just a fixed parameter).

As in Section 3.5, the assumption that the Laurent polynomial $F_{\underline{p}}(\,\cdot\,, n_0 - 1, t_{\underline{r}})$ is of full order is implied by the hypothesis that

$$\mathcal{D}_{\hat{\underline{\mu}}(n_0, t_{\underline{r}})} \in \mathcal{M}_0 \text{ for all } t_{\underline{r}} \in (t_{0,\underline{r}} - T_0, t_{0,\underline{r}} + T_0). \tag{3.298}$$

The explicit formula for $\beta(n_0, t_{\underline{r}})$ then reads (for $t_{\underline{r}} \in (t_{0,\underline{r}} - T_0, t_{0,\underline{r}} + T_0)$)

$$\begin{aligned}
&\alpha^+(n_0, t_{\underline{r}}) \beta(n_0, t_{\underline{r}}) \\
&= -\frac{1}{2} \sum_{k=1}^{q(n_0, t_{\underline{r}})} \frac{\left(d^{p_k(n_0, t_{\underline{r}})-1}\left(\zeta^{-1} y(P)\right)/d\zeta^{p_k(n_0, t_{\underline{r}})-1}\right)\big|_{P=(\zeta, \eta) = \hat{\mu}_k(n_0, t_{\underline{r}})}}{(p_k(n_0, t_{\underline{r}}) - 1)!} \\
&\quad \times \prod_{k'=1, k' \neq k}^{q(n_0, t_{\underline{r}})} (\mu_k(n_0, t_{\underline{r}}) - \mu_{k'}(n_0, t_{\underline{r}}))^{-p_k(n_0, t_{\underline{r}})} \\
&\quad + \frac{1}{2}\left(\frac{\alpha^+(n_0, t_{\underline{r}})}{\alpha(n_0, t_{\underline{r}})} + \sum_{k=1}^{q(n_0, t_{\underline{r}})} p_k(n_0, t_{\underline{r}}) \mu_k(n_0, t_{\underline{r}}) - \frac{1}{2} \sum_{m=0}^{2p+1} E_m \right). \tag{3.299}
\end{aligned}$$

With (3.285)–(3.299) in place, we can now apply the stationary formalism as summarized in Theorem 3.30, subject to the additional hypothesis (3.298), for each fixed $t_{\underline{r}} \in (t_{0,\underline{r}} - T_0, t_{0,\underline{r}} + T_0)$. This yields, in particular, the quantities

$$F_{\underline{p}}, G_{\underline{p}}, H_{\underline{p}}, \alpha, \beta, \text{ and } \hat{\underline{\mu}}, \hat{\underline{\nu}} \text{ for } (n, t_{\underline{r}}) \in \mathbb{Z} \times (t_{0,\underline{r}} - T_0, t_{0,\underline{r}} + T_0), \tag{3.300}$$

which are of the form (3.294)–(3.299), replacing the fixed $n_0 \in \mathbb{Z}$ by an arbitrary $n \in \mathbb{Z}$. In addition, one has the following result.

Lemma 3.41 *Assume Hypothesis 3.40 and condition* (3.298). *Then the following relations are valid on* $\mathbb{C} \times \mathbb{Z} \times (t_{0,\underline{r}} - T_0, t_{0,\underline{r}} + T_0)$,

$$G_{\underline{p}}^2 - F_{\underline{p}} H_{\underline{p}} = R_{\underline{p}}, \tag{3.301}$$

$$z(G_{\underline{p}}^- - G_{\underline{p}}) + z\beta F_{\underline{p}} + \alpha H_{\underline{p}}^- = 0, \tag{3.302}$$

$$z\beta F_{\underline{p}}^- + \alpha H_{\underline{p}} - G_{\underline{p}} + G_{\underline{p}}^- = 0, \tag{3.303}$$

$$-F_{\underline{p}} + zF_{\underline{p}}^- + \alpha(G_{\underline{p}} + G_{\underline{p}}^-) = 0, \tag{3.304}$$

$$z\beta(G_{\underline{p}} + G_{\underline{p}}^-) - zH_{\underline{p}} + H_{\underline{p}}^- = 0, \tag{3.305}$$

3.7 The Time-Dependent AL Algebro-Geometric IVP

and hence the stationary part, (3.195), of the algebro-geometric initial value problem holds,

$$UV_{\underline{p}} - V_{\underline{p}}^+ U = 0 \text{ on } \mathbb{C} \times \mathbb{Z} \times (t_{0,\underline{r}} - T_0, t_{0,\underline{r}} + T_0).$$

In particular, Lemmas 3.18–3.22 apply.

Lemma 3.41 now raises the following important consistency issue: On the one hand, one can solve the initial value problem (3.313), (3.314) at $n = n_0$ in some interval $t_{\underline{r}} \in (t_{0,\underline{r}} - T_0, t_{0,\underline{r}} + T_0)$, and then extend the quantities $F_{\underline{p}}, G_{\underline{p}}, H_{\underline{p}}$ to all $\mathbb{C} \times \mathbb{Z} \times (t_{0,\underline{r}} - T_0, t_{0,\underline{r}} + T_0)$ using the stationary algorithm summarized in Theorem 3.30 as just recorded in Lemma 3.41. On the other hand, one can solve the initial value problem (3.313), (3.314) at $n = n_1$, $n_1 \neq n_0$, in some interval $t_{\underline{r}} \in (t_{0,\underline{r}} - T_1, t_{0,\underline{r}} + T_1)$ with the initial condition obtained by applying the discrete algorithm to the quantities $F_{\underline{p}}, G_{\underline{p}}, H_{\underline{p}}$ starting at $(n_0, t_{0,\underline{r}})$ and ending at $(n_1, t_{0,\underline{r}})$. Consistency then requires that the two approaches yield the same result at $n = n_1$ for $t_{\underline{r}}$ in some open neighborhood of $t_{0,\underline{r}}$.

Equivalently, and pictorially speaking, envisage a vertical $t_{\underline{r}}$-axis and a horizontal n-axis. Then, consistency demands that first solving the initial value problem (3.313), (3.314) at $n = n_0$ in some $t_{\underline{r}}$-interval around $t_{0,\underline{r}}$ and using the stationary algorithm to extend $F_{\underline{p}}, G_{\underline{p}}, H_{\underline{p}}$ horizontally to $n = n_1$ and the same $t_{\underline{r}}$-interval around $t_{0,\underline{r}}$, or first applying the stationary algorithm starting at $(n_0, t_{0,\underline{r}})$ to extend $F_{\underline{p}}, G_{\underline{p}}, H_{\underline{p}}$ horizontally to $(n_1, t_{0,\underline{r}})$ and then solving the initial value problem (3.313), (3.314) at $n = n_1$ in some $t_{\underline{r}}$-interval around $t_{0,\underline{r}}$ should produce the same result at $n = n_1$ in a sufficiently small open $t_{\underline{r}}$ interval around $t_{0,\underline{r}}$.

To settle this consistency issue, we will prove the following result. To this end we find it convenient to replace the initial value problem (3.313), (3.314) by the original $t_{\underline{r}}$-dependent zero-curvature equation (3.194), $U_{t_{\underline{r}}} + U\widetilde{V}_{\underline{r}} - \widetilde{V}_{\underline{r}}^+ U = 0$ on $\mathbb{C} \times \mathbb{Z} \times (t_{0,\underline{r}} - T_0, t_{0,\underline{r}} + T_0)$.

Lemma 3.42 *Assume Hypothesis 3.40 and condition (3.298). Moreover, suppose that (3.231)–(3.233) hold on $\mathbb{C} \times \{n_0\} \times (t_{0,\underline{r}} - T_0, t_{0,\underline{r}} + T_0)$. Then (3.231)–(3.233) hold on $\mathbb{C} \times \mathbb{Z} \times (t_{0,\underline{r}} - T_0, t_{0,\underline{r}} + T_0)$, that is,*

$$F_{\underline{p},t_{\underline{r}}}(z,n,t_{\underline{r}}) = -2iG_{\underline{p}}(z,n,t_{\underline{r}})\widetilde{F}_{\underline{r}}(z,n,t_{\underline{r}})$$
$$+ i\big(\widetilde{G}_{\underline{r}}(z,n,t_{\underline{r}}) + \widetilde{K}_{\underline{r}}(z,n,t_{\underline{r}})\big)F_{\underline{p}}(z,n,t_{\underline{r}}), \qquad (3.306)$$

$$G_{\underline{p},t_{\underline{r}}}(z,n,t_{\underline{r}}) = iF_{\underline{p}}(z,n,t_{\underline{r}})\widetilde{H}_{\underline{r}}(z,n,t_{\underline{r}}) - iH_{\underline{p}}(z,n,t_{\underline{r}})\widetilde{F}_{\underline{r}}(z,n,t_{\underline{r}}), \qquad (3.307)$$

$$H_{\underline{p},t_{\underline{r}}}(z,n,t_{\underline{r}}) = 2iG_{\underline{p}}(z,n,t_{\underline{r}})\widetilde{H}_{\underline{r}}(z,n,t_{\underline{r}})$$
$$- i\big(\widetilde{G}_{\underline{r}}(z,n,t_{\underline{r}}) + \widetilde{K}_{\underline{r}}(z,n,t_{\underline{r}})\big)H_{\underline{p}}(z,n,t_{\underline{r}}), \qquad (3.308)$$

$$(z,n,t_{\underline{r}}) \in \mathbb{C} \times \mathbb{Z} \times (t_{0,\underline{r}} - T_0, t_{0,\underline{r}} + T_0).$$

Moreover,

$$\phi_{t_{\underline{r}}}(P, n, t_{\underline{r}}) = i\widetilde{F}_{\underline{r}}(z, n, t_{\underline{r}})\phi^2(P, n, t_{\underline{r}}) \qquad (3.309)$$
$$- i\big(\widetilde{G}_{\underline{r}}(z, n, t_{\underline{r}}) + \widetilde{K}_{\underline{r}}(z, n, t_{\underline{r}})\big)\phi(P, n, t_{\underline{r}}) + i\widetilde{H}_{\underline{r}}(z, n, t_{\underline{r}}),$$
$$\alpha_{t_{\underline{r}}}(n, t_{\underline{r}}) = iz\widetilde{F}_{\underline{r}}^-(z, n, t_{\underline{r}}) \qquad (3.310)$$
$$+ i\alpha(n, t_{\underline{r}})\big(\widetilde{G}_{\underline{r}}(z, n, t_{\underline{r}}) + \widetilde{K}_{\underline{r}}^-(z, n, t_{\underline{r}})\big) - i\widetilde{F}_{\underline{r}}(z, n, t_{\underline{r}}),$$
$$\beta_{t_{\underline{r}}}(n, t_{\underline{r}}) = -i\beta(n, t_{\underline{r}})\big(\widetilde{G}_{\underline{r}}^-(z, n, t_{\underline{r}}) + \widetilde{K}_{\underline{r}}(z, n, t_{\underline{r}})\big)$$
$$+ i\widetilde{H}_{\underline{r}}(z, n, t_{\underline{r}}) - iz^{-1}\widetilde{H}_{\underline{r}}^-(z, n, t_{\underline{r}}), \qquad (3.311)$$
$$(z, n, t_{\underline{r}}) \in \mathbb{C} \times \mathbb{Z} \times (t_{0,\underline{r}} - T_0, t_{0,\underline{r}} + T_0).$$

Proof By Lemma 3.41 we have (3.208), (3.209), (3.214), (3.216)–(3.218), and (3.301)–(3.305) for $(n, t_{\underline{r}}) \in \mathbb{Z} \times (t_{0,\underline{r}} - T_0, t_{0,\underline{r}} + T_0)$ at our disposal.
Differentiating (3.208) at $n = n_0$ with respect to $t_{\underline{r}}$ and inserting (3.306) and (3.307) at $n = n_0$ then yields (3.309) at $n = n_0$.
We note that the sequences $\tilde{f}_{\ell,\pm}$, $\tilde{g}_{\ell,\pm}$, $\tilde{h}_{\ell,\pm}$ satisfy the recursion relations (3.19)–(3.26) (since the homogeneous sequences satisfy these relations). Hence, to prove (3.310) and (3.311) at $n = n_0$ it remains to show

$$\alpha_{t_{\underline{r}}} = i\alpha(\tilde{g}_{r_+,+} + \tilde{g}_{r_-,-}^-) + i(\tilde{f}_{r_--1,-}^- - \tilde{f}_{r_++1,+}),$$
$$\beta_{t_{\underline{r}}} = -i\beta(\tilde{g}_{r_+,+}^- + \tilde{g}_{r_-,-}) - i(\tilde{h}_{r_++1,+}^- - \tilde{h}_{r_--1,-}). \qquad (3.312)$$

But this follows from (3.306), (3.308) at $n = n_0$ (cf. (3.284), (3.292))

$$\alpha_{t_{\underline{r}}} = i\alpha(\tilde{g}_{r_+,+} - \tilde{g}_{r_-,-}) + i(\tilde{f}_{r_-,-} - \tilde{f}_{r_++1,+}),$$
$$\beta_{t_{\underline{r}}} = i\beta(\tilde{g}_{r_+,+} - \tilde{g}_{r_-,-}) + i(\tilde{h}_{r_--1,-} - \tilde{h}_{r_+,+}).$$

Inserting now (3.25) at $\ell = r_- - 1$ and (3.22) at $\ell = r_+ - 1$ then yields (3.312).
For the step $n = n_0 \mp 1$ we differentiate (3.171)–(3.176) (which are equivalent to (3.301)–(3.305)) and insert (3.306)–(3.308), (3.198)–(3.205) at $n = n_0$. For the case $n > n_0$ we obtain $\alpha_{t_{\underline{r}}}^+$ and $\beta_{t_{\underline{r}}}^+$ from (3.306), (3.308) at $n = n_0$ as before using the other two signs in (3.284), (3.292). Iterating these arguments proves (3.306)–(3.311) for $(z, n, t_{\underline{r}}) \in \mathbb{C} \times \mathbb{Z} \times (t_{0,\underline{r}} - T_0, t_{0,\underline{r}} + T_0)$. □

We summarize Lemmas 3.41 and 3.42 next.

Theorem 3.43 *Assume Hypothesis 3.40 and condition (3.298). Moreover, suppose that*

$$f_{\ell,\pm} = f_{\ell,\pm}(n_0, t_{\underline{r}}), \quad \ell = 0, \ldots, p_\pm - 1,$$
$$g_{\ell,\pm} = g_{\ell,\pm}(n_0, t_{\underline{r}}), \quad \ell = 0, \ldots, p_\pm,$$
$$h_{\ell,\pm} = h_{\ell,\pm}(n_0, t_{\underline{r}}), \quad \ell = 0, \ldots, p_\pm - 1$$
$$\text{for all } t_{\underline{r}} \in (t_{0,\underline{r}} - T_0, t_{0,\underline{r}} + T_0),$$

3.7 The Time-Dependent AL Algebro-Geometric IVP

satisfy the autonomous first-order system of ordinary differential equations (for fixed $n = n_0$)

$$f_{\ell,\pm,t_{\underline{r}}} = \mathcal{F}_{\ell,\pm}(f_{k,-}, f_{k,+}, g_{k,-}, g_{k,+}), \quad \ell = 0, \ldots, p_\pm - 1,$$
$$g_{\ell,\pm,t_{\underline{r}}} = \mathcal{G}_{\ell,\pm}(f_{k,-}, f_{k,+}, h_{k,-}, h_{k,+}), \quad \ell = 0, \ldots, p_\pm, \quad (3.313)$$
$$h_{\ell,\pm,t_{\underline{r}}} = \mathcal{H}_{\ell,\pm}(g_{k,-}, g_{k,+}, h_{k,-}, h_{k,+}), \quad \ell = 0, \ldots, p_\pm - 1,$$

with $\mathcal{F}_{\ell,\pm}, \mathcal{G}_{\ell,\pm}, \mathcal{H}_{\ell,\pm}$ given by (3.283), (3.290), (3.291), and with initial conditions

$$f_{\ell,\pm}(n_0, t_{0,\underline{r}}), \quad \ell = 0, \ldots, p_\pm - 1,$$
$$g_{\ell,\pm}(n_0, t_{0,\underline{r}}), \quad \ell = 0, \ldots, p_\pm, \quad (3.314)$$
$$h_{\ell,\pm}(n_0, t_{0,\underline{r}}), \quad \ell = 0, \ldots, p_\pm - 1.$$

Then $F_{\underline{p}}, G_{\underline{p}}$, and $H_{\underline{p}}$ as constructed in (3.295)–(3.300) on $\mathbb{C} \times \mathbb{Z} \times (t_{0,\underline{r}} - T_0, t_{0,\underline{r}} + T_0)$ satisfy the zero-curvature equations (3.194), (3.195), and (3.234) on $\mathbb{C} \times \mathbb{Z} \times (t_{0,\underline{r}} - T_0, t_{0,\underline{r}} + T_0)$,

$$U_{t_{\underline{r}}} + U\widetilde{V}_{\underline{r}} - \widetilde{V}_{\underline{r}}^+ U = 0,$$
$$UV_{\underline{p}} - V_{\underline{p}}^+ U = 0,$$
$$V_{\underline{p},t_{\underline{r}}} - [\widetilde{V}_{\underline{r}}, V_{\underline{p}}] = 0$$

with $U, V_{\underline{p}}$, and $\widetilde{V}_{\underline{r}}$ given by (3.196). In particular, α, β satisfy (3.189) and the algebro-geometric initial value problem (3.190), (3.191) on $\mathbb{Z} \times (t_{0,\underline{r}} - T_0, t_{0,\underline{r}} + T_0)$,

$$\widetilde{AL}_{\underline{r}}(\alpha, \beta) = \begin{pmatrix} -i\alpha_{t_{\underline{r}}} - \alpha(\widetilde{g}_{r_+,+} + \widetilde{g}_{r_-,-}) + \widetilde{f}_{r_+-1,+} - \widetilde{f}_{r_--1,-} \\ -i\beta_{t_{\underline{r}}} + \beta(\widetilde{g}_{r_+,+} + \widetilde{g}_{r_-,-}) - \widetilde{h}_{r_--1,-} + \widetilde{h}_{r_+-1,+} \end{pmatrix} = 0,$$

$$(\alpha, \beta)\big|_{t=t_{0,\underline{r}}} = (\alpha^{(0)}, \beta^{(0)}),$$

$$\text{s-AL}_{\underline{p}}(\alpha^{(0)}, \beta^{(0)}) = \begin{pmatrix} -\alpha^{(0)}(g_{p_+,+} + g_{p_-,-}) + f_{p_+-1,+} - f_{p_--1,-} \\ \beta^{(0)}(g_{p_+,+} + g_{p_-,-}) - h_{p_--1,-} + h_{p_+-1,+} \end{pmatrix} = 0.$$

In addition, α, β are given by

$$\alpha^+(n, t_{\underline{r}}) = \alpha(n, t_{\underline{r}}) \left(\prod_{m=0}^{2p+1} E_m \right)^{1/2} \prod_{k=1}^{q(n,t_{\underline{r}})} \mu_k(n, t_{\underline{r}})^{-p_k(n,t_{\underline{r}})}, \quad (3.315)$$

$$\alpha^+(n, t_{\underline{r}})\beta(n, t_{\underline{r}})$$
$$= -\frac{1}{2} \sum_{k=1}^{q(n,t_{\underline{r}})} \frac{\left(d^{p_k(n,t_{\underline{r}})-1}(\zeta^{-1}y(P))/d\zeta^{p_k(n,t_{\underline{r}})-1}\right)\big|_{P=(\zeta,\eta)=\hat{\mu}_k(n,t_{\underline{r}})}}{(p_k(n, t_{\underline{r}}) - 1)!}$$

$$\times \prod_{k'=1, k' \neq k}^{q(n,t_{\underline{r}})} (\mu_k(n, t_{\underline{r}}) - \mu_{k'}(n, t_{\underline{r}}))^{-p_k(n,t_{\underline{r}})} \qquad (3.316)$$

$$+ \frac{1}{2}\left(\left(\prod_{m=0}^{2p+1} E_m\right)^{1/2} \prod_{k=1}^{q(n,t_{\underline{r}})} \mu_k(n, t_{\underline{r}})^{-p_k(n,t_{\underline{r}})} + \sum_{k=1}^{q(n,t_{\underline{r}})} p_k(n, t_{\underline{r}})\mu_k(n, t_{\underline{r}})\right.$$

$$\left. - \frac{1}{2} \sum_{m=0}^{2p+1} E_m\right), \quad (z, n, t_{\underline{r}}) \in \mathbb{Z} \times (t_{0,\underline{r}} - T_0, t_{0,\underline{r}} + T_0).$$

Moreover, Lemmas 3.18–3.22 and 3.33–3.34 apply.

As in Lemma 3.29 we now show that also in the time-dependent case, most initial divisors are well-behaved in the sense that the corresponding divisor trajectory stays away from $P_{\infty_\pm}$, $P_{0,\pm}$ for all $(n, t_{\underline{r}}) \in \mathbb{Z} \times \mathbb{R}$.

Lemma 3.44 *The set $\mathcal{M}_1$ of initial divisors $\mathcal{D}_{\hat{\underline{\mu}}(n_0,t_{0,\underline{r}})}$ for which $\mathcal{D}_{\hat{\underline{\mu}}(n,t_{\underline{r}})}$ and $\mathcal{D}_{\hat{\underline{\nu}}(n,t_{\underline{r}})}$, defined via (3.257) and (3.258), are admissible (i.e., do not contain $P_{\infty_\pm}$, $P_{0,\pm}$) and hence are nonspecial for all $(n, t_{\underline{r}}) \in \mathbb{Z} \times \mathbb{R}$, forms a dense set of full measure in the set $\mathrm{Sym}^p(\mathcal{K}_p)$ of nonnegative divisors of degree p.*

Proof Let $\mathcal{M}_{\mathrm{sing}}$ be as introduced in the proof of Lemma 3.29. Then

$$\bigcup_{t_{\underline{r}} \in \mathbb{R}} \left(\underline{\alpha}_{Q_0}(\mathcal{M}_{\mathrm{sing}}) + t_{\underline{r}} \widetilde{\underline{U}}_{\underline{r}}^{(2)}\right)$$

$$\subseteq \bigcup_{P \in \{P_{\infty_\pm}, P_{0,\pm}\}} \bigcup_{t_{\underline{r}} \in \mathbb{R}} \left(\underline{A}_{Q_0}(P) + \underline{\alpha}_{Q_0}(\mathrm{Sym}^{p-1}(\mathcal{K}_p)) + t_{\underline{r}} \widetilde{\underline{U}}_{\underline{r}}^{(2)}\right)$$

is of measure zero as well, since it is contained in the image of $\mathbb{R} \times \mathrm{Sym}^{p-1}(\mathcal{K}_p)$ which misses one real dimension in comparison to the $2p$ real dimensions of $J(\mathcal{K}_p)$. But then (cf. (3.154)),

$$\bigcup_{(n,t_{\underline{r}}) \in \mathbb{Z} \times \mathbb{R}} \left(\underline{\alpha}_{Q_0}(\mathcal{M}_{\mathrm{sing}}) + n\underline{A}_{P_{0,-}}(P_{\infty_+}) + t_{\underline{r}} \widetilde{\underline{U}}_{\underline{r}}^{(2)}\right) \qquad (3.317)$$

$$\cup \left(\bigcup_{(n,t_{\underline{r}}) \in \mathbb{Z} \times \mathbb{R}} \left(\underline{\alpha}_{Q_0}(\mathcal{M}_{\mathrm{sing}}) + n\underline{A}_{P_{0,-}}(P_{\infty_+}) + t_{\underline{r}} \widetilde{\underline{U}}_{\underline{r}}^{(2)}\right) + \underline{A}_{P_{0,-}}(P_{\infty_-})\right)$$

is also of measure zero. Applying $\underline{\alpha}_{Q_0}^{-1}$ to the complement of the set in (3.317) then yields a set $\mathcal{M}_1$ of full measure in $\mathrm{Sym}^p(\mathcal{K}_p)$. In particular, $\mathcal{M}_1$ is necessarily dense in $\mathrm{Sym}^p(\mathcal{K}_p)$. $\square$

3.8 AL Conservation Laws and the Hamiltonian Formalism   281

Theorem 3.45 *Let $\mathcal{D}_{\hat{\underline{\mu}}(n_0,t_{0,\underline{r}})} \in \mathcal{M}_1$ be an initial divisor as in Lemma 3.44. Then the sequences α, β constructed from $\hat{\underline{\mu}}(n_0, t_{0,\underline{r}})$ as described in Theorem 3.43 satisfy Hypothesis 3.32. In particular, the solution α, β of the algebro-geometric initial value problem (3.315), (3.316) is global in $(n, t_{\underline{r}}) \in \mathbb{Z} \times \mathbb{R}$.*

Proof Starting with $\mathcal{D}_{\hat{\underline{\mu}}(n_0,t_{0,\underline{r}})} \in \mathcal{M}_1$, the procedure outlined in this section and summarized in Theorem 3.43 leads to $\mathcal{D}_{\hat{\underline{\mu}}(n,t_{\underline{r}})}$ and $\mathcal{D}_{\hat{\underline{\nu}}(n,t_{\underline{r}})}$ for all $(n, t_{\underline{r}}) \in \mathbb{Z} \times (t_{0,\underline{r}} - T_0, t_{0,\underline{r}} + T_0)$ such that (3.257) and (3.258) hold. But if α, β should blow up, then $\mathcal{D}_{\hat{\underline{\mu}}(n,t_{\underline{r}})}$ or $\mathcal{D}_{\hat{\underline{\nu}}(n,t_{\underline{r}})}$ must hit one of $P_{\infty_\pm}$ or $P_{0,\pm}$, which is excluded by our choice of initial condition. □

We note, however, that in general (i.e., unless one is, e.g., in the special periodic or unitary case), $\mathcal{D}_{\hat{\underline{\mu}}(n,t_{\underline{r}})}$ will get arbitrarily close to $P_{\infty_\pm}$, $P_{0,\pm}$ since straight motions on the torus are generically dense and hence no uniform bound (and no uniform bound away from zero) on the sequences $\alpha(n, t_{\underline{r}})$, $\beta(n, t_{\underline{r}})$ exists as $(n, t_{\underline{r}})$ varies in $\mathbb{Z} \times \mathbb{R}$. In particular, these complex-valued algebro-geometric solutions of the Ablowitz–Ladik hierarchy initial value problem, in general, will not be quasi-periodic with respect to n or $t_{\underline{r}}$.

3.8 Ablowitz–Ladik Conservation Laws and the Hamiltonian Formalism

> ...trust not the computation of a single Clerk nor any other eyes than your own.
>
> *Isaac Newton*[1]

In this section we deviate from the principal theme of this book and discuss the Green's function of an $\ell^2(\mathbb{Z})$-realization of the difference expression L and systematically derive high- and low-energy expansions of solutions of an associated Riccati-type equation in connection with spatially sufficiently decaying sequences α and β, not necessarily associated with algebro-geometric coefficients. In addition, we derive local conservation laws and develop the Hamiltonian formalism for the AL hierarchy including variational derivatives and Poisson brackets.

In connection with the asymptotic expansions of various quantities we now make the following strengthened assumptions on the coefficients α and β.

Hypothesis 3.46 *Suppose that α, β satisfy*

$$\alpha, \beta \in \ell^\infty(\mathbb{Z}), \quad \alpha(n)\beta(n) \notin \{0, 1\}, \quad n \in \mathbb{Z}.$$

---
[1] Quoted in R. Westfall, *The Life of Isaac Newton*, Cambridge University Press, Cambridge, 1993, p. 226.

282   3 The Ablowitz–Ladik Hierarchy

Given Hypothesis 3.46 we introduce the $\ell^2(\mathbb{Z})$-realization $\check{L}$ of the difference expression L in (3.61) by

$$\check{L}f = Lf, \quad f \in \operatorname{dom}(\check{L}) = \ell^2(\mathbb{Z}), \tag{3.318}$$

and similarly introduce the $\ell^2(\mathbb{Z})$-realizations of the difference expression D, E, D^{-1}, and E^{-1} in (3.67)–(3.70) by

$$\check{D}f = Df, \quad f \in \operatorname{dom}(\check{D}) = \ell^2(\mathbb{Z}), \tag{3.319}$$
$$\check{E}f = Ef, \quad f \in \operatorname{dom}(\check{E}) = \ell^2(\mathbb{Z}), \tag{3.320}$$
$$\check{D}^{-1}f = D^{-1}f, \quad f \in \operatorname{dom}(\check{D}^{-1}) = \ell^2(\mathbb{Z}), \tag{3.321}$$
$$\check{E}^{-1}f = E^{-1}f, \quad f \in \operatorname{dom}(\check{E}^{-1}) = \ell^2(\mathbb{Z}). \tag{3.322}$$

The following elementary result shows that these $\ell^2(\mathbb{Z})$-realizations are meaningful; it will be used in the proof of Lemma 3.48 below.

Lemma 3.47 *Assume Hypothesis* 3.46. *Then the linear operators $\check{D}$, $\check{D}^{-1}$, $\check{E}$, $\check{E}^{-1}$, $\check{L}$, and $\check{L}^{-1}$ are bounded on $\ell^2(\mathbb{Z})$. In addition, $(\check{L} - z)^{-1}$ is norm analytic with respect to z in an open neighborhood of $z = 0$, and $(\check{L} - z)^{-1} = -z^{-1}(I - z^{-1}\check{L})^{-1}$ is analytic with respect to $1/z$ in an open neighborhood of $1/z = 0$.*

Proof By Hypothesis 3.46, $\rho^2 = 1 - \alpha\beta$, and (3.67)–(3.70), one infers that $\check{D}$, $\check{E}$, $\check{D}^{-1}$, $\check{E}^{-1}$ are bounded operators on $\ell^2(\mathbb{Z})$ whose norms are bounded by

$$\|\check{D}\|, \|\check{E}\|, \|\check{D}^{-1}\|, \|\check{E}^{-1}\| \leq 2\|\rho\|_\infty + \|\alpha\|_\infty + \|\beta\|_\infty$$
$$\leq 2(1 + \|\alpha\|_\infty + \|\beta\|_\infty).$$

Since by (3.66),

$$\check{L} = \check{D}\check{E}, \quad \check{L}^{-1} = \check{E}^{-1}\check{D}^{-1},$$

the assertions of Lemma 3.47 are evident (alternatively, one can of course invoke (3.61) and (3.64)). □

To introduce the Green's function of $\check{L}$, we need to digress a bit. Introducing the transfer matrix $T(z, \cdot)$ associated with L by

$$T(z, n) = \begin{cases} \rho(n)^{-1} \begin{pmatrix} \alpha(n) & z \\ z^{-1} & \beta(n) \end{pmatrix}, & n \text{ odd,} \\ \rho(n)^{-1} \begin{pmatrix} \beta(n) & 1 \\ 1 & \alpha(n) \end{pmatrix}, & n \text{ even,} \end{cases} \quad z \in \mathbb{C} \setminus \{0\}, \, n \in \mathbb{Z},$$

3.8 AL Conservation Laws and the Hamiltonian Formalism

recalling that $\rho = \gamma^{1/2} = (1 - \alpha\beta)^{1/2}$, one then verifies that (cf. (3.4))

$$T(z, n) = A(z, n) z^{-1/2} \rho(n)^{-1} U(z, n) A(z, n-1)^{-1}, \quad z \in \mathbb{C} \setminus \{0\}, \; n \in \mathbb{Z}. \tag{3.323}$$

Here we introduced

$$A(z, n) = \begin{cases} \begin{pmatrix} z^{1/2} & 0 \\ 0 & z^{-1/2} \end{pmatrix}, & n \text{ odd}, \\ \begin{pmatrix} 0 & 1 \\ 1 & 0 \end{pmatrix}, & n \text{ even}, \end{cases} \quad z \in \mathbb{C} \setminus \{0\}, \; n \in \mathbb{Z}. \tag{3.324}$$

Next, we consider a fundamental system of solutions

$$\Psi_\pm(z, \cdot) = \begin{pmatrix} \psi_{1,\pm}(z, \cdot) \\ \psi_{2,\pm}(z, \cdot) \end{pmatrix}$$

of

$$U(z)\Psi_\pm^-(z) = \Psi_\pm(z), \quad z \in \mathbb{C} \setminus \left(\operatorname{spec}(\check{L}) \cup \{0\}\right), \tag{3.325}$$

with $\operatorname{spec}(\check{L})$ denoting the spectrum of $\check{L}$ and U given by (3.4), such that

$$\det(\Psi_-(z), \Psi_+(z)) \neq 0. \tag{3.326}$$

The precise form of $\Psi_\pm$ will be chosen as a consequence of (3.332) below. Introducing in addition,

$$\begin{pmatrix} u_\pm(z, n) \\ v_\pm(z, n) \end{pmatrix} = C_\pm z^{-n/2} \left(\prod_{n'=1}^{n} \rho(n')^{-1} \right) A(z, n) \begin{pmatrix} \psi_{1,\pm}(z, n) \\ \psi_{2,\pm}(z, n) \end{pmatrix}, \tag{3.327}$$

$$z \in \mathbb{C} \setminus \left(\operatorname{spec}(\check{L}) \cup \{0\}\right), \; n \in \mathbb{Z},$$

for some constants $C_\pm \in \mathbb{C} \setminus \{0\}$, (3.323) and (3.327) yield

$$T(z) \begin{pmatrix} u_\pm^-(z) \\ v_\pm^-(z) \end{pmatrix} = \begin{pmatrix} u_\pm(z) \\ v_\pm(z) \end{pmatrix}. \tag{3.328}$$

Moreover, one can show (cf. the literature comments in the notes to this section for the following facts (3.329)–(3.332)) that

$$Lu_\pm(z) = zu_\pm(z), \quad L^\top v_\pm(z) = zv_\pm(z), \tag{3.329}$$

$$Dv_\pm(z) = u_\pm(z), \quad Eu_\pm(z) = zv_\pm(z), \tag{3.330}$$

where

$$L = DE, \quad L^\top = ED, \tag{3.331}$$

and hence $L^\top$ represents the difference expression associated with the transpose of the infinite matrix L (cf. (3.59)) in the standard basis of $\ell^2(\mathbb{Z})$. Next, we choose $\Psi_\pm(z)$ in such a manner that for all $n_0 \in \mathbb{Z}$,

$$\begin{pmatrix} u_\pm(z,\cdot) \\ v_\pm(z,\cdot) \end{pmatrix} \in \ell^2([n_0, \pm\infty) \cap \mathbb{Z})^2, \quad z \in \mathbb{C} \setminus \big(\operatorname{spec}\big(\check{L}\big) \cup \{0\}\big). \tag{3.332}$$

Since by hypothesis $z \in \mathbb{C} \setminus \operatorname{spec}(\check{L})$, the elements $(u_+(z,\cdot), v_+(z,\cdot))^\top$ and $(u_-(z,\cdot), v_-(z,\cdot))^\top$ are linearly independent since otherwise z would be an eigenvalue of $\check{L}$. This is of course consistent with (3.326) and (3.327).

The Green's function of $\check{L}$, the $\ell^2(\mathbb{Z})$-realization of the Lax difference expression L, is then of the form

$$\begin{aligned}
G(z, n, n') &= \big(\delta_n, (\check{L} - z)^{-1}\delta_{n'}\big) \\
&= \frac{-1}{z \det \begin{pmatrix} u_+(z,0) & u_-(z,0) \\ v_+(z,0) & v_-(z,0) \end{pmatrix}} \\
&\quad \times \begin{cases} v_-(z,n')u_+(z,n), & n' < n \text{ or } n = n' \text{ even}, \\ v_+(z,n')u_-(z,n), & n' > n \text{ or } n = n' \text{ odd}, \end{cases} \quad n, n' \in \mathbb{Z}, \\
&= -\frac{1}{4z} \frac{(1 - \phi_+(z,0))(1 - \phi_-(z,0))}{\phi_+(z,0) - \phi_-(z,0)} \\
&\quad \times \begin{cases} v_-(z,n')u_+(z,n), & n' < n \text{ or } n = n' \text{ even}, \\ v_+(z,n')u_-(z,n), & n' > n \text{ or } n = n' \text{ odd}, \end{cases} \quad n, n' \in \mathbb{Z}, \\
&\quad z \in \mathbb{C} \setminus \big(\operatorname{spec}\big(\check{L}\big) \cup \{0\}\big). \tag{3.333}
\end{aligned}$$

Introducing

$$\phi_\pm(z,n) = \frac{\psi_{2,\pm}(z,n)}{\psi_{1,\pm}(z,n)}, \quad z \in \mathbb{C} \setminus \big(\operatorname{spec}\big(\check{L}\big) \cup \{0\}\big), \ n \in \mathbb{N},$$

then (3.325) implies that $\phi_\pm$ satisfy the Riccati-type equation

$$\alpha\phi_\pm\phi_\pm^- - \phi_\pm^- + z\phi_\pm = z\beta, \tag{3.334}$$

and one introduces in addition,

$$\mathfrak{f} = \frac{2}{\phi_+ - \phi_-},$$

$$\mathfrak{g} = \frac{\phi_+ + \phi_-}{\phi_+ - \phi_-},$$

$$\mathfrak{h} = \frac{2\phi_+\phi_-}{\phi_+ - \phi_-}.$$

3.8 AL Conservation Laws and the Hamiltonian Formalism

Using the Riccati-type equation (3.334) and its consequences,

$$\alpha(\phi_+\phi_+^- - \phi_-\phi_-^-) - (\phi_+^- - \phi_-^-) + z(\phi_+ - \phi_-) = 0,$$
$$\alpha(\phi_+\phi_+^- + \phi_-\phi_-^-) - (\phi_+^- + \phi_-^-) + z(\phi_+ + \phi_-) = 2z\beta,$$

one then derives the identities

$$z(\mathfrak{g}^- - \mathfrak{g}) + z\beta\mathfrak{f} + \alpha\mathfrak{h}^- = 0, \tag{3.335}$$
$$z\beta\mathfrak{f}^- + \alpha\mathfrak{h} - \mathfrak{g} + \mathfrak{g}^- = 0, \tag{3.336}$$
$$-\mathfrak{f} + z\mathfrak{f}^- + \alpha(\mathfrak{g} + \mathfrak{g}^-) = 0, \tag{3.337}$$
$$z\beta(\mathfrak{g}^- + \mathfrak{g}) - z\mathfrak{h} + \mathfrak{h}^- = 0, \tag{3.338}$$
$$\mathfrak{g}^2 - \mathfrak{f}\mathfrak{h} = 1. \tag{3.339}$$

For the connection between $\mathfrak{f}$, $\mathfrak{g}$, and $\mathfrak{h}$ and the Green's function of L one finally obtains

$$\mathfrak{f}(z, n) = -2\alpha(n)(zG(z, n, n) + 1) - 2\rho(n)z \begin{cases} G(z, n-1, n), & n \text{ even,} \\ G(z, n, n-1), & n \text{ odd,} \end{cases}$$

$$\mathfrak{g}(z, n) = -2zG(z, n, n) - 1, \tag{3.340}$$

$$\mathfrak{h}(z, n) = -2\beta(n)zG(z, n, n) - 2\rho(n)z \begin{cases} G(z, n, n-1), & n \text{ even,} \\ G(z, n-1, n), & n \text{ odd,} \end{cases}$$

illustrating the spectral theoretic content of $\mathfrak{f}$, $\mathfrak{g}$, and $\mathfrak{h}$.

Next we digress some more and quickly prove (3.86): In the special algebro-geometric case, we may identify $\Psi_+(z, \cdot)$ with $\Psi(P, \cdot, 0)$ and $\Psi_-(z, \cdot)$ with $\Psi(P^*, \cdot, 0)$, and similarly, $\phi_+(z, \cdot)$ with $\phi(P, \cdot)$ and $\phi_-(z, \cdot)$ with $\phi(P^*, \cdot)$ (cf. also (C.72) and (C.73)). Equations (3.95) and (3.96) then imply

$$\frac{1}{2}c_{0,+}z^{-\underline{p}_-}y = F_{\underline{p}}\phi - G_{\underline{p}} = -H_{\underline{p}}\phi^{-1} + G_{\underline{p}}. \tag{3.341}$$

Introducing in analogy to (3.327),

$$\begin{pmatrix} u(P, n) \\ v(P, n) \end{pmatrix} = C(P)z^{-n/2} \left(\prod_{n'=1}^{n} \rho(n')^{-1} \right) A(z, n) \begin{pmatrix} \psi_1(P, n, 0) \\ \psi_2(P, n, 0) \end{pmatrix},$$

$$z \in \mathbb{C} \setminus \left(\operatorname{spec}(\check{L}) \cup \{0\} \right), \ n \in \mathbb{Z},$$

for some $C(P) \in \mathbb{C} \setminus \{0\}$, we may identify $u_+(z, \cdot)$ with $u(P, \cdot)$ and $u_-(z, \cdot)$ with $u(P^*, \cdot)$. Thus, one obtains from (3.76),

$$\left(P_{\underline{p}} u(P, \cdot)\right)(k) = \begin{cases} -iz F_{\underline{p}}(z, k)\left(D^{-1} u(P, \cdot)\right)(k) + i G_{\underline{p}}(z, k) u(P, k), & k \text{ odd,} \\ i H_{\underline{p}}(z, k)\left(D^{-1} u(P, \cdot)\right)(k) - i G_{\underline{p}}(z, k) u(P, k), & k \text{ even,} \end{cases}$$

$$= -\frac{i}{2} c_{0,+} z^{-p_-} y \, u(P, k), \quad P = (z, y), \, k \in \mathbb{Z}.$$

In particular,

$$L^{2p_-} P_{\underline{p}}^2 u_\pm(z, \cdot) = -\frac{1}{4} c_{0,+}^2 y^2 u_\pm(z, \cdot)$$

$$= -\frac{1}{4} c_{0,+}^2 \prod_{m=0}^{2p+1} (L - E_m) u_\pm(z, \cdot)$$

$$= -L^{2p_-} R_{\underline{p}}(L) u_\pm(z, \cdot). \tag{3.342}$$

Since $z \in \mathbb{C} \setminus \{0\}$ is arbitrary, and the difference expressions on either side of (3.342) are of finite order, one obtains (3.86). $\square$

Since we are particularly interested in the asymptotic expansion of $\phi_\pm$ in a neighborhood of the points $z = 0$ and $1/z = 0$ we now turn to this topic next.

Lemma 3.48 *Assume that α, β satisfy Hypothesis 3.46. Then $\phi_\pm$ have the following convergent expansions with respect to $1/z$ around $1/z = 0$ and with respect to z around $z = 0$,*

$$\phi_\pm(z) \underset{z \to \infty}{=} \begin{cases} \sum_{j=0}^\infty \phi_{j,+}^\infty z^{-j}, \\ \sum_{j=-1}^\infty \phi_{j,-}^\infty z^{-j}, \end{cases}$$

$$\phi_\pm(z) \underset{z \to 0}{=} \begin{cases} \sum_{j=0}^\infty \phi_{j,+}^0 z^j, \\ \sum_{j=1}^\infty \phi_{j,-}^0 z^j, \end{cases}$$

where

$$\phi_{0,+}^\infty = \beta, \quad \phi_{1,+}^\infty = \beta^- \gamma,$$

$$\phi_{j+1,+}^\infty = (\phi_{j,+}^\infty)^- - \alpha \sum_{\ell=0}^j (\phi_{j-\ell,+}^\infty)^- \phi_{\ell,+}^\infty, \quad j \in \mathbb{N}, \tag{3.343}$$

$$\phi_{-1,-}^\infty = -\frac{1}{\alpha^+}, \quad \phi_{0,-}^\infty = \frac{\alpha^{++}}{(\alpha^+)^2} \gamma^+,$$

$$\phi_{j+1,-}^\infty = -\frac{\alpha^{++}}{\alpha^+} \phi_{j,-}^\infty + \alpha^{++} \sum_{\ell=0}^j \phi_{j-\ell,-}^\infty (\phi_{\ell,-}^\infty)^+, \quad j \in \mathbb{N}_0, \tag{3.344}$$

3.8 AL Conservation Laws and the Hamiltonian Formalism

$$\phi_{0,+}^0 = \frac{1}{\alpha}, \quad \phi_{1,+}^0 = -\frac{\alpha^-}{\alpha^2}\gamma,$$

$$\phi_{j+1,+}^0 = (\phi_{j,+}^0)^+ + \alpha^+ \sum_{\ell=0}^{j+1}(\phi_{j+1-\ell,+}^0)^+ \phi_{\ell,+}^0, \quad j \in \mathbb{N}, \tag{3.345}$$

$$\phi_{1,-}^0 = -\beta^+,$$

$$\phi_{j+1,-}^0 = (\phi_{j,-}^0)^+ + \alpha^+ \sum_{\ell=1}^{j} \phi_{j+1-\ell,-}^0 (\phi_{\ell,-}^0)^+, \quad j \in \mathbb{N}. \tag{3.346}$$

Proof Since

$$\phi_\pm = \frac{\mathfrak{g} \pm 1}{\mathfrak{f}},$$

combining Lemma 3.47, (3.333) and (3.340) proves that $\phi_\pm$ has a convergent expansion with respect to z and $1/z$ in a neighborhood of $z = 0$ and $1/z = 0$, respectively. The explicit expansion coefficients $\phi_{j,\pm}^\infty$ are then readily derived by making the ansatz

$$\phi_\pm \underset{z \to \infty}{=} \sum_{j=-1}^{\infty} \phi_{j,\pm}^\infty z^{-j}, \quad \phi_{-1,+}^\infty = 0. \tag{3.347}$$

Inserting (3.347) into the Riccati-type equation (3.334) one finds

$$0 = \alpha\phi_\pm \phi_\pm^- - \phi_\pm^- + z(\phi_\pm - \beta) = \left(\alpha\phi_{-1,\pm}^\infty (\phi_{-1,\pm}^\infty)^- + \phi_{-1,\pm}^\infty\right)z^2 + O(z), \tag{3.348}$$

which yields the case distinction above and the formulas for $\phi_{j,\pm}^\infty$. The corresponding expansion coefficients $\phi_{j,\pm}^0$ are obtained analogously by making the ansatz $\phi_\pm \underset{z \to 0}{=} \sum_{j=0}^{\infty} \phi_{j,\pm}^0 z^j.$ □

For the record we list a few explicit expressions:

$$\phi_{0,+}^\infty = \beta,$$

$$\phi_{1,+}^\infty = \beta^- \gamma,$$

$$\phi_{2,+}^\infty = \gamma\big(-\alpha(\beta^-)^2 + \beta^{--}\gamma^-\big),$$

$$\phi_{-1,-}^\infty = -\frac{1}{\alpha^+},$$

$$\phi_{0,-}^\infty = \frac{\alpha^{++}}{(\alpha^+)^2}\gamma^+,$$

$$\phi_{1,-}^\infty = \frac{\gamma^+}{(\alpha^+)^3}\big(\alpha^+\alpha^{+++}\gamma^{++} - (\alpha^{++})^2\big),$$

$$\phi_{0,+}^0 = \frac{1}{\alpha},$$

$$\phi^0_{1,+} = -\frac{\alpha^-}{\alpha^2}\gamma,$$

$$\phi^0_{2,+} = \frac{\gamma}{\alpha^3}((\alpha^-)^2 - \alpha^{--}\alpha\gamma^-),$$

$$\phi^0_{1,-} = -\beta^+,$$

$$\phi^0_{2,-} = -\gamma^+\beta^{++},$$

$$\phi^0_{3,-} = \gamma^+(\alpha^+(\beta^{++})^2 - \gamma^{++}\beta^{+++}),$$

$$\phi^0_{4,-} = \gamma^+\Big(-(\alpha^+)^2(\beta^{++})^3$$
$$+ \gamma^{++}\big(2\alpha^+\beta^{++}\beta^{+++} + \alpha^{++}(\beta^{+++})^2 - \gamma^{+++}\beta^{++++}\big)\Big), \text{ etc.}$$

Later on we will also need the convergent expansions of $\ln(z + \alpha^+ \phi_\pm(z))$ with respect to z and $1/z$. We will separately provide all four expansions of $\ln(z + \alpha^+ \phi_\pm(z))$ around $1/z = 0$ and $z = 0$ and repeatedly use the general formula

$$\ln\left(1 + \sum_{j=1}^\infty \omega_j z^{\pm j}\right) = \sum_{j=1}^\infty \sigma_j z^{\pm j}, \qquad (3.349)$$

where

$$\sigma_1 = \omega_1, \qquad \sigma_j = \omega_j - \sum_{\ell=1}^{j-1} \frac{\ell}{j}\omega_{j-\ell}\sigma_\ell, \quad j \geq 2,$$

and $|z|$ as $|z| \to 0$, respectively, $1/|z|$ as $|z| \to \infty$, are assumed to be sufficiently small in (3.349). We start by expanding ϕ_+ around $1/z = 0$

$$\ln(z + \alpha^+\phi_+(z)) = \ln\left(z + \alpha^+\sum_{j=0}^\infty \phi^\infty_{j,+} z^{-j}\right)$$

$$= \ln\left(1 + \alpha^+\sum_{j=0}^\infty \phi^\infty_{j,+} z^{-j-1}\right)$$

$$= \ln\left(1 + \alpha^+\sum_{j=1}^\infty \phi^\infty_{j-1,+} z^{-j}\right)$$

$$= \sum_{j=1}^\infty \rho^\infty_{j,+} z^{-j}, \qquad (3.350)$$

where

$$\rho^\infty_{1,+} = \alpha^+\phi^\infty_{0,+}, \quad \rho^\infty_{j,+} = \alpha^+\left(\phi^\infty_{j-1,+} - \sum_{\ell=1}^{j-1}\frac{\ell}{j}\phi^\infty_{j-1-\ell,+}\rho^\infty_{\ell,+}\right), \quad j \geq 2. \quad (3.351)$$

3.8 AL Conservation Laws and the Hamiltonian Formalism

An expansion of ϕ_- around $1/z = 0$ yields

$$\ln(z + \alpha^+\phi_-(z)) = \ln\left(z + \alpha^+ \sum_{j=-1}^{\infty} \phi_{j,-}^\infty z^{-j}\right)$$

$$= \ln\left(\frac{\alpha^{++}\gamma^+}{\alpha^+}\right) + \ln\left(1 + \frac{(\alpha^+)^2}{\alpha^{++}\gamma^+} \sum_{j=1}^{\infty} \phi_{j,-}^\infty z^{-j}\right)$$

$$= \ln\left(\frac{\alpha^{++}}{\alpha^+}\right) + \ln(\gamma^+) + \ln\left(1 + \frac{(\alpha^+)^2}{\alpha^{++}\gamma^+} \sum_{j=1}^{\infty} \phi_{j,-}^\infty z^{-j}\right)$$

$$= \ln\left(\frac{\alpha^{++}}{\alpha^+}\right) + \ln(\gamma^+) + \sum_{j=1}^{\infty} \rho_{j,-}^\infty z^{-j},$$

where

$$\rho_{1,-}^\infty = \frac{(\alpha^+)^2}{\alpha^{++}\gamma^+}\phi_{1,-}^\infty, \quad \rho_{j,-}^\infty = \frac{(\alpha^+)^2}{\alpha^{++}\gamma^+}\left(\phi_{j,-}^\infty - \sum_{\ell=1}^{j-1} \frac{\ell}{j}\phi_{j-\ell,-}^\infty \rho_{\ell,-}^\infty\right), \quad j \geq 2. \tag{3.352}$$

For the expansion of ϕ_+ around $z = 0$ one gets

$$\ln(z + \alpha^+\phi_+(z)) = \ln\left(z + \alpha^+ \sum_{j=0}^{\infty} \phi_{j,+}^0 z^j\right)$$

$$= \ln\left(\frac{\alpha^+}{\alpha}\right) + \ln\left(1 + \frac{\alpha}{\alpha^+}(1 + \alpha^+\phi_{1,+}^0)z + \alpha\sum_{j=2}^{\infty} \phi_{j,+}^0 z^j\right)$$

$$= \ln\left(\frac{\alpha^+}{\alpha}\right) + \sum_{j=1}^{\infty} \rho_{j,+}^0 z^j,$$

where

$$\rho_{1,+}^0 = \frac{\alpha}{\alpha^+}(1 + \alpha^+\phi_{1,+}^0), \quad \rho_{2,+}^0 = \alpha\phi_{2,+}^0 - \frac{1}{2}(\rho_{1,+}^0)^2,$$

$$\rho_{j,+}^0 = \alpha\left(\phi_{j,+}^0 - \sum_{j=1}^{j-2} \frac{\ell}{j}\phi_{j-\ell,+}^0 \rho_{\ell,+}^0\right) - \frac{j-1}{j}\rho_{1,+}^0 \rho_{j-1,+}^0, \quad j \geq 3. \tag{3.353}$$

Finally, the expansion of ϕ_- around $z = 0$ is given by

$$\ln(z + \alpha^+\phi_-(z)) = \ln\left(z + \alpha^+ \sum_{j=1}^{\infty} \phi_{j,-}^0 z^j\right)$$

$$= \ln\left(\gamma^+ z + \alpha^+ \sum_{j=2}^{\infty} \phi_{j,-}^0 z^j\right)$$

$$= \ln(z) + \ln(\gamma^+) + \ln\left(1 + \frac{\alpha^+}{\gamma^+} \sum_{j=1}^{\infty} \phi_{j+1,-}^0 z^j\right)$$

$$= \ln(z) + \ln(\gamma^+) + \sum_{j=1}^{\infty} \rho_{j,-}^0 z^j, \qquad (3.354)$$

where

$$\rho_{1,-}^0 = \frac{\alpha^+}{\gamma^+} \phi_{2,-}^0, \quad \rho_{j,-}^0 = \frac{\alpha^+}{\gamma^+}\left(\phi_{j+1,-}^0 - \sum_{\ell=1}^{j-1} \frac{\ell}{j} \phi_{j+1-\ell,-}^0 \rho_{\ell,-}^0\right), \quad j \geq 2. \tag{3.355}$$

Explicitly, the first expansion coefficients are given by

$$\rho_{1,+}^\infty = \alpha^+\beta,$$
$$\rho_{2,+}^\infty = -\tfrac{1}{2}(\alpha^+\beta)^2 + \gamma\alpha^+\beta^-,$$
$$\rho_{3,+}^\infty = \tfrac{1}{3}(\alpha^+\beta)^3 - \gamma(\gamma^-\alpha^+\beta^{--} - (\alpha^+)^2\beta^-\beta - \alpha\alpha^+(\beta^-)^2),$$
$$\rho_{1,-}^\infty = -\alpha^{+++}\beta^{++} + (S^+ - I)\frac{\alpha^{++}}{\alpha^+},$$
$$\rho_{1,+}^0 = \alpha^-\beta + (S^+ - I)\frac{\alpha^-}{\alpha}, \tag{3.356}$$
$$\rho_{1,-}^0 = -\alpha^+\beta^{++},$$
$$\rho_{2,-}^0 = \tfrac{1}{2}(\alpha^+\beta^{++})^2 - \gamma^{++}\alpha^+\beta^{+++},$$
$$\rho_{3,-}^0 = -\tfrac{1}{3}(\alpha^+\beta^{++})^3$$
$$\qquad + \gamma^{++}\big(-\gamma^{+++}\alpha^+\beta^{++++} + (\alpha^+)^2\beta^{++}\beta^{+++} + \alpha^+\alpha^{++}(\beta^{+++})^2\big), \text{ etc.}$$

The next result shows that $\hat{g}_{j,+}$ and $\pm j\rho_{j,\pm}^\infty$, respectively $\hat{g}_{j,-}$ and $\pm j\rho_{j,\pm}^0$, are equal up to terms that are total differences, that is, are of the form $(S^+ - I)d_{j,\pm}$ for some sequence $d_{j,\pm}$. The exact form of $d_{j,\pm}$ will not be needed later. In the proof we will heavily use the equations (3.335)–(3.339).

3.8 AL Conservation Laws and the Hamiltonian Formalism 291

Lemma 3.49 *Suppose Hypothesis 3.46 holds. Then*

$$\hat{g}_{j,+} = -j\rho_{j,+}^\infty + (S^+ - I)d_{j,+} = j\rho_{j,-}^\infty + (S^+ - I)e_{j,+}, \quad j \in \mathbb{N}, \quad (3.357)$$

$$\hat{g}_{j,-} = -j\rho_{j,+}^0 + (S^+ - I)d_{j,-} = j\rho_{j,-}^0 + (S^+ - I)e_{j,-}, \quad j \in \mathbb{N}, \quad (3.358)$$

for some polynomials $d_{j,\pm}$, $e_{j,\pm}$, $j \in \mathbb{N}$, in α and β and certain shifts thereof.

Proof We consider the case for $\hat{g}_{j,+}$ first. Our aim is to show that

$$\frac{d}{dz}\ln(z + \alpha^+\phi_+) = -\frac{1}{2z}\mathfrak{g} + \frac{1}{2z} + (S^+ - I)K + (S^+ - I)M, \quad (3.359)$$

where

$$K = \frac{1}{2}\left(\mathfrak{g}\frac{\dot{\mathfrak{f}}}{\mathfrak{f}} - \dot{\mathfrak{g}}\right), \quad M = \frac{1}{2}\frac{\dot{\mathfrak{f}}}{\mathfrak{f}},$$

which implies (3.357) by (3.350). Here $\cdot$ denotes d/dz.
Since $\phi_+ = (\mathfrak{g} + 1)/\mathfrak{f}$,

$$\frac{d}{dz}\ln(z + \alpha^+\phi_+) = \frac{1 + \alpha^+\dot{\phi}_+}{z + \alpha^+\phi_+} = \frac{\mathfrak{f}^2 + \alpha^+(\mathfrak{f}\dot{\mathfrak{g}} - \dot{\mathfrak{f}}\mathfrak{g} - \dot{\mathfrak{f}})}{\mathfrak{f}(z\mathfrak{f} + \alpha^+\mathfrak{g} + \alpha^+)} \frac{z\mathfrak{f} + \alpha^+\mathfrak{g} - \alpha^+}{z\mathfrak{f} + \alpha^+\mathfrak{g} - \alpha^+}. \quad (3.360)$$

Next we treat the denominator of (3.360) using (3.335), (3.337),

$$\mathfrak{f}\big((z\mathfrak{f} + \alpha^+\mathfrak{g})^2 - (\alpha^+)^2\big) = \mathfrak{f}\big((z\mathfrak{f} + \alpha^+\mathfrak{g})^2 - (\alpha^+)^2(\mathfrak{g}^2 - \mathfrak{f}\mathfrak{h})\big)$$

$$= z\mathfrak{f}^2\left(z\mathfrak{f} + \alpha^+\mathfrak{g} + \alpha^+\mathfrak{g} + (\alpha^+)^2\frac{1}{z}\mathfrak{h}\right)$$

$$= z\mathfrak{f}^2\big(\mathfrak{f}^+ - \alpha^+\mathfrak{g}^+ + \alpha^+\mathfrak{g}^+ - \alpha^+\beta^+\mathfrak{f}^+\big)$$

$$= z\gamma^+\mathfrak{f}^2\mathfrak{f}^+.$$

Expanding the numerator in (3.360) and applying (3.335), (3.337), and their derivatives with respect to z as well as $2\mathfrak{g}\dot{\mathfrak{g}} = \dot{\mathfrak{f}}\mathfrak{h} + \mathfrak{f}\dot{\mathfrak{h}}$ yields

$$(z\mathfrak{f} + \alpha^+\mathfrak{g} - \alpha^+)(\mathfrak{f}^2 + \alpha^+(\mathfrak{f}\dot{\mathfrak{g}} - \dot{\mathfrak{f}}\mathfrak{g} - \dot{\mathfrak{f}}))$$

$$= \mathfrak{f}\big(z\mathfrak{f}^2 + z\alpha^+(\mathfrak{f}\dot{\mathfrak{g}} - \dot{\mathfrak{f}}\mathfrak{g}) + \alpha^+\mathfrak{f}\mathfrak{g} + \tfrac{1}{2}(\alpha^+)^2(\mathfrak{f}\dot{\mathfrak{h}} - \dot{\mathfrak{f}}\mathfrak{h}) - \alpha^+(\mathfrak{f} + z\dot{\mathfrak{f}} + \alpha^+\dot{\mathfrak{g}})\big)$$

$$= \frac{\mathfrak{f}}{2}\Big(2z\mathfrak{f}^2 + z\alpha^+\mathfrak{f}\dot{\mathfrak{g}} + z\mathfrak{f}(-\alpha^+\dot{\mathfrak{g}}^+ - z\dot{\mathfrak{f}} - \mathfrak{f} + \mathfrak{f}^+) - z\alpha^+\dot{\mathfrak{f}}\mathfrak{g}$$

$$+ z\dot{\mathfrak{f}}(\alpha^+\mathfrak{g}^+ + z\mathfrak{f} - \mathfrak{f}^+)$$

$$+ \alpha^+\mathfrak{f}\mathfrak{g} + \mathfrak{f}(-\alpha^+\mathfrak{g}^+ - z\mathfrak{f} + \mathfrak{f}^+)$$

$$+ \alpha^+\mathfrak{f}(-\beta^+\mathfrak{f}^+ - z\beta^+\dot{\mathfrak{f}}^+ - \mathfrak{g} + \mathfrak{g}^+ - z\dot{\mathfrak{g}} + z\dot{\mathfrak{g}}^+)$$

$$+ \alpha^+\dot{\mathfrak{f}}(z\beta^+\mathfrak{f}^+ + z\mathfrak{g} - z\mathfrak{g}^+) - 2\alpha^+(\mathfrak{f} + z\dot{\mathfrak{f}} + \alpha^+\dot{\mathfrak{g}})\Big)$$

$$= \frac{\mathfrak{f}}{2}\Big(\gamma^+\mathfrak{f}\mathfrak{f}^+ + z\gamma^+\dot{\mathfrak{f}}\mathfrak{f}^+ - z\gamma^+\mathfrak{f}\dot{\mathfrak{f}}^+ - 2\alpha^+(\mathfrak{f} + z\dot{\mathfrak{f}} + \alpha^+\dot{\mathfrak{g}})\Big).$$

292   3 The Ablowitz–Ladik Hierarchy

In summary,

$$\frac{d}{dz}\ln(z+\alpha^+\phi_+) = \frac{1}{2z} + (S^+ - I)M - \frac{\alpha^+}{z\gamma^+\mathfrak{f}\mathfrak{f}^+}\bigl(\mathfrak{f}+z\dot{\mathfrak{f}}+\alpha^+\dot{\mathfrak{g}}\bigr). \quad (3.361)$$

We multiply the numerator on the right-hand side by $-2 = -2(\mathfrak{g}^2 - \mathfrak{f}\mathfrak{h})$ and use again (3.335), (3.337), and their derivatives:

$$\begin{aligned}
&2\alpha^+\bigl(\mathfrak{f}\mathfrak{h}-\mathfrak{g}^2\bigr)\bigl(\mathfrak{f}+z\dot{\mathfrak{f}}+\alpha^+\dot{\mathfrak{g}}\bigr)\\
&= 2\alpha^+\mathfrak{f}\mathfrak{h}(\mathfrak{f}+z\dot{\mathfrak{f}}+\alpha^+\dot{\mathfrak{g}}) - 2\alpha^+\mathfrak{f}\mathfrak{g}^2 - 2z\alpha^+\dot{\mathfrak{f}}\mathfrak{g}^2 - (\alpha^+)^2\mathfrak{g}(\dot{\mathfrak{f}}\mathfrak{h}+\mathfrak{f}\dot{\mathfrak{h}})\\
&= \alpha^+\mathfrak{f}\mathfrak{h}(\mathfrak{f}+z\dot{\mathfrak{f}}) + \alpha^+\mathfrak{f}\dot{\mathfrak{g}}(-z\beta^+\mathfrak{f}^+ - z\mathfrak{g} + z\mathfrak{g}^+)\\
&\quad + \mathfrak{f}(-z\beta^+\mathfrak{f}^+ - z\mathfrak{g} + z\mathfrak{g}^+)(\dot{\mathfrak{f}}^+ - \alpha^+\dot{\mathfrak{g}}^+)\\
&\quad - \alpha^+\mathfrak{f}\mathfrak{g}^2 + \mathfrak{f}\mathfrak{g}(\alpha^+\mathfrak{g}^+ + z\mathfrak{f} - \mathfrak{f}^+) - z\alpha^+\dot{\mathfrak{f}}\mathfrak{g}^2 + z\dot{\mathfrak{f}}\mathfrak{g}(\alpha^+\mathfrak{g}^+ + z\mathfrak{f} - \mathfrak{f}^+)\\
&\quad + \alpha^+\dot{\mathfrak{f}}\mathfrak{g}(z\beta^+\mathfrak{f}^+ + z\mathfrak{g} - z\mathfrak{g}^+) + \alpha^+\mathfrak{f}\mathfrak{g}(\beta^+\mathfrak{f}^+ + z\beta^+\dot{\mathfrak{f}}^+ + \mathfrak{g} - \mathfrak{g}^+ + z\dot{\mathfrak{g}} - z\dot{\mathfrak{g}}^+)\\
&= \alpha^+\mathfrak{f}\mathfrak{h}(\mathfrak{f}+z\dot{\mathfrak{f}}) + \alpha^+\mathfrak{f}\dot{\mathfrak{g}}(-z\beta^+\mathfrak{f}^+ - z\mathfrak{g} + z\mathfrak{g}^+) + \mathfrak{f}\dot{\mathfrak{f}}^+(-z\beta^+\mathfrak{f}^+ - z\mathfrak{g} + z\mathfrak{g}^+)\\
&\quad + \alpha^+\mathfrak{f}\dot{\mathfrak{g}}^+(z\beta^+\mathfrak{f}^+ + z\mathfrak{g} - z\mathfrak{g}^+) + z\mathfrak{f}^2\mathfrak{g} - \gamma^+\mathfrak{f}\mathfrak{f}^+\mathfrak{g} + z^2\mathfrak{f}\dot{\mathfrak{f}}\mathfrak{g}\\
&\quad - z\gamma^+\dot{\mathfrak{f}}\mathfrak{f}^+\mathfrak{g} + z\beta^+\mathfrak{f}\dot{\mathfrak{f}}^+(-\alpha^+\mathfrak{g}^+ - z\mathfrak{f} + \mathfrak{f}^+) + z\alpha^+\mathfrak{f}\mathfrak{g}(\dot{\mathfrak{g}} - \dot{\mathfrak{g}}^+)\\
&= \alpha^+\mathfrak{f}\mathfrak{h}(\mathfrak{f}+z\dot{\mathfrak{f}}) - z\alpha^+\beta^+\mathfrak{f}\mathfrak{f}^+(\dot{\mathfrak{g}} - \dot{\mathfrak{g}}^+) - z\mathfrak{f}(\alpha^+\mathfrak{g} + z\mathfrak{f} - \mathfrak{f}^+)(\dot{\mathfrak{g}} - \dot{\mathfrak{g}}^+)\\
&\quad - z\mathfrak{f}\mathfrak{f}^+\mathfrak{g} + z\mathfrak{f}^2\mathfrak{g} - \gamma^+\mathfrak{f}\mathfrak{f}^+\mathfrak{g} + z^2\mathfrak{f}\dot{\mathfrak{f}}\mathfrak{g} - z\gamma^+\dot{\mathfrak{f}}\mathfrak{f}^+\mathfrak{g} + z\gamma^+\mathfrak{f}\mathfrak{f}^+\mathfrak{g}^+ - z^2\beta^+\mathfrak{f}^2\dot{\mathfrak{f}}^+\\
&= z\alpha^+\mathfrak{f}\mathfrak{f}\mathfrak{h} - z\mathfrak{f}^2(z\beta^+\dot{\mathfrak{f}}^+ + z\dot{\mathfrak{g}} - z\dot{\mathfrak{g}}^+ - \tfrac{1}{z}\alpha^+\mathfrak{h}) + z\mathfrak{f}\mathfrak{g}(\mathfrak{f}+z\dot{\mathfrak{f}}-\alpha^+\dot{\mathfrak{g}}-\mathfrak{f}^+ +\alpha^+\dot{\mathfrak{g}}^+)\\
&\quad + z\gamma^+\mathfrak{f}\mathfrak{f}^+(\dot{\mathfrak{g}}-\dot{\mathfrak{g}}^+) - \gamma^+\mathfrak{f}\mathfrak{f}^+\mathfrak{g} - z\gamma^+\dot{\mathfrak{f}}\mathfrak{f}^+\mathfrak{g} + z\gamma^+\mathfrak{f}\mathfrak{f}^+\mathfrak{g}^+\\
&= z\alpha^+\mathfrak{f}(\dot{\mathfrak{f}}\mathfrak{h}+\mathfrak{f}\dot{\mathfrak{h}} - 2\mathfrak{g}\dot{\mathfrak{g}}) + z\gamma^+\mathfrak{f}\mathfrak{f}^+(\dot{\mathfrak{g}}-\dot{\mathfrak{g}}^+) - \gamma^+\mathfrak{f}\mathfrak{f}^+\mathfrak{g} - z\gamma^+\dot{\mathfrak{f}}\mathfrak{f}^+\mathfrak{g} + z\gamma^+\mathfrak{f}\mathfrak{f}^+\mathfrak{g}^+\\
&= z\gamma^+\mathfrak{f}\mathfrak{f}^+(\dot{\mathfrak{g}}-\dot{\mathfrak{g}}^+) - \gamma^+\mathfrak{f}\mathfrak{f}^+\mathfrak{g} - z\gamma^+\dot{\mathfrak{f}}\mathfrak{f}^+\mathfrak{g} + z\gamma^+\mathfrak{f}\mathfrak{f}^+\mathfrak{g}^+.
\end{aligned}$$

Inserting this in (3.361) finally yields (3.359). The result for $\hat{g}_{j,-}$ is derived similarly starting from $\phi_- = (\mathfrak{g}-1)/\mathfrak{f}$. $\square$

Next, we turn to a derivation of the local conservation laws of the AL hierarchy and introduce the following assumption for this purpose:

Hypothesis 3.50 *Suppose that* $\alpha, \beta \colon \mathbb{Z} \times \mathbb{R} \to \mathbb{C}$ *satisfy*

$$\sup_{(n,t_{\underline{p}})\in\mathbb{Z}\times\mathbb{R}} \bigl(|\alpha(n,t_{\underline{p}})| + |\beta(n,t_{\underline{p}})|\bigr) < \infty,$$

$a(n,\cdot), b(n,\cdot) \in C^1(\mathbb{R}), \; n \in \mathbb{Z}, \quad \alpha(n,t_{\underline{p}})\beta(n,t_{\underline{p}}) \notin \{0,1\}, \; (n,t_{\underline{p}}) \in \mathbb{Z}\times\mathbb{R}.$

In accordance with the notation introduced in (3.318)–(3.322) we denote the bounded difference operator defined on $\ell^2(\mathbb{Z})$, generated by the finite difference expression $P_{\underline{p}}$ in (3.73), by the symbol $\breve{P}_{\underline{p}}$. Similarly, the bounded finite difference operator in $\ell^2(\mathbb{Z})$ generated by $P_{\underline{p}}^\top$ in (3.77) is then denoted by $\breve{P}_{\underline{p}}^\top$.

3.8 AL Conservation Laws and the Hamiltonian Formalism

We start with the following existence result.

Theorem 3.51 *Assume Hypothesis 3.50 and suppose α, β satisfy $\mathrm{AL}_{\underline{p}}(\alpha, \beta) = 0$ for some $\underline{p} \in \mathbb{N}_0^2$. In addition, let $t_{\underline{p}} \in \mathbb{R}$ and $z \in \mathbb{C} \setminus \big(\mathrm{spec}\big(\check{L}(t_{\underline{p}})\big) \cup \{0\}\big)$. Then there exist Weyl–Titchmarsh-type solutions $u_\pm = u_\pm(z, n, t_{\underline{p}})$ and $v_\pm = v_\pm(z, n, t_{\underline{p}})$ such that for all $n_0 \in \mathbb{Z}$,*

$$\begin{pmatrix} u_\pm(z,\,\cdot\,,t_{\underline{p}}) \\ v_\pm(z,\,\cdot\,,t_{\underline{p}}) \end{pmatrix} \in \ell^2([n_0, \pm\infty) \cap \mathbb{Z})^2, \quad u_\pm(z, n, \cdot), v_\pm(z, n, \cdot) \in C^1(\mathbb{R}),$$
(3.362)

and $u_\pm$ and $v_\pm$ simultaneously satisfy the following equations in the weak sense

$$\check{L}(t_{\underline{p}})u_\pm(z,\,\cdot\,,t_{\underline{p}}) = z u_\pm(z,\,\cdot\,,t_{\underline{p}}),$$
(3.363)

$$u_{\pm, t_{\underline{p}}}(z,\,\cdot\,,t_{\underline{p}}) = \check{P}_{\underline{p}}(t_{\underline{p}}) u_\pm(z,\,\cdot\,,t_{\underline{p}}),$$
(3.364)

and

$$\check{L}^\top(t_{\underline{p}}) v_\pm(z,\,\cdot\,,t_{\underline{p}}) = z v_\pm(z,\,\cdot\,,t_{\underline{p}}),$$
(3.365)

$$v_{\pm, t_{\underline{p}}}(z,\,\cdot\,,t_{\underline{p}}) = -\check{P}_{\underline{p}}^\top(t_{\underline{p}}) v_\pm(z,\,\cdot\,,t_{\underline{p}}),$$
(3.366)

respectively.

Proof Applying $\big(\check{L}(t_{\underline{p}}) - zI\big)^{-1}$ to δ_{n_0} (cf. (3.333)) yields the existence of Weyl–Titchmarsh-type solutions $\tilde{u}_\pm$ of $Lu = zu$ satisfying (3.362). Next, using the Lax commutator equation (3.79) one computes

$$z\tilde{u}_{\pm, t_{\underline{p}}} = (L\tilde{u}_\pm)_{t_{\underline{p}}} = L_{t_{\underline{p}}}\tilde{u}_\pm + L\tilde{u}_{\pm, t_{\underline{p}}} = [P_{\underline{p}}, L]\tilde{u}_\pm + L\tilde{u}_{\pm, t_{\underline{p}}}$$

$$= zP_{\underline{p}}\tilde{u}_\pm - LP_{\underline{p}}\tilde{u}_\pm + L\tilde{u}_{\pm, t_{\underline{p}}}$$

and hence

$$(L - zI)(\tilde{u}_{\pm, t_{\underline{p}}} - P_{\underline{p}}\tilde{u}_\pm) = 0.$$

Thus, $\tilde{u}_\pm$ satisfy

$$\tilde{u}_{\pm, t_{\underline{p}}} - P_{\underline{p}}\tilde{u}_\pm = C_\pm \tilde{u}_\pm + D_\pm \tilde{u}_\mp.$$

Introducing $\tilde{u}_\pm = c_\pm u_\pm$, and choosing $c_\pm$ such that $c_{\pm, t_{\underline{p}}} = C_\pm c_\pm$, one obtains

$$u_{\pm, t_{\underline{p}}} - P_{\underline{p}} u_\pm = D_\pm u_\mp.$$
(3.367)

Since $u_\pm \in \ell^2([n_0, \pm\infty) \cap \mathbb{Z})$, $n_0 \in \mathbb{Z}$, and α, β satisfy Hypothesis 3.50, (3.76) shows that $P_{\underline{p}} u_\pm \in \ell^2([n_0, \pm\infty) \cap \mathbb{Z})$. Moreover, since

$$u_\pm(z, n, t_{\underline{p}}) = d_\pm(t_{\underline{p}})\big((\check{L}(t_{\underline{p}}) - zI)^{-1}\delta_{n_0}\big)(n), \quad n \in [n_0, \pm\infty) \cap \mathbb{Z}, \quad (3.368)$$

for some $d_\pm \in C^1(\mathbb{R})$, the calculation

$$u_{\pm,t_{\underline{p}}} = d_{\pm,t_{\underline{p}}}(\check{L} - zI)^{-1}\delta_{n_0} - d_\pm(\check{L} - zI)^{-1}\check{L}_{t_{\underline{p}}}(\check{L} - zI)^{-1}\delta_{n_0}$$

also yields $u_{\pm,t_{\underline{p}}} \in \ell^2([n_0, \pm\infty) \cap \mathbb{Z})$. But then $D_\pm = 0$ in (3.367) since $u_\mp \notin \ell^2([n_0, \pm\infty) \cap \mathbb{Z})$. This proves (3.364).

Equations (3.362), (3.365), and (3.366) for $v_\pm$ are proved similarly, replacing L, $P_{\underline{p}}$ by $L^\top$, $P_{\underline{p}}^\top$ and observing that (3.79) implies

$$L_{t_{\underline{p}}}^\top(t_{\underline{p}}) + [P_{\underline{p}}^\top(t_{\underline{p}}), L^\top(t_{\underline{p}})] = 0, \quad t_{\underline{p}} \in \mathbb{R}.$$

$\square$

For the remainder of this section we will always refer to the Weyl–Titchmarsh solutions $u_\pm$, $v_\pm$ introduced in Theorem 3.51. Given $u_\pm$, $v_\pm$, we now introduce

$$\Psi_\pm(z, \cdot, t_{\underline{p}}) = \begin{pmatrix} \psi_{1,\pm}(z, \cdot, t_{\underline{p}}) \\ \psi_{2,\pm}(z, \cdot, t_{\underline{p}}) \end{pmatrix}, \quad z \in \mathbb{C}\setminus\big(\mathrm{spec}\,(\check{L}(t_{\underline{p}}))\cup\{0\}\big), \; t_{\underline{p}} \in \mathbb{R}, \quad (3.369)$$

by (cf. (3.327))

$$\begin{pmatrix} \psi_{1,\pm}(z,n,t_{\underline{p}}) \\ \psi_{2,\pm}(z,n,t_{\underline{p}}) \end{pmatrix} = D(t_{\underline{p}})z^{n/2}\bigg(\prod_{n'=1}^n \rho(n',t_{\underline{p}})\bigg)A(z,n)^{-1}\begin{pmatrix} u_\pm(z,n,t_{\underline{p}}) \\ v_\pm(z,n,t_{\underline{p}}) \end{pmatrix},$$

$$z \in \mathbb{C}\setminus\big(\mathrm{spec}\,(\check{L}(t_{\underline{p}})) \cup \{0\}\big), \; (n,t_{\underline{p}}) \in \mathbb{Z}\times\mathbb{R}, \quad (3.370)$$

with the choice of normalization

$$D(t_{\underline{p}}) = \exp\left(\frac{i}{2}\big(g_{p_+,+}(0) - g_{p_-,-}(0)\big)t_{\underline{p}}\right)D(0), \quad t_{\underline{p}} \in \mathbb{R}, \quad (3.371)$$

for some constant $D(0) \in \mathbb{C}\setminus\{0\}$.

Lemma 3.52 *Assume Hypothesis 3.50 and suppose α, β satisfy $\mathrm{AL}_{\underline{p}}(\alpha,\beta) = 0$ for some $\underline{p} \in \mathbb{N}_0^2$. In addition, let $t_{\underline{p}} \in \mathbb{R}$ and $z \in \mathbb{C}\setminus\big(\mathrm{spec}\,(\check{L}(t_{\underline{p}}))\cup\{0\}\big)$. Then $\Psi_\pm(z,\cdot,t_{\underline{p}})$ defined in (3.370) satisfy*

$$U(z,\cdot,t_{\underline{p}})\Psi_\pm^-(z,\cdot,t_{\underline{p}}) = \Psi_\pm(z,\cdot,t_{\underline{p}}), \quad (3.372)$$

$$\Psi_{\pm,t_{\underline{p}}}(z,\cdot,t_{\underline{p}}) = V_{\underline{p}}^+(z,\cdot,t_{\underline{p}})\Psi_\pm(z,\cdot,t_{\underline{p}}). \quad (3.373)$$

In addition, $\Psi_-(z,\cdot,t_{\underline{p}})$ and $\Psi_+(z,\cdot,t_{\underline{p}})$ are linearly independent.

Proof Equation (3.372) is equivalent to

$$\begin{pmatrix} \psi_{1,\pm} \\ \psi_{2,\pm} \end{pmatrix} = \begin{pmatrix} z\psi_{1,\pm}^- + \alpha\psi_{2,\pm}^- \\ z\beta\psi_{1,\pm}^- + \psi_{2,\pm}^- \end{pmatrix}. \quad (3.374)$$

3.8 AL Conservation Laws and the Hamiltonian Formalism

Using (3.324) and (3.370) one obtains

$$\begin{pmatrix} \psi_{1,\pm} \\ \psi_{2,\pm} \end{pmatrix} = D z^{n/2} \left(\prod_{n'=1}^{n} \rho(n') \right) \begin{cases} \begin{pmatrix} z^{-1/2} u_{\pm} \\ z^{1/2} v_{\pm} \end{pmatrix}, & n \text{ odd,} \\ \begin{pmatrix} v_{\pm} \\ u_{\pm} \end{pmatrix}, & n \text{ even.} \end{cases} \quad (3.375)$$

Inserting (3.375) into (3.374), one finds that (3.374) is equivalent to (3.328), thereby proving (3.372).

Equation (3.373) is equivalent to

$$\begin{pmatrix} \psi_{1,\pm,t_{\underline{p}}} \\ \psi_{2,\pm,t_{\underline{p}}} \end{pmatrix} = i \begin{pmatrix} G_{\underline{p}} \psi_{1,\pm} - F_{\underline{p}} \psi_{2,\pm} \\ H_{\underline{p}} \psi_{1,\pm} - K_{\underline{p}} \psi_{2,\pm} \end{pmatrix}. \quad (3.376)$$

We first consider the case when n is odd. Using (3.375), the right-hand side of (3.376) reads

$$i \begin{pmatrix} G_{\underline{p}} \psi_{1,\pm} - F_{\underline{p}} \psi_{2,\pm} \\ H_{\underline{p}} \psi_{1,\pm} - K_{\underline{p}} \psi_{2,\pm} \end{pmatrix} = i D z^{(n-1)/2} \left(\prod_{n'=1}^{n} \rho(n') \right) \begin{pmatrix} G_{\underline{p}} u_{\pm} - z F_{\underline{p}} v_{\pm} \\ H_{\underline{p}} u_{\pm} - z K_{\underline{p}} v_{\pm} \end{pmatrix}. \quad (3.377)$$

Equation (3.375) then implies

$$\begin{pmatrix} \psi_{1,\pm,t_{\underline{p}}} \\ \psi_{2,\pm,t_{\underline{p}}} \end{pmatrix} = D_{t_{\underline{p}}} z^{n/2} \left(\prod_{n'=1}^{n} \rho(n') \right) \begin{pmatrix} z^{-1/2} u_{\pm} \\ z^{1/2} v_{\pm} \end{pmatrix} \quad (3.378)$$

$$+ D z^{n/2} \left(\prod_{n'=1}^{n} \rho(n') \right) \begin{pmatrix} z^{-1/2} u_{\pm,t_{\underline{p}}} \\ z^{1/2} v_{\pm,t_{\underline{p}}} \end{pmatrix} + D z^{n/2} \left(\partial_{t_{\underline{p}}} \prod_{n'=1}^{n} \rho(n') \right) \begin{pmatrix} z^{-1/2} u_{\pm} \\ z^{1/2} v_{\pm} \end{pmatrix}.$$

Next, one observes that

$$\left(\partial_{t_{\underline{p}}} \prod_{n'=1}^{n} \rho(n') \right) \left(\prod_{n'=1}^{n} \rho(n') \right)^{-1} = \partial_{t_{\underline{p}}} \ln \left(\prod_{n'=1}^{n} \rho(n') \right)$$

$$= \frac{1}{2} \partial_{t_{\underline{p}}} \ln \left(\prod_{n'=1}^{n} \rho(n')^2 \right)$$

$$= \frac{1}{2} \partial_{t_{\underline{p}}} \ln \left(\prod_{n'=1}^{n} \gamma(n') \right) = \frac{1}{2} \sum_{n'=1}^{n} \frac{\gamma_{t_{\underline{p}}}(n')}{\gamma(n')}.$$

Thus, (3.378) reads

$$D^{-1}z^{-(n-1)/2}\left(\prod_{n'=1}^{n}\rho(n')\right)^{-1}\begin{pmatrix}\psi_{1,\pm,t_{\underline{p}}}\\ \psi_{2,\pm,t_{\underline{p}}}\end{pmatrix}$$
$$=\begin{pmatrix}u_{\pm,t_{\underline{p}}}\\ zv_{\pm,t_{\underline{p}}}\end{pmatrix}+\frac{1}{2}\left(\sum_{n'=1}^{n}\frac{\gamma_{t_{\underline{p}}}(n')}{\gamma(n')}\right)\begin{pmatrix}u_{\pm}\\ zv_{\pm}\end{pmatrix}+\frac{D_{t_{\underline{p}}}}{D}\begin{pmatrix}u_{\pm}\\ zv_{\pm}\end{pmatrix}. \tag{3.379}$$

Combining (3.377) and (3.379) one finds that (3.376) is equivalent to

$$\begin{pmatrix}u_{\pm,t_{\underline{p}}}\\ v_{\pm,t_{\underline{p}}}\end{pmatrix}+\frac{1}{2}\left(\sum_{n'=1}^{n}\frac{\gamma_{t_{\underline{p}}}(n')}{\gamma(n')}\right)\begin{pmatrix}u_{\pm}\\ v_{\pm}\end{pmatrix}+\frac{D_{t_{\underline{p}}}}{D}\begin{pmatrix}u_{\pm}\\ v_{\pm}\end{pmatrix}=i\begin{pmatrix}G_{\underline{p}}u_{\pm}-zF_{\underline{p}}v_{\pm}\\ z^{-1}H_{\underline{p}}u_{\pm}-K_{\underline{p}}v_{\pm}\end{pmatrix}. \tag{3.380}$$

Using (3.55), (3.16), and (3.371) we find

$$\sum_{n'=1}^{n}\frac{\gamma_{t_{\underline{p}}}(n')}{\gamma(n')}=i\sum_{n'=1}^{n}(I-S^{-})(g_{p_{+},+}-g_{p_{-},-})$$
$$=i\big((g_{p_{+},+}(n)-g_{p_{-},-}(n))-(g_{p_{+},+}(0)-g_{p_{-},-}(0))\big)$$
$$=i(G_{\underline{p}}-K_{\underline{p}})-2\frac{D_{t_{\underline{p}}}}{D}. \tag{3.381}$$

From (3.76), (3.78), (3.364), and (3.366) one obtains (we recall that n is assumed to be odd)

$$\begin{pmatrix}u_{\pm,t_{\underline{p}}}\\ v_{\pm,t_{\underline{p}}}\end{pmatrix}=i\begin{pmatrix}-zF_{\underline{p}}v_{\pm}+\frac{1}{2}(G_{\underline{p}}+K_{\underline{p}})u_{\pm}\\ +z^{-1}H_{\underline{p}}u_{\pm}-\frac{1}{2}(G_{\underline{p}}+K_{\underline{p}})v_{\pm}\end{pmatrix}, \tag{3.382}$$

using (3.330).

Inserting (3.381) into (3.380), we see that it reduces to (3.382), thereby proving (3.376) in the case when n is odd. The case with n even follows from analogous computations.

Linear independence of $\Psi_{-}(z,\cdot,t_{\underline{p}})$ and $\Psi_{+}(z,\cdot,t_{\underline{p}})$ follows from

$$\begin{pmatrix}\psi_{1,-}(z,n,t_{\underline{p}}) & \psi_{1,+}(z,n,t_{\underline{p}})\\ \psi_{2,-}(z,n,t_{\underline{p}}) & \psi_{2,+}(z,n,t_{\underline{p}})\end{pmatrix}=D(t_{\underline{p}})z^{n/2}\left(\prod_{n'=1}^{n}\rho(n',t_{\underline{p}})\right)A(z,n)^{-1}$$
$$\times\begin{pmatrix}u_{-}(z,n,t_{\underline{p}}) & u_{+}(z,n,t_{\underline{p}})\\ v_{-}(z,n,t_{\underline{p}}) & v_{+}(z,n,t_{\underline{p}})\end{pmatrix},$$

the fact that $\rho(n,t_{\underline{p}})\neq 0$, $\det(A(z,n))=(-1)^{n+1}$, and from

$$\det\left(\begin{pmatrix}u_{-}(z,n,t_{\underline{p}}) & u_{+}(z,n,t_{\underline{p}})\\ v_{-}(z,n,t_{\underline{p}}) & v_{+}(z,n,t_{\underline{p}})\end{pmatrix}\right)\neq 0, \quad (n,t_{\underline{p}})\in\mathbb{Z}\times\mathbb{R}, \tag{3.383}$$

since by hypothesis $z\in\mathbb{C}\setminus\text{spec}\big(\check{L}(t_{\underline{p}})\big)$. $\square$

3.8 AL Conservation Laws and the Hamiltonian Formalism

In the following we will always refer to the solutions $\Psi_\pm$ introduced in (3.369)–(3.371).

The next result recalls the existence of a propagator $W_{\underline{p}}$ associated with $P_{\underline{p}}$. (Below we denote by $\mathcal{B}(\mathcal{H})$ the Banach space of all bounded linear operators defined on the Hilbert space $\mathcal{H}$.)

Theorem 3.53 *Assume Hypothesis 3.50 and suppose α, β satisfy $\mathrm{AL}_{\underline{p}}(\alpha, \beta) = 0$ for some $\underline{p} \in \mathbb{N}_0^2$. Then there is a propagator $W_{\underline{p}}(s, t) \in \mathcal{B}(\ell^2(\mathbb{Z}))$, $(s, t) \in \mathbb{R}^2$, satisfying*

(i) $W_{\underline{p}}(t, t) = I, \quad t \in \mathbb{R},$ (3.384)

(ii) $W_{\underline{p}}(r, s) W_{\underline{p}}(s, t) = W_{\underline{p}}(r, t), \quad (r, s, t) \in \mathbb{R}^3,$ (3.385)

(iii) $W_{\underline{p}}(s, t)$ *is jointly strongly continuous in* $(s, t) \in \mathbb{R}^2$, (3.386)

such that for fixed $t_0 \in \mathbb{R}$, $f_0 \in \ell^2(\mathbb{Z})$,

$$f(t) = W_{\underline{p}}(t, t_0) f_0, \quad t \in \mathbb{R},$$

satisfies

$$\frac{d}{dt} f(t) = \check{P}_{\underline{p}}(t) f(t), \quad f(t_0) = f_0. \quad (3.387)$$

Moreover, $\check{L}(t)$ is similar to $\check{L}(s)$ for all $(s, t) \in \mathbb{R}^2$,

$$\check{L}(s) = W_{\underline{p}}(s, t) \check{L}(t) W_{\underline{p}}(s, t)^{-1}, \quad (s, t) \in \mathbb{R}^2. \quad (3.388)$$

This extends to appropriate functions of $\check{L}(t)$ and so, in particular, to its resolvent $(\check{L}(t) - zI)^{-1}$, $z \in \mathbb{C} \setminus \sigma(\check{L}(t))$, and hence also yields

$$\sigma(\check{L}(s)) = \sigma(\check{L}(t)), \quad (s, t) \in \mathbb{R}^2.$$

Consequently, the spectrum of $\check{L}(t)$ is independent of $t \in \mathbb{R}$.

Proof (3.384)–(3.387) are standard results which follow, for instance, from Reed and Simon (1975, Theorem X.69) under even weaker hypotheses on α, β. In particular, the propagator $W_{\underline{p}}$ admits the norm convergent Dyson series

$$W_{\underline{p}}(s, t) = I + \sum_{k \in \mathbb{N}} \int_s^t dt_1 \int_s^{t_1} dt_2 \cdots \int_s^{t_{k-1}} dt_k \, \check{P}_{\underline{p}}(t_1) \check{P}_{\underline{p}}(t_2) \cdots \check{P}_{\underline{p}}(t_k),$$

$$(s, t) \in \mathbb{R}^2.$$

3 The Ablowitz–Ladik Hierarchy

Fixing $s \in \mathbb{R}$ and introducing the operator-valued function

$$\check{K}(t) = W_{\underline{p}}(s,t)\check{L}(t)W_{\underline{p}}(s,t)^{-1}, \quad t \in \mathbb{R}, \tag{3.389}$$

one computes

$$\check{K}'(t)f = W_{\underline{p}}(s,t)\big(\check{L}'(t) - [\check{P}_{\underline{p}}(t), \check{L}(t)]\big)W_{\underline{p}}(s,t)^{-1}f = 0, \quad t \in \mathbb{R}, \ f \in \ell^2(\mathbb{Z}),$$

using the Lax commutator equation (3.79). Thus, $\check{K}$ is independent of $t \in \mathbb{R}$ and hence taking $t = s$ in (3.389) then yields $\check{K} = \check{L}(s)$ and thus proves (3.388). □

Next we briefly recall the Ablowitz–Ladik initial value problem in a setting convenient for our purpose.

Theorem 3.54 *Let $t_{0,\underline{p}} \in \mathbb{R}$ and suppose $\alpha^{(0)}, \beta^{(0)} \in \ell^q(\mathbb{Z})$ for some $q \in [1, \infty) \cup \{\infty\}$. Then the $\underline{p}$th Ablowitz–Ladik initial value problem*

$$\mathrm{AL}_{\underline{p}}(\alpha, \beta) = 0, \quad (\alpha, \beta)\big|_{t_{\underline{p}} = t_{0,\underline{p}}} = \big(\alpha^{(0)}, \beta^{(0)}\big) \tag{3.390}$$

for some $\underline{p} \in \mathbb{N}_0^2$, has a unique, local, and smooth solution in time, that is, there exists a $T_0 > 0$ such that

$$\alpha(\,\cdot\,), \beta(\,\cdot\,) \in C^\infty((t_{0,\underline{p}} - T_0, t_{0,\underline{p}} + T_0), \ell^q(\mathbb{Z})).$$

Proof Local existence and uniqueness as well as smoothness of the solution of the initial value problem (3.390) (cf. (3.54)) follow since $f_{p_{\pm}-1,\pm}$, $g_{p_{\pm},\pm}$, and $h_{p_{\pm}-1,\pm}$ depend polynomially on α, β and certain of their shifts, and the fact that the Ablowitz–Ladik flows are autonomous. □

Remark 3.55 In the special defocusing case, where $\beta = \overline{\alpha}$ and hence $\check{L}(t), t \in \mathbb{R}$, is unitary, one obtains

$$\sup_{(n,t_{\underline{p}}) \in \mathbb{N} \times (t_{0,\underline{p}} - T_0, t_{0,\underline{p}} + T_0)} |\alpha(n, t_{\underline{p}})| \leq 1$$

using $\gamma = 1 - |\alpha|^2$ and $\gamma_{t_{\underline{p}}} = i\gamma\big((g_{p_+,+} - g^-_{p_+,+}) - (g_{p_-,-} - g^-_{p_-,-})\big)$ in (3.55). This then yields a unique, global, and smooth solution of the $\underline{p}$th AL initial value problem (3.390). Moreover, if α satisfies Hypothesis 3.50 and the $\underline{p}$th AL equation $\mathrm{AL}_{\underline{p}}(\alpha, \overline{\alpha}) = 0$, then α is actually smooth with respect to $t_{\underline{p}} \in \mathbb{R}$, that is,

$$\alpha(n, \,\cdot\,) \in C^\infty(\mathbb{R}), \quad n \in \mathbb{Z}. \tag{3.391}$$

3.8 AL Conservation Laws and the Hamiltonian Formalism

Equation (3.373), that is, $\Psi_{\pm,\underline{t}_{\underline{p}}} = V_{\underline{p}}^+ \Psi_\pm$, implies that

$$\partial_{\underline{t}_{\underline{p}}} \ln\left(\frac{\psi_{1,\pm}^+}{\psi_{1,\pm}}\right) = (S^+ - I)\partial_{\underline{t}_{\underline{p}}} \ln(\psi_{1,\pm})$$

$$= (S^+ - I)\frac{\partial_{\underline{t}_{\underline{p}}} \psi_{1,\pm}}{\psi_{1,\pm}}$$

$$= i(S^+ - I)(G_{\underline{p}} - F_{\underline{p}}\phi_\pm).$$

On the other hand, equation (3.372), that is, $U\Psi_{\pm}^- = \Psi_\pm$, yields

$$\partial_{\underline{t}_{\underline{p}}} \ln\left(\frac{\psi_{1,\pm}^+}{\psi_{1,\pm}}\right) = \partial_{\underline{t}_{\underline{p}}} \ln(z + \alpha^+ \phi_\pm),$$

and thus one concludes that

$$\partial_{\underline{t}_{\underline{p}}} \ln(z + \alpha^+ \phi_\pm) = i(S^+ - I)(G_{\underline{p}} - F_{\underline{p}}\phi_\pm). \tag{3.392}$$

Below we will refer to (3.392±) according to the upper or lower sign in (3.392). Expanding (3.392±) in powers of z and $1/z$ then yields the following conserved densities:

Theorem 3.56 *Assume Hypothesis* 3.50 *and suppose* α, β *satisfy* $\mathrm{AL}_{\underline{p}}(\alpha, \beta) = 0$ *for some* $\underline{p} \in \mathbb{N}_0^2$. *Then the following infinite sequences of local conservation laws hold:*

Expansion of (3.392+) *at* $1/z = 0$:

$$\partial_{\underline{t}_{\underline{p}}} \rho_{j,+}^\infty = i(S^+ - I)\bigg(g_{p_--j,-} - \sum_{\ell=0}^{j-1} f_{p_--j+\ell,-}\phi_{\ell,+}^\infty - \sum_{\ell=0}^{p_+-1} f_{p_+-1-\ell,+}\phi_{j+\ell,+}^\infty\bigg),$$
$$j = 1, \ldots, p_-, \tag{3.393}$$

$$\partial_{\underline{t}_{\underline{p}}} \rho_{j,+}^\infty = -i(S^+ - I)\bigg(\sum_{\ell=1}^{p_-} f_{p_--\ell,-}\phi_{j-\ell,+}^\infty + \sum_{\ell=0}^{p_+-1} f_{p_+-1-\ell,+}\phi_{j+\ell,+}^\infty\bigg),$$
$$j \geq p_- + 1, \tag{3.394}$$

where $\rho_{j,+}^\infty$ *and* $\phi_{j,+}^\infty$ *are given by* (3.351) *and* (3.343).
Expansion of (3.392−) *at* $1/z = 0$:

$$\partial_{\underline{t}_{\underline{p}}} \rho_{j,-}^\infty = i(S^+ - I)\bigg(g_{p_--j,-} - \sum_{\ell=-1}^{j-1} f_{p_-+\ell-j,-}\phi_{\ell,-}^\infty$$
$$- \sum_{\ell=0}^{p_+-1} f_{p_+-1-\ell,+}\phi_{j+\ell,-}^\infty\bigg), \quad j = 1, \ldots, p_-, \tag{3.395}$$

300   3 The Ablowitz–Ladik Hierarchy

$$\partial_{t_{\underline{p}}}\rho_{j,-}^{\infty} = -i(S^+ - I)\bigg(\sum_{\ell=1}^{p_-} f_{p_--\ell,-}\phi_{j-\ell,-}^{\infty} + \sum_{\ell=0}^{p_+-1} f_{p_+-1-\ell,+}\phi_{j+\ell,-}^{\infty}\bigg),$$

$$j \geq p_- + 1, \quad (3.396)$$

where $\rho_{j,-}^{\infty}$ and $\phi_{j,-}^{\infty}$ are given by (3.352) and (3.344).

Expansion of (3.392+) at $z = 0$:

$$\partial_{t_{\underline{p}}}\rho_{j,+}^{0} = i(S^+ - I)\bigg(g_{p_+-j,+} - \sum_{\ell=1}^{p_-}\phi_{j+\ell,+}^{0} f_{p_--\ell,-} - \sum_{\ell=0}^{j}\phi_{\ell,+}^{0} f_{p_+-1-j+\ell,+}\bigg),$$

$$j = 1, \ldots, p_+ - 1,$$

$$\partial_{t_{\underline{p}}}\rho_{p_+,+}^{0} = i(S^+ - I)\bigg(g_{0,+} - \sum_{\ell=1}^{p_-}\phi_{j+\ell,+}^{0} f_{p_--\ell,-} - \sum_{\ell=0}^{p_+-1}\phi_{j+\ell-p_++1,+}^{0} f_{\ell,+}\bigg),$$

$$\partial_{t_{\underline{p}}}\rho_{j,+}^{0} = -i(S^+ - I)\bigg(\sum_{\ell=1}^{p_-}\phi_{j+\ell,+}^{0} f_{p_--\ell,-} + \sum_{\ell=0}^{p_+-1}\phi_{j+\ell-p_++1,+}^{0} f_{\ell,+}\bigg),$$

$$j \geq p_+ + 1,$$

where $\rho_{j,+}^{0}$ and $\phi_{j,+}^{0}$ are given by (3.353) and (3.345).

Expansion of (3.392−) at $z = 0$:

$$\partial_{t_{\underline{p}}}\rho_{j,-}^{0} = i(S^+ - I)\bigg(g_{p_+-j,+} - \sum_{\ell=1}^{j}\phi_{\ell,-}^{0} f_{p_+-j+\ell-1,+} - \sum_{\ell=1}^{p_-}\phi_{j+\ell,-}^{0} f_{p_--\ell,-}\bigg),$$

$$j = 1, \ldots, p_+,$$

$$\partial_{t_{\underline{p}}}\rho_{j,-}^{0} = -i(S^+ - I)\bigg(\sum_{\ell=j+1-p_+}^{j}\phi_{\ell,-}^{0} f_{p_+-j+\ell-1,+} + \sum_{\ell=1}^{p_-}\phi_{j+\ell,-}^{0} f_{p_--\ell,-}\bigg),$$

$$j \geq p_+ + 1, \quad (3.397)$$

where $\rho_{j,-}^{0}$ and $\phi_{j,-}^{0}$ are given by (3.355) and (3.346).

Proof The proof consists of expanding (3.392±) in powers of z and $1/z$ and applying (3.350)–(3.355).

Expansion of (3.392+) at $1/z = 0$: For the right-hand side of (3.392+) one finds

$$G_{\underline{p}} - F_{\underline{p}}\phi_+ = \sum_{\ell=1}^{p_-} g_{p_--\ell,-}z^{-\ell} + \sum_{\ell=0}^{p_+} g_{p_+-\ell,+}z^{\ell}$$

$$- \bigg(\sum_{\ell=1}^{p_-} f_{p_--\ell,-}z^{-\ell} + \sum_{\ell=0}^{p_+-1} f_{p_+-1-\ell,+}z^{\ell}\bigg)\sum_{j=0}^{\infty}\phi_{j,+}^{\infty}z^{-j}$$

3.8 AL Conservation Laws and the Hamiltonian Formalism

$$= g_{0,+} z^{p_+} + \sum_{j=0}^{p_+-1} \left(g_{p_+-j,+} - \sum_{\ell=0}^{p_+-j-1} f_{p_+-j-1-\ell,+} \phi^\infty_{\ell,+} \right) z^j$$

$$+ \sum_{j=1}^{p_-} \left(g_{p_--j,-} - \sum_{\ell=0}^{j-1} f_{p_--j+\ell,-} \phi^\infty_{\ell,+} - \sum_{\ell=0}^{p_+-1} f_{p_+-1-\ell,+} \phi^\infty_{j+\ell,+} \right) z^{-j}$$

$$- \sum_{j=p_-+1}^{\infty} \left(\sum_{\ell=1}^{p_-} f_{p_--\ell,-} \phi^\infty_{j-\ell,+} + \sum_{\ell=0}^{p_+-1} f_{p_+-1-\ell,+} \phi^\infty_{j+\ell,+} \right) z^{-j}.$$

Here we used the fact that all positive powers vanish because of (3.392). This yields the following additional formulas:

Conservation laws derived from ϕ_+ at $1/z = 0$:

$$(S^+ - 1) \left(g_{p_+-j,+} - \sum_{\ell=0}^{p_+-j-1} f_{p_+-j-1-\ell,+} \phi^\infty_{\ell,+} \right) = 0, \quad j = 0, \ldots, p_+ - 1,$$

$$(S^+ - 1) g_{0,+} = 0.$$

Expansion of (3.392−) at $1/z = 0$: The right-hand side of (3.392−) yields

$$G_{\underline{p}} - F_{\underline{p}} \phi_- = \sum_{\ell=1}^{p_-} g_{p_--\ell,-} z^{-\ell} + \sum_{\ell=0}^{p_+} g_{p_+-\ell,+} z^\ell$$

$$- \left(\sum_{\ell=1}^{p_-} f_{p_--\ell,-} z^{-\ell} + \sum_{\ell=0}^{p_+-1} f_{p_+-1-\ell,+} z^\ell \right) \sum_{j=-1}^{\infty} \phi^\infty_{j,-} z^{-j}$$

$$= \sum_{j=1}^{p_+} \left(g_{p_+-j,+} - \sum_{\ell=-1}^{p_+-j-1} f_{p_+-j-1-\ell,+} \phi^\infty_{\ell,-} \right) z^j$$

$$+ \left(g_{p_+,+} - \sum_{\ell=0}^{p_+-1} f_{p_+-1-\ell,+} \phi^\infty_{\ell,-} - f_{p_--1,-} \phi^\infty_{-1,-} \right)$$

$$+ \sum_{j=1}^{p_-} \left(g_{p_--j,-} - \sum_{\ell=-1}^{j-1} f_{p_-+\ell-j,-} \phi^\infty_{\ell,-} - \sum_{\ell=0}^{p_+-1} f_{p_+-1-\ell,+} \phi^\infty_{j+\ell,-} \right) z^{-j}$$

$$- \sum_{j=p_-+1}^{\infty} \left(\sum_{\ell=1}^{p_-} f_{p_--\ell,-} \phi^\infty_{j-\ell,-} + \sum_{\ell=0}^{p_+-1} f_{p_+-1-\ell,+} \phi^\infty_{j+\ell,-} \right) z^{-j}.$$

Conservation laws derived from ϕ_+ at $1/z = 0$:

$$(S^+ - I) \left(g_{p_+-j,+} - \sum_{\ell=-1}^{p_+-j-1} f_{p_+-j-1-\ell,+} \phi^\infty_{\ell,-} \right) = 0, \quad j = 1, \ldots, p_+,$$

$$i(S^+ - I)\left(g_{p_+,+} - \sum_{\ell=0}^{p_+-1} f_{p_+-1-\ell,+}\phi_{\ell,-}^\infty - f_{p_--1,-}\phi_{-1,-}^\infty\right)$$

$$= \partial_{t_{\underline{p}}} \ln\left(\frac{\alpha^{++}}{\alpha^+}\right) + \partial_{t_{\underline{p}}} \ln(\gamma^+).$$

Expansion of (3.392+) at $z = 0$: For the right-hand side of (3.392+) one finds

$$G_{\underline{p}} - F_{\underline{p}}\phi_+ = \sum_{j=1}^{p_-}\left(g_{p_--j,-} - \sum_{\ell=0}^{p_--j}\phi_{\ell,+}f_{p_--j-\ell,-}\right)z^{-j}$$

$$+ \sum_{j=0}^{p_+-1}\left(g_{p_+-j,+} - \sum_{\ell=1}^{p_-}\phi_{j+\ell,+}^0 f_{p_--\ell,-} - \sum_{\ell=0}^{j}\phi_{\ell,+}^0 f_{p_+-1-j+\ell,+}\right)z^j$$

$$+ \sum_{j=p_+}^{\infty}\left(g_{0,+}\chi_{j p_+} - \sum_{\ell=1}^{p_-}\phi_{j+\ell,+}^0 f_{p_--\ell,-} - \sum_{\ell=0}^{p_+-1}\phi_{j+\ell-p_++1,+}^0 f_{\ell,+}\right)z^j.$$

Conservation laws derived from ϕ_+ at $z = 0$:

$$(S^+ - I)\left(g_{j,-} - \sum_{\ell=0}^{j}\phi_{\ell,+}^0 f_{j-\ell,-}\right) = 0, \quad j = 1,\ldots,p_-,$$

$$(S^+ - I)\left(g_{p_+,+} - \phi_{0,+}^0 f_{p_+-1,+} - \sum_{\ell=1}^{p_-}\phi_{\ell,+}^0 f_{p_--\ell,-}\right) = \partial_{t_{\underline{p}}}\ln\left(\frac{\alpha}{\alpha^+}\right).$$

Expansion of (3.392−) at $z = 0$: For the right-hand side of (3.392−) one finds

$$G_{\underline{p}} - F_{\underline{p}}\phi_- = g_{0,-}z^{-p_-} + \sum_{j=1}^{p_--1}\left(g_{p_--j,-} - \sum_{\ell=1}^{p_--j}\phi_{\ell,-}^0 f_{p_--j-\ell,-}\right)z^{-j}$$

$$+ g_{p_+,+} - \sum_{\ell=1}^{p_-}\phi_{\ell,-}^0 f_{p_--\ell,-}$$

$$+ \sum_{j=1}^{p_+}\left(g_{p_+-j,+} - \sum_{\ell=1}^{j}\phi_{\ell,-}^0 f_{p_+-j+\ell-1,+} - \sum_{\ell=1}^{p_-}\phi_{j+\ell,-}^0 f_{p_--\ell,-}\right)z^j$$

$$- \sum_{j=p_++1}^{\infty}\left(\sum_{\ell=j+1-p_+}^{j}\phi_{\ell,-}^0 f_{p_+-j+\ell-1,+} + \sum_{\ell=1}^{p_-}\phi_{j+\ell,-}^0 f_{p_--\ell,-}\right)z^j.$$

3.8 AL Conservation Laws and the Hamiltonian Formalism

Conservation laws derived from ϕ_- at $z = 0$:

$$(S^+ - I)g_{0,-} = 0,$$

$$(S^+ - I)\left(g_{j,-} - \sum_{\ell=1}^{j} \phi^0_{\ell,-} f_{j-\ell,-}\right) = 0, \quad j = 1, \ldots, p_- - 1,$$

$$(S^+ - I)\left(g_{p_+,+} - \sum_{\ell=1}^{p_-} \phi^0_{\ell,-} f_{p_--\ell,-}\right) = \partial_{t_{\underline{p}}} \ln(\gamma^+).$$

Combining these expansions with (3.350)–(3.354) finishes the proof. $\square$

Remark 3.57 (*i*) There is a certain redundancy in the conservation laws (3.393)–(3.397) as can be observed from Lemma 3.49. Equations (3.357)–(3.358) imply

$$\rho^\infty_{j,+} = -\rho^\infty_{j,-} + \frac{1}{j}(S^+ - I)(d_{j,+} - e_{j,+}), \quad j \in \mathbb{N},$$

$$\rho^0_{j,+} = -\rho^0_{j,-} + \frac{1}{j}(S^+ - I)(d_{j,-} - e_{j,-}), \quad j \in \mathbb{N}.$$

Thus one can, for instance, transfer (3.393)–(3.394) into (3.395)–(3.396).
(*ii*) In addition to the conservation laws listed in Theorem 3.56, we recover the familiar conservation law (cf. (3.55))

$$\partial_{t_{\underline{p}}} \ln(\gamma) = i(I - S^-)(g_{p_+,+} - g_{p_-,-}), \quad \underline{p} \in \mathbb{N}_0^2. \tag{3.398}$$

(*iii*) Another consequence of Theorem 3.56 and Lemma 3.49 is that for α, β satisfying Hypothesis 3.50 and $\alpha, \beta \in C^1(\mathbb{R}, \ell^2(\mathbb{Z}))$, one has

$$\frac{d}{dt_{\underline{p}}} \sum_{n \in \mathbb{Z}} \ln(\gamma(n, t_{\underline{p}})) = 0, \quad \frac{d}{dt_{\underline{p}}} \sum_{n \in \mathbb{Z}} \hat{g}_{j,\pm}(n, t_{\underline{p}}) = 0, \quad j \in \mathbb{N}, \ \underline{p} \in \mathbb{N}_0^2. \tag{3.399}$$

Remark 3.58 The two local conservation laws coming from expansions around $z = 0$ are essentially the same since the two conserved densities, $\rho^0_{j,+}$ and $\rho^0_{j,+}$, differ by a first-order difference expression (cf. Remark 3.57). A similar argument applies to the expansions around $1/z = 0$. That there are two independent sequences of conservation laws is also clear from (3.399), which yields that $\sum_{n \in \mathbb{Z}} \hat{g}_{j,\pm}(n, t_{\underline{p}})$ are time-independent. One observes that the quantities $\hat{g}_{j,+}$, $j \in \mathbb{N}$, are related to the expansions around $1/z = 0$, that is, to $\rho^\infty_{j,\pm}$, while $\hat{g}_{j,-}$, $j \in \mathbb{N}$, are related to $\rho^0_{j,\pm}$ (cf. Lemma 3.49). In addition to the two infinite sequences of polynomial conservation laws, there is a logarithmic conservation law (cf. (3.398) and (3.399)).

The first conservation laws explicitly read as follows:
$p_+ = p_- = 1$:

$$\partial_{t_{(1,1)}}\rho_{j,\pm}^\infty = -i(S^+ - I)(f_{0,-}\phi_{j-1,\pm}^\infty + f_{0,+}\phi_{j,\pm}^\infty), \quad j \geq 1,$$

$$\partial_{t_{(1,1)}}\rho_{j,\pm}^0 = -i(S^+ - I)(f_{0,-}\phi_{j+1,\pm}^0 + f_{0,+}\phi_{j,\pm}^0), \quad j \geq 1.$$

For $j = 1$ this yields using (3.356)

$$\partial_{t_{(1,1)}}\rho_{1,+}^\infty = \partial_{t_{(1,1)}}\alpha^+\beta = i(S^+ - I)(-c_{0,-}\alpha\beta + c_{0,+}\alpha^+\beta^-\gamma),$$

$$\partial_{t_{(1,1)}}\rho_{1,-}^\infty = \partial_{t_{(1,1)}}\left(-\alpha^{+++}\beta^{++} + (S^+ - I)\frac{\alpha^{++}}{\alpha^+}\right)$$

$$= i(S^+ - I)\left(c_{0,+}\frac{\alpha^{+++}}{\alpha^+}\gamma^+\gamma^{++} - c_{0,-}\frac{\alpha\alpha^{++}}{(\alpha^+)^2}\gamma^+ - c_{0,+}\left(\frac{\alpha\alpha^{++}}{\alpha^+}\right)^2\gamma^+\right),$$

$$\partial_{t_{(1,1)}}\rho_{1,+}^0 = \partial_{t_{(1,1)}}\left(\alpha^-\beta + (S^+ - I)\frac{\alpha^-}{\alpha}\right)$$

$$= i(S^+ - I)\left(c_{0,-}\frac{\alpha^{--}}{\alpha}\gamma^-\gamma - c_{0,+}\frac{\alpha^-\alpha^+}{\alpha^2}\gamma - c_{0,-}\left(\frac{\alpha^-}{\alpha}\right)^2\gamma\right),$$

$$\partial_{t_{(1,1)}}\rho_{1,-}^0 = \partial_{t_{(1,1)}}\alpha^+\beta^{++} = i(S^+ - I)(c_{0,+}\alpha^+\beta^+ - c_{0,-}\alpha\beta^{++}\gamma^+).$$

This shows in particular that we obtain two sets of conservation laws (one from expanding near ∞ and the other from expanding near 0), where the first few equations of each set explicitly read ($p_+ = p_- = 1$):

$j = 1:$ $\partial_{t_{(1,1)}}\alpha^+\beta = i(S^+ - I)\big(-c_{0,-}\alpha\beta + c_{0,+}\alpha^+\beta^-\gamma\big),$

 $\partial_{t_{(1,1)}}\alpha\beta^+ = i(S^+ - I)\big(c_{0,+}\alpha\beta - c_{0,-}\alpha^-\beta^+\gamma\big).$

$j = 2:$ $\partial_{t_{(1,1)}}\big(-\frac{1}{2}(\alpha^+\beta)^2 + \gamma\alpha^+\beta^-\big)$

 $= i(S^+ - I)\gamma\big(-c_{0,-}\alpha\beta^- - c_{0,+}\alpha\alpha^+(\beta^-)^2 + c_{0,+}\gamma^-\alpha^+\beta^{--}\big),$

 $\partial_{t_{(1,1)}}\big(\frac{1}{2}(\alpha\beta^+)^2 - \gamma^+\alpha\beta^{++}\big)$

 $= i(S^+ - I)\gamma\big(-c_{0,+}\alpha\beta^+ - c_{0,-}\alpha^-\alpha(\beta^+)^2 + c_{0,-}\gamma^+\alpha^-\beta^{++}\big).$

Using Lemma 3.49, one observes that one can replace $\rho_{j,\pm}^{\infty,0}$ in Theorem 3.56 by $\hat{g}_{j,\pm}$ by suitably adjusting the right-hand sides in (3.393)–(3.397).

Next, we turn to a presentation of the Hamiltonian formalism for the AL hierarchy including variational derivatives, Poisson brackets, etc.

We start this section by a short review of variational derivatives for discrete systems. Consider the functional

$$\mathcal{G}: \ell^1(\mathbb{Z})^K \to \mathbb{C},$$

$$\mathcal{G}(u) = \sum_{n \in \mathbb{Z}} G\big(u(n), u^{(+1)}(n), u^{(-1)}(n), \ldots, u^{(k)}(n), u^{(-k)}(n)\big)$$

3.8 AL Conservation Laws and the Hamiltonian Formalism

for some $\kappa \in \mathbb{N}$ and $k \in \mathbb{N}_0$, where $G \colon \mathbb{Z} \times \mathbb{C}^{2r\kappa} \to \mathbb{C}$ is C^1 with respect to the $2r\kappa$ complex-valued entries and where

$$u^{(s)} = S^{(s)}u, \quad S^{(s)} = \begin{cases} (S^+)^s u & \text{if } s \geq 0, \\ (S^-)^{-s} u & \text{if } s < 0, \end{cases} \quad u \in \ell^\infty(\mathbb{Z})^\kappa.$$

For brevity we write

$$G(u(n)) = G\big(u(n), u^{(+1)}(n), u^{(-1)}(n), \ldots, u^{(k)}(n), u^{(-k)}(n)\big).$$

The functional $\mathcal{G}$ is Frechet-differentiable and one computes for any $v \in \ell^1(\mathbb{Z})^\kappa$ for the differential $d\mathcal{G}$

$$\begin{aligned}
(d\mathcal{G})_u(v) &= \frac{d}{d\epsilon}\mathcal{G}(u + \epsilon v)\Big|_{\epsilon=0} \\
&= \sum_{n\in\mathbb{Z}} \bigg(\frac{\partial G(u(n))}{\partial u} v(n) + \frac{\partial G(u(n))}{\partial u^{(+1)}} v^{(+1)}(n) + \frac{\partial G(u(n))}{\partial u^{(-1)}} v^{(-1)}(n) \\
&\quad + \cdots + \frac{\partial G(u(n))}{\partial u^{(k)}} v^{(k)}(n) + \frac{\partial G(u(n))}{\partial u^{(-k)}} v^{(-k)}(n) \bigg) \\
&= \sum_{n\in\mathbb{Z}} \bigg(\frac{\partial G(u(n))}{\partial u} + S^{(-1)} \frac{\partial G(u(n))}{\partial u^{(+1)}} + S^{(+1)} \frac{\partial G(u(n))}{\partial u^{(-1)}} \\
&\quad + \cdots + S^{(-k)} \frac{\partial G(u(n))}{\partial u^{(k)}} + S^{(k)} \frac{\partial G(u(n))}{\partial u^{(-k)}} \bigg) v(n) \\
&= \sum_{n\in\mathbb{Z}} (\nabla\mathcal{G})_u(n) v(n) = \sum_{n\in\mathbb{Z}} \frac{\delta G}{\delta u}(n) v(n),
\end{aligned}$$

where we introduced the gradient and the variational derivative of $\mathcal{G}$ by

$$\begin{aligned}
(\nabla\mathcal{G})_u &= \frac{\delta G}{\delta u} \\
&= \frac{\partial G}{\partial u} + S^{(-1)} \frac{\partial G}{\partial u^{(+1)}} + S^{(+1)} \frac{\partial G}{\partial u^{(-1)}} + \cdots + S^{(-k)} \frac{\partial G}{\partial u^{(k)}} + S^{(k)} \frac{\partial G}{\partial u^{(-k)}},
\end{aligned}$$

assuming

$$\{G(u(n))\}_{n\in\mathbb{Z}}, \left\{\frac{\partial G(u(n))}{\partial u^{(\pm j)}}\right\}_{n\in\mathbb{Z}} \in \ell^1(\mathbb{Z}), \quad j = 1, \ldots, k.$$

To establish the connection with the Ablowitz–Ladik hierarchy we make the following assumption for the remainder of this section.

Hypothesis 3.59 *Suppose*

$$\alpha, \beta \in \ell^1(\mathbb{Z}), \quad \alpha(n)\beta(n) \notin \{0, 1\}, \quad n \in \mathbb{Z}.$$

Next, let $\mathcal{G}$ be a functional of the type

$$\mathcal{G}\colon \ell^1(\mathbb{Z})^2 \to \mathbb{C},$$
$$\mathcal{G}(\alpha,\beta) = \sum_{n\in\mathbb{Z}} G(\alpha(n),\beta(n),\ldots,\alpha(n+k),\beta(n+k),\alpha(n-k),\beta(n-k))$$
$$= \sum_{n\in\mathbb{Z}} G(\alpha(n),\beta(n)),$$

where $G(\alpha,\beta)$ is polynomial in α, β and some of their shifts. The gradient $\nabla\mathcal{G}$ and symplectic gradient $\nabla_s\mathcal{G}$ of $\mathcal{G}$ are then defined by

$$(\nabla\mathcal{G})_{\alpha,\beta} = \begin{pmatrix}(\nabla\mathcal{G})_\alpha \\ (\nabla\mathcal{G})_\beta\end{pmatrix} = \begin{pmatrix}\frac{\delta\mathcal{G}}{\delta\alpha} \\ \frac{\delta\mathcal{G}}{\delta\beta}\end{pmatrix}$$

and

$$(\nabla_s\mathcal{G})_{\alpha,\beta} = \mathcal{D}(\nabla\mathcal{G})_{\alpha,\beta} = \mathcal{D}\begin{pmatrix}(\nabla\mathcal{G})_\alpha \\ (\nabla\mathcal{G})_\beta\end{pmatrix},$$

respectively. Here $\mathcal{D}$ is defined by

$$\mathcal{D} = \gamma\begin{pmatrix}0 & 1 \\ -1 & 0\end{pmatrix}, \quad \gamma = 1 - \alpha\beta.$$

In addition, we introduce the weakly nondegenerate closed 2-form

$$\Omega\colon \ell^1(\mathbb{Z})^2 \times \ell^1(\mathbb{Z})^2 \to \mathbb{C},$$
$$\Omega(u,v) = \sum_{n\in\mathbb{Z}}(\mathcal{D}^{-1}u)(n)\cdot v(n).$$

One then concludes that

$$\Omega(\mathcal{D}u,v) = \sum_{n\in\mathbb{Z}} u(n)\cdot v(n) = \sum_{n\in\mathbb{Z}}\bigl(u_1(n)v_1(n) + u_2(n)v_2(n)\bigr)$$
$$= \langle u,v\rangle_{\ell^2(\mathbb{Z})^2}, \quad u,v \in \ell^1(\mathbb{Z})^2,$$

where $\langle\,\cdot\,,\,\cdot\,\rangle_{\ell^2(\mathbb{Z})^2}$ denotes the "real" inner product in $\ell^2(\mathbb{Z})^2$, that is,

$$\langle\,\cdot\,,\,\cdot\,\rangle_{\ell^2(\mathbb{Z})^2}\colon \ell^2(\mathbb{Z})^2 \times \ell^2(\mathbb{Z})^2 \to \mathbb{C},$$
$$\langle u,v\rangle_{\ell^2(\mathbb{Z})^2} = \sum_{n\in\mathbb{Z}} u(n)\cdot v(n) = \sum_{n\in\mathbb{Z}}\bigl(u_1(n)v_1(n) + u_2(n)v_2(n)\bigr).$$

In addition, one obtains

$$(d\mathcal{G})_{\alpha,\beta}(v) = \langle(\nabla\mathcal{G})_{\alpha,\beta},v\rangle_{\ell^2(\mathbb{Z})^2} = \Omega(\mathcal{D}(\nabla\mathcal{G})_{\alpha,\beta},v) = \Omega((\nabla_s\mathcal{G})_{\alpha,\beta},v).$$

3.8 AL Conservation Laws and the Hamiltonian Formalism

Given two functionals $\mathcal{G}_1, \mathcal{G}_2$ we define their Poisson bracket by

$$\{\mathcal{G}_1, \mathcal{G}_2\} = d\mathcal{G}_1(\nabla_s \mathcal{G}_2) = \Omega(\nabla_s \mathcal{G}_1, \nabla_s \mathcal{G}_2)$$
$$= \Omega(\mathcal{D}\nabla\mathcal{G}_1, \mathcal{D}\nabla\mathcal{G}_2) = \langle \nabla\mathcal{G}_1, \mathcal{D}\nabla\mathcal{G}_2 \rangle_{\ell^2(\mathbb{Z})^2}$$
$$= \sum_{n\in\mathbb{Z}} \begin{pmatrix} \frac{\delta G_1}{\delta\alpha}(n) \\ \frac{\delta G_1}{\delta\beta}(n) \end{pmatrix} \cdot \mathcal{D} \begin{pmatrix} \frac{\delta G_2}{\delta\alpha}(n) \\ \frac{\delta G_2}{\delta\beta}(n) \end{pmatrix}.$$

Moreover, both the Jacobi identity

$$\{\{\mathcal{G}_1, \mathcal{G}_2\}, \mathcal{G}_3\} + \{\{\mathcal{G}_2, \mathcal{G}_3\}, \mathcal{G}_1\} + \{\{\mathcal{G}_3, \mathcal{G}_1\}, \mathcal{G}_2\} = 0, \quad (3.400)$$

as well as the Leibniz rule

$$\{\mathcal{G}_1, \mathcal{G}_2\mathcal{G}_3\} = \{\mathcal{G}_1, \mathcal{G}_2\}\mathcal{G}_3 + \mathcal{G}_2\{\mathcal{G}_1, \mathcal{G}_3\}, \quad (3.401)$$

hold.

If $\mathcal{G}$ is a smooth functional and (α, β) develops according to a Hamiltonian flow with Hamiltonian $\mathcal{H}$, that is,

$$\begin{pmatrix} \alpha \\ \beta \end{pmatrix}_t = (\nabla_s \mathcal{H})_{\alpha,\beta} = \mathcal{D}(\nabla \mathcal{H})_{\alpha,\beta} = \mathcal{D} \begin{pmatrix} \frac{\delta H}{\delta\alpha} \\ \frac{\delta H}{\delta\beta} \end{pmatrix},$$

then

$$\frac{d\mathcal{G}}{dt} = \frac{d}{dt} \sum_{n\in\mathbb{Z}} G(\alpha(n), \beta(n))$$
$$= \sum_{n\in\mathbb{Z}} \begin{pmatrix} \frac{\delta G}{\delta\alpha}(n) \\ \frac{\delta G}{\delta\beta}(n) \end{pmatrix} \cdot \begin{pmatrix} \alpha(n) \\ \beta(n) \end{pmatrix}_t = \sum_{n\in\mathbb{Z}} \begin{pmatrix} \frac{\delta G}{\delta\alpha}(n) \\ \frac{\delta G}{\delta\beta}(n) \end{pmatrix} \cdot \mathcal{D} \begin{pmatrix} \frac{\delta H}{\delta\alpha}(n) \\ \frac{\delta H}{\delta\beta}(n) \end{pmatrix}$$
$$= \{\mathcal{G}, \mathcal{H}\}. \quad (3.402)$$

Here, and in the remainder of this section, time-dependent equations such as (3.402) are viewed locally in time, that is, assumed to hold on some open t-interval $\mathbb{I} \subseteq \mathbb{R}$.

If a functional $\mathcal{G}$ is in involution with the Hamiltonian $\mathcal{H}$, that is,

$$\{\mathcal{G}, \mathcal{H}\} = 0,$$

then it is conserved in the sense that

$$\frac{d\mathcal{G}}{dt} = 0.$$

Next, we turn to the specifics of the AL hierarchy. We define

$$\widehat{\mathcal{G}}_{\ell,\pm} = \sum_{n\in\mathbb{Z}} \hat{g}_{\ell,\pm}(n). \quad (3.403)$$

Lemma 3.60 *Assume Hypothesis* 3.59 *and* $v \in \ell^1(\mathbb{Z})$. *Then,*

$$(d\widehat{\mathcal{G}}_{\ell,\pm})_\beta(v) = \sum_{n\in\mathbb{Z}} \frac{\delta \hat{g}_{\ell,\pm}(n)}{\delta\beta} v(n) = \pm\ell \sum_{n\in\mathbb{Z}} (\delta_n, L^{\pm\ell-1} M_\beta(v)\delta_n), \quad \ell \in \mathbb{N}, \tag{3.404}$$

$$(d\widehat{\mathcal{G}}_{\ell,\pm})_\alpha(v) = \sum_{n\in\mathbb{Z}} \frac{\delta \hat{g}_{\ell,\pm}(n)}{\delta\alpha} v(n) = \pm\ell \sum_{n\in\mathbb{Z}} (\delta_n, L^{\pm\ell-1} M_\alpha(v)\delta_n), \quad \ell \in \mathbb{N}, \tag{3.405}$$

where

$$M_\beta(v) = -v\alpha^+ + \left(\left(v^-\rho - \beta^-\frac{v\alpha}{2\rho}\right)\delta_{\text{even}} + \alpha^+\frac{v\alpha}{2\rho}\delta_{\text{odd}}\right)S^-$$
$$+ \left(\left(v\rho^+ - \beta\frac{v^+\alpha^+}{2\rho^+}\right)\delta_{\text{even}} + \alpha^{++}\frac{v^+\alpha^+}{2\rho^+}\delta_{\text{odd}}\right)S^+$$
$$- \left(\rho\frac{v^-\alpha^-}{2\rho^-} + \rho^-\frac{v\alpha}{2\rho}\right)\delta_{\text{even}}S^{--} - \left(\rho^+\frac{v^{++}\alpha^{++}}{2\rho^{++}} + \rho^{++}\frac{v^+\alpha^+}{2\rho^+}\right)\delta_{\text{odd}}S^{++},$$

$$M_\alpha(v) = -v^+\beta - \left(\left(v^+\rho - \alpha^+\frac{v\beta}{2\rho}\right)\delta_{\text{odd}} + \beta^-\frac{v\beta}{2\rho}\delta_{\text{even}}\right)S^-$$
$$- \left(\left(v^{++}\rho^+ - \alpha^{++}\frac{v^+\beta^+}{2\rho^+}\right)\delta_{\text{odd}} - \beta\frac{v^+\beta^+}{2\rho^+}\delta_{\text{even}}\right)S^+$$
$$- \left(\rho\frac{v^-\beta^-}{2\rho^-} + \rho^-\frac{v\beta}{2\rho}\right)\delta_{\text{even}}S^{--} - \left(\rho^+\frac{v^{++}\beta^{++}}{2\rho^{++}} + \rho^{++}\frac{v^+\beta^+}{2\rho^+}\right)\delta_{\text{odd}}S^{++}.$$

Proof We first consider the derivative with respect to β. By a slight abuse of notation we write $L = L(\beta)$. Using (3.403) and (3.71) one finds

$$(d\widehat{\mathcal{G}}_{\ell,\pm})_\beta v = \frac{d}{d\epsilon}\mathcal{G}(\beta + \epsilon v)\Big|_{\epsilon=0} = \sum_{n\in\mathbb{Z}} \frac{d}{d\epsilon}\hat{g}_{\ell,\pm}(\beta+\epsilon v)(n)\Big|_{\epsilon=0}$$
$$= \sum_{n\in\mathbb{Z}}(\delta_n, \frac{d}{d\epsilon}L(\beta+\epsilon v)^{\pm\ell}\delta_n)|_{\epsilon=0}. \tag{3.406}$$

Next, one considers

$$\frac{d}{d\epsilon}L(\beta+\epsilon v)^\ell|_{\epsilon=0} = \lim_{\epsilon\to 0}\frac{1}{\epsilon}\left(L(\beta+\epsilon v)^\ell - L(\beta)^\ell\right)$$
$$= \lim_{\epsilon\to 0}\frac{1}{\epsilon}\Big((L(\beta+\epsilon v) - L(\beta))L(\beta)^{\ell-1}$$
$$+ L(\beta)(L(\beta+\epsilon v) - L(\beta))L(\beta)^{\ell-2}$$
$$+ \cdots + L(\beta)^{\ell-1}(L(\beta+\epsilon v) - L(\beta))\Big) \tag{3.407}$$
$$= M_\beta L(\beta)^{\ell-1} + L(\beta)M_\beta L(\beta)^{\ell-2} + \cdots + L(\beta)^{\ell-1}M_\beta,$$

3.8 AL Conservation Laws and the Hamiltonian Formalism

where

$$M_\beta = \lim_{\epsilon \to 0} \frac{1}{\epsilon}\big(L(\beta + \epsilon v) - L(\beta)\big)$$

$$= \bigg(-v(n)\alpha(n+1)\delta_{m,n} + \Big((v(n-1)\rho(n) - \beta(n-1)\frac{v(n)\alpha(n)}{2\rho(n)}\Big)\delta_{\text{odd}}(n)$$

$$+ \alpha(n+1)\frac{v(n)\alpha(n)}{2\rho(n)}\delta_{\text{even}}(n)\Big)\delta_{m,n-1}$$

$$+ \Big((v(n)\rho(n+1) - \beta(n)\frac{v(n+1)\alpha(n+1)}{2\rho(n+1)}\Big)\delta_{\text{odd}}(n) \quad (3.408)$$

$$+ \alpha(n+2)\frac{v(n+1)\alpha(n+1)}{2\rho(n+1)}\delta_{\text{even}}(n)\Big)\delta_{m,n+1}$$

$$- \Big(\rho(n+1)\frac{v(n+2)\alpha(n+2)}{2\rho(n+2)} + \rho(n+2)\frac{v(n+1)\alpha(n+1)}{2\rho(n+1)}\Big)$$

$$\times \delta_{\text{even}}(n)\delta_{m,n+2}$$

$$- \Big(\rho(n)\frac{v(n-1)\alpha(n-1)}{2\rho(n-1)} + \rho(n-1)\frac{v(n)\alpha(n)}{2\rho(n)}\Big)\delta_{\text{odd}}(n)\delta_{m,n-2}\bigg)_{m,n\in\mathbb{Z}}.$$

Similarly one obtains

$$\frac{d}{d\epsilon}L(\beta + \epsilon v)^{-\ell}|_{\epsilon=0} \quad (3.409)$$

$$= -\Big(L(\beta)^{-1}M_\beta L(\beta)^{-\ell} + L(\beta)^{-2}M_\beta L(\beta)^{-\ell+1} + \cdots + L(\beta)^{-\ell}M_\beta L(\beta)^{-1}\Big).$$

Inserting the expression (3.407) into (3.406) one finds

$$(d\widehat{\mathcal{G}}_{\ell,+})_\beta v = \sum_{n\in\mathbb{Z}}(\delta_n, \frac{d}{d\epsilon}L(\beta + \epsilon v)^\ell \delta_n)|_{\epsilon=0}$$

$$= \sum_{n\in\mathbb{Z}}(\delta_n, \sum_{k=0}^{\ell-1} L^k M_\beta L^{\ell-1-k}\delta_n)$$

$$= \sum_{k=0}^{\ell-1}\sum_{n\in\mathbb{Z}}(\delta_n, L^k M_\beta L^{\ell-1-k}\delta_n)$$

$$= \sum_{k=0}^{\ell-1}\sum_{n\in\mathbb{Z}}(\delta_n, (L^{\ell-1}M_\beta + [L^k M_\beta, L^{\ell-1-k}])\delta_n)$$

$$= \sum_{k=0}^{\ell-1}\sum_{n\in\mathbb{Z}}\Big((\delta_n, L^{\ell-1}M_\beta\delta_n) + (\delta_n, [L^k M_\beta, L^{\ell-1-k}]\delta_n)\Big)$$

$$= \ell\sum_{n\in\mathbb{Z}}(\delta_n, L^{\ell-1}M_\beta\delta_n) + \sum_{k=0}^{\ell-1}\text{tr}\left([L^k M_\beta, L^{\ell-1-k}]\right)$$

$$= \ell\sum_{n\in\mathbb{Z}}(\delta_n, L^{\ell-1}M_\beta\delta_n).$$

Similarly, using (3.406) and (3.409), one concludes that

$$(d\widehat{\mathcal{G}}_{\ell,-})_\beta v = -\ell \sum_{n\in\mathbb{Z}} (\delta_n, L^{-\ell-1} M_\beta \delta_n).$$

For the derivative with respect to α we set $L = L(\alpha)$ and replace M_β by

$$M_\alpha = \lim_{\epsilon \to 0} \frac{1}{\epsilon}\big(L(\alpha+\epsilon v) - L(\alpha)\big)$$

$$= \bigg(-v(n+1)\beta(n)\delta_{m,n} + \Big(-\beta(n-1)\frac{v(n)\beta(n)}{2\rho(n)}\delta_{\text{odd}}(n)$$
$$- \big(v(n+1)\rho(n) - \alpha(n+1)\frac{v(n)\beta(n)}{2\rho(n)}\big)\delta_{\text{even}}(n)\Big)\delta_{m,n-1}$$
$$- \Big(\beta(n)\frac{v(n+1)\beta(n+1)}{2\rho(n+1)}\delta_{\text{odd}}(n) + \big(v(n+2)\rho(n+1) \qquad (3.410)$$
$$+ \alpha(n+2)\frac{v(n+1)\beta(n+1)}{2\rho(n+1)}\big)\delta_{\text{even}}(n)\Big)\delta_{m,n+1}$$
$$- \Big(\rho(n+1)\frac{v(n+2)\beta(n+2)}{2\rho(n+2)} + \rho(n+2)\frac{v(n+1)\beta(n+1)}{2\rho(n+1)}\Big)$$
$$\times \delta_{\text{even}}(n)\delta_{m,n+2}$$
$$- \Big(\rho(n)\frac{v(n-1)\beta(n-1)}{2\rho(n-1)} + \rho(n-1)\frac{v(n)\beta(n)}{2\rho(n)}\Big)\delta_{\text{odd}}(n)\delta_{m,n-2}\bigg)_{m,n\in\mathbb{Z}}.$$

$\square$

Lemma 3.61 *Assume Hypothesis* 3.59. *Then the following relations hold:*

$$\frac{\delta \hat{g}_{\ell,+}}{\delta \beta} = \frac{\ell}{\gamma}\big(\hat{f}_{\ell-1,+} - \alpha \hat{g}_{\ell,+}\big), \quad \ell \in \mathbb{N}, \qquad (3.411)$$

$$\frac{\delta \hat{g}_{\ell,-}}{\delta \beta} = -\frac{\ell}{\gamma}\big(\hat{f}^-_{\ell-1,-} + \alpha \hat{g}^-_{\ell,-}\big), \quad \ell \in \mathbb{N}. \qquad (3.412)$$

Proof We consider (3.411) first. By (3.71) one concludes that

$$\hat{f}_{\ell-1,+}(n) - \alpha(n)\hat{g}_{\ell,+}(n)$$
$$= (\delta_n, EL^{\ell-1}\delta_n)\delta_{\text{even}}(n) + (\delta_n, L^{\ell-1}D\delta_n)\delta_{\text{odd}}(n) - \alpha(n)(\delta_n, L^\ell\delta_n).$$

Thus one has to show that

$$\sum_{n\in\mathbb{Z}} (\delta_n, L^{\ell-1}M_\beta\delta_n) = \sum_{n\in\mathbb{Z}} \frac{v(n)}{\rho(n)^2}\big(\hat{f}_{\ell-1,+}(n) - \alpha(n)\hat{g}_{\ell,+}(n)\big),$$

since this implies (3.411), using (3.404). By (3.408), (3.67), and (3.68), and assuming

3.8 AL Conservation Laws and the Hamiltonian Formalism

$v \in \ell^1(\mathbb{Z})$ one obtains

$$\sum_{n \in \mathbb{Z}} (\delta_n, L^{\ell-1} M_\beta \delta_n)$$

$$= \sum_{n \in \mathbb{Z}} \Big(-v\alpha^+(\delta_n, L^{\ell-1}\delta_n) + v^-\rho(\delta_n, L^{\ell-1}\delta_{n-1})\delta_{\text{odd}}$$
$$+ v\rho^+(\delta_n, L^{\ell-1}\delta_{n+1})\delta_{\text{odd}}$$
$$- \frac{v\alpha}{2\rho}\big(-\alpha^+(\delta_n, L^{\ell-1}\delta_{n-1})\delta_{\text{even}} + \rho^+(\delta_{n+1}, L^{\ell-1}\delta_{n-1})\delta_{\text{even}}\big)$$
$$- \frac{v\alpha}{2\rho}\big(\beta^-(\delta_n, L^{\ell-1}\delta_{n-1})\delta_{\text{odd}} + \rho^-(\delta_n, L^{\ell-1}\delta_{n-2})\delta_{\text{odd}}\big)$$
$$- \frac{v^+\alpha^+}{2\rho^+}\big(-\alpha^{++}(\delta_n, L^{\ell-1}\delta_{n+1})\delta_{\text{even}} + \rho^{++}(\delta_n, L^{\ell-1}\delta_{n+1})\delta_{\text{even}}\big)$$
$$- \frac{v^+\alpha^+}{2\rho^+}\big(\beta(\delta_n, L^{\ell-1}\delta_{n+1})\delta_{\text{odd}} + \rho(\delta_{n-1}, L^{\ell-1}\delta_{n+1})\delta_{\text{odd}}\big)\Big)$$

$$= \sum_{n \in \mathbb{Z}} \Big(-v\alpha^+(\delta_n, L^{\ell-1}\delta_n) + v\rho^+(\delta_{n+1}, L^{\ell-1}\delta_n)\delta_{\text{even}}$$
$$+ v\rho^+(\delta_n, L^{\ell-1}\delta_{n+1})\delta_{\text{odd}}$$
$$- \frac{v\alpha}{2\rho}\big((\delta_n, EL^{\ell-1}\delta_{n-1})\delta_{\text{even}} + (\delta_n, L^{\ell-1}D\delta_{n-1})\delta_{\text{odd}}\big)$$
$$- \frac{v^+\alpha^+}{2\rho^+}\big((\delta_n, L^{\ell-1}D\delta_{n+1})\delta_{\text{even}} + (\delta_n, EL^{\ell-1}\delta_{n+1})\delta_{\text{odd}}\big)\Big)$$

$$= \sum_{n \in \mathbb{Z}} \Big(v(\delta_n, EL^{\ell-1}\delta_n)\delta_{\text{even}} + v(\delta_n, L^{\ell-1}D\delta_n)\delta_{\text{odd}}$$
$$- \frac{v\alpha}{2\rho}\big((\delta_n, EL^{\ell-1}\delta_{n-1})\delta_{\text{even}} + (\delta_n, L^{\ell-1}D\delta_{n-1})\delta_{\text{odd}}$$
$$+ (\delta_{n-1}, L^{\ell-1}D\delta_n)\delta_{\text{odd}} + (\delta_{n-1}, EL^{\ell-1}\delta_n)\delta_{\text{even}}\big)\Big)$$

$$= \sum_{n \in \mathbb{Z}} \Big(\frac{v}{\rho^2}\big(\hat{f}_{\ell-1,+} - \alpha\hat{g}_{\ell,+}\big) + \frac{v\alpha}{2\rho}\big((\delta_{n-1}, EL^{\ell-1}\delta_n)\delta_{\text{even}}$$
$$+ (\delta_n, L^{\ell-1}D\delta_{n-1})\delta_{\text{odd}} - (\delta_n, EL^{\ell-1}\delta_{n-1})\delta_{\text{even}} - (\delta_{n-1}, L^{\ell-1}D\delta_n)\delta_{\text{odd}}\big)\Big),$$

where we used (3.71) and

$$\hat{g}_{\ell,+} = (\delta_n, L^{\ell-1}DE\delta_n)$$
$$= \beta(\delta_n, L^{\ell-1}D\delta_n)\delta_{\text{odd}} + \rho(\delta_n, L^{\ell-1}D\delta_{n-1})\delta_{\text{odd}}$$
$$+ \beta(\delta_n, EL^{\ell-1}\delta_n)\delta_{\text{even}} + \rho(\delta_{n-1}, EL^{\ell-1}\delta_n)\delta_{\text{even}}$$
$$= \beta\hat{f}_{\ell-1,+} + \rho(\delta_n, L^{\ell-1}D\delta_{n-1})\delta_{\text{odd}} + \rho(\delta_{n-1}, EL^{\ell-1}\delta_n)\delta_{\text{even}}.$$

312   3 The Ablowitz–Ladik Hierarchy

Hence it remains to show that

$$(\delta_{n-1}, EL^{\ell-1}\delta_n)\delta_{\text{even}} + (\delta_n, L^{\ell-1}D\delta_{n-1})\delta_{\text{odd}}$$
$$- (\delta_n, EL^{\ell-1}\delta_{n-1})\delta_{\text{even}} - (\delta_{n-1}, L^{\ell-1}D\delta_n)\delta_{\text{odd}} = 0,$$

but this follows from $(EL^\ell)^\top = EL^\ell$ (resp., $(L^\ell D)^\top = L^\ell D$) by (3.65), (3.66). In the case (3.412) one similarly shows that

$$\sum_{n\in\mathbb{Z}}(\delta_n, L^{-\ell-1}M_\beta\delta_n) = -\sum_{n\in\mathbb{Z}}\frac{v(n+1)}{\rho(n)^2}\Big((\delta_n, D^{-1}L^{-\ell+1}\delta_n)\delta_{\text{even}}(n)$$
$$+ (\delta_n, L^{-\ell+1}E^{-1}\delta_n)\delta_{\text{odd}}(n) + \alpha(n+1)(\delta_n, L^{-\ell}\delta_n)\Big).$$

$\square$

Lemma 3.62 *Assume Hypothesis 3.59. Then the following relations hold:*

$$\frac{\delta \hat{g}_{\ell,+}}{\delta \alpha} = -\frac{\ell}{\gamma}\big(\hat{h}^-_{\ell-1,+} + \beta \hat{g}^-_{\ell,+}\big), \quad \ell \in \mathbb{N}, \tag{3.413}$$

$$\frac{\delta \hat{g}_{\ell,-}}{\delta \alpha} = \frac{\ell}{\gamma}\big(\hat{h}_{\ell-1,-} - \beta \hat{g}_{\ell,-}\big), \quad \ell \in \mathbb{N}. \tag{3.414}$$

Proof We consider (3.413) first. Using (3.405), (3.410), (3.67), and (3.68), and assuming $v \in \ell^1(\mathbb{Z})$ one obtains

$$\sum_{n\in\mathbb{Z}}(\delta_n, L^{\ell-1}M_\alpha\delta_n)$$
$$= \sum_{n\in\mathbb{Z}}\Big(-v^+\beta(\delta_n, L^{\ell-1}\delta_n) - v^+\rho(\delta_n, L^{\ell-1}\delta_{n-1})\delta_{\text{even}}$$
$$- v^{++}\rho^+(\delta_n, L^{\ell-1}\delta_{n+1})\delta_{\text{even}}$$
$$- \frac{v\beta}{2\rho}\big(\beta^-(\delta_n, L^{\ell-1}\delta_{n-1})\delta_{\text{odd}} + \rho^-(\delta_n, L^{\ell-1}\delta_{n-2})\delta_{\text{odd}}\big)$$
$$- \frac{v\beta}{2\rho}\big(-\alpha^+(\delta_n, L^{\ell-1}\delta_{n-1})\delta_{\text{even}} + \rho^+(\delta_{n+1}, L^{\ell-1}\delta_{n-1})\delta_{\text{even}}\big)$$
$$- \frac{v^+\beta^+}{2\rho^+}\big(-\alpha^{++}(\delta_n, L^{\ell-1}\delta_{n+1})\delta_{\text{even}} + \rho^{++}(\delta_n, L^{\ell-1}\delta_{n+2})\delta_{\text{even}}\big)$$
$$- \frac{v^+\beta^+}{2\rho^+}\big(\beta(\delta_n, L^{\ell-1}\delta_{n+1})\delta_{\text{odd}} + \rho(\delta_{n-1}, L^{\ell-1}\delta_{n+1})\delta_{\text{odd}}\big)\Big)$$
$$= \sum_{n\in\mathbb{Z}}\Big(-v^+\big((\delta_n, L^{\ell-1}D\delta_n)\delta_{\text{even}} + (\delta_n, EL^{\ell-1}\delta_n)\delta_{\text{odd}}\big)$$
$$- \frac{v\beta}{2\rho}\big((\delta_n, EL^{\ell-1}\delta_{n-1})\delta_{\text{even}} + (\delta_n, L^{\ell-1}D\delta_{n-1})\delta_{\text{odd}}\big)$$

3.8 AL Conservation Laws and the Hamiltonian Formalism

$$-\frac{v^+\beta^+}{2\rho^+}\left((\delta_n, L^{\ell-1}D\delta_{n+1})\delta_{\text{even}} + (\delta_n, EL^{\ell-1}\delta_{n+1})\delta_{\text{odd}}\right)$$

$$= -\sum_{n\in\mathbb{Z}} \frac{v}{\rho^2}(\hat{h}^-_{\ell-1,+} + \beta\hat{g}^-_{\ell,+}),$$

since by (3.71),

$$2\alpha\hat{h}^-_{\ell-1,+} + 2\hat{g}^-_{\ell,+} = \rho\big((\delta_n, L^{\ell-1}D\delta_{n-1})\delta_{\text{odd}} + (\delta_n, EL^{\ell-1}\delta_{n-1})\delta_{\text{even}}$$
$$+ (\delta_{n-1}, L^{\ell-1}D\delta_n)\delta_{\text{odd}} + (\delta_{n-1}, EL^{\ell-1}\delta_n)\delta_{\text{even}}\big).$$

The result (3.414) follows similarly. □

Next, we introduce the Hamiltonians

$$\widehat{\mathcal{H}}_0 = \sum_{n\in\mathbb{Z}} \ln(\gamma(n)), \quad \widehat{\mathcal{H}}_{p_\pm,\pm} = \frac{1}{p_\pm}\sum_{n\in\mathbb{Z}} \hat{g}_{p_\pm,\pm}(n), \quad p_\pm \in \mathbb{N}, \qquad (3.415)$$

$$\mathcal{H}_{\underline{p}} = \sum_{\ell=1}^{p_+} c_{p_+-\ell,+}\widehat{\mathcal{H}}_{\ell,+} + \sum_{\ell=1}^{p_-} c_{p_--\ell,-}\widehat{\mathcal{H}}_{\ell,-} + c_{\underline{p}}\widehat{\mathcal{H}}_0, \quad \underline{p} = (p_-, p_+) \in \mathbb{N}_0^2.$$

$$(3.416)$$

(We recall that $c_{\underline{p}} = (c_{p,-} + c_{p,+})/2$.)

Theorem 3.63 *Assume Hypothesis 3.59. Then the following relation holds:*

$$\text{AL}_{\underline{p}}(\alpha,\beta) = \begin{pmatrix} -i\alpha_{t_{\underline{p}}} \\ -i\beta_{t_{\underline{p}}} \end{pmatrix} + \mathcal{D}\nabla\mathcal{H}_{\underline{p}} = 0, \quad \underline{p} \in \mathbb{N}_0^2.$$

Proof This follows directly from Lemmas 3.61 and 3.62,

$$(\nabla\widehat{\mathcal{H}}_{\ell,+})_\alpha = \frac{1}{\gamma}\big(-\beta\hat{g}^-_{\ell,+} - \hat{h}^-_{\ell-1,+}\big), \quad (\nabla\widehat{\mathcal{H}}_{\ell,+})_\beta = \frac{1}{\gamma}\big(-\alpha\hat{g}_{\ell,+} + \hat{f}_{\ell-1,+}\big),$$

$$(\nabla\widehat{\mathcal{H}}_{\ell,-})_\alpha = \frac{1}{\gamma}\big(-\beta\hat{g}_{\ell,-} + \hat{h}_{\ell-1,-}\big), \quad (\nabla\widehat{\mathcal{H}}_{\ell,-})_\beta = \frac{1}{\gamma}\big(-\alpha\hat{g}^-_{\ell,-} - \hat{f}^-_{\ell-1,-}\big),$$

$$\ell \in \mathbb{N},$$

together with (3.35). □

Theorem 3.64 *Assume Hypothesis 3.50 and suppose α, β satisfy $\text{AL}_{\underline{p}}(\alpha,\beta) = 0$ for some $\underline{p} \in \mathbb{N}_0^2$. Then,*

$$\frac{d\mathcal{H}_{\underline{r}}}{dt_{\underline{p}}} = 0, \quad \underline{r} \in \mathbb{N}_0^2. \qquad (3.417)$$

Proof From Lemma 3.49 and Theorem 3.56 one obtains

$$\frac{d\hat{g}_{r_\pm,\pm}}{dt_{\underline{p}}} = (S^+ - I)J_{r_\pm,\pm}, \quad r_\pm \in \mathbb{N}_0,$$

for some $J_{r_\pm,\pm}, r_\pm \in \mathbb{N}_0$, which are polynomials in α and β and certain shifts thereof. Using definition (3.416) of $\mathcal{H}_{\underline{r}}$, the result (3.417) follows in the homogeneous case and then by linearity in the general case. □

Theorem 3.65 *Assume Hypothesis 3.59 and let $\underline{p}, \underline{r} \in \mathbb{N}_0^2$. Then,*

$$\{\mathcal{H}_{\underline{p}}, \mathcal{H}_{\underline{r}}\} = 0, \tag{3.418}$$

that is, $\mathcal{H}_{\underline{p}}$ and $\mathcal{H}_{\underline{r}}$ are in involution for all $\underline{p}, \underline{r} \in \mathbb{N}_0^2$.

Proof By Theorem 3.54, there exists $T > 0$ such that the initial value problem

$$\mathrm{AL}_{\underline{p}}(\alpha, \beta) = 0, \quad (\alpha, \beta)\big|_{t_{\underline{p}}=0} = \big(\alpha^{(0)}, \beta^{(0)}\big),$$

where $\alpha^{(0)}, \beta^{(0)}$ satisfy Hypothesis 3.59, has unique, local, and smooth solutions $\alpha(t), \beta(t)$ satisfying Hypothesis 3.59 for each $t \in [0, T)$. For this solution we know that

$$\frac{d}{dt_{\underline{p}}}\mathcal{H}_{\underline{p}}(t) = \{\mathcal{H}_{\underline{r}}(t), \mathcal{H}_{\underline{p}}(t)\} = 0.$$

Next, let $t \downarrow 0$. Then

$$0 = \{\mathcal{H}_{\underline{r}}(t), \mathcal{H}_{\underline{p}}(t)\} \underset{t\downarrow 0}{\to} \{\mathcal{H}_{\underline{r}}(0), \mathcal{H}_{\underline{p}}(0)\} = \{\mathcal{H}_{\underline{r}}, \mathcal{H}_{\underline{p}}\}\big|_{(\alpha,\beta)=(\alpha^{(0)},\beta^{(0)})}.$$

Since $\alpha^{(0)}, \beta^{(0)}$ are arbitrary coefficients satisfying Hypothesis 3.59 one concludes (3.418). □

3.9 Notes

> In science the credit goes to the man who convinces the world, not the man to whom the idea first occurs.
>
> *Sir Francis Darwin*[1]

This chapter closely follows the series of papers by Gesztesy et al. (to appear; 2007a,b; 2008a).

Section 3.1. In the mid-1970s, Ablowitz and Ladik, in a series of papers (Ablowitz and Ladik (1975; 1976a,b; 1977), see also Ablowitz (1977), Ablowitz and Clarkson

---

[1] *Eugenics Review*, April 1914.

(1991, Sect. 3.2.2), Ablowitz et al. (2004, Ch. 3), Chiu and Ladik (1977)), used inverse scattering methods to analyze certain integrable differential-difference systems. One of their integrable variants of such systems included a discretization of the celebrated AKNS-ZS system, the pair of coupled nonlinear differential-difference equations (3.1), that is,

$$-i\alpha_t - (1 - \alpha\beta)(\alpha^- + \alpha^+) + 2\alpha = 0,$$
$$-i\beta_t + (1 - \alpha\beta)(\beta^- + \beta^+) - 2\beta = 0,$$
(3.419)

which is commonly referred to as the Ablowitz–Ladik system. In particular, Ablowitz and Ladik (1976b) (see also Ablowitz et al. (2004, Ch. 3)) showed that in the defocusing case, where $\beta = \overline{\alpha}$, and in the focusing case, where $\beta = -\overline{\alpha}$, (3.419) yields the discrete analog of the nonlinear Schrödinger equation

$$-i\alpha_t - (1 \mp |\alpha|^2)(\alpha^- + \alpha^+) + 2\alpha = 0. \tag{3.420}$$

It should be noted here that the 2×2-system of differential-difference equations (3.419) is actually a special case of a more general 2×2-system introduced by Ablowitz and Ladik (1975). The latter is not studied in this monograph.

Since the mid-1970s there has been an enormous amount of activity in the area of integrable differential-difference equations. Two principal directions of research are responsible for this development: Originally, the development was driven by the theory of completely integrable systems and its applications to fields such as nonlinear optics, and more recently, it gained additional momentum due to its intimate connections with the theory of orthogonal polynomials. The more recent developments in connection with the complete integrability aspects of the AL hierarchy will naturally be discussed in the subsequent notes pertinent to each of the AL sections. Here we briefly recall some of the recent developments influenced by research on orthogonal polynomials in the following (which are not discussed in the main body of this monograph).

The connection between the Ablowitz–Ladik system (3.1) and orthogonal polynomials comes about as follows: Let $\{\alpha(n)\}_{n\in\mathbb{N}} \subset \mathbb{C}$ be a sequence of complex numbers subject to the condition $|\alpha(n)| < 1$, $n \in \mathbb{N}$, and define the transfer matrix

$$T(z) = \begin{pmatrix} z & \alpha \\ \overline{\alpha}z & 1 \end{pmatrix}, \quad z \in \mathbb{T}, \tag{3.421}$$

with spectral parameter z on the unit circle $\mathbb{T} = \{z \in \mathbb{C} \mid |z| = 1\}$. Consider the system of difference equations

$$\Phi(z, n) = T(z, n)\Phi(z, n - 1), \quad (z, n) \in \mathbb{T} \times \mathbb{N}, \tag{3.422}$$

with initial condition $\Phi(z, 0) = \begin{pmatrix} 1 \\ 1 \end{pmatrix}$, where

$$\Phi(z, n) = \begin{pmatrix} \varphi(z, n) \\ z^n \overline{\varphi(1/\bar z, n)} \end{pmatrix}, \quad (z, n) \in \mathbb{T} \times \mathbb{N}_0. \tag{3.423}$$

Then $\varphi(\,\cdot\,, n)$ are monic polynomials of degree n first introduced by Szegő in the 1920s in his seminal work on the asymptotic distribution of eigenvalues of sections of Toeplitz forms, cf. Szegő (1920; 1921), and the monograph Szegő (1978, Ch. XI). Szegő's point of departure was the trigonometric moment problem and hence the theory of orthogonal polynomials on the unit circle. Indeed, given a probability measure $d\sigma$ supported on an infinite set on the unit circle, one is interested in finding monic polynomials $\chi(\,\cdot\,, n)$ of degree $n \in \mathbb{N}_0$ in $z = e^{i\theta}, \theta \in [0, 2\pi]$, such that

$$\int_0^{2\pi} d\sigma(e^{i\theta}) \, \overline{\chi(e^{i\theta}, m)} \chi(e^{i\theta}, n) = w(n)^{-2} \delta_{m,n}, \quad m, n \in \mathbb{N}_0,$$

where $w(0)^2 = 1$, $w(n)^2 = \prod_{j=1}^n \left(1 - |\alpha(j)|^2\right)^{-1}$, $n \in \mathbb{N}$. Szegő showed that the corresponding polynomials (3.423) with φ replaced by χ satisfy the recurrence formula (3.422). Early work in this area includes important contributions by Akhiezer, Geronimus, Krein, Tomčuk, Verblunsky, Widom, and others, and is summarized in the books by Akhiezer (1965), Geronimus (1961), Szegő (1978), and especially in the two-volume treatise by Simon (2005b,c).

An important extension of (3.422) was developed by Baxter in a series of papers on Toeplitz forms (Baxter (1960; 1961a,b; 1963)). In these papers the transfer matrix T in (3.421) is replaced by the more general (complexified) transfer matrix

$$U(z) = \begin{pmatrix} z & \alpha \\ \beta z & 1 \end{pmatrix}, \tag{3.424}$$

that is, precisely the matrix U responsible for the spatial part in the Ablowitz–Ladik system in its zero-curvature formulation (3.4), (3.48). Here the sequences $\alpha = \{\alpha(n)\}_{n \in \mathbb{N}}, \beta = \{\beta(n)\}_{n \in \mathbb{N}}$ are assumed to satisfy the restriction $\alpha(n)\beta(n) \neq 1$, $n \in \mathbb{N}$. Studying the following extension of (3.422),

$$\Psi(z, n) = U(z, n)\Psi(z, n-1), \quad (z, n) \in \mathbb{T} \times \mathbb{N},$$

Baxter was led to biorthogonal polynomials on the unit circle with respect to a complex-valued measure on $\mathbb{T}$. In this context of biorthogonal Laurent polynomials we also refer to Bertola and Gekhtman (2007) and Bultheel et al. (1999).

For recent discussions of the connection between the AL hierarchy and orthogonal polynomials we refer, for instance, to Bertola and Gekhtman (2007), Deift (2007), Killip and Nenciu (2006), Li (2005), Nenciu (2005a,b; 2006), Simon (2005a–c; 2007a), and the extensive literature cited therein.

3.9 Notes

Section 3.2. The material in this section is mostly taken from Gesztesy et al. (to appear).

The first systematic discussion of the Ablowitz–Ladik hierarchy appears to be due to Schilling (1989) (cf. also Tamizhmani and Ma (2000), Vekslerchik (2002), Zeng and Rauch-Wojciechowski (1995b)).

Baxter's U matrix in (3.424) led to a new hierarchy of nonlinear difference equations, called the Szegő–Baxter (SB) hierarchy in Geronimo et al. (2005), in honor of these two pioneers of orthogonal polynomials on the unit circle. The latter hierarchy is now seen to be a special case of the AL hierarchy as pointed out in Remark 3.9 (ii).

The literature on the special case of the discrete nonlinear Schrödinger hierarchy, isolated in Remark 3.13 (i), is too voluminous to be listed here in detail. So we refer, for instance, to Ablowitz et al. (2004, Ch. 3) and the references cited therein. However, we note that several of the references in these notes cited in connection with the AL hierarchy actually refer to the discrete nonlinear Schrödinger hierarchy. The hierarchy of Schur flows discussed in Remark 3.13 (ii), on the other hand, hardly appears to have been studied at all. Instead, only its first element, (3.57), received some attention recently (cf. Ammar and Gragg (1994), Faybusovich and Gekhtman (1999; 2000), Golinskii (2006), Mukaihira and Nakamura (2002), Simon (2007b)).

Connections between the AL hierarchy and the motion of a piecewise linear curve have been established by Doliwa and Santini (1995); Bäcklund and Darboux transformations were studied by Chiu and Ladik (1977), Chowdhury and Mahato (1983), Geng (1989), Rourke (2004), and Vekslerchik (2006); the Hirota bilinear formalism, AL τ-functions, etc., were considered by Sadakane (2003), Vekslerchik (1998; 2002); connections with the discrete isotropic Heisenberg magnet are studied in Hoffmann (2000); for an application of the inverse scattering method to (3.420) we refer to Ablowitz et al. (2004, Ch. 3), Ablowitz et al. (2007), and Vekslerchik and Konotop (1992). The continuum limit of the AL hierarchy is studied in Zeng and Rauch-Wojciechowski (1995b).

Finally, we note that Remark 3.11 can be shown as in Gesztesy and Holden (2003b, Remark I.1.5).

Section 3.3. The presentation of this section closely follows some parts in Gesztesy et al. (2008a).

The half-lattice (i.e., semi-infinite) $\ell^2(\mathbb{N})$-operator realization of the difference expression L as a five-diagonal matrix was recently rediscovered by Cantero, Moral and Velazquez, see Cantero et al. (2003) (see also Cantero et al. (2005)) in their study of orthogonal polynomials on the unit circle in the special defocusing case, where $\beta = \overline{\alpha}$. Subsequently, the term CMV matrix was coined by Simon (2005b,c).

The actual history of CMV operators, however, is more intricate: The corresponding unitary semi-infinite five-diagonal matrices were first introduced by Bunse-Gerstner and Elsner (1991), and subsequently treated in detail by Watkins (1993) (cf. the recent discussion in Simon (2007a)) before they were subsequently rediscovered by Cantero et al. (2003). We also note that in a context different from orthogonal polynomials on the unit circle, Bourget et al. (2003) introduced a family of doubly infinite matrices with three sets of parameters which, for special choices of the parameters, reduces to two-sided CMV matrices on $\mathbb{Z}$. Moreover, it is possible to connect unitary block Jacobi matrices to the trigonometric moment problem (and hence to CMV matrices) as discussed by Berezansky and Dudkin (2005; 2006).

The Ablowitz–Ladik Lax pair in the special defocusing case, where $\beta = \overline{\alpha}$, in the finite-dimensional context, was recently discussed by Nenciu (2005a,b; 2006).

Finally, we note that the result of Lemma 3.14 (i.e., (3.71)) can be rewritten as follows:

Lemma 3.66 *Let* $n \in \mathbb{Z}$. *Then the homogeneous coefficients* $\{\hat{f}_{\ell,\pm}\}_{\ell \in \mathbb{N}_0}$, $\{\hat{g}_{\ell,\pm}\}_{\ell \in \mathbb{N}_0}$, *and* $\{\hat{h}_{\ell,\pm}\}_{\ell \in \mathbb{N}_0}$ *satisfy the following relations:*

$$\hat{f}_{\ell,+}(n) = \alpha(n)(\delta_n, L^{\ell+1}\delta_n) + \rho(n)(\delta_{n-1}, L^{\ell+1}\delta_n)\delta_{\text{even}}(n)$$
$$+ \rho(n)(\delta_n, L^{\ell+1}\delta_{n-1})\delta_{\text{odd}}(n), \quad \ell \in \mathbb{N}_0,$$

$$\hat{f}_{\ell,-}(n) = \alpha(n)(\delta_n, L^{-\ell}\delta_n) + \rho(n)(\delta_{n-1}, L^{-\ell}\delta_n)\delta_{\text{even}}(n)$$
$$+ \rho(n)(\delta_n, L^{-\ell}\delta_{n-1})\delta_{\text{odd}}(n), \quad \ell \in \mathbb{N}_0,$$

$$\hat{g}_{0,\pm} = \frac{1}{2}, \quad \hat{g}_{\ell,\pm}(n) = (\delta_n, L^{\pm\ell}\delta_n), \quad \ell \in \mathbb{N},$$

$$\hat{h}_{\ell,+}(n) = \beta(n)(\delta_n, L^{\ell}\delta_n) + \rho(n)(\delta_n, L^{\ell}\delta_{n-1})\delta_{\text{even}}(n)$$
$$+ \rho(n)(\delta_{n-1}, L^{\ell}\delta_n)\delta_{\text{odd}}(n), \quad \ell \in \mathbb{N}_0,$$

$$\hat{h}_{\ell,-}(n) = \beta(n)(\delta_n, L^{-\ell-1}\delta_n) + \rho(n)(\delta_n, L^{-\ell-1}\delta_{n-1})\delta_{\text{even}}(n)$$
$$+ \rho(n)(\delta_{n-1}, L^{-\ell-1}\delta_n)\delta_{\text{odd}}(n), \quad \ell \in \mathbb{N}_0.$$

Section 3.4. The material in this section is predominantly taken from Gesztesy et al. (2007a).

Since algebro-geometric solutions of the AL hierarchy are usually studied in a time-dependent context, we postpone listing pertinent references to the notes of Section 3.6.

In the defocusing case $\beta = \overline{\alpha}$, finite-arc spectral theoretic investigations of the unitary operator realization of L on the unit circle $\mathbb{T}$ were undertaken by Geronimo and Johnson (1998) in the case where the coefficients α are random variables. Under appropriate ergodicity assumptions on α and the hypothesis of a vanishing Lyapunov exponent on prescribed spectral arcs on the unit circle, Geronimo and Johnson (1996; 1998) and Geronimo and Teplyaev (1994) developed the corresponding spectral theory of the unitary operator realization of L in $\ell^2(\mathbb{Z})$. In this sense the discussion in Geronimo and Johnson (1998) is a purely stationary one and connections with a zero-curvature formalism, theta function representations, and integrable hierarchies are not made in Geronimo and Johnson (1998). More recently, the defocusing case with periodic and quasi-periodic coefficients was also studied in great detail by Bogolyubov and Prikarpatskii (1982), Deift (2007), Golinskii and Nevai (2001), Killip and Nenciu (2006), Nenciu (2005a; 2006), and Simon (2004b; 2005a,c; 2007a).

The function ϕ introduced in (3.95), which is again the fundamental object of study as everywhere else in this volume and in Volume I, is closely related to one of the variants of Weyl–Titchmarsh functions discussed in Gesztesy and Zinchenko (2006a,b), Simon (2004a) in the special defocusing case $\beta = \bar{\alpha}$ (see also, Clark et al. (2007; 2008)).

For a discussion of orthogonal polynomials on several arcs of the unit circle relevant to this section we also refer to Lukashov (2004) and the detailed list of references therein.

Section 3.5. The material of this section is taken from Gesztesy et al. (2007b).

For a pertinent discussion of Sard's theorem as needed in the proof of Lemma 3.29 we refer, for instance, to Abraham et al. (1988, Sect. 3.6).

We emphasize that the approach described in this section is not limited to the Ablowitz–Ladik hierarchy, but applies universally to the construction of stationary algebro-geometric solutions of integrable lattice hierarchies of soliton equations. In particular, it was applied to the Toda lattice hierarchy as discussed in Gesztesy et al. (2008b) and in Section 1.4 of this monograph.

We also note that while the periodic case with complex-valued α, β is of course included in our analysis, we consider throughout the more general algebro-geometric case (in which α, β need not be quasi-periodic).

Section 3.6. This section is primarily taken from Gesztesy et al. (2007a).

The first systematic and detailed treatment of algebro-geometric solutions of the AL system (3.1) was performed by Miller et al. (1995) (see also Miller (1994)) in an effort to analyze models describing oscillations in nonlinear dispersive wave systems. Related material can be found in Ahmad and Chowdhury (1987a,b), Bogolyubov and Prikarpatskii (1982) (see also Bogolyubov et al. (1981; 1982)), Chow et al. (2006), Miller (1995), and Vaninsky (2001). Algebro-geometric solutions of the AL hierarchy were also discussed in Geng et al. (2003; 2007) and Vekslerchik (1999) (by employing methods different from those in Gesztesy et al. (2007a)).

Solutions of the AL system in terms of elliptic functions (and some of their degenerations into soliton solutions) were discussed in Huang and Liu (2008).

Section 3.7. This section is based on our paper Gesztesy et al. (2007b).

Next, we mention a few references relevant at some more technical points: The fact that there exists a $T_0 > 0$, such that the first-order autonomous initial value problem (3.281), (3.287), (3.288) (with polynomial right-hand sides) and initial conditions (3.289) has a unique solution for all $t_{\underline{r}} \in (t_{0,\underline{r}} - T_0, t_{0,\underline{r}} + T_0)$, follows, for example, from Walter (1998, Sect. III.10).

In connection with the paragraph following Theorem 3.45, relevant references illustrating the fact that straight motions on the torus are generically dense are, for instance, Arnold (1989, Sect. 51) or Katok and Hasselblatt (1995, Sects. 1.4, 1.5). In addition, for the definition of quasi-periodic functions, one can consult, for example, Pastur and Figotin (1992, p. 31).

We emphasize that the approach described in this section is not limited to the Ablowitz–Ladik hierarchy but applies universally to constructing algebro-geometric solutions of $(1+1)$-dimensional integrable soliton equations. In particular, it applies to the Toda lattice hierarchy as discussed in Gesztesy et al. (2008b) and in Chapter 1 of this monograph. Moreover, the principal idea of replacing Dubrovin-type equations by a first-order system of the type (3.281), (3.287), and (3.288) is also relevant in the context of general non-normal Lax operators for the continuous models in $(1+1)$-dimensions. In particular, the models studied in detail in Gesztesy and Holden (2003b) can be revisited from this point of view. However, the fact that the set in (3.317) is of measure zero relies on the fact that n varies in the countable set $\mathbb{Z}$ and hence is not applicable to continuous models in $(1+1)$-dimensions.

We also note that while the periodic case with complex-valued α, β is of course included in our analysis, we consider throughout the more general algebro-geometric case (in which α, β need not be quasi-periodic).

Although the following is not treated in this volume, the interested reader might want to notice that the initial value problem for half-infinite discrete linear Schrödinger equations and the Schur flow were discussed by Common (1992) (see also Common and Hafez (1990)) using a continued fraction approach. The corresponding nonabelian cases on a finite interval were studied by Gekhtman (1993).

Section 3.8. This section is based on some of the material in Gesztesy et al. (2008a).

Infinitely many conservation laws are discussed, for instance, in Ablowitz and Ladik (1976b), Ablowitz et al. (2004, Ch. 3), Ding et al. (2006), Zhang and Chen (2002a), and Zhang et al. (2006); the bi-Hamiltonian structure of the AL hierarchy

3.9 Notes

is considered by Ercolani and Lozano (2006), Hydon (2005), and Lozano (2004); multi-Hamiltonian structures for the defocusing AL hierarchy were studied by Faybusovich and Gekhtman (2000), Gekhtman and Nenciu (to appear), Zeng and Rauch-Wojciechowski (1995b), and Zhang and Chen (2002b); Poisson brackets for orthogonal polynomials on the unit circle relevant to the case of the defocusing AL hierarchy (where $\beta = \overline{\alpha}$) have been studied by Cantero and Simon (to appear), Killip and Nenciu (2006), and Nenciu (2007); Lenard recursions and the Hamiltonian formalism were discussed in Geng and Dai (2007), Geng et al. (2007), Suris (2003), Tang et al. (2007), and Vekslerchik (1993). Quantum aspects of the AL equation are discussed, for instance, in Enolskii et al. (1992).

In connection with Theorem 3.56 we note that Zhang and Chen (2002a) study local conservation laws for the full 4×4 Ablowitz–Ladik system in a similar way to the one employed here. However, they only expand their equation around a point that corresponds to $1/z = 0$.

Next we mention some references supporting more technical points in this section: Equations (3.329)–(3.332) are discussed in Gesztesy and Zinchenko (2006b) in the special defocusing case, where $\beta = \overline{\alpha}$. While (3.329)–(3.331) are of an algebraic nature and hence immediately extend to the case where $\beta \neq \overline{\alpha}$, the existence of solutions $u_\pm$ and $v_\pm$ satisfying $u_\pm, v_\pm \in \ell^2([n_0, \pm\infty) \cap \mathbb{Z})$ can be inferred from applying the resolvents of $\check{L}$ and $\check{L}^\top$ to the element δ_{n_0},

$$u_\pm(n) = c_\pm\big((\check{L} - z)^{-1}\delta_{n_0}\big)(n), \quad n \in [n_0, \pm\infty) \cap \mathbb{Z},$$
$$v_\pm(n) = d_\pm\big((\check{L}^\top - z)^{-1}\delta_{n_0}\big)(n), \quad n \in [n_0, \pm\infty) \cap \mathbb{Z},$$

for some constants $c_\pm, d_\pm \in \mathbb{C} \setminus \{0\}$.

The existence of the propagator $W_{\underline{p}}(\,\cdot\,,\,\cdot\,)$ satisfying equations (3.384)–(3.387) is a standard result which follows, for instance, from Theorem X.69 of Reed and Simon (1975) (under even weaker hypotheses on α, β).

Local existence and uniqueness as well as smoothness of the solution of the initial value problem (3.390) (cf. (3.54)) in Theorem 3.54 follows from Abraham et al. (1988, Theorem 4.1.5) since $f_{p_\pm-1,\pm}$, $g_{p_\pm,\pm}$, and $h_{p_\pm-1,\pm}$ depend polynomially on α, β and certain of their shifts, and the fact that the Ablowitz–Ladik flows are autonomous.

An application of Abraham et al. (1988, Proposition 4.1.22) yields the unique, global, and smooth solution of the $\underline{p}$th AL initial value problem [4] (3.390) in Remark 3.55, and the same argument also shows that if α satisfies Hypothesis 3.50 and the $\underline{p}$th AL equation $\mathrm{AL}_{\underline{p}}(\alpha, \overline{\alpha}) = 0$, then α is actually smooth with respect to $t_{\underline{p}} \in \mathbb{R}$ and satisfies (3.391).

The Jacobi identity (3.400) and the Leibniz rule (3.401) follow since $\Omega(\cdot,\cdot)$ is a weakly nondegenerate closed 2-form as discussed in Kriegl and Michor (1997, Theorem 48.8).

Finally, we briefly add some comments further illustrating the notion of conserved densities and local conservation laws: Consider sequences

$$\{\alpha(n,t),\beta(n,t)\}_{n\in\mathbb{Z}}\in\ell^1(\mathbb{Z})$$

(satisfying some additional assumptions as in Section 3.8), parametrized by the deformation (time) parameter $t\in\mathbb{R}$, that are solutions of the Ablowitz–Ladik equations

$$\begin{pmatrix}-i\alpha_t-(1-\alpha\beta)(\alpha^-+\alpha^+)+2\alpha\\-i\beta_t+(1-\alpha\beta)(\beta^-+\beta^+)-2\beta\end{pmatrix}=0.$$

Then clearly

$$\partial_t\sum_{n\in\mathbb{Z}}\alpha^+(n,t)\beta(n,t)=\partial_t\sum_{n\in\mathbb{Z}}\alpha(n,t)\beta^+(n,t)=0.$$

Additional calculations also show that

$$\partial_t\sum_{n\in\mathbb{Z}}\left(\tfrac{1}{2}(\alpha^+(n,t)\beta(n,t))^2-\gamma(n,t)\alpha^+(n,t)\beta^-(n,t)\right)=0,$$

$$\partial_t\sum_{n\in\mathbb{Z}}\left(\tfrac{1}{2}(\alpha(n,t)\beta^+(n,t))^2-\gamma^+(n,t)\alpha(n,t)\beta^{++}(n,t)\right)=0.$$

Indeed, as we proved in Section 3.8, there exists an infinite sequence $\{\rho_{j,\pm}\}_{j\in\mathbb{N}}$ of polynomials of α,β and certain shifts thereof, with the property that the lattice sum is time-independent,

$$\partial_t\sum_{n\in\mathbb{Z}}\rho_{j,\pm}(n,t)=0,\quad j\in\mathbb{N}.$$

This result is obtained by deriving local conservation laws of the type

$$\partial_t\rho_{j,\pm}+(S^+-I)J_{j,\pm}=0,\quad j\in\mathbb{N},$$

for certain polynomials $J_{j,\pm}$ of α,β and certain shifts thereof. The polynomials $J_{j,\pm}$ have been constructed in Section 3.8 via an explicit recursion relation.

The above analysis extends to the full Ablowitz–Ladik hierarchy as follows: Given $\underline{p}=(p_-,p_+)\in\mathbb{N}_0^2$ and the $\underline{p}$th AL equation $\mathrm{AL}_{\underline{p}}(\alpha,\beta)=0$, the associated conserved densities $\rho_{j,\pm}$ are independent of the equation in the hierarchy while the currents $J_{\underline{p},j,\pm}$ depend on $\underline{p}$ and one finds (cf. Theorem 3.56)

$$\partial_{t_{\underline{p}}}\rho_{j,\pm}+(S^+-I)J_{\underline{p},j,\pm}=0,\quad j\in\mathbb{N},\ \underline{p}\in\mathbb{N}_0^2.$$

For $\alpha, \beta \in \ell^1(\mathbb{Z})$ it then follows that

$$\frac{d}{dt_{\underline{p}}} \sum_{n \in \mathbb{Z}} \rho_{j,\pm}(n, t_{\underline{p}}) = 0, \quad t_{\underline{p}} \in \mathbb{R}, \ j \in \mathbb{N}, \ \underline{p} \in \mathbb{N}_0^2.$$

By showing that $\rho_{j,\pm}$ equals $\hat{g}_{j,\pm}$ up to a first-order difference expression (cf. Lemma 3.49), and by investigating the time-dependence of $\gamma = 1 - \alpha\beta$, one concludes (cf. Remark 3.57) that

$$\frac{d}{dt_{\underline{p}}} \sum_{n \in \mathbb{Z}} \ln(\gamma(n, t_{\underline{p}})) = 0, \quad \frac{d}{dt_{\underline{p}}} \sum_{n \in \mathbb{Z}} \hat{g}_{j,\pm}(n, t_{\underline{p}}) = 0, \quad t_{\underline{p}} \in \mathbb{R}, \ j \in \mathbb{N}, \ \underline{p} \in \mathbb{N}_0^2,$$

represent the two infinite sequences of AL conservation laws.

We emphasize that our recursive and systematic approach to local conservation laws of the Ablowitz–Ladik hierarchy appears to be new. Moreover, our treatment of Poisson brackets and variational derivatives, and their connections with the diagonal Green's function of the underlying Lax operator, now puts the AL hierarchy on precisely the same level as the Toda and KdV hierarchy with respect to these particular aspects of the Hamiltonian formalism (cf. Gesztesy and Holden (2003b, Ch. 1)).

Appendix A
Algebraic Curves and Their Theta Functions in a Nutshell

> You know my methods. Apply them.
> *Sherlock Holmes*[1]

This appendix treats some of the basic aspects of complex algebraic curves and their theta functions as used at various places in this monograph. The material below is standard, and we include it for two major reasons: On the one hand it allows us to summarize a variety of facts and explicit formulas needed in connection with the construction of algebro-geometric solutions of completely integrable equations, and, on the other hand, it will simultaneously enable us to introduce a large part of the notation used throughout this volume. We emphasize that the summary presented in this appendix is not intended as a substitute for textbook consultations. Relevant literature in this context is mentioned in the notes to this appendix.

Definition A.1 *An affine plane (complex) algebraic curve $\mathcal{K}$ is the locus of zeros in $\mathbb{C}^2$ of a (nonconstant) polynomial $\mathcal{F}$ in two variables. The polynomial $\mathcal{F}$ is called nonsingular at a root (z_0, y_0) if*

$$\nabla \mathcal{F}(z_0, y_0) = (\mathcal{F}_z(z_0, y_0), \mathcal{F}_y(z_0, y_0)) \neq 0.$$

The affine plane curve $\mathcal{K}$ of roots of $\mathcal{F}$ is called nonsingular at $P_0 = (z_0, y_0)$ if $\mathcal{F}$ is nonsingular at P_0. The curve $\mathcal{K}$ is called nonsingular, or smooth, if it is nonsingular at each of its points (otherwise, it is called singular).

The implicit function theorem allows one to conclude that a smooth affine curve $\mathcal{K}$ is locally a graph and to introduce complex charts on $\mathcal{K}$ as follows. If $\mathcal{F}(P_0) = 0$ with $\mathcal{F}_y(P_0) \neq 0$, there is a holomorphic function g_{P_0} such that in a neighborhood U_{P_0} of P_0, the curve $\mathcal{K}$ is characterized by the graph $y = g_{P_0}(z)$. Hence the projection

$$\tilde{\pi}_z \colon U_{P_0} \to \tilde{\pi}_z(U_{P_0}) \subset \mathbb{C}, \quad (z, y) \mapsto z, \tag{A.1}$$

---
[1] In Sir Arthur Conan Doyle, *The Hound of the Baskervilles*, The Strand Magazine, vol. XXII, 1901, p. 4.

yields a complex chart on $\mathcal{K}$. If, on the other hand, $\mathcal{F}(P_0) = 0$ with $\mathcal{F}_z(P_0) \neq 0$, then the projection

$$\tilde{\pi}_y \colon U_{P_0} \to \tilde{\pi}_y(U_{P_0}) \subset \mathbb{C}, \quad (z, y) \mapsto y, \tag{A.2}$$

defines a chart on $\mathcal{K}$. In this way, as long as $\mathcal{K}$ is nonsingular, one arrives at a complex atlas on $\mathcal{K}$. The space $\mathcal{K} \subset \mathbb{C}^2$ is second countable and Hausdorff. In order to obtain a Riemann surface one needs connectedness of $\mathcal{K}$ which is implied by adding the assumption of irreducibility[1] of the polynomial $\mathcal{F}$. Thus, $\mathcal{K}$ equipped with charts (A.1) and (A.2) is a Riemann surface if $\mathcal{F}$ is nonsingular and irreducible. Affine plane curves $\mathcal{K}$ are unbounded as subsets of $\mathbb{C}^2$, and hence noncompact. The compactification of $\mathcal{K}$ is conveniently described in terms of the projective plane $\mathbb{CP}^2$, the set of all one-dimensional (complex) subspaces of $\mathbb{C}^3$.

To simplify notations, we temporarily abbreviate $x_1 = y$ and $x_2 = z$. Moreover, we denote the linear span of $(x_2, x_1, x_0) \in \mathbb{C}^3 \setminus \{0\}$ by $[x_2 : x_1 : x_0]$. Since the homogeneous coordinates $[x_2 : x_1 : x_0]$ satisfy

$$[x_2 : x_1 : x_0] = [cx_2 : cx_1 : cx_0], \quad c \in \mathbb{C} \setminus \{0\},$$

the space $\mathbb{CP}^2$ can be viewed as the quotient space of $\mathbb{C}^3 \setminus \{0\}$ by the multiplicative action of $\mathbb{C} \setminus \{0\}$, that is, $\mathbb{CP}^2 = (\mathbb{C}^3 \setminus \{0\})/(\mathbb{C} \setminus \{0\})$, and hence $\mathbb{CP}^2$ inherits a Hausdorff topology which is the quotient topology induced by the natural map

$$\iota \colon \mathbb{C}^3 \setminus \{0\} \to \mathbb{CP}^2, \quad (x_2, x_1, x_0) \mapsto [x_2 : x_1 : x_0].$$

Next, define the open sets

$$U^m = \{[x_2 : x_1 : x_0] \in \mathbb{CP}^2 \mid x_m \neq 0\}, \quad m = 0, 1, 2.$$

Then

$$f^0 \colon U^0 \to \mathbb{C}^2, \quad [x_2 : x_1 : x_0] \mapsto (x_2/x_0, x_1/x_0)$$

with inverse

$$(f^0)^{-1} \colon \mathbb{C}^2 \to U^0, \quad (x_2, x_1) \mapsto [x_2 : x_1 : 1],$$

and analogously for functions f^1 and f^2 (relative to sets U^1 and U^2, respectively), are homeomorphisms. In particular, U^0, U^1, and U^2 together cover $\mathbb{CP}^2$. Thus, $\mathbb{CP}^2$ is compact since it is covered by the closed unit (poly)disks in U^0, U^1, and U^2. The element $[x_2 : x_1 : 0] \in \mathbb{CP}^2$ represents the point at infinity along the direction $x_2 : x_1$ in $\mathbb{C}^2$ (identifying $[x_2 : x_1 : 0] \in \mathbb{CP}^2$ and $[x_2 : x_1] \in \mathbb{CP}^1$). The set of all such elements then represents the line at infinity, $L_\infty = \{[x_2 : x_1 : x_0] \in \mathbb{CP}^2 \mid x_0 = 0\}$,

---

[1] The polynomial $\mathcal{F}$ in two variables is called irreducible if it cannot be factored into $\mathcal{F} = \mathcal{F}_1 \mathcal{F}_2$ with $\mathcal{F}_1$ and $\mathcal{F}_2$ both nonconstant polynomials in two variables.

and yields the compactification $\mathbb{CP}^2$ of $\mathbb{C}^2$. In other words, $\mathbb{CP}^2 \cong \mathbb{C}^2 \cup L_\infty$, $\mathbb{CP}^1 \cong \mathbb{C}_\infty$, and $L_\infty \cong \mathbb{CP}^1$.

Let $\mathcal{P}$ be a (nonconstant) homogeneous polynomial of degree d in (x_2, x_1, x_0), that is,

$$\mathcal{P}(cx_2, cx_1, cx_0) = c^d \mathcal{P}(x_2, x_1, x_0),$$

and introduce

$$\overline{\mathcal{K}} = \{[x_2 : x_1 : x_0] \in \mathbb{CP}^2 \mid \mathcal{P}(x_2, x_1, x_0) = 0\}.$$

The set $\overline{\mathcal{K}}$ is well-defined (even though $\mathcal{P}(u, v, w)$ is not for $[u : v : w] \in \mathbb{CP}^2$) and closed in $\mathbb{CP}^2$. The intersections,

$$\mathcal{K}^m = \overline{\mathcal{K}} \cap U^m, \quad m = 0, 1, 2,$$

are affine plane curves when transported to $\mathbb{C}^2$. In particular,

$$\mathcal{K}^0 \cong \{(x_2, x_1) \in \mathbb{C}^2 \mid \mathcal{P}(x_2, x_1, 1) = 0\}$$

represents the affine curve $\mathcal{K}$ defined by $\mathcal{F}(z, y) = 0$, where $\mathcal{F}(x_2, x_1) = \mathcal{P}(x_2, x_1, 1)$, that is, $\mathcal{K}^0$ represents the affine part of $\overline{\mathcal{K}}$. ($\mathcal{F}$ has degree d provided x_0 is not a factor of $\mathcal{P}$, i.e., provided $\overline{\mathcal{K}}$ does not contain the projective line L_∞.)

Conversely, given the affine curve $\mathcal{K}$ defined by

$$\mathcal{F}(x_2, x_1) = \sum_{\substack{r,s=0 \\ r+s \leq d}}^{d} a_{r,s} z^r y^s = 0,$$

with $\mathcal{F}$ of degree d, the associated homogeneous polynomial $\mathcal{P}$ of degree d can be obtained from

$$\mathcal{P}(x_2, x_1, x_0) = x_0^d \mathcal{F}(x_2/x_0, x_1/x_0).$$

The affine curve $\mathcal{K}$ is then the intersection of the projective curve $\overline{\mathcal{K}}$ defined by $\mathcal{P}(x_2, x_1, x_0) = 0$ with U^0, that is, $\mathcal{K} \cong \overline{\mathcal{K}} \cap U^0 = \mathcal{K}^0$. The intersection of $\overline{\mathcal{K}}$ with L_∞, the line at infinity, then consists of the finite set of points

$$\overline{\mathcal{K}} \setminus \mathcal{K} = \left\{ [x_2 : x_1 : 0] \in \mathbb{CP}^2 \,\middle|\, \sum_{r=0}^{d} a_{r,d-r} z^r y^{d-r} = 0 \right\}.$$

Definition A.2 *A projective plane (complex) algebraic curve $\overline{\mathcal{K}}$ is the locus of zeros in $\mathbb{CP}^2$ of a homogeneous polynomial $\mathcal{P}$ in three variables. A homogeneous (nonconstant) polynomial $\mathcal{P}$ in three variables is called nonsingular if there are no common*

Algebraic Curves and Their Theta Functions in a Nutshell

solutions $(x_{2,0}, x_{1,0}, x_{0,0}) \in \mathbb{C}^3 \setminus \{0\}$ of

$$\mathcal{P}(x_{2,0}, x_{1,0}, x_{0,0}) = 0,$$
$$\nabla \mathcal{P}(x_{2,0}, x_{1,0}, x_{0,0}) = (\mathcal{P}_{x_2}, \mathcal{P}_{x_1}, \mathcal{P}_{x_0})(x_{2,0}, x_{1,0}, x_{0,0}) = 0.$$

The set $\overline{\mathcal{K}}$ is called a smooth (or nonsingular) projective plane curve (of degree $d \in \mathbb{N}$) if $\mathcal{P}$ is nonsingular (and of degree $d \in \mathbb{N}$).

If $x_{0,0} \neq 0$, then $[x_{2,0} : x_{1,0} : x_{0,0}] \in \mathbb{CP}^2$ is a nonsingular point of the projective curve $\overline{\mathcal{K}}$ (defined by $\mathcal{P}(x_2, x_1, x_0) = 0$) if and only if $(x_{2,0}/x_{0,0}, x_{1,0}/x_{0,0}) \in \mathbb{C}^2$ is a nonsingular point of the affine curve $\mathcal{K}$ (defined by $\mathcal{P}(x_2, x_1, 1) = 0$).

One verifies that the homogeneous polynomial $\mathcal{P}$ is nonsingular if and only if each $\mathcal{K}^m$ is a smooth affine plane curve in $\mathbb{C}^2$. Moreover, any nonsingular homogeneous polynomial $\mathcal{P}$ is irreducible and consequently each $\mathcal{K}^m$ is a Riemann surface for $m = 0, 1, 2$. The coordinate charts on each $\mathcal{K}^m$ are simply the projections, that is, x_2/x_0 and x_1/x_0 for $\mathcal{K}^0$, x_2/x_1 and x_0/x_1 for $\mathcal{K}^1$, and finally, x_1/x_2 and x_0/x_2 for $\mathcal{K}^2$. These separate complex structures on $\mathcal{K}^m$ are compatible on $\overline{\mathcal{K}}$ and hence induce a complex structure on $\overline{\mathcal{K}}$.

The zero locus of a nonsingular homogeneous polynomial $\mathcal{P}(x_2, x_1, x_0)$ in $\mathbb{CP}^2$ defines a smooth projective plane curve $\overline{\mathcal{K}}$ which is a compact Riemann surface. Topologically, this Riemann surface is a sphere with g handles where

$$g = (d-1)(d-2)/2, \tag{A.3}$$

with d the degree of $\mathcal{P}$. In particular, $\overline{\mathcal{K}}$ has topological genus g and we indicate this by writing $\overline{\mathcal{K}}_g$. However, for notational convenience we shall use the symbol $\mathcal{K}_g$ instead (i.e., $\mathcal{K}_g$ always denotes the corresponding compact Riemann surface). In general, the projective curve $\mathcal{K}_g$ can be singular even though the associated affine curve $\mathcal{K}_g^0$ is nonsingular. In this case one has to account for the singularities at infinity and properly amend the genus formula (A.3) according to results of Clebsch, M. Noether, and Plücker.

Next, let $\mathcal{K}_g$ be a smooth projective curve not containing the point $[0, 1, 0]$ associated with the homogeneous polynomial $\mathcal{P}$ of degree d. Then

$$\mathcal{P}(z, y, 1) = 0$$

defines y as a multi-valued function of z such that away from ramification points there correspond precisely d values of y to each value of $z \in \mathbb{C}$. The set of finite ramification points of $\mathcal{K}_g$ is given by

$$\{[z : y : 1] \in \mathbb{CP}^2 \mid \mathcal{P}(z, y, 1) = \mathcal{P}_y(z, y, 1) = 0\}. \tag{A.4}$$

Similarly, ramification points at infinity are defined by

$$\{[1 : y : 0] \in \mathbb{CP}^2 \mid \mathcal{P}(1, y, 0) = \mathcal{P}_y(1, y, 0) = 0\}. \tag{A.5}$$

The set of ramification points of $\mathcal{K}_g$ is then the union of points in (A.4) and (A.5). Given the set of ramification points $\{P_1, \ldots, P_r\}$ one can cut the complex plane along smooth nonintersecting arcs $\mathcal{C}_q$ (e.g., straight lines if $P_1, \ldots, P_r$ are suitably situated) connecting P_q and P_{q+1} for $q = 1, \ldots, r-1$, and define holomorphic functions $f_1, \ldots, f_d$ on the cut plane $\Pi = \mathbb{C} \setminus \bigcup_{q=1}^{r-1} \mathcal{C}_q$ such that

$$\mathcal{P}(z, y, 1) = 0 \text{ for } y \in \Pi \text{ if and only if } y = f_j(z) \text{ for some } j \in \{1, \ldots, d\}.$$

This yields a topological construction of $\mathcal{K}_g$ by appropriately gluing together d copies of the cut plane Π, the result being a sphere with g handles (g depending on the order of the ramification points). If $\mathcal{K}_g$ is singular, this procedure requires appropriate modifications.

There is an alternative description of $\mathcal{K}_g$ as a (branched) covering surface of the Riemann sphere $\mathcal{K}_0 = \mathbb{C}_\infty$, which naturally leads to the notion of branch points. To begin with, we briefly consider the case of a general (not necessarily compact) Riemann surface. Starting from a (real) two-dimensional connected C^0-manifold[1] $(\mathcal{M}, \mathcal{A} = (U_\alpha, z_\alpha)_{\alpha \in I})$ and a nonconstant map $F: \mathcal{M} \to \mathbb{C}_\infty$ such that

$$F \circ z_\alpha^{-1}: z_\alpha(U_\alpha) \to \mathbb{C}_\infty \text{ is nonconstant and holomorphic for all } \alpha \in I, \quad \text{(A.6)}$$

one defines a maximal atlas $\mathcal{A}(F)$ compatible with (A.6). The Riemann surface $\mathcal{R}_F = (\mathcal{M}, \mathcal{A}(F))$ is then a covering surface[2] of $\mathbb{C}_\infty$ and branch points on $\mathcal{R}_F$ are identified with those of F. More precisely, if $z = F(P)$ has a k-fold z_0-point[3] at $P = P_0$, $P_0 \in \mathcal{R}_F$ for some $k \in \mathbb{N}$, then P_0 is called unbranched (unramified) for $k = 1$ and has a branch point (respectively ramification point)[4] of order $k - 1$ (respectively k) for $k \geq 2$. The set of branch points of $\mathcal{R}_F$ will be denoted by $\mathcal{B}(\mathcal{R}_F)$. Depending on the branching behavior of $P \in \mathcal{R}_F$, one then introduces the following system of charts on $\mathcal{R}_F$.

---

[1] $(\mathcal{M}, \tau)$ is a (real) two-dimensional connected C^0-manifold if $(\mathcal{M}, \tau)$ is a second countable connected Hausdorff topological space with topology τ, $\mathcal{M} = \bigcup_{\alpha \in I} U_\alpha$, $U_\alpha \in \tau$, $z_\alpha: U_\alpha \to \mathbb{C}$ are homeomorphisms, $z_\alpha(U_\alpha)$ is open in $\mathbb{C}$, $\alpha \in I$ (an index set), and $\mathcal{A}$ is a maximal C^0-atlas on $\mathcal{M}$.

[2] In this monograph we only deal with covering surfaces of the Riemann sphere $\mathcal{K}_0 = \mathbb{C}_\infty$. The study of special elliptic algebro-geometric solutions of integrable hierarchies, however, is most naturally connected with covers of the torus $\mathcal{K}_1$.

[3] F has a k-fold z_0-point at $P = P_0$, if for some chart (U_{P_0}, ζ_{P_0}) on $\mathcal{R}_F$ at P_0 with $\zeta_{P_0}(P_0) = 0$ and some chart (V_{P_0}, w_{P_0}) on $\mathbb{C}_\infty$ at $z_0 = F(P_0)$ with $w_{P_0}(F(P_0)) = 0$, $(w_{P_0} \circ F \circ \zeta_{P_0}^{-1})(\zeta) = \zeta^k$ for all $\zeta \in \zeta_{P_0}(U_{P_0})$. This includes, of course, the possibility that $z_0 = \infty$.

[4] This definition of branch points is not universally adopted. Many monographs distinguish ramification and branch points in the sense that a branch point is the image of a ramification point under the covering map. In this monograph we found it convenient to follow the convention used in Farkas and Kra (1992, Sect. I.2).

(i) $F(P_0) = z_0 \in \mathbb{C}$: One defines for appropriate $C_0 > 0$

$$U_{P_0} = \{P \in \mathcal{R}_F \mid |z - z_0| < C_0\}, \quad V_{P_0} = \{\zeta \in \mathbb{C} \mid |\zeta| < C_0^{1/k}\},$$
$$\zeta_{P_0} : U_{P_0} \to V_{P_0}, \quad P \mapsto (z - z_0)^{1/k},$$
$$\zeta_{P_0}^{-1} : V_{P_0} \to U_{P_0}, \quad \zeta \mapsto z_0 + \zeta^k.$$

(ii) $F(P_0) = z_0 = \infty$: One defines for appropriate $C_\infty > 0$

$$U_{P_0} = \{P \in \mathcal{R}_F \mid |z| > C_\infty\}, \quad V_{P_0} = \{\zeta \in \mathbb{C} \mid |\zeta| < C_\infty^{-1/k}\},$$
$$\zeta_{P_0} : U_{P_0} \to V_{P_0}, \quad P \mapsto z^{-1/k},$$
$$\zeta_{P_0}^{-1} : V_{P_0} \to U_{P_0}, \quad \zeta \mapsto \zeta^{-k}.$$

Next, consider an open nonempty and connected subset $\mathcal{S}$ of $\mathcal{R}_F$ over $\mathbb{C}_\infty$ such that F is univalent in $\mathcal{S}$, that is, F is analytic in $\mathcal{S}$ and takes distinct values at distinct points of $\mathcal{S}$ (thus, F maps $\mathcal{S}$ onto a subset $F(\mathcal{S})$ of $\mathbb{C}_\infty$ in a one-to-one fashion). If $\mathcal{S}$ is maximal with respect to this property (i.e., $\mathcal{S}$ cannot be extended to $\widetilde{\mathcal{S}} \supsetneq \mathcal{S}$ with F univalent on $\widetilde{\mathcal{S}}$), $\mathcal{S}$ is called a sheet of $\mathcal{R}_F$. In this manner $\mathcal{R}_F$ can be pictured as consisting of finitely many or countably infinitely many sheets over $\mathbb{C}_\infty$ which are connected along branch cuts in such a way that $\mathcal{R}_F$ can be covered locally by disks (if $k = 1$ for the center of such disks) and by k-fold disks (if the center of the disk is a branch point of order $k - 1$, $k \geq 2$). The choice of branch cuts is largely arbitrary as long as they connect branch points and are non-self-intersecting. In the special case of compact Riemann surfaces the total number of sheets, branch cuts, and branch points is finite.

An important aspect is the possibility of analytic continuation of a given (circular) function element in $\mathbb{C}_\infty$ (i.e., a convergent power series expansion in some disk) along all possible continuous paths on a (covering) Riemann surface $\mathcal{R}_F$ in such a way that the resulting function f has algebroidal[1] behavior at any point of $\mathcal{R}_F$. More precisely, $\mathcal{R}_F$ and a function $f : \mathcal{R}_F \to \mathbb{C}_\infty$ are said to correspond to each other if f is meromorphic on $\mathcal{R}_F$, two function elements of f associated with two different points on $\mathcal{R}_F$ over the same point $z \in \mathbb{C}_\infty$ are distinct, and $\mathcal{R}_F$ is maximal in the sense that there is no $\widetilde{\mathcal{R}}_F \supsetneq \mathcal{R}_F$ such that these properties hold with $\mathcal{R}_F$ replaced by $\widetilde{\mathcal{R}}_F$. The basic fact concerning analytic functions and corresponding covering Riemann surfaces is then the following: To every function element φ in $\mathbb{C}_\infty$ there exists a covering Riemann surface $\mathcal{R}_F$, such that $\mathcal{R}_F$ and the analytic function f, obtained by analytic continuation of φ along any possible continuous path on $\mathcal{R}_F$, correspond to each other.

---

[1] f has algebroidal behavior at $P \in \mathcal{R}_F$ if $f(P) = \sum_{n=-p}^{\infty} c_n (z - z_0)^{n/k}$ for $z_0 \in \mathbb{C}$ and $f(P) = \sum_{n=-q}^{\infty} d_n z^{-n/k}$ for $z_0 = \infty$.

Finally we briefly consider the special case of compact Riemann surfaces, the case at hand in this monograph. We recall that $w = f(z)$ is called algebraic if there exists an irreducible polynomial $\mathcal{P}$ in two variables such that $\mathcal{P}(z, w) = 0$ for all $z \in \mathbb{C}$. The fundamental connection between compact Riemann surfaces and algebraic functions then reads as follows: f corresponds to a compact Riemann surface if and only if f is algebraic.

Next, we consider the notion of the meromorphic function field $\mathcal{M}(\mathcal{R}_F)$ of $\mathcal{R}_F$, which by definition consists of all analytic maps $f \colon \mathcal{R}_F \to \mathbb{C}_\infty$. If $\mathcal{R}_F$ corresponds to an algebraic function in the sense described above, then $g \colon \mathcal{R}_F \to \mathbb{C}_\infty$ belongs to $\mathcal{M}(\mathcal{R}_F)$, that is, $g \in \mathcal{M}(\mathcal{R}_F)$, if and only if $g(z)$ is a rational function in the two variables z and $f(z)$. In addition, if $N \in \mathbb{N}$ denotes the number of sheets of $\mathcal{R}_F$, $1, f, f^2, \ldots, f^{N-1}$ forms a basis in $\mathcal{M}(\mathcal{R}_F)$ and g can be uniquely represented as

$$g(z) = r_0(z) + r_1(z) f(z) + \cdots + r_{N-1}(z) f(z)^{N-1},$$

with r_j, $j = 0, \ldots, N - 1$ rational functions. Moreover, if $f_1, f_2 \in \mathcal{M}(\mathcal{R}_F)$, then $\mathcal{Q}(f_1(z), f_2(z)) = 0$ for all $z \in \mathbb{C}$ for some irreducible polynomial $\mathcal{Q}$ in two variables.

We also mention an alternative to (A.3) for computing the topological genus of $\mathcal{R}_F$ covering $\mathbb{C}_\infty$. Denote by N the number of sheets of $\mathcal{R}_F$, by B the total branching number of $\mathcal{R}_F$,

$$B = \sum_{P \in \mathcal{R}_F} (k(P) - 1) = \sum_{P \in \mathcal{B}(\mathcal{R}_F)} (k(P) - 1),$$

where $k(P) - 1$ denotes the branching order of $P \in \mathcal{R}_F$ (of course $k(P) = 1$ for all but finitely many $P \in \mathcal{R}_F$), and by $\mathcal{B}(\mathcal{R}_F)$ the set of branch points of $\mathcal{R}_F$. Then the topological genus g of $\mathcal{R}_F$ is given by the Riemann–Hurwitz formula

$$g = 1 - N + (B/2).$$

Since hyperelliptic Riemann surfaces are of particular importance to the main body of this monograph, we end this informal introduction with a precise definition of this special case.

Definition A.3 *A compact Riemann surface is called hyperelliptic if it admits a meromorphic function of degree two (i.e., a nonconstant meromorphic function with precisely two poles counting multiplicity).*

We will describe hyperelliptic Riemann surfaces $\mathcal{K}_g$ of genus $g \in \mathbb{N}$ as two-sheeted coverings of the Riemann sphere $\mathbb{C}_\infty$ branched at $2g+2$ points in great detail in Appendix B. The meromorphic function of degree two alluded to in Definition A.3 is then given by the projection $\tilde{\pi}$ as defined in (B.24). Here we just add one more

Algebraic Curves and Their Theta Functions in a Nutshell 331

brief comment on hyperelliptic curves, the principal object in the main body of this text. The projective curve

$$x_1^2 x_0^{k-2} = \prod_{\ell=1}^{k}(x_2 - e_\ell x_0), \quad k \in \mathbb{N}, \tag{A.7}$$

in $\mathbb{CP}^2$ of degree k, with $e_1, \ldots, e_k$ distinct complex numbers, is called elliptic if $k = 3, 4$ and hyperelliptic if $k \geq 5$. However, to simplify matters, all curves in (A.7) are usually called hyperelliptic and this convention is adopted in Definition A.3. The projective curve (A.7) is smooth (nonsingular) if and only if $1 \leq k \leq 3$; if $k \geq 4$ it has the unique singular point $[0, 1, 0]$.

For most of the remainder of Appendix A we suppose that $\mathcal{K}_g$ is a compact Riemann surface of genus $g \in \mathbb{N}$ and choose a homology basis $\{a_j, b_j\}_{j=1}^{g}$ on $\mathcal{K}_g$ in such a way that the intersection matrix of the cycles satisfies

$$a_j \circ b_k = \delta_{j,k}, \quad a_j \circ a_k = 0, \quad b_j \circ b_k = 0, \quad j, k = 1, \ldots, g \tag{A.8}$$

(with a_j and b_k intersecting to form a right-handed coordinate system, cf. Figures A.1, A.2). In particular, the first homology group of $\mathcal{K}_g$ with integer coefficients, $H_1(\mathcal{K}_g, \mathbb{Z})$, is the free abelian group on the generators $[a_j], [b_j], j = 1, \ldots, g$, where $[c]$ denotes the homology class of the cycle c.

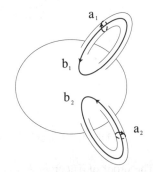

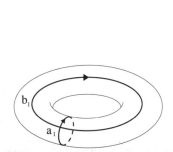

Fig. A.1. Genus $g = 1$.   Fig. A.2. Genus $g = 2$. $(a_j \circ b_k = \delta_{jk})$

Unless explicitly stated otherwise, it will be assumed that $g \geq 1$ for the remainder of this appendix.

Turning briefly to meromorphic differentials (1-forms) on $\mathcal{K}_g$, we state the following result.

Theorem A.4 (Riemann's period relations) *Suppose ω and ν are closed C^1 meromorphic differentials (1-forms) on $\mathcal{K}_g$. Then,*

(i)
$$\iint_{\mathcal{K}_g} \omega \wedge v = \sum_{j=1}^{g} \left(\left(\int_{a_j} \omega \right) \left(\int_{b_j} v \right) - \left(\int_{b_j} \omega \right) \left(\int_{a_j} v \right) \right). \quad (A.9)$$

If, in addition ω and v are holomorphic 1-forms on $\mathcal{K}_g$, then

$$\sum_{j=1}^{g} \left(\left(\int_{a_j} \omega \right) \left(\int_{b_j} v \right) - \left(\int_{b_j} \omega \right) \left(\int_{a_j} v \right) \right) = 0. \quad (A.10)$$

(ii) If ω is a nonzero holomorphic 1-form on $\mathcal{K}_g$, then

$$\operatorname{Im} \left(\sum_{j=1}^{g} \left(\int_{a_j} \omega \right) \left(\int_{b_j} \overline{\omega} \right) \right) > 0. \quad (A.11)$$

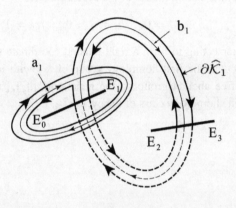

Fig. A.3. Canonical dissection, genus $g = 1$ ($\partial \widehat{\mathcal{K}}_1 = a_1 b_1 a_1^{-1} b_1^{-1}$).

The proof of Theorem A.4 is usually based on Stokes' theorem and a canonical dissection of (cf. Figure A.3) $\mathcal{K}_g$ along its cycles yielding the simply connected interior $\widehat{\mathcal{K}}_g$ of the fundamental polygon $\partial \widehat{\mathcal{K}}_g$ given by

$$\partial \widehat{\mathcal{K}}_g = a_1 b_1 a_1^{-1} b_1^{-1} a_2 b_2 a_2^{-1} b_2^{-1} \ldots a_g^{-1} b_g^{-1}. \quad (A.12)$$

Given the cycles $\{a_j, b_j\}_{j=1}^g$, we denote by $\{\omega_j\}_{j=1}^g$ the corresponding normalized basis of the space of holomorphic differentials (also called abelian differentials of the first kind) on $\mathcal{K}_g$, that is,

$$\int_{a_k} \omega_j = \delta_{j,k}, \quad j, k = 1, \ldots, g. \quad (A.13)$$

Algebraic Curves and Their Theta Functions in a Nutshell   333

The b-periods of ω_j are then defined by

$$\tau_{j,k} = \int_{b_k} \omega_j, \quad j,k = 1,\ldots,g. \tag{A.14}$$

Theorem A.4 then implies the following result.

Theorem A.5 *The $g \times g$ matrix $\tau = (\tau_{j,k})_{j,k=1,\ldots,g}$ is symmetric, that is,*

$$\tau_{j,k} = \tau_{k,j}, \quad j,k = 1,\ldots,g,$$

with a positive definite imaginary part,

$$\mathrm{Im}(\tau) = \frac{1}{2i}(\tau - \tau^*) > 0. \tag{A.15}$$

Next we briefly study some consequences of a change of homology basis. Let

$$\{a_1,\ldots,a_g,b_1,\ldots,b_g\} \tag{A.16}$$

be a canonical homology basis on $\mathcal{K}_g$ with intersection matrix satisfying (A.8) and

$$\{a'_1,\ldots,a'_g,b'_1,\ldots,b'_g\} \tag{A.17}$$

a homology basis on $\mathcal{K}_g$ related to each other by

$$\begin{pmatrix} \underline{a'}^\top \\ \underline{b'}^\top \end{pmatrix} = X \begin{pmatrix} \underline{a}^\top \\ \underline{b}^\top \end{pmatrix},$$

where

$$\underline{a}^\top = (a_1,\ldots,a_g)^\top, \quad \underline{b}^\top = (b_1,\ldots,b_g)^\top,$$
$$\underline{a'}^\top = (a'_1,\ldots,a'_g)^\top, \quad \underline{b'}^\top = (b'_1,\ldots,b'_g)^\top,$$
$$X = \begin{pmatrix} A & B \\ C & D \end{pmatrix},$$

with $A, B, C,$ and D being $g \times g$ matrices with integer entries. Then (A.17) is also a canonical homology basis on $\mathcal{K}_g$ with intersection matrix satisfying (A.8) if and only if

$$X \in \mathrm{Sp}(g,\mathbb{Z}),$$

where

$$\mathrm{Sp}(g,\mathbb{Z}) = \left\{ X = \begin{pmatrix} A & B \\ C & D \end{pmatrix} \,\bigg|\, X \begin{pmatrix} 0 & I_g \\ -I_g & 0 \end{pmatrix} X^\top = \begin{pmatrix} 0 & I_g \\ -I_g & 0 \end{pmatrix},\, \det(X) = 1 \right\}$$

denotes the symplectic modular group of genus $g \in \mathbb{N}$ (here A, B, C, D in X are again $g \times g$ matrices with integer entries). If $\{\omega_j\}_{j=1}^g$ and $\{\omega'_j\}_{j=1}^g$ are the normalized bases of holomorphic differentials corresponding to the canonical homology bases

(A.16) and (A.17), with τ and τ' the associated b and b'-periods of $\omega_1, \ldots, \omega_g$ and $\omega'_1, \ldots, \omega'_g$, respectively, one computes

$$\underline{\omega}' = \underline{\omega}(A + B\tau)^{-1}, \quad \tau' = (C + D\tau)(A + B\tau)^{-1}, \tag{A.18}$$

where $\underline{\omega} = (\omega_1, \ldots, \omega_g)$, $\underline{\omega}' = (\omega'_1, \ldots, \omega'_g)$.

Abelian differentials of the second kind, $\omega^{(2)}$, are characterized by the property that all their residues vanish. They will usually be normalized by the vanishing of all their a-periods (this is achieved by adding a suitable linear combination of differentials of the first kind)

$$\int_{a_j} \omega^{(2)} = 0, \quad j = 1, \ldots, g.$$

We may add in this context that the sum of the residues of any meromorphic differential ν on $\mathcal{K}_g$ vanishes, the residue at a pole $Q_0 \in \mathcal{K}_g$ of ν being defined by

$$\mathrm{res}_{Q_0}(\nu) = \frac{1}{2\pi i} \int_{\gamma_{Q_0}} \nu,$$

where γ_{Q_0} is a smooth, counterclockwise oriented, simple, closed contour, encircling Q_0, but no other pole of ν.

Theorem A.6 *Assume* $\omega^{(2)}_{Q_1,n}$ *to be a differential of the second kind on* $\mathcal{K}_g$, *whose only pole is* $Q_1 \in \widehat{\mathcal{K}}_g$ *with principal part* $\zeta_{Q_1}^{-n-2} d\zeta_{Q_1}$ *for some* $n \in \mathbb{N}_0$ *and* $\omega^{(1)}$ *a differential of the first kind on* $\mathcal{K}_g$ *of the type*

$$\omega^{(1)} = \left(\sum_{m=0}^{\infty} c_m(Q_1) \zeta_{Q_1}^m \right) d\zeta_{Q_1}$$

near Q_1. *Then,*

$$\frac{1}{2\pi i} \sum_{j=1}^{g} \left(\left(\int_{a_j} \omega^{(1)} \right) \left(\int_{b_j} \omega^{(2)}_{Q_1,n} \right) - \left(\int_{a_j} \omega^{(2)}_{Q_1,n} \right) \left(\int_{b_j} \omega^{(1)} \right) \right) = \frac{c_n(Q_1)}{n+1},$$

$$n \in \mathbb{N}_0.$$

In particular, if $\omega^{(2)}_{Q_1,n}$ *is normalized and*

$$\omega^{(1)} = \omega_j = \left(\sum_{m=0}^{\infty} c_{j,m}(Q_1) \zeta_{Q_1}^m \right) d\zeta_{Q_1},$$

Algebraic Curves and Their Theta Functions in a Nutshell 335

then the vector of b-periods of $\omega_{Q_1,n}^{(2)}/(2\pi i)$, denoted by $\underline{U}_n^{(2)}$, reads

$$\underline{U}_n^{(2)} = (U_{n,1}^{(2)}, \ldots, U_{n,g}^{(2)}), \quad U_{n,j}^{(2)} = \frac{1}{2\pi i} \int_{b_j} \omega_{Q_1,n}^{(2)} = \frac{c_{j,n}(Q_1)}{n+1}, \quad (A.19)$$

$$n \in \mathbb{N}_0, \quad j = 1, \ldots, g.$$

Any meromorphic differential $\omega^{(3)}$ on $\mathcal{K}_g$ not of the first or second kind is said to be of the third kind. It is common to normalize $\omega^{(3)}$ by the vanishing of its a-periods, that is, by

$$\int_{a_j} \omega^{(3)} = 0, \quad j = 1, \ldots, g. \quad (A.20)$$

A normal differential of the third kind, denoted $\omega_{Q_1,Q_2}^{(3)}$, associated with two distinct points $Q_1, Q_2 \in \widehat{\mathcal{K}}_g$, by definition, is holomorphic on $\mathcal{K}_g \setminus \{Q_1, Q_2\}$ and has simple poles at Q_ℓ with residues $(-1)^{\ell+1}$, $\ell = 1, 2$, and vanishing a-periods.

Theorem A.7 *Suppose $\omega^{(3)}$ to be a differential of the third kind on $\mathcal{K}_g$ whose only singularities are simple poles at $Q_n \in \widehat{\mathcal{K}}_g$ with residues c_n, $n = 1, \ldots, N$. Denote by $\omega^{(1)}$ a differential of the first kind on $\mathcal{K}_g$. Then*

$$\frac{1}{2\pi i} \sum_{j=1}^{g} \left(\left(\int_{a_j} \omega^{(1)} \right) \left(\int_{b_j} \omega^{(3)} \right) - \left(\int_{b_j} \omega^{(1)} \right) \left(\int_{a_j} \omega^{(3)} \right) \right)$$

$$= \sum_{n=1}^{N} c_n \int_{Q_0}^{Q_n} \omega^{(1)}, \quad (A.21)$$

where $Q_0 \in \widehat{\mathcal{K}}_g$ is any fixed base point. In particular, if $\omega^{(3)}$ is normalized and $\omega^{(1)} = \omega_j$, then,

$$\frac{1}{2\pi i} \int_{b_j} \omega^{(3)} = \sum_{n=1}^{N} c_n \int_{Q_0}^{Q_n} \omega_j, \quad j = 1, \ldots, g. \quad (A.22)$$

Moreover, if $\omega_{Q_1,Q_2}^{(3)}$ is a normal differential of the third kind on $\mathcal{K}_g$ holomorphic on $\mathcal{K}_g \setminus \{Q_1, Q_2\}$, then

$$\frac{1}{2\pi i} \int_{b_j} \omega_{Q_1,Q_2}^{(3)} = \int_{Q_2}^{Q_1} \omega_j, \quad j = 1, \ldots, g. \quad (A.23)$$

Theorem A.8 *Let $P, Q \in \widehat{\mathcal{K}}_g$, $P \neq Q$, and $\omega_{P,Q}^{(3)}$ a normal differential of the third kind on $\mathcal{K}_g$ holomorphic on $\mathcal{K}_g \setminus \{P, Q\}$. Pick a base point $Q_0 \in \partial \widehat{\mathcal{K}}_g$ and connect P and Q_0 and Q and Q_0 with two smooth paths $\gamma_{P,Q_0} \in \widehat{\mathcal{K}}_g$ and $\gamma_{Q,Q_0} \in \widehat{\mathcal{K}}_g$, respectively. Define $\widehat{\widehat{\mathcal{K}}}_g = \widehat{\mathcal{K}}_g \setminus \{\gamma_{P,Q_0} \cup \gamma_{Q,Q_0}\}$ and pick $R, S \in \widehat{\widehat{\mathcal{K}}}_g$, $R \neq S$. Finally,* let $\omega_{R,S}^{(3)}$ be a normal differential of the third kind on $\mathcal{K}_g$ holomorphic on $\mathcal{K}_g \setminus \{R, S\}$ and connect R and Q_0 and S and Q_0 with two smooth paths $\gamma_{R,Q_0} \in \widehat{\widehat{\mathcal{K}}}_g$ and $\gamma_{S,Q_0} \in \widehat{\widehat{\mathcal{K}}}_g$, respectively. The paths $\gamma_{P,Q_0}, \gamma_{Q,Q_0}, \gamma_{R,Q_0}$, and γ_{S,Q_0} are assumed to be mutually nonintersecting. Then,

$$\int_S^R \omega_{P,Q}^{(3)} = \int_Q^P \omega_{R,S}^{(3)},$$

where the paths from S to R and from Q to P lie in the simply connected region $\widehat{\mathcal{K}}_g \setminus \{\gamma_{P,Q_0} \cup \gamma_{Q,Q_0} \cup \gamma_{R,Q_0} \cup \gamma_{S,Q_0}\}$.

We shall always assume (without loss of generality) that all poles of differentials of the second and third kind on $\mathcal{K}_g$ lie on $\widehat{\mathcal{K}}_g$ (i.e., not on $\partial \widehat{\mathcal{K}}_g$). This can always be achieved by an appropriate choice of the cycles a_j and b_j. Moreover, we assume that all integration paths on the right-hand sides of (A.21)–(A.23) stay away from the cycles a_j and b_k.

Next, we turn to divisors on $\mathcal{K}_g$ and the Jacobi variety $J(\mathcal{K}_g)$ of $\mathcal{K}_g$. Let $\mathcal{M}(\mathcal{K}_g)$ and $\mathcal{M}^1(\mathcal{K}_g)$ denote the set of meromorphic functions (0-forms) and meromorphic 1-forms on $\mathcal{K}_g$, respectively, for some $g \in \mathbb{N}_0$.

Definition A.9 Let $g \in \mathbb{N}_0$. Suppose $f \in \mathcal{M}(\mathcal{K}_g)$, $\omega = h(\zeta_{Q_0}) d\zeta_{Q_0} \in \mathcal{M}^1(\mathcal{K}_g)$, and (U_{Q_0}, ζ_{Q_0}) is a chart near some point $Q_0 \in \mathcal{K}_g$.
(i) If $(f \circ \zeta_{Q_0}^{-1})(\zeta) = \sum_{n=m_0}^\infty c_n(Q_0) \zeta^n$ for some $m_0 \in \mathbb{Z}$ (which turns out to be independent of the chosen chart), the order $\nu_f(Q_0)$ of f at Q_0 is defined by

$$\nu_f(Q_0) = m_0.$$

One defines $\nu_f(P) = \infty$ for all $P \in \mathcal{K}_g$ if f is identically zero on $\mathcal{K}_g$.
(ii) If $h_{Q_0}(\zeta_{Q_0}) = \sum_{n=m_0}^\infty d_n(Q_0) \zeta_{Q_0}^n$ for some $m_0 \in \mathbb{Z}$ (which again is independent of the chart chosen), the order $\nu_\omega(Q_0)$ of ω at Q_0 is defined by

$$\nu_\omega(Q_0) = m_0.$$

Next, we turn to divisors and introduce some structure on the set of all divisors.

Definition A.10 Let $g \in \mathbb{N}_0$.
(i) A divisor $\mathcal{D}$ on $\mathcal{K}_g$ is a map $\mathcal{D}: \mathcal{K}_g \to \mathbb{Z}$, where $\mathcal{D}(P) \neq 0$ for only finitely many $P \in \mathcal{K}_g$. On the set of all divisors $\mathrm{Div}(\mathcal{K}_g)$ on $\mathcal{K}_g$ one introduces the partial ordering

$$\mathcal{D} \geq \mathcal{E} \text{ if } \mathcal{D}(P) \geq \mathcal{E}(P), \quad P \in \mathcal{K}_g.$$

(ii) The degree $\deg(\mathcal{D})$ of $\mathcal{D} \in \mathrm{Div}(\mathcal{K}_g)$ is defined by

$$\deg(\mathcal{D}) = \sum_{P \in \mathcal{K}_g} \mathcal{D}(P).$$

(iii) $\mathcal{D} \in \mathrm{Div}(\mathcal{K}_g)$ *is called nonnegative (or effective) if*

$$\mathcal{D} \geq 0,$$

where 0 *denotes the zero divisor* $0(P) = 0$ *for all* $P \in \mathcal{K}_g$.
(iv) Let $\mathcal{D}, \mathcal{E} \in \mathrm{Div}(\mathcal{K}_g)$. *Then* $\mathcal{D}$ *is called a multiple of* $\mathcal{E}$ *if*

$$\mathcal{D} \geq \mathcal{E}.$$

$\mathcal{D}$ *and* $\mathcal{E}$ *are called relatively prime if*

$$\mathcal{D}(P)\mathcal{E}(P) = 0, \quad P \in \mathcal{K}_g.$$

(v) If $f \in \mathcal{M}(\mathcal{K}_g) \setminus \{0\}$ *and* $\omega \in \mathcal{M}^1(\mathcal{K}_g) \setminus \{0\}$, *then the divisor* (f) *of* f *is defined by*

$$(f) \colon \mathcal{K}_g \to \mathbb{Z}, \quad P \mapsto \nu_f(P)$$

(thus, f *is holomorphic if and only if* $(f) \geq 0$), *and the divisor of* ω *is defined by*

$$(\omega) \colon \mathcal{K}_g \to \mathbb{Z}, \quad P \mapsto \nu_\omega(P)$$

(thus, ω *is a differential of the first kind if and only if* $(\omega) \geq 0$). *The divisor* (f) *is called a principal divisor, and* (ω) *a canonical divisor.*
(vi) The divisors $\mathcal{D}, \mathcal{E} \in \mathrm{Div}(\mathcal{K}_g)$ *are called equivalent, written* $\mathcal{D} \sim \mathcal{E}$, *if*

$$\mathcal{D} - \mathcal{E} = (f)$$

for some $f \in \mathcal{M}(\mathcal{K}_g) \setminus \{0\}$. *The divisor class* $[\mathcal{D}]$ *of* $\mathcal{D}$ *is defined by*

$$[\mathcal{D}] = \{\mathcal{E} \in \mathrm{Div}(\mathcal{K}_g) \mid \mathcal{E} \sim \mathcal{D}\}. \tag{A.24}$$

Lemma A.11 *Let* $g \in \mathbb{N}_0$. *Suppose* $f \in \mathcal{M}(\mathcal{K}_g)$ *and* $\omega \in \mathcal{M}^1(\mathcal{K}_g)$. *Then,*

$$\deg((f)) = 0$$

and

$$\deg((\omega)) = 2(g - 1).$$

Clearly, $\mathrm{Div}(\mathcal{K}_g)$ forms an abelian group with respect to addition of divisors. The principal divisors form a subgroup $\mathrm{Div}_P(\mathcal{K}_g)$ of $\mathrm{Div}(\mathcal{K}_g)$. The quotient group $\mathrm{Div}(\mathcal{K}_g)/\mathrm{Div}_P(\mathcal{K}_g)$ consists of the cosets of divisors, the divisor classes defined in (A.24). Also the set of divisors of degree zero, $\mathrm{Div}_0(\mathcal{K}_g)$, forms a subgroup of $\mathrm{Div}(\mathcal{K}_g)$. Since $\mathrm{Div}_P(\mathcal{K}_g) \subset \mathrm{Div}_0(\mathcal{K}_g)$, one can introduce the quotient group $\mathrm{Pic}(\mathcal{K}_g) = \mathrm{Div}_0(\mathcal{K}_g)/\mathrm{Div}_P(\mathcal{K}_g)$, called the Picard group of $\mathcal{K}_g$.

Definition A.12 *Let* $g \in \mathbb{N}_0$, *and define*

$$\mathcal{L}(\mathcal{D}) = \{f \in \mathcal{M}(\mathcal{K}_g) \mid f = 0 \text{ or } (f) \geq \mathcal{D}\},$$
$$\mathcal{L}^1(\mathcal{D}) = \{\omega \in \mathcal{M}^1(\mathcal{K}_g) \mid \omega = 0 \text{ or } (\omega) \geq \mathcal{D}\}.$$

Both $\mathcal{L}(\mathcal{D})$ and $\mathcal{L}^1(\mathcal{D})$ are linear spaces over $\mathbb{C}$. We denote their (complex) dimensions by

$$r(\mathcal{D}) = \dim \mathcal{L}(\mathcal{D}), \tag{A.25}$$

$$i(\mathcal{D}) = \dim \mathcal{L}^1(\mathcal{D}). \tag{A.26}$$

$i(\mathcal{D})$ is also called the index of specialty of $\mathcal{D}$. $\mathcal{D}$ is called special (resp., nonspecial) if $i(\mathcal{D}) \geq 1$ (resp., $i(\mathcal{D}) = 0$).

Lemma A.13 *Let $g \in \mathbb{N}_0$ and $\mathcal{D} \in \mathrm{Div}(\mathcal{K}_g)$. Then $\deg(\mathcal{D})$, $r(\mathcal{D})$, and $i(\mathcal{D})$ only depend on the divisor class $[\mathcal{D}]$ of $\mathcal{D}$ (and not on the particular representative $\mathcal{D}$). Moreover, for $\omega \in \mathcal{M}^1(\mathcal{K}_g) \setminus \{0\}$ one infers*

$$i(\mathcal{D}) = r(\mathcal{D} - (\omega)), \quad \mathcal{D} \in \mathrm{Div}(\mathcal{K}_g).$$

Theorem A.14 (Riemann–Roch) *Let $g \in \mathbb{N}_0$ and $\mathcal{D} \in \mathrm{Div}(\mathcal{K}_g)$. Then $r(-\mathcal{D})$ and $i(\mathcal{D})$ are finite and*

$$r(-\mathcal{D}) = \deg(\mathcal{D}) + i(\mathcal{D}) - g + 1.$$

In particular, Riemann's inequality

$$r(-\mathcal{D}) \geq \deg(\mathcal{D}) - g + 1$$

holds.

Next we turn to Jacobi varieties and the Abel map.

Definition A.15 *Define the period lattice L_g in $\mathbb{C}^g$ by*

$$L_g = \{\underline{z} \in \mathbb{C}^g \mid \underline{z} = \underline{n} + \underline{m}\tau, \ \underline{n}, \underline{m} \in \mathbb{Z}^g\}. \tag{A.27}$$

Then the Jacobi variety $J(\mathcal{K}_g)$ of $\mathcal{K}_g$ is defined by

$$J(\mathcal{K}_g) = \mathbb{C}^g / L_g, \tag{A.28}$$

and the Abel maps are defined by

$$\underline{A}_{Q_0} \colon \mathcal{K}_g \to J(\mathcal{K}_g), \quad P \mapsto \underline{A}_{Q_0}(P) = (A_{Q_0,1}(P), \ldots, A_{Q_0,g}(P)) \tag{A.29}$$

$$= \left(\int_{Q_0}^P \omega_1, \ldots, \int_{Q_0}^P \omega_g \right) \pmod{L_g}$$

and

$$\underline{\alpha}_{Q_0} \colon \mathrm{Div}(\mathcal{K}_g) \to J(\mathcal{K}_g), \quad \mathcal{D} \mapsto \underline{\alpha}_{Q_0}(\mathcal{D}) = \sum_{P \in \mathcal{K}_g} \mathcal{D}(P) \underline{A}_{Q_0}(P), \tag{A.30}$$

Algebraic Curves and Their Theta Functions in a Nutshell

where $Q_0 \in \mathcal{K}_g$ is a fixed base point and (for convenience only) the same path is chosen from Q_0 to P for all $j = 1, \dots, g$ in (A.29) and (A.30).[1]

Clearly, $\underline{A}_{Q_0}$ is well-defined since changing the path from Q_0 to P amounts to adding a closed cycle whose contribution in the integral (A.29) consists of adding a vector in L_g. Moreover, $\underline{\alpha}_{Q_0}$ is a group homomorphism and $J(\mathcal{K}_g)$ is a complex torus of (complex) dimension g that depends on the choice of the homology basis $\{a_j, b_j\}_{j=1}^g$. However, different homology bases yield isomorphic Jacobians.

Theorem A.16 (Abel's theorem) *A divisor $\mathcal{D} \in \text{Div}(\mathcal{K}_g)$ is principal if and only if*

$$\deg(\mathcal{D}) = 0 \text{ and } \underline{\alpha}_{Q_0}(\mathcal{D}) = 0. \tag{A.31}$$

We note that the apparent base point dependence of the Abel map on Q_0 in (A.31) disappears if $\deg(\mathcal{D}) = 0$.

Remark A.17 The preceding results, starting with Definition A.10, are of considerable relevance in connection with the linearizing property of the Abel map of auxiliary divisors (see, e.g., (1.125), (1.259), (3.138), (3.139), (3.257), and (3.258)) in the following sense: The Abel map is injective on the set of nonnegative nonspecial divisors of degree g. To illustrate this fact we denote by $\mathcal{D}$ a nonnegative divisor of degree g and by $|\mathcal{D}|$ the set of nonnegative divisors linearly equivalent to $\mathcal{D}$. By the Riemann–Roch theorem, Theorem A.14,

$$r(-\mathcal{D}) - 1 = i(\mathcal{D}). \tag{A.32}$$

By Abel's theorem, Theorem A.16,

$$\underline{\alpha}_{Q_0}^{-1}(\underline{\alpha}_{Q_0}(\mathcal{D})) = |\mathcal{D}|,$$

with $\underline{\alpha}_{Q_0}^{-1}(\cdot)$ denoting the inverse image of $\underline{\alpha}_{Q_0}$. If $\mathcal{D}$ is nonspecial, that is, if $i(\mathcal{D}) = 0$, (A.32) yields $|\mathcal{D}| = \mathcal{D}$ and hence the asserted injectivity property since $\dim(\underline{\alpha}_{Q_0}^{-1}(\underline{\alpha}_{Q_0}(\mathcal{D}))) = r(-\mathcal{D}) - 1$.

Next, we turn to Riemann theta functions and a constructive approach to the Jacobi inversion problem. We assume $g \in \mathbb{N}$ for the remainder of this appendix.

Given the Riemann surface $\mathcal{K}_g$, the homology basis $\{a_j, b_j\}_{j=1}^g$, and the matrix τ of b-periods of the differentials of the first kind, $\{\omega_j\}_{j=1}^g$ (cf. (A.14)), the Riemann theta function associated with $\mathcal{K}_g$ and the homology basis is defined as

$$\theta(\underline{z}) = \sum_{\underline{n} \in \mathbb{Z}^g} \exp\left(2\pi i (\underline{n}, \underline{z}) + \pi i (\underline{n}, \underline{n}\tau)\right), \quad \underline{z} \in \mathbb{C}^g, \tag{A.33}$$

---

[1] This convention allows one to avoid the multiplicative version of the Riemann–Roch theorem at various places in this monograph.

where $(\underline{u},\underline{v}) = \overline{\underline{u}}\,\underline{v}^\top = \sum_{j=1}^g \overline{u_j} v_j$ denotes the scalar product in $\mathbb{C}^g$. Because of (A.15), θ is well-defined and represents an entire function on $\mathbb{C}^g$. Elementary properties of θ are, for instance,

$$\theta(z_1,\ldots,z_{j-1},-z_j,z_{j+1},\ldots,z_n) = \theta(\underline{z}), \quad \underline{z} = (z_1,\ldots,z_g) \in \mathbb{C}^g, \tag{A.34}$$

$$\theta(\underline{z} + \underline{m} + \underline{n}\tau) = \theta(\underline{z})\exp\big(-2\pi i(\underline{n},\underline{z}) - \pi i(\underline{n},\underline{n}\tau)\big), \quad \underline{m},\underline{n} \in \mathbb{Z}^g, \underline{z} \in \mathbb{C}^g. \tag{A.35}$$

Lemma A.18 *Let $\underline{\xi} \in \mathbb{C}^g$ and define*

$$\widehat{F} \colon \widehat{\mathcal{K}}_g \to \mathbb{C}, \quad P \mapsto \theta(\underline{\xi} - \underline{\widehat{A}}_{Q_0}(P)), \tag{A.36}$$

where

$$\underline{\widehat{A}}_{Q_0} \colon \widehat{\mathcal{K}}_g \to \mathbb{C}^g, \tag{A.37}$$

$$P \mapsto \underline{\widehat{A}}_{Q_0}(P) = \big(\widehat{A}_{Q_0,1}(P),\ldots,\widehat{A}_{Q_0,g}(P)\big) = \left(\int_{Q_0}^P \omega_1,\ldots,\int_{Q_0}^P \omega_g\right).$$

Suppose $\widehat{F}$ is not identically zero on $\widehat{\mathcal{K}}_g$, that is, $\widehat{F} \not\equiv 0$. Then $\widehat{F}$ has precisely g zeros on $\widehat{\mathcal{K}}_g$ counting multiplicities.

Lemma A.18 can be proven by integrating $d\ln(\widehat{F})$ along $\partial\widehat{\mathcal{K}}_g$.

For subsequent use in Remark A.30 we also introduce

$$\underline{\hat{\alpha}}_{Q_0} \colon \mathrm{Div}(\widehat{\mathcal{K}}_g) \to \mathbb{C}^g, \quad \mathcal{D} \mapsto \underline{\hat{\alpha}}_{Q_0}(\mathcal{D}) = \sum_{P \in \widehat{\mathcal{K}}_g} \mathcal{D}(P)\underline{\widehat{A}}_{Q_0}(P), \tag{A.38}$$

in addition to $\underline{\widehat{A}}_{Q_0}$ in (A.37).

Theorem A.19 *Let $\underline{\xi} \in \mathbb{C}^g$ and define $\widehat{F}$ as in (A.36). Assume that $\widehat{F}$ is not identically zero on $\widehat{\mathcal{K}}_g$, and let $Q_1,\ldots,Q_g \in \mathcal{K}_g$ be the zeros of $\widehat{F}$ (multiplicities included) given by Lemma A.18. Define the corresponding positive divisor $\mathcal{D}_{\underline{Q}}$ of degree g on $\mathcal{K}_g$ by*

$$\mathcal{D}_{\underline{Q}} \colon \mathcal{K}_g \to \mathbb{N}_0, \quad P \mapsto \mathcal{D}_{\underline{Q}}(P) = \begin{cases} m & \text{if } P \text{ occurs } m \text{ times in } \{Q_1,\ldots,Q_g\}, \\ 0 & \text{if } P \notin \{Q_1,\ldots,Q_g\}, \end{cases}$$

$$\underline{Q} = \{Q_1,\ldots,Q_g\} \in \mathrm{Sym}^g(\mathcal{K}_g), \tag{A.39}$$

and recall the Abel map $\underline{\alpha}_{Q_0}$ in (A.30). Then there exists a vector $\underline{\Xi}_{Q_0} \in \mathbb{C}^g$, the vector of Riemann constants, such that

$$\underline{\alpha}_{Q_0}(\mathcal{D}_{\underline{Q}}) = (\underline{\xi} - \underline{\Xi}_{Q_0}) \pmod{L_g}. \tag{A.40}$$

Algebraic Curves and Their Theta Functions in a Nutshell 341

The vector $\underline{\Xi}_{Q_0} = (\Xi_{Q_0,1}, \ldots, \Xi_{Q_0,g})$ is given by

$$\Xi_{Q_0,j} = \frac{1}{2}(1 + \tau_{j,j}) - \sum_{\substack{\ell=1 \\ \ell \neq j}}^{g} \int_{a_\ell} \omega_\ell(P) \int_{Q_0}^{P} \omega_j, \quad j = 1, \ldots, g. \tag{A.41}$$

For the proof of Theorem A.19 one integrates $\widehat{A}_{P_0,j}(P) d \ln(\widehat{F}(P))$ along $\partial \widehat{\mathcal{K}}_g$.
Clearly, $\underline{\Xi}_{Q_0}$ depends on the base point Q_0 and on the choice of the homology basis $\{a_j, b_j\}_{j=1}^{g}$. In the special hyperelliptic case where $\mathcal{K}_g$ is derived from $\mathcal{F}(z, y) = y^2 - \prod_{m=0}^{2g}(z - E_m) = 0$ and $\{E_m\}_{m=0,\ldots,2g}$ are $2g + 1$ distinct points in $\mathbb{C}$, equation (A.41) simplifies to

$$\Xi_{P_\infty,j} = \frac{1}{2}\left(j + \sum_{k=1}^{g} \tau_{j,k}\right), \quad j = 1, \ldots, g,$$

where the base point Q_0 has been chosen to be the unique point P_∞ of $\mathcal{K}_g$ at infinity.

Remark A.20 Theorem A.19 yields a partial solution of Jacobi's inversion problem which can be stated as follows: Given $\underline{\xi} \in \mathbb{C}^g$, find a divisor $\mathcal{D}_{\underline{Q}} \in \text{Div}(\mathcal{K}_g)$ such that

$$\underline{\alpha}_{Q_0}(\mathcal{D}_{\underline{Q}}) = \underline{\xi} \pmod{L_g}.$$

Indeed, if $\widetilde{F}(\cdot) = \theta(\underline{\Xi}_{Q_0} - \underline{\widehat{A}}_{Q_0}(\cdot) + \underline{\xi}) \not\equiv 0$ on $\widehat{\mathcal{K}}_g$, the zeros $Q_1, \ldots, Q_g \in \widehat{\mathcal{K}}_g$ of $\widetilde{F}$ (guaranteed by Lemma A.18) satisfy Jacobi's inversion problem by (A.40). Thus it remains to specify conditions such that $\widetilde{F} \not\equiv 0$ on $\widehat{\mathcal{K}}_g$.

Theorem A.21 *Let $\mathcal{D} \in \text{Div}(\mathcal{K}_g)$ be of degree $2(g - 1)$, $g \in \mathbb{N}$. Then $\mathcal{D}$ is a canonical divisor (i.e., the divisor of a meromorphic differential on $\mathcal{K}_g$) if and only if*

$$\underline{\alpha}_{Q_0}(\mathcal{D}) = -2\underline{\Xi}_{Q_0}.$$

Remark A.22 While θ is well-defined (in fact, entire) on $\mathbb{C}^g$, it is not well-defined on $J(\mathcal{K}_g) = \mathbb{C}^g/L_g$ because of (A.35). Nevertheless, θ is a "multiplicative function" on $J(\mathcal{K}_g)$ since the multipliers in (A.35) cannot vanish. In particular, if $\underline{z}_1 = \underline{z}_2$ (mod L_g), then $\theta(\underline{z}_1) = 0$ if and only if $\theta(\underline{z}_2) = 0$. Hence it is meaningful to state that θ vanishes at points of $J(\mathcal{K}_g)$. Since the Abel map $\underline{A}_{Q_0}$ maps $\mathcal{K}_g$ into $J(\mathcal{K}_g)$, the function $\theta(\underline{\xi} - \underline{A}_{Q_0}(P))$ for $\underline{\xi} \in \mathbb{C}^g$, becomes a multiplicative function on $\mathcal{K}_g$. Again it makes sense to say that $\theta(\underline{\xi} - \underline{A}_{Q_0}(\cdot))$ vanishes at points of $\mathcal{K}_g$. In particular, Lemma A.18 and Theorem A.19 extend to the case where $\widetilde{F}$ is replaced by $F: \mathcal{K}_g \to \mathbb{C}, P \mapsto \theta(\underline{\xi} - \underline{A}_{Q_0}(P))$.

In the following we use the obvious notation

$$X + Y = \{(\underline{x} + \underline{y}) \in J(\mathcal{K}_g) \mid \underline{x} \in X, \underline{y} \in Y\},$$
$$-X = \{-\underline{x} \in J(\mathcal{K}_g) \mid \underline{x} \in X\}, \tag{A.42}$$
$$X + \underline{z} = \{(\underline{x} + \underline{z}) \in J(\mathcal{K}_g) \mid \underline{x} \in X\},$$

for $X, Y \subset J(\mathcal{K}_g)$ and $\underline{z} \in J(\mathcal{K}_g)$. Furthermore, we identify the mth symmetric power of $\mathcal{K}_g$, denoted $\mathrm{Sym}^m(\mathcal{K}_g)$, with the set of nonnegative divisors of degree $m \in \mathbb{N}$ on $\mathcal{K}_g$. For notational convenience, the restriction of the Abel map to $\mathrm{Sym}^m(\mathcal{K}_g)$ (i.e., its restriction to the set of nonnegative divisors of degree m) will be denoted by the same symbol $\underline{\alpha}_{Q_0}$. Moreover, we introduce the convenient notation

$$\mathcal{D}_{Q_0 \underline{Q}} = \mathcal{D}_{Q_0} + \mathcal{D}_{\underline{Q}}, \quad \mathcal{D}_{\underline{Q}} = \mathcal{D}_{Q_1} + \cdots + \mathcal{D}_{Q_m},$$
$$\underline{Q} = \{Q_1, \ldots, Q_m\} \in \mathrm{Sym}^m(\mathcal{K}_g), \quad Q_0 \in \mathcal{K}_g, \ m \in \mathbb{N}, \tag{A.43}$$

where for any $Q \in \mathcal{K}_g$,

$$\mathcal{D}_Q \colon \mathcal{K}_g \to \mathbb{N}_0, \quad P \mapsto \mathcal{D}_Q(P) = \begin{cases} 1 & \text{for } P = Q, \\ 0 & \text{for } P \in \mathcal{K}_g \setminus \{Q\}. \end{cases} \tag{A.44}$$

Definition A.23
(i) Define

$$\underline{W}_0 = \{0\} \subset J(\mathcal{K}_g), \quad \underline{W}_{n, Q_0} = \underline{\alpha}_{Q_0}(\mathrm{Sym}^n(\mathcal{K}_g)), \quad n \in \mathbb{N}.$$

(ii) $Q \in \mathcal{K}_g$ is called a Weierstrass point of $\mathcal{K}_g$ if $i(g\mathcal{D}_Q) \geq 1$, where $g\mathcal{D}_Q = \sum_{j=1}^g \mathcal{D}_Q$.

Remark A.24
(i) Since $i(\mathcal{D}_P) = 0$ for all $P \in \mathcal{K}_1$, $\mathcal{K}_1$ has no Weierstrass points.
(ii) If $g \geq 2$ and $\mathcal{K}_g$ is hyperelliptic, $\mathcal{K}_g$ has precisely $2g + 2$ Weierstrass points. In particular, if $\mathcal{K}_g$ is given as a double cover of the Riemann sphere $\mathbb{C}_\infty$, then the Weierstrass points of $\mathcal{K}_g$ are precisely its $2g + 2$ branch points.
(iii) If $g \geq 3$ and $\mathcal{K}_g$ is not hyperelliptic, then the number N of Weierstrass points of $\mathcal{K}_g$ satisfies the inequality

$$2g + 2 \leq N \leq g^3 - g.$$

(iv) Special divisors $\mathcal{D}_{\underline{Q}}$ with $\deg(\underline{Q}) = N \geq g$ and $\underline{Q} = \{Q_1, \ldots, Q_N\} \in \mathrm{Sym}^N(\mathcal{K}_g)$ are the critical points of the Abel map $\underline{\alpha}_{Q_0} \colon \mathrm{Sym}^N(\mathcal{K}_g) \to J(\mathcal{K}_g)$, that is, the set of points $\mathcal{D}$ at which the rank of the differential $d\underline{\alpha}_{Q_0}$ is less than g.
(v) While $\mathrm{Sym}^m(\mathcal{K}_g) \not\subset \mathrm{Sym}^n(\mathcal{K}_g)$ for $m < n$, one has $\underline{W}_{m, Q_0} \subseteq \underline{W}_{n, Q_0}$ for $m < n$. Thus $\underline{W}_{n, Q_0} = J(\mathcal{K}_g)$ for $n \geq g$ by Theorem A.27 below.

Theorem A.25 *The set* $\Theta = \underline{W}_{g-1,Q_0} + \underline{\Xi}_{Q_0} \subset J(\mathcal{K}_g)$, *the so-called theta divisor, is the complete set of zeros of* θ *on* $J(\mathcal{K}_g)$, *that is,*

$$\theta(\underline{X}) = 0 \text{ if and only if } \underline{X} \in \Theta \tag{A.45}$$

(i.e., if and only if $\underline{X} = \left(\underline{\alpha}_{Q_0}(\mathcal{D}) + \underline{\Xi}_{Q_0}\right)$ *(mod* L_g*) for some* $\mathcal{D} \in \mathrm{Sym}^{g-1}(\mathcal{K}_g)$*). The theta divisor* Θ *has (complex) dimension* $g - 1$ *and is independent of the base point* Q_0.

Theorem A.26 (Riemann's vanishing theorem) *Let* $\underline{\xi} \in \mathbb{C}^g$.
(i) If $\theta(\underline{\xi}) \neq 0$, *then there exists a unique* $\mathcal{D} \in \mathrm{Sym}^g(\mathcal{K}_g)$ *such that*

$$\underline{\xi} = \left(\underline{\alpha}_{Q_0}(\mathcal{D}) + \underline{\Xi}_{Q_0}\right) \text{ (mod } L_g) \tag{A.46}$$

and

$$i(\mathcal{D}) = 0.$$

(ii) If $\theta(\underline{\xi}) = 0$ *and* $g = 1$, *then,*

$$\underline{\xi} = \underline{\Xi} \text{ (mod } L_1) = (1 + \tau)/2 \text{ (mod } L_1), \quad L_1 = \mathbb{Z} + \tau\mathbb{Z}, \quad -i\tau > 0.$$

(iii) Assume $\theta(\underline{\xi}) = 0$ *and* $g \geq 2$. *Let* $s \in \mathbb{N}$ *with* $s \leq g - 1$ *be the smallest integer such that* $\theta(\underline{W}_{s,Q_0} - \underline{W}_{s,Q_0} - \underline{\xi}) \neq 0$ *(i.e., there exist* $\mathcal{E}, \mathcal{F} \in \mathrm{Sym}^s(\mathcal{K}_g)$ *with* $\mathcal{E} \neq \mathcal{F}$ *such that* $\theta(\underline{\alpha}_{Q_0}(\mathcal{E}) - \underline{\alpha}_{Q_0}(\mathcal{F}) - \underline{\xi}) \neq 0$*). Then there exists a* $\mathcal{D} \in \mathrm{Sym}^{g-1}(\mathcal{K}_g)$ *such that*

$$\underline{\xi} = \left(\underline{\alpha}_{Q_0}(\mathcal{D}) + \underline{\Xi}_{Q_0}\right) \text{ (mod } L_g) \tag{A.47}$$

and

$$i(\mathcal{D}) = s.$$

All partial derivatives of θ *with respect to* $A_{Q_0,j}$ *for* $j = 1, \ldots, g$ *of order strictly less than* s *vanish at* $\underline{\xi}$, *whereas at least one partial derivative of* θ *of order* s *is nonzero at* $\underline{\xi}$. *Moreover,* $s \leq (g+1)/2$ *and the integer* s *is the same for* $\underline{\xi}$ *and* $-\underline{\xi}$.

Note that there is no explicit reference to the base point Q_0 in the formulation of Theorem A.26 since the set $\underline{W}_{s,Q_0} - \underline{W}_{s,Q_0} \subset J(\mathcal{K}_g)$ (cf. (A.42)) is independent of the base point Q_0, while $\underline{W}_{s,Q_0}$ alone is obviously not.

Theorem A.27 (Jacobi's inversion theorem) *The Abel map restricted to the set of nonnegative divisors,* $\underline{\alpha}_{Q_0} \colon \mathrm{Sym}^g(\mathcal{K}_g) \to J(\mathcal{K}_g)$, *is surjective. More precisely, given* $\underline{\tilde{\xi}} = (\underline{\xi} + \underline{\Xi}_{Q_0}) \in \mathbb{C}^g$, *the divisors* $\mathcal{D}$ *in (A.46) and (A.47) (respectively* $\mathcal{D} = \mathcal{D}_{Q_0}$ *if* $g = 1$*) solve the Jacobi inversion problem for* $\underline{\xi} \in \mathbb{C}^g$.

A special case of this analysis can be summarized as follows. Consider the function

$$G(\cdot) = \theta\big(\underline{\Xi}_{Q_0} - \underline{A}_{Q_0}(\cdot) + \sum_{j=1}^{g} \underline{A}_{Q_0}(Q_j)\big), \quad Q_j \in \mathcal{K}_g, \quad j = 1, \ldots, g$$

on $\mathcal{K}_g$. Then

$$\begin{aligned} G(Q_k) &= \theta\big(\underline{\Xi}_{Q_0} + \sum_{\substack{j=1 \\ j \neq k}}^{g} \underline{A}_{Q_0}(Q_j)\big) \\ &= \theta\big(\underline{\Xi}_{Q_0} + \underline{\alpha}_{Q_0}(\mathcal{D}_{(Q_1,\ldots,Q_{k-1},Q_{k+1},\ldots,Q_g)})\big) = 0, \quad k = 1, \ldots, g \end{aligned}$$

by Theorem A.25. Moreover, by Lemma A.18, Remark A.22, and Theorem A.26, the points $Q_1, \ldots, Q_g$ are the only zeros of G on $\mathcal{K}_g$ if and only if $\mathcal{D}_{\underline{Q}}$ is nonspecial, that is, if and only if

$$i(\mathcal{D}_{\underline{Q}}) = 0, \quad \underline{Q} = \{Q_1, \ldots, Q_g\} \in \mathrm{Sym}^g(\mathcal{K}_g).$$

Conversely, $G \equiv 0$ on $\mathcal{K}_g$ if and only if $\mathcal{D}_{\underline{Q}}$ is special, that is, if and only if $i(\mathcal{D}_{\underline{Q}}) \geq 1$. Thus, one obtains the following fact.

Theorem A.28 *Let $\underline{Q} = \{Q_1, \ldots, Q_g\} \in \mathrm{Sym}^g(\mathcal{K}_g)$ and assume $\mathcal{D}_{\underline{Q}}$ to be nonspecial, that is, $i(\mathcal{D}_{\underline{Q}}) = 0$. Then,*

$$\theta(\underline{\Xi}_{Q_0} - \underline{A}_{Q_0}(P) + \underline{\alpha}_{Q_0}(\mathcal{D}_{\underline{Q}})) = 0 \text{ if and only if } P \in \{Q_1, \ldots, Q_g\}.$$

We also mention the elementary change in the Abel map and in Riemann's vector if one changes the base point,

$$\underline{A}_{Q_1} = \big(\underline{A}_{Q_0} - \underline{A}_{Q_0}(Q_1)\big) \pmod{L_g}, \tag{A.48}$$

$$\underline{\Xi}_{Q_1} = \big(\underline{\Xi}_{Q_0} + (g-1)\underline{A}_{Q_0}(Q_1)\big) \pmod{L_g}, \quad Q_0, Q_1 \in \mathcal{K}_g. \tag{A.49}$$

Remark A.29 The L_g quasi-periodic holomorphic function θ on $\mathbb{C}^g$ can be used to construct L_g periodic, meromorphic functions f on $\mathbb{C}^g$ as follows. Either
(i)

$$f(\underline{z}) = \prod_{j=1}^{N} \frac{\theta(\underline{z} + \underline{c}_j)}{\theta(\underline{z} + \underline{d}_j)}, \quad \underline{z}, \underline{c}_j, \underline{d}_j \in \mathbb{C}^g, \quad j = 1, \ldots, N,$$

where

$$\sum_{j=1}^{N} \underline{c}_j = \sum_{j=1}^{N} \underline{d}_j \pmod{\mathbb{Z}^g},$$

or
(ii)
$$f(\underline{z}) = \partial_{z_j} \ln\left(\frac{\theta(\underline{z}+\underline{e})}{\theta(\underline{z}+\underline{h})}\right), \quad j = 1,\ldots,g, \quad \underline{z}, \underline{e}, \underline{h} \in \mathbb{C}^g,$$

or
(iii)
$$f(\underline{z}) = \partial^2_{z_j z_k} \ln \theta(\underline{z}), \quad \underline{z} \in \mathbb{C}^g, \quad j, k = 1,\ldots,g.$$

Then indeed in all cases (i)–(iii),
$$f(\underline{z} + \underline{m} + \underline{n}\tau) = f(\underline{z}), \quad \underline{z} \in \mathbb{C}^g, \quad \underline{m}, \underline{n} \in \mathbb{Z}^g$$

holds by (A.35).

Remark A.30 In the main text we frequently deal with theta function expressions of the type

$$\phi(P) = \frac{\theta(\underline{\Xi}_{Q_0} - \underline{A}_{Q_0}(P) + \underline{\alpha}_{Q_0}(\mathcal{D}_1))}{\theta(\underline{\Xi}_{Q_0} - \underline{A}_{Q_0}(P) + \underline{\alpha}_{Q_0}(\mathcal{D}_2))} \exp\left(\int_{Q_0}^{P} \omega^{(3)}_{Q_1, Q_2}\right), \quad P \in \mathcal{K}_g \quad \text{(A.50)}$$

and

$$\psi(P) = \frac{\theta(\underline{\Xi}_{Q_0} - \underline{A}_{Q_0}(P) + \underline{\alpha}_{Q_0}(\mathcal{D}_1))}{\theta(\underline{\Xi}_{Q_0} - \underline{A}_{Q_0}(P) + \underline{\alpha}_{Q_0}(\mathcal{D}_2))} \exp\left(-c \int_{Q_0}^{P} \Omega^{(2)}\right), \quad P \in \mathcal{K}_g, \quad \text{(A.51)}$$

where $\mathcal{D}_j \in \mathrm{Sym}^g(\mathcal{K}_g)$, $j = 1, 2$, are nonspecial positive divisors of degree g, $c \in \mathbb{C}$ is a constant, $Q_j \in \mathcal{K}_g \setminus \{Q_0\}$, $j = 1, 2$, $Q_1 \neq Q_2$, $\omega^{(3)}_{Q_1, Q_2}$ is a normal differential of the third kind, and $\Omega^{(2)}$ a normalized differential of the second kind with singularities contained in $\{P_1, \ldots, P_N\} \subset \mathcal{K}_g \setminus \{Q_0\}$, $P_k \neq P_\ell$, for $k \neq \ell$, $k, \ell = 1, \ldots, N$, for some $N \in \mathbb{N}$. In particular, one has

$$\int_{a_j} \omega^{(3)}_{Q_1, Q_2} = \int_{a_j} \Omega^{(2)} = 0, \quad j = 1, \ldots, g. \quad \text{(A.52)}$$

Even though we agree always to choose identical paths of integration from Q_0 to P in all abelian integrals (A.50) and (A.51), this is not sufficient to render ϕ and ψ single-valued on $\mathcal{K}_g$. To achieve single-valuedness one needs to replace $\mathcal{K}_g$ by its simply connected canonical dissection $\widehat{\mathcal{K}}_g$ and then replace $\underline{A}_{Q_0}$, $\underline{\alpha}_{Q_0}$ in (A.50) and (A.51) with $\underline{\widehat{A}}_{Q_0}$, $\underline{\hat{\alpha}}_{Q_0}$, as introduced in (A.37) and (A.38). In particular, one regards a_j, b_j as curves (being a part of $\partial \widehat{\mathcal{K}}_g$, cf. (A.12)) and not as homology classes $[a_j]$, $[b_j]$ in $H_1(\mathcal{K}_g, \mathbb{Z})$. Similarly, one then replaces $\underline{\Xi}_{Q_0}$ by $\underline{\widehat{\Xi}}_{Q_0}$ (replacing $\underline{A}_{Q_0}$ by $\underline{\widehat{A}}_{Q_0}$ in (A.41), etc.). Moreover, to render ϕ single-valued on $\widehat{\mathcal{K}}_g$ one needs to assume in addition that

$$\underline{\hat{\alpha}}_{Q_0}(\mathcal{D}_1) - \underline{\hat{\alpha}}_{Q_0}(\mathcal{D}_2) = 0 \quad \text{(A.53)}$$

(as opposed to merely $\underline{\alpha}_{Q_0}(\mathcal{D}_1) - \underline{\alpha}_{Q_0}(\mathcal{D}_2) = 0 \pmod{L_g}$). Similarly, in connection with ψ, one introduces the vector of b-periods $\underline{U}^{(2)}$ of $\Omega^{(2)}$ by

$$\underline{U}^{(2)} = (U_1^{(2)}, \ldots, U_g^{(2)}), \quad U_j^{(2)} = \frac{1}{2\pi i} \int_{b_j} \Omega^{(2)}, \quad j = 1, \ldots, g, \quad (A.54)$$

and then renders ψ single-valued on $\widehat{\mathcal{K}}_g$ by requiring

$$\hat{\underline{\alpha}}_{Q_0}(\mathcal{D}_1) - \hat{\underline{\alpha}}_{Q_0}(\mathcal{D}_2) = c\,\underline{U}^{(2)} \quad (A.55)$$

(as opposed to merely $\underline{\alpha}_{Q_0}(\mathcal{D}_1) - \underline{\alpha}_{Q_0}(\mathcal{D}_2) = c\,\underline{U}^{(2)} \pmod{L_g}$). These statements easily follow from (A.23) and (A.35) in the case of ϕ and simply from (A.35) in the case of ψ. In fact, by (A.35),

$$\hat{\underline{\alpha}}_{Q_0}(\mathcal{D}_1 + \mathcal{D}_{Q_1}) - \hat{\underline{\alpha}}_{Q_0}(\mathcal{D}_2 + \mathcal{D}_{Q_2}) \in \mathbb{Z}^g, \quad (A.56)$$

respectively,

$$\hat{\underline{\alpha}}_{Q_0}(\mathcal{D}_1) - \hat{\underline{\alpha}}_{Q_0}(\mathcal{D}_2) - c\,\underline{U}^{(2)} \in \mathbb{Z}^g, \quad (A.57)$$

suffice to guarantee single-valuedness of ϕ, respectively, ψ on $\widehat{\mathcal{K}}_g$. Without the replacement of $\underline{A}_{Q_0}$ and $\underline{\alpha}_{Q_0}$ by $\hat{\underline{A}}_{Q_0}$ and $\hat{\underline{\alpha}}_{Q_0}$ in (A.50) and (A.51) and the assumptions (A.53) and (A.55) (or (A.56) and (A.57)), ϕ and ψ are multiplicative (multi-valued) functions on $\mathcal{K}_g$, and then most effectively discussed by introducing the notion of characters on $\mathcal{K}_g$. For simplicity, we decided to avoid the latter possibility and throughout this text will tacitly always assume (A.53) and (A.55) without particularly emphasizing this convention each time it is used.

Remark A.31 Let $\underline{\xi} \in J(\mathcal{K}_g)$ be given, assume that $\theta(\underline{\Xi}_{Q_0} - \underline{A}_{Q_0}(\cdot) + \underline{\xi}) \not\equiv 0$ on $\mathcal{K}_g$ and suppose that $\underline{\alpha}_{Q_0}^{-1}(\underline{\xi}) = \{Q_1, \ldots, Q_g\} \in \operatorname{Sym}^g(\mathcal{K}_g)$ is the unique solution of Jacobi's inversion problem. Let $f \in \mathcal{M}(\mathcal{K}_g) \setminus \{0\}$ and suppose $f(Q_j) \neq \infty$ for $j = 1, \ldots, g$. Then $\underline{\xi}$ uniquely determines the values $f(Q_1), \ldots, f(Q_g)$. Moreover, any symmetric function of these values is a single-valued meromorphic function of $\underline{\xi} \in J(\mathcal{K}_g)$, that is, an abelian function on $J(\mathcal{K}_g)$. Any such meromorphic function on $J(\mathcal{K}_g)$ can be expressed in terms of the Riemann theta function on $\mathcal{K}_g$. For instance, for the elementary symmetric functions of the second kind (Newton polynomials) one obtains from the residue theorem in analogy to the proof of Lemma A.18 that

$$\sum_{j=1}^{g} f(Q_j)^n = \sum_{j=1}^{g} \int_{a_j} f(P)^n \omega_j(P)$$
$$- \sum_{\substack{P_r \in \mathcal{K}_g \\ f(P_r) = \infty}} \operatorname*{res}_{P = P_r} \left(f(P)^n d \ln \left(\theta\big(\underline{\Xi}_{Q_0} - \underline{A}_{Q_0}(P) + \underline{\alpha}_{Q_0}(\mathcal{D}_{\underline{Q}})\big) \right) \right),$$

$$\underline{Q} = \{Q_1, \ldots, Q_n\} \in \operatorname{Sym}^n(\mathcal{K}_g). \quad (A.58)$$

Algebraic Curves and Their Theta Functions in a Nutshell

Here an appropriate homology basis $\{a_j, b_j\}_{j=1}^g$, avoiding $\{Q_1, \ldots, Q_g\}$ and the poles $\{P_r\}$ of f, has been chosen.

In the special case of hyperelliptic Riemann surfaces, special divisors are characterized as follows. Denote by $*: \mathcal{K}_g \to \mathcal{K}_g$ the sheet exchange map (involution)

$$*: \mathcal{K}_g \to \mathcal{K}_g, \quad [x_2 : x_1 : x_0] \mapsto [x_2 : -x_1 : x_0].$$

Theorem A.32 *Suppose $\mathcal{K}_g$ is a hyperelliptic Riemann surface of genus $g \in \mathbb{N}$ and $\mathcal{D}_{\underline{Q}} \in \mathrm{Sym}^g(\mathcal{K}_g)$ with $\underline{Q} = \{Q_1, \ldots, Q_g\} \in \mathrm{Sym}^g(\mathcal{K}_g)$. Then,*

$$1 \leq i(\mathcal{D}_{\underline{Q}}) = s$$

if and only if $\{Q_1, \ldots, Q_g\}$ contains s pairings of the type $\{P, P^*\}$. (This includes, of course, branch points of $\mathcal{K}_g$ for which $P = P^*$.) One has $s \leq g/2$.

We add one more result in connection with hyperelliptic Riemann surfaces.

Theorem A.33 *Suppose $\mathcal{K}_g$ is a hyperelliptic Riemann surface of genus $g \in \mathbb{N}$, $\mathcal{D}_{\underline{\hat{\mu}}} \in \mathrm{Sym}^g(\mathcal{K}_g)$ is nonspecial, $\underline{\hat{\mu}} = \{\hat{\mu}_1, \ldots, \hat{\mu}_g\}$, and $\hat{\mu}_{g+1} \in \mathcal{K}_g$ with $\hat{\mu}_{g+1}^* \notin \{\hat{\mu}_1, \ldots, \hat{\mu}_g\}$. Let $\{\hat{\lambda}_1, \ldots, \hat{\lambda}_{g+1}\} \subset \mathcal{K}_g$ with $\mathcal{D}_{\underline{\hat{\lambda}}\hat{\lambda}_{g+1}} \sim \mathcal{D}_{\underline{\hat{\mu}}\hat{\mu}_{g+1}}$ (i.e., $\mathcal{D}_{\underline{\hat{\lambda}}\hat{\lambda}_{g+1}} \in [\mathcal{D}_{\underline{\hat{\mu}}\hat{\mu}_{g+1}}]$). Then any g points $\hat{\nu}_j \in \{\hat{\lambda}_1, \ldots, \hat{\lambda}_{g+1}\}$, $j = 1, \ldots, g$, define a nonspecial divisor $\mathcal{D}_{\underline{\hat{\nu}}} \in \mathrm{Sym}^g(\mathcal{K}_g)$, $\underline{\hat{\nu}} = \{\hat{\nu}_1, \ldots, \hat{\nu}_g\}$.*

Proof Since $i(\mathcal{D}_P) = 0$ for all $P \in \mathcal{K}_1$, there is nothing to be proven in the special case $g = 1$. Hence we assume $g \geq 2$. Let $Q_0 \in \mathcal{B}(\mathcal{K}_g)$ be a fixed branch point of $\mathcal{K}_g$ and suppose that $\mathcal{D}_{\underline{\hat{\nu}}}$ is special. Then by Theorem A.32 there is a pair $\{\hat{\nu}, \hat{\nu}^*\} \subset \{\hat{\nu}_1, \ldots, \hat{\nu}_g\}$ such that

$$\underline{\alpha}_{Q_0}(\mathcal{D}_{\underline{\hat{\nu}}}) = \underline{\alpha}_{Q_0}(\mathcal{D}_{\underline{\hat{\nu}}}),$$

where $\underline{\hat{\underline{\nu}}} = \{\hat{\nu}_1, \ldots, \hat{\nu}_g\} \setminus \{\hat{\nu}, \hat{\nu}^*\} \in \mathrm{Sym}^{g-2}(\mathcal{K}_g)$. Let $\hat{\nu}_{g+1} \in \{\hat{\lambda}_1, \ldots, \hat{\lambda}_{g+1}\} \setminus \{\hat{\nu}_1, \ldots, \hat{\nu}_g\}$ so that $\{\hat{\nu}_1, \ldots, \hat{\nu}_{g+1}\} = \{\hat{\lambda}_1, \ldots, \hat{\lambda}_{g+1}\} \in \mathrm{Sym}^{g+1}(\mathcal{K}_g)$. Then

$$\underline{\alpha}_{Q_0}(\mathcal{D}_{\underline{\hat{\underline{\nu}}}\hat{\nu}_{g+1}}) = \underline{\alpha}_{Q_0}(\mathcal{D}_{\underline{\hat{\nu}}\hat{\nu}_{g+1}}) = \underline{\alpha}_{Q_0}(\mathcal{D}_{\underline{\hat{\lambda}}\hat{\lambda}_{g+1}})$$

$$= \underline{\alpha}_{Q_0}(\mathcal{D}_{\underline{\hat{\mu}}\hat{\mu}_{g+1}}) = -\underline{A}_{Q_0}(\hat{\mu}_{g+1}^*) + \underline{\alpha}_{Q_0}(\mathcal{D}_{\underline{\hat{\mu}}}), \tag{A.59}$$

and hence by Theorem A.25 and (A.59),

$$0 = \theta(\underline{\Xi}_{Q_0} + \underline{\alpha}_{Q_0}(\mathcal{D}_{\underline{\hat{\underline{\nu}}}\hat{\nu}_{g+1}})) = \theta(\underline{\Xi}_{Q_0} - \underline{A}_{Q_0}(\hat{\mu}_{g+1}^*) + \underline{\alpha}_{Q_0}(\mathcal{D}_{\underline{\hat{\mu}}})). \tag{A.60}$$

Since by hypothesis $\mathcal{D}_{\underline{\hat{\mu}}}$ is nonspecial and $\hat{\mu}_{g+1}^* \notin \{\hat{\mu}_1, \ldots, \hat{\mu}_g\}$, (A.60) contradicts Theorem A.28. Thus, $\mathcal{D}_{\underline{\hat{\nu}}}$ is nonspecial. □

We conclude this appendix with a brief summary of Riemann surfaces with symmetries, which is a topic of fundamental importance in characterizing real-valued algebro-geometric solutions of the Toda hierarchy.

Assuming $\mathcal{K}_g$ to be a compact Riemann surface of genus g, let

$$\rho: \mathcal{K}_g \to \mathcal{K}_g$$

be an antiholomorphic involution on $\mathcal{K}_g$ (i.e., $\rho^2 = \mathrm{id}\,|_{\mathcal{K}_g}$). Moreover, let

$$\mathcal{R} = \{P \in \mathcal{K}_g \,|\, \rho(P) = P\}$$

be the set of fixed points of ρ (sometimes called the set of "real" points of $\mathcal{K}_g$) and denote by r the number of nontrivial connected components of $\mathcal{R}$. Topologically, these connected components are circles.

Theorem A.34 *Let ρ be an antiholomorphic involution on $\mathcal{K}_g$. Then either $\mathcal{K}_g \setminus \mathcal{R}$ is connected (and then the quotient space (the Klein surface) $\mathcal{K}_g/\rho$ is nonorientable) or $\mathcal{K}_g \setminus \mathcal{R}$ consists precisely of two connected components (and then $\mathcal{K}_g/\rho$ is orientable). In the latter case, if $\mathcal{R} \neq \emptyset$, $\mathcal{K}_g \cup \mathcal{R}$ is a bordered Riemann surface and $(\mathcal{K}_g, \rho)$ is the complex double of $\mathcal{K}_g \cup \mathcal{R}$.*

Definition A.35 *Suppose ρ is an antiholomorphic involution on $\mathcal{K}_g$. Define*

$$\varepsilon = \begin{cases} + & \text{if } \mathcal{K}_g \setminus \mathcal{R} \text{ is disconnected,} \\ - & \text{if } \mathcal{K}_g \setminus \mathcal{R} \text{ is connected.} \end{cases}$$

The pair $(\mathcal{K}_g, \rho)$ is called a symmetric Riemann surface, the triple (g, r, ε) denotes the type of $(\mathcal{K}_g, \mathcal{R})$. If $\varepsilon = +$, $(\mathcal{K}_g, \rho)$ is of dividing (separating) type; if $\varepsilon = -$, $(\mathcal{K}_g, \rho)$ is of nondividing (nonseparating) type.
If $r = g + 1$, then $\mathcal{K}_g$ is called an M*-curve.*

Theorem A.36 *Assume ρ is an antiholomorphic involution on $\mathcal{K}_g$.*
(i) If $\varepsilon = +$, then $1 \leq r \leq g+1$, $r \equiv g+1 \pmod 2$, $g = r - 1 + 2k$, $0 \leq k \leq (g+1-r)/2$.
(ii) If $\varepsilon = -$, then $0 \leq r \leq g$.
(iii) If $r = 0$, then $\varepsilon = -$. If $r = g + 1$, then $\varepsilon = +$.

Example A.37
(i) Consider the hyperelliptic Riemann surface $\mathcal{K}_g: y^2 = \prod_{m=0}^{2g+1}(z - E_m)$ with E_m, $m = 0, \ldots, 2g+1$, grouped into k real and ℓ complex conjugate pairs, $k + \ell = g + 1$. Define the antiholomorphic involution $\rho_+: (z, y) \mapsto (\bar{z}, \bar{y})$ on $\mathcal{K}_g$. If $(\mathcal{K}_g, \rho_+)$ is of dividing type $(g, r, +)$, then either $r = g + 1 = k$ and $\ell = 0$ (if $E_m \in \mathbb{R}$, $m = 0, \ldots, 2g+1$), or else, $r = 1$ if g is even and $r = 2$ if g is odd if none of the E_m are real (and hence only occur in complex conjugate pairs). In particular, if

$\prod_{m=0}^{2g+1}(z - E_m)$ contains $2r > 0$ real roots, then $(\mathcal{K}_g, \rho_+)$ is of type $(g, r, +)$ if and only if $r = g + 1$ and of type $(g, r, -)$ if and only if $1 \leq r \leq g$. If $(\mathcal{K}_g, \rho_+)$ is of nondividing type, then $r = k$ (and of course $1 \leq r \leq g$).
(ii) The hyperelliptic Riemann surfaces $\mathcal{K}_g \colon y^2 = \pm \prod_{m=0}^{g} |z - \tilde{E}_m|^2$ are of the type $(g, 0, -)$ with respect to the antiholomorphic involutions $\rho_\pm \colon (z, y) \mapsto (\bar{z}, \pm \bar{y})$, respectively, since $\mathcal{R} = \emptyset, r = 0$ in either case.
(iii) Consider the hyperelliptic Riemann surface $\mathcal{K}_g \colon y^2 = \prod_{m=0}^{2g}(z - E_m)$ with $E_0 \in \mathbb{R}$ and $E_m, m = 1, \ldots, 2g$, grouped into k real and ℓ complex conjugate pairs, $k + \ell = g$. In addition, define the antiholomorphic involution $\rho_+ \colon (z, y) \mapsto (\bar{z}, \bar{y})$ on $\mathcal{K}_g$. Then $(\mathcal{K}_g, \rho_+)$ is of dividing type $(g, r, +)$ if and only if $r = g + 1 = k + 1$ and $\ell = 0$ (and hence $E_m \in \mathbb{R}, m = 0, \ldots, 2g$). If $(\mathcal{K}_g, \rho_+)$ is of nondividing type, then $r = k + 1$ (and of course $1 \leq r \leq g$).

In the following it is convenient to abbreviate

$$\underline{\mathrm{diag}}(M) = (M_{1,1}, \ldots, M_{g,g})$$

for a $g \times g$ matrix M with entries in $\mathbb{C}$.

Theorem A.38 *Let $(\mathcal{K}_g, \rho)$ be a symmetric Riemann surface.*
(i) There exists a canonical homology basis $\{a_j, b_j\}_{j=1}^{g}$ on $\mathcal{K}_g$ with intersection matrix (A.8) and a symmetric $g \times g$ matrix R such that the $2g \times 2g$ matrix S of complex conjugation of the action of ρ on $\mathrm{H}_1(\mathcal{K}_g, \mathbb{Z})$ in this basis is given by

$$S = \begin{pmatrix} I_g & R \\ 0 & -I_g \end{pmatrix}, \quad R^\top = R,$$

that is,

$$(\underline{\rho(a)}, \underline{\rho(b)}) = (\underline{a}, \underline{b}) \begin{pmatrix} I_g & R \\ 0 & -I_g \end{pmatrix} = (\underline{a}, \underline{a}R - \underline{b}),$$

$$\underline{a} = (a_1, \ldots, a_g), \quad \underline{b} = (b_1, \ldots, b_g),$$

$$\underline{\rho(a)} = (\rho(a_1), \ldots, \rho(a_g)), \quad \underline{\rho(b)} = (\rho(b_1), \ldots, \rho(b_g)),$$

where R is of the following form.[1]

---

[1] Blank entries are representing zeros in the matrices (A.61)–(A.64). Moreover, 0 in (A.61) and (A.64) denotes a 1×1 element whereas 0 in (A.62) represents an $(r - 1) \times (r - 1)$ block matrix (which is absent for $r = 1$).

If $\mathcal{R} \neq \emptyset$ and $\mathcal{K}_g \setminus \mathcal{R}$ is disconnected,

$$R = \begin{pmatrix} \sigma_1 & & & & \\ & \ddots & & & \\ & & \sigma_1 & & \\ & & & 0 & \\ & & & & \ddots \\ & & & & & 0 \end{pmatrix}, \quad \operatorname{rank}(R) = g + 1 - r \tag{A.61}$$

(*in particular, $R = 0$ if $r = g + 1$*).
If $\mathcal{R} \neq \emptyset$ and $\mathcal{K}_g \setminus \mathcal{R}$ is connected,

$$R = \begin{pmatrix} I_{g+1-r} & \\ & 0 \end{pmatrix}, \quad \operatorname{rank}(R) = g + 1 - r \tag{A.62}$$

(*in particular, $R = 0$ if $r = g + 1$*).
If $\mathcal{R} = \emptyset$ and g is even,

$$R = \begin{pmatrix} \sigma_1 & & \\ & \ddots & \\ & & \sigma_1 \end{pmatrix}, \quad \operatorname{rank}(R) = g \tag{A.63}$$

(*in particular, $R = 0$ if $g = 0$*).
If $\mathcal{R} = \emptyset$ and g is odd,

$$R = \begin{pmatrix} \sigma_1 & & & \\ & \ddots & & \\ & & \sigma_1 & \\ & & & 0 \end{pmatrix}, \quad \operatorname{rank}(R) = g - 1 \tag{A.64}$$

(*in particular, $R = 0$ if $g = 1$*).
Here σ_1 denotes the 2×2 Pauli matrix

$$\sigma_1 = \begin{pmatrix} 0 & 1 \\ 1 & 0 \end{pmatrix}.$$

(*ii*) *Given the basis of cycles $\{a_j, b_j\}_{j=1}^g$ of item (i), introduce the corresponding basis $\{\omega_j\}_{j=1}^g$ of normalized holomorphic differentials satisfying (A.13) and define the associated matrix τ of b-periods as in (A.14) and the Riemann theta function θ as in (A.33). Then, ω_j, $j = 1, \ldots, g$, are ρ-real, that is,*

$$\rho^* \omega_j = \overline{\omega_j}, \quad j = 1, \ldots, g, \tag{A.65}$$

where $\rho^*\omega$ denotes the pull back[1] of a meromorphic differential ω by the involution ρ. Moreover,

$$\overline{\tau} = R - \tau, \quad \operatorname{Re}(\tau) = \frac{1}{2}R, \tag{A.66}$$

$$\overline{\theta(\underline{z})} = \theta\left(\overline{\underline{z}} + \frac{1}{2}\operatorname{diag}(R)\right), \quad \underline{z} \in \mathbb{C}^g, \tag{A.67}$$

$$\overline{\Xi}_{Q_0} = \underline{\Xi}_{Q_0} + \frac{1}{2}\operatorname{diag}(R) + (g-1)\underline{\alpha}_{Q_0}(\rho(Q_0)). \tag{A.68}$$

Finally, assume $\mathcal{R} \neq \emptyset$, $\mathcal{K}_g \setminus \mathcal{R}$ is disconnected (cf. (A.61)), $\underline{x} \in \mathbb{R}^g$, $\chi_m \in \{0, 1\}$, $m = \ell + 1, \ldots, g$, $\ell = \operatorname{rank}(R)$. Then

$$\theta\left(i\underline{x} + \frac{1}{2}(0, \ldots, 0, \chi_{\ell+1}, \ldots, \chi_g)\right) \neq 0$$

if and only if $\chi_m = 0$, $m = \ell + 1, \ldots, g$.

Assuming P_0 to be a Weierstrass point of $\mathcal{K}_g$ (implying $g \geq 2$), $\underline{\Xi}_{P_0}$ is a half-period, that is,

$$\underline{\Xi}_{P_0} = -\underline{\Xi}_{P_0} \pmod{L_g}, \tag{A.69}$$

$$\underline{\Xi}_{P_0} = \underline{\beta} + \underline{\gamma}\tau, \quad \underline{\beta}, \underline{\gamma} \in \frac{1}{2}\mathbb{Z}^g. \tag{A.70}$$

In this case (A.66) implies in addition to (A.68) and (A.69),

$$\overline{\Xi}_{P_0} = -\underline{\Xi}_{P_0} \pmod{\mathbb{Z}^g}. \tag{A.71}$$

In the context of the hyperelliptic Riemann surfaces described as two-sheeted covers of the Riemann sphere $\mathbb{C}_\infty$ in Appendix B, equation (A.71) remains valid if P_0 is any of the associated $2g + 2$ branch points, $g \in \mathbb{N}$.

Notes

This appendix, apart from some minor corrections and additions, is nearly identical to Appendix A in Gesztesy and Holden (2003b).

The bulk of the material of this appendix is standard and taken from textbooks. The one we relied on most was Farkas and Kra (1992). Moreover, we used material from Mumford (1983, Ch. II) and Mumford (1984, Ch. IIIa) (concerning divisors on hyperelliptic Riemann surfaces constructed originally in Jacobi (1846)), and Behnke and Sommer (1965) (in connection with covering Riemann surfaces). In addition, the following well-known sources make for great collateral reading on various topics

---

[1] If $\omega = f(\zeta)d\zeta$, then $\rho^*\omega = f(\rho(\zeta))d\rho(\zeta)$ and $\int_\gamma \rho^*\omega = \int_{\rho(\gamma)} \omega$, $\gamma \in H_1(\mathcal{K}_g, \mathbb{Z})$. In particular, if ω is ρ-real, that is, $\rho^*(\omega) = \overline{\omega}$, then $\overline{\int_\gamma \omega} = \int_{\rho(\gamma)} \omega$, $\gamma \in H_1(\mathcal{K}_g, \mathbb{Z})$.

relevant to the applications discussed in this monograph: Belokolos et al. (1994, Ch. 2), Fay (1973), Forster (1999), Forsyth (1965), Griffiths (1989), Griffiths and Harris (1978), Gunning (1972), Hofmann (1888), Kirwan (1992), Markushevich (1992), Miranda (1995), Narasimhan (1992), Rauch and Farkas (1974), Reyssat (1989), Rodin (1988), Schlichenmaier (1989), Shokurov (1994), Siegel (1988), Springer (1981), and the recently reprinted classical treatise by Baker (1995). Finally, there are various reviews on compact Riemann surfaces and their associated theta functions. A masterpiece in this connection and still of great relevance is Dubrovin (1981). In addition, we call attention to the following reviews, Bost (1992), Korotkin (1998), Krichever (1983), Krichever and Novikov (1999), Lewittes (1964), Matveev (1976; 2008), Rodin (1987), Smith (1989), and Taimanov (1997).

The fact that different homology bases yield isomorphic Jacobians, as alluded to after Definition A.15, is discussed, for instance, in Farkas and Kra (1992, p. 137) and Gunning (1966, Sect. 8(b)). For a detailed discussion of multiplicative (multi-valued) functions in connection with Remark A.30 (a topic we circumvent in this monograph) we refer to Farkas and Kra (1992, Sects. III.9, VI.2). Theorem A.32 can be found in Krazer (1970, Sect. X.3).

In connection with Remark A.17 we refer to Farkas and Kra (1992, pp. 73, 150, 152, 309).

Finally, the material on symmetric Riemann surfaces is mainly taken from Gross and Harris (1981) and Vinnikov (1993). In particular, Theorem A.38 is proved in Vinnikov (1993) (compare Alpay and Vinnikov (2002) for the proof of (A.67)). The case $\mathcal{R} \neq \emptyset$ and $\mathcal{K}_g \setminus \mathcal{R}$ disconnected is treated in Fay (1973, Ch. VI). Classical sources for this material are Comessatti (1924), Harnack (1876), Klein (1893), and Weichold (1883). For modern treatments of this subject we refer to Bujalance et al. (2001), Fay (1973, Ch. VI), Gross and Harris (1981), Natanzon (1980; 1990), Silhol (1982), and Wilson (1978); applications to algebro-geometric solutions can be found, for instance, in Date (1982), Dubrovin (1982; 1983), Dubrovin and Natanzon (1982), Natanzon (1995), Taimanov (1990), and Zhivkov (1989; 1994).

Appendix B
Hyperelliptic Curves of the Toda-Type

...and he that contemneth small things,
shall fall by little and little.

Ecclesiasticus, 19:1

We briefly summarize some of the basic facts on hyperelliptic Toda-type curves (i.e., those not branched at infinity) as employed in this volume. We freely use the notation and results collected in Appendix A.

Fix $p \in \mathbb{N}_0$. We are going to construct the hyperelliptic Riemann surface $\mathcal{K}_p$ of genus p associated with the Toda-type curve,

$$\mathcal{F}_p(z, y) = y^2 - R_{2p+2}(z) = 0,$$

$$R_{2p+2}(z) = \prod_{m=0}^{2p+1} (z - E_m), \quad \{E_m\}_{m=0,\ldots,2p+1} \subset \mathbb{C}, \qquad (B.1)$$

$$E_m \neq E_{m'} \text{ for } m \neq m', \ m, m' = 0, \ldots, 2p+1.$$

Introducing an appropriate set of $p + 1$ (nonintersecting) cuts $\mathcal{C}_j$ joining $E_{m(j)}$ and $E_{m'(j)}$ we denote

$$\mathcal{C} = \bigcup_{j \in J} \mathcal{C}_j, \quad \mathcal{C}_j \cap \mathcal{C}_k = \emptyset, \quad j \neq k,$$

where the finite index set $J \subseteq \{1, \ldots, p + 1\}$ has cardinality $p + 1$. Define the cut plane Π,

$$\Pi = \mathbb{C} \setminus \mathcal{C},$$

and introduce the holomorphic function

$$R_{2p+2}(\cdot)^{1/2} \colon \Pi \to \mathbb{C}, \quad z \mapsto \left(\prod_{m=0}^{2p+1} (z - E_m) \right)^{1/2} \qquad (B.2)$$

353

on Π with an appropriate choice of the square root branch (B.2). Next, define

$$\mathcal{M}_p = \{(z, \sigma R_{2p+2}(z)^{1/2}) \mid z \in \mathbb{C}, \sigma \in \{1, -1\}\} \cup \{P_{\infty_+}, P_{\infty_-}\},$$

by extending $R_{2p+2}(\cdot)^{1/2}$ to $\mathcal{C}$ and joining P_{∞_+} and P_{∞_-}, $P_{\infty_+} \neq P_{\infty_-}$, the two points at infinity. To describe charts on $\mathcal{M}_p$ let $Q_0 \in \mathcal{M}_p$, $U_{Q_0} \subset \mathcal{M}_p$ a neighborhood of Q_0, $\zeta_{Q_0} : U_{Q_0} \to V_{Q_0} \subset \mathbb{C}$ a homeomorphism defined below, and write

$$Q_0 = (z_0, \sigma_0 R_{2p+2}(z_0)^{1/2}) \text{ or } Q_0 = P_{\infty_\pm},$$
$$Q = (z, \sigma R_{2p+2}(z)^{1/2}) \in U_{Q_0} \subset \mathcal{M}_p, \quad V_{Q_0} = \zeta_{Q_0}(U_{Q_0}) \subset \mathbb{C}.$$

Branch points on $\mathcal{M}_p$ are defined by

$$\mathcal{B}_s(\mathcal{K}_p) = \begin{cases} \{(E_m, 0)\}_{m=0,1}, & p = 0, \\ \{(E_m, 0)\}_{m=0,\ldots,2p+1} \cup \{P_{\infty_+}, P_{\infty_-}\}, & p \in \mathbb{N}. \end{cases}$$

While $P_{\infty_\pm}$ are never branch points, $P_{\infty_\pm}$ are nonsingular for $p = 0$ and singular for $p \in \mathbb{N}$.

Charts on $\mathcal{M}_p$ are now introduced distinguishing three cases, (i) $Q_0 \in \mathcal{M}_p \setminus (\mathcal{B}_s(\mathcal{K}_p) \cup \{P_{\infty_+}, P_{\infty_-}\})$, (ii) $Q_0 = P_{\infty_\pm}$, and (iii) $Q_0 = (E_m, 0)$ for some $m = 0, \ldots, 2p + 1$.

(i) $Q_0 \in \mathcal{M}_p \setminus (\mathcal{B}_s(\mathcal{K}_p) \cup \{P_{\infty_+}, P_{\infty_-}\})$: Then one defines

$$U_{Q_0} = \{Q \in \mathcal{M}_p \mid |z - z_0| < C_0\}, \quad C_0 = \min_{m=0,\ldots,2p+1} |z_0 - E_m|, \tag{B.3}$$

where $\sigma R_{2p+2}(z)^{1/2}$ is the branch obtained by straight line analytic continuation starting from z_0,

$$V_{Q_0} = \{\zeta \in \mathbb{C} \mid |\zeta| < C_0\}, \tag{B.4}$$

and

$$\zeta_{Q_0} : U_{Q_0} \to V_{Q_0}, \quad Q \mapsto (z - z_0) \tag{B.5}$$

with inverse

$$\zeta_{Q_0}^{-1} : V_{Q_0} \to U_{Q_0}, \quad \zeta \mapsto (z_0 + \zeta, \sigma R_{2p+2}(z_0 + \zeta)^{1/2}). \tag{B.6}$$

(ii) Let $Q_0 = P_{\infty_\pm}$: Then one introduces

$$U_{P_{\infty_\pm}} = \{Q \in \mathcal{M}_p \mid |z| > C_\infty\}, \quad C_\infty = \max_{m=0,\ldots,2p+1} |E_m|, \tag{B.7}$$

$$V_{P_{\infty_\pm}} = \{\zeta \in \mathbb{C} \mid |\zeta| < C_\infty^{-1}\}, \tag{B.8}$$

and

$$\zeta_{P_{\infty_\pm}} : U_{P_{\infty_\pm}} \to V_{P_{\infty_\pm}}, \quad Q \mapsto z^{-1}, \quad P_{\infty_\pm} \mapsto 0 \tag{B.9}$$

with inverse

$$\zeta_{P_{\infty\pm}}^{-1} : V_{P_{\infty\pm}} \to U_{P_{\infty\pm}},$$

$$\zeta \mapsto \left(\zeta^{-1}, \mp\left(\prod_{m=0}^{2p+1}(1-E_m\zeta)\right)^{1/2}\zeta^{-p-1}\right), \quad 0 \mapsto P_{\infty\pm}, \quad \text{(B.10)}$$

where the square root is chosen such that

$$\left(\prod_{m=0}^{2p+1}(1-E_m\zeta)\right)^{1/2} = 1 - \frac{1}{2}\left(\sum_{m=0}^{2p+1} E_m\right)\zeta + O(\zeta^2). \quad \text{(B.11)}$$

(iii) $Q_0 = (E_m, 0)$: Then one defines

$$U_{Q_0} = \{Q \in \mathcal{M}_p \mid |z - E_{m_0}| < C_{m_0}\}, \quad C_{m_0} = \min_{m=0,\dots,2p+1} |E_m - E_{m_0}|, \quad \text{(B.12)}$$

$$V_{Q_0} = \{\zeta \in \mathbb{C} \mid |\zeta| < C_{m_0}^{1/2}\}, \quad \text{(B.13)}$$

and

$$\zeta_{Q_0} : U_{Q_0} \to V_{Q_0}, \quad Q \mapsto \sigma(z - E_{m_0})^{1/2}, \quad \sigma \in \{1, -1\} \quad \text{(B.14)}$$

with inverse

$$\zeta_{Q_0}^{-1} : V_{Q_0} \to U_{Q_0}, \quad \zeta \mapsto \left(E_{m_0} + \zeta^2, \left(\prod_{\substack{m=0 \\ m \neq m_0}}^{2p+1}(E_{m_0} - E_m + \zeta^2)\right)^{1/2}\zeta\right), \quad \text{(B.15)}$$

where the square root branches are chosen in order to yield compatibility with the charts in (B.3)–(B.6) and (B.7)–(B.11).

The set $\mathcal{M}_p$ and the complex structure (B.3)–(B.15) just defined, then yield a compact Riemann surface of topological genus p which we denoted by $\overline{\mathcal{K}}_p$ in Appendix A. For simplicity of notation we use the symbol $\mathcal{K}_p$ to denote both the affine curve (B.1) and its projective closure $\overline{\mathcal{K}}_p$ throughout major parts of this monograph. The construction of $\mathcal{K}_p$ is sketched in Figure B.1 in the genus $p = 1$ case. A typical homology basis on $\mathcal{K}_p$ in the genus $p = 3$ case is depicted in Figure B.2 (it differs from that shown in Figure B.1).

Next, for the reader's convenience, we provide a detailed treatment of branch points for the two most frequently occurring situations: the self-adjoint case, where $\{E_m\}_{m=0,\dots,2p+1} \subset \mathbb{R}$, and the case in which $\{E_m\}_{m=0,\dots,2p+1} = \{\widetilde{E}_\ell, \overline{\widetilde{E}}_\ell\}_{\ell=0,\dots,p}$ consists of complex conjugate pairs.

Let us first consider the case with real and distinct roots, that is,

$$\{E_m\}_{m=0,\dots,2p+1} \subset \mathbb{R}, \quad E_0 < E_1 < \cdots < E_{2p+1}. \quad \text{(B.16)}$$

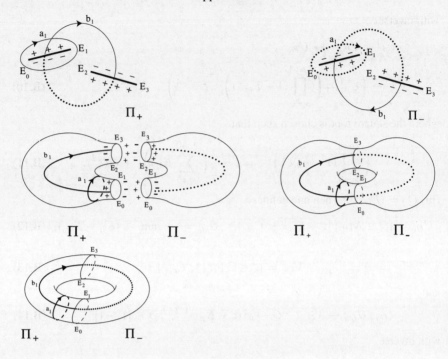

Fig. B.1. Genus $p = 1$.

In this case
$$\mathcal{C}_j = [E_{2j}, E_{2j+1}], \quad j = 0, \ldots, p,$$

and the square root branch in (B.2) is chosen according to
$$R_{2p+2}(\lambda)^{1/2} = \lim_{\varepsilon \downarrow 0} R_{2p+2}(\lambda + i\varepsilon)^{1/2}, \quad \lambda \in \mathcal{C}, \tag{B.17}$$

and

$$R_{2p+2}(\lambda)^{1/2} = |R_{2p+2}(\lambda)^{1/2}| \begin{cases} -1, & \lambda \in (E_{2p+1}, \infty), \\ (-1)^{p+j}, & \lambda \in (E_{2j-1}, E_{2j}), \ j = 1, \ldots, p, \\ (-1)^p, & \lambda \in (-\infty, E_0), \\ i(-1)^{p+j+1}, & \lambda \in (E_{2j}, E_{2j+1}), \ j = 0, \ldots, p, \end{cases}$$
$$\lambda \in \mathbb{R}. \tag{B.18}$$

The square root branches in (B.14) and (B.15) are defined by

$$(z - E_{m_0})^{1/2} = |(z - E_{m_0})^{1/2}| \exp\big((i/2) \arg(z - E_{m_0})\big),$$

$$\arg(z - E_{m_0}) \in \begin{cases} [0, 2\pi), & m_0 \text{ even,} \\ (-\pi, \pi], & m_0 \text{ odd,} \end{cases}$$

and

$$\left(\prod_{\substack{m=0 \\ m \neq m_0}}^{2p+1} (E_{m_0} - E_m + \zeta^2) \right)^{1/2} = (-1)^p i^{-m_0-1} \left| \left(\prod_{\substack{m=0 \\ m \neq m_0}}^{2p+1} (E_{m_0} - E_m) \right)^{1/2} \right|$$

$$\times \left(1 + \frac{1}{2} \left(\sum_{\substack{m=0 \\ m \neq m_0}}^{2p+1} (E_{m_0} - E_m)^{-1} \right) \zeta^2 + O(\zeta^4) \right), \quad (B.19)$$

in order to guarantee compatibility of all charts.

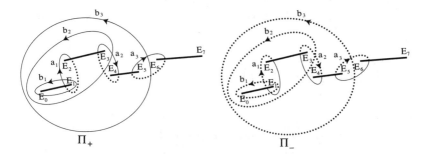

Fig. B.2. Genus $p = 3$.

Next we turn to the case where the roots form complex conjugate pairs,

$$\{E_m\}_{m=0,\ldots,2p+1} = \{\widetilde{E}_\ell, \overline{\widetilde{E}}_\ell\}_{\ell=0,\ldots,p},$$

where we assume

$$\operatorname{Re}(\widetilde{E}_\ell) < \operatorname{Re}(\widetilde{E}_{\ell+1}), \quad \ell = 0, \ldots, p-1, \quad \operatorname{Im}(\widetilde{E}_\ell) < \operatorname{Im}(\overline{\widetilde{E}}_\ell), \quad \ell = 0, \ldots, p.$$

In this case

$$\mathcal{C}_\ell = \{z \in \mathbb{C} \mid z = \widetilde{E}_\ell + t(\overline{\widetilde{E}}_\ell - \widetilde{E}_\ell), \ 0 \leq t \leq 1\}, \quad \ell = 0, \ldots, p,$$

and the square root branch in (B.2) is chosen according to

$$R_{2p+2}(z)^{1/2} = \lim_{\varepsilon \downarrow 0} R_{2p+2}(z + (-1)^{p+\ell}\varepsilon)^{1/2}, \quad z \in \mathcal{C}_\ell, \quad \ell = 0, \ldots, p, \quad (B.20)$$

where

$$R_{2p+2}(\lambda)^{1/2} = |R_{2p+2}(\lambda)^{1/2}|$$
$$\times \begin{cases} -1, & \text{Re}(\lambda) \in (\widetilde{E}_p, \infty), \\ (-1)^{p+\ell+1}, & \lambda \in (\text{Re}(\widetilde{E}_\ell), \text{Re}(\widetilde{E}_{\ell+1})), \ \ell = 0, \ldots, p-1, \\ (-1)^p, & \lambda \in (-\infty, \text{Re}(\widetilde{E}_0)). \end{cases} \quad (B.21)$$

The square root branches in (B.14) and (B.15) then are defined by

$$(z - E_{m_0})^{1/2} = |(z - E_{m_0})^{1/2}| \exp\big((i/2)\arg(z - E_{m_0})\big),$$

where, for p even,

$$\arg(z - \widetilde{E}_\ell) \in \begin{cases} (\frac{\pi}{2}, \frac{5\pi}{2}], & \ell \text{ even}, \\ [\frac{\pi}{2}, \frac{5\pi}{2}), & \ell \text{ odd}, \end{cases} \quad \arg(z - \overline{\widetilde{E}}_\ell) \in \begin{cases} [-\frac{\pi}{2}, \frac{3\pi}{2}), & \ell \text{ even}, \\ (-\frac{\pi}{2}, \frac{3\pi}{2}], & \ell \text{ odd}, \end{cases}$$

and for p odd,

$$\arg(z - \widetilde{E}_\ell) \in \begin{cases} [\frac{\pi}{2}, \frac{5\pi}{2}), & \ell \text{ even}, \\ (\frac{\pi}{2}, \frac{5\pi}{2}], & \ell \text{ odd}, \end{cases} \quad \arg(z - \overline{\widetilde{E}}_\ell) \in \begin{cases} (-\frac{\pi}{2}, \frac{3\pi}{2}], & \ell \text{ even}, \\ [-\frac{\pi}{2}, \frac{3\pi}{2}), & \ell \text{ odd}. \end{cases}$$

Here

$$\bigg(\prod_{\substack{m=0 \\ m \neq m_0}}^{2p+1} (E_{m_0} - E_m + \zeta^2)\bigg)^{1/2}$$

$$= \exp\bigg((i/2) \sum_{\substack{m=0 \\ m \neq m_0}}^{2p+1} \arg(E_{m_0} - E_m)\bigg) \bigg|\bigg(\prod_{\substack{m=0 \\ m \neq m_0}}^{2p+1} (E_{m_0} - E_m)\bigg)^{1/2}\bigg|$$

$$\times \bigg(1 + \frac{1}{2}\bigg(\sum_{\substack{m=0 \\ m \neq m_0}}^{2p+1} (E_{m_0} - E_m)^{-1}\bigg)\zeta^2 + O(\zeta^4)\bigg), \quad (B.22)$$

where $\exp(\frac{i}{2} \sum_{\substack{m=0 \\ m \neq m_0}}^{2p+1} \arg(E_{m_0} - E_m))$ is determined by analytic continuation in (B.21).

In the following we return to the general case (B.1). Points $P \in \mathcal{K}_p \setminus \{P_{\infty_-}, P_{\infty_+}\}$ are denoted by

$$P = (z, \sigma R_{2p+2}(z)^{1/2}) = (z, y), \quad P \in \mathcal{K}_p \setminus \{P_{\infty_-}, P_{\infty_+}\},$$

that is, we abbreviate $y(P) = \sigma R_{2p+2}(z)^{1/2}$. Moreover, we introduce the holomorphic sheet exchange map (involution)

$$*: \mathcal{K}_p \to \mathcal{K}_p, \quad P = (z, y) \mapsto P^* = (z, -y), \quad P_{\infty_\pm} \mapsto P^*_{\infty_\pm} = P_{\infty_\mp}, \quad (B.23)$$

and the two meromorphic projection maps

$$\tilde{\pi}: \mathcal{K}_p \to \mathbb{C} \cup \{\infty\}, \quad P = (z, y) \mapsto z, \quad P_{\infty_\pm} \mapsto \infty, \tag{B.24}$$

and

$$y: \mathcal{K}_p \to \mathbb{C} \cup \{\infty\}, \quad P = (z, y) \mapsto y, \quad P_{\infty_\pm} \mapsto \infty. \tag{B.25}$$

The map $\tilde{\pi}$ has poles of order 1 at $P_{\infty_\pm}$, and y has poles of order $p + 1$ at $P_{\infty_\pm}$. Moreover,

$$\tilde{\pi}(P^*) = \tilde{\pi}(P), \quad y(P^*) = -y(P), \quad P \in \mathcal{K}_p. \tag{B.26}$$

Thus $\mathcal{K}_p$ is a two-sheeted branched covering of the Riemann sphere $\mathbb{CP}^1$ ($\cong \mathbb{C} \cup \{\infty\}$) branched at the $2p + 2$ points $\{(E_m, 0)\}_{m=0,\ldots,2p+1}$. Moreover, $\mathcal{K}_p$ is compact (since $\tilde{\pi}$ is open and $\mathbb{CP}^1$ is compact), and $\mathcal{K}_p$ is hyperelliptic (since it admits the meromorphic function $\tilde{\pi}$ of degree two). In this context we denote the set of branch points of $\mathcal{K}_p$ by $\mathcal{B}(\mathcal{K}_p)$. Topologically, $\mathcal{K}_p$ is a sphere with p handles and hence has (topological) genus p.

The projection $\tilde{\pi}$ has two simple zeros at $(0, \pm R_{2p+2}(0)^{1/2})$ if $R_{2p+2}(0) \neq 0$ or a double zero at $(0, 0)$ if $R_{2p+2}(0)^{1/2} = 0$ (i.e., if $0 \in \{E_m\}_{m=0,\ldots,2p+1}$) and y has $2p + 2$ simple zeros at $(E_m, 0)$ for $m = 0, \ldots, 2p + 1$.

For the rest of this appendix we assume that $p \in \mathbb{N}$.

We introduce the upper and lower sheets by

$$\Pi_\pm = \{(z, \pm R_{2p+2}(z)^{1/2}) \in \mathcal{M}_p \mid z \in \Pi\} \tag{B.27}$$

and the associated charts

$$\zeta_\pm: \Pi_\pm \to \Pi, \quad P \mapsto z. \tag{B.28}$$

In particular, the charts in (B.3)–(B.15) are chosen to be compatible with $\zeta_\pm$ wherever they overlap.

Using the local chart near $P_{\infty_\pm}$ one verifies that dz/y is a holomorphic differential on $\mathcal{K}_p$ with zeros of order $p - 1$ at $P_{\infty_\pm}$ and hence

$$\eta_j = \frac{z^{j-1} dz}{y}, \quad j = 1, \ldots, p, \tag{B.29}$$

form a basis for the space of holomorphic differentials on $\mathcal{K}_p$. Assume that $\{a_j, b_j\}_{j=1,\ldots,p}$ is a homology basis for $\mathcal{K}_p$ with intersection properties (A.8). Introducing the invertible matrix C in $\mathbb{C}^p$,

$$C = (C_{j,k})_{j,k=1,\ldots,p}, \quad C_{j,k} = \int_{a_k} \eta_j, \tag{B.30}$$

$$\underline{c}(k) = (c_1(k), \ldots, c_p(k)), \quad c_j(k) = (C^{-1})_{j,k}, \quad j, k = 1, \ldots, p, \tag{B.31}$$

the normalized holomorphic differentials ω_j for $j = 1, \ldots, p$ (cf. (A.13)),

$$\omega_j = \sum_{\ell=1}^{p} c_j(\ell)\eta_\ell, \quad \int_{a_k} \omega_j = \delta_{j,k}, \quad j,k = 1, \ldots, p, \tag{B.32}$$

form a canonical basis for the space of holomorphic differentials on $\mathcal{K}_p$.

In the charts $(U_{P_{\infty\pm}}, \zeta_{P_{\infty\pm}})$ induced by $1/\tilde{\pi}$ near $P_{\infty\pm}$ one infers

$$y(P) \underset{\zeta \to 0}{=} \mp \zeta^{-p-1} \sum_{k=0}^{\infty} c_k(\underline{E}) \zeta^k$$

$$\underset{\zeta \to 0}{=} \mp \left(1 - \frac{1}{2}\left(\sum_{m=0}^{2p+1} E_m\right)\zeta + O(\zeta^2)\right)\zeta^{-p-1} \text{ as } P \to P_{\infty\pm}, \zeta = 1/z.$$

Here $\underline{E} = (E_0, \ldots, E_{2p+1})$ and we used

$$\left(\prod_{m=0}^{2p+1}(1 - E_m\zeta)\right)^{1/2} = \sum_{k=0}^{\infty} c_k(\underline{E})\zeta^k,$$

for $\zeta \in \mathbb{C}$ such that $|\zeta|^{-1} > \max\{|E_0|, \ldots, |E_{2p+1}|\}$, where $c_k(\underline{E})$, $k \in \mathbb{N}_0$, are defined in (C.5).

In addition, one obtains the expansion

$$\underline{\omega} = (\omega_1, \ldots, \omega_p) = \pm \sum_{j=1}^{p} \underline{c}(j) \frac{\zeta^{p-j} d\zeta}{(\prod_{m=0}^{2p+1}(1 - E_m\zeta))^{1/2}}$$

$$\underset{\zeta \to 0}{=} \pm \left(\sum_{q=0}^{\infty} \sum_{k=1}^{p} \underline{c}(k)\hat{c}_{k-p+q}(\underline{E})\zeta^q\right)d\zeta \tag{B.33}$$

$$\underset{\zeta \to 0}{=} \pm \left(\underline{c}(p) + \left(\frac{1}{2}\underline{c}(p)\sum_{m=0}^{2p+1} E_m + \underline{c}(p-1)\right)\zeta + O(\zeta^2)\right)d\zeta, \quad \zeta = 1/z,$$

where we used

$$\left(\prod_{m=0}^{2p+1}(1 - E_m\zeta)\right)^{-1/2} = \sum_{k=0}^{\infty} \hat{c}_k(\underline{E})\zeta^k$$

for $\zeta \in \mathbb{C}$ such that $|\zeta|^{-1} > \max\{|E_0|, \ldots, |E_{2p+1}|\}$ with $\hat{c}_k(\underline{E})$, $k \in \mathbb{N}_0$, defined in (C.2), and we agreed to set

$$\hat{c}_{-k}(\underline{E}) = 0, \quad k \in \mathbb{N}.$$

Combining (A.19) and (B.33) one computes for the vector $\underline{U}^{(2)}_{\pm,q}$ of b-periods of

$\omega^{(2)}_{P_{\infty_\pm},q}/(2\pi i)$, the normalized differential of the second kind, holomorphic on $\mathcal{K}_p \setminus \{P_{\infty_\pm}\}$, with principal part $\zeta^{-q-2}d\zeta/(2\pi i)$,

$$\underline{U}^{(2)}_{\pm,q} = (U^{(2)}_{\pm,q,1}, \ldots, U^{(2)}_{\pm,q,p}), \tag{B.34}$$

$$U^{(2)}_{\pm,q,j} = \frac{1}{2\pi i} \int_{b_j} \omega^{(2)}_{P_{\infty_\pm},q} = \frac{\pm 1}{q+1} \sum_{k=1}^{p} c_j(k)\hat{c}_{k-p+q}(\underline{E}), \quad j = 1, \ldots, p, \ q \in \mathbb{N}_0.$$

We note in passing that $\omega^{(2)}_{P_{\infty_+},q} - \omega^{(2)}_{P_{\infty_-},q}$ has the explicit form

$$\omega^{(2)}_{P_{\infty_+},q} - \omega^{(2)}_{P_{\infty_-},q} = \frac{z^p}{y} \sum_{k=0}^{q+1} c_{q+1-k}(\underline{E}) z^k dz + \frac{\lambda'_p}{y} \prod_{j=1}^{p-1} (z - \lambda'_j) dz, \quad q \in \mathbb{N}_0,$$

where $c_k(\underline{E})$, $k \in \mathbb{N}_0$, are defined in (C.5) and λ'_j, $j = 1, \ldots, p$, are uniquely determined by the normalization

$$\int_{a_j} \left(\omega^{(2)}_{P_{\infty_+},q} - \omega^{(2)}_{P_{\infty_-},q}\right) = 0, \quad j = 1, \ldots, p.$$

In the special self-adjoint case (B.16), the matrix τ of b-periods satisfies in addition to (A.15),

$$\tau = iT, \quad T > 0, \tag{B.35}$$

since

$$C_{j,k} = \int_{a_k} \eta_j = 2 \int_{E_{2k-1}}^{E_{2k}} \frac{x^{j-1}dx}{R_{2p+2}(x)^{1/2}} \in \mathbb{R}, \quad j,k = 1, \ldots, p, \tag{B.36}$$

and

$$\int_{b_k} \eta_j = -2 \sum_{\ell=1}^{k} \int_{E_{2\ell-2}}^{E_{2\ell-1}} \frac{x^{j-1}dx}{R_{2p+2}(x+i0)^{1/2}} \in i\mathbb{R}, \quad j,k = 1, \ldots, p, \tag{B.37}$$

using a homology basis of the type described in Figure B.2 for $p = 3$.

Next, assuming $0 \notin \{E_m\}_{m=0,\ldots,2p+1}$, one then computes in the charts $(U_{P_{0,\pm}}, \zeta_{P_{0,\pm}})$ induced by $\tilde{\pi}$ near $P_{0,\pm} = (0, y(P_{0,\pm}))$,

$$\underline{\omega} \underset{\zeta \to 0}{=} \pm \frac{1}{y_{0,+}} \left(\underline{c}(1) + \left(\frac{1}{2}\underline{c}(1) \sum_{m=0}^{2p+1} E_m^{-1} + \underline{c}(2)\right)\zeta + O(\zeta^2)\right)d\zeta, \quad \zeta = z,$$

$$y(P_{0,+}) = y_{0,+}, \quad y_{0,+}^2 = \prod_{m=0}^{2p+1} E_m.$$

using

$$y(P) \underset{\zeta \to 0}{=} \pm y_{0,+} + O(\zeta) \text{ as } P \to P_{0,\pm}, \ \zeta = z,$$

with the sign of $y_{0,+}$ determined appropriately. In particular, $P_{0,\pm}$ and $P_{\infty_\pm}$ are not necessarily on the same sheet.

Finally, if $E_0 = 0$, $E_m \neq 0$, $m = 1, \ldots, 2p+1$, one computes in the chart (U_{P_0}, ζ_{P_0}) induced by $\tilde{\pi}^{1/2}$ near $P_0 = (0, 0)$,

$$\omega \underset{\zeta \to 0}{=} -2i\left(\frac{c(1)}{\tilde{y}_1} + O(\zeta^2)\right)d\zeta \text{ as } P \to P_0, \ \tilde{y}_1 = \left(\prod_{m=1}^{2p+1} E_m\right)^{1/2}, \quad \text{(B.38)}$$

$$\zeta = \sigma z^{1/2}, \ \sigma \in \{1, -1\},$$

using

$$y(P) \underset{\zeta \to 0}{=} i\tilde{y}_1 \zeta + O(\zeta^3) \text{ as } P \to P_0, \ \zeta = \sigma z^{1/2}, \ \sigma \in \{1, -1\}, \quad \text{(B.39)}$$

with the sign of $\tilde{y}_1$ determined by the compatibility of charts.

Explicit formulas for normal differentials of the third kind, $\omega^{(3)}_{Q_1, Q_2}$, with simple poles at Q_1 and Q_2, corresponding residues $+1$ and -1, vanishing a-periods, and holomorphic on $\mathcal{K}_p \setminus \{Q_1, Q_2\}$, can easily be found. One obtains

$$\omega^{(3)}_{P_{\infty_+}, P_{\infty_-}} = \frac{z^p \, dz}{y} + \sum_{j=1}^{p} \gamma_j \omega_j = \frac{1}{y} \prod_{j=1}^{p}(z - \lambda_j) \, dz, \quad \text{(B.40)}$$

$$\omega^{(3)}_{P_1, P_{\infty_\pm}} = \frac{y + y_1}{z - z_1} \frac{dz}{2y} \mp \frac{1}{2y} \prod_{j=1}^{p}(z - \lambda_{1, \pm, j}) \, dz, \quad \text{(B.41)}$$

$$\omega^{(3)}_{P_1, P_2} = \left(\frac{y + y_1}{z - z_1} - \frac{y + y_2}{z - z_2}\right)\frac{dz}{2y} + \frac{\lambda''_p}{y} \prod_{j=1}^{p-1}(z - \lambda''_j) \, dz, \quad \text{(B.42)}$$

$$P_1, P_2 \in \mathcal{K}_p \setminus \{P_{\infty_+}, P_{\infty_-}\},$$

where $\gamma_j, \lambda_j, \lambda_{1,\pm,j}, \lambda''_j, j = 1, \ldots, p$, are uniquely determined by the requirement of vanishing a-periods and we abbreviated $P_j = (z_j, y_j)$, $j = 1, 2$. (If $p = 0$ in (B.40) or (B.41) and $p = 1$ in (B.42), we use the standard conventions that products and sums over empty index sets are replaced by 1 and 0, respectively; if $p = 0$, the product in (B.42) is replaced by 0.) We also note the expansion

$$\omega^{(3)}_{P_{\infty_+}, P_{\infty_-}} \underset{\zeta \to 0}{=} \pm \zeta^{-1}\left(1 + \left(\frac{1}{2}\sum_{m=0}^{2p+1} E_m - \sum_{j=1}^{p}\lambda_j\right)\zeta + O(\zeta^2)\right)d\zeta \quad \text{(B.43)}$$

$$\text{as } P \to P_{\infty_\pm}, \ \zeta = 1/z.$$

Next, we turn to the theta function representation of symmetric functions of values of a meromorphic function as discussed in Remark A.31 in the current case of Toda-type hyperelliptic Riemann surfaces. The choice $f = \tilde{\pi}$ in (A.58) then yields, after a standard residue calculation at $P_{\infty_\pm}$,

$$\sum_{j=1}^{p} \mu_j = \sum_{j=1}^{p} \int_{a_j} \tilde{\pi}\omega_j$$
$$+ \sum_{j=1}^{p} U^{(2)}_{+,0,j} \partial_{w_j} \ln\left(\frac{\theta\big(\underline{\Xi}_{Q_0} - \underline{A}_{Q_0}(P_{\infty_+}) + \underline{\alpha}_{Q_0}(\mathcal{D}_{\underline{\hat\mu}}) + \underline{w}\big)}{\theta\big(\underline{\Xi}_{Q_0} - \underline{A}_{Q_0}(P_{\infty_-}) + \underline{\alpha}_{Q_0}(\mathcal{D}_{\underline{\hat\mu}}) + \underline{w}\big)}\right)\bigg|_{\underline{w}=0},$$

where $\underline{\hat\mu} = \{\hat\mu_1, \ldots, \hat\mu_p\}$, $\hat\mu_j = (\mu_j, y(\hat\mu_j)) \in \mathcal{K}_p$, $j = 1, \ldots, p$, assuming $\mathcal{D}_{\underline{\hat\mu}} \in \mathrm{Sym}^p(\mathcal{K}_p)$ to be nonspecial and using

$$\underline{A}_{Q_0}(P) - \underline{A}_{Q_0}(P_{\infty_\pm}) \underset{\zeta\to 0}{=} \pm \underline{U}^{(2)}_{+,0}\zeta + O(\zeta^2) \text{ as } P \to P_{\infty_\pm}, \quad \underline{U}^{(2)}_{+,0} = \underline{c}(p)$$

according to (B.33)–(B.34). Here $Q_0 \in \mathcal{K}_p \setminus \{P_{\infty_+}, P_{\infty_-}\}$ denotes an appropriate base point. In the present hyperelliptic context, the constant $\sum_{j=1}^{p} \int_{a_j} \tilde{\pi}\omega_j$ can be related to the zeros $\{\lambda_j\}_{j=1,\ldots,p}$ of the normal differential of the third kind, $\omega^{(3)}_{P_{\infty_+},P_{\infty_-}}$, as follows,

$$\sum_{j=1}^{p} \int_{a_j} \tilde{\pi}\omega_j = \sum_{j=1}^{p} \lambda_j.$$

This is proven as in (F.68) and (F.70) of Volume I. Hence, one finally obtains

$$\sum_{j=1}^{p} \mu_j = \sum_{j=1}^{p} \lambda_j$$
$$+ \sum_{j=1}^{p} U^{(2)}_{+,0,j} \partial_{w_j} \ln\left(\frac{\theta\big(\underline{\Xi}_{Q_0} - \underline{A}_{Q_0}(P_{\infty_+}) + \underline{\alpha}_{Q_0}(\mathcal{D}_{\underline{\hat\mu}}) + \underline{w}\big)}{\theta\big(\underline{\Xi}_{Q_0} - \underline{A}_{Q_0}(P_{\infty_-}) + \underline{\alpha}_{Q_0}(\mathcal{D}_{\underline{\hat\mu}}) + \underline{w}\big)}\right)\bigg|_{\underline{w}=0}.$$

Finally, we formulate the following Riemann–Roch-type uniqueness results for the Baker–Akhiezer functions needed in Chapters 1 and 3. In the following, $Q_0 \in \mathcal{K}_p \setminus \{P_{\infty_\pm}\}$ in the Toda case, and $Q_0 \in \mathcal{K}_p \setminus \{P_{\infty_\pm}, P_{0,\pm}\}$ in the AL case, is an appropriate base point.

Lemma B.1 *Let* $(n, t), (n_0, t_0) \in \Omega$ *for some* $\Omega \subseteq \mathbb{Z} \times \mathbb{R}$. *Assume* $\psi(\,\cdot\,, n, t)$ *to be meromorphic on* $\mathcal{K}_n \setminus \{P_{\infty_\pm}\}$ *with possible essential singularities at* $P_{\infty_\pm}$ *in the Toda case, and meromorphic on* $\mathcal{K}_p \setminus \{P_{\infty_\pm}, P_{0,\pm}\}$ *with possible essential singularities at* $P_{\infty_\pm}, P_{0,\pm}$ *in the AL case, such that* $\tilde\psi(\,\cdot\,, n, t)$ *defined by*

$$\tilde\psi(P, n, t) = \psi(P, n, t) \exp\left((t - t_0)\int_{Q_0}^{P} \widetilde\Omega^{(2)}\right)$$

is meromorphic on $\mathcal{K}_p$ and its divisor satisfies

$$(\tilde{\psi}(\,\cdot\,,n,t)) \geq -\mathcal{D}_{\hat{\underline{\mu}}(n_0,t_0)} + (n-n_0)(\mathcal{D}_{P_{\infty_+}} - \mathcal{D}_{P_{\infty_-}})$$

in the Toda case and

$$(\tilde{\psi}(\,\cdot\,,n,t)) \geq -\mathcal{D}_{\hat{\underline{\mu}}(n_0,t_0)} + (n-n_0)(\mathcal{D}_{P_{0,-}} - \mathcal{D}_{P_{\infty_+}})$$

in the AL case, for some positive divisor $\mathcal{D}_{\hat{\underline{\mu}}(n_0,t_0)}$ of degree p. Here $\tilde{\Omega}^{(2)}$ is defined as in (1.247) in the Toda case, and as in (3.251) in the AL case. In addition, the path of integration is chosen identical to that in the Abel maps[1] (A.29) and (A.30). Define a divisor $\mathcal{D}_0(n,t)$ by

$$(\tilde{\psi}(\,\cdot\,,n,t)) = \mathcal{D}_0(n,t) - \mathcal{D}_{\hat{\underline{\mu}}(n_0,t_0)} + (n-n_0)(\mathcal{D}_{P_{\infty_+}} - \mathcal{D}_{P_{\infty_-}})$$

in the Toda case and by

$$(\tilde{\psi}(\,\cdot\,,n,t)) = \mathcal{D}_0(n,t) - \mathcal{D}_{\hat{\underline{\mu}}(n_0,t_0)} + (n-n_0)(\mathcal{D}_{P_{0,-}} - \mathcal{D}_{P_{\infty_+}})$$

in the AL case. Then

$$\mathcal{D}_0(n,t) \in \operatorname{Sym}^p \mathcal{K}_p, \quad \mathcal{D}_0(n,t) > 0, \quad \deg(\mathcal{D}_0(n,t)) = p.$$

Moreover, if $\mathcal{D}_0(n,t)$ is nonspecial for all $(n,t) \in \Omega$, that is, if

$$i(\mathcal{D}_0(n,t)) = 0, \quad (n,t) \in \Omega,$$

then $\psi(\,\cdot\,,n,t)$ is unique up to a constant multiple (which may depend on the parameters (n,t), $(n_0,t) \in \Omega$).

The proof of Lemma B.1 is analogous to that of Lemma B.2 in Volume I.

Notes

This appendix essentially coincides with Appendix C in Gesztesy and Holden (2003b) except, for simplicity, we restrict ourselves to hyperelliptic curves with nonsingular affine parts only. In addition, we appropriately adapted Lemma B.1 to the Toda and AL cases.

---

[1] This is to avoid multi-valued expressions and hence the use of the multiplicative Riemann–Roch theorem in the proof of Lemma B.1.

Appendix C
Asymptotic Spectral Parameter Expansions and Nonlinear Recursion Relations

> It has long been an axiom of mine that the little things are infinitely the most important.
>
> Sherlock Holmes[1]

In this appendix we discuss asymptotic spectral parameter expansions related to the basic polynomials and Laurent polynomials in the context of the Toda and Ablowitz–Ladik hierarchies. In addition, we discuss nonlinear recursion relations for the corresponding homogeneous recursion coefficients $\hat{f}_\ell$, $\hat{g}_\ell$ (resp., $\hat{h}_\ell$) in the Toda and Ablowitz–Ladik cases.

Before discussing each completely integrable system separately, we start with the following elementary results (which are consequences of the binomial expansion and have already been used in Appendix B). Let

$$\{E_m\}_{m=0,\ldots,2p+1} \subset \mathbb{C} \text{ for some } p \in \mathbb{N}_0$$

and $\eta \in \mathbb{C}$ such that $|\eta| < \min\{|E_0|^{-1}, \ldots, |E_{2p+1}|^{-1}\}$,

and abbreviate

$$\underline{E} = (E_0, \ldots, E_{2p+1}), \quad \underline{E}^{-1} = (E_0^{-1}, \ldots, E_{2p+1}^{-1}).$$

Then

$$\left(\prod_{m=0}^{2p+1} (1 - E_m \eta)\right)^{-1/2} = \sum_{k=0}^{\infty} \hat{c}_k(\underline{E}) \eta^k, \tag{C.1}$$

where

$$\hat{c}_0(\underline{E}) = 1,$$

$$\hat{c}_k(\underline{E}) = \sum_{\substack{j_0,\ldots,j_{2p+1}=0 \\ j_0+\cdots+j_{2p+1}=k}}^{k} \frac{(2j_0)! \cdots (2j_{2p+1})!}{2^{2k}(j_0!)^2 \cdots (j_{2p+1}!)^2} E_0^{j_0} \cdots E_{2p+1}^{j_{2p+1}}, \quad k \in \mathbb{N}. \tag{C.2}$$

---

[1] In Sir Arthur Conan Doyle, *A Case of Identity*, The Strand Magazine, 1891.

The first few coefficients explicitly read

$$\hat{c}_0(\underline{E}) = 1, \quad \hat{c}_1(\underline{E}) = \frac{1}{2}\sum_{m=0}^{2p+1} E_m,$$

$$\hat{c}_2(\underline{E}) = \frac{1}{4}\sum_{\substack{m_1,m_2=0 \\ m_1<m_2}}^{2p+1} E_{m_1}E_{m_2} + \frac{3}{8}\sum_{m=0}^{2p+1} E_m^2, \quad \text{etc.} \quad (C.3)$$

For notational convenience in connection with Theorem C.2 we also introduce

$$\hat{c}_{-k}(\underline{E}) = 0, \ k \in \mathbb{N}.$$

Similarly, one has

$$\left(\prod_{m=0}^{2p+1}(1-E_m\eta)\right)^{1/2} = \sum_{k=0}^{\infty} c_k(\underline{E})\eta^k, \quad (C.4)$$

where

$$c_0(\underline{E}) = 1, \quad (C.5)$$

$$c_k(\underline{E}) = \sum_{\substack{j_0,\ldots,j_{2p+1}=0 \\ j_0+\cdots+j_{2p+1}=k}}^{k} \frac{(2j_0)!\cdots(2j_{2p+1})!\, E_0^{j_0}\cdots E_{2p+1}^{j_{2p+1}}}{2^{2k}(j_0!)^2\cdots(j_{2p+1}!)^2(2j_0-1)\cdots(2j_{2p+1}-1)}, \quad k \in \mathbb{N}.$$

The first few coefficients are given explicitly by

$$c_0(\underline{E}) = 1, \quad c_1(\underline{E}) = -\frac{1}{2}\sum_{m=0}^{2p+1} E_m,$$

$$c_2(\underline{E}) = \frac{1}{4}\sum_{\substack{m_1,m_2=0 \\ m_1<m_2}}^{2p+1} E_{m_1}E_{m_2} - \frac{1}{8}\sum_{m=0}^{2p+1} E_m^2, \quad \text{etc.}$$

Multiplying (C.1) and (C.4) and comparing coefficients of η^k one finds

$$\sum_{\ell=0}^{k}\hat{c}_{k-\ell}(\underline{E})c_\ell(\underline{E}) = \delta_{k,0}, \quad k \in \mathbb{N}_0. \quad (C.6)$$

Next, we turn to asymptotic expansions of various quantities in the case of the Toda hierarchy. We start with some general results in connection with the corresponding Lax difference expression $L = aS^+ + a^-S^- + b$, assuming $a, b \in \mathbb{C}^{\mathbb{Z}}$, $a(n) \neq 0$, $n \in \mathbb{Z}$. Consider a fundamental system of solutions $\psi_\pm(z, \cdot)$ of $L\psi(z) = z\psi(z)$ for $z \in \mathbb{C}$ (or in some subdomain of $\mathbb{C}$), such that

$$W(\psi_-(z), \psi_+(z)) \neq 0,$$

Asymptotic Spectral Parameter Expansions 367

where $W(f, g) = a(fg^+ - f^+g)$ denotes the Wronskian of f and g for $f, g \in \mathbb{C}^{\mathbb{Z}}$. Introducing

$$\phi_\pm(z, n) = \frac{\psi_\pm^+(z, n)}{\psi_\pm(z, n)}, \quad z \in \mathbb{C}, \ n \in \mathbb{N}, \tag{C.7}$$

then $\phi_\pm$ satisfy the Riccati-type equation

$$a\phi_\pm + a^-(\phi_\pm^-)^{-1} = z - b, \tag{C.8}$$

and one introduces in addition,

$$\mathfrak{f} = \frac{1}{a(\phi_+ - \phi_-)}, \tag{C.9}$$

$$\mathfrak{g} = -\frac{\phi_+ + \phi_-}{\phi_+ - \phi_-}. \tag{C.10}$$

Using the Riccati-type equation (C.8) and its consequences,

$$(z-b)\phi_\pm^- - a\phi_\pm \phi_\pm^- = a^-,$$
$$(z-b)(\phi_+^- - \phi_-^-) - a(\phi_+\phi_+^- - \phi_-\phi_-^-) = 0,$$

one derives the identities

$$\mathfrak{f}^+ = \frac{\phi_+ \phi_-}{a(\phi_+ - \phi_-)}, \tag{C.11}$$

$$2(z-b)\mathfrak{f} + \mathfrak{g} + \mathfrak{g}^- = 0, \tag{C.12}$$

$$(z-b)^2\mathfrak{f} + (z-b)\mathfrak{g} + a^2\mathfrak{f}^+ - (a^-)^2\mathfrak{f}^- = 0, \tag{C.13}$$

$$\mathfrak{g}^2 - 4a^2\mathfrak{f}\mathfrak{f}^+ = 1. \tag{C.14}$$

Moreover, computing $\mathfrak{f}$ in terms of $\mathfrak{g}$ and vice versa, using (C.12) and (C.13), and inserting the respective results into (C.14) also yields

$$(z-b)^4\mathfrak{f}^2 - 2a^2(z-b)^2\mathfrak{f}\mathfrak{f}^+ - 2(a^-)^2(z-b)^2\mathfrak{f}\mathfrak{f}^- + a^4(\mathfrak{f}^+)^2$$
$$+ (a^-)^4(\mathfrak{f}^-)^2 - 2a^2(a^-)^2\mathfrak{f}^+\mathfrak{f}^- = (z-b)^2, \tag{C.15}$$

$$(z-b)(z-b^+)\mathfrak{g}^2 - a^2(\mathfrak{g}^- + \mathfrak{g})(\mathfrak{g} + \mathfrak{g}^+) = (z-b)(z-b^+). \tag{C.16}$$

Formally, that is, ignoring possible boundary conditions at $+\infty$ and/or $-\infty$, the Green's function of an $\ell^2(\mathbb{Z})$-realization of the Lax difference expression L is of the form

$$G(z, n, n') = (L-z)^{-1}(n, n')$$

$$= \frac{1}{W(\psi_-(z), \psi_+(z))} \begin{cases} \psi_-(z, n')\psi_+(z, n), & n' \leq n, \\ \psi_+(z, n')\psi_-(z, n), & n' \geq n, \end{cases} \quad n, n' \in \mathbb{Z},$$

and hence one obtains

$$\begin{align}
\mathfrak{f}(z, n) &= G(z, n, n), \\
\mathfrak{f}^+(z, n) &= G(z, n+1, n+1), \\
\mathfrak{g}(z, n) &= 1 - 2a(n)G(z, n, n+1) = 1 - 2a(n)G(z, n+1, n),
\end{align} \tag{C.17}$$

illustrating the spectral theoretic content of $\mathfrak{f}$ and $\mathfrak{g}$.

Next, assuming the existence of asymptotic expansions of $\mathfrak{f}$ and $\mathfrak{g}$,

$$\mathfrak{f}(z) \underset{\substack{|z| \to \infty \\ z \in C_R}}{=} -\sum_{\ell=0}^{\infty} \hat{\mathfrak{f}}_\ell z^{-\ell-1}, \quad \mathfrak{g}(z) \underset{\substack{|z| \to \infty \\ z \in C_R}}{=} -\sum_{\ell=-1}^{\infty} \hat{\mathfrak{g}}_\ell z^{-\ell-1}, \tag{C.18}$$

for z in some cone C_R with apex at $z = 0$ and some opening angle in $(0, 2\pi]$, exterior to a disk centered at $z = 0$ of sufficiently large radius $R > 0$, for some set of coefficients $\hat{\mathfrak{f}}_\ell, \ell \in \mathbb{N}_0$ and $\hat{\mathfrak{g}}_\ell, \ell \in \mathbb{N}_0 \cup \{-1\}$, one can prove the following result.

Theorem C.1 *Assume $a, b \in \mathbb{C}^\mathbb{Z}$, $a(n) \neq 0$, $n \in \mathbb{Z}$, and the existence of the asymptotic expansions (C.18). Then $\mathfrak{f}$ and $\mathfrak{g}$ have the following asymptotic expansions as $|z| \to \infty$, $z \in C_R$,*

$$\mathfrak{f}(z) \underset{\substack{|z| \to \infty \\ z \in C_R}}{=} -\sum_{\ell=0}^{\infty} \hat{f}_\ell z^{-\ell-1}, \quad \mathfrak{g}(z) \underset{\substack{|z| \to \infty \\ z \in C_R}}{=} -\sum_{\ell=-1}^{\infty} \hat{g}_\ell z^{-\ell-1}, \tag{C.19}$$

where $\hat{f}_\ell$ and $\hat{g}_\ell$ are the homogeneous versions of the coefficients f_ℓ and g_ℓ defined in (1.7) and (1.8). In particular, $\hat{f}_\ell$ and $\hat{g}_\ell$ can be computed from the nonlinear recursion relations

$$\hat{f}_0 = 1, \quad \hat{f}_1 = b, \quad \hat{f}_2 = a^2 + (a^-)^2 + b^2,$$

$$\begin{aligned}
\hat{f}_{\ell+2} = &-\frac{1}{2}\sum_{k=1}^{\ell+1} \hat{f}_{\ell+2-k}\hat{f}_k + 2b\sum_{k=0}^{\ell+1} \hat{f}_{\ell+1-k}\hat{f}_k \\
&+ \sum_{k=0}^{\ell} \left(-3b^2 \hat{f}_{\ell-k}\hat{f}_k + a^2 \hat{f}^+_{\ell-k}\hat{f}_k + (a^-)^2 \hat{f}_{\ell-k}\hat{f}^-_k \right) \\
&+ \sum_{k=0}^{\ell-1} \left(2b^3 \hat{f}_{\ell-1-k}\hat{f}_k - 2a^2 b \hat{f}^+_{\ell-1-k}\hat{f}_k - 2(a^-)^2 b \hat{f}_{\ell-1-k}\hat{f}^-_k \right) \\
&+ \sum_{k=0}^{\ell-2} \left(a^2 b^2 \hat{f}^+_{\ell-2-k}\hat{f}_k + (a^-)^2 b^2 \hat{f}_{\ell-2-k}\hat{f}^-_k + a^2(a^-)^2 \hat{f}^+_{\ell-2-k}\hat{f}^-_k \right. \\
&\left. \quad -\frac{1}{2}a^4 \hat{f}^+_{\ell-2-k}\hat{f}^+_k - \frac{1}{2}(a^-)^4 \hat{f}^-_{\ell-2-k}\hat{f}^-_k \right), \quad \ell \in \mathbb{N},
\end{aligned} \tag{C.20}$$

Asymptotic Spectral Parameter Expansions

and

$$\hat{g}_{-1} = -1, \quad \hat{g}_0 = 0, \quad \hat{g}_1 = -2a^2,$$

$$\hat{g}_{\ell+1} = -\frac{1}{2}\sum_{k=-1}^{\ell}(b+b^+)\hat{g}_{\ell-1-k}\hat{g}_k + \frac{1}{2}\sum_{k=0}^{\ell}\hat{g}_{\ell-k}\hat{g}_k \qquad (C.21)$$

$$+ \frac{1}{2}\sum_{k=-1}^{\ell-1}\left(bb^+\hat{g}_{\ell-2-k}\hat{g}_k - a^2(\hat{g}^-_{\ell-2-k} + \hat{g}_{\ell-2-k})(\hat{g}_k + \hat{g}^+_k)\right), \quad \ell \in \mathbb{N}.$$

Proof Inserting expansion (C.18) for $\mathfrak{f}$ into (C.15) and expansion (C.18) for $\mathfrak{g}$ into (C.16) then yields the nonlinear recursion relations (C.20) and (C.21), but with $\hat{f}_\ell$ and $\hat{g}_\ell$ replaced by $\hat{\mathfrak{f}}_\ell$ and $\hat{\mathfrak{g}}_\ell$, respectively. More precisely, one first obtains $|\hat{\mathfrak{f}}_0| = |\hat{\mathfrak{g}}_{-1}| = 1$ and upon choosing the signs of $\hat{\mathfrak{f}}_0$ and $\hat{\mathfrak{g}}_{-1}$ such that $\hat{\mathfrak{f}}_0 = \hat{f}_0 = 1$ and $\hat{\mathfrak{g}}_{-1} = -1$ one obtains (C.20) and (C.21) (still with $\hat{f}_\ell$ and $\hat{g}_\ell$ replaced by $\hat{\mathfrak{f}}_\ell$ and $\hat{\mathfrak{g}}_\ell$). Next, inserting the expansions (C.18) for $\mathfrak{f}$ and $\mathfrak{g}$ into (C.12) and (C.13), and comparing powers of $z^{-\ell}$ as $|z| \to \infty$, $z \in C_R$, one infers that $\hat{\mathfrak{f}}_\ell$ and $\hat{\mathfrak{g}}_\ell$ satisfy the linear recursion relations (1.3)–(1.5). Hence one concludes that

$$\hat{\mathfrak{f}}_\ell = f_\ell, \quad \hat{\mathfrak{g}}_\ell = g_\ell, \quad \ell \in \mathbb{N}_0,$$

for certain values of the summation constants c_ℓ. To show that actually $\hat{\mathfrak{f}}_\ell = \hat{f}_\ell$, $\hat{\mathfrak{g}}_\ell = \hat{g}_\ell$, and hence all c_ℓ, $\ell \in \mathbb{N}$, vanish, we now rely on the notion of degree as introduced in Remark 1.3. To this end we recall that

$$\deg\left(\hat{f}_\ell\right) = \ell, \quad \deg\left(\hat{g}_\ell\right) = \ell + 1, \quad \ell \in \mathbb{N},$$

(cf. (1.12)). Similarly, the nonlinear recursion relations (C.20) and (C.21) yield inductively that

$$\deg\left(\hat{\mathfrak{f}}_\ell\right) = \ell, \quad \deg\left(\hat{\mathfrak{g}}_\ell\right) = \ell + 1, \quad \ell \in \mathbb{N}.$$

Hence one concludes that

$$\hat{\mathfrak{f}}_\ell = \hat{f}_\ell, \quad \hat{\mathfrak{g}}_\ell = \hat{g}_\ell, \quad \ell \in \mathbb{N}_0.$$

$\square$

Given this general result on asymptotic (Green's function) expansions for Jacobi-type difference expressions, we now specialize to the algebro-geometric case at hand. We recall our convention $y(P) = \mp(\zeta^{-p-1} + O(\zeta^{-p}))$ for P near $P_{\infty\pm}$ (where $\zeta = 1/z$).

Theorem C.2 *Assume* (1.1), $\text{s-Tl}_p(a,b) = 0$, *and suppose* $P = (z,y) \in \mathcal{K}_p \setminus \{P_{\infty_+}, P_{\infty_-}\}$. *Then* F_p/y *and* G_{p+1}/y *have the following convergent expansions as* $P \to P_{\infty_\pm}$,

$$\frac{F_p(z)}{y} = \mp \sum_{\ell=0}^{\infty} \hat{f}_\ell \zeta^{\ell+1}, \quad \frac{G_{p+1}(z)}{y} = \mp \sum_{\ell=-1}^{\infty} \hat{g}_\ell \zeta^{\ell+1}, \quad \text{(C.22)}$$

where $\zeta = 1/z$ *is the local coordinate near* $P_{\infty_\pm}$ *described in* (B.7)–(B.11) *and* $\hat{f}_\ell$ *and* $\hat{g}_\ell$ *are the homogeneous versions[1] of the coefficients* f_ℓ *and* g_ℓ *as introduced in* (1.7) *and* (1.8). *Moreover, one infers for the* E_m-*dependent summation constants* c_ℓ, $\ell = 0, \ldots, p+1$, *in* F_p *and* G_{p+1} *that*

$$c_\ell = c_\ell(\underline{E}), \quad \ell = 0, \ldots, p+1. \quad \text{(C.23)}$$

In addition, one has the following relations between the homogeneous and nonhomogeneous recursion coefficients,[2]

$$f_\ell = \sum_{k=0}^{\ell} c_{\ell-k}(\underline{E}) \hat{f}_k, \quad \ell = 0, \ldots, p,$$

$$g_\ell = \sum_{k=1}^{\ell} c_{\ell-k}(\underline{E}) \hat{g}_k - c_{\ell+1}(\underline{E}), \quad \ell = 0, \ldots, p-1, \quad \text{(C.24)}$$

$$g_p = \sum_{k=0}^{p} c_{p-k}(\underline{E}) \hat{g}_k - c_{p+1}(\underline{E}) - f_{p+1},$$

$$\hat{f}_\ell = \sum_{k=0}^{\ell \wedge p} \hat{c}_{\ell-k}(\underline{E}) f_k, \quad \ell \in \mathbb{N}_0,$$

$$\hat{g}_\ell = \sum_{k=0}^{\ell \wedge p} \hat{c}_{\ell-k}(\underline{E}) g_k - \hat{c}_{\ell+1}(\underline{E}) + \hat{c}_{\ell-p}(\underline{E}) f_{p+1}, \quad \ell \in \mathbb{N}_0. \quad \text{(C.25)}$$

Proof Identifying

$$\psi_+(z, \cdot) \text{ with } \psi(P, \cdot, 0) \text{ and } \psi_-(z, \cdot) \text{ with } \psi(P^*, \cdot, 0),$$

recalling $W(\psi(P, \cdot, 0), \psi(P^*, \cdot, 0)) = -y F_p(z, 0)^{-1}$ (cf. (1.84)), and similarly, identifying

$$\phi_+(z, \cdot) \text{ with } \phi(P, \cdot) \text{ and } \phi_-(z, \cdot) \text{ with } \phi(P^*, \cdot),$$

---

[1] Strictly speaking, the coefficients $\hat{f}_\ell$ and $\hat{g}_\ell$ in (C.22) no longer have a well-defined degree and hence represent a slight abuse of notation since we assumed that $\text{s-Tl}_p(a,b) = 0$. At any rate, they are explicitly given by (C.25).

[2] $m \wedge n = \min\{m, n\}$.

Asymptotic Spectral Parameter Expansions 371

a comparison of (C.7)–(C.10), (C.11) and the results of Lemmas 1.12 and 1.17 show that we may also identify

$$\mathfrak{f} \text{ with } \pm F_p/y \text{ and } \mathfrak{g} \text{ with } \pm G_{p+1}/y,$$

the sign depending on whether P tends to $P_{\infty\pm}$. In particular, (C.12)–(C.16) then correspond to (1.38)–(1.40), (1.40), (1.42), and (1.43), respectively. Since F_p/y and G_{p+1}/y clearly have an asymptotic (in fact, even convergent) expansion as $|z| \to \infty$ around $1/z = 0$, the results of Theorem C.1 apply and using (C.1), (1.30), and (1.31) one obtains

$$\frac{F_p(z)}{y} \underset{\zeta \to 0}{=} \mp \left(\sum_{k=0}^{\infty} \hat{c}_k(\underline{E}) \zeta^k \right) \left(\sum_{\ell=0}^{p} f_\ell \zeta^{\ell+1} \right)$$

$$\underset{\zeta \to 0}{=} \mp \sum_{\ell=0}^{\infty} \hat{f}_\ell \zeta^{\ell+1}, \tag{C.26}$$

$$\frac{G_{p+1}(z)}{y} \underset{\zeta \to 0}{=} \mp \left(\sum_{k=0}^{\infty} \hat{c}_k(\underline{E}) \zeta^k \right) \left(-1 + \sum_{\ell=0}^{p} g_\ell \zeta^{\ell+1} + f_{p+1} \zeta^{p+1} \right)$$

$$\underset{\zeta \to 0}{=} \mp \sum_{\ell=-1}^{\infty} \hat{g}_\ell \zeta^{\ell+1}, \tag{C.27}$$

and hence (C.22).
A comparison of coefficients in (C.26) proves (C.25) for $\hat{f}_\ell$. Similarly, we use (C.27) to establish (C.25) for $\hat{g}_\ell$. Next, multiplying (C.1) and (C.4), a comparison of coefficients of η^k yields

$$\sum_{\ell=0}^{k} \hat{c}_{k-\ell}(\underline{E}) c_\ell(\underline{E}) = \delta_{k,0}, \quad k \in \mathbb{N}_0. \tag{C.28}$$

Thus, one computes

$$\sum_{m=0}^{\ell} c_{\ell-m}(\underline{E}) \hat{f}_m = \sum_{m=0}^{\ell} \sum_{k=0}^{m} c_{\ell-m}(\underline{E}) \hat{c}_{m-k}(\underline{E}) f_k = \sum_{k=0}^{\ell} \sum_{q=k}^{\ell} c_{\ell-q}(\underline{E}) \hat{c}_{q-k}(\underline{E}) f_k$$

$$= \sum_{k=0}^{\ell} \left(\sum_{m=0}^{\ell-k} c_{\ell-k-m}(\underline{E}) \hat{c}_m(\underline{E}) \right) f_k = f_\ell, \quad \ell = 0, \ldots, p,$$

applying (C.28). Hence one obtains (C.24) for f_ℓ and thus (C.23) (cf. (1.9), (1.10)). The corresponding proof of (C.24) for g_ℓ is similar to that of f_ℓ. □

Next, we turn to asymptotic expansions of various quantities in the case of the Ablowitz–Ladik hierarchy. We start with some general results which eventually will be connected with the corresponding Lax difference expression L introduced in

(3.61), assuming $\alpha, \beta \in \mathbb{C}^{\mathbb{Z}}$, $\alpha(n)\beta(n) \notin \{0, 1\}$, $n \in \mathbb{Z}$. Consider a fundamental system of solutions $\Psi_\pm(z, \cdot) = (\psi_{1,\pm}(z, \cdot), \psi_{2,\pm}(z, \cdot))^\top$ of $U(z)\Psi_\pm^-(z) = \Psi_\pm(z)$ for $z \in \mathbb{C}$ (or in some subdomain of $\mathbb{C}$), with U given by (3.4), such that

$$\det(\Psi_-(z), \Psi_+(z)) \neq 0.$$

Introducing

$$\phi_\pm(z, n) = \frac{\psi_{2,\pm}(z, n)}{\psi_{1,\pm}(z, n)}, \quad z \in \mathbb{C}, \, n \in \mathbb{N}, \tag{C.29}$$

then $\phi_\pm$ satisfy the Riccati-type equation

$$\alpha \phi_\pm \phi_\pm^- - \phi_\pm^- + z\phi_\pm = z\beta, \tag{C.30}$$

and one introduces in addition,

$$\mathfrak{f} = \frac{2}{\phi_+ - \phi_-}, \tag{C.31}$$

$$\mathfrak{g} = \frac{\phi_+ + \phi_-}{\phi_+ - \phi_-}, \tag{C.32}$$

$$\mathfrak{h} = \frac{2\phi_+ \phi_-}{\phi_+ - \phi_-}. \tag{C.33}$$

Using the Riccati-type equation (C.30) and its consequences,

$$\alpha(\phi_+ \phi_+^- - \phi_- \phi_-^-) - (\phi_+^- - \phi_-^-) + z(\phi_+ - \phi_-) = 0,$$
$$\alpha(\phi_+ \phi_+^- + \phi_- \phi_-^-) - (\phi_+^- + \phi_-^-) + z(\phi_+ + \phi_-) = 2z\beta,$$

one derives the identities

$$z(\mathfrak{g}^- - \mathfrak{g}) + z\beta\mathfrak{f} + \alpha\mathfrak{h}^- = 0, \tag{C.34}$$

$$z\beta\mathfrak{f}^- + \alpha\mathfrak{h} - \mathfrak{g} + \mathfrak{g}^- = 0, \tag{C.35}$$

$$-\mathfrak{f} + z\mathfrak{f}^- + \alpha(\mathfrak{g} + \mathfrak{g}^-) = 0, \tag{C.36}$$

$$z\beta(\mathfrak{g}^- + \mathfrak{g}) - z\mathfrak{h} + \mathfrak{h}^- = 0, \tag{C.37}$$

$$\mathfrak{g}^2 - \mathfrak{f}\mathfrak{h} = 1. \tag{C.38}$$

Moreover, (C.34)–(C.37) and (C.38) also permit one to derive nonlinear difference equations for $\mathfrak{f}$, $\mathfrak{g}$, and $\mathfrak{h}$ separately, and one obtains

$$\left((\alpha^+ + z\alpha)^2 \mathfrak{f} - z(\alpha^+)^2 \gamma \mathfrak{f}^-\right)^2 - 2z\alpha^2 \gamma^+ \left((\alpha^+ + z\alpha)^2 \mathfrak{f} + z(\alpha^+)^2 \gamma \mathfrak{f}^-\right) \mathfrak{f}^+$$
$$+ z^2 \alpha^4 (\gamma^+)^2 (\mathfrak{f}^+)^2 = 4(\alpha\alpha^+)^2 (\alpha^+ + \alpha z)^2, \tag{C.39}$$

$$(\alpha^+ + z\alpha)(\beta + z\beta^+)(z + \alpha^+\beta)(1 + z\alpha\beta^+)\mathfrak{g}^2$$
$$+ z(\alpha^+ \gamma \mathfrak{g}^- + z\alpha\gamma^+ \mathfrak{g}^+)(z\beta^+ \gamma \mathfrak{g}^- + \beta\gamma^+ \mathfrak{g}^+)$$
$$- z\gamma\left((\alpha^+\beta + z^2\alpha\beta^+)(2 - \gamma^+) + 2z(1 - \gamma^+)(2 - \gamma)\right)\mathfrak{g}^- \mathfrak{g}$$

$$-z\gamma^+\big(2z(1-\gamma)(2-\gamma^+) + (\alpha^+\beta + z^2\alpha\beta^+)(2-\gamma)\big)\mathfrak{g}^+\mathfrak{g}$$
$$= (\alpha^+\beta - z^2\alpha\beta^+)^2, \tag{C.40}$$

$$z^2\big((\beta^+)^2\gamma\mathfrak{h}^- - \beta^2\gamma^+\mathfrak{h}^+\big)^2 - 2z(\beta + z\beta^+)^2\big((\beta^+)^2\gamma\mathfrak{h}^- + \beta^2\gamma^+\mathfrak{h}^+\big)\mathfrak{h}$$
$$+ (\beta + z\beta^+)^4\mathfrak{h}^2 = 4z^2(\beta\beta^+)^2(\beta + \beta^+z)^2. \tag{C.41}$$

To make the connection with L and its Green's function, we need to digress a bit. Introducing the transfer matrix $T(z, \cdot)$ associated with L by

$$T(z,n) = \begin{cases} \rho(n)^{-1}\begin{pmatrix} \alpha(n) & z \\ z^{-1} & \beta(n) \end{pmatrix}, & n \text{ odd,} \\ \rho(n)^{-1}\begin{pmatrix} \beta(n) & 1 \\ 1 & \alpha(n) \end{pmatrix}, & n \text{ even,} \end{cases} \quad n \in \mathbb{Z}, \tag{C.42}$$

recalling that $\rho = \gamma^{1/2} = (1 - \alpha\beta)^{1/2}$, one then verifies that

$$T(z, n) = A(z, n)z^{-1/2}\rho(n)^{-1}U(z, n)A(z, n-1)^{-1}. \tag{C.43}$$

Here we introduced

$$A(z, n) = \begin{cases} \begin{pmatrix} z^{1/2} & 0 \\ 0 & z^{-1/2} \end{pmatrix}, & n \text{ odd,} \\ \begin{pmatrix} 0 & 1 \\ 1 & 0 \end{pmatrix}, & n \text{ even,} \end{cases} \quad n \in \mathbb{Z}.$$

Introducing in addition

$$\begin{pmatrix} u_\pm(z,n) \\ v_\pm(z,n) \end{pmatrix} = C_\pm z^{-n/2}\left(\prod_{n'=1}^n \rho(n')^{-1}\right) A(z,n) \begin{pmatrix} \psi_{1,\pm}(z,n) \\ \psi_{2,\pm}(z,n) \end{pmatrix}, \quad n \in \mathbb{Z}, \tag{C.44}$$

for some constants $C_\pm \in \mathbb{C} \setminus \{0\}$, (C.43) and (C.44) yield

$$T(z)\begin{pmatrix} u_\pm^-(z) \\ v_\pm^-(z) \end{pmatrix} = \begin{pmatrix} u_\pm(z) \\ v_\pm(z) \end{pmatrix}.$$

Moreover, one can show that

$$Lu_\pm(z) = zu_\pm(z), \quad L^\top v_\pm(z) = zv_\pm(z),$$
$$Dv_\pm(z) = u_\pm(z), \quad Eu_\pm(z) = zv_\pm(z),$$

where $L^\top$ denotes the difference expression associated with the transpose of the infinite matrix L (cf. (3.59)) in the standard basis of $\ell^2(\mathbb{Z})$.

Again, formally, that is, ignoring possible boundary conditions at $+\infty$ and/or $-\infty$, the Green's function of an $\ell^2(\mathbb{Z})$-realization of the Lax difference expression L is then of the form

$$G(z, n, n') = (L - z)^{-1}(n, n') = (\delta_n, (L - z)^{-1}\delta_{n'})$$

$$= \frac{-1}{z \det \begin{pmatrix} u_+(z,0) & u_-(z,0) \\ v_+(z,0) & v_-(z,0) \end{pmatrix}}$$

$$\times \begin{cases} v_-(z, n')u_+(z, n), & n' < n \text{ and } n = n' \text{ even,} \\ v_+(z, n')u_-(z, n), & n' > n \text{ and } n = n' \text{ odd,} \end{cases} \quad n, n' \in \mathbb{Z},$$

$$= -\frac{1}{4z} \frac{(1 - \phi_+(z,0))(1 - \phi_-(z,0))}{\phi_+(z,0) - \phi_-(z,0)}$$

$$\times \begin{cases} v_-(z, n')u_+(z, n), & n' < n \text{ and } n = n' \text{ even,} \\ v_+(z, n')u_-(z, n), & n' > n \text{ and } n = n' \text{ odd,} \end{cases} \quad n, n' \in \mathbb{Z}.$$

Here $\Psi_\pm(z, \cdot)$ in (C.44) have to be chosen such that for all $n_0 \in \mathbb{Z}$,

$$\begin{pmatrix} u_\pm(z, \cdot) \\ v_\pm(z, \cdot) \end{pmatrix} \in \ell^2([n_0, \pm\infty) \cap \mathbb{Z})^2. \tag{C.45}$$

For the connection between $\mathfrak{f}$, $\mathfrak{g}$, and $\mathfrak{h}$ and the Green's function of L one then obtains

$$\mathfrak{f}(z, n) = -2\alpha(n)(zG(z, n, n) + 1) - 2\rho(n)z \begin{cases} G(z, n-1, n), & n \text{ even,} \\ G(z, n, n-1), & n \text{ odd,} \end{cases}$$

$$\mathfrak{g}(z, n) = -2zG(z, n, n) - 1,$$

$$\mathfrak{h}(z, n) = -2\beta(n)zG(z, n, n) - 2\rho(n)z \begin{cases} G(z, n, n-1), & n \text{ even,} \\ G(z, n-1, n), & n \text{ odd,} \end{cases}$$

illustrating the spectral theoretic content of $\mathfrak{f}$, $\mathfrak{g}$, and $\mathfrak{h}$.

Next, we assume the existence of the following asymptotic expansions of $\mathfrak{f}$, $\mathfrak{g}$, and $\mathfrak{h}$ near $1/z = 0$ and $z = 0$. More precisely, near $1/z = 0$ we assume that

$$\mathfrak{f}(z) \underset{\substack{|z| \to \infty \\ z \in C_R}}{=} -\sum_{\ell=0}^\infty \hat{\mathfrak{f}}_{\ell,+} z^{-\ell-1}, \quad \mathfrak{g}(z) \underset{\substack{|z| \to \infty \\ z \in C_R}}{=} -\sum_{\ell=0}^\infty \hat{\mathfrak{g}}_{\ell,+} z^{-\ell},$$

$$\mathfrak{h}(z) \underset{\substack{|z| \to \infty \\ z \in C_R}}{=} -\sum_{\ell=0}^\infty \hat{\mathfrak{h}}_{\ell,+} z^{-\ell}, \tag{C.46}$$

for z in some cone C_R with apex at $z = 0$ and some opening angle in $(0, 2\pi]$, exterior to a disk centered at $z = 0$ of sufficiently large radius $R > 0$, for some set of coefficients $\hat{\mathfrak{f}}_{\ell,+}$, $\hat{\mathfrak{g}}_{\ell,+}$, and $\hat{\mathfrak{h}}_{\ell,+}$, $\ell \in \mathbb{N}_0$. Similarly, near $z = 0$ we assume that

$$\mathfrak{f}(z) \underset{\substack{|z|\to 0 \\ z\in C_r}}{=} -\sum_{\ell=0}^{\infty} \hat{\mathfrak{f}}_{\ell,-} z^{\ell}, \quad \mathfrak{g}(z) \underset{\substack{|z|\to 0 \\ z\in C_r}}{=} -\sum_{\ell=0}^{\infty} \hat{\mathfrak{g}}_{\ell,-} z^{\ell},$$
$$\mathfrak{h}(z) \underset{\substack{|z|\to 0 \\ z\in C_r}}{=} -\sum_{\ell=0}^{\infty} \hat{\mathfrak{h}}_{\ell,-} z^{\ell+1},$$
(C.47)

for z in some cone C_r with apex at $z = 0$ and some opening angle in $(0, 2\pi]$, interior to a disk centered at $z = 0$ of sufficiently small radius $r > 0$, for some set of coefficients $\hat{\mathfrak{f}}_{\ell,-}$, $\hat{\mathfrak{g}}_{\ell,-}$, and $\hat{\mathfrak{h}}_{\ell,-}$, $\ell \in \mathbb{N}_0$. Then one can prove the following result.

Theorem C.3 *Assume $\alpha, \beta \in \mathbb{C}^{\mathbb{Z}}$, $\alpha(n)\beta(n) \notin \{0, 1\}$, $n \in \mathbb{Z}$, and the existence of the asymptotic expansions* (C.46) *and* (C.47). *Then $\mathfrak{f}$, $\mathfrak{g}$, and $\mathfrak{h}$ have the following asymptotic expansions as $|z| \to \infty$, $z \in C_R$, respectively, $|z| \to 0$, $z \in C_r$,*

$$\mathfrak{f}(z) \underset{\substack{|z|\to \infty \\ z\in C_R}}{=} -\sum_{\ell=0}^{\infty} \hat{f}_{\ell,+} z^{-\ell-1}, \quad \mathfrak{g}(z) \underset{\substack{|z|\to \infty \\ z\in C_R}}{=} -\sum_{\ell=0}^{\infty} \hat{g}_{\ell,+} z^{-\ell},$$
$$\mathfrak{h}(z) \underset{\substack{|z|\to \infty \\ z\in C_R}}{=} -\sum_{\ell=0}^{\infty} \hat{h}_{\ell,+} z^{-\ell},$$
(C.48)

and

$$\mathfrak{f}(z) \underset{\substack{|z|\to 0 \\ z\in C_r}}{=} -\sum_{\ell=0}^{\infty} \hat{f}_{\ell,-} z^{\ell}, \quad \mathfrak{g}(z) \underset{\substack{|z|\to 0 \\ z\in C_r}}{=} -\sum_{\ell=0}^{\infty} \hat{g}_{\ell,-} z^{\ell},$$
$$\mathfrak{h}(z) \underset{\substack{|z|\to 0 \\ z\in C_r}}{=} -\sum_{\ell=0}^{\infty} \hat{h}_{\ell,-} z^{\ell+1},$$
(C.49)

where $\hat{f}_{\ell,\pm}$, $\hat{g}_{\ell,\pm}$, and $\hat{h}_{\ell,\pm}$ are the homogeneous versions of the coefficients $f_{\ell,\pm}$, $g_{\ell,\pm}$, and $h_{\ell,\pm}$ defined in (3.32)–(3.34). *In particular, $\hat{f}_{\ell,\pm}$, $\hat{g}_{\ell,\pm}$, and $\hat{h}_{\ell,\pm}$ can be computed from the following nonlinear recursion relations*[1]

$$\hat{f}_{0,+} = -\alpha^+, \quad \hat{f}_{1,+} = (\alpha^+)^2 \beta - \gamma^+ \alpha^{++},$$
$$\hat{f}_{2,+} = -(\alpha^+)^3 \beta^2 + (\alpha^+)^2 \beta^+ - \alpha^{+++} \gamma^+ \gamma^{++}$$
$$+ (\alpha^+)^2 \big(\beta^- \gamma - 2\alpha^{++}\beta\beta^+\big) + \alpha^+ \alpha^{++}\big(2\beta - \alpha^{++}(\beta^+)^2\big),$$

---

[1] Here sums with upper limits strictly less than their lower limits are interpreted as zero.

$$\alpha^4 \alpha^+ \hat{f}_{\ell,+} = \frac{1}{2}\bigg((\alpha^+)^4 \sum_{m=0}^{\ell-4} \hat{f}_{m,+}\hat{f}_{\ell-m-4,+} + \alpha^4 \sum_{m=1}^{\ell-1} \hat{f}_{m,+}\hat{f}_{\ell-m,+}$$

$$- 2(\alpha^+)^2 \sum_{m=0}^{\ell-3} \hat{f}_{m,+}\big(-2\alpha\alpha^+ \hat{f}_{\ell-m-3,+} + (\alpha^+)^2 \gamma \hat{f}^-_{\ell-m-3,+}m$$

$$+ \alpha^2 \gamma^+ \hat{f}^+_{\ell-m-3,+}\big)$$

$$+ \sum_{m=0}^{\ell-2} \big(\alpha^4 (\gamma^+)^2 \hat{f}^+_{m,+}\hat{f}^+_{\ell-m-2,+} + (\alpha^+)^2 \gamma \hat{f}^-_{m,+}((\alpha^+)^2 \gamma \hat{f}^-_{\ell-m-2,+}$$

$$- 2\alpha^2 \gamma^+ \hat{f}^+_{\ell-m-2,+})$$

$$- 2\alpha\alpha^+ \hat{f}_{m,+}(-3\alpha\alpha^+ \hat{f}_{\ell-m-2,+} + 2(\alpha^+)^2 \gamma \hat{f}^-_{\ell-m-2,+} + 2\alpha^2 \gamma^+ \hat{f}^+_{\ell-m-2,+}))$$

$$- 2\alpha^2 \sum_{m=0}^{\ell-1} \hat{f}_{m,+}\big(-2\alpha\alpha^+ \hat{f}_{\ell-m-1,+} + (\alpha^+)^2 \gamma \hat{f}^-_{\ell-m-1,+}m + \alpha^2 \gamma^+ \hat{f}^+_{\ell-m-1,+}\big)\bigg),$$

$$\ell \geq 3, \qquad \text{(C.50)}$$

$$\hat{f}_{0,-} = \alpha, \quad \hat{f}_{1,-} = \gamma\alpha^- - \alpha^2 \beta^+,$$

$$\hat{f}_{2,-} = \alpha^{--}\gamma^-\gamma - (\alpha^-)^2\beta\gamma - 2\alpha^-\alpha\gamma\beta^+ + \alpha^2(\alpha(\beta^+)^2 - \gamma^+)\beta^{++},$$

$$\alpha(\alpha^+)^4 \hat{f}_{\ell,-} = -\frac{1}{2}\bigg(\alpha^4 \sum_{m=0}^{\ell-4} \hat{f}_{m,-}\hat{f}_{\ell-m-4,-} + (\alpha^+)^4 \sum_{m=1}^{\ell-1} \hat{f}_{m,-}\hat{f}_{\ell-m,-}$$

$$- 2\alpha^2 \sum_{m=0}^{\ell-3} \hat{f}_{m,-}\big(-2\alpha\alpha^+ \hat{f}_{\ell-m-3,-} + (\alpha^+)^2 \gamma \hat{f}^-_{\ell-m-3,-} + \alpha^2 \gamma^+ \hat{f}^+_{\ell-m-3,-}\big)$$

$$+ \sum_{m=0}^{\ell-2} \big(\alpha^4 (\gamma^+)^2 \hat{f}^+_{m,-}\hat{f}^+_{\ell-m-2,-}$$

$$+ (\alpha^+)^2 \gamma \hat{f}^-_{m,-}((\alpha^+)^2 \gamma \hat{f}^-_{\ell-m-2,-} - 2\alpha^2 \gamma^+ \hat{f}^+_{\ell-m-2,-})$$

$$- 2\alpha\alpha^+ \hat{f}_{m,-}(-3\alpha\alpha^+ \hat{f}_{\ell-m-2,-} + 2(\alpha^+)^2 \gamma \hat{f}^-_{\ell-m-2,-} + 2\alpha^2 \gamma^+ \hat{f}^+_{\ell-m-2,-}))$$

$$- 2(\alpha^+)^2 \sum_{m=0}^{\ell-1} \hat{f}_{m,-}\big(-2\alpha\alpha^+ \hat{f}_{\ell-m-1,-} + (\alpha^+)^2 \gamma \hat{f}^-_{\ell-m-1,-}$$

$$+ \alpha^2 \gamma^+ \hat{f}^+_{\ell-m-1,-}\big)\bigg), \quad \ell \geq 3, \qquad \text{(C.51)}$$

$$\hat{g}_{0,+} = \frac{1}{2}, \quad \hat{g}_{1,+} = -\alpha^+ \beta,$$

$$\hat{g}_{2,+} = -\alpha^{++}\beta + (\alpha^+)^2 \beta^2 + \alpha^+(-\beta^-\gamma + \alpha^{++}\beta\beta^+),$$

Asymptotic Spectral Parameter Expansions

$$(\alpha\beta^+)^2 \hat{g}_{\ell,+} = -\bigg((\alpha^+)^2\beta^2 \sum_{m=0}^{\ell-4} \hat{g}_{m,+}\hat{g}_{\ell-m-4,+} + \alpha^2(\beta^+)^2 \sum_{m=1}^{\ell-1} \hat{g}_{m,+}\hat{g}_{\ell-m,+}$$

$$+ \alpha^+\beta \sum_{m=0}^{\ell-3} \big(\gamma\gamma^+ \hat{g}^-_{m,+}\hat{g}^+_{\ell-m-3,+} + \hat{g}_{m,+}((1+\alpha\beta)(1+\alpha^+\beta^+)\hat{g}_{\ell-m-3,+}$$

$$- (\gamma + \alpha^+\beta^+\gamma)\hat{g}^-_{\ell-m-3,+} + (-2+\gamma)\gamma^+ \hat{g}^+_{\ell-m-3,+})\big)$$

$$+ \sum_{m=0}^{\ell-2} \big(\alpha^+\beta^+\gamma^2 \hat{g}^-_{m,+}\hat{g}^-_{\ell-m-2,+} + \alpha\beta(\gamma^+)^2 \hat{g}^+_{m,+}\hat{g}^+_{\ell-m-2,+}$$

$$+ \hat{g}_{m,+}((\alpha^+\beta^+ + \alpha^2\alpha^+\beta^2\beta^+ + \alpha\beta(1+\alpha^+\beta^+)^2)\hat{g}_{\ell-m-2,+}$$

$$- 2(\alpha^+(1+\alpha\beta)\beta^+\gamma \hat{g}^-_{\ell-m-2,+} + \alpha\beta(1+\alpha^+\beta^+)\gamma^+ \hat{g}^+_{\ell-m-2,+}))\big)$$

$$+ \alpha\beta^+ \sum_{m=0}^{\ell-1} \big(\gamma\gamma^+ \hat{g}^-_{m,+}\hat{g}^+_{\ell-m-1,+} + \hat{g}_{m,+}((1+\alpha\beta)(1+\alpha^+\beta^+)\hat{g}_{\ell-m-1,+}$$

$$- (\gamma + \alpha^+\beta^+\gamma)\hat{g}^-_{\ell-m-1,+} + (-2+\gamma)\gamma^+ \hat{g}^+_{\ell-m-1,+})\big)\bigg), \quad \ell \geq 3,$$

(C.52)

$$\hat{g}_{0,-} = \frac{1}{2}, \quad \hat{g}_{1,-} = -\alpha\beta^+,$$

$$\hat{g}_{2,-} = -\alpha^-\gamma\beta^+ + \alpha(\alpha(\beta^+)^2 - \gamma^+)\beta^{++},$$

$$(\alpha^+)^2\beta^2 \hat{g}_{\ell,-} = -\bigg(\alpha^2(\beta^+)^2 \sum_{m=0}^{\ell-4} \hat{g}_{m,-}\hat{g}_{\ell-m-4,-} + (\alpha^+)^2\beta^2 \sum_{m=1}^{\ell-1} \hat{g}_{m,-}\hat{g}_{\ell-m,-}$$

$$+ \alpha\beta^+ \sum_{m=0}^{\ell-3} \big(\gamma\gamma^+ \hat{g}^-_{m,-}\hat{g}^+_{\ell-m-3,-} + \hat{g}_{m,-}((1+\alpha\beta)(1+\alpha^+\beta^+)\hat{g}_{\ell-m-3,-}$$

$$- (\gamma + \alpha^+\beta^+\gamma)\hat{g}^-_{\ell-m-3,-} + (-2+\gamma)\gamma^+ \hat{g}^+_{\ell-m-3,-})\big)$$

$$+ \sum_{m=0}^{\ell-2} \big(\alpha^+\beta^+\gamma^2 \hat{g}^-_{m,-}\hat{g}^-_{\ell-m-2,-} + \alpha\beta(\gamma^+)^2 \hat{g}^+_{m,-}\hat{g}^+_{\ell-m-2,-}$$

$$+ \hat{g}_{m,-}((\alpha^+\beta^+ + \alpha^2\alpha^+\beta^2\beta^+ + \alpha\beta(1+\alpha^+\beta^+)^2)\hat{g}_{\ell-m-2,-}$$

$$- 2(\alpha^+(1+\alpha\beta)\beta^+\gamma \hat{g}^-_{\ell-m-2,-} + \alpha\beta(1+\alpha^+\beta^+)\gamma^+ \hat{g}^+_{\ell-m-2,-}))\big)$$

$$+ \alpha^+\beta \sum_{m=0}^{\ell-1} \big(\gamma\gamma^+ \hat{g}^-_{m,-}\hat{g}^+_{\ell-m-1,-} + \hat{g}_{m,-}((1+\alpha\beta)(1+\alpha^+\beta^+)\hat{g}_{\ell-m-1,-}$$

$$- (\gamma + \alpha^+\beta^+\gamma)\hat{g}^-_{\ell-m-1,-} + (-2+\gamma)\gamma^+ \hat{g}^+_{\ell-m-1,-})\big)\bigg), \quad \ell \geq 3,$$

(C.53)

$$\hat{h}_{0,+} = \beta, \quad \hat{h}_{1,+} = \gamma\beta^- - \alpha^+\beta^2,$$
$$\hat{h}_{2,+} = \alpha^2(\beta^-)^2\beta + \beta^{--}\gamma^-\gamma - \alpha\beta^-(\beta^- - 2\alpha^+\beta^2)$$
$$+ \beta\big(-\alpha^{++}\beta + (\alpha^+)^2\beta^2 + \alpha^+(-2\beta^- + \alpha^{++}\beta\beta^+)\big),$$

$$\beta(\beta^+)^4 \hat{h}_{\ell,+} = -\frac{1}{2}\bigg(\beta^4 \sum_{m=0}^{\ell-4} \hat{h}_{m,+}\hat{h}_{\ell-m-4,+} + (\beta^+)^4 \sum_{m=1}^{\ell-1} \hat{h}_{m,+}\hat{h}_{\ell-m,+}$$

$$- 2\beta^2 \sum_{m=0}^{\ell-3} \hat{h}_{m,+}\big(-2\beta\beta^+\hat{h}_{\ell-m-3,+} + (\beta^+)^2\gamma\hat{h}^-_{\ell-m-3,+} + \beta^2\gamma^+\hat{h}^+_{\ell-m-3,+}\big)$$

$$+ \sum_{m=0}^{\ell-2} \big(\beta^4(\gamma^+)^2 \hat{h}^+_{m,+}\hat{h}^+_{\ell-m-2,+}$$

$$+ (\beta^+)^2\gamma\hat{h}^-_{m,+}((\beta^+)^2\gamma\hat{h}^-_{\ell-m-2,+} - 2\beta^2\gamma^+\hat{h}^+_{\ell-m-2,+})$$

$$- 2\beta\beta^+\hat{h}_{m,+}(-3\beta\beta^+\hat{h}_{\ell-m-2,+} + 2(\beta^+)^2\gamma\hat{h}^-_{\ell-m-2,+} + 2\beta^2\gamma^+\hat{h}^+_{\ell-m-2,+})\big)$$

$$- 2(\beta^+)^2 \sum_{m=0}^{\ell-1} \hat{h}_{m,+}\big(-2\beta\beta^+\hat{h}_{\ell-m-1,+} + (\beta^+)^2\gamma\hat{h}^-_{\ell-m-1,+}$$

$$+ \beta^2\gamma^+\hat{h}^+_{\ell-m-1,+}\big)\bigg), \quad \ell \geq 3, \tag{C.54}$$

$$\hat{h}_{0,-} = -\beta^+, \quad \hat{h}_{1,-} = -\gamma^+\beta^{++} + \alpha(\beta^+)^2,$$
$$\hat{h}_{2,-} = \alpha^-\gamma(\beta^+)^2 - \alpha^2(\beta^+)^3 + 2\alpha\beta^+\gamma^+\beta^{++} + \gamma^+\big(\alpha^+(\beta^{++})^2 - \gamma^{++}\big)\beta^{+++},$$

$$\beta^+\beta^4\hat{h}_{\ell,-} = \frac{1}{2}\bigg((\beta^+)^4 \sum_{m=0}^{\ell-4} \hat{h}_{m,-}\hat{h}_{\ell-m-4,-} + \beta^4 \sum_{m=1}^{\ell-1} \hat{h}_{m,-}\hat{h}_{\ell-m,-}$$

$$- 2(\beta^+)^2 \sum_{m=0}^{\ell-3} \hat{h}_{m,-}\big(-2\beta\beta^+\hat{h}_{\ell-m-3,-} + (\beta^+)^2\gamma\hat{h}^-_{\ell-m-3,-}$$

$$+ \beta^2\gamma^+\hat{h}^+_{\ell-m-3,-}\big)$$

$$+ \sum_{m=0}^{\ell-2} \big(\beta^4(\gamma^+)^2 \hat{h}^+_{m,-}\hat{h}^+_{\ell-m-2,-}$$

$$+ (\beta^+)^2\gamma\hat{h}^-_{m,-}((\beta^+)^2\gamma\hat{h}^-_{\ell-m-2,-} - 2\beta^2\gamma^+\hat{h}^+_{\ell-m-2,-})$$

$$- 2\beta\beta^+\hat{h}_{m,-}(-3\beta\beta^+\hat{h}_{\ell-m-2,-} + 2(\beta^+)^2\gamma\hat{h}^-_{\ell-m-2,-}$$

$$+ 2\beta^2\gamma^+\hat{h}^+_{\ell-m-2,-})\big)$$

$$- 2\beta^2 \sum_{m=0}^{\ell-1} \hat{h}_{m,-}\big(-2\beta\beta^+\hat{h}_{\ell-m-1,-} + (\beta^+)^2\gamma\hat{h}^-_{\ell-m-1,-}$$

$$+ \beta^2\gamma^+\hat{h}^+_{\ell-m-1,-}\big)\bigg), \quad \ell \geq 3. \tag{C.55}$$

Asymptotic Spectral Parameter Expansions

Proof We first consider the expansions (C.48) near $1/z = 0$ and the nonlinear recursion relations (C.50), (C.52), and (C.54) in detail. Inserting expansion (C.46) for $\mathfrak{f}$ into (C.39), the expansion (C.46) for $\mathfrak{g}$ into (C.40), and the expansion (C.46) for $\mathfrak{h}$ into (C.41), then yields the nonlinear recursion relations (C.50), (C.52), and (C.54), but with $\hat{f}_{\ell,+}$, $\hat{g}_{\ell,+}$, and $\hat{h}_{\ell,+}$ replaced by $\hat{\mathfrak{f}}_{\ell,+}$, $\hat{\mathfrak{g}}_{\ell,+}$, and $\hat{\mathfrak{h}}_{\ell,+}$, respectively. From the leading asymptotic behavior one finds that $\hat{\mathfrak{f}}_{0,+} = -\alpha^+$, $\hat{\mathfrak{g}}_{0,+} = \frac{1}{2}$, and $\hat{\mathfrak{h}}_{0,+} = \beta$. Next, inserting the expansions (C.46) for $\mathfrak{f}$, $\mathfrak{g}$, and $\mathfrak{h}$ into (C.34)–(C.37), and comparing powers of $z^{-\ell}$ as $|z| \to \infty$, $z \in C_R$, one infers that $\mathfrak{f}_{\ell,+}$, $\mathfrak{g}_{\ell,+}$, and $\mathfrak{h}_{\ell,+}$ satisfy the linear recursion relations (3.19)–(3.22). Here we have used (3.16). The coefficients $\hat{\mathfrak{f}}_{0,+}$, $\hat{\mathfrak{g}}_{0,+}$, and $\hat{\mathfrak{h}}_{0,+}$ are consistent with (3.19) for $c_{0,+} = 1$. Hence one concludes that

$$\hat{\mathfrak{f}}_{\ell,+} = f_{\ell,+}, \quad \hat{\mathfrak{g}}_{\ell,+} = g_{\ell,+}, \quad \hat{\mathfrak{h}}_{\ell,+} = h_{\ell,+}, \quad \ell \in \mathbb{N}_0,$$

for certain values of the summation constants $c_{\ell,+}$. To conclude that actually, $\hat{\mathfrak{f}}_{\ell,+} = \hat{f}_{\ell,+}$, $\hat{\mathfrak{g}}_{\ell,+} = \hat{g}_{\ell,+}$, $\hat{\mathfrak{h}}_{\ell,+} = \hat{h}_{\ell,+}$, $\ell \in \mathbb{N}_0$, and hence all $c_{\ell,+}$, $\ell \in \mathbb{N}$, vanish, we now rely on the notion of degree as introduced in Remark 3.6. To this end we recall that

$$\deg\left(\hat{f}_{\ell,+}\right) = \ell + 1, \quad \deg\left(\hat{g}_{\ell,+}\right) = \ell, \quad \deg\left(\hat{h}_{\ell,+}\right) = \ell, \quad \ell \in \mathbb{N}_0,$$

(cf. (3.36)). Similarly, the nonlinear recursion relations (C.50), (C.52), and (C.54) yield inductively that

$$\deg\left(\hat{\mathfrak{f}}_{\ell,+}\right) = \ell + 1, \quad \deg\left(\hat{\mathfrak{g}}_{\ell,+}\right) = \ell, \quad \deg\left(\hat{\mathfrak{h}}_{\ell,+}\right) = \ell, \quad \ell \in \mathbb{N}_0.$$

Hence one concludes

$$\hat{\mathfrak{f}}_{\ell,+} = \hat{f}_{\ell,+}, \quad \hat{\mathfrak{g}}_{\ell,+} = \hat{g}_{\ell,+}, \quad \hat{\mathfrak{h}}_{\ell,+} = \hat{h}_{\ell,+}, \quad \ell \in \mathbb{N}_0.$$

The proof of the corresponding asymptotic expansion (C.49) and the nonlinear recursion relations (C.51), (C.53), and (C.55) follows precisely the same strategy and is hence omitted. □

Given this general result on asymptotic (Green's function) expansions for Ablowitz–Ladik Lax difference expressions of the type (3.61), we now specialize to the algebro-geometric case at hand. We recall our conventions $y(P) = \mp(\zeta^{-p-1} + O(\zeta^{-p}))$ for P near $P_{\infty\pm}$ (where $\zeta = 1/z$) and $y(P) = \pm((c_{0,-}/c_{0,+}) + O(\zeta))$ for P near $P_{0,\pm}$ (where $\zeta = z$).

Theorem C.4 *Assume* (3.87), s-AL$_{\underline{p}}(\alpha, \beta) = 0$, *and suppose* $P = (z, y) \in \mathcal{K}_p \setminus \{P_{\infty_+}, P_{\infty_-}\}$. *Then* $z^{p-}F_{\underline{p}}/y$, $z^{p-}G_{\underline{p}}/y$, *and* $z^{p-}H_{\underline{p}}/y$ *have the following convergent expansions as* $P \to P_{\infty\pm}$, *respectively*, $P \to P_{0,\pm}$,

$$\frac{z^{p-}}{c_{0,+}} \frac{F_{\underline{p}}(z)}{y} = \begin{cases} \mp \sum_{\ell=0}^{\infty} \hat{f}_{\ell,+} \zeta^{\ell+1}, & P \to P_{\infty\pm}, \quad \zeta = 1/z, \\ \pm \sum_{\ell=0}^{\infty} \hat{f}_{\ell,-} \zeta^{\ell}, & P \to P_{0,\pm}, \quad \zeta = z, \end{cases} \quad \text{(C.56)}$$

$$\frac{z^{p_-}}{c_{0,+}}\frac{G_{\underline{p}}(z)}{y} = \begin{cases} \mp \sum_{\ell=0}^{\infty} \hat{g}_{\ell,+} \zeta^{\ell}, & P \to P_{\infty_{\pm}}, \quad \zeta = 1/z, \\ \pm \sum_{\ell=0}^{\infty} \hat{g}_{\ell,-} \zeta^{\ell}, & P \to P_{0,\pm}, \quad \zeta = z, \end{cases} \quad (C.57)$$

$$\frac{z^{p_-}}{c_{0,+}}\frac{H_{\underline{p}}(z)}{y} = \begin{cases} \mp \sum_{\ell=0}^{\infty} \hat{h}_{\ell,+} \zeta^{\ell}, & P \to P_{\infty_{\pm}}, \quad \zeta = 1/z, \\ \pm \sum_{\ell=0}^{\infty} \hat{h}_{\ell,-} \zeta^{\ell+1}, & P \to P_{0,\pm}, \quad \zeta = z, \end{cases} \quad (C.58)$$

where $\zeta = 1/z$ (resp., $\zeta = z$) is the local coordinate near $P_{\infty_{\pm}}$ (resp., $P_{0,\pm}$) and $\hat{f}_{\ell,\pm}$, $\hat{g}_{\ell,\pm}$, and $\hat{h}_{\ell,\pm}$ are the homogeneous versions[1] of the coefficients $f_{\ell,\pm}$, $g_{\ell,\pm}$, and $h_{\ell,\pm}$ as introduced in (3.32)–(3.34). Moreover, one infers for the E_m-dependent summation constants $c_{\ell,\pm}$, $\ell = 0, \ldots, p_{\pm}$, in $F_{\underline{p}}$, $G_{\underline{p}}$, and $H_{\underline{p}}$ that

$$c_{\ell,\pm} = c_{0,\pm} c_{\ell}(\underline{E}^{\pm 1}), \quad \ell = 0, \ldots, p_{\pm}. \quad (C.59)$$

In addition, one has the following relations between the homogeneous and nonhomogeneous recursion coefficients:

$$f_{\ell,\pm} = c_{0,\pm} \sum_{k=0}^{\ell} c_{\ell-k}(\underline{E}^{\pm 1}) \hat{f}_{k,\pm}, \quad \ell = 0, \ldots, p_{\pm}, \quad (C.60)$$

$$g_{\ell,\pm} = c_{0,\pm} \sum_{k=0}^{\ell} c_{\ell-k}(\underline{E}^{\pm 1}) \hat{g}_{k,\pm}, \quad \ell = 0, \ldots, p_{\pm}, \quad (C.61)$$

$$h_{\ell,\pm} = c_{0,\pm} \sum_{k=0}^{\ell} c_{\ell-k}(\underline{E}^{\pm 1}) h_{k,\pm}, \quad \ell = 0, \ldots, p_{\pm}. \quad (C.62)$$

Furthermore, one has

$$c_{0,\pm} \hat{f}_{\ell,\pm} = \sum_{k=0}^{\ell} \hat{c}_{\ell-k}(\underline{E}^{\pm 1}) f_{k,\pm}, \quad \ell = 0, \ldots, p_{\pm} - 1, \quad (C.63)$$

$$c_{0,\pm} \hat{f}_{p_{\pm},\pm} = \sum_{k=0}^{p_{\pm}-1} \hat{c}_{p_{\pm}-k}(\underline{E}^{\pm 1}) f_{k,\pm} + \hat{c}_0(\underline{E}^{\pm 1}) f_{p_{\mp}-1,\mp}, \quad (C.64)$$

$$c_{0,\pm} \hat{g}_{\ell,\pm} = \sum_{k=0}^{\ell} \hat{c}_{\ell-k}(\underline{E}^{\pm 1}) g_{k,\pm}, \quad \ell = 0, \ldots, p_{\pm} - 1, \quad (C.65)$$

$$c_{0,\pm} \hat{g}_{p_{\pm},\pm} = \sum_{k=0}^{p_{\pm}-1} \hat{c}_{p_{\pm}-k}(\underline{E}^{\pm 1}) g_{k,\pm} + \hat{c}_0(\underline{E}^{\pm 1}) g_{p_{\mp},\mp}, \quad (C.66)$$

---

[1] Strictly speaking, the coefficients $\hat{f}_{\ell,\pm}$, $\hat{g}_{\ell,\pm}$, and $\hat{h}_{\ell,\pm}$ in (C.56)–(C.58) no longer have a well-defined degree and hence represent a slight abuse of notation since we assumed that s-AL$_{\underline{p}}(\alpha, \beta) = 0$. At any rate, they are explicitly given by (C.69)–(C.71).

$$c_{0,\pm}\hat{h}_{\ell,\pm} = \sum_{k=0}^{\ell} \hat{c}_{\ell-k}(\underline{E}^{\pm 1})h_{k,\pm}, \quad \ell = 0, \ldots, p_{\pm} - 1, \tag{C.67}$$

$$c_{0,\pm}\hat{h}_{p_{\pm},\pm} = \sum_{k=0}^{p_{\pm}-1} \hat{c}_{p_{\pm}-k}(\underline{E}^{\pm 1})h_{k,\pm} + \hat{c}_0(\underline{E}^{\pm 1})h_{p_{\mp}-1,\mp}. \tag{C.68}$$

For general ℓ (not restricted to $\ell \leq p_{\pm}$) one has[1]

$$c_{0,\pm}\hat{f}_{\ell,\pm} = \begin{cases} \sum_{k=0}^{\ell} \hat{c}_{\ell-k}(\underline{E}^{\pm 1}) f_{k,\pm}, & \ell = 0, \ldots, p_{\pm} - 1, \\ \sum_{k=0}^{p_{\pm}-1} \hat{c}_{\ell-k}(\underline{E}^{\pm 1}) f_{k,\pm} \\ \quad + \sum_{k=(p-\ell)\vee 0}^{p_{\mp}-1} \hat{c}_{\ell+k-p}(\underline{E}^{\pm 1}) f_{k,\mp}, & \ell \geq p_{\pm}, \end{cases} \tag{C.69}$$

$$c_{0,\pm}\hat{g}_{\ell,\pm} = \begin{cases} \sum_{k=0}^{\ell} \hat{c}_{\ell-k}(\underline{E}^{\pm 1}) g_{k,\pm}, & \ell = 0, \ldots, p_{\pm} - \delta_{\pm}, \\ \sum_{k=0}^{p_{\pm}-\delta_{\pm}} \hat{c}_{\ell-k}(\underline{E}^{\pm 1}) g_{k,\pm} \\ \quad + \sum_{k=(p-\ell)\vee 0}^{p_{\mp}-\delta_{\pm}} \hat{c}_{\ell+k-p}(\underline{E}^{\pm 1}) g_{k,\mp}, & \ell \geq p_{\pm} - \delta_{\pm} + 1, \end{cases} \tag{C.70}$$

$$c_{0,\pm}\hat{h}_{\ell,\pm} = \begin{cases} \sum_{k=0}^{\ell} \hat{c}_{\ell-k}(\underline{E}^{\pm 1}) h_{k,\pm}, & \ell = 0, \ldots, p_{\pm} - 1, \\ \sum_{k=0}^{p_{\pm}-1} \hat{c}_{\ell-k}(\underline{E}^{\pm 1}) h_{k,\pm} \\ \quad + \sum_{k=(p-\ell)\vee 0}^{p_{\mp}-1} \hat{c}_{\ell+k-p}(\underline{E}^{\pm 1}) h_{k,\mp}, & \ell \geq p_{\pm}. \end{cases} \tag{C.71}$$

Here we used the convention

$$\delta_{\pm} = \begin{cases} 0, & +, \\ 1, & -. \end{cases}$$

Proof Identifying

$$\Psi_+(z, \cdot) \text{ with } \Psi(P, \cdot, 0) \text{ and } \Psi_-(z, \cdot) \text{ with } \Psi(P^*, \cdot, 0), \tag{C.72}$$

recalling that $W(\Psi(P, \cdot, 0), \Psi(P^*, \cdot, 0)) = -c_{0,+} z^{n-n_0-p-y} F_{\underline{p}}(z, 0)^{-1} \Gamma(n, n_0)$
(cf. (3.109)), and similarly, identifying

$$\phi_+(z, \cdot) \text{ with } \phi(P, \cdot) \text{ and } \phi_-(z, \cdot) \text{ with } \phi(P^*, \cdot), \tag{C.73}$$

a comparison of (C.29)–(C.33) and the results of Lemmas 3.18 and 3.21 shows that we may also identify

$$\mathfrak{f} \text{ with } \mp \frac{2F_{\underline{p}}}{c_{0,+}z^{-p}-y}, \quad \mathfrak{g} \text{ with } \mp \frac{2G_{\underline{p}}}{c_{0,+}z^{-p}-y}, \text{ and } \mathfrak{h} \text{ with } \mp \frac{2H_{\underline{p}}}{c_{0,+}z^{-p}-y},$$

the sign depending on whether P tends to $P_{\infty_{\pm}}$ or to $P_{0,\pm}$. In particular, (C.34)–(C.41) then correspond to (3.8)–(3.11), (3.39), (3.43)–(3.45), respectively. Since

---

[1] $m \vee n = \max\{m, n\}$.

$z^{p_-}F_{\underline{p}}/y$, $z^{p_-}G_{\underline{p}}/y$, and $z^{p_-}H_{\underline{p}}/y$ clearly have asymptotic (in fact, even convergent) expansions as $|z| \to \infty$ and as $|z| \to 0$, the results of Theorem C.3 apply. Thus, as $P \to P_{\infty\pm}$, one obtains the following expansions using (C.1) and (3.13)–(3.15):

$$\frac{z^{p_-}}{c_{0,+}} \frac{F_{\underline{p}}(z)}{y}$$

$$\underset{\zeta \to 0}{=} \mp \frac{1}{c_{0,+}} \bigg(\sum_{k=0}^{\infty} \hat{c}_k(\underline{E}) \zeta^k \bigg) \bigg(\sum_{\ell=1}^{p_-} f_{p_--\ell,-} \zeta^{p_++\ell} + \sum_{\ell=0}^{p_+-1} f_{p_+-1-\ell,+} \zeta^{p_+-\ell} \bigg)$$

$$\underset{\zeta \to 0}{=} \mp \sum_{\ell=0}^{\infty} \hat{f}_{\ell,+} \zeta^{\ell+1}, \qquad (C.74)$$

$$\frac{z^{p_-}}{c_{0,+}} \frac{G_{\underline{p}}(z)}{y}$$

$$\underset{\zeta \to 0}{=} \mp \frac{1}{c_{0,+}} \bigg(\sum_{k=0}^{\infty} \hat{c}_k(\underline{E}) \zeta^k \bigg) \bigg(\sum_{\ell=1}^{p_-} g_{p_--\ell,-} \zeta^{p_++\ell} + \sum_{\ell=0}^{p_+} g_{p_+-\ell,+} \zeta^{p_+-\ell} \bigg)$$

$$\underset{\zeta \to 0}{=} \mp \sum_{\ell=0}^{\infty} \hat{g}_{\ell,+} \zeta^{\ell}, \qquad (C.75)$$

$$\frac{z^{p_-}}{c_{0,+}} \frac{H_{\underline{p}}(z)}{y}$$

$$\underset{\zeta \to 0}{=} \mp \frac{1}{c_{0,+}} \bigg(\sum_{k=0}^{\infty} \hat{c}_k(\underline{E}) \zeta^k \bigg) \bigg(\sum_{\ell=0}^{p_--1} h_{p_--1-\ell,-} \zeta^{p_++\ell} + \sum_{\ell=1}^{p_+} h_{p_+-\ell,+} \zeta^{p_+-\ell} \bigg)$$

$$\underset{\zeta \to 0}{=} \mp \sum_{\ell=0}^{\infty} \hat{h}_{\ell,+} \zeta^{\ell}. \qquad (C.76)$$

This implies (C.56)–(C.58) as $P \to P_{\infty\pm}$.

Similarly, as $P \to P_{0,\pm}$, (C.1) and (3.13)–(3.15), and (3.41) imply

$$\frac{z^{p_-}}{c_{0,+}} \frac{F_{\underline{p}}(z)}{y} \underset{\zeta \to 0}{=} \pm \frac{1}{c_{0,-}} \bigg(\sum_{k=0}^{\infty} \hat{c}_k(\underline{E}^{-1}) \zeta^k \bigg)$$

$$\times \bigg(\sum_{\ell=1}^{p_-} f_{p_--\ell,-} \zeta^{p_+-\ell} + \sum_{\ell=0}^{p_+-1} f_{p_+-1-\ell,+} \zeta^{p_++\ell} \bigg)$$

$$\underset{\zeta \to 0}{=} \pm \sum_{\ell=0}^{\infty} \hat{f}_{\ell,-} \zeta^{\ell},$$

$$\frac{z^{p_-}}{c_{0,+}}\frac{G_{\underline{p}}(z)}{y} \underset{\zeta\to 0}{=} \pm\frac{1}{c_{0,-}}\left(\sum_{k=0}^{\infty}\hat{c}_k(\underline{E}^{-1})\zeta^k\right)$$

$$\times\left(\sum_{\ell=1}^{p_-}g_{p_--\ell,-}\zeta^{p_+-\ell}+\sum_{\ell=0}^{p_+}g_{p_+-\ell,+}\zeta^{p_++\ell}\right)$$

$$\underset{\zeta\to 0}{=}\pm\sum_{\ell=0}^{\infty}\hat{g}_{\ell,-}\zeta^{\ell},$$

$$\frac{z^{p_-}}{c_{0,+}}\frac{H_{\underline{p}}(z)}{y} \underset{\zeta\to 0}{=} \pm\frac{1}{c_{0,-}}\left(\sum_{k=0}^{\infty}\hat{c}_k(\underline{E}^{-1})\zeta^k\right)$$

$$\times\left(\sum_{\ell=0}^{p_--1}h_{p_--1-\ell,-}\zeta^{p_+-\ell}+\sum_{\ell=1}^{p_+}h_{p_+-\ell,+}\zeta^{p_++\ell}\right)$$

$$\underset{\zeta\to 0}{=}\pm\sum_{\ell=0}^{\infty}\hat{h}_{\ell,-}\zeta^{\ell+1}.$$

Thus, (C.56)–(C.58) hold as $\underline{P}\to \underline{P}_{0,\pm}$.

Next, comparing powers of ζ in the second and third term of (C.74), formula (C.63) follows (and hence (C.69) as well). Formulas (C.65) and (C.67) follow by using (C.75) and (C.76), respectively.

To prove (C.60) one uses (C.6) and finds

$$c_{0,\pm}\sum_{m=0}^{\ell}c_{\ell-m}(\underline{E}^{\pm 1})\hat{f}_{m,\pm} = \sum_{m=0}^{\ell}c_{\ell-m}(\underline{E})\sum_{k=0}^{m}\hat{c}_{m-k}(\underline{E}^{\pm 1})f_{k,\pm} = f_{\ell,\pm}.$$

The proofs of (C.61) and (C.62) and those of (C.70) and (C.71) are analogous. □

Finally, we also mention the following system of recursion relations for the homogeneous coefficients $\hat{f}_{\ell,\pm}$, $\hat{g}_{\ell,\pm}$, and $\hat{h}_{\ell,\pm}$.

Lemma C.5 *The homogeneous coefficients $\hat{f}_{\ell,\pm}$, $\hat{g}_{\ell,\pm}$, and $\hat{h}_{\ell,\pm}$ are uniquely defined by the following recursion relations:*

$$\hat{g}_{0,+} = \frac{1}{2}, \quad \hat{f}_{0,+} = -\alpha^+, \quad \hat{h}_{0,+} = \beta,$$

$$\hat{g}_{\ell+1,+} = \sum_{k=0}^{\ell}\hat{f}_{\ell-k,+}\hat{h}_{k,+} - \sum_{k=1}^{\ell}\hat{g}_{\ell+1-k,+}\hat{g}_{k,+}, \qquad \text{(C.77)}$$

$$\hat{f}_{\ell+1,+}^- = \hat{f}_{\ell,+} - \alpha(\hat{g}_{\ell+1,+} + \hat{g}_{\ell+1,+}^-),$$

$$\hat{h}_{\ell+1,+} = \hat{h}_{\ell,+}^- + \beta(\hat{g}_{\ell+1,+} + \hat{g}_{\ell+1,+}^-),$$

and

$$\hat{g}_{0,-} = \frac{1}{2}, \quad \hat{f}_{0,-} = \alpha, \quad \hat{h}_{0,-} = -\beta^+,$$

$$\hat{g}_{\ell+1,-} = \sum_{k=0}^{\ell} \hat{f}_{\ell-k,-}\hat{h}_{k,-} - \sum_{k=1}^{\ell} \hat{g}_{\ell+1-k,-}\hat{g}_{k,-},$$

$$\hat{f}_{\ell+1,-} = \hat{f}_{\ell,-}^- + \alpha(\hat{g}_{\ell+1,-} + \hat{g}_{\ell+1,-}^-),$$

$$\hat{h}_{\ell+1,-}^- = \hat{h}_{\ell,-} - \beta(\hat{g}_{\ell+1,-} + \hat{g}_{\ell+1,-}^-).$$

Proof One verifies that the coefficients defined via these recursion relations satisfy (3.19)–(3.22) (resp., (3.23)–(3.26)). Since they are homogeneous of the required degree this completes the proof. □

Notes

This appendix is patterned after Appendix D in Gesztesy and Holden (2003b).

High-energy expansions for the Toda system were studied in Section 4 of Bulla et al. (1998) (see also Teschl (2000, Sect. 12.3)).

The high-energy expansion for the AL-system is taken from Gesztesy et al. (to appear).

For more details on (C.42)–(C.45) we refer to Gesztesy and Zinchenko (2006b) and Gesztesy et al. (2008a).

Appendix D
Lagrange Interpolation

Tvertimot.
Henrik Ibsen's very last word[1]

In this appendix we briefly review essentials of the standard Lagrange interpolation formalism. Near the end we briefly turn to the general situation where the interpolating polynomial need not have distinct zeros.

Assuming $p \in \mathbb{N}$ to be fixed and introducing

$$\mathcal{S}_k = \{\underline{\ell} = (\ell_1, \ldots, \ell_k) \in \mathbb{N}^k \mid 1 \leq \ell_1 < \cdots < \ell_k \leq p\}, \quad k = 1, \ldots, p,$$
$$\mathcal{I}_k^{(j)} = \{\underline{\ell} = (\ell_1, \ldots, \ell_k) \in \mathcal{S}_k \mid \ell_m \neq j, \, m = 1, \ldots, k\},$$
$$k = 1, \ldots, p-1, \; j = 1, \ldots, p,$$

one defines the symmetric functions

$$\Psi_0(\underline{\mu}) = 1, \quad \Psi_k(\underline{\mu}) = (-1)^k \sum_{\underline{\ell} \in \mathcal{S}_k} \mu_{\ell_1} \cdots \mu_{\ell_k}, \quad k = 1, \ldots, p,$$

$$\Phi_0^{(j)}(\underline{\mu}) = 1,$$
$$\Phi_k^{(j)}(\underline{\mu}) = (-1)^k \sum_{\underline{\ell} \in \mathcal{I}_k^{(j)}} \mu_{\ell_1} \cdots \mu_{\ell_k}, \quad k = 1, \ldots, p-1, \; j = 1, \ldots, p,$$
$$\Phi_p^{(j)}(\underline{\mu}) = 0, \quad j = 1, \ldots, p,$$

where $\underline{\mu} = (\mu_1, \ldots, \mu_p) \in \mathbb{C}^p$. Explicitly, one verifies

$$\Psi_1(\underline{\mu}) = -\sum_{\ell=1}^p \mu_\ell, \quad \Psi_2(\underline{\mu}) = \sum_{\substack{\ell_1, \ell_2 = 1 \\ \ell_1 < \ell_2}}^p \mu_{\ell_1} \mu_{\ell_2}, \text{ etc.},$$

[1] "On the contrary."

$$\Phi_1^{(j)}(\underline{\mu}) = -\sum_{\substack{\ell=1 \\ \ell \neq j}}^{p} \mu_\ell, \quad \Phi_2^{(j)}(\underline{\mu}) = \sum_{\substack{\ell_1,\ell_2=1 \\ \ell_1,\ell_2 \neq j \\ \ell_1 < \ell_2}}^{p} \mu_{\ell_1}\mu_{\ell_2}, \text{ etc.}$$

Introducing

$$F_p(z) = \prod_{j=1}^{p}(z - \mu_j) = \sum_{\ell=0}^{p} \Psi_{p-\ell}(\underline{\mu})z^\ell, \quad z \in \mathbb{C}, \tag{D.1}$$

one infers

$$F_{p,z}(\mu_k) = \prod_{\substack{j=1 \\ j \neq k}}^{p}(\mu_k - \mu_j).$$

The general form of Lagrange's interpolation theorem then reads as follows.

Theorem D.1 *Assume that $\mu_1, \ldots, \mu_p$ are p distinct complex numbers. Then,*

$$\sum_{j=1}^{p} \frac{\mu_j^{m-1}}{F_{p,z}(\mu_j)} \Phi_k^{(j)}(\underline{\mu}) = \delta_{m,p-k} - \Psi_{k+1}(\underline{\mu})\delta_{m,p+1}, \tag{D.2}$$

$$m = 1, \ldots, p+1, \quad k = 0, \ldots, p-1.$$

Proof Let C_R be a circle with center at the origin and radius R that contains the zeros μ_j of the polynomial F_p and that is oriented counterclockwise. Cauchy's theorem then yields

$$\frac{1}{2\pi i} \oint_{C_R} d\zeta \, \frac{\zeta^{m-1}}{F_p(\zeta)(\zeta - z)} = \frac{z^{m-1}}{F_p(z)} + \sum_{k=1}^{p} \frac{\mu_k^{m-1}}{F_{p,z}(\mu_j)(\mu_j - z)},$$

$$z \neq \mu_1, \ldots, \mu_p, \quad m = 1, \ldots, p+1.$$

However, by letting $R \to \infty$, we infer that

$$\frac{1}{2\pi i} \oint_{C_R} d\zeta \, \frac{\zeta^{m-1}}{F_p(\zeta)(\zeta - z)} = \lim_{R \to \infty} \frac{R^{m-1}}{F_p(R)} = \delta_{m,p+1}, \quad m = 1, \ldots, p+1,$$

which implies

$$z^{m-1} - \sum_{k=1}^{p} \frac{\mu_k^{m-1} F_p(z)}{F_{p,z}(\mu_j)(z - \mu_j)} = F_p(z)\delta_{m,p+1}. \tag{D.3}$$

Using the symmetric functions Ψ_j, we may write

$$F_p(z) = \sum_{j=0}^{p} \Psi_j(\underline{\mu}) z^{p-j} \tag{D.4}$$

and
$$\frac{F_p(z)}{z - \mu_j} = \sum_{k=0}^{p-1} \Phi_k^{(j)}(\underline{\mu}) z^{p-1-k}. \tag{D.5}$$

Expanding both sides of equation (D.3) in powers in z, using (D.4) on the right-hand side and (D.5) on the left-hand side, proves (D.2). □

The simplest Lagrange interpolation formula reads in the case $k = 0$,
$$\sum_{j=1}^{p} \frac{\mu_j^{m-1}}{F_{p,z}(\mu_j)} = \delta_{m,p}, \quad m = 1, \ldots, p.$$

As a consequence, if Q_{p-1} denotes a polynomial of degree $p - 1$, then
$$Q_{p-1}(z) = F_p(z) \sum_{j=1}^{p} \frac{Q_{p-1}(\mu_j)}{F_{p,z}(\mu_j)(z - \mu_j)}$$
$$= \sum_{j=1}^{p} Q_{p-1}(\mu_j) \prod_{\substack{k=1 \\ k \neq j}}^{p} \frac{z - \mu_k}{\mu_j - \mu_k}, \quad z \in \mathbb{C}, \tag{D.6}$$

assuming $\mu_1, \ldots, \mu_p$ to be pairwise distinct.

For use in the main text we finally observe the following results.

Lemma D.2 *Assume that $\mu_1, \ldots, \mu_p$ are p distinct complex numbers. Then,*

(i) $\Psi_{k+1}(\underline{\mu}) + \mu_j \Phi_k^{(j)}(\underline{\mu}) = \Phi_{k+1}^{(j)}(\underline{\mu}), \quad k = 0, \ldots, p-1, \; j = 1, \ldots, p.$

$$\tag{D.7}$$

(ii) $\sum_{\ell=0}^{k} \Psi_{k-\ell}(\underline{\mu}) \mu_j^\ell = \Phi_k^{(j)}(\underline{\mu}), \quad k = 0, \ldots, p, \; j = 1, \ldots, p.$ (D.8)

(iii) $\sum_{\ell=0}^{k-1} \Phi_{k-1-\ell}^{(j)}(\underline{\mu}) z^\ell = \frac{1}{z - \mu_j} \left(\sum_{\ell=0}^{k} \Psi_{k-\ell}(\underline{\mu}) z^\ell - \Phi_k^{(j)}(\underline{\mu}) \right),$ (D.9)

$$k = 0, \ldots, p, \; j = 1, \ldots, p.$$

Proof (i) Adding (D.4) to μ_j times (D.5), one finds
$$F_p(z) + \mu_j \frac{F_p(z)}{z - \mu_j} = \sum_{k=0}^{p-1} \left(\Psi_{k+1} + \mu_j \Phi_k^{(j)} \right) z^{p-k-1} + z^p.$$

However, one also has

$$F_p(z) + \mu_j \frac{F_p(z)}{z - \mu_j} = z\frac{F_p(z)}{z - \mu_j} = \sum_{k=0}^{p-1} \Phi_{k+1}^{(j)} z^{p-k-1} + z^p,$$

using (D.5) and recalling $\Phi_p^{(j)} = 0$. Thus, (D.7) holds.

(ii) We prove (D.8) by induction on k. Equation (D.8) clearly holds for $k = 0$; next assume that

$$\sum_{\ell=0}^{k-1} \Psi_{k-1-\ell} \mu_j^\ell = \Phi_{k-1}^{(j)}$$

holds. Then one finds that

$$\sum_{\ell=0}^{k} \Psi_{k-\ell} \mu_j^\ell = \Psi_k + \mu_j \sum_{\ell=1}^{k} \Psi_{k-\ell} \mu_j^{\ell-1}$$

$$= \Psi_k + \mu_j \sum_{\ell=0}^{k-1} \Psi_{k-1-\ell} \mu_j^\ell = \Psi_k + \mu_j \Phi_{k-1}^{(j)} = \Phi_k^{(j)},$$

using first the induction hypothesis and then (D.7).

(iii) Using (D.7) and $\Psi_0(\underline{\mu}) = \Phi_0^{(j)}(\underline{\mu}) = 1$ one computes

$$(z - \mu_j) \sum_{\ell=0}^{k-1} \Phi_{k-1-\ell}^{(j)}(\underline{\mu}) z^\ell = \sum_{m=1}^{k} \Phi_{k-m}^{(j)}(\underline{\mu}) z^m - \sum_{\ell=0}^{k-1} \mu_j \Phi_{k-1-\ell}^{(j)}(\underline{\mu}) z^\ell$$

$$= \sum_{m=1}^{k} \Phi_{k-m}^{(j)}(\underline{\mu}) z^m + \sum_{\ell=0}^{k-1} \Psi_{k-\ell}(\underline{\mu}) z^\ell - \sum_{\ell=0}^{k-1} \Phi_{k-\ell}^{(j)}(\underline{\mu}) z^\ell$$

$$= \sum_{\ell=0}^{k-1} \Psi_{k-\ell}(\underline{\mu}) z^\ell + \Phi_0^{(j)}(\underline{\mu}) z^k - \Phi_k^{(j)}(\underline{\mu})$$

$$= \sum_{\ell=0}^{k} \Psi_{k-\ell}(\underline{\mu}) z^\ell - \Phi_k^{(j)}(\underline{\mu}).$$

□

Next, assuming $\mu_j \neq \mu_{j'}$ for $j \neq j'$, we introduce the $p \times p$ matrix $U_p(\underline{\mu})$ by

$$U_1(\underline{\mu}) = 1, \quad U_p(\underline{\mu}) = \left(\frac{\mu_k^{j-1}}{\prod_{\substack{m=1 \\ m \neq k}}^{p} (\mu_k - \mu_m)} \right)_{j,k=1}^{p}, \quad (D.10)$$

where $\underline{\mu} = (\mu_1, \ldots, \mu_p) \in \mathbb{C}^p$.

Lagrange Interpolation   389

Lemma D.3 *Suppose $\mu_j \in \mathbb{C}$, $j = 1, \ldots, p$, are p distinct complex numbers. Then,*

$$U_p(\underline{\mu})^{-1} = \left(\Phi^{(j)}_{p-k}(\underline{\mu})\right)^p_{j,k=1}. \quad (D.11)$$

Proof One observes that (D.10) may be written as

$$U_p(\underline{\mu}) = \left(\frac{\mu_k^{j-1}}{F_{p,z}(\mu_k)}\right)^p_{j,k=1}. \quad (D.12)$$

Using Lagrange's interpolation result, Theorem D.1 (replacing k by $p - k$ in (D.2)), then proves the result. □

In the next result we derive some formulas useful in connection with the Toda hierarchy. More precisely, we express f_ℓ and $\widetilde{F}_r$ in terms of elementary symmetric functions of $\mu_1, \ldots, \mu_p$.

Lemma D.4 *Let $\hat{c}_k(\underline{E})$ be defined as in (C.2). Then one infers the following result for the homogeneous coefficients $\hat{f}_\ell$ introduced in (1.7) in connection with the Toda hierarchy,*[1]

$$\hat{f}_\ell = \sum_{k=0}^{\ell \wedge p} \hat{c}_{\ell-k}(\underline{E})\Psi_k(\underline{\mu}), \quad \ell \in \mathbb{N}_0. \quad (D.13)$$

Moreover, let $r \in \mathbb{N}_0$, then[2]

$$\widehat{F}_r(\mu_j) = \sum_{s=(r-p)\vee 0}^{r} \hat{c}_s(\underline{E})\Phi^{(j)}_{r-s}(\underline{\mu}).$$

Proof It suffices to refer to (1.30), (1.64) and to note

$$F_p(z) = \sum_{\ell=0}^{p} f_{p-\ell}z^\ell = \prod_{j=1}^{p}(z - \mu_j) = \sum_{\ell=0}^{p} \Psi_{p-\ell}(\underline{\mu})z^\ell,$$

that is, $f_\ell = \Psi_\ell(\underline{\mu})$, $\ell = 0, \ldots, p$. Equation (D.13) then follows from (C.25). By definition,

$$\widehat{F}_r(z) = \sum_{\ell=0}^{r} \hat{f}_{r-\ell}z^\ell = \sum_{\ell=0}^{r} z^\ell \sum_{m=0}^{(r-\ell)\wedge p} \hat{c}_{r-\ell-m}(\underline{E})\Psi_m(\underline{\mu}).$$

---
[1] $m \wedge n = \min\{m, n\}$.
[2] $m \vee n = \max\{m, n\}$.

Consider first the case $r \leq p$. Then

$$\widehat{F}_r(z) = \sum_{s=0}^{r} \hat{c}_s(\underline{E}) \sum_{\ell=0}^{r-s} \Psi_{r-\ell-s}(\underline{\mu}) z^\ell$$

and hence

$$\widehat{F}_r(\mu_j) = \sum_{s=0}^{r} \hat{c}_s(\underline{E}) \Phi_{r-s}^{(j)}(\underline{\mu}),$$

using (D.8). In the case $r \geq p + 1$, we find, applying (D.1),

$$\widehat{F}_r(z) = \sum_{m=0}^{p} \Psi_m(\underline{\mu}) \sum_{s=0}^{r-m} \hat{c}_s(\underline{E}) z^{r-m-s}$$

$$= \sum_{s=0}^{r-p} \hat{c}_s(\underline{E}) \bigg(\sum_{\ell=0}^{p} \Psi_\ell(\underline{\mu}) z^{p-\ell} \bigg) z^{r-p-s} + \sum_{s=r-p+1}^{r} \hat{c}_s(\underline{E}) \sum_{\ell=0}^{r-s} \Psi_\ell(\underline{\mu}) z^{r-s-\ell}$$

$$= F_p(z) \sum_{s=0}^{r-p} \hat{c}_s(\underline{E}) z^{r-p-s} + \sum_{s=r-p+1}^{r} \hat{c}_s(\underline{E}) \sum_{\ell=0}^{r-s} \Psi_\ell(\underline{\mu}) z^{r-s-\ell}$$

$$= F_p(z) \sum_{s=0}^{r-p} \hat{c}_s(\underline{E}) z^{r-p-s} + \sum_{s=r-p+1}^{r} \hat{c}_s(\underline{E}) \sum_{\ell=0}^{r-s} \Psi_{r-s-\ell}(\underline{\mu}) z^\ell.$$

Hence,

$$\widehat{F}_r(\mu_j) = \sum_{s=r-p+1}^{r} \hat{c}_s(\underline{E}) \Phi_{r-s}^{(j)}(\underline{\mu}),$$

using (D.8) again. □

Introducing

$$d_{\ell,k}(\underline{E}) = \sum_{m=0}^{\ell-k} c_{\ell-k-m}(\underline{E}) \hat{c}_m(\underline{E}), \quad k = 0, \ldots, \ell, \; \ell = 0, \ldots, p, \tag{D.14}$$

$$\tilde{d}_{r,k}(\underline{E}) = \sum_{s=0}^{r-k} \tilde{c}_{r-k-s} \hat{c}_s(\underline{E}), \quad k = 0, \ldots, r \wedge p, \tag{D.15}$$

for a given set of constants $\{\tilde{c}_s\}_{s=1,\ldots,r} \subset \mathbb{C}$, the corresponding nonhomogeneous quantities f_ℓ, $F_p(\mu_j)$, and $\widetilde{F}_r(\mu_j)$ in the Toda case are then given by

$$f_\ell = \sum_{k=0}^{\ell} c_{\ell-k}(\underline{E}) \hat{f}_k = \sum_{k=0}^{\ell} d_{\ell,k}(\underline{E}) \Psi_k(\underline{\mu}), \quad \ell = 0, \ldots, p, \tag{D.16}$$

$$F_p(\mu_j) = \sum_{\ell=0}^{p} c_{p-\ell}(\underline{E})\widehat{F}_\ell(\mu_j) = \sum_{\ell=0}^{p} d_{p,\ell}(\underline{E})\Phi_\ell^{(j)}(\underline{\mu}), \quad c_0 = 1, \tag{D.17}$$

$$\widetilde{F}_r(\mu_j) = \sum_{s=0}^{r} \tilde{c}_{r-s}\widehat{F}_s(\mu_j) = \sum_{k=0}^{r \wedge p} \tilde{d}_{r,k}(\underline{E})\Phi_k^{(j)}(\underline{\mu}), \quad r \in \mathbb{N}_0, \ \tilde{c}_0 = 1. \tag{D.18}$$

Here $c_k(\underline{E})$, $k \in \mathbb{N}_0$, are defined by (C.5).

In the remainder of this appendix we recall a useful interpolation formula which goes beyond the standard Lagrange interpolation formula for polynomials in the sense that the zeros of the interpolating polynomial need not be distinct.

Lemma D.5 *Let $p \in \mathbb{N}$ and S_{p-1} be a polynomial of degree $p - 1$. In addition, let F_p be a monic polynomial of degree p of the form*

$$F_p(z) = \prod_{k=1}^{q}(z - \mu_k)^{p_k}, \quad p_j \in \mathbb{N}, \ \mu_j \in \mathbb{C}, \ j = 1, \dots, q, \quad \sum_{k=1}^{q} p_k = p.$$

Then,

$$S_{p-1}(z) = F_p(z) \sum_{k=1}^{q} \sum_{\ell=0}^{p_k-1} \frac{S_{p-1}^{(\ell)}(\mu_k)}{\ell!(p_k - \ell - 1)!} \tag{D.19}$$

$$\times \left(\frac{d^{p_k-\ell-1}}{d\zeta^{p_k-\ell-1}} \left((z-\zeta)^{-1} \prod_{k'=1, k' \neq k}^{q} (\zeta - \mu_{k'})^{-p_{k'}} \right) \right) \bigg|_{\zeta = \mu_k}, \quad z \in \mathbb{C}.$$

In particular, S_{p-1} is uniquely determined by prescribing the p values

$$S_{p-1}(\mu_k), S'_{p-1}(\mu_k), \dots, S_{p-1}^{(p_k-1)}(\mu_k), \quad k = 1, \dots, q,$$

at the given points $\mu_1, \dots, \mu_q$.
Conversely, prescribing the p complex numbers

$$\alpha_k^{(0)}, \alpha_k^{(1)}, \dots, \alpha_k^{(p_k-1)}, \quad k = 1, \dots, q,$$

there exists a unique polynomial T_{p-1} of degree $p - 1$,

$$T_{p-1}(z) = F_p(z) \sum_{k=1}^{q} \sum_{\ell=0}^{p_k-1} \frac{\alpha_k^{(\ell)}}{\ell!(p_k - \ell - 1)!} \tag{D.20}$$

$$\times \left(\frac{d^{p_k-\ell-1}}{d\zeta^{p_k-\ell-1}} \left((z-\zeta)^{-1} \prod_{k'=1, k' \neq k}^{q} (\zeta - \mu_{k'})^{-p_{k'}} \right) \right) \bigg|_{\zeta = \mu_k}, \quad z \in \mathbb{C},$$

such that

$$T_{p-1}(\mu_k) = \alpha_k^{(0)}, T'_{p-1}(\mu_k) = \alpha_k^{(1)}, \dots, T_{p-1}^{(p_k-1)}(\mu_k) = \alpha_k^{(p_k-1)}, \quad k = 1, \dots, q.$$

Proof Our starting point for proving (D.19) is the following formula,

$$S_{p-1}(z) = \frac{1}{2\pi i} \oint_\Gamma \frac{d\zeta \, S_{p-1}(\zeta)}{F_p(\zeta)} \frac{F_p(\zeta) - F_p(z)}{\zeta - z}, \quad z \in \mathbb{C}, \tag{D.21}$$

where Γ is a simple, smooth, counterclockwise oriented curve encircling the points $\mu_1, \ldots, \mu_q$. Since the integrand in (D.21) is analytic at the point $\zeta = z$, we may, without loss of generality, assume that Γ does not encircle z. With this assumption one obtains

$$\frac{1}{2\pi i} \oint_\Gamma \frac{d\zeta \, S_{p-1}(\zeta)}{\zeta - z} = 0$$

and hence deforming Γ into sufficiently small counterclockwise oriented circles Γ_k with center at μ_k, $k = 1, \ldots, q$, such that no $\mu_{k'}$, $k' \neq k$, is encircled by Γ_k, one obtains

$$\begin{aligned}
S_{p-1}(z) &= -\frac{F_p(z)}{2\pi i} \oint_\Gamma \frac{d\zeta \, S_{p-1}(\zeta)}{F_p(\zeta)(\zeta - z)} \\
&= -\frac{F_p(z)}{2\pi i} \sum_{k=1}^{q} \oint_{\Gamma_k} \frac{d\zeta \, S_{p-1}(\zeta)}{F_p(\zeta)(\zeta - z)} \\
&= -\frac{F_p(z)}{2\pi i} \sum_{k=1}^{q} \sum_{\ell=0}^{p-1} \frac{S_{p-1}^{(\ell)}(\mu_k)}{\ell!} \oint_{\Gamma_k} \frac{d\zeta \, (\zeta - \mu_k)^\ell}{F_p(\zeta)(\zeta - z)} \\
&= -\frac{F_p(z)}{2\pi i} \sum_{k=1}^{q} \sum_{\ell=0}^{p-1} \frac{S_{p-1}^{(\ell)}(\mu_k)}{\ell!} \oint_{\Gamma_k} \frac{d\zeta \, (\zeta - \mu_k)^\ell}{(\zeta - z) \prod_{k'=1}^{q} (\zeta - \mu_{k'})^{p_{k'}}} \\
&= -\frac{F_p(z)}{2\pi i} \sum_{k=1}^{q} \sum_{\ell=0}^{p-1} \frac{S_{p-1}^{(\ell)}(\mu_k)}{\ell!} \oint_{\Gamma_k} \frac{d\zeta \, (\zeta - \mu_k)^{\ell - p_k}}{(\zeta - z) \prod_{\substack{k'=1 \\ k' \neq k}}^{q} (\zeta - \mu_{k'})^{p_{k'}}} \\
&= -\frac{F_p(z)}{2\pi i} \sum_{k=1}^{q} \sum_{\ell=0}^{p_k - 1} \frac{S_{p-1}^{(\ell)}(\mu_k)}{\ell!} \oint_{\Gamma_k} \frac{d\zeta \, (\zeta - \mu_k)^{\ell - p_k}}{(\zeta - z) \prod_{\substack{k'=1 \\ k' \neq k}}^{q} (\zeta - \mu_{k'})^{p_{k'}}}, \tag{D.22}
\end{aligned}$$

where we used

$$\oint_{\Gamma_k} d\zeta \, (\zeta - \mu_k)^{\ell - p_k} f(\zeta) = 0 \text{ for } \ell \geq p_k, \, \ell \in \mathbb{N},$$

for any function f analytic in a neighborhood of the disk D_k with boundary Γ_k, $k = 1, \ldots, q$, to arrive at the last line of (D.22). An application of Cauchy's formula for derivatives of analytic functions to (D.22) then yields

$$S_{p-1}(z) = -F_p(z) \sum_{k=1}^{q} \sum_{\ell=0}^{p_k-1} \frac{S_{p-1}^{(\ell)}(\mu_k)}{\ell!}$$

$$\times \frac{1}{2\pi i} \oint_{\Gamma_k} d\zeta \frac{1}{(\zeta - \mu_k)^{(p_k-\ell-1)+1}} \frac{1}{(\zeta - z) \prod_{k'=1, k' \neq k}^{q} (\zeta - \mu_{k'})^{p_{k'}}}$$

$$= F_p(z) \sum_{k=1}^{q} \sum_{\ell=0}^{p_k-1} \frac{S_{p-1}^{(\ell)}(\mu_k)}{\ell!(p_k - \ell - 1)!}$$

$$\times \left(\frac{d^{p_k-\ell-1}}{d\zeta^{p_k-\ell-1}} \left(\frac{1}{(z-\zeta) \prod_{k'=1, k' \neq k}^{q} (\zeta - \mu_{k'})^{p_{k'}}} \right) \right) \bigg|_{\zeta=\mu_k}, \quad z \in \mathbb{C},$$

and hence (D.19). Conversely, a linear algebraic argument shows that any polynomial T_{p-1} of degree $p-1$ is uniquely determined by data of the type

$$T_{p-1}(\mu_k), T'_{p-1}(\mu_k), \ldots, T_{p-1}^{(p_k-1)}(\mu_k), \quad k = 1, \ldots, q.$$

Uniqueness of the representation (D.19) then proves (D.20). □

We briefly mention two special cases of (D.19). First, assume the generic case where all zeros of F_p are distinct, that is,

$$q = p, \quad p_k = 1, \quad \mu_k \neq \mu_{k'} \text{ for } k \neq k', \ k, k' = 1, \ldots, p.$$

In this case (D.19) reduces to the classical Lagrange interpolation formula (cf. (D.6))

$$S_{p-1}(z) = F_p(z) \sum_{k=1}^{p} \frac{S_{p-1}(\mu_k)}{((dF_p(\zeta)/d\zeta)|_{\zeta=\mu_k})(z - \mu_k)}, \quad z \in \mathbb{C}.$$

Second, we consider the other extreme case where all zeros of F_p coincide, that is,

$$q = 1, \quad p_1 = p, \quad F_p(z) = (z - \mu_1)^p, \quad z \in \mathbb{C}.$$

In this case (D.19) reduces of course to the Taylor expansion of S_{p-1} around $z = \mu_1$,

$$S_{p-1}(z) = \sum_{\ell=0}^{p-1} \frac{S_{p-1}^{(\ell)}(\mu_1)}{\ell!} (z - \mu_1)^\ell, \quad z \in \mathbb{C}.$$

Notes

The material of the first part of this appendix (up to Lemma D.4), where the zeros of the polynomial F_p are assumed to be distinct, is mostly taken from Gesztesy and Holden (2002; 2003a) (see also Gesztesy and Holden (2003b, Appendix E,

Lemma F.2)). A proof of Lagrange's interpolation result in the simplest case $k = 0$ can be found, for example, in Toda (1989b, Appendix E).

Lemma D.5 in the final part of this appendix, where some (or even all) the zeros of the polynomial F_p are permitted to coincide, is taken from Gesztesy et al. (2008b) (see also Gesztesy et al. (2007b)).

Formula (D.21) is derived, for instance, in Markushevich (1985, Part 2, Sect. 2.11, p. 68).

List of Symbols

> Algebra is rich in structure, but weak in meaning.
> *René Thom*[1]

$\mathbb{N}$, the natural numbers
$\mathbb{N}_0 = \mathbb{N} \cup \{0\}$, the nonnegative integers
$\mathbb{Z}$, the integers
$\mathbb{R}$, the real numbers
$\mathbb{T}$, the one-dimensional torus (homeomorphic to the circle S^1)
$\mathbb{C}$, the complex numbers
$\mathrm{Re}(z), \mathrm{Im}(z)$, the real and imaginary part of $z \in \mathbb{C}$
$\arg(z)$, the argument of $z \in \mathbb{C}$
$\bar{z}$, the complex conjugate of z
$\mathbb{C}_\infty = \mathbb{C} \cup \{\infty\} \cong \mathbb{CP}^1$, the Riemann sphere
$\mathbb{CP}^2 = (\mathbb{C}^3 \setminus \{0\})/(\mathbb{C} \setminus \{0\})$, the projective plane p. 325
$\mathbb{C}_\pm = \{z \in \mathbb{C} \mid \mathrm{Im}(z) \gtrless 0\}$, the open upper (lower)
  complex half-plane
$\lfloor x \rfloor = \sup\{n \in \mathbb{Z} \mid n \leq x\}$, the largest integer not exceeding x
$p \vee q$, the maximum of p and q
$p \wedge q$, the minimum of p and q
I_m, the identity matrix in $\mathbb{C}^m$, $m \geq 2$
$\underline{a} = (a_1, \ldots, a_m)$, a row vector in $\mathbb{C}^m$, $\underline{a}^\top$, a column vector in $\mathbb{C}^m$
$M^\top$, the transpose of the matrix M
M^*, the adjoint (conjugate transpose) of the matrix M
$\underline{\mathrm{diag}}(M) = (M_{1,1}, \ldots, M_{m,m})$, a row vector built of the diagonal
  terms of an $m \times m$ matrix M
$\mathrm{dom}(T)$, the domain of an operator T

---

[1] As quoted on http://www-groups.dcs.st-and.ac.uk/~history/Mathematicians/Thom.html

List of Symbols

$\ker(T)$, the kernel (null space) of a linear operator T
$\operatorname{spec}(T)$, the spectrum of a closed linear operator T
$\mathcal{B}(\mathcal{H})$, the Banach space of all bounded linear operators
　defined on the Hilbert space $\mathcal{H}$
$\operatorname{tr}(A)$, the trace of a trace-class operator A
$[A, B] = AB - BA$, the commutator of A and B
$\int^{\oplus}$, the direct integral of Hilbert spaces or linear operators p. 63
$\check{L}$, the Jacobi operator p. 118
$G(z, n, n')$, the Green's function of $\check{L}$ pp. 119, 284
$\check{L}^D_{\pm,n_0}$, the half-line Dirichlet operator p. 118
$G^D_{\pm,n_0}$, the Green's function of the half-line
　Dirichlet operator $\check{L}^D_{\pm,n_0}$ p. 120
$\ell^p(M)$, with $p \in \mathbb{N}$ and M is $\mathbb{N}$, $\mathbb{N}_0$, or $\mathbb{Z}$, etc.,
　the space of p-summable complex-valued
　sequences indexed by M
$\ell^\infty(M)$, where M is $\mathbb{N}$, $\mathbb{N}_0$, $\mathbb{Z}$, etc.,
　the set of complex-valued bounded sequences
　indexed by M
$(f, g) = \sum_{n \in M} \overline{f(n)} g(n)$, the scalar product
　in the Hilbert space $\ell^2(M)$
$C^k(\Omega)$, the set of all k times continuously differentiable
　functions on an open subset $\Omega \subseteq \mathbb{R}$
$C^\infty(\Omega, \mathcal{K})$, the set of all infinitely differentiable
　functions on Ω taking values in $\mathcal{K}$
$M_n(\mathbb{C})$, the set of all $n \times n$ matrices with complex-valued entries
$g = O(f)$ as $x \to x_0$, ("big-Oh") if g/f is bounded
　in a neighborhood of x_0
$g = o(f)$, ("little-Oh") if $g(x)/f(x) \to 0$ as $x \to x_0$
$\partial_w = \frac{\partial}{\partial w}$, the (partial) derivative with respect to w
$\partial_w^m = \frac{\partial^m}{\partial w^m}, m \in \mathbb{N}$, $\partial^2_{w_1 w_2} = \frac{\partial^2}{\partial w_1 \partial w_2}$
$f^\pm(n) = S^\pm f(n) = f(n \pm 1), n \in \mathbb{Z}$
$f^{(r)} = S^{(r)} f$, $S^{(r)} = (S^+)^r$ if $r \geq 0$; $S^{(r)} = (S^-)^{-r}$ if $r < 0$ p. 28
$W(f, g)(n) = a(fg^+ - f^+ g)$, the Wronskian of f and g
　for Jacobi difference expressions p. 43
$\operatorname{Sym}^n(X) = \{\{x_1, \ldots, x_n\} \mid x_j \in X, \ j = 1, \ldots, n\}$,
　the nth symmetric product of X
$\Psi_k(\underline{\mu})$, elementary symmetric functions, $\Phi^{(j)}_k(\underline{\mu})$ p. 385
$\mathcal{K}_g$, a compact Riemann surface of genus g p. 327
$\mathcal{B}(\mathcal{K}_n)$, the set of branch points of $\mathcal{K}_n$
　in the hyperelliptic case pp. 328, 359

List of Symbols

$\partial \widehat{\mathcal{K}}_g = a_1 b_1 a_1^{-1} b_1^{-1} \ldots a_g b_g a_g^{-1} b_g^{-1}$,
the fundamental polygon of $\mathcal{K}_g$ p. 332
$\widehat{\mathcal{K}}_g$, the simply connected interior
    of the fundamental polygon $\partial \widehat{\mathcal{K}}_g$ p. 345
$\tilde{\pi}_z, \tilde{\pi}_y$, projections   p. 325
$[x_2 : x_1 : x_0]$, homogeneous coordinates   p. 325
$\{a_j, b_j\}_{j=1}^g$, a homology basis on $\mathcal{K}_g$ p. 331
ω, Abelian differential of the first kind   p. 332
$\omega^{(2)}, \Omega^{(2)}$, Abelian differentials of the second kind   p. 334
$\omega^{(3)}, \Omega^{(3)}$, Abelian differentials of the third kind   p. 335
$\mathrm{res}_{P=Q} f(P)$, the residue of a meromorphic function f
    on a Riemann surface $\mathcal{K}_g$ at $Q \in \mathcal{K}_g$
$\mathcal{M}(\mathcal{K}_g)$, the set of meromorphic functions (0-forms) on $\mathcal{K}_g$ p. 336
$\mathcal{M}^1(\mathcal{K}_g)$, the set of meromorphic 1-forms on $\mathcal{K}_g$ p. 336
L_g, the period lattice   p. 338
$J(\mathcal{K}_g) = \mathbb{C}^g/L_g$, the Jacobi variety of $\mathcal{K}_g$ p. 338
$\mathcal{D}$, a divisor   p. 336
$\mathrm{Div}(\mathcal{K})$, the set of divisors   p. 336
$\deg(\mathcal{D})$, the degree of a divisor $\mathcal{D}$ p. 336
(f), the divisor of $f \in \mathcal{M}(\mathcal{K}_g) \setminus \{0\}$ p. 337
(ω), the divisor of $\omega \in \mathcal{M}^1(\mathcal{K}_g) \setminus \{0\}$ p. 337
$\mathcal{D} \sim \mathcal{E}$, the equivalence of divisors $\mathcal{D}$ and $\mathcal{E}$ p. 337
$[\mathcal{D}]$, the divisor class of the divisor $\mathcal{D}$ p. 337
$i(\mathcal{D}) = \dim\{\omega \in \mathcal{M}^1(\mathcal{K}) \mid (\omega) \geq \mathcal{D}\}$,
    the index of specialty of the divisor $\mathcal{D}$ p. 338
$r(\mathcal{D}) = \dim\{f \in \mathcal{M}(\mathcal{K}) \mid (f) \geq \mathcal{D}\}$ p. 338
$\mathcal{D}_{\underline{Q}}$, a nonnegative divisor of
    degree $m \in \mathbb{N}$, $\underline{Q} = \{Q_1, \ldots, Q_m\} \in \mathrm{Sym}^m(\mathcal{K}_g)$ p. 340
$\underline{A}_Q, \underline{\alpha}_Q, \underline{\hat{A}}_Q, \underline{\hat{\alpha}}_Q$, Abel maps   pp. 338, 340
θ, Riemann's theta function   p. 339
$\underline{\Xi}_Q$, the vector of Riemann constants   p. 340
$*: \mathcal{K}_n \to \mathcal{K}_n$, the hyperelliptic involution
    (sheet exchange map)   pp. 347, 358
$\Pi_\pm$, the upper and lower sheets of $\mathcal{K}_n$
    in the hyperelliptic case   p. 359
sn, the Jacobian elliptic function   p. 71
$\psi(P, n, n_0), \psi(P, n, n_0, t_r, t_{0,r})$, Baker–Akhiezer functions for
    the Toda hierarchy   pp. 43, 87
$\Psi(P, n, n_0), \Psi(P, n, n_0, t_r, t_{0,r})$, Baker–Akhiezer functions for
    the Ablowitz–Ladik hierarchy   pp. 223, 252

Bibliography[1]

> This paper contains much that is new and much that is true.
> Unfortunately, that which is true is not new and that which is
> new is not true.
>
> Howard W. Eves[2]

ABLOWITZ, M. J. 1977. Nonlinear evolution equations — continuous and discrete. *SIAM Rev.*, **19**, 663–684.

ABLOWITZ, M. J., BIONDINI, G., AND PRINARI, B. 2007. Inverse scattering transform for the integrable discrete nonlinear Schrödinger equation with nonvanishing boundary conditions. *Inverse Problems*, **23**, 1711–1758.

ABLOWITZ, M. J. AND CLARKSON, P. A. 1991. *Solitons, Nonlinear Evolution Equations and Inverse Scattering*. Cambridge: Cambridge University Press.

ABLOWITZ, M. J., KAUP, D. J., NEWELL, A. C., AND SEGUR, H. 1973a. Method for solving the sine-Gordon equation. *Phys. Rev. Lett.*, **30**, 1262–1264.

ABLOWITZ, M. J., KAUP, D. J., NEWELL, A. C., AND SEGUR, H. 1973b. Nonlinear-evolution equations of physical significance. *Phys. Rev. Lett.*, **31**, 125–127.

ABLOWITZ, M. J., KAUP, D. J., NEWELL, A. C., AND SEGUR, H. 1974. The inverse scattering transform – Fourier analysis for nonlinear problems. *Stud. Appl. Math.*, **53**, 249–315.

ABLOWITZ, M. J. AND LADIK, J. F. 1975. Nonlinear differential-difference equations. *J. Math. Phys.*, **16**, 598–603.

ABLOWITZ, M. J. AND LADIK, J. F. 1976a. A nonlinear difference scheme and inverse scattering. *Stud. Appl. Math.*, **55**, 213–229.

ABLOWITZ, M. J. AND LADIK, J. F. 1976b. Nonlinear differential-difference equations and Fourier analysis. *J. Math. Phys.*, **17**, 1011–1018.

ABLOWITZ, M. J. AND LADIK, J. F. 1977. On the solution of a class of nonlinear partial difference equations. *Stud. Appl. Math.*, **57**, 1–12.

ABLOWITZ, M. J., PRINARI, B., AND TRUBATCH, A. D. 2004. *Discrete and Continuous Nonlinear Schrödinger Systems*. Cambridge: Cambridge University Press.

---

[1] Publications with three or more authors are abbreviated with "First author et al. (year)" in the text. If more than one publication yields the same abbreviation, latin letters a,b,c, etc., are added after the year. Publications are alphabetically ordered using all authors' names and year of publication.

[2] Quoted in *Return to Mathematical Circles*, Boston: Prindle, Weber & Schmidt, 1988, p. 158.

Bibliography

ABRAHAM, R., MARSDEN, J. E., AND RATIU, T. 1988. *Manifolds, Tensor Analysis, and Applications*. Second edn. New York: Springer.

ADLER, M. 1979. On the trace functional for formal pseudo-differential operators and the symplectic structure of the Korteweg–Devries type equations. *Invent. Math.*, **50**, 219–248.

ADLER, M. 1981. On the Bäcklund transformation for the Gel'fand–Dickey equations. *Comm. Math. Phys.*, **80**, 517–527.

ADLER, M., HAINE, L., AND VAN MOERBEKE, P. 1993. Limit matrices for the Toda flow and periodic flags for loop groups. *Math. Ann.*, **296**, 1–33.

ADLER, M. AND VAN MOERBEKE, P. 1980a. Completely integrable systems, Euclidean Lie algebras, and curves. *Adv. in Math.*, **38**, 267–317.

ADLER, M. AND VAN MOERBEKE, P. 1980b. Linearization of Hamiltonian systems, Jacobi varieties and representation theory. *Adv. in Math.*, **38**, 318–379.

ADLER, M. AND VAN MOERBEKE, P. 1982. Kowalewski's asymptotic method, Kac–Moody Lie algebras and regularization. *Comm. Math. Phys.*, **83**, 83–106.

ADLER, M. AND VAN MOERBEKE, P. 1991. The Toda lattice, Dynkin diagrams, singularities and Abelian varieties. *Invent. Math.*, **103**, 223–278.

ADLER, M. AND VAN MOERBEKE, P. 1995. Matrix integrals, Toda symmetries, Virasoro constraints, and orthogonal polynomials. *Duke Math. J.*, **80**, 863–911.

AGROTIS, M. A. AND DAMIANOU, P. A. 2007. Volterra's realization of the KM-system. *J. Math. Anal. Appl.*, **325**, 157–165.

AHMAD, S. AND CHOWDHURY, A. ROY. 1987a. On the quasi-periodic solutions to the discrete non-linear Schrödinger equation. *J. Phys. A*, **20**, 293–303.

AHMAD, S. AND CHOWDHURY, A. ROY. 1987b. The quasi-periodic solutions to the discrete non-linear Schrödinger equation. *J. Math. Phys.*, **28**, 134–1137.

AKHIEZER, N. I. 1965. *The Classical Moment Problem*. London: Oliver and Boyd.

ALBER, M. S., CAMASSA, R., AND GEKHTMAN, M. 2000. Billiard weak solutions of nonlinear PDE's and Toda flows. Pages 1–11 of: LEVI, D. AND RAGNISCO, O. (eds.), *SIDE III–Symmetries and Integrability of Difference Equations*. CRM Proceedings and Lecture Notes, vol. 25. Providence, RI: Amer. Math. Soc.

ALBER, S. J. 1989. On finite-zone solutions of relativistic Toda lattices. *Lett. Math. Phys.*, **17**, 149–155.

ALBER, S. J. 1991a. Associated integrable systems. *J. Math. Phys.*, **32**, 916–922.

ALBER, S. J. 1991b. Hamiltonian systems on the Jacobi varieties. Pages 23–32 of: RATIU, T. (ed.), *The Geometry of Hamiltonian Systems*. New York: Springer.

ALPAY, D. AND VINNIKOV, V. 2002. Finite dimensional de Branges spaces on Riemann surfaces. *J. Funct. Anal.*, **189**, 283–335.

AMMAR, G. S. AND GRAGG, W. B. 1994. Schur flows for orthogonal Hessenberg matrices. Pages 27–34 of: BLOCH, A. (ed.), *Hamiltonian and Gradient Flows, Algorithms and Control*. Fields Inst. Commun., vol. 3. Providence, RI: Amer. Math. Soc.

ANTONY, A. J. AND KRISHNA, M. 1992. Almost periodicity for some Jacobi matrices. *Proc. Indian Acad. Sci. Math. Sci.*, **102**, 175–188.

ANTONY, A. J. AND KRISHNA, M. 1994. Inverse spectral theory for Jacobi matrices and their almost periodicity. *Proc. Indian Acad. Sci. Math. Sci.*, **104**, 777–818.

APTEKAREV, A. I. 1986. Asymptotic properties of polynomials orthogonal on a system of contours, and periodic motions of Toda lattices. *Math. USSR-Sb.*, **53**, 233–260.

ARNOLD, V. I. 1989. *Mathematical Methods of Classical Mechanics*. Second edn. New York: Springer.

ATLEE JACKSON, E., PASTA, J. R., AND WATERS, J. F. 1968. Thermal conductivity of one-dimensional lattices. *J. Comput. Phys.*, **2**, 207–227.

BABELON, O., BERNARD, D., AND TALON, M. 2003. *Introduction to Classical Integrable Systems*. Cambridge: Cambridge University Press.

BABICH, M. V. 1996. Willmore surfaces, 4-particle Toda lattice and double coverings of hyperelliptic surfaces. *Amer. Math. Soc. Transl. Ser. 2*, **174**, 143–168.

BAKER, H. F. 1995. *Abelian Functions. Abel's Theorem and the Allied Theory of Theta Functions*. Cambridge: Cambridge University Press.

BATCHENKO, V. AND GESZTESY, F. 2005a. On the spectrum of Jacobi operators with quasi-periodic algebro-geometric coefficients. *IMRP Int. Math. Res. Pap.*, 511–563.

BATCHENKO, V. AND GESZTESY, F. 2005b. On the spectrum of Schrödinger operators with quasi-periodic algebro-geometric KdV potentials. *J. Anal. Math.*, **95**, 333–387.

BÄTTIG, D., BLOCH, A. M., GUILLOT, J. C., AND KAPPELER, T. 1995. On the symplectic structure of the phase space for periodic KdV, Toda, and defocusing NLS. *Duke Math. J.*, **79**, 549–604.

BÄTTIG, D., GRÉBERT, B., GUILLOT, J. C., AND KAPPELER, T. 1993. Fibration of the phase space of the periodic Toda lattice. *J. Math. Pures Appl. (9)*, **72**, 553–565.

BAXTER, G. 1960. Polynomials defined by a difference system. *Bull. Amer. Math. Soc. (N.S.)*, **66**, 187–190.

BAXTER, G. 1961a. A convergence equivalence related to polynomials orthogonal on the unit circle. *Trans. Amer. Math. Soc.*, **99**, 471–487.

BAXTER, G. 1961b. Polynomials defined by a difference system. *J. Math. Anal. Appl.*, **2**, 223–263.

BAXTER, G. 1963. A norm inequality for a "finite-section" Wiener–Hopf equation. *Illinois J. Math.*, **7**, 97–103.

BEALS, R., SATTINGER, D. H., AND SZMIGIELSKI, J. 2001. Peakons, strings, and the finite Toda lattice. *Comm. Pure Appl. Math.*, **54**, 91–106.

BEFFA, G. M. 1997. The second Hamiltonian structure for the periodic Toda lattice. *Proc. Roy. Soc. Edinburgh Sect. A*, **127**, 547–566.

BEHNKE, H. AND SOMMER, F. 1965. *Theorie der analytischen Funktionen einer komplexen Veränderlichen*. Third edn. Grundlehren, vol. 77. Berlin: Springer.

BELOKOLOS, E. D., BOBENKO, A. I., ENOL'SKII, V. Z., ITS, A. R., AND MATVEEV, V. B. 1994. *Algebro-Geometric Approach to Nonlinear Integrable Equations*. Berlin: Springer.

BEREZANSKI, YU. M. 1986. The integration of semi-infinite Toda chain by means of inverse spectral problem. *Rep. Math. Phys.*, **24**, 21–47.

BEREZANSKII, JU. M. 1968. *Expansions in Eigenfunctions of Self-Adjoint Operators*. Transl. Math. Monographs, vol. 17. Providence, RI: Amer. Math. Soc.

BEREZANSKIĬ, YU. M. 1985. Integration of nonlinear difference equations by the inverse spectral problem method. *Soviet Math. Dokl.*, **31**, 264–267.

BEREZANSKII, YU. M. AND GEKHTMAN, M. I. 1990. Inverse problem of the spectral analysis and non-abelian chains of nonlinear equations. *Ukrainian Math. J.*, **42**, 645–658.

BEREZANSKII, YU. M., GEKHTMAN, M. I., AND SMOISH, M. E. 1986. Integration of some chains of nonlinear difference equations by the method of the inverse spectral problem. *Ukrainian Math. J.*, **38**, 74–78.

BEREZANSKII, YU. M. AND SMOISH, M. E. 1990. Nonisospectral nonlinear difference equations. *Ukrainian Math. J.*, **42**, 492–495.

BEREZANSKII, YU. M. AND SMOISH, M. E. 1994. Nonisospectral flows on semi-infinite Jacobi matrices. *J. Nonlinear Math. Phys.*, **1**, 116–146.

BEREZANSKY, YU. M. 1990. Inverse problems of spectral analysis and the integration of nonlinear equations. Pages 56–63 of: CARILLO, S. AND RAGNISCO, O. (eds.), *Nonlinear Evolution Equations and Dynamical Systems*. Berlin: Springer.

BEREZANSKY, YU. M. 1996. Integration of nonlinear nonisospectral difference-differential equations by means of the inverse spectral problem. Pages 11–20 of: ALFINITO, E., BOITI, M., MARTINA, L., AND PEMPINELLI, F. (eds.), *Nonlinear Physics. Theory and Experiment*. Singapore: World Scientific.

BEREZANSKY, YU. M. AND DUDKIN, M. E. 2005. The direct and inverse spectral problems for the block Jacobi type unitary matrices. *Meth. Funct. Anal. Top.*, **11**, 327–345.

BEREZANSKY, YU. M. AND DUDKIN, M. E. 2006. The complex moment problem and direct and inverse spectral problems for the block Jacobi type bounded normal matrices. *Meth. Funct. Anal. Top.*, **12**, 1–31.

BERTOLA, M. AND GEKHTMAN, M. 2007. Biorthogonal Laurent polynomials, Töplitz determinants, minimal Toda orbits and isomonodromic tau functions. *Constr. Approx.*, **26**, 383–430.

BLOCH, A. M., BROCKETT, R. W., AND RATIU, T. 1992. Completely integrable gradient flows. *Comm. Math. Phys.*, **147**, 57–74.

BLOCH, A. M., FLASCHKA, H., AND RATIU, T. 1990. A convexity theorem for isospectral manifolds of Jacobi matrices in a compact Lie algebra. *Duke Math. J.*, **61**, 41–65.

BLOCH, A. M. AND GEKHTMAN, M. I. 1998. Hamiltonian and gradient structures in the Toda flows. *J. Geom. Phys.*, **27**, 230–248.

BLOCH, A. M. AND GEKHTMAN, M. I. 2007. Lie algebraic aspects of the finite nonperiodic Toda flows. *J. Comput. Appl. Math.*, **202**, 3–25.

BOGOLYUBOV, N. N. AND PRIKARPATSKII, A. K. 1982. The inverse periodic problem for a discrete approximation of a nonlinear Schrödinger equation. *Soviet Phys. Dokl.*, **27**, 113–116.

BOGOLYUBOV, N. N., PRIKARPATSKII, A. K., AND SAMOILENKO, V. G. 1981. Discrete periodic problem for the modified nonlinear Korteweg–de Vries equation. *Soviet Phys. Dokl.*, **26**, 490–492.

BOGOLYUBOV, N. N., PRIKARPATSKII, A. K., AND SAMOILENKO, V. G. 1982. Discrete periodic problem for the modified nonlinear Korteweg–de Vries equation. *Theoret. and Math. Phys.*, **50**, 75–81.

BOGOYAVLENSKII, O. I. 1991. Algebraic constructions of integrable dynamical systems–extensions of the Volterra system. *Russian Math. Surveys*, **46** (3), 1–64.

BOGOYAVLENSKY, O. I. 1988. Integrable discretizations of the KdV equation. *Phys. Lett. A*, **134**, 34–38.

BOGOYAVLENSKY, O. I. 1990. On integrable generalizations of Volterra systems. Pages 165–175 of: RABINOWITZ, P. H. AND ZEHNDER, E. (eds.), *Analysis, et cetera*. Boston: Academic Press.

BOLEY, D. AND GOLUB, G. H. 1984. A modified method for reconstructing periodic Jacobi matrices. *Math. Comp.*, **42**, 143–150.

BOLEY, D. AND GOLUB, G. H. 1987. A survey of matrix inverse eigenvalue problems. *Inverse Problems*, **3**, 595–622.

BOST, J. B. 1992. Introduction to compact Riemann surfaces, Jacobians, and abelian varieties. Pages 64–211 of: WALDSCHMIDT, M., MOUSSA, P., LUCK, J. M., AND ITZYKSON, C. (eds.), *From Number Theory to Physics*. Berlin: Springer.

BOURGET, O., HOWLAND, J. S., AND JOYE, A. 2003. Spectral analysis of unitary band matrices. *Comm. Math. Phys.*, **234**, 191–227.

BRAZOVSKII, S. A., DZYALOSHINSKII, N. E., AND KRICHEVER, I. M. 1982. Discrete Peierls models with exact solutions. *Soviet Phys. JETP*, **56**, 212–225.

BUJALANCE, E., CIRRE, F. J., GAMBOA, J. M., AND GROMADZKI, G. 2001. *Symmetry Types of Hyperelliptic Riemann Surfaces.* Mem. Soc. Math. France, vol. 86. Paris: Soc. Math. France.

BULLA, W., GESZTESY, F., HOLDEN, H., AND TESCHL, G. 1998. Algebro-geometric quasi-periodic finite-gap solutions of the Toda and Kac–van Moerbeke hierarchy. *Mem. Amer. Math. Soc.*, **135** (641), 1–79.

BULTHEEL, A., GONZÁLEZ-VERA, P., HENDRIKSEN, E., AND NJÅSTAD, O. 1999. *Orthogonal Rational Functions.* Cambridge: Cambridge University Press.

BUNSE-GERSTNER, A. AND ELSNER, L. 1991. Schur parameter pencils for the solution of unitary eigenproblem. *Linear Algebra Appl.*, **154/156**, 741–778.

BUYS, M. AND FINKEL, A. 1984. The inverse periodic problem for Hill's equation with a finite-gap potential. *J. Differential Equations*, **55**, 257–275.

BYRD, P. F. AND FRIEDMAN, M. D. 1971. *Handbook of Elliptic Integrals for Engineers and Physicists.* Second revised edn. Berlin: Springer.

CANTERO, M. J., MORAL, L., AND VELÁZQUEZ, L. 2003. Five-diagonal matrices and zeros of orthogonal polynomials on the unit circle. *Linear Algebra Appl.*, **362**, 29–56.

CANTERO, M. J., MORAL, L., AND VELÁZQUEZ, L. 2005. Minimal representations of unitary operators and orthogonal polynomials on the unit circle. *Linear Algebra Appl.*, **408**, 40–65.

CANTERO, M. J. AND SIMON, B. (To appear.) Poisson brackets of orthogonal polynomials. *J. Approx. Theory.*

CARMONA, R. AND KOTANI, S. 1987. Inverse spectral theory for random Jacobi matrices. *J. Statist. Phys.*, **46**, 1091–1114.

CARMONA, R. AND LACROIX, J. 1990. *Spectral Theory of Random Schrödinger Operators.* Boston: Birkhäuser.

CASIAN, L. AND KODAMA, Y. 2006. Toda lattice, cohomology of compact Lie groups and finite Chevalley groups. *Invent. Math.*, **165**, 163–208.

CASIAN, L. AND KODAMA, Y. 2007. Singular structure of Toda lattices and cohomology of certain compact Lie groups. *J. Comput. Appl. Math.*, **202**, 56–79.

CHIU, S. C. AND LADIK, J. F. 1977. Generating exactly soluble nonlinear discrete evolution equations by a generalized Wronskian technique. *J. Math. Phys.*, **18**, 690–700.

CHOW, K. W., CONTE, R., AND XU, N. 2006. Analytic doubly periodic wave patterns for the integrable discrete nonlinear Schrödinger (Ablowitz–Ladik) model. *Phys. Lett. A*, **349**, 422–429.

CHOWDHURY, A. R. AND MAHATO, G. 1983. A Darboux–Bäcklund transformation associated with a discrete nonlinear Schrödinger equation. *Lett. Math. Phys.*, **7**, 313–317.

CHU, M. T. 1985. Asymptotic analysis of Toda lattice on diagonalizable matrices. *Nonlinear Anal.*, **9**, 193–201.

CHU, M. T. 1994. A list of matrix flows with applications. Pages 87–97 of: BLOCH, A. (ed.), *Hamiltonian and Gradient Flows, Algorithms and Control.* Fields Institute Communication Series, vol. 3. Providence, RI: Amer. Math. Soc.

CHU, M. T. AND NORRIS, L. K. 1988. Isospectral flows and abstract matrix factorizations. *SIAM J. Numer. Anal.*, **25**, 1383–1391.

CLARK, S., GESZTESY, F., AND RENGER, W. 2005. Trace formulas and Borg-type theorems for matrix-valued Jacobi and Dirac finite difference operators. *J. Differential Equations*, **219**, 144–182.

CLARK, S., GESZTESY, F., AND ZINCHENKO, M. 2007. Weyl–Titchmarsh and Borg–Marchenko-type uniqueness results for CMV operators with matrix-valued Verblunsky coefficients. *Oper. Matrices*, **1**.

CLARK, S., GESZTESY, F., AND ZINCHENKO, M. 2008. Borg–Marchenko-type uniqueness results for CMV operators. Preprint.

COMESSATTI, A. 1924. Sulle varietà abeliane reali. *Ann. Mat. Pura Appl. (4)*, **2**, 67–106.

COMMON, A. K. 1992. A solution of the initial value problem for half-infinite integrable lattice systems. *Inverse Problems*, **8**, 393–408.

COMMON, A. K. AND HAFEZ, S. T. 1990. Continued-fraction solutions to the Riccati equation and integrable lattice systems. *J. Phys. A*, **23**, 455–466.

CORDUNEANU, C. 1989. *Almost Periodic Functions*. New York: Chelsea.

DAI, H. H. AND GENG, X. 2003. Decomposition of a 2 + 1-dimensional Volterra type lattice and its quasi-periodic solutions. *Chaos Solitons Fractals*, **18**, 1031–1044.

DAMIANOU, P. A. 1993. Symmetries of Toda equations. *J. Phys. A*, **26**, 3791–3796.

DAMIANOU, P. A. 2000. Multiple Hamiltonian structures for Toda systems of type A–B–C. *Regul. Chaotic Dyn.*, **5** (1), 17–32.

DAMIANOU, P. A. 2004. Multiple Hamiltonian structures of Bogoyavlensky–Toda lattices. *Rev. Math. Phys.*, **16**, 175–241.

DAMIANOU, P. A. AND FERNANDES, R. L. 2002. From the Toda lattice to the Volterra lattice and back. *Rep. Math. Phys.*, **50**, 361–378.

DATE, E. 1982. Multi-soliton solutions and quasi-periodic solutions of nonlinear equations of sine-Gordon type. *Osaka J. Math.*, **19**, 125–158.

DATE, E. AND TANAKA, S. 1976a. Analogue of inverse scattering theory for the discrete Hill's equation and exact solutions for the periodic Toda lattice. *Progr. Theoret. Phys.*, **55**, 457–465.

DATE, E. AND TANAKA, S. 1976b. Periodic multi-soliton solutions of Korteweg–de Vries equation and Toda lattice. *Progr. Theoret. Phys. Suppl.*, **59**, 107–125.

DAUXOIS, T. 2008. *Fermi, Pasta, Ulam and a mysterious lady*. arXiv:physics.hist/0801.1590.

DAVIS, M. W. 1987. Some aspherical manifolds. *Duke Math. J.*, **55**, 105–139.

DEIFT, P. A. 1978. Applications of a commutation formula. *Duke Math. J.*, **45**, 267–310.

DEIFT, P. [A]. 1999. *Orthogonal Polynomials and Random Matrices: A Riemann–Hilbert Approach*. Courant Lecture Notes in Mathematics, vol. 3. New York: Courant Institute of Mathematical Sciences.

DEIFT, P. [A]. 2007. Riemann–Hilbert methods in the theory of orthogonal polynomials. Pages 715–740 of: GESZTESY, F., DEIFT, P. [A.], GALVEZ, C., PERRY, P., AND SCHLAG, W. (eds.), *Spectral Theory and Mathematical Physics: A Festschrift in Honor of Barry Simon's 60th Birthday. Ergodic Schrödinger Operators, Singular Spectrum, Orthogonal Polynomials, and Inverse Spectral Theory*. Providence, RI: Amer. Math. Soc.

DEIFT, P. [A.], KRIECHERBAUER, T., AND VENAKIDES, S. 1995a. Forced lattice vibrations: Part I. *Comm. Pure Appl. Math.*, **48**, 1187–1250.

DEIFT, P. [A.], KRIECHERBAUER, T., AND VENAKIDES, S. 1995b. Forced lattice vibrations: Part II. *Comm. Pure Appl. Math.*, **48**, 1251–1298.

DEIFT, P. [A.], KRIECHERBAUER, T., AND VENAKIDES, S. 1996. Forced lattice vibrations. Pages 377–389 of: ALFINITO, E., BOITI, M., MARTINA, L., AND PEMPINELLI., F. (eds.), *Nonlinear Physics: Theory and Experiment*. River Edge, NJ: World Scientific.

DEIFT, P. A. AND LI, L. C. 1989. Generalized affine Lie algebras and the solution of a class of flows associated with the QR eigenvalue algorithm. *Comm. Pure Appl. Math.*, **42**, 963–991.

DEIFT, P. [A.] AND LI, L. C. 1991. Poisson geometry of the analog of the Miura maps and Bäcklund–Darboux transformations for equations of Toda type and periodic Toda flows. *Comm. Math. Phys.*, **143**, 201–214.

DEIFT, P. [A.], LI, L. C., NANDA, T., AND TOMEI, C. 1986. The Toda flow on a generic orbit is integrable. *Comm. Pure Appl. Math.*, **39**, 183–232.

DEIFT, P. [A.], LI, L. C., AND TOMEI, C. 1985. Toda flows with infinitely many variables. *J. Funct. Anal.*, **64**, 358–402.

DEIFT, P. [A.], LI, L. C., AND TOMEI, C. 1989. Matrix factorizations and integrable systems. *Comm. Pure Appl. Math.*, **42**, 443–521.

DEIFT, P. [A.], LI, L. C., AND TOMEI, C. 1993. Symplectic aspects of some eigenvalue algorithms. Pages 511–536 of: FOKAS, A. S. AND ZAKHAROV, V. E. (eds.), *Important Developments in Soliton Theory*. Berlin: Springer.

DEIFT, P. [A.] AND MCLAUGHLIN, K. T. R. 1998. A continuum limit of the Toda lattice. *Mem. Amer. Math. Soc.*, **131** (624), 1–216.

DEIFT, P. [A.] AND TRUBOWITZ, E. 1981. A continuum limit of matrix inverse problems. *SIAM J. Math. Anal.*, **12**, 799–818.

DING, H. Y., SUN, Y. P., AND XU, X. X. 2006. A hierarchy of nonlinear lattice soliton equations, its integrable coupling systems and infinitely many conservation laws. *Chaos Solitons Fractals*, **30**, 227–234.

DOLIWA, A. AND SANTINI, P. M. 1995. Integrable dynamics of a discrete curve and the Ablowitz–Ladik hierarchy. *J. Math. Phys.*, **36**, 1259–1273.

DRIESSEL, K. R. 1986. On isospectral gradient flows — solving matrix eigenproblems using differential equations. Pages 69–91 of: CANNON, J. R. AND HORNUNG, U. (eds.), *Inverse Problems*. Basel: Birkhäuser.

DUBROVIN, B. A. 1981. Theta functions and non-linear equations. *Russian Math. Surveys*, **36** (2), 11–92.

DUBROVIN, B. A. 1982. Multidimensional theta functions and their application to the integration of nonlinear equations. Pages 83–150 of: NOVIKOV, S. P. (ed.), *Mathematical Physics Reviews*. Soviet Scientific Reviews, Section C, vol. 3. Chur: Harwood.

DUBROVIN, B. A. 1983. Matrix finite-zone operators. *Revs. Sci. Tech.*, **23**, 20–50.

DUBROVIN, B. A., KRICHEVER, I. M., AND NOVIKOV, S. P. 1990. Integrable systems. I. Pages 173–280 of: ARNOLD, V. I. AND NOVIKOV, S. P. (eds.), *Dynamical Systems IV*. Berlin: Springer.

DUBROVIN, B. A., MATVEEV, V. B., AND NOVIKOV, S. P. 1976. Non-linear equations of the Korteweg–de Vries type, finite-zone linear operators and Abelian varieties. *Russian Math. Surveys*, **31** (1), 59–146. Reprinted in Novikov et al. (1981), pp. 53–140.

DUBROVIN, B. A. AND NATANZON, S. M. 1982. Real two-zone solutions of the sine-Gordon equations. *Functional Anal. Appl.*, **16**, 21–33.

DZYALOSHINSKII, I. E. AND KRICHEVER, I. M. 1983. Sound and charge-density wave in the discrete Peierls model. *Soviet Phys. JETP*, **58**, 1031–1040.

EGOROVA, I., MICHOR, J., AND TESCHL, G. (To appear.) Soliton solutions of the Toda hierarchy on quasi-periodic background revisited. *Math. Nachr.*

EGOROVA, I., MICHOR, J., AND TESCHL, G. 2006. Scattering theory of Jacobi operators with quasi-periodic background. *Comm. Math. Phys.*, **264**, 811–842.

EGOROVA, I., MICHOR, J., AND TESCHL, G. 2007a. Inverse scattering transform for the Toda hierarchy with quasi-periodic background. *Proc. Amer. Math. Soc.*, **135**, 1817–1827.

EGOROVA, I., MICHOR, J., AND TESCHL, G. 2007b. Scattering theory for Jacobi operators with a steplike quasi-periodic background. *Inverse Problems*, **23**, 905–918.

EGOROVA, I., MICHOR, J., AND TESCHL, G. 2008. Scattering theory for Jacobi operators with general steplike quasi-periodic background. *J. Math. Phys. Anal. Geom.*, **4**, 33–62.

EILENBERGER, G. 1983. *Solitons*. Berlin: Springer. Second corrected printing.

ENOLSKII, V. Z., SALERNO, M., SCOTT, A. C., AND EILBECK, J. C. 1992. There's more than one way to skin Schrödinger's cat. *Phys. D*, **59**, 1–24.

ERCOLANI, N. M., FLASCHKA, H., AND SINGER, S. 1993. The geometry of the full Kostant–Toda lattice. Pages 181–225 of: BABELON, O., CARTIER, P., AND KOSMANN SCHWARZBACH, Y. (eds.), *Integrable Systems. The Verdier Memorial Conference*. Progress in Mathematics, vol. 115. Boston: Birkhäuser.

ERCOLANI, N. [M.] AND LOZANO, G. I. 2006. A bi-Hamiltonian structure for the integrable, discrete non-linear Schrödinger system. *Phys. D*, **218**, 105–121.

FADDEEV, L. AND TAKHTAJAN, L. 1987. *Hamiltonian Methods in the Theory of Solitons*. Berlin: Springer.

FARKAS, H. M. AND KRA, I. 1992. *Riemann Surfaces*. Second edn. New York: Springer.

FAY, J. D. 1973. *Theta Functions on Riemann Surfaces*. Lecture Notes in Mathematics, vol. 352. Berlin: Springer.

FAYBUSOVICH, L. 1992. Toda flows and isospectral manifolds. *Proc. Amer. Math. Soc.*, **115**, 837–847.

FAYBUSOVICH, L. 1994. Rational functions, Toda flows, and LR-like algorithms. *Linear Algebra Appl.*, **203–204**, 359–381.

FAYBUSOVICH, L. AND GEKHTMAN, M. 1999. On Schur flows. *J. Phys. A*, **32**, 4671–4680.

FAYBUSOVICH, L. AND GEKHTMAN, M. 2000. Elementary Toda orbits and integrable lattices. *J. Math. Phys.*, **41**, 2905–2921.

FAYBUSOVICH, L. AND GEKHTMAN, M. 2001. Inverse moment problem for elementary co-adjoint orbits. *Inverse Problems*, **17**, 1295–1306.

FERGUSON, W. E. 1980. The construction of Jacobi and periodic Jacobi matrices with prescribed spectra. *Math. Comp.*, **35**, 1203–1220.

FERMI, E., PASTA, J., AND ULAM, S. M. 1955. Studies in nonlinear problems. Tech. Rep. Los Alamos Sci. Lab. Reprinted in Newell (1974), pp. 143–156 and Mattis (1993), pp. 851–870.

FERNANDES, R. L. 1993. On the master symmetries and bi-Hamiltonian structure of the Toda lattice. *J. Phys. A*, **26**, 3797–3803.

FERNANDES, R. L. AND VANHAECKE, P. 2001. Hyperelliptic Prym varieties and integrable systems. *Comm. Math. Phys.*, **221**, 169–196.

FINKEL, A., ISAACSON, E., AND TRUBOWITZ, E. 1987. An explicit solution of the inverse periodic problem for Hill's equation. *SIAM J. Math. Anal.*, **18**, 46–53.

FLASCHKA, H. 1974a. On the Toda lattice. II. *Progr. Theoret. Phys.*, **51**, 703–716.

FLASCHKA, H. 1974b. On the Toda lattice. II. Existence of integrals. *Phys. Rev. B (3)*, **9**, 1924–1925.

FLASCHKA, H. 1975. Discrete and periodic illustrations of some aspects of the inverse method. Pages 441–466 of: MOSER, J. (ed.), *Dynamical Systems, Theory and Applications*. Lecture Notes in Physics, vol. 38. Berlin: Springer.

FLASCHKA, H. 1988. The Toda lattice in the complex domain. Pages 141–154 of: KASHIWARA, M. AND KAWAI, T. (eds.), *Algebraic Analysis, Vol. I*. Boston, MA: Academic Press.

FLASCHKA, H. 1994. Integrable systems and torus actions. Pages 43–101 of: BABELON, O., CARTIER, P., AND KOSMANN SCHWARZBACH, Y. (eds.), *Lectures on Integrable Systems. In Memory of Jean-Louis Verdier*. Singapore: World Scientific.

FLASCHKA, H. AND HAINE, L. 1991. Variétés de drapeaux et réseaux de Toda. *Math. Z.*, **208**, 545–556.

FLASCHKA, H. AND MCLAUGHLIN, D. W. 1976a. Canonically conjugate variables for the Korteweg–de Vries equation and the Toda lattice with periodic boundary conditions. *Progr. Theoret. Phys.*, **55**, 438–456.

FLASCHKA, H. AND MCLAUGHLIN, D. W. 1976b. Some comments on Bäcklund transformations, canonical transformations, and the inverse scattering method. Pages 252–295 of: MIURA, R. M. (ed.), *Bäcklund Transformations, the Inverse Scattering Method, Solitons, and their Applications*. Lecture Notes in Mathematics, vol. 515. Berlin: Springer.

FORD, J. 1961. Equipartition of energy for nonlinear systems. *J. Math. Phys.*, **2**, 387–393.

FORD, J. AND WATERS, J. 1964. Computer studies of energy sharing and ergodicity for nonlinear oscillator systems. *J. Math. Phys.*, **4**, 1293–1306.

FORSTER, O. 1999. *Lectures on Riemann Surfaces*. New York: Springer. Fourth corr. printing.

FORSYTH, A. R. 1965. *Theory of Functions of a Complex Variable I*. Third edn. New York: Dover.

FRIED, D. 1986. The cohomology of an isospectral flow. *Proc. Amer. Math. Soc.*, **98**, 363–368.

GARDNER, C. S., GREENE, J. M., KRUSKAL, M. D., AND MIURA, R. M. 1967. Method for solving the Korteweg–de Vries equation. *Phys. Rev. Lett.*, **19**, 1095–1097.

GARDNER, C. S., GREENE, J. M., KRUSKAL, M. D., AND MIURA, R. M. 1974. Korteweg–de Vries equation and generalizations. VI. Methods for exact solution. *Comm. Pure Appl. Math.*, **27**, 97–133.

GEKHTMAN, M. [I]. 1990. Integration of non-abelian Toda-type chains. *Functional Anal. Appl.*, **24**, 231–233.

GEKHTMAN, M. [I]. 1991a. On the integration of the infinite Toda lattice. Pages 29–31 of: MAKHANKOV, V. G. AND PASHAEV, O. K. (eds.), *Nonlinear Evolution Equations and Dynamical Systems*. Berlin: Springer.

GEKHTMAN, M. [I]. 1991b. Solution of infinite Toda chains. *Functional Anal. Appl.*, **25**, 230–232.

GEKHTMAN, M. [I]. 1993. Non-abelian nonlinear lattice equations on finite interval. *J. Phys. A*, **26**, 6303–6317.

GEKHTMAN, M. [I]. 1998. Hamiltonian structure of non-abelian Toda lattice. *Lett. Math. Phys.*, **46**, 189–205.

GEKHTMAN, M. [I.] AND NENCIU, I. (To appear.) Multi-hamiltonian structure for the finite defocusing Ablowitz–Ladik equation. *Comm. Pure Appl. Math.*

GEKHTMAN, M. [I.] AND SHAPIRO, M. Z. 1997. Completeness of real Toda flows and totally positive matrices. *Math. Z.*, **226**, 51–66.

GEKHTMAN, M. [I.] AND SHAPIRO, M. Z. 1999. Noncommutative and commutative integrability of generic Toda flows in simple Lie algebras. *Comm. Pure Appl. Math.*, **52**, 53–84.

GEL'FAND, I. M. AND DIKII, L. A. 1975. Asymptotic behaviour of the resolvent of Sturm–Liouville equations and the algebra of the Korteweg–de Vries equations. *Russian Math. Surveys*, **30** (5), 77–113.

GEL'FAND, I. M. AND ZAKHAREVICH, I. 2000. Webs, Lenard schemes, and the local geometry of bi-Hamiltonian Toda and Lax structures. *Selecta Math. (N.S.)*, **6**, 131–183.

GENG, X. 1989. Darboux transformation of the discrete Ablowitz–Ladik eigenvalue problem. *Acta Math. Sci.*, **9**, 21–26.

GENG, X. AND DAI, H. H. 2007. Nonlinearization of the Lax pairs for discrete Ablowitz–Ladik hierarchy. *J. Math. Anal. Appl.*, **327**, 829–853.

GENG, X., DAI, H. H., AND CAO, C. 2003. Algebro-geometric constructions of the discrete Ablowitz–Ladik flows and applications. *J. Math. Phys.*, **44**, 4573–4588.

GENG, X., DAI, H. H., AND ZHU, J. 2007. Decompositions of the discrete Ablowitz–Ladik hierarchy. *Stud. Appl. Math.*, **118**, 281–312.

GERONIMO, J. S., GESZTESY, F., AND HOLDEN, H. 2005. Algebro-geometric solutions of the Baxter–Szegő difference equation. *Comm. Math. Phys.*, **258**, 149–177.

GERONIMO, J. S. AND JOHNSON, R. 1996. Rotation number associated with difference equations satisfied by polynomials orthogonal on the unit circle. *J. Differential Equations*, **132**, 140–178.

GERONIMO, J. S. AND JOHNSON, R. 1998. An inverse problem associated with polynomials orthogonal on the unit circle. *Comm. Math. Phys.*, **193**, 125–150.

GERONIMO, J. S. AND TEPLYAEV, A. 1994. A difference equation arising from the trigonometric moment problem having random reflection coefficients—an operator theoretic approach. *J. Funct. Anal.*, **123**, 12–45.

GERONIMUS, YA. L. 1961. *Orthogonal Polynomials*. New York: Consultants Bureau.

GESZTESY, F. 1989. Some applications of commutation methods. Pages 93–117 of: HOLDEN, H. AND JENSEN, A. (eds.), *Schrödinger Operators*. Lecture Notes in Physics, vol. 345. Berlin: Springer.

GESZTESY, F. 1992. Quasi-periodic, finite-gap solutions of the modified Korteweg–de Vries equation. Pages 428–471 of: ALBEVERIO, S., FENSTAD, J. E., HOLDEN, H., AND LINDSTRØM, T. (eds.), *Ideas and Methods in Mathematical Analysis, Stochastics, and Applications*. Cambridge: Cambridge University Press.

GESZTESY, F. AND HOLDEN, H. 1994. Trace formulas and conservation laws for nonlinear evolution equations. *Rev. Math. Phys.*, **6**, 51–95.

GESZTESY, F. AND HOLDEN, H. 2002. Dubrovin equations and integrable systems on hyperelliptic curves. *Math. Scand.*, **91**, 91–126.

GESZTESY, F. AND HOLDEN, H. 2003a. Algebro-geometric solutions of the Camassa–Holm hierarchy. *Rev. Mat. Iberoamericana*, **19**, 73–142.

GESZTESY, F. AND HOLDEN, H. 2003b. *Soliton Equations and Their Algebro-Geometric Solutions. Vol. I. $(1+1)$-Dimensional Continuous Models*. Vol. 1. Cambridge: Cambridge University Press.

GESZTESY, F. AND HOLDEN, H. 2006. Local conservation laws and the Hamiltonian formalism for the Toda hierarchy revisited. *Roy. Norw. Soc. Sci. Lett. Trans.*, **2006** (3), 1–30.

GESZTESY, F., HOLDEN, H., MICHOR, J., AND TESCHL, G. (To appear.) The Ablowitz–Ladik hierarchy revisited. In: JANAS, J., KURASOV, P., LAPTEV, A., NABOKO, S., AND STOLZ, G. (eds.), *Operator Theory, Analysis in Mathematical Physics*. Operator Theory, Advances and Applications. Basel: Birkhäuser.

GESZTESY, F., HOLDEN, H., MICHOR, J., AND TESCHL, G. 2007a. Algebro-geometric finite-band solutions of the Ablowitz–Ladik hierarchy. *Int. Math. Res. Not. IMRN*, **2007**, 1–55. article ID rnm082, doi:10.1093/imrn/rnm082.

GESZTESY, F., HOLDEN, H., MICHOR, J., AND TESCHL, G. 2007b. *The algebro-geometric initial value problem for the Ablowitz–Ladik hierarchy.* arXiv:nlin/07063370.

GESZTESY, F., HOLDEN, H., MICHOR, J., AND TESCHL, G. 2008a. Local conservation laws and the Hamiltonian formalism for the Ablowitz–Ladik hierarchy. *Stud. Appl. Math.*, **120**, 361–423.

GESZTESY, F., HOLDEN, H., SIMON, B., AND ZHAO, Z. 1993. On the Toda and Kac–van Moerbeke systems. *Trans. Amer. Math. Soc.*, **339**, 849–868.

GESZTESY, F., HOLDEN, H., AND TESCHL, G. 2008b. The algebro-geometric Toda hierarchy initial value problem for complex-valued initial data. *Rev. Mat. Iberoamericana*, **24**, 117–182.

GESZTESY, F., KRISHNA, M., AND TESCHL, G. 1996a. On isospectral sets of Jacobi operators. *Comm. Math. Phys.*, **181**, 631–645.

GESZTESY, F., RACE, D., UNTERKOFLER, K., AND WEIKARD, R. 1994. On Gelfand–Dickey and Drinfeld–Sokolov systems. *Rev. Math. Phys.*, **6**, 227–276.

GESZTESY, F. AND RENGER, W. 1997. New classes of Toda soliton solutions. *Comm. Math. Phys.*, **184**, 27–50.

GESZTESY, F., SCHWEIGER, W., AND SIMON, B. 1991. Commutation methods applied to the mKdV-equation. *Trans. Amer. Math. Soc.*, **324**, 465–525.

GESZTESY, F. AND SIMON, B. 1996. The xi function. *Acta Math.*, **176**, 49–71.

GESZTESY, F. AND SIMON, B. 1997. m-functions and inverse spectral analysis for finite and semi-infinite Jacobi matrices. *J. Anal. Math.*, **73**, 267–297.

GESZTESY, F., SIMON, B., AND TESCHL, G. 1996b. Spectral deformation of one-dimensional Schrödinger operators. *J. Anal. Math.*, **70**, 267–324.

GESZTESY, F. AND SVIRSKY, R. 1995. (m)KdV solitons on the background of quasi-periodic finite-gap solutions. *Mem. Amer. Math. Soc.*, **118** (563), 1–88.

GESZTESY, F. AND TESCHL, G. 1996. Commutation methods for Jacobi operators. *J. Differential Equations*, **128**, 252–299.

GESZTESY, F. AND YUDITSKII, P. 2006. Spectral properties of a class of reflectionless Schrödinger operators. *J. Funct. Anal.*, **241**, 486–527.

GESZTESY, F. AND ZINCHENKO, M. (To appear.) Local spectral properties of reflectionless Jacobi, CMV, and Schrödinger operators. *J. Differential Equations*.

GESZTESY, F. AND ZINCHENKO, M. 2006a. A Borg-type theorem associated with orthogonal polynomials on the unit circle. *J. London Math. Soc. (2)*, **74**, 757–777.

GESZTESY, F. AND ZINCHENKO, M. 2006b. Weyl–Titchmarsh theory for CMV operators associated with orthogonal polynomials on the unit circle. *J. Approx. Theory*, **139**, 179–213.

GIBSON, P. C. 2002. *Spectral distributions and isospectral sets of tridiagonal matrices.* arXiv:math/0207041.

GIESEKER, D. 1996. The Toda hierarchy and the KdV hierarchy. *Comm. Math. Phys.*, **181**, 587–603.

GINDIKIN, S. G., GUILLEMIN, V. W., KIRILLOV, A. A., KOSTANT, B., AND STERNBERG, S. (eds.). 1987. *Izrail M. Gelfand. Collected Papers*. Berlin: Springer.

GLAZMAN, I. M. 1965. *Direct Methods of Qualitative Spectral Analysis of Singular Differential Operators*. Jerusalem: Israel Program for Scientific Translations.

GOLINSKII, L. 2006. Schur flows and orthogonal polynomials on the unit circle. *Sbornik Math.*, **197**, 1145–1165.

GOLINSKII, L. AND NEVAI, P. 2001. Szegő difference equations, transfer matrices and orthogonal polynomials on the unit circle. *Comm. Math. Phys.*, **223**, 223–259.

GRIFFITHS, P. A. 1989. *Introduction to Algebraic Curves*. Providence, RI: Amer. Math. Soc.

GRIFFITHS, P. A. AND HARRIS, J. 1978. *Principles of Algebraic Geometry*. New York: Wiley.

GROSS, B. H. AND HARRIS, J. 1981. Real algebraic curves. *Ann. Sci. École Norm. Sup. (4)*, **14**, 157–182.

GUEST, M. A. 1997. *Harmonic Maps, Loop Groups, and Integrable Systems*. London Mathematical Society Student Texts, vol. 38. Cambridge: Cambridge University Press.

GUILLOT, J. C. 1994. Fibration of the phase space for the periodic non-linear Schrödinger equation and for the periodic Toda lattice equations. Pages 15–31 of: IKAWA, M. (ed.), *Spectral and Scattering Theory*. New York: Marcel Dekker.

GUNNING, R. C. 1966. *Lectures on Riemann Surfaces*. Princeton Mathematical Notes. Princeton: Princeton University Press.

GUNNING, R. C. 1972. *Lectures on Riemann Surfaces: Jacobi Varieties*. Princeton Mathematical Notes. Princeton: Princeton University Press.

GUSEINOV, I. M. AND KHANMAMEDOV, A. KH. 1999. The $t \to \infty$ asymptotic regime of the Cauchy problem solution for the Toda chain with threshold-type initial data. *Theoret. and Math. Phys.*, **119**, 739–749.

GUSTAFSON, K. E. AND RAO, D. K. M. 1997. *Numerical Range. The Field of Values of Linear Operators and Matrices*. New York: Springer.

HARNACK, A. 1876. Ueber die Vieltheiligkeit der ebenen algebraischen Curven. *Math. Ann.*, **10**, 189–209.

HÉNON, M. 1974. Integrals of the Toda lattice. *Phys. Rev. B (3)*, **9**, 1921–1923.

HENRICI, A. AND KAPPELER, T. 2008. Results on normal forms for FPU chains. *Comm. Math. Phys.*, **278**, 145–177.

HIROTA, R. 1975. Nonlinear transformations among differential-difference equations that exhibit solitons. *Progr. Theoret. Phys.*, **54**, 288.

HIROTA, R. AND SATSUMA, J. 1976a. N-soliton solutions of nonlinear network equations describing a Volterra system. *J. Phys. Soc. Japan*, **40**, 891–900.

HIROTA, R. AND SATSUMA, J. 1976b. A variety of nonlinear network equations generated from the Bäcklund transformation for the Toda lattice. *Progr. Theoret. Phys. Suppl.*, **59**, 64–100.

HOFFMANN, T. 2000. On the equivalence of the discrete nonlinear Schrödinger equation and the discrete isotropic Heisenberg magnet. *Phys. Lett. A*, **265**, 62–67.

HOFFMANN, T. AND KUTZ, N. 2004. Discrete curves in $\mathbb{CP}^1$ and the Toda lattice. *Stud. Appl. Math.*, **113**, 31–55.

HOFMANN, F. 1888. *Methodik der stetigen Deformation von zweiblättrigen Riemann'schen Flächen*. Halle: Verlag von L. Nebert.

HUANG, W. AND LIU, Y. 2008. Doubly periodic wave solutions and soliton solutions of Ablowitz–Ladik lattice system. *Int. J. Theoret. Phys.*, **47**, 338–349.

HYDON, P. E. 2005. Multisymplectic conservation laws for differential and differential-difference equations. *Proc. Roy. Soc. London Ser. A*, **461**, 1627–1637.

IGUCHI, K. 1992a. Exact wave functions of an electron on a quasiperiodic lattice: Definition of an infinite-dimensional Riemann theta function. *J. Math. Phys.*, **33**, 3938–3947.

IGUCHI, K. 1992b. Quasiperiodic systems without Cantor-set-like energy bands. *J. Math. Phys.*, **33**, 3736–3739.

ILIEV, P. 2007. Heat kernel expansions on the integers and the Toda lattice hierarchy. *Selecta Math. (N.S.)*, **13**, 497–530.

INOUE, R. 2003. The extended Lotka–Volterra lattice and affine Jacobi varieties of spectral curves. *J. Math. Phys.*, **44**, 338–351.

INOUE, R. 2004. The matrix realization of affine Jacobi varieties and the extended Lotka–Volterra lattice. *J. Phys. A*, **37**, 1277–1298.

INOUE, R. AND WADATI, M. 1997. Hungry Volterra model and a new hierarchy of discrete models. *J. Phys. Soc. Japan*, **66**, 1291–1293.

ITS, A. R. AND MATVEEV, V. B. 1975a. Hill's operator with finitely many gaps. *Functional Anal. Appl.*, **9**, 65–66.

ITS, A. R. AND MATVEEV, V. B. 1975b. Schrödinger operators with finite-gap spectrum and N-soliton solutions of the Korteweg–de Vries equation. *Theoret. and Math. Phys.*, **23**, 343–355.

JACOBI, C. G. T. 1846. Über eine neue Methode zur Integration der hyperelliptischen Differentialgleichung und über die rationale Form ihrer vollständigen algebraischen Integralgleichungen. *J. Reine Angew. Math.*, **32**, 220–226.

JOHNSON, R. AND MOSER, J. 1982. The rotation number for almost periodic potentials. *Comm. Math. Phys.*, **84**, 403–438. Erratum, *ibid.* **90**, 317–318 (1983).

KAC, M. AND VAN MOERBEKE, P. 1975a. A complete solution of the periodic Toda problem. *Proc. Nat. Acad. Sci. U.S.A.*, **72**, 2879–2880.

KAC, M. AND VAN MOERBEKE, P. 1975b. On an explicitly soluble system of nonlinear differential equations related to certain Toda lattices. *Adv. in Math.*, **16**, 160–169.

KAC, M. AND VAN MOERBEKE, P. 1975c. On some periodic Toda lattices. *Proc. Nat. Acad. Sci. U.S.A.*, **72**, 1627–1629.

KAJIWARA, K., MAZZOCCO, M., AND OHTA, Y. 2007. A remark on the Hankel determinant formula for solutions of the Toda equation. *J. Phys. A*, **40**, 12661–12675.

KAMVISSIS, S. 1993. On the long time behavior of the doubly infinite Toda lattice under initial data decaying at infinity. *Comm. Math. Phys.*, **153**, 479–519.

KAMVISSIS, S. AND TESCHL, G. 2007a. Stability of periodic soliton equations under short range perturbations. *Phys. Lett. A*, **364**, 480–483.

KAMVISSIS, S. AND TESCHL, G. 2007b. *Stability of the periodic Toda lattice under short range perturbations*. arXiv:0705.0346 v1 2 May 2007.

KATO, Y. 1983. On the spectral density of periodic Jacobi matrices. Pages 153–181 of: JIMBO, M. AND MIWA, T. (eds.), *Non-linear Integrable Systems–Classical Theory and Quantum Theory*. Singapore: World Scientific.

KATOK, A. AND HASSELBLATT, B. 1995. *Introduction to the Modern Theory of Dynamical Systems*. Cambridge: Cambridge University Press.

KHANMAMEDOV, AG. KH. 2005a. On the integration of an initial-boundary value problem for the Volterra lattice. *Differ. Equ.*, **41**, 1192–1195.

KHANMAMEDOV, AG. KH. 2005b. Rapidly decreasing solution of the initial boundary-value problem for the Toda lattice. *Ukrainian Math. J.*, **57**, 1350–1359.

KILLIP, R. AND NENCIU, I. 2006. CMV: The unitary analogue of Jacobi matrices. *Comm. Pure Appl. Math.*, **59**, 1–41.

KIRWAN, F. 1992. *Complex Algebraic Curves*. Cambridge: Cambridge University Press.

KLEIN, F. 1893. Ueber Realitätsverhältnisse bei der einem beliebigen Geschlechte zugehörigen Normalcurve der φ. *Math. Ann.*, **42**, 1–29.

KNILL, O. 1993a. Factorization of random Jacobi operators and Bäcklund transformations. *Comm. Math. Phys.*, **151**, 589–605.

KNILL, O. 1993b. Isospectral deformations of random Jacobi operators. *Comm. Math. Phys.*, **151**, 403–426.

KNILL, O. 1993c. Renormalization of random Jacobi operators. *Comm. Math. Phys.*, **164**, 195–215.

KODAMA, Y. 2002. Topology of the real part of the hyperelliptic Jacobian associated with the periodic Toda lattice. *Theoret. and Math. Phys.*, **133**, 1692–1711.

KODAMA, Y. AND YE, J. 1996a. Iso-spectral deformations of general matrix and their reductions on Lie algebras. *Comm. Math. Phys.*, **178**, 765–788.

KODAMA, Y. AND YE, J. 1996b. Toda hierarchy with indefinite metric. *Phys. D*, **91**, 321–339.

KODAMA, Y. AND YE, J. 1998. Toda lattices with indefinite metric II: Topology of the iso-spectral manifolds. *Phys. D*, **121**, 89–108.

KOELLING, M. E., BLOCH, A. M., AND GEKHTMAN, M. 2004. Qualitative behavior of non-Abelian Toda-like flows. *Phys. D*, **199**, 317–338.

KOROTKIN, D. A. 1998. Introduction to the functions on compact Riemann surfaces and theta-functions. Pages 109–139 of: WÓJCIK, D. AND CIEŚLIŃSKI, J. (eds.), *Nonlinearity & Geometry*. Warsaw: PWN.

KOSTANT, B. 1979. The solution to a generalized Toda lattice and representation theory. *Adv. in Math.*, **34**, 195–338.

KOTANI, S. AND SIMON, B. 1988. Stochastic Schrödinger operators and Jacobi matrices on the strip. *Comm. Math. Phys.*, **119**, 403–429.

KRAZER, A. 1970. *Lehrbuch der Thetafunktionen*. New York: Chelsea.

KRIČEVER, I. M. 1982. Algebro-geometric spectral theory of the Schrödinger difference operator and the Peierls model. *Soviet Math. Dokl.*, **26**, 194–198.

KRICHEVER, I. M. 1978. Algebraic curves and non-linear difference equations. *Russian Math. Surveys*, **33** (4), 255–256. Reprinted in Novikov et al. (1981), pp. 170–171.

KRICHEVER, I. M. 1982. The Peierls model. *Functional Anal. Appl.*, **16**, 248–263.

KRICHEVER, I. M. 1983. Nonlinear equations and elliptic curves. *Revs. Sci. Tech.*, **23**, 51–90.

KRICHEVER, I. M. 2006. Integrable linear equations and the Riemann–Schottky problem. Pages 497–514 of: *Algebraic Geometry and Number Theory*. Progr. Math., vol. 253. Boston, MA: Birkhäuser Boston.

KRICHEVER, I. M. AND NOVIKOV, S. P. 1999. Periodic and almost periodic potentials in inverse problems. *Inverse Problems*, **15**, R117–R144.

KRIEGL, A. AND MICHOR, P. W. 1997. *The Convenient Setting of Global Analysis*. Mathematical Surveys and Monographs, vol. 53. Providence, RI: Amer. Math. Soc.

KRÜGER, H. AND TESCHL, G. 2007. Long-time asymptotics for the Toda lattice in the soliton region. arXiv:0711.2793v2 [nlin.SI].

KUDRYAVTSEV, M. 2002. The Cauchy problem for the Toda lattice with a class of non-stabilized initial data. Pages 209–214 of: WEDER, R., EXNER, P., AND GRÉBERT, B. (eds.), *Mathematical Results in Quantum Mechanics*. Contemp. Math., vol. 307. Providence, RI: Amer. Math. Soc.

KUIJLAARS, A. B. J. 2000. On the finite-gap ansatz in the continuum limit of the Toda lattice. *Duke Math. J.*, **104**, 433–462.

KUIJLAARS, A. B. J. AND MCLAUGHLIN, K. T. R. 2001. Long time behavior of the continuum limit of the Toda lattice, and the generation of infinitely many gaps from the C^∞ initial data. *Comm. Math. Phys.*, **221**, 305–333.

KUPERSHMIDT, B. A. 1985. *Discrete Lax equations and differential-difference calculus*. Vol. 123. Astérisque.

KUTZ, N. 2003. Tri-Hamiltonian Toda lattice and a canonical bracket for closed discrete curves. *Lett. Math. Phys.*, **64**, 229–234.

LADIK, J. F. AND CHIU, S. C. 1977. Solutions of nonlinear network equations by the inverse method. *J. Math. Phys.*, **18**, 701–704.

LAKSHMANAN, M. AND RAJASEKAR, S. 2003. *Nonlinear Dynamics. Integrability, Chaos and Patterns*. New York: Springer.

LAX, P. D. 1968. Integrals of nonlinear equations of evolution and solitary waves. *Comm. Pure Appl. Math.*, **21**, 467–490.

LEITE, R. S., SALDANHA, N. C., AND TOMEI, C. 2008. An atlas for trigonal isospectral manifolds. *Linear Algebra Appl.*, **429**, 387–402.

LEVITAN, B. M. AND ZHIKOV, V. V. 1982. *Almost Periodic Functions and Differential Equations*. Cambridge: Cambridge University Press.

LEWITTES, J. 1964. Riemann surfaces and the theta function. *Acta Math.*, **111**, 37–61.

LEZNOV, A. N. AND SAVELIEV, M. V. 1992. *Group-Theoretical Methods for Integration of Nonlinear Dynamical Systems*. Basel: Birkhäuser.

LI, L. C. 1987. Long time behaviour of an infinite particle system. *Comm. Math. Phys.*, **110**, 617–623.

LI, L. C. 1997. On the complete integrability of some Lax equations on a periodic lattice. *Trans. Amer. Math. Soc.*, **349**, 331–372.

LI, L. C. 2005. Some remarks on CMV matrices and dressing orbits. *Int. Math. Res. Not. IMRN*, **40**, 2437–2446.

LOZANO, G. I. 2004. *Poisson Geometry of the Ablowitz–Ladik Equations*. Ph.D. thesis, University of Arizona.

LUKASHOV, A. L. 2004. Circular parameters of polynomials orthogonal on several arcs of the unit circle. *Sbornik Math.*, **195**, 1639–1663.

MANAKOV, S. V. 1975. Complete integrability and stochastization of discrete dynamical systems. *Soviet Phys. JETP*, **40**, 269–274.

MARČENKO, V. A. 1974a. A periodic Korteweg–de Vries problem. *Soviet Math. Dokl.*, **15**, 1052–1056.

MARČENKO, V. A. 1974b. The periodic Korteweg–de Vries problem. *Math. USSR-Sb.*, **24**, 319–344.

MARKUSHEVICH, A. I. 1985. *Theory of Functions of a Complex Variable*. Second edn. New York: Chelsea.

MARKUSHEVICH, A. I. 1992. *Introduction to the Classical Theory of Abelian Functions*. Translations of Mathematical Monographs, vol. 96. Providence, RI: Amer. Math. Soc.

MATTIS, D. C. (ed.). 1993. *The Many-Body Problem*. Singapore: World Scientific.

MATVEEV, V. B. 1976. *Abelian functions and solitons.* Univ. of Wroclaw, Preprint, no 373.
MATVEEV, V. B. 2008. 30 years of finite-gap integration theory. *Philos. Trans. Roy. Soc. London Ser. A*, **366**, 837–875.
MATVEEV, V. B. AND SALLE, M. A. 1991. *Darboux Transformations and Solitons.* Berlin: Springer.
MCDANIEL, A. AND SMOLINSKY, L. 1992. A Lie-theoretic Galois theory for the spectral curves of an integrable system. I. *Comm. Math. Phys.*, **149**, 127–148.
MCDANIEL, A. AND SMOLINSKY, L. 1997. A Lie-theoretic Galois theory for the spectral curves of an integrable system. II. *Trans. Amer. Math. Soc.*, **349**, 713–746.
MCDANIEL, A. AND SMOLINSKY, L. 1998. Lax equations, weight lattices, and Prym–Tjurin varieties. *Acta Math.*, **181**, 283–305.
MCKEAN, H. P. 1979. Integrable systems and algebraic curves. Pages 83–200 of: GRMELA, M. AND MARSDEN, J. E. (eds.), *Global Analysis.* Lecture Notes in Mathematics, vol. 755. Berlin: Springer.
MCKEAN, H. P. 1985. Variation on a theme of Jacobi. *Comm. Pure Appl. Math.*, **38**, 669–678.
MCKEAN, H. P. AND VAN MOERBEKE, P. 1980. Hill and Toda curves. *Comm. Pure Appl. Math.*, **33**, 23–42.
MCLAUGHLIN, D. W. 1975. Four examples of the inverse method as a canonical transformation. *J. Math. Phys.*, **16**, 96–99, erratum: p. 1704.
MCLAUGHLIN, D. W. 1989. *The Periodic Toda Chain: Lecture Notes.*
MICHOR, J. 2005. *Scattering Theory for Jacobi Operators and Applications to Completely Integrable Systems.* Ph.D. thesis, University of Vienna, Austria.
MICHOR, J. AND TESCHL, G. 2007a. On the equivalence of different Lax pairs for the Kac–van Moerbeke hierarchy. arXiv:0710.2184 nonlin.SI.
MICHOR, J. AND TESCHL, G. 2007b. Trace formulas for Jacobi operators in connection with scattering theory for quasi-periodic background. Pages 69–76 of: JANAS, J., KURASOV, P., LAPTEV, A., NABOKO, S., AND STOLZ, G. (eds.), *Operator Theory, Analysis and Mathematical Physics.* Operator Theory: Advances and Applications, vol. 174. Basel: Birkhäuser.
MIKHAILOV, A. V., OLSHANETSKY, M. A., AND PERELOMOV, A. M. 1981. Two-dimensional generalized Toda lattice. *Comm. Math. Phys.*, **79**, 473–488.
MILLER, P. D. 1994. *Macroscopic Lattice Dynamics.* Ph.D. thesis, University of Arizona, Arizona, USA.
MILLER, P. D. 1995. Macroscopic behavior in the Ablowitz–Ladik equations. Pages 158–167 of: MAKHANKOV, V. G., BISHOP, A. R., AND HOLM, D. D. (eds.), *Nonlinear Evolution Equations & Dynamical Systems.* River Edge, NJ: World Scientific.
MILLER, P. D., ERCOLANI, N. M., KRICHEVER, I. M., AND LEVERMORE, C. D. 1995. Finite genus solutions to the Ablowitz–Ladik equations. *Comm. Pure Appl. Math.*, **48**, 1369–1440.
MIRANDA, R. 1995. *Algebraic Curves and Riemann Surfaces.* Graduate Texts in Mathematics, vol. 5. Providence, RI: Amer. Math. Soc.
MOSER, J. 1975a. Finitely many mass points on the line under the influence of an exponential potential – an integrable system. Pages 467–497 of: MOSER, J. (ed.), *Dynamical Systems, Theory and Applications.* Lecture Notes in Physics, vol. 38. Berlin: Springer.
MOSER, J. 1975b. Three integrable Hamiltonian systems connected with isospectral deformations. *Adv. in Math.*, **16**, 197–220.

MUKAIHIRA, A. AND NAKAMURA, Y. 2002. Schur flow for orthogonal polynomials on the unit circle and its integrable discretization. *J. Comput. Appl. Math.*, **139**, 75–94.

MUMFORD, D. 1978. An algebro-geometric construction of commuting operators and of solutions to the Toda lattice equation, Korteweg deVries equation and related non-linear equations. *Proceedings of the International Symposium on Algebraic Geometry (Kyoto Univ., Kyoto, 1977)*, 115–153.

MUMFORD, D. 1983. *Tata Lectures on Theta I*. Progress in Mathematics, vol. 28. Boston: Birkhäuser.

MUMFORD, D. 1984. *Tata Lectures on Theta II*. Progress in Mathematics, vol. 43. Boston: Birkhäuser.

NAĬMAN, P. B. 1962. On the theory of periodic and limit-periodic Jacobian matrices. *Soviet Math. Dokl.*, **3**, 383–385.

NAĬMAN, P. B. 1964. On the spectral theory of non-symmetric periodic Jacobi matrices. *Zap. Meh.-Mat. Fak. Harprime kov. Gos. Univ. i Harprime kov. Mat. Obšč.*, **30** (4), 138–151. Russian.

NAKAMURA, Y. AND KODAMA, Y. 1995. Moment problem of Hamburger, hierarchies of integrable systems, and the positivity of tau-functions. *Acta Appl. Math.*, **39**, 435–443.

NARASIMHAN, R. 1992. *Compact Riemann Surfaces*. Basel: Birkhäuser.

NATANZON, S. M. 1980. Moduli spaces of real curves. *Trans. Moscow Math. Soc.*, **37**, 233–272.

NATANZON, S. M. 1990. Klein surfaces. *Russian Math. Surveys*, **45** (6), 53–108.

NATANZON, S. M. 1995. Real nonsingular finite zone solutions of soliton equations. *Amer. Math. Soc. Transl. Ser. 2*, **170**, 153–183.

NENCIU, I. 2005a. *Lax Pairs for the Ablowitz–Ladik System via Orthogonal Polynomials on the Unit Circle*. Ph.D. thesis, California Institute of Technology, Pasadena, CA, USA.

NENCIU, I. 2005b. Lax pairs for the Ablowitz–Ladik system via orthogonal polynomials on the unit circle. *Int. Math. Res. Not. IMRN*, **11**, 647–686.

NENCIU, I. 2006. CMV matrices in random matrix theory and integrable systems: a survey. *J. Phys. A*, **39**, 8811–8822.

NENCIU, I. 2007. *Poisson brackets for orthogonal polynomials on the unit circle*. Preprint.

NEWELL, A. C. (ed.). 1974. *Nonlinear Wave Motion*. Lectures in Applied Mathematics, vol. 15. Providence, RI: Amer. Math. Soc.

NIJHOFF, F. W. 2000. Discrete Dubrovin equations and separation of variables for discrete systems. *Chaos Solitons Fractals*, **11**, 19–28.

NOVIKOV, S. P. 1974. The periodic problem for the Korteweg–de Vries equation. *Functional Anal. Appl.*, **8**, 236–246.

NOVIKOV, S. [P.], MANAKOV, S. V., PITAEVSKII, L. P., AND ZAKHAROV, V. E. 1984. *Theory of Solitons*. New York: Consultants Bureau.

NOVIKOV, S. P., MATVEEV, V. B., GELFAND, I. M., KRICHEVER, I. M., DIKII, L. A., VINOGRADOV, A. M., YUSIN, B. V., KUPERSHMIDT, B. A., DUBROVIN, B. A., AND KRASILSHCHIK, I. S. (eds.). 1981. *Integrable Systems*. London Mathematical Society Lecture Notes Series, vol. 60. Cambridge: Cambridge University Press.

OLSHANETSKY, M. A. AND PERELOMOV, A. M. 1979. Explicit solutions of classical generalized Toda models. *Invent. Math.*, **54**, 261–269.

OLSHANETSKY, M. A. AND PERELOMOV, A. M. 1981. Classical integrable finite-dimensional systems related to Lie algebras. *Phys. Rep.*, **71**, 313–400.

Bibliography 415

OLSHANETSKY, M. A. AND PERELOMOV, A. M. 1994. Integrable systems and finite-dimensional Lie algebras. Pages 87–116 of: ARNOL'D, V. I. AND NOVIKOV, S. P. (eds.), *Dynamical Systems VII. Integrable Systems, Nonholonomic Dynamical Systems.* Encyclopedia of Mathematical Sciences, vol. 16. Berlin: Springer.

ORLOV, A. YU. 2006. Hypergeometric functions as infinite-soliton tau functions. *Theoret. and Math. Phys.*, **146**, 183–206.

OSIPOV, A. S. 1997. Integration of non-abelian Langmuir type lattices by the inverse spectral problem method. *Functional Anal. Appl.*, **31**, 67–70.

PASTUR, L. A. 2006. From random matrices to quasi-periodic Jacobi matrices via orthogonal polynomials. *J. Approx. Theory*, **139**, 269–292.

PASTUR, L. [A.] AND FIGOTIN, A. 1992. *Spectra of Random and Almost-Periodic Operators*. Grundlehren der mathematischen Wissenschaften, vol. 297. Berlin: Springer.

PAYTON, D. N., RICH, M., AND VISSCHER, W. M. 1967. Lattice thermal conductivity in disordered harmonic and anharmonic crystal models. *Phys. Rev.*, **160**, 706–711.

PAYTON, D. N. AND VISSCHER, W. M. 1967a. Dynamics of disordered harmonic lattices. I. Normal-mode spectra for randomly disordered isotopic binary lattices. *Phys. Rev.*, **154**, 802–811.

PAYTON, D. N. AND VISSCHER, W. M. 1967b. Dynamics of disordered harmonic lattices. II. Normal modes of isotopically disordered binary lattices. *Phys. Rev.*, **156**, 1032–1038.

PAYTON, D. N. AND VISSCHER, W. M. 1968. Dynamics of disordered harmonic lattices. III. Normal-mode spectra for abnormal arrays. *Phys. Rev.*, **175**, 1201–1207.

PEHERSTORFER, F. 1995. Elliptic orthogonal and extremal polynomials. *Proc. London Math. Soc. (3)*, **70**, 605–624.

PEHERSTORFER, F. 2001. On Toda lattices and orthogonal polynomials. Pages 519–534 of: *Proceedings of the Fifth International Symposium on Orthogonal Polynomials, Special Functions and their Applications (Patras, 1999)*, vol. 133.

PEHERSTORFER, F., SPIRIDONOV, V. P., AND ZHEDANOV, A. S. 2007. Toda chain, Stieltjes function, and orthogonal polynomials. *Theoret. and Math. Phys.*, **151**, 505–528.

PENSKOI, A. V. 1998a. Canonically conjugate variables for the Volterra lattice with periodic boundary conditions. *Math. Notes*, **64**, 98–109.

PENSKOI, A. V. 1998b. The Volterra lattice as a gradient flow. *Regul. Chaotic Dyn.*, **3** (1), 76–77.

PENSKOI, A. V. 2007. The Volterra system and the topology of the isospectral variety of zero-diagonal Jacobi matrices. *Russian Math. Surveys*, **62** (3), 626–628.

PENSKOI, A. V. 2008. Integrable systems and the topology of isospectral manifolds. *Theoret. and Math. Phys.*, **155** (1), 627–632.

PERELOMOV, A. M. 1990. *Integrable Systems of Classical Mechanics and Lie Algebras*. Vol. I. Basel: Birkhäuser.

RAGNISCO, O. AND BRUSCHI, M. 1996. Peakons, r-matrix and Toda lattices. *Phys. A*, **228**, 150–159.

RAUCH, H. E. AND FARKAS, H. M. 1974. *Theta Functions with Applications to Riemann Surfaces*. Baltimore: The Williams & Wilkins Co.

RAZUMOV, A. V. AND SAVELIEV, M. V. 1997. *Lie Algebras, Geometry, and Toda-type Systems*. Cambridge: Cambridge University Press.

REED, M. AND SIMON, B. 1975. *Methods of Modern Mathematical Physics III. Scattering Theory*. New York: Academic Press.

RENGER, W. 1999. Toda soliton limits on general backgrounds. *J. Differential Equations*, **151**, 191–230.

REYMAN, A. G. AND SEMENOV-TIAN-SHANSKY, M. A. 1979. Reduction of Hamiltonian systems, affine Lie algebras and Lax equations. *Invent. Math.*, **54**, 81–100.

REYMAN, A. G. AND SEMENOV-TIAN-SHANSKY, M. A. 1981. Reduction of Hamiltonian systems, affine Lie algebras and Lax equations. II. *Invent. Math.*, **63**, 423–432.

REYMAN, A. G. AND SEMENOV-TIAN-SHANSKY, M. A. 1994. Group-theoretical methods in the theory of finite-dimensional integrable systems. Pages 116–225 of: ARNOL'D, V. I. AND NOVIKOV, S. P. (eds.), *Dynamical Systems VII. Integrable Systems, Nonholonomic Dynamical Systems*. Encyclopedia of Mathematical Sciences, vol. 16. Berlin: Springer.

REYSSAT, E. 1989. *Quelques Aspects des Surfaces de Riemann*. Progress in Mathematics, vol. 77. Boston: Birkhäuser.

RODIN, YU. L. 1987. The Riemann boundary problem on closed Riemann surfaces and integrable systems. *Phys. D*, **24**, 1–53.

RODIN, YU. L. 1988. *The Riemann Boundary Problem on Riemann Surfaces*. Dordrecht: Reidel.

ROLANÍA, D. BARRIOS AND HEREDERO, R. HERNÁNDEZ. 2006. On the relation between the complex Toda and Volterra lattices. arXiv:nlin.SI/0610010.

ROURKE, D. E. 2004. Elementary Bäcklund transformations for a discrete Ablowitz–Ladik eigenvalue problem. *J. Phys. A*, **37**, 2693–2708.

SADAKANE, T. 2003. Ablowitz–Ladik hierarchy and two-component Toda lattice hierarchy. *J. Phys. A*, **36**, 87–97.

SAKHNOVICH, L. A. 1989. Study of the semi-infinite Toda chain. *Theoret. and Math. Phys.*, **81**, 1018–1026.

SCHIEBOLD, C. 1998. An operator theoretic approach to the Toda lattice equation. *Phys. D*, **122**, 37–61.

SCHIEBOLD, C. 2005. *Integrable Systems and Operator Equations*. Habilitationsschrift, Friedrich-Schiller-Universität, Jena, Germany.

SCHILLING, R. J. 1989. A systematic approach to the soliton equations of a discrete eigenvalue problem. *J. Math. Phys.*, **30**, 1487–1501.

SCHLICHENMAIER, M. 1989. *An Introduction to Riemann Surfaces, Algebraic Curves and Moduli Spaces*. Lecture Notes in Physics, vol. 322. Berlin: Springer.

SHARIPOV, R. A. 1990. Integration of Bogoyavlenskii chains. *Math. Notes*, **47**, 101–103.

SHARIPOV, R. A. 1991. Minimal tori in the five-dimensional sphere in $\mathbb{C}^3$. *Theoret. and Math. Phys.*, **87**, 363–369.

SHIOTA, T. 1986. Characterization of Jacobian varieties in terms of soliton equations. *Invent. Math.*, **83**, 333–382.

SHIOTA, T. 1990. The KP equation and the Schottky problem. *Sugaku Expositions*, **3**, 183–211.

SHMOISH, M. E. 1989. Nonisospectral deformations of Jacobi matrices and nonlinear difference equations. *Ukrainian Math. J.*, **41**, 254–257.

SHOKUROV, V. V. 1994. I. Riemann surfaces and algebraic curves. Pages 1–166 of: SHAFAREVICH, I. R. (ed.), *Algebraic Geometry I*. Encyclopaedia of Mathematical Sciences, vol. 23. Berlin: Springer.

SIEGEL, C. L. 1988. *Topics in Complex Functions II. Automorphic Functions and Abelian Integrals*. New York: Wiley.

SILHOL, R. 1982. Real Abelian varieties and the theory of Comessatti. *Math. Z.*, **181**, 345–364.

SIMON, B. 2004a. Analogs of the *m*-function in the theory of orthogonal polynomials on the unit circle. *J. Comput. Appl. Math.*, **171**, 411–424.

SIMON, B. 2004b. Orthogonal polynomials on the unit circle: New results. *Int. Math. Res. Not. IMRN*, **53**, 2837–2880.

SIMON, B. 2005a. OPUC on one foot. *Bull. Amer. Math. Soc. (N.S.)*, **42**, 431–460.

SIMON, B. 2005b. *Orthogonal Polynomials on the Unit Circle, Part 1: Classical Theory.* Providence, RI: Amer. Math. Soc.

SIMON, B. 2005c. *Orthogonal Polynomials on the Unit Circle, Part 2: Spectral Theory.* Providence, RI: Amer. Math. Soc.

SIMON, B. 2005d. *Trace Ideals and their Applications.* Second edn. Providence, RI: Amer. Math. Soc.

SIMON, B. 2007a. CMV matrices: Five years after. *J. Comput. Appl. Math.*, **208**, 120–154.

SIMON, B. 2007b. Zeros of OPUC and long time asymptotics of Schur and related flows. *Inverse Probl. Imaging*, **1**, 189–215.

SMIRNOV, A. O. 1989. Finite-gap solutions of Abelian Toda chain of genus 4 and 5 in elliptic functions. *Theoret. and Math. Phys.*, **78**, 6–13.

SMITH, R. 1989. The Jacobian variety of a Riemann surface and its theta geometry. Pages 350–427 of: CORNALBA, M., GOMEZ MONT, X., AND VERJOVSKY, A. (eds.), *Lectures on Riemann Surfaces.* Singapore: World Scientific.

SODIN, M. AND YUDITSKIĬ, P. 1994. Infinite-zone Jacobi matrices with pseudo-extendible Weyl functions and homogeneous spectrum. *Russian Acad. Sci. Dokl. Math.*, **49**, 364–368.

SODIN, M. AND YUDITSKIĬ, P. 1997. Almost periodic Jacobi matrices with homogeneous spectrum, infinite dimensional Jacobi inversion, and Hardy classes of character-automorphic functions. *J. Geom. Anal.*, **7**, 387–435.

SPRINGER, G. 1981. *Introduction to Riemann Surfaces.* Second edn. New York: Chelsea.

SUN, M. N., DENG, S. F., AND CHEN, D. Y. 2005. The Bäcklund transformation and novel solutions for the Toda lattice. *Chaos Solitons Fractals*, **23**, 1169–1175.

SURIS, Y. B. 2003. *The Problem of Integrable Discretization: Hamiltonian Approach.* Basel: Birkhäuser.

SURIS, Y. B. 2006. Toda Lattices. Pages 235–244 of: FRANCOISE, J. P., NABER, G., AND TSOU, S. T. (eds.), *Encyclopedia of Mathematical Physics, Vol. 5.* Amsterdam: Academic Press, Elsevier.

SYMES, W. W. 1980a. Hamiltonian group actions and integrable systems. *Phys. D*, **1**, 339–374.

SYMES, W. W. 1980b. Systems of Toda type, inverse spectral problems, and representation theory. *Invent. Math.*, **59**, 13–51.

SZEGŐ, G. 1920. Beiträge zur Theorie der Toeplitzschen Formen I. *Math. Z.*, **6**, 167–202.

SZEGŐ, G. 1921. Beiträge zur Theorie der Toeplitzschen Formen II. *Math. Z.*, **9**, 167–190.

SZEGŐ, G. 1978. *Orthogonal Polynomials.* Amer. Math. Soc. Colloq. Publ., vol. 23. Providence, RI: Amer. Math. Soc.

TAIMANOV, I. A. 1990. Smooth real finite-zonal solutions of sine-Gordon equations. *Math. Notes*, **47**, 94–100.

TAIMANOV, I. A. 1997. Secants of Abelian varieties, theta functions, and soliton equations. *Russian Math. Surveys*, **52** (1), 147–218.

TAKASAKI, K. 1984. Initial value problem for the Toda lattice hierarchy. Pages 139–163 of: OKAMOTO, K. (ed.), *Advanced Studies in Pure Mathematics*, vol. 4. Amsterdam: North-Holland.

TAKEBE, T. 1990. Toda lattice hierarchy and conservation laws. *Comm. Math. Phys.*, **129**, 281–318.

TAMIZHMANI, K. M. AND MA, WEN XIU. 2000. Master symmetries from Lax operators for certain lattice soliton hierarchies. *J. Phys. Soc. Japan*, **69**, 351–361.

TANG, Y., CAO, J., LIU, X., AND SUN, Y. 2007. Symplectic methods for the Ablowitz–Ladik discrete nonlinear Schrödinger equation. *J. Phys. A*, **40**, 2425–2437.

TESCHL, G. 1995. *Spectral Theory for Jacobi Operators*. Ph.D. thesis, University of Missouri, Columbia, MO, USA.

TESCHL, G. 1997. Spectral deformations of Jacobi operators. *J. Reine Angew. Math.*, **491**, 1–15.

TESCHL, G. 1998. Trace formulas and inverse spectral theory for Jacobi operators. *Comm. Math. Phys.*, **196**, 175–202.

TESCHL, G. 1999a. Inverse scattering transform for the Toda hierarchy. *Math. Nachr.*, **202**, 163–171.

TESCHL, G. 1999b. On the Toda and Kac–van Moerbeke hierarchies. *Math. Z.*, **231**, 325–344.

TESCHL, G. 2000. *Jacobi Operators and Completely Integrable Nonlinear Lattices*. Mathematical Surveys and Monographs, vol. 72. Providence, RI: Amer. Math. Soc.

TESCHL, G. 2001. Almost everything you always wanted to know about the Toda equation. *Jahresber. Deutsch. Math.-Verein.*, **103**, 149–162.

TESCHL, G. 2007. Algebro-geometric constraints on solitons with respect to quasi-periodic backgrounds. *Bull. London Math. Soc.*, **39**, 677–684.

TODA, M. 1967a. Vibration of a chain with nonlinear interaction. *J. Phys. Soc. Japan*, **22**, 431–436.

TODA, M. 1967b. Wave propagation in anharmonic lattices. *J. Phys. Soc. Japan*, **23**, 501–506.

TODA, M. 1970. Waves in nonlinear lattice. *Progr. Theoret. Phys. Suppl.*, **45**, 174–200.

TODA, M. 1975. Studies of a non-linear lattice. *Phys. Rep.*, **18**, 1–124.

TODA, M. 1976. Development of the theory of a nonlinear lattice. *Progr. Theoret. Phys. Suppl.*, **59**, 1–35.

TODA, M. 1989a. *Nonlinear Waves and Solitons*. Dordrecht: Kluwer.

TODA, M. 1989b. *Theory of Nonlinear Lattices*. Second enlarged edn. Solid-State Sciences, vol. 20. Berlin: Springer.

TODA, M. 1993. *Selected Papers of Morikazu Toda*. River Edge, USA: World Scientific. M. Wadati (ed.).

TOMEI, C. 1984. The topology of isospectral manifolds of tridiagonal matrices. *Duke Math. J.*, **51**, 981–996.

TSIGANOV, A. V. 2007. On two different bi-Hamiltonian structures for the Toda lattice. *J. Phys. A*, **40**, 6395–6406.

UENO, K. AND TAKASAKI, K. 1984. Toda lattice hierarchy. Pages 1–95 of: OKAMOTO, K. (ed.), *Advanced Studies in Pure Mathematics*, vol. 4. Amsterdam: North-Holland.

ULAM, S. M. 1991. *Adventures of a Mathematician*. Berkeley: University of California Press.

VAN MOERBEKE, P. 1976. The spectrum of Jacobi matrices. *Invent. Math.*, **37**, 45–81.

VAN MOERBEKE, P. 1979. About isospectral deformations of discrete Laplacians. Pages 313–370 of: GRMELA, M. AND MARSDEN, J. E. (eds.), *Global Analysis*. Lecture Notes in Mathematics, vol. 755. Berlin: Springer.

VAN MOERBEKE, P. 1993. The boundary of isospectral manifolds, Bäcklund transforms and regularization. Pages 163–177 of: BABELON, O., CARTIER, P., AND KOSMANN SCHWARZBACH, Y. (eds.), *Integrable Systems: The Verdier Memorial Conference*. Boston: Birkhäuser.

VAN MOERBEKE, P. AND MUMFORD, D. 1979. The spectrum of difference operators and algebraic curves. *Acta Math.*, **143**, 93–154.

VANINSKY, K. L. 2001. An additional Gibbs' state for the cubic Schrödinger equation on the circle. *Comm. Pure Appl. Math.*, **54**, 537–582.

VANINSKY, K. L. 2003. The Atiyah–Hitchin bracket and the open Toda lattice. *J. Geom. Phys.*, **46**, 283–307.

VEKSLERCHIK, V. E. 1993. Finite nonlinear Schrödinger chain. *Phys. Lett. A*, **174**, 285–288.

VEKSLERCHIK, V. E. 1998. Functional representation of the Ablowitz–Ladik hierarchy. *J. Phys. A*, **31**, 1087–1099.

VEKSLERCHIK, V. E. 1999. Finite genus solutions for the Ablowitz–Ladik hierarchy. *J. Phys. A*, **32**, 4983–4994.

VEKSLERCHIK, V. E. 2002. Functional representation of the Ablowitz–Ladik hierarchy. II. *J. Nonlinear Math. Phys.*, **9**, 157–180.

VEKSLERCHIK, V. E. 2005. Functional representation of the Volterra hierarchy. *J. Nonlinear Math. Phys.*, **12**, 409–431.

VEKSLERCHIK, V. E. 2006. Implementation of the Bäcklund transformations for the Ablowitz–Ladik hierarchy. *J. Phys. A*, **39**, 6933–6953.

VEKSLERCHIK, V. E. AND KONOTOP, V. V. 1992. Discrete nonlinear Schrödinger equation under non-vanishing boundary conditions. *Inverse Problems*, **8**, 889–909.

VESELOV, A. P. 1991. Integrable maps. *Russian Math. Surveys*, **46:5**, 1–51.

VESELOV, A. P. AND PENSKOI, A. V. 1998. On algebro-geometric Poisson brackets for the Volterra lattice. *Regul. Chaotic Dyn.*, **3** (2), 3–9.

VESELOV, A. P. AND PENSKOI, A. V. 1999. Algebraic-geometric Poisson brackets for difference operators and Volterra systems. *Dokl. Math.*, **59**, 391–394.

VINNIKOV, V. 1993. Self-adjoint determinantal representations of real plane curves. *Math. Ann.*, **296**, 453–479.

VINNIKOV, V. AND YUDITSKII, P. 2002. Functional models for almost periodic Jacobi matrices and the Toda hierarchy. *Mat. Fiz. Anal. Geom.*, **9**, 206–219.

VOLKOV, A. YU. 1988. Miura transformation on a lattice. *Theoret. and Math. Phys.*, **74**, 96–99.

WADATI, M. 1976. Transformation theories for nonlinear discrete systems. *Progr. Theoret. Phys. Suppl.*, **59**, 36–63.

WADATI, M. AND WATANABE, M. 1977. Conservation laws of a Volterra system and nonlinear self-dual network equation. *Progr. Theoret. Phys.*, **57**, 808–811.

WALTER, W. 1998. *Ordinary Differential Equations*. New York: Springer.

WATERS, J. AND FORD, J. 1964. A method of solution for resonant nonlinear coupled oscillator systems. *J. Math. Phys.*, **7**, 399–403.

WATKINS, D. S. 1984. Isospectral flows. *SIAM Rev.*, **26**, 379–391.

WATKINS, D. S. 1993. Some perspectives on the eigenvalue problem. *SIAM Rev.*, **35**, 430–471.

WEICHOLD, G. 1883. Ueber symmetrische Riemann'sche Flächen und die Periodicitätsmoduln der zugehörigen Abel'schen Normalintegrale erster Gattung. *Z. Math. Phys.*, **28**, 321–351.

WEIDMANN, J. 1980. *Linear Operators in Hilbert Spaces*. New York: Springer.

WIDOM, H. 1997. Some classes of solutions to the Toda lattice hierarchy. *Comm. Math. Phys.*, **184**, 653–667.

WILSON, G. 1978. Hilbert's sixteenth problem. *Topology*, **17**, 53–73.

YAMAZAKI, S. 1987. On the system of non-linear differential equations $\dot{y}_k = y_k(y_{k+1} - y_{k-1})$. *J. Phys. A*, **20**, 6237–6241.

YAMAZAKI, S. 1989. On an exponential-type dynamical system: $\dot{x}_k = (-1/2)[\exp(x_{k-1} - x_k) + \exp(x_k - x_{k+1})]$. *J. Phys. A*, **22**, 611–616.

YAMAZAKI, S. 1990. The semi-infinite system of nonlinear differential equations $\dot{A}_k = 2A_k(A_{k+1} - A_{k-1})$; methods of integration and asymptotic time behaviours. *Nonlinear Anal.*, **3**, 653–676.

YAMAZAKI, S. 1992. On specification of a class of infinite Jacobi matrices. *J. Phys. A*, **25**, L403–L408.

YURKO, V. A. 1995. Integrations of nonlinear dynamic systems with the method of inverse spectral problems. *Math. Notes*, **57**, 672–675.

ZABUSKY, N. J. AND KRUSKAL, M. D. 1965. Interactions of "solitons" in a collisionless plasma and the recurrence of initial states. *Phys. Rev. Lett.*, **15**, 240–243.

ZAKHAROV, V. E. (ed.). 1991. *What is Integrability?* Berlin: Springer.

ZAKHAROV, V. E., MUSHER, S. L., AND RUBENCHIK, A. M. 1974. Nonlinear stage of parametric wave excitation in a plasma. *Soviet Phys. JETP Lett.*, **19**, 151–152.

ZAKHAROV, V. E. AND SHABAT, A. B. 1972. Exact theory of two-dimensional self-focusing and one-dimensional self-modulation of waves in nonlinear media. *Soviet Phys. JETP*, **34**, 62–69.

ZAKHAROV, V. E. AND SHABAT, A. B. 1973. Interaction between solitons in a stable medium. *Soviet Phys. JETP*, **37**, 823–828.

ZAKHAROV, V. E. AND SHABAT, A. B. 1974. A scheme for integrating the nonlinear equations of mathematical physics by the method of inverse scattering transform. I. *Functional Anal. Appl.*, **8**, 226–235.

ZENG, Y. AND RAUCH-WOJCIECHOWSKI, S. 1995a. Continuous limits for the Kac–Van Moerbeke hierarchy and for their restricted flows. *J. Phys. A*, **28**, 3825–3840.

ZENG, Y. AND RAUCH-WOJCIECHOWSKI, S. 1995b. Restricted flows of the Ablowitz–Ladik hierarchy and their continuous limits. *J. Phys. A*, **28**, 113–134.

ZHANG, D. J. AND CHEN, D. Y. 2002a. The conservation laws of some discrete soliton systems. *Chaos Solitons Fractals*, **14**, 573–579.

ZHANG, D. J. AND CHEN, D. Y. 2002b. Hamiltonian structure of discrete soliton systems. *J. Phys. A*, **35**, 7225–7241.

ZHANG, D. J., NING, T. K., BI, J. B., AND CHEN, D. Y. 2006. New symmetries for the Ablowitz–Ladik hierarchies. *Phys. Lett. A*, **359**, 458–466.

ZHEDANOV, A. 1990. The Toda chain: Solutions with dynamical symmetry and classical orthogonal polynomials. *Theoret. and Math. Phys.*, **82**, 6–11.

ZHERNAKOV, N. V. 1986. Direct and inverse problems for a periodic Jacobi matrix. *Ukrainian Math. J.*, **38**, 665–668.

ZHERNAKOV, N. V. 1987. Integration of Toda chains in the class of Hilbert–Schmidt operators. *Ukrainian Math. J.*, **39**, 527–530.

ZHIVKOV, A. I. 1989. Isospectral classes of matrix-valued finite-gap operators with symmetries. *Russian Math. Surveys*, **44** (1), 269–270.

ZHIVKOV, A. I. 1994. Finite-gap matrix potentials with one and two involutions. *Bull. Sci. Math.*, **118**, 403–440.

Index

Abel maps, 338, 340
Abel's theorem, 339
Abelian differentials
  of the first kind, 332
  of the second kind, 334
  of the third kind, 335
Ablowitz–Ladik hierarchy
  Baker–Akhiezer vector
    stationary, 223
    time-dependent, 252
  conservation laws (local), 299
  constant solution, 236
  degree, 194
  Dubrovin equations, 257
  Hamiltonian formalism, 310
  homogeneous
    coefficients, 193
    stationary, 194
    time-dependent, 201
  hyperelliptic curve, 198, 221
  Lax relation
    stationary, 211
    time-dependent, 211
  recursion
    linear, 191
    nonlinear, 375
  Riccati-type equation
    stationary, 223
    time-dependent, 253
  stationary (definition), 192
  theta function representation
    Baker–Akhiezer function (stationary), 232
    Baker–Akhiezer function (time-dependent), 263
    ϕ (stationary), 232
    ϕ (time-dependent), 263
  time-dependent (definition), 200

trace formulas
  stationary, 225
  time-dependent, 258
zero-curvature formalism
  stationary, 189
  time-dependent, 199
algebraic curve
  affine plane curve
    definition, 324
    nonsingular, 324
    singular, 324
    smooth, 324
  charts, 324
  projections, 324
  projective plane curve
    definition, 326
    smooth, 327
algebraic function, 330
algebroidal, 329

Baker–Akhiezer function
  Toda hierarchy, 14, 20, 43, 87
Baker–Akhiezer vector
  Ablowitz–Ladik hierarchy, 223, 252
  Toda hierarchy, 43, 87
Baxter–Szegő hierarchy, 197
branch (ramified) point, 328
Burchnall–Chaundy curve
  Toda hierarchy, 41
Burchnall–Chaundy Laurent polynomial
  Ablowitz–Ladik hierarchy, 220
Burchnall–Chaundy polynomial
  Kac–van Moerbeke hierarchy, 165
  Toda hierarchy, 36
Burchnall–Chaundy theorem
  Toda hierarchy, 11

Index

conservation laws (local)
  Ablowitz–Ladik hierarchy, 299
  Toda hierarchy, 130

degree
  Ablowitz–Ladik hierarchy, 194
  Toda hierarchy, 28
Dirichlet data, 60
Dirichlet divisor
  Toda hierarchy, 51
discrete nonlinear Schrödinger equation, 201
  defocusing, 201
  focusing, 201
divisor, $\mathcal{D}$
  canonical, 337
  definition, 336
  degree, deg($\mathcal{D}$), 336
  index of specialty, $i(\mathcal{D})$, 338
  nonnegative, 337
  nonspecial, 17, 338
  principal, 337
  special, 338
  zero, 337
divisors
  equivalent, 337
  partial ordering, 336
  relatively prime, 337
  the set of all, 336
Dubrovin equations
  Ablowitz–Ladik hierarchy, 257
  Toda hierarchy, 21, 90, 151

Flaschka's variables, 3, 6, 146
function element, 329

genus
  topological, 355

Hamiltonian formalism
  Ablowitz–Ladik hierarchy, 310
  Toda hierarchy, 140
homogeneous coordinates
  definition, 325
homology basis, 331
hyperelliptic curve
  Ablowitz–Ladik hierarchy, 198
  Toda hierarchy, 36
hyperelliptic Riemann surface
  definition, 330

intersection matrix, 331

Jacobi variety, $J(\mathcal{K}_g)$, 336, 338
Jacobi's inversion theorem, 343

Kac–van Moerbeke hierarchy
  Burchnall–Chaundy polynomial, 165
  constant solution, 178
  Lax relation
    stationary, 164
    time-dependent, 166
  stationary (definition), 165
  theta function representation
    Baker–Akhiezer function (stationary), 177
    Baker–Akhiezer function (time-dependent), 180
    ρ (stationary), 177
    ρ (time-dependent), 180
  time-dependent (definition), 166
Klein surface, 348

Lagrange's interpolation theorem, 386
Lax pair
  Ablowitz–Ladik hierarchy, 211
  Kac–van Moerbeke hierarchy, 163
  Toda hierarchy, 31
Lax relation
  Ablowitz–Ladik hierarchy, 211
  Kac–van Moerbeke hierarchy, 164, 166
  Toda hierarchy, 6, 8
line at infinity, 326

maximal atlas, 328
meromorphic differentials (1-forms), 331
meromorphic function field, 330
Miura-type identity, 167

nonsingular curve, *see* algebraic curve

period lattice, L_g, 338
Picard group, 337
projective curve
  ramification point, 327
projective plane, $\mathbb{CP}^2$, 325

quasi-periodic sequence
  definition, 60

Riccati-type equation
  Ablowitz–Ladik hierarchy, 223, 253
  Toda hierarchy, 20, 43, 88
Riemann sphere, $\mathbb{C}_\infty$, 328
Riemann surface, 328
  definition, 325
  genus, 327
  sheet, 329
  symmetric, 348
Riemann theta function
  definition, 339
  properties, 340
Riemann's inequality, 338
Riemann's period relations, 331

Index

Riemann's vanishing theorem, 343
Riemann–Roch theorem, 338

Schur flow, 201
sheet exchange map (involution), 347
singular curve, *see* algebraic curve
solitons
  Toda hierarchy
    stationary, 71
    time-dependent, 101
symmetric functions
  definition of, 385

theta divisor, 343
theta function representation
  Ablowitz–Ladik hierarchy
    Baker–Akhiezer function (stationary), 232
    Baker–Akhiezer function (time-dependent), 263
    ϕ (stationary), 232
    ϕ (time-dependent), 263
  Kac–van Moerbeke hierarchy
    Baker–Akhiezer function (stationary), 177
    Baker–Akhiezer function (time-dependent), 180
    ρ (stationary), 177
    ρ (time-dependent), 180
  Toda hierarchy
    Baker–Akhiezer function (stationary), 55
    Baker–Akhiezer function (time-dependent), 95
    ϕ (stationary), 55
    ϕ (time-dependent), 95
Toda hierarchy
  algebro-geometric solutions, 11, 32
  Baker–Akhiezer function
    stationary, 14, 43
    time-dependent, 20, 87
  Baker–Akhiezer vector
    stationary, 43
    time-dependent, 87
  Burchnall–Chaundy curve, 41
  Burchnall–Chaundy polynomial, 36
  Burchnall–Chaundy theorem, 11
  conservation laws (local), 130
  constant solution, 68
  curve, 353
  degree, 28
  Dirichlet divisor, 51
  Dubrovin equations
    stationary, 151
    time-dependent, 21, 90
  Hamiltonian formalism, 140
  homogeneous
    coefficients, 29
    stationary, 32
    time-dependent, 38
  hyperelliptic curve, 12, 36
  Its–Matveev formula
    stationary, 17
    time-dependent, 22
  Lax pair, 8, 31
  Lax relation, 6, 8
  periodic solution, 61
  potentials, 11
  recursion
    linear, 7, 27
    nonlinear, 368
  Riccati-type equation, 20
    stationary, 43
    time-dependent, 88
  solitons
    stationary, 71
    time-dependent, 101
  stationary (definition), 10, 31
  theta function representation
    Baker–Akhiezer function (stationary), 55
    Baker–Akhiezer function (time-dependent), 95
    ϕ (stationary), 55
    ϕ (time-dependent), 95
    potential (stationary), 17
    solution (time-dependent), 22
  time-dependent (definition), 8, 37
  trace formulas
    stationary, 15, 48
    time-dependent, 91
  zero-curvature formalism
    stationary, 40
    time-dependent, 9, 40
trace formulas
  Ablowitz–Ladik hierarchy, 225, 258
  Toda hierarchy, 15, 48, 91

vector of Riemann constants, 340

Weierstrass point
  definition, 342
  properties, 342

zero-curvature formalism
  Ablowitz–Ladik hierarchy, 189, 199
  Toda hierarchy, 9, 40

Errata and Addenda for Volume I[1]

Changes appear in grey. Line $k+$ (resp., line $k-$) denotes the kth line from the top (resp., the bottom) of a page.

INTRODUCTION

Page 10. In equation (0.34), it should read:

$$P_{2n+1}$$

Page 15. In line 3+, it should read:

$$P_{2n+1}$$

CHAPTER 1

Page 28. In line 8+, the second term should read:

$$\bigl(u_{t_n}+(L-z)P_{2n+1}\bigr)\big|_{\ker(L-z)} = 0,$$

Page 32. In equation (1.46) and in line 4−, it should read:

$$P_{2n+1}$$

Page 32. In equation (1.52), it should read:

$$F_{\underline{n}}(z, x')$$

Page 32. In lines 3− and 2−, it should read: Equation (1.48) follows by combining (1.41) and (1.44).

Page 35. In line 9+, it should read:

$$\underline{\hat{\lambda}}^{\beta,\ell} = \{\hat{\lambda}_0^\beta, \hat{\lambda}_1^\beta, \ldots, \hat{\lambda}_{\ell-1}^\beta, \hat{\lambda}_{\ell+1}^\beta, \ldots, \hat{\lambda}_n^\beta\}, \quad \ell = 2, \ldots, n-1,$$

---

[1] An updated list of errata and addenda is to be found at www.math.ntnu.no/~holden/solitons

Page 39. Equation (1.79) should have only one factor of i on the right-hand side, so the i to the right of $\sum_{j=1}^{n}$ should be stricken, that is, equation (1.79) should read

$$F_{n,x}(z) = 2i \sum_{j=1}^{n} y(\hat{\mu}_j) \prod_{\substack{k=1 \\ k \neq j}}^{n} (z - \mu_k)(\mu_j - \mu_k)^{-1},$$

Page 40. Equation (1.80) should have $+F_{n,x}(z)$ on the right-hand side, that is, it should read

$$\partial_\beta K_{n+1}^\beta(z) = F_{n,x}(z) + 2\beta F_n(z)$$

Similarly, the first line of (1.81) should have $+F_{n,x}(\lambda_\ell^\beta)$ on the right-hand side.

Page 41. In line 2+, it should read:
 ...with (1.5), (1.11), and (1.16) taken into account.

Page 41. In line 9+, it should read:
 ...Relations (1.85) and (1.86)...

Page 41. In line 10+, it should read:
 ...for K_{n+1}^β with (1.16) and (1.56) taken into account.

Page 41. In Lemma 1.18 it suffices to assume $u \in C^\infty(\mathbb{R})$. The additional assumption $u \in L^\infty(\mathbb{R})$ is superfluous since for each fixed $x \in \mathbb{R}$, $|\mu_j(x)|$ is finite and hence $\hat{\mu}_j(x)$ cannot coincide with the branch point P_∞ (a fact used in the last part of the proof on p. 42). Analogously, it suffices to assume that u satisfies Hypothesis 1.33 in the time-dependent setting.

Page 42. In line 12+, it should read:

$$\lim_{x \to x_0} \hat{\mu}_{j_p}(x) = (\mu_0, -(i/2)F_{n,x}(\mu_0, x_0)), \quad p = 1, \ldots, N;$$

Page 44. Line 5− should read:
 One infers from (1.98) that

Page 44. Equation (1.100) should read:

$$\int_{Q_0}^{P} \omega_{P_\infty,0}^{(2)} \underset{\zeta \to 0}{=} -\zeta^{-1} + e_{0,0} + O(\zeta) \text{ as } P \to P_\infty$$

for some $e_{0,0} \in \mathbb{C}$.

Page 44. In lines 1−, 2−, and 3−, strike the sentences
 since by (1.98),...sheets $\Pi_\pm$. Thus, ...contains no constant term.

Page 45. In equation (1.105), it should read:

$$\cdots \exp\left(-i(x - x_0)\left(\int_{Q_0}^{P} \omega_{P_\infty,0}^{(2)} - e_{0,0}\right)\right)$$

Page 52. In equation (1.128) and in the line following it, replace $\mathbb{Z}^n \setminus \{0\}$ by $(\mathbb{Z}^n \setminus \{0\})\tau$

Page 52. In the second line following equation (1.128), replace $i\Omega U^{(2)}_{0,j} = m_j$ by
$$i\Omega U^{(2)}_{0,j} = m_1 \tau_{1,j} + \sum_{k=2}^{n}(m_k - m_{k-1})\tau_{k,j}$$
Page 52. In the fifth line following equation (1.128), it should read:
... interval of length Ω, $x \in [x_0, x_0 + \Omega]$, for some $x_0 \in \mathbb{R}$.
Pages 57–63. There is a systematic error in Examples 1.30–1.32: All formulas for y_j should be replaced by iy_j, $j = 1, 2$. In Example 1.30, the quantities $\mathcal{F}_k(z, y)$ should therefore be of the form $\mathcal{F}_k(z, y) = y^2 - R_{2k+1}(z)$ for $k = 1, 2, 3$, and $k = n$, for monic polynomials R_{2k+1} of the form z^3, z^5, z^7, and z^{2n+1}. In Example 1.31, $\mathcal{F}_k(z, y)$ should be of the form

$$\mathcal{F}_n(z, y) = y^2 - z \prod_{j=1}^{n}(z + \kappa_j^2)^2 = 0$$

and in Example 1.32 it should read

$$\mathcal{F}_1(z, y) = y^2 - \left(z^3 - \frac{g_2}{4}z + \frac{g_3}{4}\right) = 0$$

and

$$\mathcal{F}_2(z, y) = y^2 - \left(z^5 - \frac{21}{4}g_2 z^3 - \frac{27}{4}g_3 z^2 + \frac{27}{4}g_2^2 z + \frac{81}{4}g_2 g_3\right)$$
$$= y^2 - (z^2 - 3g_2)\left(z^3 - \frac{9}{4}g_2 z - \frac{27}{4}g_3\right) = 0$$

Pages 61. In Example 1.31 assume $c_j, \kappa_j \in \mathbb{C} \setminus \{0\}$, $j = 1, \ldots, n$.
Pages 61. In Example 1.31 replace s-$\widehat{\text{KdV}}_n(u_n) = 0$ by

$$\text{s-KdV}_n(u_n) = 0$$

for an appropriate set of integration constants $\{c_\ell\}_{\ell=1,\ldots,n} \subset \mathbb{C}$ (cf. (1.15)).
Pages 66. Add the following to the end of the sentence following (1.164):
... except at collisions of certain μ_j (respectively, ν_ℓ), where one can only assert continuity of μ_j (respectively, ν_ℓ) with respect to (x, t_r).
Page 76. Equation (1.207) should read:

$$\int_{Q_0}^{P} \widetilde{\Omega}^{(2)}_{P_\infty, 2r} \underset{\zeta \to 0}{=} -\sum_{q=0}^{r} \tilde{c}_{r-q} \zeta^{-2q-1} + \tilde{e}_{r,0} + O(\zeta) \text{ as } P \to P_\infty$$

for some $\tilde{e}_{r,0} \in \mathbb{C}$.
Page 77. In equation (1.211), it should read:

$$\cdots \exp\left(-i(x - x_0)\left(\int_{Q_0}^{P} \omega^{(2)}_{P_\infty, 0} - e_{0,0}\right) - i(t_r - t_{0,r})\left(\int_{Q_0}^{P} \widetilde{\Omega}^{(2)}_{P_\infty, 2r} - \tilde{e}_{r,0}\right)\right)$$

Page 79. In line 9−, replace $-\widetilde{\underline{U}}_{2r}^{(2)}$ by $i\widetilde{\underline{U}}_{2r}^{(2)}$
Page 81. In lines 9− and 10−, it should read:

$$\cdots \exp\Bigg(-i(x-x_0)\bigg(\int_{Q_0}^P \omega_{P_\infty,0}^{(2)} - e_{0,0}\bigg)$$
$$-i(t_r - t_{0,r})\bigg(\int_{Q_0}^P \widetilde{\Omega}_{P_\infty,2r}^{(2)} - \tilde{e}_{r,0}\bigg) - \int_{Q_0}^P \omega_{\hat{\lambda}_0^\beta(x_0,t_{0,r}),\hat{\lambda}_0^\beta(x,t_r)}^{(3)}\bigg)$$

Page 81. In line 4− replace ... purely imaginary ... by ... real ...
Pages 86. In Example 1.52 assume $c_j, \kappa_j \in \mathbb{C} \setminus \{0\}, j = 1, \ldots, n$.
Pages 86. Line 1− should read:

$$\text{s-KdV}_n(u_n) = 0, \quad \text{KdV}_r(u_n) = 0$$

for appropriate sets of integration constants $\{c_\ell\}_{\ell=1,\ldots,n} \subset \mathbb{C}$ (cf. (1.15)) and $\{\tilde{c}_s\}_{s=1,\ldots,r} \subset \mathbb{C}$.

Page 101. Line 2+ should start with: genus n. ...
Page 102. In the last line of equation (1.282) and in equation (1.283), the remainder term can be replaced by:

$$O(|z|^{-n-(3/2)})$$

Page 102. In line 2 of equation (1.282), it should read:

$$\bigg(\prod_{m=0}^{2n}(1-(E_m/z))\bigg)^{-1/2}$$

Page 102. In the last line of equation (1.284), the remainder term can be replaced by:

$$O(|z|^{-n-(5/2)})$$

Page 102. Line 4− should read:

$$\frac{\delta \hat{f}_\ell}{\delta u} = \frac{2\ell - 1}{2}\hat{f}_{\ell-1}, \quad \ell = 1, \ldots, n.$$

Page 103. The last line of equation (1.287) should read:

$$= u_{t_n} - \partial_x(\nabla \mathcal{H}_n)_u,$$

Page 103. In line 2 of Theorem 1.62, it should read:

$$\ldots \ell = 1, \ldots, n \text{ for } n \in \mathbb{N} \ldots$$

Page 117. In line 11+ replace $\tau \mathbb{Z}^n$ by $\mathbb{Z}^n \tau$.

CHAPTER 2

Page 141. Equation (2.106) should read:
$$\int_{Q_0}^{P} \omega^{(2)}_{P_\infty, 0} \underset{\zeta \to 0}{=} -\zeta^{-1} + e_{0,0} + O(\zeta) \text{ as } P \to P_\infty$$
for some $e_{0,0} \in \mathbb{C}$.

Page 142. In equation (2.111), it should read:
$$\cdots \exp\left(-i(x-x_0)\left(\int_{Q_0}^{P} \omega^{(2)}_{P_\infty, 0} - e_{0,0}\right)\right)$$

Page 142. In equation (2.112), it should read:
$$\cdots \exp\left(-\int_{Q_0}^{P} \omega^{(3)}_{P_\infty, P_0} + \tfrac{1}{2}\ln(E_{m_0}) - i(x-x_0)\left(\int_{Q_0}^{P} \omega^{(2)}_{P_\infty, 0} - e_{0,0}\right)\right)$$

Pages 151. Add the following to the end of the sentence following (2.159):
...except at collisions of certain μ_j (respectively, ν_k), where one can only assert continuity of μ_j (respectively, ν_k) with respect to (x, t_r).

Page 159. Equation (2.197) should read:
$$\int_{Q_0}^{P} \widetilde{\Omega}^{(2)}_{P_\infty, r} \underset{\zeta \to 0}{=} \begin{cases} -\sum_{q=0}^{r-1} \tilde{c}_{r-1-q} \zeta^{-2q-1} + \tilde{e}_{r,0} + O(\zeta), & r \in \mathbb{N}, \\ 0, & r = 0, \end{cases} \text{ as } P \to P_\infty$$
for some $\tilde{e}_{r,0} \in \mathbb{C}$.

Page 159. Equation (2.201) should read:
$$\frac{\tilde{\alpha}}{\alpha} Q^{1/2} \int_{Q_0}^{P} \omega^{(2)}_{P_0, 0} \underset{\zeta \to 0}{=} -\frac{\tilde{\alpha}}{\alpha} Q^{1/2} \zeta^{-1} + d_0 + O(\zeta) \text{ as } P \to P_0$$
for some $d_0 \in \mathbb{C}$.

Page 159. In lines 8−, 9−, 10− and 11−, strike the sentence
Since by (2.200), ... no constant term.

Page 160. In equation (2.204), it should read:
$$\cdots \exp\left(-i(x-x_0)\left(\int_{Q_0}^{P} \omega^{(2)}_{P_\infty, 0} - e_{0,0}\right)\right.$$
$$\left. + (t_r - t_{0,r})\left(\frac{\tilde{\alpha}}{\alpha} Q^{1/2} \int_{Q_0}^{P} \omega^{(2)}_{P_0, 0} - d_0 + \int_{Q_0}^{P} \widetilde{\Omega}^{(2)}_{P_\infty, r} - \tilde{e}_{r,0}\right)\right)$$

Page 160. In equation (2.205), it should read:
$$\cdots \exp\left(-\int_{Q_0}^{P} \omega^{(3)}_{P_\infty, P_0} + (1/2)\ln(E_{m_0}) - i(x-x_0)\left(\int_{Q_0}^{P} \omega^{(2)}_{P_\infty, 0} - e_{0,0}\right)\right.$$
$$\left. + (t_r - t_{0,r})\left(\frac{\tilde{\alpha}}{\alpha} Q^{1/2} \int_{Q_0}^{P} \omega^{(2)}_{P_0, 0} - d_0 + \int_{Q_0}^{P} \widetilde{\Omega}^{(2)}_{P_\infty, r} - \tilde{e}_{r,0}\right)\right)$$

Page 168. The second paragraph should read as follows:

If, in addition, $\ell = 0$ and one is interested in spatially periodic solutions u with a real period $\Omega > 0$, the additional periodicity constraints

$$i\Omega \underline{U}_0^{(2)} \in (\mathbb{Z}^n \setminus \{0\})\tau$$

must be imposed. (By (B.45) this is equivalent to $2i\Omega \underline{c}(n) \in (\mathbb{Z}^n \setminus \{0\})\tau$.)

CHAPTER 3

Page 180. Line 1−, should read:

$$Q_{n+1} = \sum_{\ell=0}^{n+1} c_{n+1-\ell}\widehat{Q}_\ell, \quad \widehat{Q}_0 = i\begin{pmatrix} -1 & 0 \\ 0 & 1 \end{pmatrix}.$$

Page 181. In line 15+, replace KdV by AKNS.
Page 189. In line 4+, it should read $\breve{h}_{n-\ell} = Ah_{n-\ell}$.
Page 199. Delete τ at the end of equation (3.108).
Page 204. In line 4−, it should read:
   with non-self-adjoint Dirac-type operators ...
Page 212. In equation (3.146), it should read:

$$p(x, \cdot), q(x, \cdot) \in C^1(\mathbb{R})$$

Page 214. In line 8−, it should read:
   ...properties of $F_n, G_{n+1}, H_n, \ldots$
Pages 214. Add the following to the end of the sentence following (3.169):
   ...except at collisions of certain μ_j (respectively, ν_k), where one can only assert continuity of μ_j (respectively, ν_k) with respect to (x, t_r).
Page 222. In line 9−, replace $\widehat{F}_r$ by $\widetilde{F}_r$ twice.

CHAPTER 5

Page 302. In line 3+, replace P_∞ by P_{∞_+}.
Page 310. In line 7−, it should read:
   ...properties of $F_n, G_n, H_n, \ldots$
Pages 310. Add the following to the end of the sentence following (5.132):
   ...except at collisions of certain μ_j (respectively, ν_k), where one can only assert continuity of μ_j (respectively, ν_k) with respect to (x, t_r).

APPENDIX A

Page 329. In line 9−, it should read: ...The intersection of ...
Page 330. Line 10− should read:

$$\mathcal{P}(z, y, 1) = 0$$

Page 331. In line 6+, it should read:
... the ramification points). If ...
Page 337. In line 4+, it should read: ... is a smooth simple, ...
Page 347. In lines 7− and 6−, it should read:
... where $\{P_{\infty_1}, \ldots, P_{\infty_N}\}$ (typically, $N \in \{1, 2\}$ in the main text), denotes the set of ...
Page 347. In line 2−, P_0 should be replaced by Q_0.
Page 354. In line 7−, it should read:
The case $\mathcal{R} \neq 0$, and ...

APPENDIX B

Page 363. In (B.39) replace $\int_{E_{2k-1}}^{E_{2k}} \frac{z^{j-1} dz}{R_{2n+1}(z)^{1/2}}$ by $\sum_{\ell=k}^{n} \int_{E_{2\ell-1}}^{E_{2\ell}} \frac{x^{j-1} dx}{R_{2n+1}(x)^{1/2}}$
and refer to the homology basis described on top of p. 360, recalling the ordering $E_0 < E_1 < \cdots < E_{2n}$.

APPENDIX C

Page 375. In (C.39) replace $\int_{E_{2k-2}}^{E_{2k-1}} \frac{z^{j-1} dz}{R_{2n+2}(z)^{1/2}}$ by $\int_{E_{2k-1}}^{E_{2k}} \frac{x^{j-1} dx}{R_{2n+2}(x)^{1/2}}$
Page 375. In (C.40) replace $\int_{E_{2k-1}}^{E_{2k}} \frac{z^{j-1} dz}{R_{2n+2}(z)^{1/2}}$ by $-\sum_{\ell=1}^{k} \int_{E_{2\ell-2}}^{E_{2\ell-1}} \frac{x^{j-1} dx}{R_{2n+2}(x+i0)^{1/2}}$
and refer to the homology basis indicated in Fig. C.2 on p. 37, changing all E_m into real position with the ordering $E_0 < E_1 < \cdots < E_{2n+1}$.

APPENDIX D

Page 383. In line 2+ it should read:
... coefficients of η^k yields
Page 385. In line 8−, it should read:
... $c_\ell, \ell = 0, \ldots, n+1$,

APPENDIX E

Page 399. In line 2−, replace (E.14) by (E.13)

APPENDIX F

Page 404. In equation (F.18), it should read:
$$\sum_{\ell=0}^{n} d_{n,\ell}(\underline{E}) \Phi_\ell^{(j)}(\underline{\mu})$$

LIST OF SYMBOLS

Page 468. In line 3+, it should read
$$\partial \widehat{\mathcal{K}}_g = a_1 b_1 a_1^{-1} b_1^{-1} \ldots a_g b_g a_g^{-1} b_g^{-1}$$

BIBLIOGRAPHY

Page 473. Belokolos, E. D., Gesztesy, F., Makarov, K. A., and Sakhnovich, L. A. (To appear), appeared in *Evolution Equations*, G. Ruiz Goldstein, R. Nagel, and S. Romanelli (eds.), Lecture Notes in Pure and Applied Mathematics, Vol. 234, Marcel Dekker, New York, 2003, p. 1–34.
Page 473. Bobenko, A. I. (To appear), appeared in *arXiv:math/9909003 v1 1 Sep 1999*.
Page 481. Gesztesy, F. and Holden, H. (To appear, a.) appeared in *Rev. Mat. Iberoamericana* **19**, 73–142 (2003).
Page 481. Gesztesy, F. and Holden, H. (To appear, b.) appeared in *Phil. Trans. Roy. Soc. London Ser. A* **366**, 1025–1054 (2008).
Page 482. Gesztesy, F. and Sakhnovich, L. A. (To appear), appeared in *Reproducing Kernel Hilbert Spaces, Positivity, System Theory and Related Topics*, D. Alpay (ed.), Operator Theory: Advances and Applications, Vol. 143, Birkhäuser, Basel, 2003, p. 213–253.
Page 483. Gesztesy, F. Unterkofler, K. and Weikard, R. 2003, appeared in *Trans. Amer. Math. Soc.* **358**, 603–656 (2006).
Page 485. Holod, I. P. and Prikarpatsky, A. K. 1978, appeared in Prikarpatskiĭ, A. K. and Golod, P. Ī. 1979. A periodic problem for the classical two-dimensional Thirring model. *Ukrain. Mat. Zh.* **31**, no. 4, 454–459, 479. In Russian.
Page 485. Holm, D. D. and Qiao, Z. 2002, appeared in Z. Qiao. 2003. The Camassa–Holm hierarchy, N-dimensional integrable systems, and algebro-geometric solution on a symplectic submanifold. *Comm. Math. Phys.* **239**, 309–341.
Page 492. Replace Novikov, S. P. 1999 by Novikov, D. P. 1999

Addenda

Page 68. Lemma 1.35 is equivalent to:

$$-V_{n+1,t_r} + \left[\widetilde{V}_{r+1}, V_{n+1}\right] = 0.$$

Page 104. The following considerations are relevant in connection with KdV conservation laws: Assuming that $u = u(x, t_n)$ satisfies $u(\,\cdot\,, t_n) \in \mathcal{S}_\mathbb{R}(\mathbb{R})$, $t_n \in \mathbb{R}$ (for simplicity), we recall that the one-dimensional Schrödinger equation

$$L(t_n)\psi(z, \,\cdot\,, t_n) = z\psi(z, \,\cdot\,, t_n), \quad z \in \mathbb{C} \setminus \mathbb{R}, \ t_n \in \mathbb{R}$$

has unique (up to constant multiples) Weyl–Titchmarsh solutions $\psi_\pm(z, \,\cdot\,, t_n)$ satisfying for all $R \in \mathbb{R}$,

$$\psi_\pm(z, \,\cdot\,, t_n) \in L^2([R, \pm\infty)), \quad z \in \mathbb{C} \setminus \mathbb{R}, \ t_n \in \mathbb{R}.$$

The corresponding Weyl–Titchmarsh functions $m_\pm(z, x, t_n)$ are then defined by

$$m_\pm(z, x, t_n) = \partial_x \ln(\psi_\pm(z, x, t_n)).$$

One obtains an asymptotic spectral parameter expansion of $m_\pm(\,\cdot\,, x, t_n)$ as $z \to i\infty$ of the type (cf. (J.29), (J.30))

$$m_\pm(z, x, t_n) \underset{z \to i\infty}{=} \sum_{j=-1}^{\infty} m_{\pm,j}(x, t_n) z^{-j/2}, \tag{D.23}$$

and the Riccati equation (J.26) for $m_\pm(z, x, t_n)$ then implies the following recursion relations for the expansion coefficients $m_{\pm,j}$,

$$\begin{aligned}
m_{\pm,-1} &= \pm i, \quad m_{\pm,0} = 0, \quad m_{\pm,1} = \mp \frac{i}{2} u, \quad m_{\pm,2} = \frac{1}{4} u_x, \\
m_{\pm,j+1} &= \pm \frac{i}{2}\left(m_{\pm,j,x} + \sum_{\ell=1}^{j-1} m_{\pm,\ell} m_{\pm,j-\ell}\right), \quad j = 2, 3, \ldots
\end{aligned} \tag{D.24}$$

Moreover, we recall

$$m_{-,j} = (-1)^j m_{+,j}, \quad j \in \{-1\} \cup \mathbb{N}_0.$$

Theorem D.6 *Suppose that $u \in \mathcal{S}_\mathbb{R}(\mathbb{R})$ satisfies the nth KdV equation (1.287) (for some set of integration constants c_ℓ, $\ell = 1, \ldots, n$, $n \in \mathbb{N}$). Then, the infinite sequence of KdV conservation laws takes on the form*

$$\partial_{t_n} m_{\pm,2\ell+1} = \partial_x \left(\sum_{k=0}^{n} c_{n-k} \sum_{p=0}^{k} \hat{f}_{k-p} m_{\pm, 2\ell+1+2p} \right), \quad \ell \in \mathbb{N}_0. \tag{D.25}$$

Here $\hat{f}_\ell$ are the homogeneous coefficients (1.6). Similarly to the recursion relation (D.24) for the coefficients $m_{\pm,j}$, the coefficients $\hat{f}_\ell$ can be computed recursively from (1.4) (putting all integration constants equal to zero) or directly from the nonlinear recursion relation (D.8).

Proof The key to the derivation of (D.25) is the innocent looking identity

$$\partial_{t_n}(\partial_x \ln(\psi_\pm(z, x, t_n,))) = \partial_x(\partial_{t_n} \ln(\psi_\pm(z, x, t_n))),$$

or equivalently,

$$m_{\pm, t_n} = ((\ln(\psi_\pm))_{t_n})_x.$$

Assuming that $\psi_\pm(z, x, t_n)$ are chosen so that

$$\psi_{\pm, t_n} = P_{2n+1} \psi_{\pm, t_n} = F_n \psi_{\pm, x} - (1/2) F_{n,x} \psi_\pm \tag{D.26}$$

(cf. (1.18)) one obtains

$$m_{\pm,t_n} = \partial_x(F_n m_\pm - (1/2)F_{n,x}).$$

(That $\psi_\pm$ can be chosen to satisfy (D.26) has been discussed in [2].) Since by (1.11), $F_n = \sum_{k=0}^n c_{n-k} \sum_{p=0}^k \hat{f}_{k-p} z^p$, it suffices to consider the homogeneous case $\widehat{F}_n = \sum_{k=0}^n \hat{f}_{n-k} z^k$. One then obtains

$$m_{\pm,t_n} = \sum_{j=1}^\infty m_{\pm,j,t_n} z^{-j/2}$$

$$= \partial_x(F_n m_\pm - (1/2)F_{n,x})$$

$$= \partial_x\left(\widehat{F}_n \sum_{j=-1}^\infty m_{\pm,j} z^{-j/2} - (1/2)\widehat{F}_{n,x}\right)$$

$$= \partial_x\left(\sum_{k=0}^n \hat{f}_{n-k} z^k \sum_{j=-1}^\infty m_{\pm,j} z^{-j/2} - (1/2)\widehat{F}_{n,x}\right).$$

A comparison of powers of $z^{-j/2}$ then yields

$$m_{\pm,j,t_n} = \partial_x\left(\sum_{k=0}^n \hat{f}_{n-k} m_{\pm,j+2k}\right), \quad j \in \mathbb{N}.$$

Since every even order coefficient $m_{\pm,2\ell}$ is known to be a total derivative (i.e., $m_{\pm,2\ell} = \partial_x(\dots)$), the conservation laws associated with $m_{\pm,2\ell}$, $\ell \in \mathbb{N}$, are all trivial. The odd order coefficients $m_{\pm,2\ell+1}$ lead to a nontrivial infinite sequence of conservation laws. (Of course, by (D.26), $m_{+,2\ell+1}$ and $m_{-,2\ell+1}$, $\ell \in \mathbb{N}_0$, yield the same infinite sequence of KdV conservation laws.) □

The basic KdV functionals $\widehat{\mathcal{I}}_\ell = \widehat{\mathcal{I}}_\ell(u, u_x, u_{xx}, \dots, \partial_x^k u)$, are then given in terms of $\hat{m}_{+,2\ell+1}$ and $\hat{f}_{\ell+1}$ by

$$\widehat{\mathcal{I}}_\ell = i \int_\mathbb{R} dx\, \hat{m}_{+,2\ell+1}(x) = \frac{1}{2\ell+1} \int_\mathbb{R} dx\, \hat{f}_{\ell+1}(x), \quad \ell \in \mathbb{N}_0 \qquad (D.27)$$

(cf. (1.268) and (1.285)). Equation (D.25) yields a direct proof of

$$\frac{d\widehat{\mathcal{I}}_\ell}{dt_n} = 0, \quad \ell \in \mathbb{N}_0,\ n \in \mathbb{N}$$

(cf. Theorem 1.62).

Real-valuedness of u is not essential for these considerations and can be dropped. Moreover, the decay assumptions on u as $|x| \to \infty$ can be considerably relaxed and replaced by the finiteness of certain moments of u.

This representation of the KdV conservation laws is perhaps simpler for computational purposes than the traditional one relying on the Lenard recursion operator. For a new twist to conserved KdV polynomials we refer to [4] (see also [1]).

Next, we supplement this particular addendum on KdV conservation laws by briefly sketching the extension of the Hamiltonian formalism to Bohr almost periodic KdV solutions in the space variable as discussed by Johnson and Moser [3]:

We start by noting that if f denotes a Bohr (uniformly) almost periodic function on $\mathbb{R}$, its ergodic mean $\langle f \rangle$ is given by

$$\langle f \rangle = \lim_{R \uparrow \infty} \frac{1}{2R} \int_{-R}^{R} dx\, f(x).$$

Suppose that u has the frequency module $\mathcal{M}(u)$. Then given a density F as on p. 97, one has

$$\mathcal{F}(u) = \lim_{R \uparrow \infty} \int_{-R}^{R} dx\, F(u, u_x, u_{xx}, \ldots, \partial_x^m u) = \langle F(u) \rangle,$$

and assuming that the frequency module $\mathcal{M}(v)$ of v satisfies $\mathcal{M}(v) \subseteq \mathcal{M}(u)$, one obtains

$$(d\mathcal{F})_u(v) = \frac{d}{d\epsilon} \mathcal{F}(u + \epsilon v)\Big|_{\epsilon=0}$$

$$= \lim_{R \uparrow \infty} \int_{-R}^{R} dx \left(\sum_{k=0}^{m} (-\partial_x)^k \partial_{u^{(k)}} F\, v \right)(x)$$

$$= \langle (\nabla \mathcal{F})_u v \rangle = \left\langle \frac{\delta F}{\delta u} v \right\rangle.$$

In analogy to p. 98, the Poisson brackets of two functionals $\mathcal{F}_1, \mathcal{F}_2$ are then given by

$$\{\mathcal{F}_1, \mathcal{F}_2\} = \left\langle \frac{\delta F_1}{\delta u} \left(\partial_x \frac{\delta F_2}{\delta u} \right) \right\rangle.$$

Again one verifies that both the Jacobi identity as well as the Leibniz rule hold in this case. Moreover, if $\mathcal{F}$ is a smooth functional and u develops according to a Hamiltonian flow with Hamiltonian $\mathcal{H}$, that is,

$$u_t = (\nabla_s \mathcal{H})_u = \partial_x (\nabla \mathcal{H})_u = \partial_x \frac{\delta H}{\delta u},$$

then

$$\frac{d\mathcal{F}}{dt} = \frac{d}{dt} \langle F(u) \rangle = \{\mathcal{F}, \mathcal{H}\}.$$

Next, one introduces the fundamental function w by

$$w(z) = -\frac{1}{2} \left\langle \frac{1}{g(z, \cdot)} \right\rangle$$

for $|z|$ sufficiently large. Since

$$w'(z) = \langle g(z, \cdot)\rangle, \quad z \in \mathbb{C} \setminus \operatorname{spec}(H)$$

(cf. (1.266)), w extends analytically to $z \in \mathbb{C} \setminus \operatorname{spec}(H)$. One infers from $g = 1/(m_- - m_+)$ that

$$w(z) = \pm\langle m_\pm(z, \cdot)\rangle, \quad z \in \mathbb{C} \setminus \operatorname{spec}(H).$$

Here $m_\pm$ denote the half-line Weyl–Titchmarsh functions associated with H. The asymptotic expansion of $m_\pm$ as $|z| \to \infty$ has been recorded in (D.23), and its connection with the homogeneous coefficients $\hat{f}_\ell$ and hence with the KdV functionals $\widehat{\mathcal{I}}_\ell$ has been noted in (D.27). In particular, introducing

$$\widehat{\mathcal{I}}_\ell = i\langle \hat{m}_{+,2\ell+1}\rangle = \frac{1}{2\ell+1}\langle \hat{f}_{\ell+1}\rangle, \quad \mathcal{I}_\ell = \sum_{k=0}^{\ell} c_{\ell-k}\widehat{\mathcal{I}}_k, \quad \ell \in \mathbb{N}_0,$$

the KdV equations again take on the form

$$\operatorname{KdV}_n(u) = u_{t_n} - 4\partial_x(\nabla I_{n+1})_u = 0, \quad n \in \mathbb{N}_0.$$

Finally, one can show that $w(z_1)$ and $w(z_2)$ are in involution for arbitrary $z_1, z_2 \in \mathbb{C} \setminus \operatorname{spec}(H)$, and hence obtains

$$\{w(z_1), w(z_2)\} = 0, \quad z_1, z_2 \in \mathbb{C} \setminus \operatorname{spec}(H), \tag{D.28}$$
$$\{w(z), I_p\} = 0, \quad \{I_p, I_r\} = 0, \quad z \in \mathbb{C} \setminus \operatorname{spec}(H), \; p, r \in \mathbb{N}_0. \tag{D.29}$$

Naturally, these considerations apply to the special periodic case in which $\langle f\rangle$ for a periodic function f on $\mathbb{R}$ is to be interpreted as the periodic mean value.

Page 122. An extension of formula (1.299) already appeared on p. 428 in [3].

Page 145. Line 4– can be more effectively replaced by:
By equation (2.94) one concludes that

Page 182. Line 7+: It would have been more natural to write equation (3.18) as:

$$G_{n+1}(z) = \sum_{\ell=0}^{n+1} g_{n+1-\ell} z^\ell = \sum_{\ell=0}^{n+1} c_{n+1-\ell} \widehat{G}_\ell(z),$$

Page 198. We note that $\Omega_0^{(2)}$ in equation (3.96) has the explicit form

$$\Omega_0^{(2)} = \omega_{P_{\infty+},0}^{(2)} - \omega_{P_{\infty-},0}^{(2)} = \frac{z^n}{y}\sum_{k=0}^{1} c_{1-k}(\underline{E})z^k dz + \frac{\tilde{\lambda}_n}{y}\prod_{j=1}^{n-1}(z - \tilde{\lambda}_j)dz,$$

where $c_k(\underline{E})$, $k \in \mathbb{N}_0$, are defined in (D.5) and $\tilde{\lambda}_j$, $j = 1, \ldots, n$, are uniquely determined by the normalization

$$\int_{a_j} \Omega_0^{(2)} = 0, \quad j = 1, \ldots, n.$$

This comment also applies to $\Omega_{\infty,0}^{(2)}$ in (4.215).

Pages 200 and 224. One uses the equality

$$\underline{z}(P_{\infty_-}, \underline{\hat{\nu}}) = \underline{z}(P_{\infty_+}, \underline{\hat{\mu}})$$

(an elementary consequence of (3.58)) to compute the constant C in equations (3.113) and (3.224).

Page 220. We note that $\omega_{P_{\infty_+},q}^{(2)} - \omega_{P_{\infty_-},q}^{(2)}$ in equation (3.207) has the explicit form

$$\omega_{P_{\infty_+},q}^{(2)} - \omega_{P_{\infty_-},q}^{(2)} = \frac{z^n}{y} \sum_{k=0}^{q+1} c_{q+1-k}(\underline{E}) z^k dz + \frac{\tilde{\lambda}_n}{y} \prod_{j=1}^{n-1} (z - \tilde{\lambda}_j) dz, \quad q \in \mathbb{N}_0,$$

where $c_k(\underline{E})$, $k \in \mathbb{N}_0$, are defined in (D.5) and $\tilde{\lambda}_j$, $j = 1, \ldots, n$, are uniquely determined by the normalization

$$\int_{a_j} \left(\omega_{P_{\infty_+},q}^{(2)} - \omega_{P_{\infty_-},q}^{(2)} \right) = 0, \quad j = 1, \ldots, n.$$

This comment is also relevant in connection with (C.37).

BIBLIOGRAPHY

[1] L. A. Dickey, On a generalization of the Tréves criterion for the first integrals of the KdV hierarchy to higher GD hierarchies, *Lett. Math. Phys.* **65**, 187–197 (2003).
[2] F. Gesztesy and B. Simon, Constructing solutions of the mKdV-equation, *J. Funct. Anal.* **89**, 53–60 (1990).
[3] R. Johnson and J. Moser, The rotation number for almost periodic potentials, *Comm. Math. Phys.* **84**, 403–438 (1982).
[4] F. Tréves, An algebraic characterization of the Korteweg–de Vries hierarchy. *Duke Math. J.* **201**, 251–294 (2001).